Geophysical Monograph Series

Including

IUGG Volumes

Maurice Ewing Volumes

Mineral Physics Volumes

Geophysical Monograph 105

New Perspectives on the Earth's Magnetotail

A. Nishida
D. N. Baker
S. W. H. Cowley
Editors

American Geophysical Union
Washington, DC

Published under the aegis of the AGU Books Board

Library of Congress Cataloging-in-Publication Data

New perspectives on the earth's magnetotail / A. Nishida, D. N. Baker, and S. W. H. Cowley, editors.
p. cm. -- (Geophysical monograph series ; 105)
Includes bibliographical references.
ISBN 0-87590-088-7
1. Magnetotails. I. Nishida, A. (Atsuhiro), 1936-
II. Baker, D. N. III. Cowley, S. W. H. IV. Series.
QC809.M35N48 1998
538'.76--dc21 98-38668
CIP

ISBN 0-87590-088-7
ISSN 0065-8448

Printed in the United States of America.

CONTENTS

CONTENTS

PREFACE

On the nightside of the Earth, a long magnetic tail is formed by the tangential stress that is exerted by the solar wind as it flows by the planet. The magnetotail is the nightside extension of the Earth's magnetosphere in which the geomagnetic field is confined by the solar wind, and its framework is formed by the field lines emanating from the polar caps. The magnetotail plays a pivotal role in magnetospheric dynamics as the reservoir of the energy that produces a wide variety of phenomena in the Earth's magnetosphere. The interaction of the magnetotail with the ambient solar wind provides virtually all of the energy to drive auroral and geomagnetic processes.

Earth-orbiting spacecraft discovered the Earth's magnetotail more than three decades ago, and early interplanetary missions hinted at its great size. The revelation that thin current sheets and highly ordered magnetic field structures exist on such vast scales fundamentally changed our view of how astrophysical plasmas are arranged. Indeed, the Earth's magnetotail has been a cosmic laboratory for studying plasma convection, particle acceleration, and large-scale current systems. The magnetotail where collisionless plasma is embedded in a relatively simple magnetic field configuration is ideal for understanding the basic processes in space plasmas such as magnetic reconnection and slow shock formation.

Space physicists have made great strides in understanding the physics of the Earth's magnetotail because of significant advances in satellite observations and theoretical investigations. This monograph offers a new perspective of the Earth's magnetotail made possible by recent studies. The first section deals with structure of the magnetotail in terms of both the magnetic field and the resident plasma. The strong influence of the interplanetary magnetic field (IMF) on the tail structure demonstrates that the magnetic reconnection between the IMF and the geomagnetic field controls the formation and filling of the magnetotail. The reconnection process operates under both southward and northward IMF conditions. Observation of the low-latitude boundary layer suggests that diffusive entry of the solar wind plasma is also operative.

Heating and acceleration processes of plasma that occur inside the magnetotail are discussed in the second section. These processes are governed by magnetic reconnection and the accompanying slow shocks. In the collisionless plasma of the magnetotail, kinetic behaviors of individual ions and electrons produce non-Maxwellian and non-isotropic distribution functions in phase space which reflect the history of the energization undergone by the particles.

The third section discusses the relationship of tail dynamics to magnetospheric substorms. The onset of magnetic reconnection in the near-Earth region of the tail leads to the substorm expansion phase; the flux rope that is produced by this reconnection propagates tailward through the distant magnetotail. On the earthward side the plasma penetrates toward the inner magnetosphere and undergoes both adiabatic and non-adiabatic heating.

The fourth section deals with waves and turbulence in the magnetotail. A great variety of plasma waves is produced from non-equilibrium in the velocity distribution functions of the tail plasma; the nature of these waves has been clarified by highly time-resolved observations and computer simulations. The magnetic field in the plasma sheet is often highly variable so that the generation of the turbulence is also an important subject of magnetotail research.

This monograph grew out of the presentations and discussions at the Chapman Conference held in Kanazawa, Japan, November 5-9, 1996. Much of the new material which is presented here has been obtained from the International Solar Terrestrial Physics (ISTP) program, which has promoted the advanced understanding of geospace. The monograph is dedicated to the memory of two of the early leaders of the ISTP program, F.L. Scarf and S. D. Shawhan, and also to the memory of T. Yamamoto, who played a major role in the magnetometer experiment on GEOTAIL.

A. Nishida
The Institute of Space and Astronomical Science, Japan
D. N. Baker
University of Colorado, Boulder, Colorado
S. W. H. Cowley
University of Leicester, Leicester, U.K.

Editors

The Distant Magnetotail: Its Structure, IMF Dependence, and Thermal Properties

K. Maezawa and T. Hori

Department of Physics, Nagoya University, Nagoya, Aichi, 464-01, Japan

Magnetic and thermal structures of the distant magnetotail as surveyed recently by the GEOTAIL spacecraft are reviewed. An analysis of the distant (-220 Re$<x<-150$ Re) tail structure is synthesized based on a coordinate system whose x-axis is taken to be parallel to the hourly direction of the solar wind measured in the upstream region. In this coordinate system, the distant magnetotail is found to be cylindrical, with almost the same dimension (about 50~55 Re) in the y and z directions for average IMF conditions. The most prominent feature of the distant magnetotail is that the tail current sheet is twisted under the presence of IMF B_y. The twist angle is larger for northward IMF than for southward IMF, showing that significant amounts of magnetospheric field lines are reconnected with the solar wind field lines even when the IMF is directed northward. The plasma mantle which is known to be formed in the high-latitude magnetosphere near the Earth is twisted in the distant magnetotail and forced against the plasma sheet owing to the large torque exerted by the IMF. Examination of individual ion distribution functions near the boundary suggests that the plasma mantle is often located adjacent to the plasma sheet. Sometimes an accelerated ion beam is superposed on the mantle plasma, suggesting that mantle field lines are reconnected at the plasma sheet boundary. An analysis of the magnetic field component normal to the twisted neutral sheet supports the view that reconnection occurs within the distant neutral sheet even during quiet times when the IMF is directed northward.

1. INTRODUCTION

Study of the Earth's magnetotail is an important key to clarify the interaction mechanism between the solar wind and the magnetosphere. The formation of the magnetotail itself is a direct consequence of the tangential stress operating on the interface between the solar wind and the Earth's magnetic field. Without this stress, the Earth's magnetosphere cannot have such a long tail but should be confined to a finite dimension determined by the normal (dynamic plus static) pressure of the solar wind. Thus the actual structure of the magnetotail will provide information about the mechanisms producing this tangential stress.

The most promising candidate for the mechanism providing this tangential stress is magnetic field reconnection, which develops a magnetic field component normal to the boundary, thus producing a magnetically open magnetosphere. A strong piece of evidence favoring the open magnetosphere comes from the confirmation of the prediction that, owing to the tangential stress produced by the normal component, newly opened-up field lines should be transported from the dayside magnetosphere to the magnetotail. Such transport of magnetic flux leads to the magnetic energy build-up in the tail lobes during intervals of southward-directed IMF. The magnetic field energy stored in the tail by this mechanism is released during the expansion phase of the substorm.

New Perspectives on the Earth's Magnetotail
Geophysical Monograph 105

During intervals of northward IMF, reconnection would be able to occur on magnetospheric field lines tailward of the dayside cusp region [Dungey, 1961]. However, it has been difficult to prove existence and consequences of reconnection for northward IMF as compared to the case for southward IMF. One reason for this is that the high-latitude magnetosphere near the cusp has not yet been fully surveyed as compared to the equatorial magnetosphere [see, however, Kessell et al., 1996; Gosling et al., 1991]. The second reason is that the reconnection with northward IMF does not necessarily produce a net transfer of field lines from the dayside magnetosphere to the magnetotail, and hence would not result in a large-scale reconfiguration of the magnetosphere. In fact, evidence for reconnection with northward IMF has principally been obtained from ground or low-altitude magnetospheric signatures confined to the polar cap [Maezawa, 1976; Burke et al., 1979; Reiff and Burch, 1985].

When the IMF has a finite y component (B_y), the tangential stress resulting from the JxB force operating on the reconnected field lines has a dawn-dusk component. In other words, the reconnected solar wind field lines will exert a torque on the open field line portions of the magnetosphere. *Cowley* [1979, 1981] proposed that the magnetotail would become twisted because of this torque. Thus the IMF B_y-associated twist of the magnetotail constitutes another test to see whether the magnetospheric field lines are really connected with the solar wind field lines. It should be noted that such twists have been detected for both the near-Earth and distant magnetotail observations. However, the average twist angles suggested from the observations are diverse. Further, the dependence of the strength of the twist on the north-south component of the IMF has not been clarified yet, with the exception of the work of *Owen et al.* [1995], who suggested that the twist angle is larger for northward IMF than for southward IMF.

Since 1992, the GEOTAIL spacecraft has been making a detailed survey of the Earth's magnetotail. In this paper we principally review the results of our analysis of the structure of the distant magnetotail, where the effect of the tangential stress due to reconnection is accumulated and most clearly seen. We are particularly interested in the IMF dependence of the distant magnetotail structure. We will confirm that the twist angle of the tail is larger for northward IMF than for southward IMF, indicating that a significant amount of magnetospheric field lines are reconnected with the IMF field lines in the case of northward IMF.

Another important subject dealt with in this paper is the mechanism of solar wind plasma supply to the magnetosphere. Although the transfer mechanism of magnetic field lines to the magnetotail by dayside reconnection has been widely accepted, the mechanism of plasma (mass) supply to the magnetotail has not been fully understood. Since the direction of plasma convection in the near-Earth magnetosphere is sunward, the magnetotail should play an important role as a plasma source for the near-earth magnetosphere. Particularly, the plasma in the plasma sheet is constantly lost by earthward convection, and the plasma sheet should be replenished either from the ionosphere or from the solar wind through the boundaries of the magnetotail.

The ISEE-3 spacecraft confirmed a B_y-dependent asymmetry in the low-energy electron distribution within the tail lobe, supporting the view that the solar wind plasma supply is controlled by the reconnection with the IMF. However, details of the mechanism of plasma entry have not yet been clarified, partly because thermal ion measurements were not available from the ISEE-3 spacecraft. In this paper, on the basis of GEOTAIL observations, we discuss the locations where the solar wind plasma can gain access to the distant plasma sheet for different IMF conditions. It is concluded that the plasma mantle, whose twisted location in the distant tail is adjacent to the plasma sheet, plays a significant role in the plasma supply to the plasma sheet. Finally, by examining individual examples of ion distribution functions, we obtain a clue of the nature of the processes by which the mantle plasma is heated and assimilates with the plasma sheet plasma.

2. OBSERVATIONS PRIOR TO THE GEOTAIL MISSION

The ISEE-3 mission to the distant tail region provided us with basic knowledge of the structure of the distant magnetotail prior to the GEOTAIL mission. ISEE-3 observations indicated that the tail at distances of x=-200Re has the same two-lobe structure as observed nearer to the Earth [*Tsurutani et al.,* 1986 *and references therein*]. An IMF B_y-associated asymmetry of the lobe plasma density was confirmed to occur deep in the regions adjacent to the current sheet [*Gosling et al.,* 1985]. The tilt of the tail current sheet was confirmed [*Sibeck et al.,* 1985; *Sibeck et al.,* 1986a, *Macwan,* 1992; *Owen et al.,* 1995]. However, the estimates of the tilt angle varied among the studies of different authors. For example, *Sibeck et al.* [1986a] derived an average tilt of 18° for a fixed polarity of IMF B_y. *Macwan* [1992] suggested much larger twist on the basis of the current sheet normals determined for selected events. *Owen et al.* [1995] obtained smaller values of 5~12° for the southward IMF cases, and somewhat larger values of 13~24° for the northward IMF cases. Some of the discrepancies among the authors may be attributable to the different magnitudes of IMF B_y that occurred for individual examples, and to different methods of analysis.

It should be noted that the study of the distant tail structure must be made taking into account the possible excursions of the tail axis from its nominal position in response to the changing solar wind direction. Thus, although the satellite path is usually confined to low latitude regions of the GSM yz plane, the satellite may actually encounter various portions of the magnetotail cross section, as they are brought to the satellite location by the shift of the tail axis. To deal with this situation correctly, *Maezawa et al.* [1997], hereafter referred to as Paper 1, used a coordinate system whose x-axis is aligned with the solar wind direction (in the Earth's rest frame) as observed simultaneously by another satellite in the upstream region. The sign convention for the x-axis is such that it is positive sunward. All the other definitions for the y and z-axis are exactly the same as those of the GSM coordinate system once the x-axis is specified; the y-axis is parallel to the vector product of the x direction and the Earth's dipole moment, and the z-axis completes the right handed Cartesian coordinates. In this paper we use this coordinate system extensively, called the 'Geocentric Solar-Wind Magnetospheric coordinates' (GSW coordinates for short).

3. GEOTAIL DATA

In the framework of the above coordinate system, we now discuss the average tilt of the distant tail current sheet and the overall cross sectional shape of the distant tail beyond x=-150 Re. The GEOTAIL data used for the analysis are the bulk plasma parameters (ion density N, ion temperature T, and the three components of the bulk velocity V_x, V_y, and V_z) obtained from the low energy plasma (LEP) experiment, and the three magnetic field components (B_x, B_y, and B_z) obtained from the magnetic field (MGF) experiment. Details of the instrumental characteristics have been given elsewhere [*Mukai et al.*, 1994; *Kokubun et al.*, 1994]. Solar wind and IMF data used in the analysis are the IMP-8 hourly values for which proper time correction is made for the solar wind propagation time delay between the IMP-8 and GEOTAIL satellites.

3.1. *Identification of the Magnetotail*

A difficult but essential step for the statistical analysis of the distant magnetotail is to identify the tail region correctly. We will see that bulk parameters (density, temperature, and bulk speed) of ions provide a useful way to distinguish the magnetotail plasma from the magnetosheath plasma, particularly if they are normalized by the corresponding quantities observed in the upstream solar wind region. We will see that the tail plasma regime is composed of two major sub-regimes, the plasma sheet and the tail lobe, though there is a certain transition state as will be shown later. Of these two sub-regimes, the plasma sheet can be easily distinguished from the magnetosheat h due to its higher ion temperature (T> 300 eV) as compared to the magnetosheath ion temperature (T << 300 eV). The rest of the tail region occupied by lower temperature plasmas will be termed "lobe" in this paper. Since the lobe and magnetosheath ion temperatures are often similar, the ion temperature cannot be used as a standard to distinguish the lobe and magnetosheath plasmas. Hence a more sophisticated method has been used in Paper 1, by normalizing the x component V_x of the ion bulk speed by the simultaneously observed solar wind speed. We note that the normalized V_x is actually a very useful parameter to distinguish the tail and the magnetosheath plasma, provided that the plasma sheet data for which V_x is highly variable have already been removed. Specifically, the data points having a normalized value of V_x greater than -0.8 can be judged to belong to the tail lobe (see the next paragraph for the reasoning). Note that in the GSW coordinate system, the upstream solar wind has V_x = -1. (The negative sign represents the anti-sunward direction.) Then a V_x value greater than -0.8 means that the speed of the plasma has been significantly reduced from that of the solar wind. (Note that the sign convention for V_x in this paper is opposite to that in Paper 1. Thus the condition V_x> -0.8 in this paper is expressed as V_x < 0.8 in Paper 1.)

Of course, these procedures used for defining the lobe and the plasma sheet plasmas are not straightforward and need verification from other plasma and magnetic field observations. For further discussions of this matter, the reader is referred to Paper 1, in which we showed that the distant magnetosheath plasma usually has a V_x value close to -1, and that the plasma populations satisfying the two conditions V_x > -0.8 and T < $3x10^6$ K are generally accompanied by magnetic field azimuthal angles close to 0° or 180°, representing a tail lobe field structure.

In passing we note that the lobe as defined in the present context includes a variety of plasma densities, ranging over more than two orders of magnitude in the normalized ion density. It will be shown, however, that the density and speed of the lobe plasma has a unique relationship, indicating that the lobe plasma population can be viewed as consisting of a single component entering the magnetosphere from the plasma mantle.

3.2. *Average Cross Section of the Distant Magnetotail*

We begin by examining the average size and shape of the cross section of the distant magnetotail. To do this all the data points (one minute averaged values) obtained when GEOTAIL was located with x < -150 Re were first classified as either belonging to the magnetosheath or to the magnetotail following the criteria introduced above. Then we calculated the probability of the satellite sitting in the magnetotail for each 4 x 4 Re^2 bin in the GSW y-z plane, by dividing the number of data points identified as "belonging to the magnetotail" for each bin by the total number of data points for each bin. Figure 1a shows the resultant contour map, seen looking towards the Earth, of the occurrence probability of the tail in the GSW y-z plane. The color coding in Figure 1a is such that red to blue tones represent occurrence probabilities greater than 50 per cent, while greenish tones represent smaller probabilities of observing the tail. The thick white line represents the 50

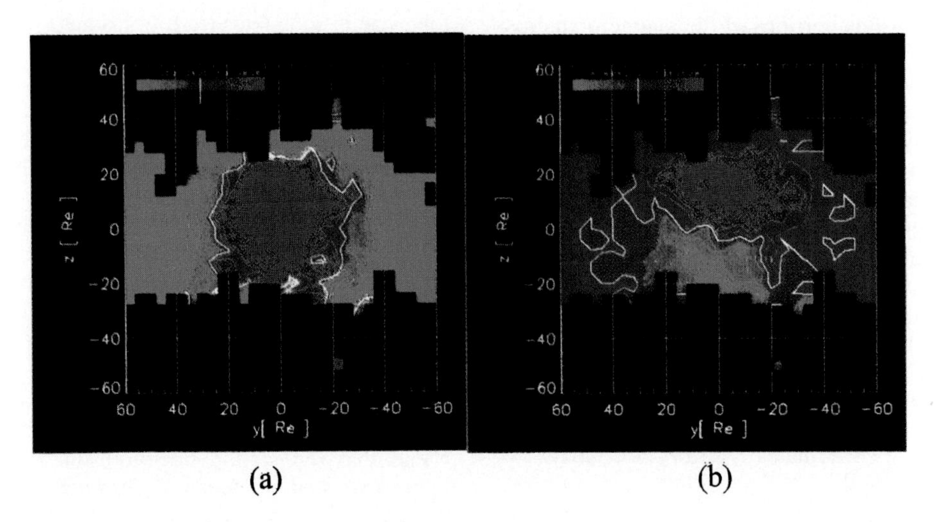

Figure 1. (a) Contour map of the probability of observing the magnetotail for each 4 x 4 Re bin in the GSW y-z plane, as seen looking towards the Earth. The color code is such that reddish colors represent occurrence probabilities greater than 50 per cent, while greenish colors represent smaller probabilities. The white thick line represents the 50 per cent contour line, which may be taken as the average location of the magnetopause. This figure represents the case for positive IMF B_y (see text for details). (b) Contour map of the difference (Pn - Ps) between the probabilities of observing a northern tail magnetic polarity and a southern tail magnetic polarity. Reddish colors represent the case where the northern tail is observed more frequently than the southern tail and greenish colors represent the opposite case. The thick white line represents the zero contour line, which may be taken as the average location of the neutral sheet. This figure represents the case for positive IMF B_y.

per cent contour line. We take this 50 per cent contour line as representing the average boundary location of the tail at x<-150 Re.

Special care was taken in the construction of Figure 1a in order to represent properly the possible dependence of the tail structure on IMF B_y. Adopting the reasonable assumption that the tail configuration for negative IMF B_y is simply a mirror image (about the z-axis) of that for positive IMF B_y, we inverted the GSW y position of the satellite for all the data points that occurred with IMF $B_y <$ 0. Thus Figure 1a represents the case for positive IMF B_y, though actually both positive and negative IMF B_y cases contributed to this figure, as explained above.

Several things are clear from Figure 1a. First, regarding the size of the tail, the 50 per cent contour line shows that the y dimension of the tail is about 55 Re, measured where the dimension is maximum, while the z dimension of the tail is also about 55 Re. The z dimension of the tail may actually be slightly larger because a part of the southern and northern high latitude contour lines is missing due to the lack of observation points. There is no indication of a major flattening of the tail in the y direction as proposed by some authors [*Sibeck et al.*, 1986b]. Elongation of the tail in the z direction has recently been pointed out by Nakamura and Kokubun [1997] for storm intervals. Second, the shape of the 80 per cent contour line (the highest contour level in the figure) is elongated in the z direction showing that the boundary location is more variable at lower latitudes. Third, there seems to be no significant tilt in the overall silhouette of the tail. This is also suggested from the fact that the 80 per cent contour line is elongated roughly in the z direction.

3.3. *Tilt of the Distant Neutral Sheet*

We now address ourselves to the twist of the tail as represented by the tilt angle of the tail current sheet. The location of the current sheet is obtained as the demarcation line between the regions of northern and southern lobe field polarities, as statistically determined in the y-z plane. We will discuss the overall tilt of the demarcation line between the northern and southern tail lobes, in order to avoid influences from local and/or transient effects.

The position of the current sheet was obtained as follows. Let Pn be the probability of observing the tail region with a positive B_x polarity. Similarly, let Ps be the probability of observing the tail region with a negative B_x polarity. (We note that the sum P=Pn+Ps represents the total probability of observing the tail, plotted in Figure 1a.) Then statistically the difference D=(Pn – Ps) should vanish at the position of the current sheet. In Figure 1b we show a contour map of the difference D= Pn - Ps in the GSW y-z plane obtained in Paper 1. The thick white line shows the contour line D=0, which is expected to be the average position of the current sheet. It is clear that the current sheet is tilted toward dawn, consistent with theoretical expectations. (Remember that the figure is drawn for positive IMF B_y, with the switch of y for IMF By < 0.) By

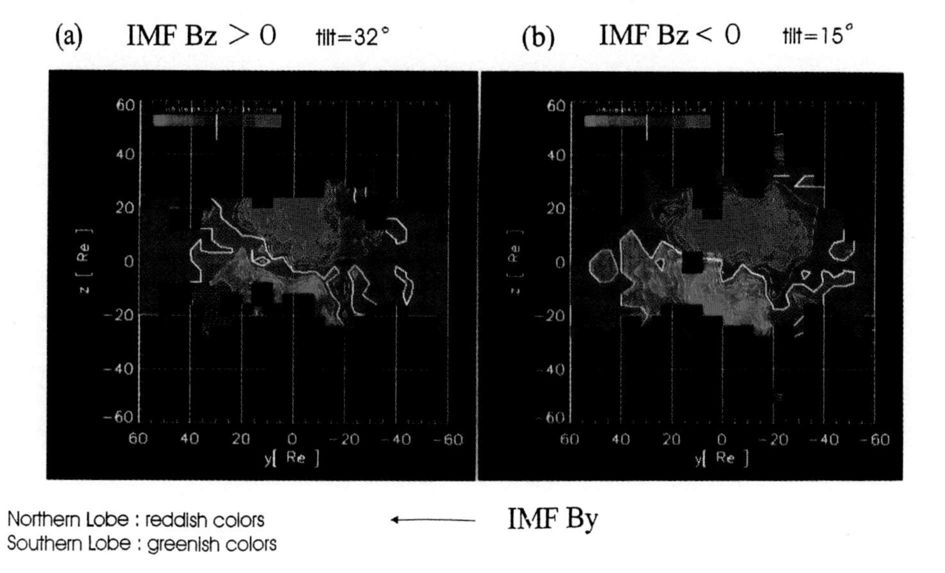

Figure 2. Same as Figure 1b but for the case of (a) northward IMF, and (b) southward IMF.

making a least squares linear fit to the portion of the D=0 contour inside the 50 percentage probability contour given in Figure 1a, the tilt angle is obtained as 20.0°.

Figure 1b provides us with information about the field structure within the lobe as well. The region where D is close to 1.0 is the region where the north tail lobe is most stably observed. By the same token, the region where D is close to -1.0 is the region where the southern lobe is most stably observed. (Of course, the absolute value of D is small near the current sheet.) We arbitrarily call the region where the absolute value of D exceeds 0.8 the 'core lobe', since it is interpreted to be the most stable part of the tail as far as the B_x polarity is concerned.

It is apparent from Figure 1b that the centers of the north and south core lobes are not tilted from the z-axis. Therefore, the current sheet tilt induced by IMF B_y does not lead to an overall rotation of the tail structure. This is because the major magnetic stress exerted by the reconnected field lines operates at the mantle, which we will show later is located at relatively low latitudes. On the other hand, the core lobes are located at higher latitudes and may be free from magnetic stress.

It is interesting to make analyses of the above effect separately for northward and southward polarities of the IMF, in order to obtain an indication of the difference in the solar wind-magnetosphere interaction mechanism for the different z polarities of IMF. Figures 2a and 2b show the same statistics as given in Figure 1b, but obtained separately for the cases IMF $B_z > 0$ (northward IMF) and IMF $B_z < 0$ (southward IMF). It is immediately apparent that the current sheet tilt for northward IMF given in Figure 2a is larger than that for southward IMF given in Figure 2b. The least squares fit shows that the tilt angle is 32° for northward IMF. On the other hand, the tilt angle seen in Figure 2b is much smaller, and the least square fit gives 15 ° for the southward IMF case. Therefore, the tilt angle for northward IMF (with an average B_z = +2 nT) is more than twice as large as that for southward IMF (with an average B_z = -2 nT).

Generally speaking, the twist angles obtained here are larger than the values obtained for the regions nearer to the Earth and are not inconsistent with some of the values suggested from ISEE-3 observations listed in Introduction. The most important point emerging from the GEOTAIL analysis is that the twist angle for northward IMF is definitely larger than that for southward IMF. As far as we know, there has been only one paper suggesting this tendency, which was based on the statistics of the orientation of the current sheet normal inferred from energetic particle anisotropy measurements at individual neutral sheet crossings. On the other hand, our statistics show the global presence of this tendency, as evidenced in the form of the global map of magnetic field polarities.

We believe that the larger twist angle of the magnetotail for northward IMF indicates that the solar wind is strongly interacting with the magnetosphere even when the IMF is directed northward [*Maezawa*,1976]. Further, the primary mode of this interaction should be the same for both the northward and southward IMF cases, since a similar IMF B_y effect is observed, differing only in magnitude. The most promising candidate is, of course, magnetic reconnection with the IMF, and we would like to obtain additional supporting evidence for the occurrence of magnetic reconnection during northward IMF conditions. In the

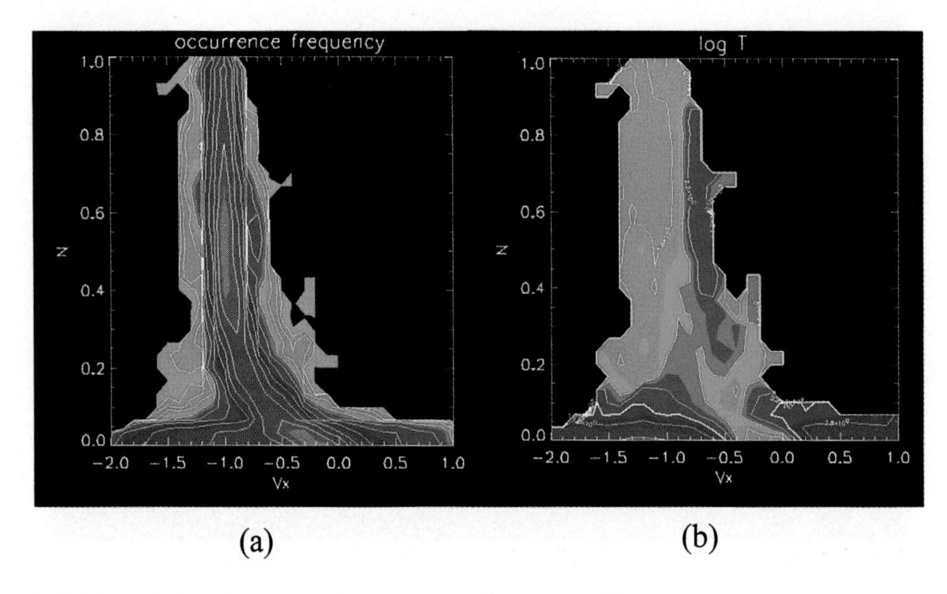

Figure 3. (a) Color-coded contour map of the occurrence frequency of data points in the parameter space spanned by the normalized x component of the plasma bulk speed and the normalized number density. Occurrence frequency is color-coded in such a way that it increases from green to pink. Magnetosheath data are included in the map. (b) Distribution of ion temperature for the same parameter space. The color code is such that the logarithm of ion temperature increases from green to pink. Contour levels of two adjacent contour lines differ by a factor of 2. The thick white line represents the 300 eV contour line.

next section we will show, based on an analysis of the thermal properties of the tail plasma, that the mantle plasma that is observed for southward IMF can be equally observed for northward IMF conditions. Indeed, we will show that the plasma mantle, which has been suggested to be the product of dayside reconnection, occupies a significant portion of the distant tail cross section regardless of the north-south polarity of the IMF.

4. PLASMA MANTLE AS A MAJOR SOURCE OF TAIL PLASMA

The statistics presented in the previous section confirm that the distant magnetotail is under the influence of IMF B_y regardless of the north-south polarity of the IMF. A reasonable way to interpret this behavior is to assume that a significant amount of magnetospheric field lines are reconnected with the IMF during periods of northward IMF. In this section, we study this possibility using a different approach, by analyzing the thermal properties of the lobe plasma.

4.1. *Identification of the Mantle Plasma*

We stated previously that the "lobe" plasma as defined in this paper has a variety of normalized density values, but that they are likely to originate from a single component entering the magnetotail via the polar mantle. Strong evidence for this assertion comes from the unique relation found between the ion speed and the ion density observed in the "lobe" region. Figure 3a shows this relationship in the form of an occurrence frequency map of data points in a parameter space spanned by the normalized V_x (x component of the bulk velocity) and the normalized number density N of ions. (V_x and N are normalized by simultaneous solar wind values as stated previously). In this figure, all the observations are presented, including magnetosheath data points. The color-coding is such that the occurrence frequency increases from green to pink. It is seen that, except for a flat, horizontal distribution seen at the bottom, almost all the population falls on a single curve which runs almost vertically from the top and approaches the origin in a slant way at the bottom. A careful examination of this curve shows that its upper portion (i. e. higher density portion) is on the line V_x=-1, but that the curve begins to deviate from the line V_x=-1 at the middle, and approaches the origin (i. e. vanishingly small density and speed) at the bottom. In other words, the upper portion of the curve represents a high density plasma that is flowing with almost the same speed as the upstream solar wind, while the lower portion represents plasmas that have been decelerated and rarefied. It is not unreasonable to assume that the high-density plasma having the solar wind speed (V_x = -1) is the magnetosheath plasma (it has V_x = -1 at these distances), while the rest of the curve represents the plasma whose speed has been modified by various means after entering the magnetotail. (Recall that we have included all the data points measured by GEOTAIL in the region x<-150 Re.)

The basic question is of course why there should be a unique relationship between the density and the speed of plasmas that have entered the magnetotail. A hint to this question comes from the work of Siscoe et al. [1994] which is based on their earlier theoretical work [*Siscoe et al.*, 1970, *Siscoe and Sanchez*, 1987]. The authors calculated the expected relationship between the density and the speed of the mantle plasma, assuming that they are determined by the action of rarefaction waves, which propagate along the reconnected field lines. They made a comparison between theory and observation for several individual magnetopause crossings made by GEOTAIL, and concluded that the theory and observation were consistent. The thick white curve in Figure 4 gives the typical density-speed relationship (called a "template" in their paper) obtained by *Siscoe et al.* [1994], superposed on the density-speed contour map we obtained in Figure 3a. In superposing the two curves, we varied our density scale to obtain the best result, since the upstream solar wind density is not given (but assumed) for their observation. Using IMP-8 solar wind data to translate their density to our normalized density, we found that the scale obtained is only 10 to 15 % different from the best fit one calculated without the solar wind data. It is clear that the two results are in reasonable agreement with each other, indicating that the majority of plasma sitting on the curve discussed in Figure 3a can be interpreted as originating in the plasma mantle. Then, according to the theory of Siscoe et al., the plasmas sitting on the portion of the curve near the line V_x =-1 in Figure 3a represent the ones which have just crossed the magnetopause. The plasmas sitting closer to the origin represent the ones observed after they have convected deeper into the magnetotail. Since the action of the rarefaction wave continues after the solar wind plasmas enter the magnetotail along the reconnected field lines, the plasmas are continually decelerated and rarefied, leading to the monotonic decrease of density and speed from the solar wind ones.

The only plasma population that does not belong to the curve discussed above is the one which is distributed horizontally along the bottom portion of the Figure 3a. The implication of this distribution is clear from Figure 3b, where we have plotted a contour map of the average ion temperature in the V_x-N parameter space. The color-coding is such that the temperature increases from green to pink. It is seen that the plasma population distributed along the bottom has an ion temperature much higher than the rest of population, showing that it represents plasma sheet ions. We note that the plasma sheet ions have a wide range of V_x values, which can easily exceed the solar wind speed in magnitude. This is the reason why we could not use the V_x values for distinguishing the plasma sheet plasma from the magnetosheath plasma. On the other hand, the monotonic decrease of V_x from the magnetosheath values to the core lobe values helped us to distinguish the magnetosheath and lobe plasmas on the basis of the V_x values.

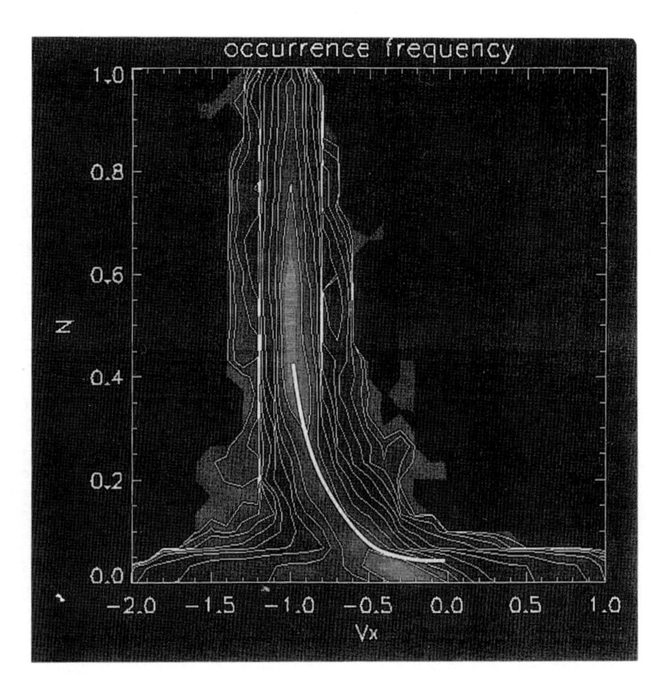

Figure 4. A thick white curve, representing the speed-density relationship obtained by Siscoe et al. (1994) for the mantle plasma, is superposed on the occurrence frequency map of data points obtained in Figure 3a.

4.2. *Phase Diagram of the Distant Tail Plasma*

We conclude that the distant plasma population is basically made up of two major components, the plasma sheet and lobe plasmas. Since the lobe component can be interpreted as originating from the plasma mantle, it may alternatively be called the mantle plasma, and both names will be used interchangeably in the rest of the paper.

Since the origin of the plasma sitting on the unique curve in the V_x-N plane is now interpreted to be the magnetosheath plasma entering the magnetosphere from the mantle, the remaining question is the origin of the plasma sheet plasma. Further, since the plasma sheet ions have a temperature much higher than that of the magnetosheath ions, an efficient heating process must operate somewhere in the tail. In order to obtain a clue about the thermal evolution of ions within the magnetotail, we have constructed a kind of "phase diagram" in which we examine which parts of parameter space can be occupied by the ions resident in the magnetotail. As is done for the HR diagram representing the evolution of stars in astrophysics, we assume that the ions move along a certain path in parameter space as their thermal state evolves in the magnetotail, and that the occurrence rate of the ions at a certain position in the diagram is proportional to the time they spend at that position.

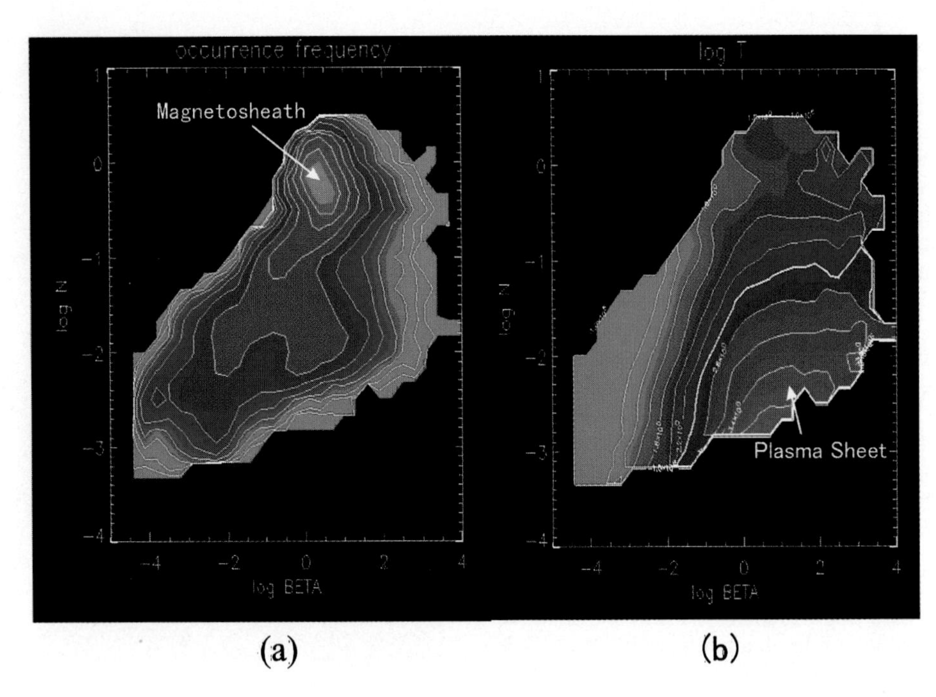

Figure 5. (a) Color-coded occurrence frequency map of the data points for the parameter space spanned by the ion beta value (not normalized) and the normalized ion density. All the one-minute values are used including magnetosheath data. (b) The distribution of ion temperature for the same parameter space. Color-coding is such that the log of ion temperature increases from green to pink. Contour levels are the same as in Figure 3b. Thick white curve represents the 300 eV contour line.

Figure 5a shows an example of such a diagram. The parameter space chosen is spanned by the normalized plasma density and the plasma beta value (not normalized). Although not shown here, we have tested many pairs of plasma bulk parameters (density, temperature, pressure, velocity, magnetic field magnitudes, etc.), and the combination presented here represents one of the best pairs, in that the magnetosheath, lobe, and plasma sheet plasmas occupy distinct positions in the two-dimensional space, and the interconnection among the three can be studied. (For simplicity we have limited our parameter spaces to two-dimensional ones)

The color-coded contour map in Figure 5a represents the occurrence frequency of data points. It is clear that the occurrence frequency is peaked around the position beta = 1 and N =1 (log N = 0). This position (the most reddish area in Figure 5a) represents the magnetosheath population. There is a ridge of occurrence frequencies running from the magnetosheath point diagonally toward the bottom left of the figure. It is also clear that there is another ridge that branches off from the middle of the first one and runs also diagonally towards lower right. We call the first one the "mantle-lobe ridge" and the second one the "plasma sheet ridge". What these two ridges represent becomes clear if we examine Figure 5b where the average ion temperature is given in the form of a color-coded map in the same parameter space. It is seen that the mantle-lobe ridge is accompanied by relatively cold temperatures (of the order of, or less than that of the magnetosheath ions), while the plasma sheet ridge is accompanied by much higher temperatures exceeding 300 eV. (The 300 eV contour line is shown by a thick white curve.) Therefore, according to the definition given in the preceding sections, the mantle-lobe ridge represents the mantle (lobe) plasma and the plasma sheet ridge represents the plasma sheet plasma.

The meaning of the diagonal trend of the mantle-lobe ridge is then clear from the theoretical model of Siscoe et al. It represents the temporal (and spatial) evolution of mantle plasma as it is convected towards the inner part of the magnetotail. The action of the rarefaction wave (actually a slow expansion fan) gradually reduces the plasma density as well as the plasma beta value as the mantle plasma is convected into the inner region (core lobe) of the magnetotail where the magnetic field magnitude is higher than the boundary region.

It is interesting that the plasma sheet ridge representing the plasma sheet plasma connects to the mantle-lobe ridge. It looks as if the plasma sheet ridge branched off from the mantle-lobe ridge at the middle of the figure. In other words, there is a spectrum of thermal states that smoothly connects the mantle-lobe and plasma sheet plasmas. We tentatively take this fact as evidence that part of the mantle plasma evolves into the plasma sheet plasma. In other words, the mantle plasma can get access to the plasma sheet where it has a chance to be heated before it is significantly rarefied to the very low density value of the core lobe.

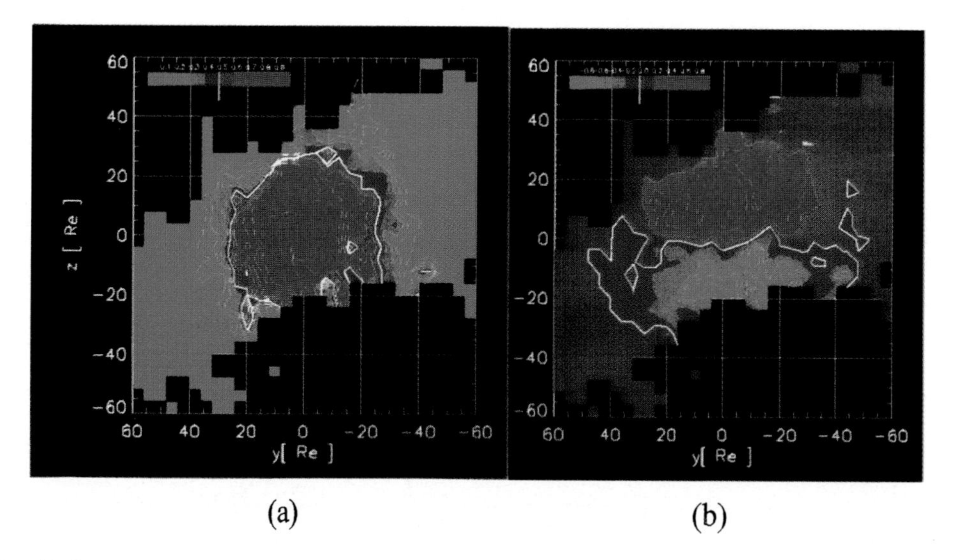

Figure 6. (a) Same as Figure 1a but presented in neutral sheet coordinates. The y axis is defined to be the empirically obtained direction of the current sheet (see text for details). (b) Same as Figure 1b but presented in neutral sheet coordinates. The zero contour line is parallel to the y axis as expected.

We will present in the Discussion section individual examples of ion distribution functions, and confirm that such heating of the mantle plasma is really occurring in the distant magnetotail. Finally, we note that the gross features presented in Figures 5a and 5b are seen both for northward and southward IMF cases (Details will be published elsewhere).

4.3. *Location of the Mantle Relative to the Plasma Sheet*

In this section we examine the location of the mantle relative to the plasma sheet in order to confirm the possibility that the former can indeed be a direct plasma source for the latter. Since we know from our statistics that the tail neutral sheet is twisted by the action of the IMF B_y, and that the twist angle is a function of IMF B_z, we adopt a coordinate system in which the y-axis is rotated around the x-axis in such a way that the new y-axis becomes parallel to the empirical direction of the neutral sheet in the y-z plane. In this new coordinate system, which we hereafter call neutral sheet coordinates, the average neutral sheet is expected to lie on the y-axis, and the z-axis is perpendicular to the neutral sheet. For the actual coordinate transformation we adopt the rotation angles obtained from observations, i.e., 32° and 15° for northward and southward IMF conditions, respectively.

Figures 6a and 6b show the occurrence probability map of the magnetotail data points in the y-z plane and the distribution of the polarity of B_x, respectively, in the same format as in Figures 1a and 1b but in neutral sheet coordinates. The overall elliptical shape of the tail magnetopause is apparent in Figure 6a except for the part of the boundary where observations are missing. Further, Figure 6b confirms that the neutral sheet (zero contour line of B_x) is oriented horizontally in the new coordinate system as expected.

Figures 7a and 7b show distributions of the temperature (not normalized) and the normalized density of the magnetotail plasma in the neutral sheet coordinate system. Magnetosheath data have been excluded according to the criteria introduced earlier. Otherwise, all the data are shown. In Figure 7a, regions having an ion temperature higher than 300 eV are coded by reddish colors in order to indicate the position of the plasma sheet. The central high temperature belt representing the plasma sheet is rather thick. There is an indication that the boundaries of the plasma sheet are slightly inclined to the y-axis, in spite of the fact that the neutral sheet is located horizontally (Figure 6b). This would mean that the location of the plasma sheet boundary has a y-dependent north-south asymmetry with respect to the neutral sheet, being probably induced by the twist effect. In other words, the northern (southern) portion of the plasma sheet seems to be thicker (thinner) on the dawn side for positive IMF B_y. However, these features are not closely related to the main theme of this paper and we do not wish to discuss it further here.

The distribution of the normalized density of plasma is shown in Figure 7b in order to delineate the region occupied by the mantle plasma. For this purpose, regions of normalized density higher than 0.03 are indicated by reddish colors. Two broad, higher density regions, one in each hemisphere, which can be identified as the mantle, occur in inclined positions along the magnetopause. Surprisingly, the north and south mantles are consistently inclined to the y-axis even in the coordinate system for which the tilt of the neutral sheet has already been corrected. This means that the plasma mantle is much more severely twisted than the neutral sheet. As a matter of fact, a

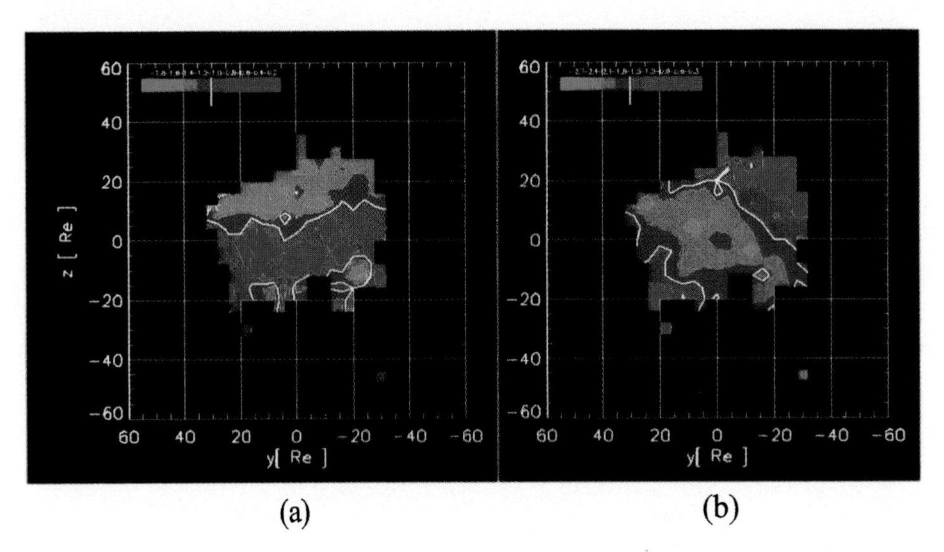

Figure 7. (a) Distribution of ion temperature in the y-z plane plotted in neutral sheet coordinates. The regions where the average ion temperature exceeds 300eV are coded by reddish colors to represent the position of the plasma sheet. (b) Distribution of normalized ion density in the y-z plane in neutral sheet coordinates. The regions where the average normalized density exceeds 0.03 are coded by reddish colors to represent the position of the plasma mantle.

careful comparison of Figures 7a and 7b shows that, because of the stronger twist of the mantle, the plasma mantle is located in juxtaposition with the plasma sheet, or even partly overlaps with the plasma sheet. Thus, as far as the relative locations are concerned, there exists the possibility that the mantle plasma can get access to the plasma sheet. The relative locations of the mantle and the plasma sheet are schematically drawn in Figure 8.

5. INDIVIDUAL EXAMPLES OF PLASMA MIXING AT THE INTERFACE BETWEEN THE PLASMA MANTLE AND THE PLASMA SHEET

Let us examine the mantle-plasma sheet transition in more detail on the basis of individual examples of the ion distribution function. Figure 9 gives magnetic field and plasma data (one minute averages) for the 4 hour period 1500 to 1900 UT on 9 December 1993, when GEOTAIL spacecraft was located near the dawn-side boundary of the distant magnetotail. The IMF was directed northward ($B_z \approx 2$ nT) throughout the period and the Kp index was low (0 and 1). The location of the satellite in GSW coordinates is given at the bottom of the figure. The plotted variables are, from top to bottom, the logarithm of the plasma density, the bulk velocity vectors projected onto the GSM xy plane, the logarithm of the ion temperature, the projections of magnetic field vectors on the GSM xy plane, and the latitudinal angle of the magnetic field. In the velocity and magnetic field vector diagrams, the sun is upward and dusk is to the left. The upstream solar wind speed was around 390 km s^{-1} as measured by the IMF 8 spacecraft. From the density and velocity vectors shown in the top two panels, we see that the satellite was located in the magnetotail throughout this period except for two short time intervals from 1550 to 1615 UT and from 1820 to 1823 UT. During these periods the satellite was probably outside the magnetotail because the observed plasma speed approached the upstream solar wind value. Indeed if we apply the threshold value V_x = -0.8 introduced in Section 3.1 to the present data, these time intervals are identified as the magnetosheath. For all the other times GEOTAIL was either in the plasma sheet or in the lobe/mantle region.

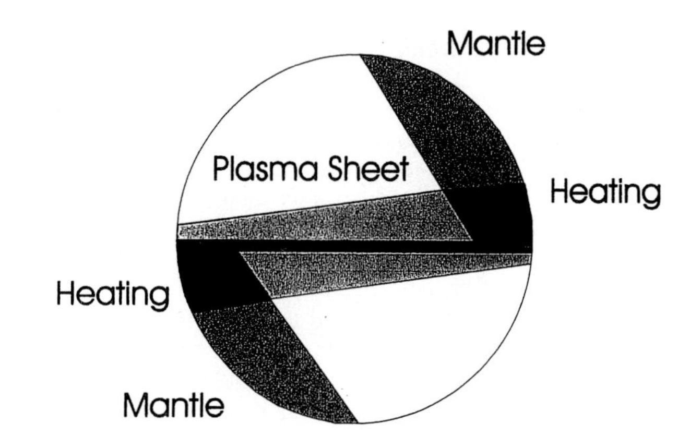

Figure 8. Schematic drawing of the tail cross section showing the average positions of the mantle and the plasma sheet in neutral sheet coordinates. The plasma mantle locations are severely inclined even in the coordinates in which the neutral sheet is oriented horizontally. The location of the mantle is in juxtaposition with the plasma sheet or probably overlaps with the plasma sheet. Heating of the mantle plasma at the mantle/plasma-sheet interface is suggested.

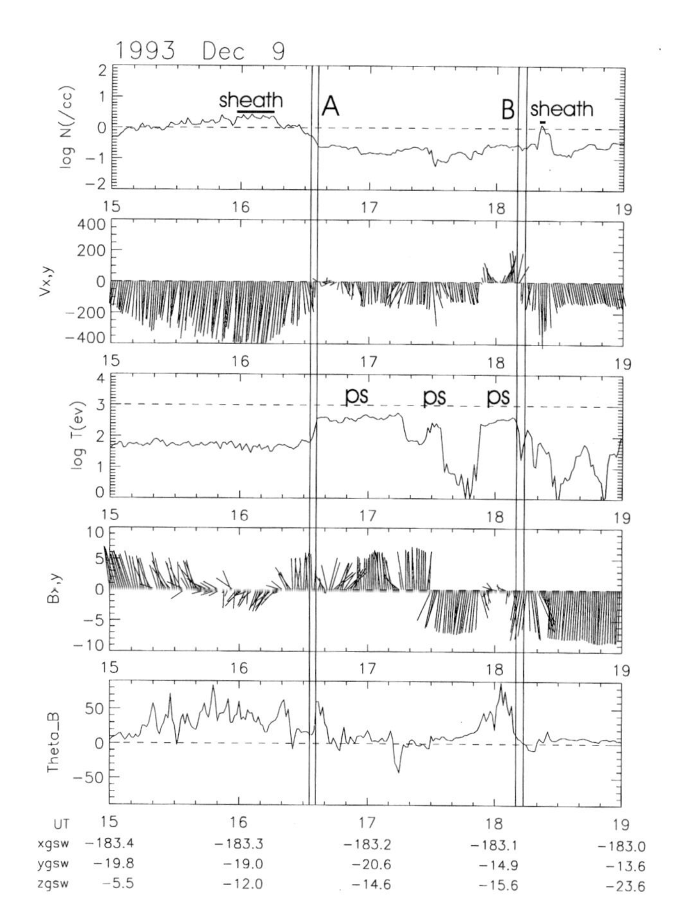

Figure 9. A 4-hour plot of plasma and magnetic field data for the period 1500 to 1900 UT on 9 December 1997. Plotted from top to bottom are the ion number density, the ion flow vectors in the GSM x-y plane, the ion temperature, the magnetic field vectors in the GSM x-y plane, and the latitude angle of the magnetic field. All the data are one-minute averages. Possible magnetosheath intervals are indicated in the top panel. Time intervals corresponding to the plasma sheet are labeled 'ps' in the middle panel. The satellite was in the mantle for the rest of the time. The two intervals of mantle/plasma-sheet interface observations discussed in the text are indicated by the pairs of vertical lines labeled A and B, respectively. The position of the satellite in GSW coordinates is given at the bottom.

Particularly, judging from the density and the velocity data, GEOTAIL was in the mantle near the magnetopause until 1630 UT with the exception of the sheath interval (1550 to 1615 UT) mentioned earlier. The magnetic field rotation on either side of the magnetosheath interval may indicate the presence of a rotational discontinuity, which is consistent with the presence of the mantle plasma. Except for the magnetosheath intervals, the magnetic field vectors plotted in the fourth panel show that the magnetic field direction was consistently sunward until 1730 UT. It then underwent a transition from the sunward to the antisunward direction around 1730 UT, and stayed generally in the antisunward direction thereafter. This transition suggests that the satellite was first located in the northern half of the tail, and was then engulfed by the southern half of the tail, presumably due to a northward movement of the current sheet. Note that the time variation of the z position of the satellite in GSW coordinates supports this view. Note also that the occurrence of the current sheet at negative z values is consistent with the expected tilt of the current sheet for the IMF away sector. (IMF B_y was about 3 nT throughout the interval plotted).

The ion temperature plot in the middle panel shows that there are at least three time intervals, 1636 to 1715 UT, 1729 to 1733 UT, and 1752 to 1810 UT, when the satellite was within the plasma sheet. These intervals are labeled as 'PS'. Actually the ion temperature for these intervals was lower than the typical temperature (~1 keV) of the plasma sheet at this distance, and was almost equal to the threshold value (300 eV) that we set to define the plasma sheet. We note that this temperature roughly corresponds to the position in Figure 5a where the plasma sheet ridge branched off from the mantle-lobe ridge.

In order to see what kind of heating mechanism is operating at the interface between the mantle and the plasma sheet during quiet times, we take two examples of the passing of the satellite across the mantle/plasma-sheet boundary. The time intervals corresponding to these two examples are each marked by a pair of vertical lines in Figure 9.

The first example occurred in the interval 16:32 – 16:35 UT (labeled A). The ion temperature data in Figure 9 suggests that GEOTAIL entered the plasma sheet from the mantle. Five snapshots of the ion distribution function during this interval are shown in Figure 10. Time increases from left to right in the upper row and similarly in the bottom row. The coordinates adopted are the BCE coordinates [*Fujimoto et al.,* 1996], which are useful to analyze a convecting plasma having anisotropic distributions relative to the magnetic field direction in velocity space. Specifically, the coordinates are defined by three axes, B, C, and E. B is parallel to the magnetic field, C is parallel to the convection (electric field drift) direction (that is, the ExB direction) , and E is parallel to the convection electric field. If we are in the tail lobe, B is roughly the (positive or negative) x direction in GSM, C is normally directed in the negative (positive) z direction in GSM for the northern (southern) tail lobe, and E is directed in the positive y direction in GSM if the electric field is directed from dawn to dusk. We present cuts of the ion distribution function in a plane that is parallel to the B-E plane (i.e. approximately parallel to the xy plane) and that contains the peak of the distribution function. In Figure 10, and in all the subsequent similar figures, the abscissa is always parallel to the B axis, but its orientation is such that the Sun always lies to the left of the figure. Thus the mantle flow is always directed from left to right regardless of the sign of the B_x component.

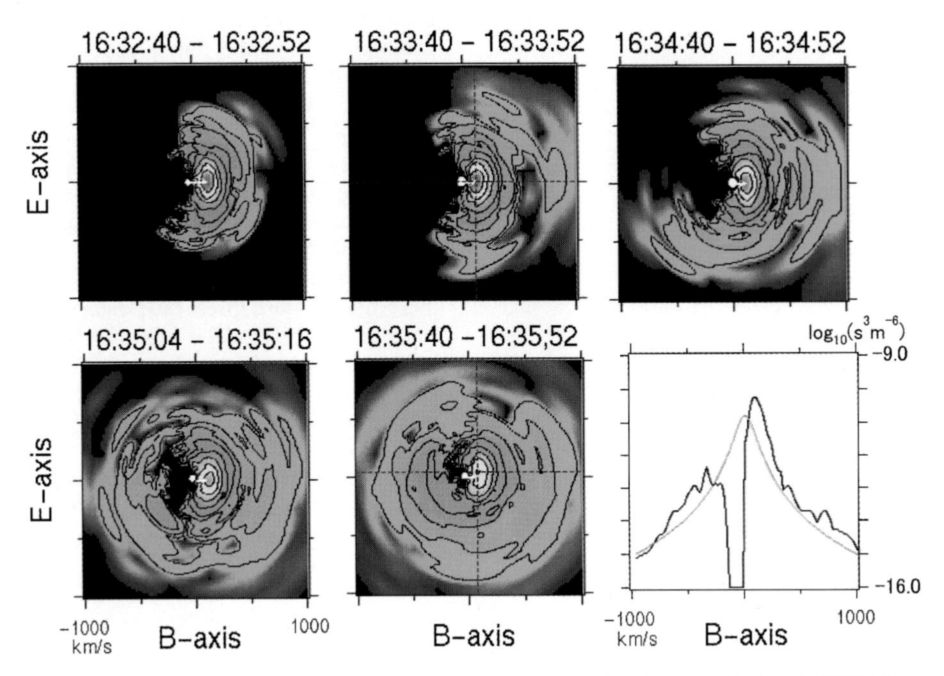

Figure 10. Evolution of the ion distribution function in velocity space for the period when GEOTAIL was crossing the mantle/plasma-sheet interface from 16:32 to 16:35 UT on 9 December 1993. Time passes from left to right in the upper row and similarly in the bottom row, except for the last diagram. The last diagram gives a plot of the ion distribution function in the final phase, expressed as a plot of the logarithm of the phase-space density (mks unit) along a line parallel to the B axis.

The top left panel in Figure 10 shows the color-coded contour map of the ion distribution function measured at 16:32:40 UT when GEOTAIL was located in the mantle. Since this is a typical example of a mantle-type ion distribution function, two other orthogonal cuts are given in Figure 11, to obtain three-dimensional information. The plots of the phase space density along lines parallel to the B and C axes passing through the density peak are also given at the bottom of the figure. The green lines indicate the background count level. White areas are out of sight for the detector. The mantle ions have two salient features. First, the distribution in the velocity space is roughly hemispherical and there are practically no ions going sunward. This is consistent with the following interpretation: (1) mantle ions entered the magnetosphere far upstream, (2) they are now flowing anti-sunward, and (3) they have never been reflected back because the mantle field lines are open. The second important characteristic of mantle ions is the temperature anisotropy; the mantle ions are generally gyrotropic in the E-C plane, but in the B-E (and B-C) plane they are highly anisotropic with the perpendicular temperature appreciably higher than the parallel temperature. This anisotropy is clearly seen in the color-coded contour maps, and in the bottom two panels of Figure 11, where we give the cuts of ion distribution function along the B and C axes, respectively.

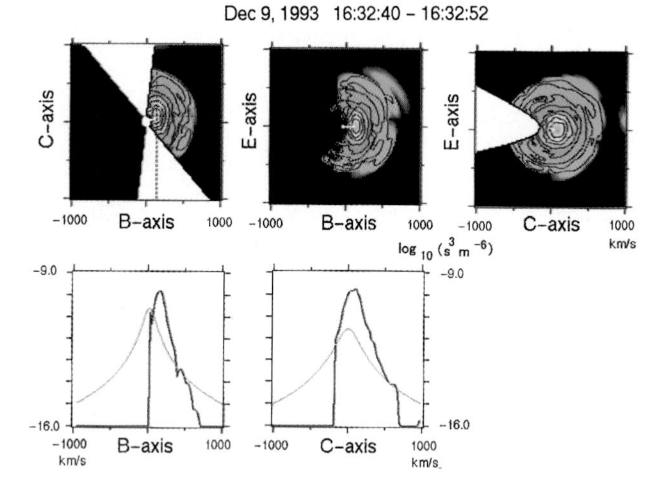

Figure 11. Three orthogonal cuts of the ion distribution function in velocity space for the period from 16:32:40 to 16:32:52 UT (corresponding to the first panel of Figure 10). The three panels show cuts perpendicular to the E, C, and B axes, respectively. All the cuts contain the peak of the distribution function. The sunward direction is pointing to the left. White areas are out of the field of view of the instrument. The profiles of the distribution function along the lines passing through the location of the peak in the directions parallel to B and C are shown at the bottom. It is seen that the ion temperature perpendicular B is appreciably higher than that parallel to B.

The magnetic field and plasma density plots in Figure 9 show that the satellite entered the plasma sheet from the mantle during this time interval. We now study how the transition in the ion bulk parameters is reflected in the time sequence of ion velocity space-distribution maps. Comparison of the five panels given in Figure 10 shows that the transition started by the appearance of anti-sunward moving ions that have energies appreciably higher than the mantle ions (in the middle panel in the upper row). Then these more energetic (>1 keV) ions become gradually more isotropic forming ring-type distributions. At the same time, the peak of the distribution formed by mantle-type ions become less prominent, and the hemispherical distribution of lower energy mantle ions disintegrates to form a more isotropic distribution at middle energies. In other words, the mantle ions are mixing with higher energy ions to form a new unified distribution, which has a temperature higher than that of the original mantle plasma. It is important to notice that (1) the mantle ions are still recognizable at the peak of the new distribution, (2) the mantle plasma is heated during the process, and (3) the sunward hemisphere of the distribution in velocity space, which was initially void, is filled with both high and middle energy ions in the last sample (middle bottom). The cut through the peak of the distribution along the B axis for the last sample is shown at the bottom right.

These examples show that the mantle plasma provides a significant mass supply to the distant plasma sheet at the mantle-plasma sheet interface. We believe that this accounts for the connection of the mantle-lobe and plasma sheet ridges seen in the phase space diagram given earlier.

There remains a question regarding the origin of the higher energy ions which become mixed with the mantle ions at the mantle-plasma sheet interface. Although we do not have a general answer to this question, there are cases where their origin can be identified with reasonable certainty. The second series of samples shown in Figure 12 gives such an example. These samples were measured when GEOTAIL was exiting the plasma sheet during the time interval 18:10 to 18:14 UT (labeled B in Figure 9). The order of the presentation of samples is opposite to the actual time sequence to facilitate a direct comparison with Figure 10 for which the satellite was entering the plasma sheet. Otherwise the format of the presentation is exactly the same as for the previous case shown in Figure 10.

It is seen in the top left panel of Figure 12 that initially the mantle ions had a velocity space distribution very similar to that of the mantle ions displayed earlier in Figures 10 ad 11. They are characterized by a hemispherical distribution with a perpendicular temperature higher than the parallel temperature. In the present case, however, the transition from the mantle to the plasma sheet begins with the sudden appearance of an earthward-directed ion beam, which has energies appreciably higher than that of mantle ions. The speed of the ion beam is estimated to be 600 - 700 km s^{-1} earthward. The combination of the tailward moving mantle population and

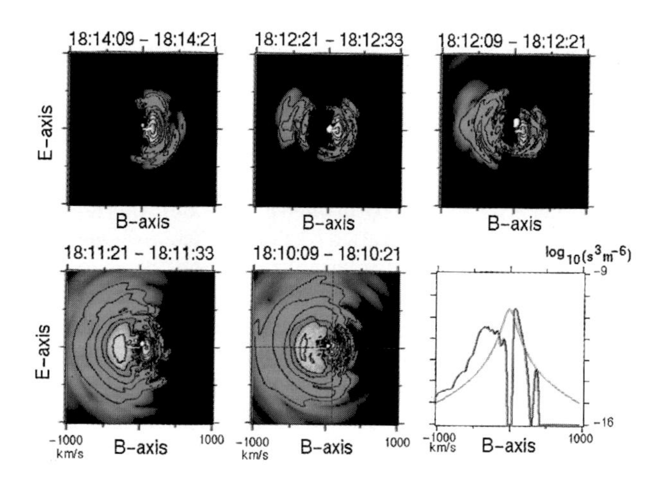

Figure 12. Evolution of the ion distribution function in velocity space is presented in reverse time order for the period 18:14 to 18:10 UT on 9 December 1993, when GEOTAIL was crossing the mantle/plasma-sheet interface. All the colored diagrams show a cut along the E-B plane passing through the distribution peak.

the earthward directed ion beam constitutes an counter streaming signature in velocity space (middle and right panels in the upper row). This kind of counter streaming ion signature is often observed in the near-earth and middle tail regions [*Hoshino et al.,* 1996; *Kawano et al.,* 1996], and has been interpreted to be the signature of magnetic reconnection. As a matter of fact, magnetic reconnection would be the only mechanism capable of producing 600-700 km s^{-1} ions jetting along the field line.

The distribution function obtained in the mixing phase (bottom two panels to the left) is also somewhat different from the previous case. Now the earthward-moving ions occupy a significant part of the distribution, though the distribution peak is still at the point originally occupied by mantle ions. In fact, the anti-sunward hemisphere in velocity space is not totally filled. The bulk velocity calculated for the last sample is about 200 km s^{-1} earthward, showing that the mixed population as a whole is moving earthward at this distance, x=-183 Re downtail.

If we assume that the earthward-directed ion beam measured at the mantle-plasma sheet interface was produced by magnetic reconnection, two important conclusions can be deduced from the observation. First, magnetic reconnection can occur in the magnetotail even during magnetically quiet times for which the IMF is directed northward. Second, judging from the direction of the beam, an active neutral line is present tailward of the position x=-183 Re at least for certain cases. We will support these conclusions on a more statistical basis in the discussion section. Here we only note that, if the magnetic reconnection is actually occurring, it provides an easy way to account for the transition in the ion thermal state from that of mantle-like ions to plasma-sheet like ions.

Mixing of the local mantle plasma with the accelerated ion beams will occur if the mantle magnetic field lines are reconnected in the distant plasma sheet.

6. DISCUSSION

The material presented in this paper confirms that the magnetic field tension exerted by the reconnected IMF field lines plays an important role in determining the shape and structure of the distant magnetotail. Particularly, the GEOTAIL results clearly show that the Y component of the IMF causes a twist of the tail current sheet. We stress that the current sheet position was obtained globally as a demarcation line between the regions of north and south tail field polarities. Therefore, the results are free from short time-scale and/or local variations in the current sheet orientation, which might have affected other methods, though the basic results are in agreement with the previous ones.

The large tilt angle found for northward IMF shows that the IMF strongly interacts with the Earth's magnetosphere even when it is directed northward. Further, the B_y-dependent torque and the presence of the mantle plasma suggest that the plasma mantle extends into the distant magnetotail, forming open field lines loaded with tailward moving plasma. In this case, lobe field lines do not have both ends in the ionosphere as predicted by some merging models for northward IMF. Our results are basically consistent with the open lobe reconnection model first suggested by *Russell*[1972], which has been expanded and applied to ground observations by *Maezawa* [1976], and later supported by low-altitude satellite observations [e.g. *Burke et al.*, 1979; for related theoretical models, see *Crooker*, 1979; *Reiff and Burch*, 1985]. It is noted that the closed field line reconnection models for the northward IMF such as the one first introduced by *Dungey* [1961] require almost due northward IMF. A due northward IMF may occur for individual cases, but statistically the probability of having a due northward IMF is much lower than that for having a less strictly northward IMF. Therefore, it is expected that the effect of Dungey type (closed lobe field line) reconnection processes would not show up in our statistics of northward IMF even though it may exist.

We do not have a definite answer to the question of why the tilt angle of the current sheet should be larger for northward IMF than for southward IMF. A possible reasoning would be that the tilt angle is determined by the total torque exerted by the reconnected IMF, and this torque should be integrated over the tail surface. Now the integrated torque is a function of at least two factors, i.e. the total amount of reconnected field lines, and the force a unit amount of reconnected field lines exert on the tail surface in the direction tangent to the tail boundary. The former quantity, which is proportional to the normal component of the magnetic field, B_n, integrated over the tail surface, is believed to be larger for southward IMF than for northward IMF. However, the latter quantity (the tangential force for a unit number of field lines) is dependent on the clock angle of the IMF. To fix ideas, let us take the case of positive IMF B_y. For the north dawn and south dusk quadrants where larger numbers of reconnected field lines are being convected, the draped IMF has a larger component B_t tangential to the magnetopause for northward IMF as compared to southward IMF. Therefore, when B_n is fixed, the tangential stress (B_nB_t/μ_0) is larger when the IMF is directed northward. Thus the two effects (B_n and B_t) compete with each other, and the observation indicates that the latter effect actually dominates.

Our statistics showed that the distant magnetotail plasma consists of plasma sheet-type and mantle-type plasmas. From the normalized density map in the tail cross section, we showed that the distant mantle occupies a significant portion of the tail lobe. This means that the major portion of the lobe plasma originates from the plasma mantle. The severely twisted position of the mantle confirms that the mantle really represents the region where the reconnected IMF field lines are exerting a torque on the magnetotail. The twisted position of the plasma mantle has an important implication for the supply mechanism of plasma to the distant plasma sheet. In fact, from the phase diagram and the analysis of individual ion distribution functions, we showed that part of the mantle plasma can get direct access to the plasma sheet, be mixed with the plasma sheet plasma, and be heated. Since we showed that magnetic connection with the IMF is an almost permanent feature of the magnetotail regardless of the north-south polarity of the IMF, we expect that the same supply mechanism is working both for northward and southward IMF conditions.

It is interesting to see how the mixture of the mantle and the plasma sheet plasmas is realized. We propose here that such mixing is produced by magnetic reconnection which is occurring in the distant plasma sheet even when the IMF is directed northward, i.e. even during magnetically quiet times. Since the mantle location is in juxtaposition with or even overlaps with the current sheet, as shown schematically in Figure 8, the mantle field lines are involved in the reconnection process and the mantle plasma can be accelerated and heated. Individual examples of ion distribution functions suggesting the occurrence of such acceleration/heating on the field lines threading the mantle has been shown in Section 5.

In a global, steady-state reconnection model of the magnetosphere, it has been customary to assume the presence of a distant neutral line. At this distant neutral line, open field lines of the north and south lobes will be reconnected, and the resulting closed field lines will return to the dayside magnetosphere, in order to balance the loss of dayside closed field lines due to dayside reconnection. We note that this idea works for the case of southward IMF, when magnetic reconnection occurs with dayside closed field lines. On the other hand, the idea may not be applicable to the case of northward IMF where magnetic reconnection may occur on the open field lines tailward of the cusp, and the amount of dayside closed field lines will not change. Thus it is interesting to see if the distant neutral line exists during northward IMF. The distant neutral line has been inferred to exist at least for active times [*Slavin et al.*, 1985; *Nishida et al.*, 1996]. Specifically, *Nishida et al.* [1996] estimated that the distant neutral line is located at about x =-140 Re for active times (K_p=4 on average). However, its presence during the periods of northward IMF has not been clarified yet. Both ISEE-3 [*Slavin et al.*, 1987; *Heikilla*, 1988] and GEOTAIL [*Yamamoto et al.*, 1994; *Nishida et al.* 1995] observations indicate that the average value of the north-south component B_z of the magnetic field at the tail current sheet is positive for prolonged periods of quiet geomagnetic activity. These observations might suggest that, on average, no neutral line exists at least up to a distance of 200 Re during prolonged quiet periods. They noted further that positive B_z is frequently accompanied by a tailward movement of plasma, indicating a net tailward loss of Earth's closed field lines during quiet times, which is of course puzzling.

In the above discussion, the average value of B_z at the current sheet has played an important role. The net magnetic flux crossing the current sheet is inferred from the average value of B_z at the current sheet. If the average B_z is negative, it is taken to indicate that a magnetic neutral line exists earthward of the position where B_z is measured, and the field lines there are not connected to the Earth.

The topology of field lines in the tail has thus been discussed mainly on the basis of the statistics of B_z in the GSM coordinate system. However, when the current sheet is twisted, the magnetic field component perpendicular to the actual current sheet should be used instead of B_z [*Maezawa et al.*, 1997; *Nishida et al*, 1998]. In fact, when the current sheet is tilted, a simple coordinate transformation yields

$$B\perp = B_z \cos\theta - B_y \sin\theta \quad (1)$$

where $B\perp$ is the magnetic field component perpendicular to the current sheet, B_y and B_z are the y and z field components in GSM coordinates, respectively, and θ represents the tilt angle of the current sheet (positive clockwise seen looking towards the sun). $B\perp$ in equation (1) is just the z component of the magnetic field expressed in the neutral sheet coordinate system defined earlier. $B\perp$ tends to be smaller than B_z since both the tilt angle θ and B_y in the tail are positively correlated with IMF B_y [*Maezawa et al.*, 1997; see also *Nishida et al*, 1998]. In the following, we calculate $B\perp$ using the average values of the twist angle obtained for the northward and southward IMF cases separately.

A comment is needed for the zero level of magnetic field measurements made with the MGF instrument. The uncertainty in the zero level is at most 0.5 nT for the magnetic field component parallel to the satellite spin axis, which is almost parallel to the GSE z axis [*Kokubun et al.*, 1994]. Uncertainties in the zero level of other components are negligibly small. In order to estimate the zero level of the GSE z component measured by GEOTAIL, we made a calibration analysis by statistically comparing hourly averaged values of the GSE z component of the IMF measured by GEOTAIL and by IMP-8, for the periods when both satellites were upstream of the bow shock. We found that the GSE z component measured by GEOTAIL is consistently larger than that measured by IMP-8 by 0.33 nT. Considering the fact that a vanishingly small (0.07 nT) long-term (> 20 years) average has been obtained for IMF B_z values measured by IMP-8, suggesting that the offset on the part of IMP-8 is very small (*see also Ahluwalia et al.*, 1996), we subtracted 0.33 nT from the GSE B_z values measured by GEOTAIL. It is stressed, however, that the present estimate is still preliminary and a more official calibration analysis is being made by the GEOTAIL MGF team.

Figure 13 shows a scatter plot of $B\perp$ versus the x component, V_x, of the ion bulk velocity for all the data points belonging to the plasma sheet at x<-150 Re. Small plus signs represent the individual data points (1 minute values) and large open circles represent the bin averages for 50 km s^{-1} wide bins. Several things are clear from this figure. First, at x<-150 Re, most data points have negative V_x, showing that the flow in the plasma sheet is predominantly anti-sunward [*Slavin et al.*, 1985]. Second, the anti-sunward velocity is weaker on average for the data points with positive $B\perp$ than for the ones with negative $B\perp$. In fact, although there is much scatter in the data, the bin averages of V_x show a step function-like (or arctangent-function like) behavior around $B\perp$=0. Averages of V_x for large negative values of $B\perp$ are close to -570 km s^{-1}, while averages of V_x for large positive values of $B\perp$ are about -220 km s^{-1}, a transition smoothly occurring in between. This kind of step function-like behavior of V_x is just what we would expect for flows accelerated in the reconnection process. We only have to recall the fact that magnetic reconnection generally produces plasma jets in opposite directions, one with positive $B\perp$ and the other with negative $B\perp$. The speed of the jets would be approximately the Alfven speed in the tail lobe, in either direction. Therefore, if we randomly sample the resultant flows along the x-axis we will observe the step-function like dependence of V_x on $B\perp$. This interpretation of the observed relationship between V_x and $B\perp$ has two important implications. First, since the overall average of V_x is negative (-395 km s^{-1}), the reconnection line itself should be moving in the anti-sunward direction with the speed of the order of 400 km s^{-1} relative to the Earth. This point will shortly be discussed. Second, we note that the lobe Alfven speed of $\pm$175 km s^{-1} is not unreasonable if the reconnected lobe field lines are filled with mantle plasma with number density ~1 cm^{-3}. Third, the fact that both positive and negative $B\perp$ can be observed at x<-150 Re means that the reconnection line

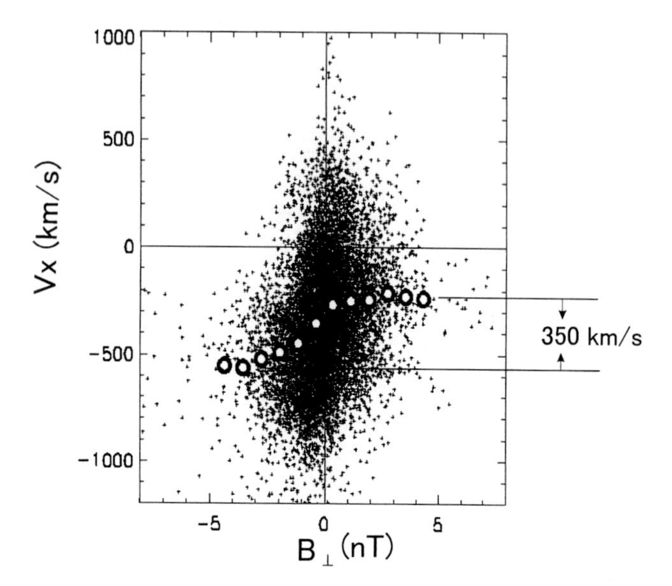

Figure 13. Scatter plot of the x component, V_x of the ion bulk velocity versus the magnetic field component $B\perp$ perpendicular to the current sheet, for all the data points belonging to the plasma sheet at x<-150 Re. Open circles represent bin averages for 0.5 nT bins.

may either be earthward or tailward of the satellite at this distance. In other words, a model of a distant neutral line located at a certain fixed position in the tail does not fit these observations. This last point is not so surprising since at least for geomagnetically active times, we know that neutral lines can move in the x direction when plasmoids are formed. Thus the data for southward IMF may include both plasmoid and distant neutral line effects.

As we stated earlier, we are interested in the field line topology for periods of northward IMF since from the theoretical point of view there is then no need for the presence of a distant neutral line if magnetic reconnection of dayside close field lines ceases. Further, contamination from plasmoids would be practically negligible for quiet times. The scatter plot of V_x versus $B\perp$ for the cases of $Kp \leq 1$ is shown in Figure 14. Although the number of data points is small, the overall distribution is very similar to that for Figure 13, with the same step function-like dependence of V_x on $B\perp$. The difference between the bin averages of V_x for positive and negative values of $B\perp$ is about 330 km s^{-1}, which is comparable to 395 km s^{-1} obtained in Figure 13 for all Kp (The difference may be attributable to higher solar wind speeds generally related to disturbed conditions). The dependence of V_x on $B\perp$ strongly suggests that magnetic reconnection is occurring in the distant plasma sheet even during quiet times. Since the effects of plasmoids are minimal, we conclude that magnetic reconnection is occurring locally in the distant plasma sheet. Further, the same reasoning as stated earlier suggests that the reconnection line is moving tailward and that it can occur either earthward or tailward of the satellite location examined in this paper (the x range of the satellite location examined in this paper is -150 > x > -220 Re).

We note that the overall average of $B\perp$ for the entire set of plasma sheet data shown in Figure 13 is –0.05 nT, while that for quiet times is –0.06 nT. Thus the average $B\perp$ is slightly negative for both data sets, and its absolute value is very small, while positive B_z values in GSM coordinates were reported for quiet times [*Slavin et al.*, 1987; *Yamamoto et al.*, 1994; *Nishida et al.*, 1994; *Nishida et al.*, 1995]. The small absolute values of averaged $B\perp$ may either be due to the random location of the neutral line or to the absence of the reconnection process itself. Since we have seen evidence for reconnection for individual cases, we interpret the negligibly small $B\perp$ value as being due to the random location of the neutral line relative to the satellite for the case of northward IMF. Another clue concerning magnetic reconnection can be obtained from data of the V x B electric field. Figure 15 shows the statistics of the y component (E_y) of the V x B electric field in neutral sheet coordinates as a function of the GSM B_z component of the IMF. Only plasma sheet data points are plotted. Open circles represent bin average for every 0.5 nT bin. It is seen that E_y is positive on average practically for all ranges of IMF B_z, though E_y for northward IMF is much less than that for southward IMF. Positive values of E_y show that the electric field in neutral sheet coordinates is directed from dawn to dusk, so that in a steady state the closed field lines produced by reconnection will return to the dayside magnetosphere for practically all ranges of IMF Bz (in the average sense). (We note, however, that this is not necessarily the case for the non-steady-state model that

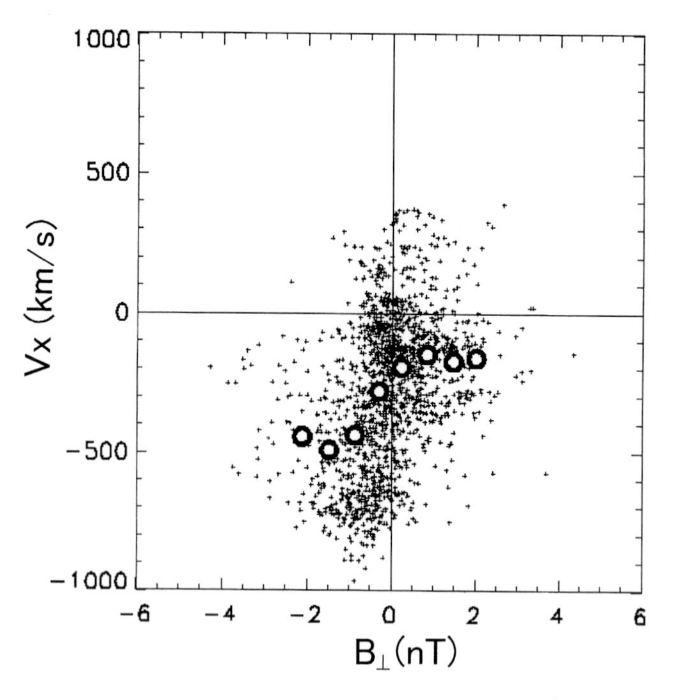

Figure 14. Same as Figure 13 but for the case of $Kp \leq 1$.

will be discussed in the next paragraph. In such cases, the electric field is not uniform in the rest frame of the Earth. See discussion in the next paragraph.)

A model of the distant neutral line which is consistent with all of these observed features is shown in Figure 16. The most important point of this model is that reconnection occurs on the field lines threading the mantle, so that the field lines to be reconnected are significantly loaded by mantle plasma moving tailward. In this case, the reconnection line would not be able to stay in a fixed position relative to the Earth but would drift tailward together with the mantle plasma since it is the only frame of reference for the reconnected plasma. The movement of the distant neutral line is schematically shown in time sequence in panels (a) to (c). We speculate that when the reconnection line has moved to a sufficiently tailward point, a new reconnection line is formed (panel d), and a new cycle begins with panel (a). It should be noted that in this model most of the reconnected field lines are convected tailward in the rest frame of the Earth, though part of the closed field lines produced by reconnection may convect earthward if the lobe Alfven speed exceeds the speed of the mantle plasma. (A detailed examination of Figures 13 and 14 suggests that such earthward motion in the rest frame of the Earth does occur for a limited number of individual cases.) This limited possibility of an earthward flow in the rest frame of the Earth is indicated by dotted arrows in Figure 16. Thus this distant tail model is consistent with the open lobe reconnection model for the northward IMF, in that the amount of closed magnetic flux in the dayside magnetosphere is almost conserved during northward IMF. The large scatter in V_x noted in Figure 13 may be explained by diversity in the speed of the mantle plasma at the

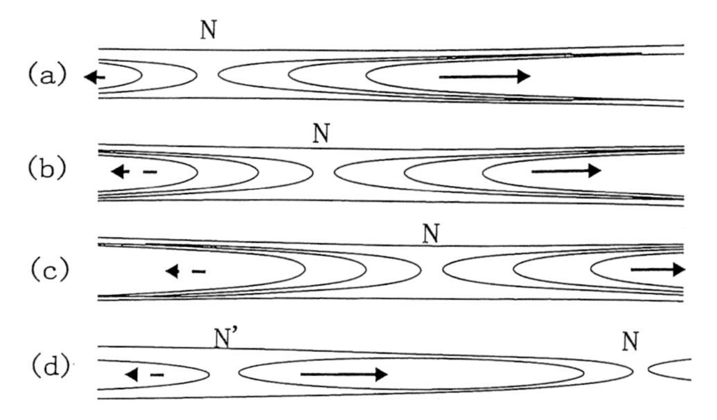

Figure 16. A model of the drifting distant neutral line. The figure represents a cut parallel to the xz plane in neutral sheet coordinates. Since magnetic field lines to be reconnected are loaded with mantle plasma, the reconnection line is not able to stay in a fixed position relative to the Earth but drifts tailward together with the mantle plasma. The movement of the distant neutral line is schematically shown in time sequence in panels (a) to (c). We speculate that when the reconnection line has moved to a sufficiently tailward point, a new reconnection line is formed (panel d), and a new cycle begins with panel (a).

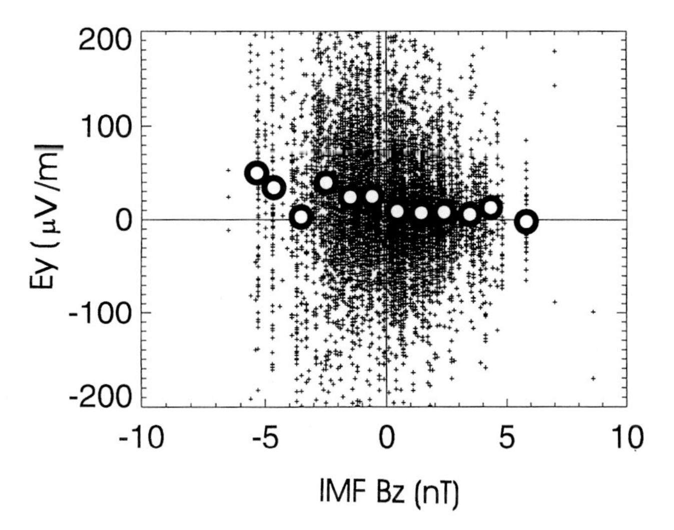

Figure 15. Scatter plot of the y component (E_y) of the V x B electric field in neutral sheet coordinates versus the GSM B_z component of the IMF. Only plasma sheet data for x <-50 Re are shown. Open circles represent bin averages for 0.5 nT bins.

mantle/plasma sheet interface, and a diversity in the lobe (mantle) Alfven speed. Finally, we note that the average electric field should be directed from dawn to dusk (in neutral sheet coordinates) in this model. This is because the tailward movement of the entire reconnection geometry produces negative and positive Ey on the earth and tail sides of the reconnection line, respectively, but these oppositely directed fields tend to cancel each other when they are time averaged at a fixed spatial point. (We assume randomness of the relative location of the neutral line with respect to the satellite). As a result, only the dawn-to-dusk electric field produced by reconnection should stand out in neutral sheet coordinates, consistent with observations.

We note that the difficulty raised earlier concerning the tailward loss of closed field lines from the Earth during quiet times [*Nishida et al.,* 1995] has been removed by the appropriate use of the neutral sheet coordinate system. This point has been discussed by *Nishida et al.* [1998] and *Nishida and Ogino* [1998] from somewhat different viewpoints.

Our model predicts that, at least in the distant tail, density and temperature of plasma sheet plasma is maintained by reconnection of field lines loaded by mantle plasma during quiet times. In this connection, our model may have an important implication for the plasma supply to the near-Earth plasma sheet during quiet times. *Terasawa et al.* [1997] recently showed that during prolonged quiet periods with northward IMF polarity, the near-Earth plasma sheet tends to be filled by a dense, cold plasma. We note that the ions ejected earthward from the tailward-moving neutral line shown in Figure 16 will have relatively low energy since the earthward momentum gained through reconnection is canceled to a large extent by the initial tailward motion of the mantle plasma. Further, the VxB electric field observed in the distant tail is not inconsistent with the idea that closed magnetic field lines produced there by reconnection during quiet times can return to the Earth under limited conditions. Further, *Nishida et al.* [1998] recently reported that the plasma flows in the *near-Earth* neutral sheet ($-19 > x > -29$ Re) are earthward on average during quiet times. Thus if the ions reach the near-earth plasma sheet with their low earthward speeds (in directions both parallel and perpendicular to B), the near-earth plasma sheet will be gradually filled with low-energy ions. The plasma sheet ion density for such periods could be larger than that for the disturbed times, because the twisted mantle position is closer to the plasma sheet during northward IMF, and therefore the low-latitude mantle field lines would contain denser ions. Thus our model is consistent with the observation of *Terasawa et al* [1997]. Finally, we note that the solar wind ion access to the near-Earth plasma sheet along the reconnected field lines has been discussed by *Fujimoto et al.* [1996] in terms of the formation mechanism of the outer LLBL. The relationship between the formation mechanism of the mantle and the LLBL will be an interesting topic to be pursued in future.

7. SUMMARY

The principal characteristics of the distant magnetotail structure as revealed by GEOTAIL observations can be summarized as follows:

(1) The distant tail current sheet ($x< -150$ Re) tilts towards the dawn (dusk) side when IMF B_y is positive (negative) both for northward and southward IMF cases. This is consistent with the fact that the torque on the tail is determined basically by the sign of IMF B_y,

(2) The tilt of the current sheet for northward IMF is more than twice as large as that for southward IMF. The average tilt angle for IMF $B_z >0$ is 32°, while the average for IMF $B_z <0$ is 15°.

(3) The distant tail plasma can be roughly divided into lobe/mantle and plasma sheet components according to their temperature. The plasma sheet component is characterized by higher temperatures and occurs along a belt containing the tilted current sheet.

(4) The mantle component, characterized by a cool, dense plasma of solar wind origin, occurs in a wide region adjacent to the magnetopause. The mantle occurs at low latitudes because of the torque exerted by the reconnected field lines.

(5) The mantle is in juxtaposition to the plasma sheet particularly when the IMF is directed northward. This location enables the mantle plasma to mix with the plasma sheet plasma and thereby be heated.

(6) The magnetic field component crossing the current sheet is correlated well with the x component of the plasma velocity, showing that magnetic field reconnection is the driving force for a large part of the plasma flow in the distant magnetotail.

(7) Magnetic reconnection occurs even during geomagnetically quiet times when the IMF is directed northward. Individual examples suggest that reconnection causes the mixing and heating of the mantle plasma at the mantle/plasma-sheet boundary.

Acknowledgements We thank A. Nishida for valuable comments on the manuscript.

REFERENCES

Ahluwalia, H. S., S. S. Xue, and M. M. Fikani, Long term variability of the solar wind speed, in Solar Wind Eight, AIP Conference Proceedings 382, AIP Press, 1996

Burke, W.J., M.C. Kelley, R. C. Sagalyn, M. Smiddy, and S. T. Lai, Polar cap electric field structures with northward interplanetary magnetic field, Geophys. Res. Lett., 6, 21, 1979.

Cowley, S. W. H., On the distribution of B_y in the geomagnetic tail, Planet. Space Sci., 27, 769, 1979.

Cowley, S. W. H., Magnetospheric asymmetries associated with the Y component of the IMF, Planet. Space Sci., 29, 79, 1981.

Crooker, N. U., Dayside merging and cusp geometry, J. Geophys. Res., 84, 951, 1979.

Fujimoto, M., A. Nishida, T. Mukai, Y. Saito, T. Yamamoto, and S. Kokubun, Plasma entry from the flanks of the near-earth magnetotail: GEOTAIL observations in the dawnside LLBL and plasma sheet, J. Geomag.Geoelectr., 48, 711, 1996.

Fujimoto, M., T. Mukai, H. Kawano, M. Nakamura, A. Nishida, Y. Saito, T. Yamomoto, and S. Kokubun, A GEOTAIL observation of low-latitude boundary layer, Adv. Space Res., 20, 813, 1997.

Gosling, J.T., D. N. Baker, S. J. Bame, W. C. Feldman, R. D. Zwickl, and E. J. Smith, North-south and dawn-dusk plasma asymmetries in the distant tail lobes: ISEE 3, J. Geophys. Res., 90, 6354, 1985.

Gosling, J. T., M.F. Thomsen, S. J. Bame, abd R. C. Elphic, Observation of reconnection of interplanetary and lobe magnetic field lines at the high-latitude magnetosphere, J. Geophys. Res., 96, 14,097, 1991.

Heikkila, W. J., Current sheet crossings in the distant magnetotail, Geophys Res. Lett., 15, 299, 1988.

Hoshino, M., T. Mukai, A. Nishida, Y. Saito, T. Yamamoto, and S. Kokubun, Evidence of two active reconnection sites in the distant magnetotail, J. Geomag. Geoelectr., 48, 515, 1996.

Kawano, H., A. Nishida, M. Fujimoto, T. Mukai, S. Kokubun, T. Yamamoto, T. Terasawa, M. Hirahara, Y. Saito, S. Machida, K. Yumoto, H. Matsumoto, and T. Murata, A quasi-stagnant plasmoid observed with Geotail on October 15, 1993, J. Geomag. Geoelectr., 48, 525, 1996.

Kessell, R. L., S. -H. Chen, J. L. Green, S. F. Fung, S. S. Boardsen, L. C. Tan, T. E. Eastman, J. D. Craven, and L. A. Frank., Geophys. Res. Lett, 23, 583, 1996.

Maezawa, K, Magnetospheric convection induced by the positive and negative z components of the interplanetary magentic field: Quantitative analyis using polar cap magnetic records, J. Geophys. Res., 81, 2289, 1976.

Maezawa, K. T. Hori, T. Mukai, Y. Saito, T. Yamamoto, S Kokubun, and A. Nishida, Structure of the distant magnetotail and its dependence on the IMF B_y component: GEOTAIL observations, Adv. Space Res., 20, 949,1997.

Macwan, S. E., A determination of twisting of the Earth's magnetotail at distances 115-220 Re: ISEE 3, J. Geophys. Res., 97, 19239, 1992.

Mukai, T, S. Machida, Y. Saito, M. Hirahara, T. Terasawa, N. Kaya, T. Obara, M. Ejiri, and A. Nishida, The low energy particle (LEP) experiment onboard the GEOTAIL satellite, J. Geomag. Geoelectr., 46, 669, 1994

Nakamura, R., and S. Kokubun, Changes in the distant tail configuration during geomagnetic storms, J. Geophys. Res, 102, 9587, 1997..

Nishida, A., T. Yamamoto, K. Tsuruda, H. Hayakawa, A. Matsuoka, S. Kokubun, and M. Nakamura, Structure of the neutral sheet in the distant tail (x=-210 Re) in geomagnetically quiet times, Geophys. Res. Lett., 21, 2951, 1994.

Nishida, A., T. Mukai, T. Yamamoto, Y. Saito, S. Kokubun, and K. Maezawa, GEOTAIL observation of magnetospheric convection in the distant tail at 200 RE in quiet times, J. Geophys. Res., 100, 23663, 1995.

Nishida, A., T. Mukai, T. Yamamoto, Y Saito, and S. Kokubun, Magnetotail convection in geomagnetically active times 1. distance to the neutral line, J. Geomag. Geoelectr., 48, 489, 1996.

Nishida, A., T. Mukai, T. Yamamoto, S. Kokubun and K. Maezawa, A unified model of the magnetotail convection in geomagnetically quiet and active times,J. Geophys Res., in press, 1998.

Nishida, A. and T. Ogino, Convection and reconnectin in the Earth's magnetotail, this volume, 1998.

Owen, C. J., J. A. Slavin, I. G. Richardson, N. Murphy, and R. J. Hynds, Average motion, structure, and orientation of the distant magnetotail determined from remote sensing of the edge of the plasma sheet boundary layer with E> 35 keV ions, J. Geophys. Res., 100, 185, 1995.

Reiff, P. H., and J. L. Burch, IMF B_y-dependent plasma flow and Birkelandcurrents in the dayside magnetosphere 2, A global model for northward and southward IMF, J. Geophys. Res., 90, 1595, 1985.

Russell, C. T., The configuration of the magnetosphere, in Critical Problems of Magnetospheric Phyiscs, Proceedings of the joint COSPAR/IAGA/URSI symposium, IUCSTP Secretariat c/o National Academy of Sciences, 1972.

Sibeck, D. G., G. L. Siscoe, J. A. Slavin, E. J. Smith, B. T. Tsurutani, and R. P. Lepping, The distant magnetotail's response to a strong interplanetary magnetic field B_y: twisting, flattening, and field line bending, J. Geophys. Res., 90, 4011, 1985.

Sibeck, D. G., J. A. Slavin, E. J. Smith, and B. T. Tsurutani, Twisting of the geomagnetic tail in Solar Wind-Magnetosphere Coupling, Terra Publishing Company, 731-738, 1986a.

Sibeck, D. G., G. L. Siscoe, J. A. Slavin, and R. P. Lepping, Major flattening of the distant gomagnetic tail, J. Geophys. Res., 91, 4223, 1986b.

Siscoe, G. L., F. L. Scalf, D. S. Intriligator, J. H. Wolfe, J. H. Binsack, H. S. Bridge, and V. M. Vasyliunas, Evidence for a geomagnetic wake at 500 Earth radii, J. Geophys Res., 75, 5319, 1970.

Siscoe, G. L., and E. Sanchez, An MHD model for the complete open magnetotail boundary, J. Geophys. Res., 29, 7405, 1987.

Siscoe, G. L., L. A. Frank, K. L. Ackerson, and W. R. Paterson, Properties of the mantle-like magnetotail boundary layer: GEOTAIL data compared with a mantle model, Geophys. Res. Lett, 21, 2975, 1994.

Slavin, J. A., E. J. Smith, D. G. Sibeck, D. N. Baker, R. D. Zwickl, and S. -I. Akasofu, An ISEE 3 study of average and substorm conditions in the distant magnetotail, J. Geophys. Res., 90, 10875, 1985.

Slavin, J. A., P. W. Daly, E. J. Smith, T. R. Sanderson, K. -P. Wenzel, R. P. Lepping, and H. W. Kroehl, Magnetic configuration of the distant plasma sheet: ISEE 3 observations, in Magnetotail Physics (ed. A. T. Y. Lui), The John Hopkins University Press, 59, 1987.

Terasawa, T., M. Fujimoto, T. Mukai, I. Shinohara, Y. Saito, T. Yamamoto, S. Machida, S. Kokubun, A. J. Lazarus, J. T. Steinberg, and R. P. Lepping, Solar wind control of density and temperature in the near-Earth plasma sheet: WIND/GEOTAIL collaboration, Geophys. Res. Lett, 24, 935, 1997.

Tsurutani, B. T., B. E. Goldstein, M. E. Burton and D. E. Jones, A review of the ISEE-3 geotail magnetic field results, Planet. Space Sci., 34, 931, 1986.

Yamamoto, T., A. Matsuoka, K. Tsuruda, H. Hayakawa, A. Nishida, M Nakamura, and S. Kokubun, Dense plasmas in the distant magnetotail as observed by GEOTAIL, Geophys. Res. Lett., 21, 2879, 1994.

K. Maezawa, and T. Hori, Department of Physics, Nagoya University, Furo-cho, Chikusa-ku, Nagoya 464-01 Japan

Large-Scale Structure of the Magnetosphere

D. N. Baker and T. I. Pulkkinen[1]

Laboratory for Atmospheric and Space Physics, University of Colorado, Boulder, CO

[1]Permanently at: Finnish Meteorological Institute, Helsinki, Finland

This paper reviews the structure of the magnetotail in terms of the most recent large-scale magnetic field models, with special emphasis on the distant tail properties. Recent observations have revealed that tailward flowing, singly charged ionospheric ions are often present in the distant magnetotail lobes. The velocities of these beamlike populations vary from relatively low ($V \sim 50$ km/s) to quite energetic ($V \geq 500$ km/s). Models of magnetic and convection electric fields together with drift calculations are used to show that the low-energy ion populations tend to reach the distant tail lobes only under relatively weak electric fields. These results suggest that the large-scale cross-tail electric field may be non-uniformly distributed, the strongest fields being in the region where the tail is locally open. On the other hand, during disturbed geomagnetic conditions, the outflowing polar cap and cleft ions can be strongly accelerated parallel to the open magnetic field lines by electric fields set up by ambipolar motion of suprathermal electrons and magnetosheath ions entering the distant tail. This mechanism would allow the commingling of cold sheath ions, bidirectional electrons of solar wind origin, and ionospheric ions of exceptionally high tailward-directed velocity.

1. INTRODUCTION

Understanding the large-scale magnetotail magnetic and electric fields as well as the sources of the ambient plasmas throughout the tail are crucial problems. It is widely accepted that the energy to drive magnetospheric processes comes overwhelmingly from the incident solar wind flow [*Cowley*, 1986]. However, even given this fact, there continues to be debate about the specific mechanisms by which solar wind energy is imparted to the magnetosphere-ionosphere system. The best evidence seems to be that dayside reconnection and magnetic flux transport to the nightside accounts for much – if not most – of the energy to drive magnetospheric dynamics (see *Rostoker et al.* [1987] and references therein). However, an important role of viscous interactions has also been suggested (see *Lundin et al.* [1991] and references therein).

In contrast to the clear consensus for driving of magnetospheric processes by solar wind energy, there is much more debate about primary plasma sources. *Chappell et al.* [1987] have argued that the ionosphere can more than adequately supply all of the ions making up the main particle populations, i.e., the plasma sheet, the plasmasphere, and the tail lobes. Other researchers have also argued for a significant role of ionospheric plasma sources [e.g., *Gloeckler and Hamilton*, 1987; *Daglis et al.*, 1994]. On the other hand, there are clear observational and theoretical arguments that the plasma mantle is of quite direct solar wind origin, and that the mantle supplies considerable mass flux to the plasma sheet and the tail lobes [*Rosenbauer et al.*, 1975].

The magnetic field structure within the magnetospheric cavity is highly time variable, being shaped both by the incident solar wind flow carrying the interplanetary magnetic field (IMF) and by the internal dynamic processes. Magnetospheric missions have shown that the magnetosphere has a coherent structure with two well-defined lobes separated by a current-carrying plasma sheet at least out to ~250 R_E in the anti-solar direction [*Slavin et al.*, 1985; *Yamamoto et al.*, 1994]. Whereas several models describe the statistical average of the inner magnetospheric configuration [*Tsyganenko*, 1987; 1989], only a few models include a description also of the more distant magnetotail [*Pulkkinen et al.*, 1996; *Tsyganenko*, 1995].

As the magnetotail lobes are almost completely current free, the charged particle populations present must be of

New Perspectives on the Earth's Magnetotail
Geophysical Monograph 105

relatively low energy and low density. However, measurements both from the ISEE-3 deep tail mission in 1982-83 and GEOTAIL distant tail passages during 1993-1995 have revealed distinct electron and ion populations that indicate structuring of the tail lobes. The GEOTAIL measurements showed relatively low-energy ion beams of both solar wind and ionospheric origin within the tail lobes [*Mukai et al.*, 1994; *Hirahara et al.*, 1995]. Furthermore, the GEOTAIL observations confirmed the earlier findings by ISEE-3 that the lobes are often populated by low-energy electrons with strong enough field-aligned anisotropies to create a clear bidirectional distribution [*Fairfield and Scudder*, 1985; *Baker et al.*, 1986; 1987; 1997].

This paper gives an overview of the large-scale tail magnetic field structure by comparing and contrasting several of the recently developed field models. Observations of both electrons and ions in the tail lobes are reviewed and compared with corresponding measurements in the polar ionosphere. The processes giving rise to these ion and electron populations are discussed; it is argued that for a large fraction of time these populations must be present simultaneously. A drift model is used to compute trajectories of the cold ion population in the distant regions of the magnetotail, and the results are used to discuss tail electric field structuring.

2. MAGNETIC FIELD OBSERVATIONS AND MODELS

Since the structure of the magnetosphere is strongly delineated by the magnetic fields, knowledge of the instantaneous large-scale magnetic field configuration is important for global studies. However, even in the ISTP era, studies are limited to having a few measurement points at widely separated locations in the magnetosphere. To overcome this difficulty, statistical databases have been collected containing magnetic field measurements of most regions of the magnetosphere, and models have been developed to represent the statistical average of the magnetic field configuration.

A portion of the available magnetic field measurements have been carefully analyzed, averaged over about 30 min or over 0.5 R_E of the satellite trajectory, and tagged with magnetic activity indices, solar wind parameters, and IMF values when available. This database, which now contains more than 79000 data points mostly from inside ~40 R_E [*Fairfield et al.*, 1994], has been used to develop empirical models for the magnetic field [*Tsyganenko*, 1987; 1989; 1995]. The T87 and T89 models give the field values as a function of the dipole tilt and magnetic activity (Kp). The T95 model includes several new field sources (an explicit representation of the magnetopause, the region 1 and region 2 Birkeland current systems, and IMF penetration through the magnetopause) and accepts the tilt angle, IMF B_Y and B_Z, solar wind pressure, Dst, and AE as input parameters. A still newer version (T96) somewhat simplifies the input by leaving out the AE-dependence.

To improve the accuracy of field-aligned mappings and to study the inner-tail current sheet structure during substorms, *Pulkkinen et al.* [1992; 1994] introduced time-varying modifications to the T89 field model. Scaling the intensity of the inner tail currents as well as adjusting the thickness of the tail current sheet based on multipoint measurements in a given substorm event allowed *Pulkkinen et al.* to obtain a large-scale field configuration that accurately represented the temporal evolution during the substorm growth and recovery phases. The model predicted strong thinning of the inner tail current sheet during the substorm growth phase, as well as strong steepening of the radial gradient of the equatorial magnetic field as a function of downtail distance. Both these features affect the stability and dynamics of the current sheet, in addition to changing the magnetic mapping between the ionosphere and magnetosphere.

The left panels of Figure 1 show field lines in the noon-midnight cross-section of the magnetotail for the three models: T89 (top), T96 (middle) and modified T89 for conditions at the end of the growth phase (bottom). The explicit magnetopause surface in the T96 model confines the tail field inside a cylindrical volume with ~30 R_E radius, in the T89 model the field flaring angle is larger than observations would suggest [*Pulkkinen and Tsyganenko*, 1996].

The right panels show the magnetic field normal component at the current sheet center, the lobe field at Z=10 R_E and the current sheet thickness along the tail axis. The T89 model is shown with open diamonds, the T96 model with open triangles, and the growth phase model (modified T89) with solid circles. The B_n profile for the growth phase model indicates that under disturbed conditions the gradient in the X direction is often very large, and that there is a field minimum in the inner magnetotail region. This very rapid transition from quasidipolar to taillike field $\partial B_Z/\partial X$ larger than predicted by the statistical models) is quite frequently observed [*Sergeev et al.*, 1993].

The lobe field variations reflect variability of the magnetic activity; the energy loading during the growth phase increases the inner tail lobe field especially. The current sheet thickness in the T96 model is constant along the tail, whereas in the T89 model it decreases from a larger inner tail value. The growth phase model includes a strong thinning of the current sheet near the inner edge of the plasma sheet [*Pulkkinen et al.*, 1992], and illustrates the significant variations from the statistical average that may occur during active periods. The field line plot of the growth phase model clearly shows the effects of the small D and B_n in the inner tail: the field lines are very stretched and compressed compared to the statistical average.

Magnetic field observations from the distant magnetotail come almost solely from two missions, the ISEE-3 deep tail era during 1982-1983, and the GEOTAIL distant tail passages during 1993-1995. The ISEE-3 observations confirmed that the magnetotail is well-ordered at least up to 240 R_E from the Earth, with distinct lobes separated by a current/plasma sheet [*Slavin et al.*, 1985]. The average lobe field intensity was found to decrease slowly as a function of downtail distance Earthward of about 100 R_E, beyond which it remains relatively constant. The observed plasma sheet field was more variable and the field magnitude was smaller at the greater distances. The B_Z component was shown to reach a value very close to zero at large radial distances. These signatures together with the observations of the flow direction reversal at X ~ -100 R_E led to the interpretation that there was a Y shaped neutral line in that region, which accelerated reconnecting flux tubes both Earthward and tailward from the neutral line [*Scholer et al.*, 1984].

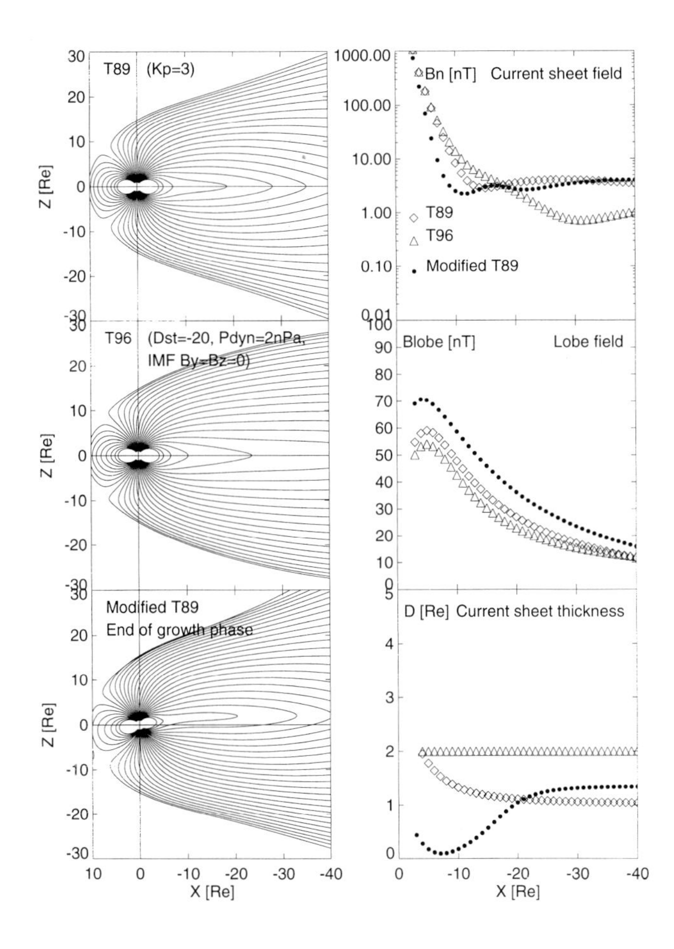

Figure 1. Comparison of models for the inner magnetosphere. Magnetic field lines in the noon-midnight meridian for (a) T89 model, (b) T96 model, and (c) modified T89 model. (d) Magnetic field normal to the current sheet at the current sheet center and at Y=0. (e) Lobe magnetic field at Y=0 and Z=10 R_E. (f) Current sheet thickness at Y=0. The three different curves in (d)-(f) correspond to T89 (open diamonds), T96 (open triangles), and modified T89 (solid circles).

Pulkkinen et al. [1996] used the ISEE-3 measurements to construct an empirical model analogous to the Tsyganenko models in the inner tail. Statistical analyses showed that B_Y is sufficiently small to allow for a two-dimensional field model. The field was modeled by a cross-tail current sheet similar in form to the tail current in T87 model, balanced by return currents at the magnetopause (top panel of Figure 2). The current sheet structure was examined using energetic ion measurements to determine the satellite velocity relative to the current sheet at times when ISEE-3 passed through the entire current sheet in a relatively short period of time [*Pulkkinen et al.,* 1993]. This allowed the estimation of the current sheet thickness, which was found to be quite thin with an average thickness of about 2.5 R_E.

The two bottom panels of Figure 2 show the field at the current sheet center and in the lobe for the T89, T96, and distant tail field models (open diamonds, open triangles, and solid squares, respectively). It is evident that the lobe field in T89 decreases much faster than the two other models; the distant tail model giving the largest lobe field. The normal components of both T89 and distant tail field model remain positive below 1 nT at all distances, whereas the T96 model includes a distant neutral line at about X=-180 R_E.

The more recent GEOTAIL observations largely confirmed the *Slavin et al.* [1985] results: The radial profile of the lobe field was found to be consistent with that found earlier [*Yamamoto et al.,* 1994; *Pulkkinen et al.,* 1996]. The GEOTAIL measurements also confirmed the rather surprising result that B_Z averaged a small but consistently positive value at all radial distances, in apparent contradiction with the *Dungey* [1961] picture of the convecting magnetosphere [*Nishida et al.,* 1996].

Nishida et al. [1997] addressed the issue of the convecting magnetosphere by arguing that reconnection in a magnetotail field twisted by the presence of the IMF B_Y component can account for the northward polarity of the field lines: In the case of northward IMF B_Z, the tailward convecting field lines tailward of the distant neutral line actually do not connect to the Earth, but are attached to the solar wind from both ends. In this case, positive B_Z and tailward V_X would not carry closed magnetic flux tailward as had been suggested before. Under southward IMF conditions, the tail field is less twisted and the B_Z is much more negative, and the plasma flow carries mostly southward

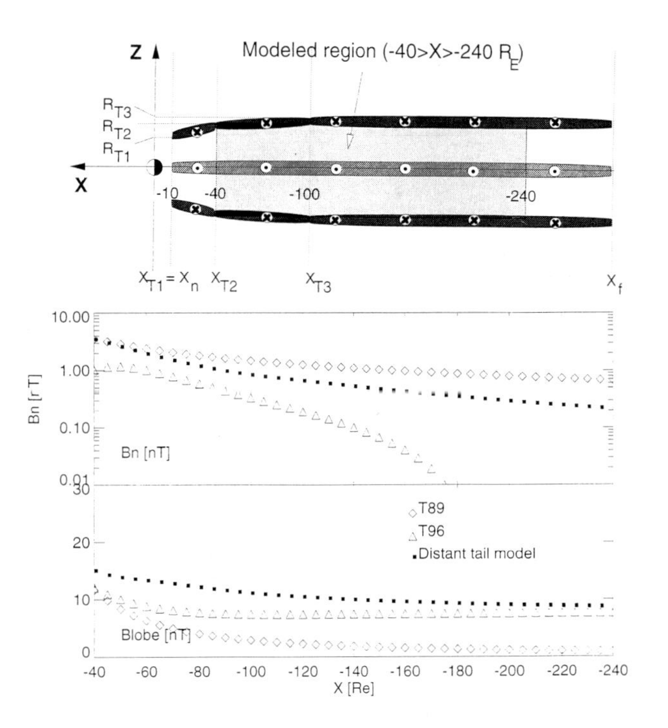

Figure 2. (a) Schematic of the distant tail model. (b) Magnetic field normal to the current sheet at the current sheet center and at Y=0. (c) Lobe magnetic field at Y=0 and Z=10 R_E. The three different curves in (b)-(c) correspond to T89 (open diamonds), T96 (open triangles), and distant tail model (solid squares).

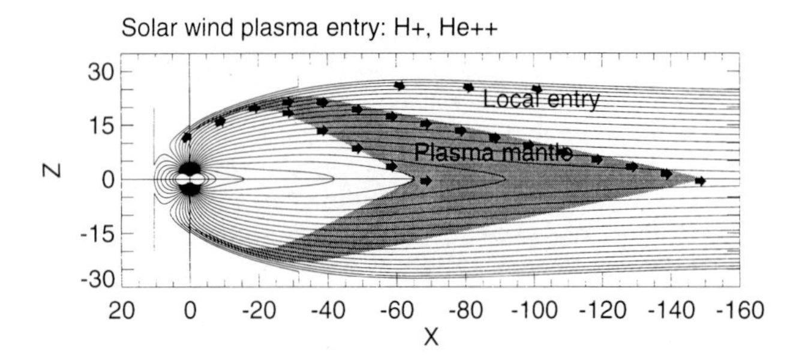

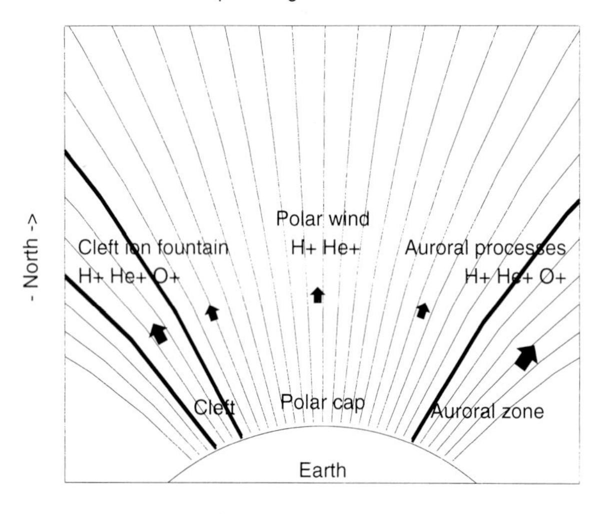

Figure 3. (top) Schematic of the ion entry from the solar wind into the plasma mantle. (bottom) Upflow of ionospheric ions from the cleft ion fountain, the polar cap, and the auroral zone.

flux tailward. *Nishida et al.* concluded that the observations are consistent with the Dungey picture, when it is suitably generalized to allow three-dimensional convective motion of the field lines and twisting of the tail field. Furthermore, this analysis would suggest that the convection reversal region at X ~ -100 R_E is also the location of the distant-tail neutral line.

3. ION POPULATIONS IN THE TAIL LOBES

3.1 *Plasma Mantle*

The plasma mantle [*Rosenbauer et al.*, 1975] in the high-latitude magnetotail is formed by magnetosheath ions which enter into the magnetosphere along reconnected field lines: The solar wind flow drags these field lines poleward and into the nightside magnetosphere to form the (open) tail lobes, and the mantle plasmas are convected along with the field lines in the tailward direction. The top panel of Figure 3 shows the plasma mantle in a schematic diagram of the noon-midnight cross-section of the magnetosphere. The thickening, or expansion, of the plasma mantle in the north-south direction can be understood in terms of the **E** x **B** drift of mantle plasmas (and associated magnetic field lines) toward the central plane of the tail. Simple order-of-magnitude estimation suggests that the largest distance the mantle ions can reach before becoming part of the plasma sheet population is ~150 R_E [*Baker et al.,* 1996].

In addition to the ion entry through dayside reconnection, the magnetosheath ions can also enter the magnetosphere through the locally open magnetopause. The large momentum of the entering ions maintains their mostly antisunward motion. This tailward moving ion (and accompanying electron) population accounts for the much larger average lobe densities in the distant magnetotail than are observed in the inner magnetosphere lobes.

As illustrated by Figure 3, the mantle plasmas are hypothesized to enter the polar magnetosphere near the cusp boundary. Since the directed mantle flow speed is large, the mantle ions would be rather energetic (200 km/s implies a kinetic energy of ~210 eV for protons). According to the above calculations, lower speed protons (or other accompanying ions) would **E** x **B** drift to the tail central plane even closer to the Earth than the above estimates for the bulk of the mantle particles. In particular, any cold ions from the ionospheric cleft ion fountain poleward of the cusp [*Chappell et al.*, 1987] or from the polar cap would be expected to drift to the plasma sheet much closer to the Earth than would the mantle population.

3.2 *Ionospheric outflow*

As reported by *Mukai et al.* [1994] and Hirahara et al. [1996], the GEOTAIL spacecraft detected tailward-flowing cold ions in the tail lobes at essentially all explored distances. The ion flow speeds vary widely, ranging from 50 km/s to 500 km/s (for protons, this corresponds to directed kinetic energies from ~13 eV to ~1.3 keV). The energy-time spectra often show two or three energy/charge components, which *Mukai et al.* [1994] identified as different singly-charged ion species. The energy ratios suggested that the lowest energy/charge component was H^+, and that the two higher energy/charge components were He^+ and O^+.

Figure 4a shows an example of cold tailward ion flow on Set 20, 1993, (1700-1900 UT) when GEOTAIL was at X_{GSM} ~ -60 R_E near the center of the tail and relatively close to the plasma sheet boundary (Z_{GSM} ~ 4 R_E) (from *Hirahara et al.* [1996]). The top two panels show the measured magnetic field intensity and polar angle, showing that the satellite was in the northern tail lobe throughout the period of interest. The bottom panels show the omnidirectional electron (c) and ion (d) data as a color-coded energy-time (E-t) spectrogram from the energy-per-charge (EA-ion) analyzers of the Low Energy Particle (LEP) instrument onboard GEOTAIL (see *Mukai et al.* [1994] for instrumental details). Examination of the data in different look directions (not shown) reveals that the two ion populations observed between 1745 and 1850 UT were flowing tailward. The high counting rates at ~0.1 keV in the fourth panel from the top is indicative of dense, low-energy proton flow [see *Hirahara et al.*, 1996]. *Mukai et al.* [1994] and *Hirahara et al.* [1996] found from moment calculations that the tailward-directed beams had temperatures of at most

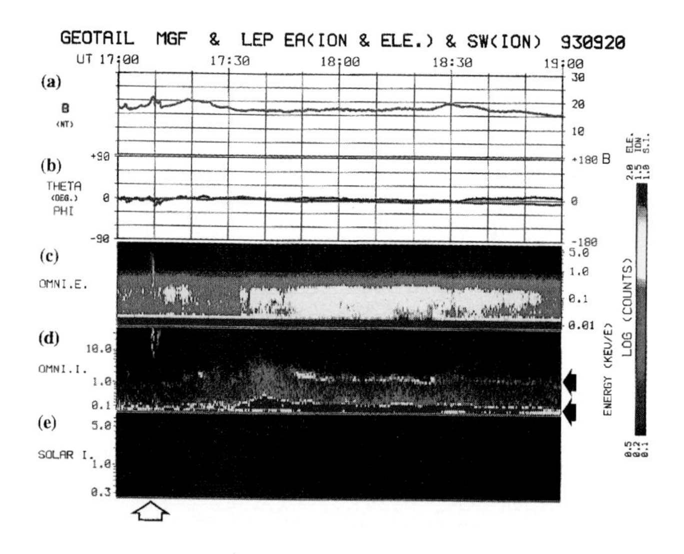

Figure 4a. GEOTAIL observations from Sept. 20, 1993. (a) Magnetic field magnitude, (b) magnetic field polar angle, (c) omnidirectional electron flux, (d) omnidirectional ion flux, and (e) solar wind detector (from *Hirahara et al.* [1996]).

several tens of eV (i.e., the beams were cold). Furthermore, when present, the higher energy trace (at ~1-3 keV) was consistently at an energy 16 times higher than the lower energy trace, indicative of an oxygen population with the same flow velocity.

Figure 4b shows the magnetic activity covering the time period of Figure 4a: The top three panels show the IMF components during most of the day as measured by IMP-8. IMF B_Z was negative continuously after about 0800 UT, B_Y was strongly negative, and B_X was usually small. The next panel shows a quasi-AL index created from the CANOPUS magnetometer chain, and the bottom panel shows geostationary electron fluxes from the Los Alamos National Laboratory instrument onboard spacecraft 1984-129. All data sets show that the event took place during a relatively disturbed period: Several substorms occurred after about 0800 UT, powered by the continuously negative IMF B_Z.

Given the singly-charged character of the cold ion flows as seen in Figure 4a, it is natural to identify the plasma source as the Earth's ionosphere (see the lower panel of Figure 3). This identification would be clearest for He^+ or O^+, but protons can also come from the ionosphere. Such events can be found at various distances from the Earth, ranging from the relatively near-Earth tail to close to the GEOTAIL apogee. Furthermore, the ion energies can also vary by an order of magnitude: proton energies ranging from 0.1 keV to above 1 keV have been observed. It is also typical to find these events during relatively disturbed periods, at times when IMF is predominantly southward and the motional electric field in the tail could be assumed to be large.

The source of these ions, the polar regions, have been extensively studied using low altitude polar orbiting satellites. The polar wind constitutes a very low-energy (few eV) ion population accelerated upward by ambipolar electric fields from the polar cap [*Hoffman and Dodson*, 1980] (see bottom panel of Figure 3). Because the accelerating field is very small, the polar wind has a mixture of H^+ and He^+ ions, heavier ions being gravitationally bound to the ionosphere. However, *Shelley et al.* [1982] reported also a more energetic population (10-100 eV) found throughout the polar cap, with large enough densities to constitute a significant source for the plasma sheet. *Lockwood et al.* [1985] and *Moore et al.* [1986] located the source of these ions in the dayside cleft region, and argued that transverse ion heating below 1 R_E altitude is a major contributor to the upward acceleration of these ions. A strong dependence of the ion outflow on geomagnetic activity was also established by these studies. More energetic ions, including ~1 keV O^+, have been found to originate from the auroral latitudes [e.g., *Wahlund and Opgenoorth*, 1989]. However, these ions are directly injected to the plasma sheet and hence do not enter the tail lobe open field lines. The acceleration mechanisms that lead to the ≥1 keV ion populations in the tail lobes are as yet not well understood.

3.3 *Tail configuration and trajectory calculations*

In order to examine the drifts of the ionospheric ions, calculations were performed for a variety of initial field-aligned energies and cross-tail electric field values. The drift

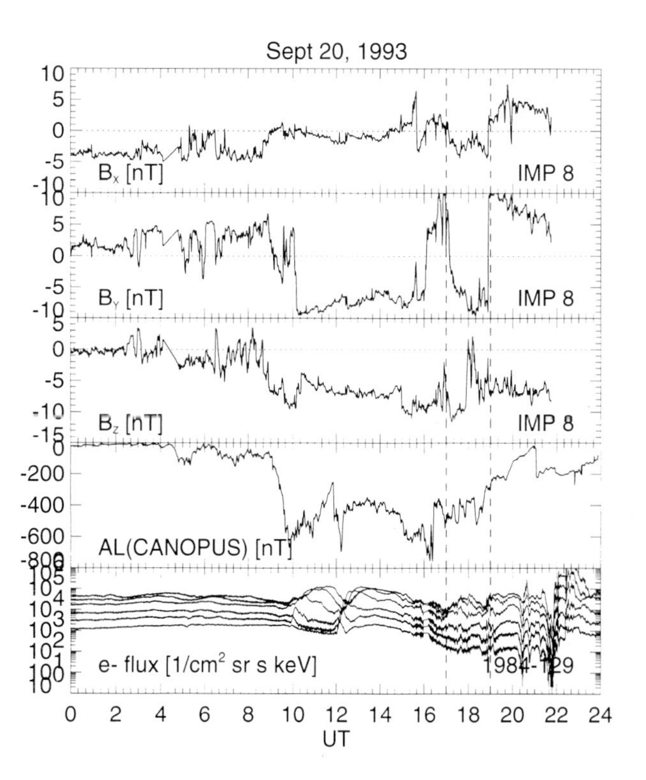

Figure 4b. Magnetospheric activity on Sept. 20, 1993. (a)-(c) IMF components in GSM coordinates from IMP-8. (d) Quasi-AL index from CANOPUS magnetometer chain. (d) Selected magnetograms from IMAGE magnetometer chain. (e) Energetic electron fluxes from LANL instrument onboard s/c 1984-129.

model developed by *Toivanen* [1997], which can utilize any magnetic field model and any simple electric field model, is employed for the calculations.

Accurate computations of ion trajectories in the distant magnetotail requires a model that can reproduce the cylindrical shape of the magnetotail and the almost constant lobe field values at large distances. As these computations are relatively time consuming, the most recent T96 model, which would have these properties, is too complex (it is ~10 times slower than the earlier T87 and T89 models). In order to get a fast model that would have these distant tail properties, the T87 (Kp = 3) model was modified to include a much stronger tail field. The modifications included were simple changes in the parameter values: $B_0 = 4.5$, $B_1 = -750.0$, $B_2 = -6500.0$, $X_n = -1.0$, $X_1 = 2.0$, $X_2 = 11.0$, and $R_T = 25.0$ were used instead of the tabulated values. The modified T87 model predicts a rather strong inner tail field; both B_n and B_{LOBE} are somewhat larger than the other models predict. Furthermore, the minimum in the field is shifted to larger distances, to around X=-60 R_E. Although the inner tail representation is not the best possible, this model is assumed to be sufficiently accurate for the purposes in this paper, i.e., ion trajectory calculations from the high-latitude polar cap to the distant magnetotail.

To illustrate the modeling results, we show here calculated particle trajectories for ions having a field-aligned velocity of 120 km/s, corresponding to oxygen energy of 1.2 keV or proton energy of 75 eV. Two different values for the cross-tail electric field were assumed: 0.75 mV/m representing strong electric field during strongly southward IMF, and 0.1 mV/m representing relatively weak electric field during weak solar wind coupling. Figure 5 shows calculated particle trajectories for the two electric field values. We allowed the ions to emerge from the ionosphere at all latitudes poleward of the cleft ion fountain: trajectories for the latitudes ranging from 77° on the dayside to 70° on the nightside (all in the noon-midnight meridian) are shown in the plot.

Note that when the electric field is weak, the outflowing ions from the cleft and polar cap fill much of the distant tail lobe volume. Even relatively low-energy oxygen from the nightside polar cap, under the conditions of very weak cross-tail electric field, can make it to relatively large distances before drifting to the central plane of the tail. In contrast, when the electric field is strong, the ions from all launching positions drift to the plasma sheet much more rapidly.

It is quite evident from the modeling results shown here that for weak convection electric fields in the tail (E_Y ~ 0.1 mV/m), even quite low energy ionospheric ions could emerge from the polar cap or from the cleft ion fountain and still survive as a focused beam of tailward streaming particles relatively far down the tail lobes ($X_{GSM} < -50$ R_E). However, for more substantial (and perhaps typical) values of the average cross-tail potential (E_Y ~ 0.5-1.0 mV/m), one would expect most of the outflowing ionospheric ions to have drifted into the plasma sheet around X_{GSM} = -100 R_E.

The above model calculations are time stationary, and thus illustrate what would happen when a relative steady-state had been attained. How changing solar wind coupling

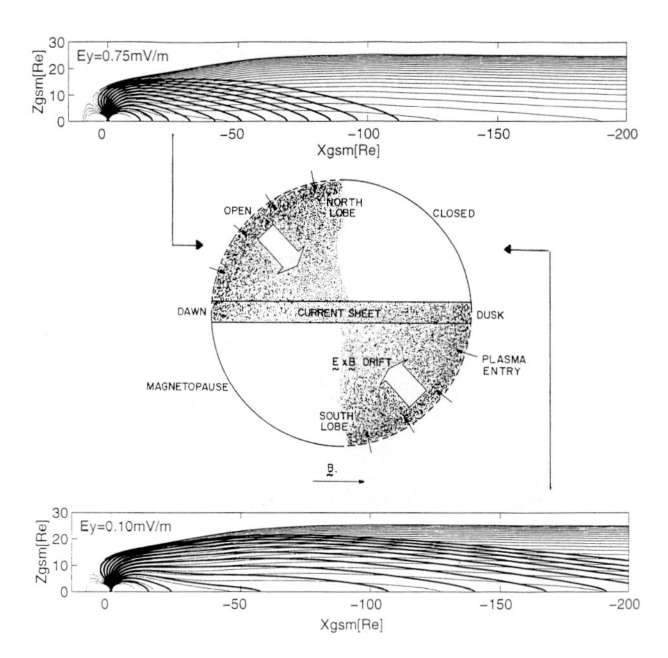

Figure 5. Calculated ion trajectories in the northern magnetotail lobe. The field lines shown in the GSM X-Z plane are computed from the model used in the calculations (see text). (top) E_Y = 0.75 mV/m. (bottom) E_Y = 0.1 mV/m. (middle) A schematic (from *Gosling et al.* [1985]) showing the local entry of plasma under IMF away sector and positive B_Y.

and configuration changes caused by substorms affect the ion trajectories is still an unresolved question. As the ion beams are predominantly found during magnetically active periods [see *Seki et al.*, 1997], one would assume that they are associated with relatively strong average cross-tail electric fields.

An important possibility to consider is that the magnetotail electric field may not be uniformly distributed across the tail cross-section. In the center of Figure 5 is a reproduction of a diagram from *Gosling et al.* [1985], which illustrates the nature of plasma entry and tail lobe convection inferred from ISEE-3 observations. The picture shows the magnetotail cross-section at a relatively distant location (~100 R_E) as viewed from the Earth. The dayside reconnection region is asymmetric depending on the strengths of the IMF B_Y and B_Z components: When IMF B_Y is positive, the magnetopause is locally open in the northern dawn boundary and southern dusk boundary (i.e., there the normal component of B is finite at the magnetopause). For negative IMF B_Y, the opposite quadrants are open for plasma entry (see also *Baker et al.* [1996]). This can lead to faster plasma convection toward the tail center on those sides of the tail where most of the plasma and flux are entering.

We would argue from results such as those shown in Figure 5 that even for southward IMF, it would be possible to have weak convection electric fields on large portions of the lobe field lines. In the sectors of the tail lobe where the magnetopause is locally closed, the drift velocities can be very small, and ionospheric ion beams may propagate to great geocentric distances before convecting down to the plasma sheet. On the other hand, in the regions of the tail where the magnetopause is locally open, one would expect to see much stronger convection electric fields, and it would be likely that low-energy ionospheric ion beams would drift into the plasma sheet relatively close to the Earth. *Seki et al.* [1997], however, have shown that O^+ beams are seen during disturbed geomagnetic conditions, primarily in the "loaded" lobe sector [see *Gosling et al.,* 1985]. Thus, in the majority of cases, the ions of ionospheric origin must be strongly accelerated parallel to the magnetic field to have any chance of making it to the distant tail lobes.

4. BIDIRECTIONAL ELECTRONS IN THE TAIL LOBES

4.1 *Local entry of solar wind electrons*

Earlier observations in the magnetotail lobes showed that counterstreaming, low-energy electrons were often present in the magnetotail lobes [*Fairfield and Scudder,* 1985; *Baker et al.,* 1986]. As these electrons were observed only in the hemisphere where the interconnecting field lines were attached to the Sun (i.e., north lobe for IMF away sectors and south lobe for IMF toward sectors), it was clear that the electrons were of solar origin. Furthermore, they were assumed to enter the magnetosphere through the locally open boundary, preferentially under southward IMF conditions (see Figure 6). Having low mass and high speed, these electrons would follow the magnetic field lines to the ionosphere where many would be reflected, thus forming the bidirectional population. The top panel of Figure 6 illustrates the local entry of these electrons into the north lobe during an IMF away sector.

Baker et al. [1997] examined GEOTAIL data in order to assess the occurrence and properties of the bidirectional electrons in the tail lobes. Figure 7a shows data from the LEP instrument in the form of energy-time spectrograms recorded on Oct. 5, 1994 1200-1800 UT. In this illustrative case, GEOTAIL was continually within the aberrated magnetotail; the satellite was at (-153.9, -14.0, 8.3) R_E in aberrated GSM coordinates. The top four panels of Figure 7 summarize the electron properties, while the next four panels show ion features. For both the electron and ion panels, the E-t spectrograms are color coded according to the log of the measured counts per sample. Characteristics of the plasmas in the sunward, duskward, tailward, and dawnward flow quadrants are shown separately for the electrons and ions.

The top panels of Figure 7a reveal that the sunward and tailward quadrants register higher counting rates than the dawnward and duskward flow quadrants. The effect is clearest prior to 1520 UT and after 1640 UT. Such an asymmetry is indicative of bidirectional electron anisotropies. Concurrent magnetic field data from the GEOTAIL magnetometer make clear that the spacecraft was mostly in the tail lobe during this time with the magnetic field directed sunward. Thus, the bidirectional electrons were field-aligned in analogy with the cases discussed previously for ISEE-3 [e.g., *Baker et al.,* 1987].

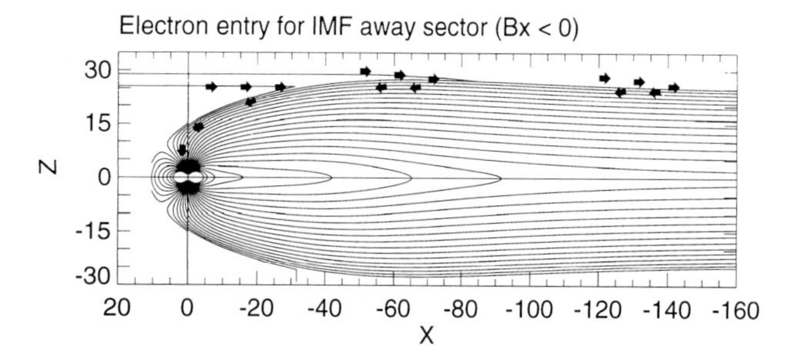

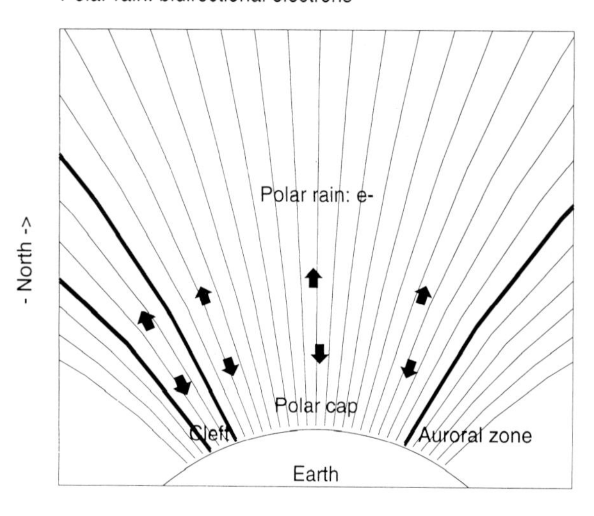

Figure 6. (top) Schematic of the electron entry from the solar wind into the plasma mantle. (bottom) Polar rain electrons at low-altitude open field lines.

The top panels of Figure 7b shows three components of the IMF for this period as measured by IMP-8. The spacecraft was at (-4.7, 18.8, -26.6) R_E in GSM coordinates, near the dusk-side bowshock. The field longitude at times when IMP-8 was not in the magnetosheath was about 135°, which is the classic Parker spiral angle for an IMF away sector. This would favor bidirectional electrons in the north lobe of the tail, as was observed. The next panel shows the magnetic activity as measured by the CANOPUS magnetometer chain; the bottom panel shows the geostationary electron fluxes from the LANL instrument. All the activity indicators show extended substorm activity from about 0800 UT until 2000 UT.

Early low-altitude polar-orbiting satellites discovered "polar rain" low-energy electrons which were present only within the region of open field lines, i.e., the polar cap [*Winningham and Heikkila*, 1974; *Fennell et al*., 1975]. *Baker et al.* [1987] argued that the solar wind heat flux electrons, when entering the magnetosphere, constitute a suprathermal (≥50 eV) population, which can move relatively freely as a test-particle population, just as they do in the solar wind. These electrons can enter the distant tail

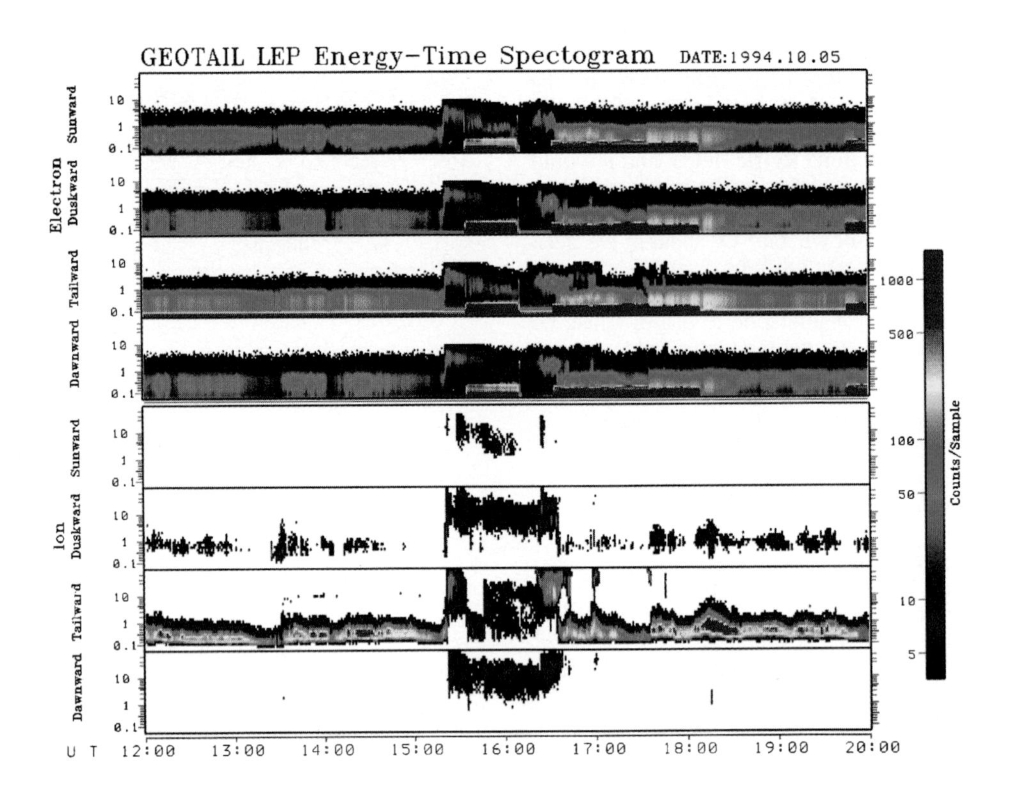

Figure 7a. GEOTAIL/LEP observations from Oct. 5, 1994. Ion (top four panels) and electron (bottom four panels) E-t spectrograms in color coding are shown for the dawnward, sunward, duskward, and tailward quadrants.

along locally open field lines and move rapidly down to low altitudes. The most narrowly field-aligned part of this suprathermal population then constitutes the (50-500 eV) polar rain. The bottom panel of Figure 6 illustrates the location of the polar rain electrons.

Figure 8 illustrates that the ISEE 3 deep-tail measurements reveal two different electron lobe populations. The thermal population is quite dense and controls the local tail conditions; this population arises principally from local sheath plasma entry. These thermal electrons are bound to the locally entering sheath ions which are strongly tailward flowing in the distant tail, and, as a consequence, most thermal electrons never reach the low-altitude polar cap. In contrast, the solar wind heat flux electrons move rapidly down to low altitudes to constitute the suprathermal polar rain.

4.2 *Connection between bidirectional electrons and cold ions*

The dense plasma population in the distant tail observed both by ISEE-3 and GEOTAIL is interpreted as being due to local entry of sheath plasma. Both ions and electrons enter the tail at these large distances, thus maintaining local charge neutrality. This dense thermal plasma population also controls the dynamics of the distant tail region. As shown by comparison of total electron densities in the distant tail lobes with those at low altitude, most of the distant tail electrons never enter the low-altitude polar cap; rather they are electrostatically bound to the thermal protons which flow tailward due to their large tailward momentum [*Baker et al.*, 1987; 1997]. This feature accounts for the much higher bulk plasma densities seen in the deep tail compared to the polar rain densities seen in the polar cap.

The event on Oct. 5, 1994, (Figure 7a) shows both bidirectional electron signatures as well as the ion beams, which indicates that the two processes may be coupled. The ion data show a strong, mostly tailward flowing cold ion population extending up to ~1-2 keV. There is also occasionally a thin ion trace near 10 keV which is identified as tailward streaming ionospheric O^+ ions [*Hirahara et al.*, 1996; *Baker et al.*, 1996]. Thus, there is the commingling of magnetosheath and ionospheric ions accompanying the bidirectional electrons. The ion populations have relatively small thermal spreads and hence are rather beamlike, and they flow generally tailward. The electrons, on the other hand, are flowing both along and opposite to the field lines.

When the bidirectional electrons were enhanced (as at ~1330 UT or ~1740 UT on Oct. 5, 1994) there was a concurrent strong intensification of the ion fluxes, especially in the tailward flow direction. There is also a (weaker) increase in the duskward flow direction. The intensities are increased in the ion energy range from perhaps 200 eV to somewhat above 1 keV. These ion enhancements decay on the same time scale as the electron events. When the bidirectional electrons are weak, or absent, as from ~1310 to ~1320 UT on Oct. 5, 1994, the tailward plasma ion flow is hardly detectable. As this pattern has been found to be quite

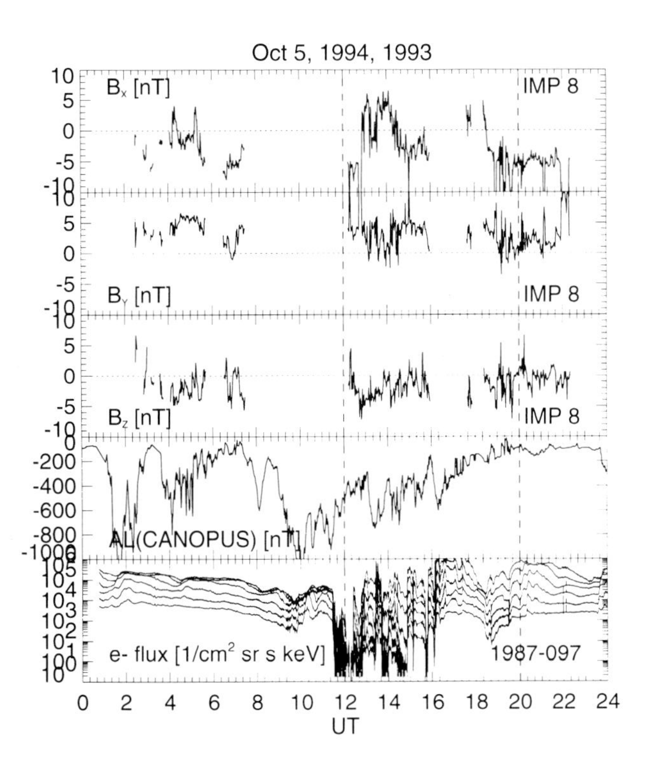

Figure 7b. Magnetospheric activity on Oct. 5, 1993. (a)-(c) IMF components in GSM coordinates from IMP-8. (d) Quasi-AL index from CANOPUS magnetometer chain. (d) Selected magnetograms from IMAGE magnetometer chain. (e) Energetic electron fluxes from LANL instrument onboard s/c 1987-097.

regular, the strong bidirectional electrons in the tail lobes seem regularly to be accompanied by dense, tailward flowing cold ions [see also *Seki et al.,* 1997].

Baker et al. [1987] speculated that both the heat flux electrons and the magnetosheath ions enter through the locally open magnetopause. However, because of their large momentum, the ions maintain their antisunward motion while the electrons can move back upstream toward the Earth. The relative motion of the electrons as they run away from the tailward flowing ions may then set up parallel electric fields which help to accelerate ionospheric ions (both H^+ and O^+) seen in the tail lobe field lines at surprisingly large distances. This acceleration leads to directional kinetic energies of the ions far in excess of those expected for upflowing ions from the cleft ion fountain region.

5. DISCUSSION AND SUMMARY

This paper reviews magnetic field and plasma observations in the magnetotail lobes. It is evident that even though the tail lobes are relatively current free, they often host highly structured populations of both electrons and ions at relatively low energies. Whereas the ions are observed to stream mainly tailward, the lighter electrons create a bidirectional distribution flowing both Earthward and tailward. Using the charge state of the ions to infer their origin, it was concluded that in many cases the ions seen in the distant tail come from the ionosphere, although in other cases clear indications of solar wind origin were seen as well [*Mukai et al.,* 1994; *Hirahara et al.,* 1996].

The observations of these ion beams at large geocentric distances were surprising in two ways: First, how could the low-energy (heavy) ions (below 1 keV) make their way to the distant magnetotail without convecting across the field lines to the plasma sheet? This is even more puzzling when it becomes apparent that these events are mostly associated with strong geomagnetic activity, which would indicate strong cross-tail electric fields and hence fast cross-tail drift speeds [*Baker et al.,* 1996]. This paper discussed the possibility that the lobe electric field is not uniformly distributed across the magnetotail, but rather is localized to the quadrant which is locally open for the solar wind entry.

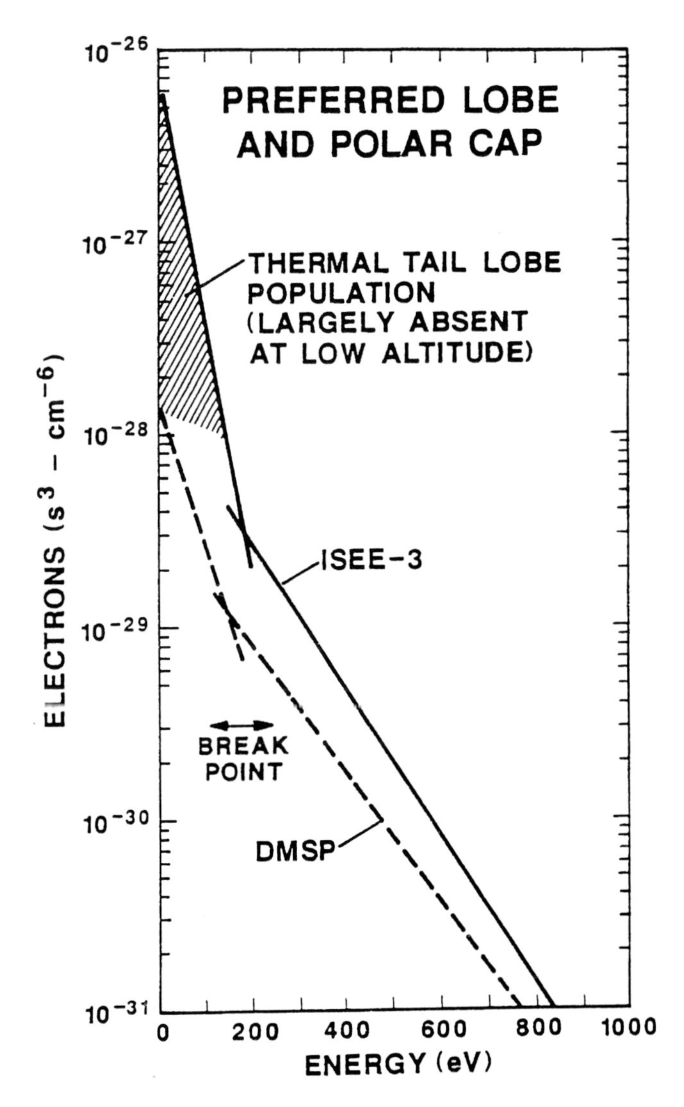

Figure 8. A schematic summary of the typical electron distribution function features seen by ISEE 3 and DMSP in the preferred lobe and polar cap regions (from *Baker et al.,* [1987]).

The second unexpected feature is the high energies (1-10 keV) that the ionospheric ions sometimes have. What is the acceleration mechanism capable of accelerating ionospheric ions up to these energies? It was suggested here that different flow directions between the ions and electrons entering through the magnetopause in the distant magnetotail may set up ambipolar electric fields, which could then accelerate ions from the polar cap and cleft regions to these large distances. Unfortunately, there are no quantitative estimates of the magnitude of such electric fields. Other acceleration mechanisms have been suggested that involve magnetic field curvature and convection electric field [*Cladis*, 1986], the centrifugal force due to the magnetic field curvature [*Horwitz et al.*, 1994], or transverse heating at low altitude [*Moore et al.*, 1986]. However, none of these mechanisms seem to be able to accelerate the ions to the required energies. A recent statistical study correlating the ion beam occurrences with the IMF direction and geomagnetic activity shows that cold ion beams occur mostly in the loaded lobe region [*Seki et al.*, 1997].

The present day magnetic field models give quite an accurate representation of the average large-scale magnetic field configuration in the magnetosphere [*Tsyganenko*, 1995], although large variations from the statistical average occur [*Pulkkinen et al.*, 1992]. However, the electric field structure is much more poorly known. Most often the large-scale magnetospheric electric field is assumed to be a constant E_Y induced by the solar wind interaction, or that obtained by field-aligned mapping of the ionospheric potential patterns. However, *Toivanen* [1997] has shown that even slow time variability is sufficient to provide inductive fields (with parallel components) that can dominate over the mapped or cross-tail electric field contribution.

The complex coupling of the fields and particle populations suggest that a self-consistent model is required for accurate description of the interactions between the polar ionosphere, the magnetotail, and the solar wind. The only present-day models capable of describing each of these regions simultaneously are the global MHD models [e.g., *Mobarry et al.*, 1996]. However, many of the dynamic processes discussed here involve inherently non-MHD physics: bidirectional electrons, monoenergetic ion beams, or charge-separation electric fields. Furthermore, the vast range of parameter space covered as one moves from the high field, high density ionosphere to the weak field and low density plasmas in the lobes make the accurate representation of the entire region still a challenge for the future. One might envision that one of the great tasks that would be possible with the ISTP fleet of spacecraft would be to compile a statistical database of the plasma parameters in the magnetosphere to be used together with the magnetic field database and statistical field models.

Acknowledgments. The authors thank Dr. T. Mukai for providing the GEOTAIL/LEP data, and P. Toivanen for the ion trajectory calculations. Dr. T. Hughes and the Canadian Space Agency are acknowledged for providing the CANOPUS magnetometer data, Dr. R. Lepping for providing the IMP-8 magnetometer data, and Dr. G. Reeves and Dr. R. Belian for providing the Los Alamos energetic particle data. The work of TP was supported by the Finnish Fulbright Association. This research was supported by NASA grant NAS 5-97140 and GC 124827.

REFERENCES

Baker, D. N., S. J. Bame, W. C. Feldman, J. T. Gosling, R. D. Zwickl, J. A. Slavin, and E. J. Smith, Strong electron bidirectional anisotropies in the distant tail: ISEE 3 observations of polar rain, *J. Geophys. Res., 91*, 5637, 1986.

Baker, D. N., S. J. Bame, J. T. Gosling, and M. S. Gussenhoven, Observations of polar rain at low and high altitudes, *J. Geophys. Res., 92*, 13457, 1987.

Baker, D. N., T. I. Pulkkinen, P. Toivanen, M. Hesse, and R. L. McPherron, A possible interpretation of cold ion beams in the Earth's tail lobes, *J. Geomag. Geoelectr., 48*, 699, 1996.

Baker, D. N., A. Nishida, T. Mukai, T. Yamamoto, Y. Saito, Y. Matsuno, S. Kokubun, and T. I. Pulkkinen, Observations of bidirectional electrons in the distant tail lobes: GEOTAIL results, *Geophys. Res. Lett., 24*, 959, 1997.

Chappell, C. R., T. E. Moore, and J. H. Waite, Jr., The ionosphere as a fully adequate source of plasma for the interactions, *J. Geomag. Geoelectr., 38*, 1223, 1987.

Cladis, J. B., Parallel acceleration and transport of ions from polar ionosphere to plasma sheet, *Geophys. Res. Lett., 13*, 893, 1986.

Cowley, S. W. H., The impact of recent observations on theoretical understanding of solar wind-magnetosphere interactions, *J. Geomag. Geoelectr., 38*, 1223, 1986.

Daglis, I. A., S. Livi, E. T. Sarris, and B. Wilken, Energy density of ionospheric and solar wind origin ions in the near-Earth magnetotail during substorms, *J. Geophys. Res., 99*, 5691, 1994.

Dungey, J. R., Interplanetary magnetic field and the auroral zones, Phys. Rev. Lett., 6, 47, 1961. Fairfield, D. H., and J. D. Scudder, Polar rain: solar coronal electrons in the Earth's magnetosphere, *J. Geophys. Res., 90*, 4055, 1985.

Fairfield, D. H., N. A. Tsyganenko, A. V. Usmanov, and M. V. Malkov, A large magnetosphere magnetic field database, *J. Geophys. Res., 99*, 11319, 1994.

Fennell, J. F., P. F. Mizera, and D. R. Croley, Jr., Low energy polar cap electrons during quiet times, *Conf. Pap. Int. Cosmic Ray Conf. 14th*, 4 (MG8-3), 1267, 1975.

Gloeckler, G., and D. C. Hamilton, AMPTE ion composition results, *Phys. Scr., T18*, 73, 1987.

Gosling, J. T., D. N. Baker, S. J. Bame, W. C. Feldman, R. D. Zwickl, and E. J. Smith, North-south and dawn-dusk plasma asymmetries in the distant tail lobes: ISEE 3, *J. Geophys. Res., 90*, 6354, 1985.

Hirahara, M., T. Mukai, T. Terasawa, S. Machida, Y. Saito, T. Yamamoto, and S. Kokubun, Cold dense ion flows with multiple components observed in the distant tail lobe by GEOTAIL, *J. Geophys. Res., 101*, 7769, 1996.

Hoffman, J. H., and W. H. Dodson, Light ion concentrations and fluxes in the polar regions during magnetically quiet times, *J. Geophys. Res., 85*, 626, 1980.

Horwitz, J. L., C. W. Ho, H. D. Scarbro, G. R. Wilson, and T. E. Moore, Centrifugal acceleration of the polar wind, *J. Geophys. Res., 99*, 15051, 1994.

Mobarry, C. M., J. A. Fedder, and J. G. Lyon, Equatorial plasma convection from global simulations of the Earth's magnetosphere, *J. Geophys. Res., 101*, 7859, 1996.

Nishida, A., T. Mukai, T. Yamamoto, Y. Saito, and S. Kokubun, Magnetotail convection in geomagnetically active times. 1. Distance to the neutral lines, *J. Geomagn. Geoelectr. 48*, 489, 1996.

Nishida, A., T. Mukai, T. Yamamoto, S. Kokubun, and K. Maezawa, A unified model of the magnetotail convection in geomagnetically quiet and active times, *J. Geophys. Res.*, in press, 1997.

Lockwood, M., M. O. Chandler, J. L. Horwitz, J. H. Waite, Jr., T. E. Moore, and C. R. Chappell, The Cleft ion fountain, *J. Geophys. Res., 90*, 9736, 1985.

Lundin, R., I. Sandahl, J. Woch, and R. Elphinstone, The contributions of the boundary layer EMF to magnetospheric substorms, in *Geophysical Monograph 64*, p. 355, American Geophys. Union, Washington, D.C., 1991.

Moore, T. E., M. Lockwood, M. O. Chandler, J. H. Waite, Jr., C. R. Chappell, A. Persoon, and M. Sugiura, Upwelling O+ ion source characteristics, *J. Geophys. Res., 91*, 7019, 1986.

Mukai, T., M. Hirahara, S. Machida, Y. Saito, T. Terasawa, and A. Nishida, GEOTAIL observations of cold ion streams in the medium distance magnetotail lobe in the course of a substorm, *Geophys. Res. Lett., 21*, 1023, 1994.

Pulkkinen, T. I., D. N. Baker, R. J. Pellinen, J. Büchner, H. E. J. Koskinen, R. E. Lopez, R. L. Dyson, and L. A. Frank, Particle scattering and current sheet stability in the geomagnetic tail during the substorm growth phase, *J. Geophys. Res., 97*, 19,283, 1992.

Pulkkinen, T. I., D. N. Baker, C. J. Owen, J. T. Gosling, and N. Murphy, Thin current sheets in the deep geomagnetic tail, *Geophys. Res. Lett., 20*, 2427, 1993.

Pulkkinen, T. I., D. N. Baker, P. K. Toivanen, R. J. Pellinen, R. H. W. Friedel, and A. Korth, Magnetospheric field and current distributions during the substorm recovery phase, *J. Geophys. Res., 99*, 10955, 1994.

Pulkkinen, T. I., D. N. Baker, C. J. Owen, and J. A. Slavin, A model for the distant tail field: ISEE 3 revisited, *J. Geomagn. Geoelectr., 48*, 455, 1996.

Rosenbauer, H., H. Grunwaldt, M. D. Montgomery, G. Paschmann, and N. Sckopke, HEOS/2 plasma observations in the distant polar magnetosphere: the plasma mantle, *J. Geophys. Res., 80*, 2723, 1975.

Rostoker, G., D. N. Baker, J. Lemaire, and V. Vasyliunas, Dialogue on the relative roles of reconnection and the "viscous" interaction in providing solar-wind energy to the magnetosphere, in *Magnetotail Physics*, edited by A. T. Y. Lui, p. 409, JHU Press, Baltimore, MD, 1987.

Scholer, M., G. Gloeckler, B. Klecker, F. M. Ipavich, D. Hovestadt, and E. J. Smith, Fast moving plasma structures in the distant magnetotail, *J. Geophys. Res., 89*, 6717, 1984.

Seki, K., M. Hirahara, T. Terasawa, T. Mukai, Y. Saito, S. Machida, T. Yamamoto, and S. Kokubun, Statistical properties and possible supply mechanisms of tailward cold O^+ beams in the lobe/mantle regions, *J. Geophys. Res.*, submitted, 1997.

Sergeev, V. A., M. Malkov, and K. Mursula, Testing the isotropic boundary algorithm method to evaluate the magnetic field configuration in the tail, *J. Geophys. Res., 98*, 7609, 1993.

Shelley, E. G., W. K. Peterson, A. G Ghielmetti, and J. Geiss, The polar ionosphere as a source of energetic magnetospheric plasma *Geophys. Res. Lett., 9*, 941, 1982.

Slavin, J. A., E. J. Smith, D. G. Sibeck, D. N. Baker, R. D. Zwickl, and S.-I. Akasofu, An ISEE-3 study of average and substorm conditions in the distant magnetotail, *J. Geophys. Res., 90*, 10,875, 1985.

Toivanen, P. K., Effects of the large-scale electric field on particle drifts in the near-Earth tail, *J. Geophys. Res., 102*, 2405, 1997.

Tsyganenko, N. A., Global quantitative models of the geomagnetic field in the cislunar magnetosphere for different disturbance levels, *Planet. Space Sci., 35*, 1347, 1987.

Tsyganenko, N. A., Magnetospheric magnetic field model with a warped tail current sheet, *Planet. Space Sci., 37*, 5, 1989.

Tsyganenko, N. A., Modeling the Earth's magnetospheric magnetic field confined within a realistic magnetopause, *J. Geophys. Res., 100*, 5599, 1995.

Winningham, J. D., and W. J. Heikkila, Polar cap auroral electron fluxes observed with Isis 1, J. Geophys. Res., 79, 949, 1974. Yamamoto, T., K. Shiokawa, and S. Kokubun, Magnetic field structures of the magnetotail as observed by GEOTAIL, *Geophys. Res. Lett., 21*, 2875, 1994.

D. N. Baker and T. I. Pulkkinen, Laboratory for Atmospheric and Space Physics, University of Colorado, 1234 Innovation Drive, Boulder, CO 80303, USA (Internet.tuija.pulkkinen@lasp.colorado.edu)

The Low-Latitude Boundary Layer in the Tail-Flanks

M. Fujimoto

Dept. Earth and Planet. Sci., Tokyo Inst.of Tech., Tokyo, Japan

T. Terasawa

Dept. of Earth and Planet. Phys., Univ. of Tokyo, Tokyo, Japan

T. Mukai

ISAS, Sagamihara, Japan

Geotail observations of the low-latitude boundary layer (LLBL) in the tail-flanks show that it is where the cold-dense ions appear with stagnant flow signatures accompanied by bi-directional thermal electrons (< 300 eV). It is concluded from these findings that the tail-LLBL is the site of capturing the cold-dense plasma of the magnetosheath origin on to the closed field lines of the magnetosphere. There are also cases suggesting that the cold-dense plasma entry from the flanks can be significant to fill a substantial part of the magnetotail. In such cases, which are detected mostly during northward IMF intervals, the cold-dense plasma is not spatially restricted to a layer attached to the magnetopause (LLBL) but continues to well inside the magnetotail, constituting the cold-dense plasma sheet. The continuity of the cold-dense plasma all the way from the magnetospheric boundary strongly supports the idea that the magnetosheath ions are directly supplied into the cold-dense plasma sheet from the flank. This cold-dense plasma in the near-Earth region shows significant contrast with the nominal hot-tenuous plasma presumably transported from the distant tail. A statistical study showing significant control on the near-Earth plasma sheet status by the IMF B_z component supports the idea that the cold-dense ion supply to the near-Earth tail from the flanks dominates over the hot-tenuous ion transport from the distant tail during northward IMF periods. We suggest that the formation processes of the plasma sheet differ according to the IMF B_z component.

1. INTRODUCTION

It is widely accepted that plasma from the ionosphere and that of the solar wind origin are the two sources for the hot plasma trapped in the plasma sheet. As for the solar wind plasma, which is considered to supply the majority of the plasma sheet protons, two entering processes have been considered: The entries via high-latitude lobes and via low-latitude flanks.

Protons in the near-Earth plasma sheet usually have temperature of several keV exceeding a typical kinetic energy of solar wind protons [*Baumjohann et al.*, 1989], meaning that some heating process should be operative while they are transported from the solar wind to

New Perspectives on the Earth's Magnetotail
Geophysical Monograph 105

the near-Earth part. During southward IMF periods, a part of solar wind plasma and energy are considered to flow into the magnetosphere through two magnetic reconnection processes, one at the dayside magnetopause that opens the field lines, and the other in the distant tail (or in the near-Earth part during substorm periods) that closes them. The energy stored in the magnetotail is released in the course, heating the plasma sheet ions up to ~ several keV or more as they convect to the near-Earth part [e.g., *Baumjohann*, 1993].

In contrast to the vast knowledge on the entry process via high-latitude lobes, whose very origin goes back to the first paper on the open magnetosphere by *Dungey* [1961], considerably less is known about the low-latitude entry into the plasma sheet. During strongly northward IMF periods, the high-latitude entry is said to become less effective. Several authors [e.g., Baumjohann et al., 1989] have noted that the plasma sheet becomes colder and denser during northward IMF periods than during southward IMF periods. While this could be taken as indicating the switch in the dominating transport process, there seems to have been no general agreement on this because of the little knowledge on the low-latitude entry.

One of the reasons for this is the theoretical difficulty in understanding efficient plasma transport across the low-latitude boundary. This is even the case for the extensively studied dayside cases. Observations in the dayside low-latitude boundary layer (LLBL) showing mixture of magnetospheric and magnetosheath ions on closed field lines have been known for some time [e.g., *Sckopke et al.*, 1981]. Studies following these early works have clarified the properties of the dayside-LLBL, such as bi-directional anisotropy of thermal electrons [e.g., *Ogilvie et al.*, 1984; *Hall et al.*, 1991; *Phan et al.*, 1997; *Fujimoto et al.*, 1998a]. However, the mechanism of this mixing, that is, the mixing that seems to take place without a change in the field line topology, is still under debate. Attempts to explain this mixing by local diffusive process seem to conclude that the wave amplitude in the high frequency range, through resonant interactions with which plasma are supposed to diffuse, are usually too small [*Winske et al.*, 1995; *Treumann et al.*, 1995]. *Terasawa et al.* [1992], *Thomas and Winske* [1993], and *Fujimoto and Terasawa* [1994] propose the Kelvin-Helmholtz (K-H) vortex, a low frequency phenomena that is not taken into account in the above approach, could be a vital driver of the mixing. *Song and Russell* [1992] and *Le et al.* [1996] propose that the field line closing by re-reconnection of open field lines having one of their ends in the ionosphere above the cusp latitudes is the formation mechanism of the closed LLBL for northward IMF.

The other reason for less knowledge on the low-latitude entry into the magnetotail has been the limitation of the data available. *Williams et al.*[1985] have shown that the LLBL in the tail consists of two layers, as in the dayside. *Mitchell et al.* [1987] and *Traver et al.* [1991] reported that the topology of the field lines in this region is controlled by the IMF B_z component, closed (open) when northward (southward). Observations of the tail-LLBL beyond the terminator meridian by ISEE 1 and 2 satellites are, however, quite limited so that studying the physics of the tail-LLBL with satellite data of an up-to-date performance has not been possible until the Geotail data become available.

Leaving aside how the magnetosheath plasma penetrate to the closed field line region of the plasma sheet, there have been some studies on the effects of the low-latitude entry. Direct supply of plasma from the tail flanks has been implicitly assumed in calculating the particle flux across LLBL into the magnetosphere in *Eastman et al.* [1976], and it has been concluded that the ions occupying low energy parts of the velocity space of the plasma sheet ions are the unheated population diffusively transported from the LLBL [*Eastman et al.*, 1985]. *Lennartsson* [1992] worked on the plasma sheet composition statistics and found that the plasma sheet population becomes very solar wind-like during northward IMF periods. From this finding, *Lennartsson* proposed a flank entry model for northward IMF periods. *Zwolakowska and Popielawska* [1992] and *Popielawska et al.* [1996] have also discussed the significance of the flank entry by mapping their data at high-latitude down to the magnetic equator. *Spence and Kivelson* [1994] have developed a convection model of the magnetotail including the dawn flank as one of the plasma sources. In their model, the plasma sheet is assumed to be filled with ions coming from the distant tail and drifting from the dawnside flank (Thus, the duskside boundary cannot operate as a plasma source in their modeling.) It was shown that the contribution from the flank can make the resultant plasma sheet much more closely resemble the observations, compared to the situation when only the distant tail source is taken into account.

In this paper, we review our recent studies on the unresolved problem of the low-latitude entry [*Fujimoto et al.*, 1996, 1998b; *Terasawa et al.*, 1997]. The GEOTAIL satellite, owning to its well-designed orbit, provides useful information on LLBL at $X_{GSM} > -30R_E$ as it skims the flanks of the magnetotail. Moreover, now that 3D particle data with much higher time resolution (12 seconds) are available, we are able to uncover some processes that were previously unidentified. In accordance with previous studies but with a more complete data set, it is shown that low-latitude flanks of the tail is the site where a part of the cold-dense magnetosheath ions flowing past is captured on the closed field lines of the magnetosphere, and that bi-directional thermal electrons are seen to accompany the cold-dense plasma. It is also shown that such captured cold-dense plasma fill a substantial part of the magnetotail at times. In such cases, since the plasma is not only stagnant but also is occupying a large area not restricted to the proximity of the magnetopause, we propose that the region occupied by the cold-dense plasma is better be termed as the cold-dense plasma sheet rather than the LLBL. Inspired by the fact that these remarkable cases are found for northward IMF, a statistical study on the status of the near-Earth plasma sheet has also been made. We also review the results showing that the plasma sheet becomes significantly colder and denser than during southward IMF periods when the northward IMF continues, and that this low-temperature/high-density status appears most prominently near the dawn and dusk flanks. These are consistent with the idea of direct supply (with little subsequent heating) of the cold-dense magnetosheath ions from the flanks to the near-Earth plasma sheet. Here we note that not only the Geotail data, but also concurrent solar wind/IMF conditions available from WIND have been essential in making these studies.

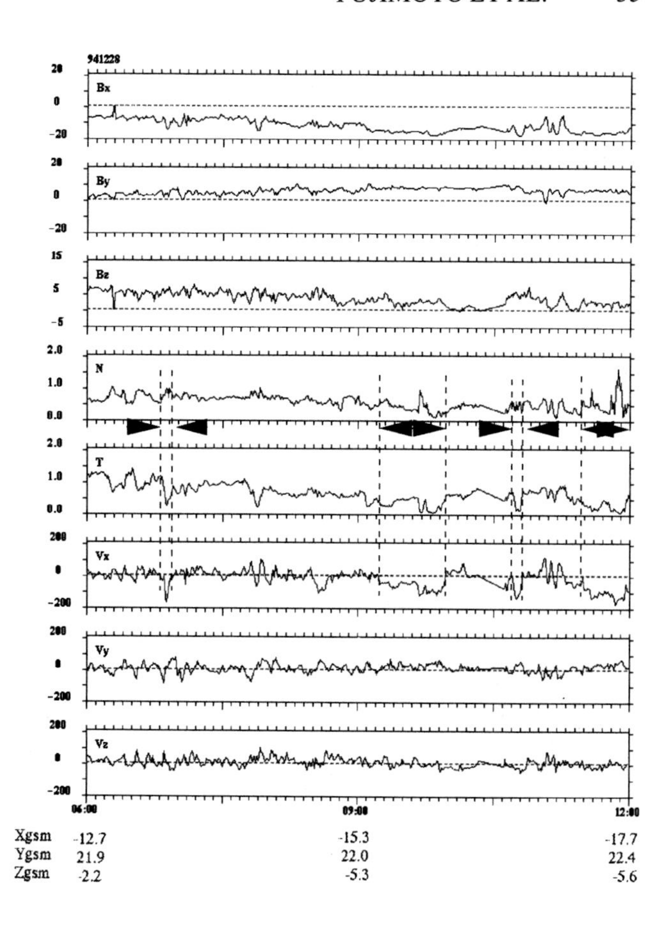

Figure 1. Overview of the data in the duskside tail-LLBL.

2. TAIL-LLBL ON THE DUSKSIDE

In this section, we will review observations of a tail-LLBL on the duskside, along an orbit that is more or less parallel to the magnetopause. Figure 1 shows the data obtained around $(X_{GSM}, Y_{GSM}, Z_{GSM})$=(-14, 22 , -4) R_E (at 07:30 UT). Shown are the three components of the magnetic field (nT), the ion density (cm^{-3}) and temperature (keV), and the three components of the ion bulk flow (km/s). GSM coordinates are used. The key parameter data from WIND show that IMF was weakly northward. The data can be divided into the magnetosheath-like intervals characterized by enhanced tailward flow (between the vertical bars) and the intervals in a stagnant region. One may expect the stagnant region to be the plasma sheet filled with hot-tenuous plasma. What is interesting, however, is that ions with (T, n) $\sim$ (1 keV, 1 cm^{-3}), that is cold and dense compared to a nominal plasma sheet value, occupy the latter.

Figure 2 expands the interval 06:30-08:30 UT. The E-t diagram for omni-directional ions shows that the stagnant ions are composed of two distinct populations, one at several keV and the other at $<$ 1keV. With this plot, the cold-dense feature can be understood to be due to the presence of the latter. While the E-t diagram appears to be similar to those seen at the plasma sheet boundary layer, where a low energy lobe component and an energetic beam leaking from the plasma sheet along the field line are superposed [e.g., *DeCoster and Frank*, 1979], an inspection of the distribution function shows that they are totally different. Figure 3a shows the ion distribution function at 07:10:57 UT (marked by a triangle in Figure 2). Shown on the left is the slice of the 3D distribution by the plane including the magnetic field and the ion bulk flow vectors (the BC plane). The B and C axes are aligned with the magnetic field and the perpendicular flow, respectively. Shown on the right is the cut of this distribution along the dashed line in the left (field aligned spectrum in the field line rest frame). From the profile on the right, the two count rate peaks in Figure 2 are understood to be a manifestation of two populations having different average energy. The colder component sitting close to the origin of the plot looks similar to the magnetosheath ions, and is responsible for the cold-dense feature. Around 2.2 keV (V_B = $\pm$ 600 km/s), the energy spectrum shape shows a transition to a hotter component, which looks similar

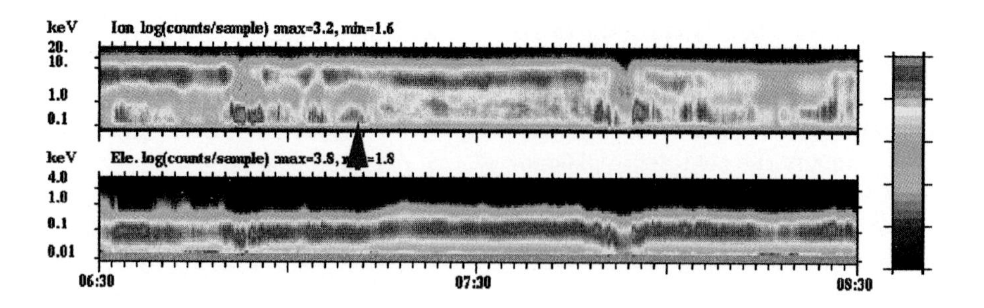

Figure 2. The interval 06:30 - 08:30 UT expanded. The E-t diagram for omnidirectional ions clearly depicts the mixing feature.

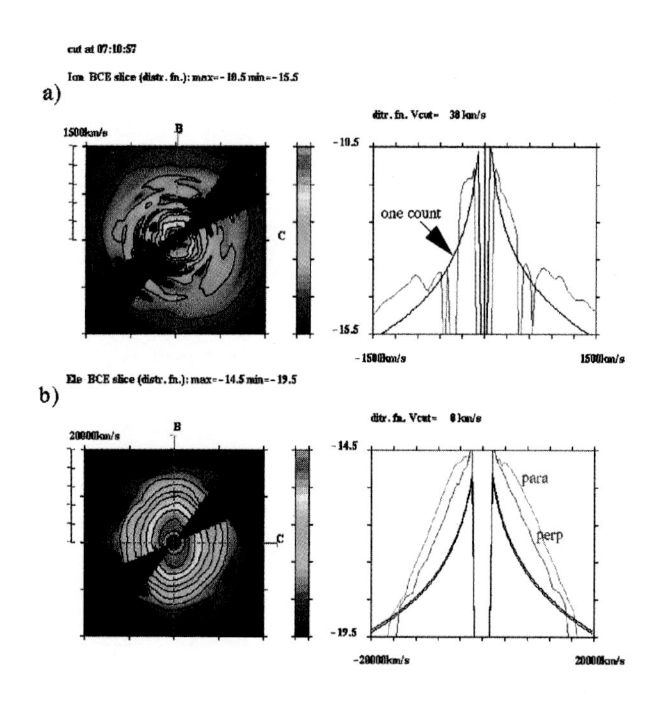

Figure 3. An example of the (a) ion and (b) electron distribution function in the duskside tail-LLBL. The slices of 3D distribution data by the plane including the magnetic field and the ion bulk velocity vector (the BC plane) are shown on the left. Shown on the right are the cuts along the dashed lines in the left.

to magnetospheric ions. Figure 3b depicts the electron distributions obtained at the same time. It can be seen that that the stagnant cold-dense ions are accompanied by thermal electrons having a football-shaped distribution. In terms of pitch angle anisotropy, the electrons are characterized by bi-directional anisotropy with the parallel-to-perpendicular flux ratio of $\sim$ 10 at 100-200 eV (< 10,000 km/s) energy ranges. This anisotropy is extended up to, but are gradually lost at, 300 eV and becomes more or less isotropic at higher energies.

The stagnant region is characterized not only by the mixing of stagnant cold-dense ions with the magnetospheric but also by the bi-directional 100-200 eV electrons having balanced flux in both the directions. We have also found that the region is highly more rich in isotropic energetic electrons (> 300 eV) than the magnetosheath. Summing all the evidence together, we interpret that the stagnant region is on closed field lines, and thus, is a region in the magnetosphere: Hereafter, the region will be called the tail-LLBL. This idea of closed topology naturally explains not only the electron characteristics but also those of the ion flow. In this picture, LLBL electrons are visualized as the mixture of those from the magnetosheath that are heated up to hundreds eV in the parallel direction and those from the magnetosphere responsible for the more energetic isotropic part. The former show balanced bi-directionality and the latter remain trapped, because of the closed topology. Ions are also from both sources, and the different characteristic energies of the sources are reflected in the two-component distribution function of the ions in the LLBL. Ions are convecting only slowly in the tail-LLBL. Inspection of the distribution function show that the slowing down is not due to the co-existence of two populations counter-streaming to each other, but because both the components are stagnant. This stagnant feature is compatible with the closed topology.

Despite the vast difference in the parallel spectra from the magnetosheath population, we consider thermal bi-directional component of the tail-LLBL electrons to be of the magnetosheath origin. Since the adjacent magnetosheath is the reservoir that the LLBL has the easiest access, we consider that the electrons make an accompanying entry to the closed field lines in order to satisfy the charge neutrality. In contrast to the relatively unheated entry of the ions, the electrons experience parallel heating upon crossing the border.

It is interesting to note that these plasma features of the duskside tail-LLBL is quite similar to the dayside (inner-) LLBL (the part that seems to be on closed field lines [*Fujimoto et al.*, 1998]). Just as the formation mechanism of this extensively studied region remains open, the formation of the tail-LLBL is an open question. It is also noted that we find the magnetospheric ions to be less on the dawnside tail-LLBL. This dawn-dusk asymmetry is presumably because magnetospheric ions drift (curvature or grad-B) from dawn-to-dusk and cannot access the region located on the dawnside edge of the magnetotail.

3. COLD-DENSE PLASMA SHEET: A CASE STUDY

The above observations showing capturing of cold-dense ions from the solar wind onto closed field lines in the tail-LLBL is representative of a group of data from the tail-LLBL at $X_{GSM} > -30R_E$ obtained by Geotail (There is another group characterized by less evident capturing feature. IMF B_z possibly has a control on the capturing efficiency as will be discussed later.). While the above example does show evidence for occasional solar wind plasma capturing at the tail-flanks, a shortcoming is that the observations are made more or less along the magnetopause so that its influence to the status of the plasma sheet further inside the magnetotail cannot be known. In this section, we will show an example strongly suggesting that the cold-dense plasma entry from the flanks can indeed be significant at times, and is likely to be responsible for the formation of the cold-dense plasma sheet.

Figure 4 shows the Geotail orbit on March 24, 1995, along which the data we will discuss has been obtained. As is seen, the orbit goes across the tail-LLBL towards the center of the tail. WIND data show that IMF was northward, and the solar wind dynamic pressure had only small variations during the interval of interest. Figure 5 shows the Geotail data. With the spacecraft's motion into the magnetotail from the duskside magnetosheath, fast tailward flow terminates around 0900 UT ($(X_{GSM}, Y_{GSM}) \sim$ (-15.3, 18.5) R_E). While this seems to mark the spacecraft's last crossing of the magnetopause, the stagnant ions further inside the boundary stays cold ($<$ 1 keV) and dense ($>$ 1 cm^{-3}) compared to nominal values, and this cold-dense status continues to the end of the plot. As shown in Figure 6, ions of $<$ 1 keV continues to show the largest count rates exceeding that of the magnetospheric component (several keV) throughout the interval, being responsible for the cold-dense status. Magnetospheric ions are less than

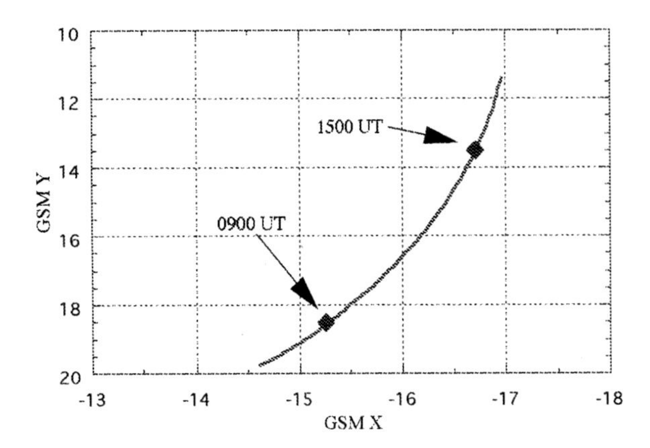

Figure 4. The orbit on March 24, 1995.

the previous example, such as not to show a second peak in the E-t diagram. This is presumably because the magnetoshere was in the "colder" state: This topic will be touched in the next subsection. Since the absolute value of the B_x component remains relatively small (and even crosses the neutral sheet several times) until the jump around 1500 UT, the cold-dense plasma is likely to be situated at a low-latitude part of the magnetotail. If we conservatively take the time of the B_x jump as the end of the low-latitude region survey, at which time Geotail was locted at (X_{GSM}, Y_{GSM}) = (-16.7, 13.5) R_E, the distance in Y_{GSM} from this inner limit to the last tailward flow position sums up to 5 R_E (see Figure 4). Since the solar wind dynamic pressure showed only small variations, we may reasonably assume that the magnetopause position did not change significantly during the interval. Then, this value is a rough estimate of the size of the region that the cold-dense plasma occupies at low-latitude. Thus, the cold-dense plasma occupies a substantial part of the magnetotail for this case. With this finding that the cold-dense ions are not spatially restricted to a thin layer attached to the magnetopause, we think that it is better to term the low-latitude region filled with the cold-dense stagnant ions as the plasma sheet in a cold-dense status, or the cold-dense plasma sheet, rather than the boundary layer located at low-latitude (LLBL). A similar case is reported by *Sauvaud et al.* [1997].

The ion temperature can be seen to increase in the innermost part of the cold-dense plasma sheet (13-15 UT) in Figure 5. Figure 6 shows that the temperature increase is due to some heating of the cold-dense component, rather than the increased contribution from the magnetospheric component. After the last tailward flow around 9 UT, thermal electrons are persistently seen to

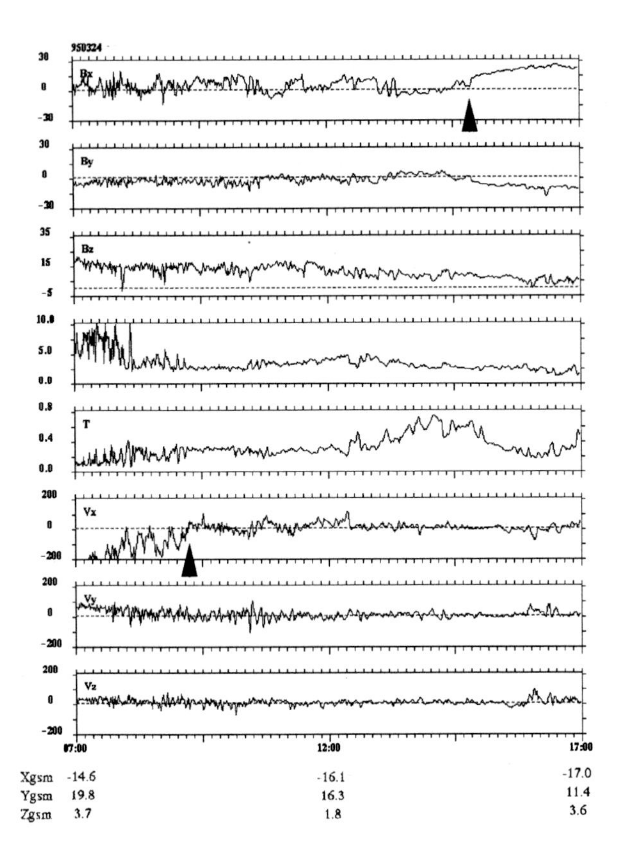

Figure 5. Overview of the data obtained along the orbit (March 24, 1995) from the duskside magnetosheath to inside the magnetotail. The cold-dense status of ions continues throughout the interval.

show clear bi-directional anisotropy up to 300 eV energy range (not shown). This feature is seen to fade out after 13:00 UT simultaneous with the ion temperature increase in the innermost part. The observations of the fading out of the bi-directional anisotropy and the heating of the cold-dense ions suggest that the plasma from the flanks tends to lose its distinguishing characteristics as they reside in the inner part of the magnetotail being subject to some internal processes.

Figure 6 also indicates that several keV magnetospheric ions exist even on tailward convecting flux tubes detected earlier than 09:00 UT. In Figure 7, we show an example of particle distributions in these tailward flux tubes. In Panel a), which shows the ion distribution in the BC plane (left) and the perpendicular spectrum (right), the magnetospheric contribution is seen as a hotter component superposed on a cold one. It can also be seen that the hotter component has a slight pan-cake anisotropy while the colder one has a weak dumbbell anisotropy. Panel b) shows the electrons. The distribution in the BC plane is shown on the left, and the parallel (red) and perpendicular (blue) spectra are compared on the right. The parallel spectrum on the right (red) shows a flat-top feature at $|V_B| < 4000$ km/s, which may give us a hint on the parallel heating process. These distribution characteristics of ions and electrons are essentially the same as those detected in the stagnant region encountered after 09:00 UT. The only difference is the presence/absence of the shift of the center of the ion distribution from the origin, which reflects the difference in the bulk flow. As we interpret the stagnant region to be on closed field lines, this similarity in the distribution functions suggests that we should interpret the tailward flowing flux tubes filled with mixed ions/bi-directional electrons to be closed as well. Significant tailward velocities with these flux tubes suggest that they are in the course to become stagnant, that is, they are to become members of the cold-dense plasma sheet flux tubes.

4. THE COLD-DENSE PLASMA SHEET: A STATISTICAL STUDY

A control by IMF B_z on the thickness of the boundary layer has been reported by *Mitchell et al.* [1987]. Our preliminary survey of the tail-LLBL structure (which still needs to be refined because of its complexity) also suggests the tendency that the capturing process at the boundary becomes more efficient for northward IMF. This tendency makes us expect that the plasma sheet changes to a cold-dense status as IMF becomes northward. This is what has recently been shown by *Terasawa et al.* [1997].

The data used are from Geotail in the plasma sheet and from WIND monitoring the solar wind/IMF conditions. Time lags estimated from dividing the geocentric distances of WIND by the observed solar wind velocities are used in synchronizing the data from both spacecrafts. The plasma sheet dataset used in the present study is constructed as follows: For the interval Nov. 1994 - Dec. 1995, 1034 1-hour segments of observations are available from the near-Earth magnetotail ($-50 < X_{GSM'} < -15R_E$, $|Y_{GSM'}| < 25R_E$: Here, GSM' is a modified GSM coordinates where an aberration angle of 4° is taken into account.). After excluding those obtained in the close proximity of the magnetopause to avoid contamination from the boundary layer, steady plasma sheet intervals are selected by requiring (1) 1-hour average plasma β (ion thermal pressure divided by magnetic pressure; hereinafter, electron contribution is neglected) is larger than 1, and, (2) standard deviation of the total pressure (ion thermal plus magnetic pressure) is less than 20 % of its 1-hour average value. The condition (2) rejects data obtained

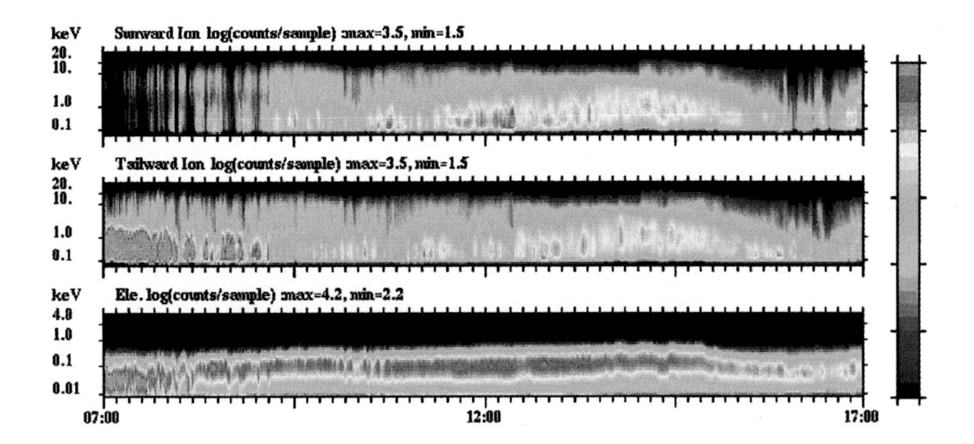

Figure 6. The E-t diagrams for sunward/tailward ions. Ions of < 1 keV show the largest counts throughout.

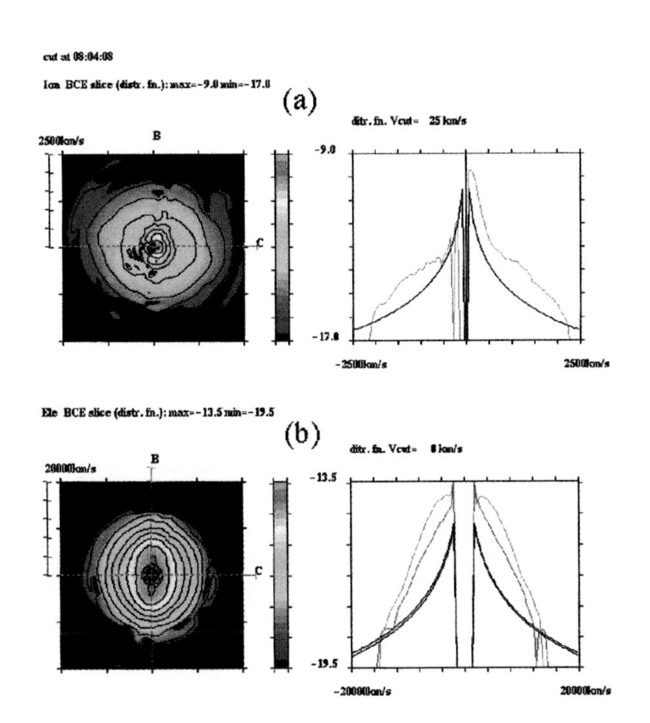

Figure 7. An example of particle distribution functions in a tailward flowing mixing flux tube. (a) Ions showing two-component feature. (b) Bi-directional electrons showing a football shaped distribution.

during highly active intervals, such as those including high speed flows. Each of these 1-hour segments contains data obtained at a more or less fixed position in the $(X_{GSM'}, Y_{GSM'})$-plane, but includes dependence on the distance from the neutral sheet. To eliminate this, we made linear regression analysis on B_x versus ion density for each of these 1-hour segments. After obtaining a linear regression line from data points in a 1-hour segment,

$$n = a \cdot \mathrm{B_x} + b$$

the ion density at the neutral sheet for the 1-hour segment is estimated by substituting $\mathrm{B_x} = 0$ in the above equation ($n = b$). The same procedure is used for estimating ion temperatures at the neutral sheet. These neutral sheet values will be used in the following analyses. It should be mentioned that since the linear dependence of ion density/temperature on B_x is not necessarily guaranteed, we check the validity of this assumption by comparing the ion thermal pressure at the neutral sheet calculated from the estimated density and temperature with the 1-hour averaged total pressure. By requiring both to agree within a 20 % error, 497 plasma sheet data points are selected.

Figure 8a and 8b show the plasma sheet temperature versus the solar wind kinetic energy and the plasma sheet density versus the solar wind density, respectively. Color of a data point depicts IMF latitudinal angle. Here, solar wind/IMF parameters are averaged for the same 1-hour interval as the plasma sheet data. While both panels show positive correlations, which indicates the fact that the state of the plasma sheet is controlled by the state of the solar wind, there are large scatter of data points. Blued points (IMF B_z positive) are more or less evenly distributed. On the other hand, reddish data points (IMF B_z negative) tend to be clustered at high temperature-low density (hot-tenuous) region. Figure 8c presents normalized plasma sheet density versus IMF latitudinal angle (theta). Color depicts the normalized plasma sheet temperature (plasma sheet temperature divided by the solar wind kinetic energy). To confirm the above statement from a reversed angle, it is evident

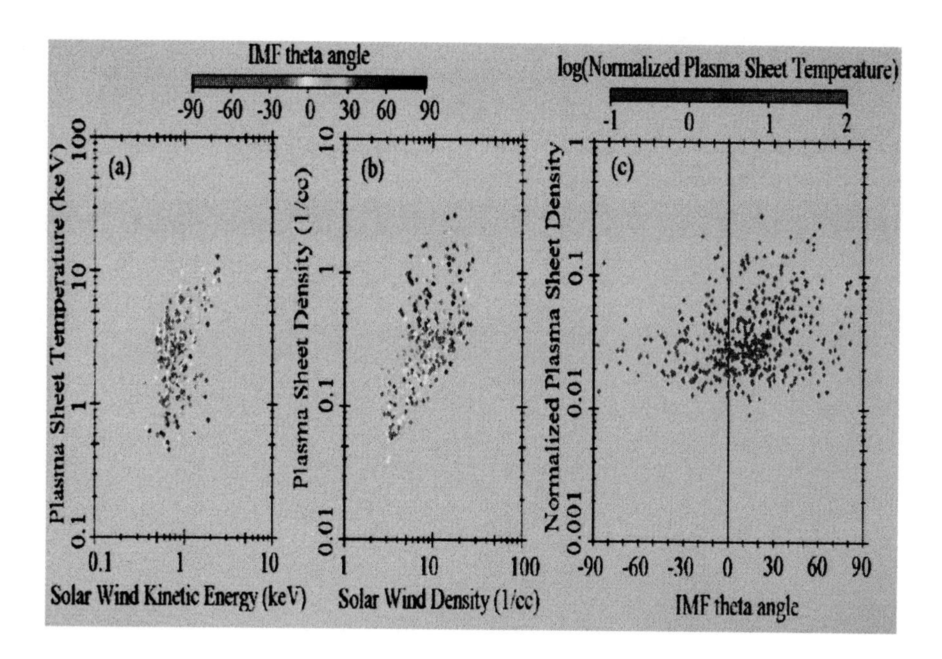

Figure 8. (a) Estimated ion temperature at the neutral sheet plotted against solar wind kinetic energy. (b) Estimated ion density at the neutral sheet plotted against the solar wind density. (c) Normalized plasma sheet density plotted against IMF latitudinal angle. The solar wind parameters are averaged for the same 1-hour interval as the plasma sheet data.

that the low temperature-high density (cold-dense) status is obtained only by the data points sampled during northward IMF periods. It is clear at this point that the cold-dense plasma sheet occasionally appears when IMF is northward, but very unlikely when IMF is southward.

To see if there is better correlation of the plasma sheet status with the IMF B_z component, we have made various trials. It is found that the large scatter for northward IMF cases can be reduced significantly by averaging the solar wind parameters over longer periods. The normalizing factors in the previous analysis has been the solar wind parameters averaged for the same 1-hour interval as the plasma sheet data. Instead, for a plasma sheet data obtained during a 1-hour interval (t-1, t), we have tried with the solar wind parameters averaged over the interval (t-N, t) (N=1 in the previous analysis). By calculating the correlation coefficients for northward IMF subsets with N=1,2,3,...48, we have found that the normalized plasma sheet density becomes most correlated with IMF latitudinal angle (theta) for N=6 – 12, with a broad peak at N=9 (Figure 4 of *Terasawa et al.* [1997]). The normalized plasma sheet temperature is also seen to be most anti-correlated at mostly the same N value. Figure 9 shows the same as Figure 8 but for 9-hours averaged solar wind parameters (N=9). There are clearer separations between blued and reddish data points in Panels (a) and (b). Panel (c) shows a clear positive correlation between the normalized plasma sheet density and the IMF theta angle for theta > 0. Also in Panel (c), blued data points appear mostly only when IMF is northward. Since these cases (normalized temperature ≤ 1) are indicative of little subsequent heating within the plasma sheet, it is suggested that direct supply of cold-dense plasma to the plasma sheet is dominantly operative during prolonged northward IMF periods. In contrast, some heating must intervene for southward IMF periods to make the plasma sheet hot and tenuous.

Plotting the normalized density/temperature for N=9 against $Y_{GSM'}$ (Figure 10), it can be seen that the plasma sheet becomes denser and colder toward the dawn and dusk edge of the plasma sheet for northward IMF (Sparse data does not allow us any conclusion for southward cases.). This is consistent with the idea that the direct supply of cold-dense plasma during northward IMF periods is from the flanks. While the meaning of the best correlation at N=9 still remains to be discussed, which time scale is considerably longer than the characteristic convection time scale of 1 – 2 hours during a southward IMF period [e.g., *McPherron*, 1991], a possible interpretation is that it reflects the slower

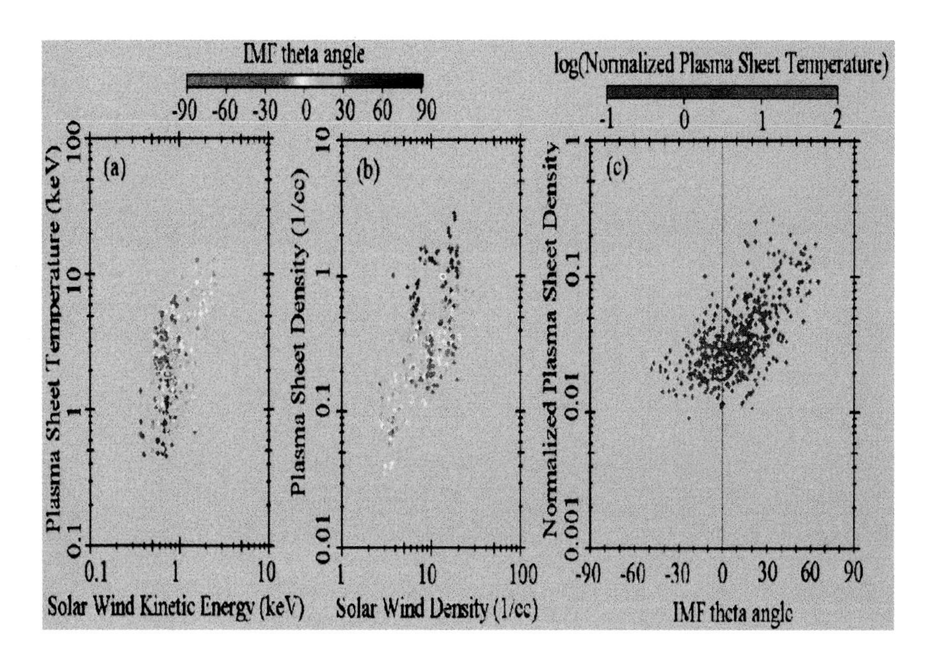

Figure 9. The same as Figure 8 but with the solar wind parameters averaged over 9 hours period.

transport from the flanks to the plasma sheet during a northward IMF period.

5. DISCUSSION

The structure of the tail-LLBL [e.g., *Mitchell et al.*, 1987; *Traver et al.*, 1991] and the impact of the plasma transport across it to the status of the plasma sheet [e.g., *Eatsman et al.*, 1985; *Lennartsson*, 1992] has been a long standing issue in the magnetospheric physics. In this paper, this unresolved problem is attacked by the study on dataset from Geotail.

The observations of the tail-LLBL can be summarized as follows: It is where the cold-dense plasma appears with stagnant flow signatures. The cold-dense ions have only slow tailward convection velocity, and are even flowing sunward in some parts. The mixing of energetic magnetospheric ions with the cold-dense ions from the magnetosheath is clear on the duskside, with the two components showing two distinct peaks in count rates resulting in a double-band structure in the Et diagram. Thermal electrons (< 300 eV) are enhanced in the field-aligned direction, with fluxes in both the parallel/anti-parallel directions being balanced. Higher energy electrons are found to be isotropic in pitch angles. Slow convection and these electron characteristics suggest the closed topology of the field lines.

We have also shown a case that strongly suggests that the cold-dense plasma entry from the flanks can be significant to fill a substantial part of the magnetotail. In this case, the cold-dense plasma is not spatially restricted to a layer attached to the magnetopause but is forming the cold-dense plasma sheet. In agreement with our preliminary survey suggesting higher efficiency of the flank entry for northward IMF, the fact that the plasma sheet changes to a cold-dense status as IMF becomes northward is shown by a statistical study. It is further shown that the cold-dense regime occurs most prominently near the dawn and dusk flanks, which is consistent with the idea of directly supplying the cold-dense ions from the flanks.

The formation mechanism of the tail-LLBL (the cold-dense plasma sheet) is the first question that one would come up with. Observations that have relevance to this topic are those of closed flux tubes flowing tailward. To remind the readers, it is shown that there are tailward flowing ($\sim$ -100 km/s) flux tubes containing mixed ions and bi-directional electrons, which hold essentially the same characteristics as those in the cold-dense plasma sheet (Figure 7). It is discussed that this similarity suggests the closed topology of the field lines, and that the flux tube are to become stagnant. It is also noted that this event was observed during strongly northward IMF period.

One possible mechanism of capturing the cold-dense plasma is the re-reconnection of open field lines which have one of their ends in the ionosphere and on which

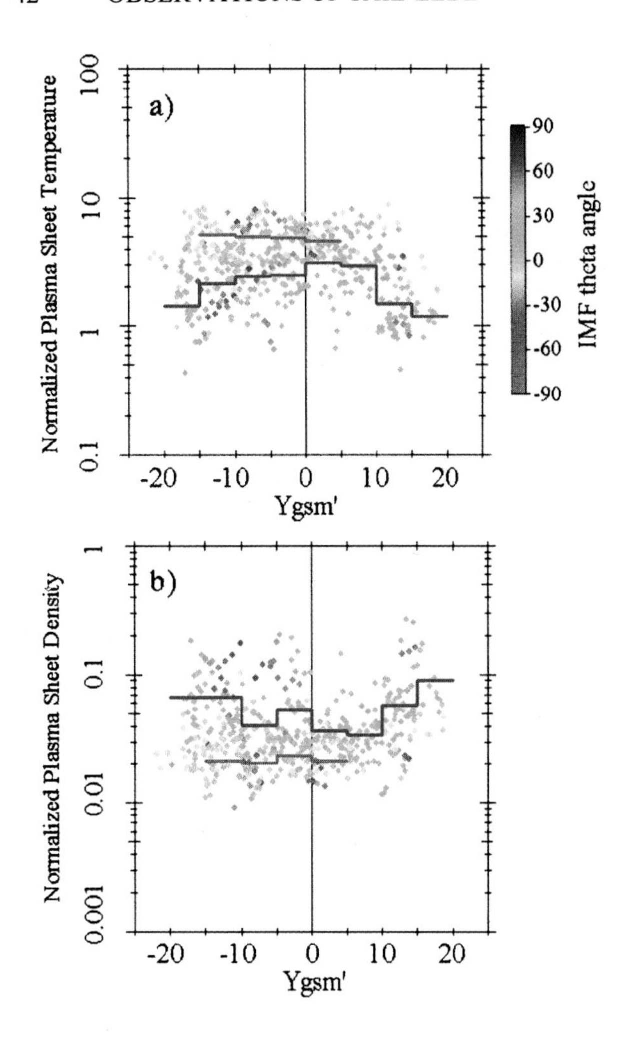

Figure 10. (a) Normalized temperature and (b) normalized density plotted against Y_{GSM}'. 9 hours averaged solar wind parameters are used. Color of the points depicts the IMF latitudinal angle. Blue solid lines shows the averages in 5 R_E bins for northward IMF (theta > 15°). The cold-dense status for northward IMF is most prominent at the dawn/dusk edge.

the magnetosheath plasma resides [e.g., *Song and Russell,*, 1992]. In the tail-LLBL, ions that used to be flowing tailward with an open field line find the re-reconnection to change the topology of the accompanying field line. They will be slowed down due to the field line tension [e.g., *Phan et al.*,1989; *Drakou et al.*, 1994]. Reconnection might also explain the electron's anisotropy: Electrons are heated in the parallel direction at the current layer and the bi-directionality results from their bounce motions along the closed field lines [e.g., *Fuselier et al.*, 1995] (Note however that *Fuselier et al.* considered this mechanism for open field lines). According to this model, detection of the tailward convecting closed field lines at a certain position would imply that such re-reconnection is already taking place at locations earthward, but not so far away as to leave enough time for full decceleration of the flow. The energetic ions on these field lines may have come from the neighboring magnetosphere by magnetic drifts after the closing of the field lines. It is worth noting that *Gosling et al.* [1986; 1991] report evidence for reconnection tailward of the dawn-dusk terminator. *Ogino et al.* [1994] and *Raeder et al.* [1995] report that a thick tail-LLBL (but with tailward flow) does form by this mechanism in their global MHD simulation with northward IMF. Direct evidence for reconnection closing the field lines at high-latitude is desired to prove the model, which may become available from INTERBALL data.

The above explains that the tail-LLBL is formed as closed flux tubes are added at the flanks of the magnetotail. Alternatively, the boundary layer may be formed as magnetosheath plasma is added to the closed flux tubes of the magnetotail. The ion mixing driven by the K-H instability [e.g., *Fujimoto and Terasawa*, 1994] is a possibility (Mixing and bi-directional heating of electrons are beyond the scope of the model that is based on hybrid simulation results). On the basis of a detailed analysis on the data before 9 UT shown in Figure 5 and on those from the interval a few hours prior, *Farifield et al.* [1996] proposes the K-H instability to be destabilized at the magnetospheric boundary for this day.

Although the mechanism of the low-latitude entry into the plasma sheet remains open, from the studies on the Geotail data, we are able to conclude that, in addition to the hot-tenuous plasma supply presumably via high-latitude lobes and the distant tail region, there clearly is direct supply of cold-dense ions from the flanks to the near-Earth plasma sheet. The latter becomes visible during prolonged northward IMF periods. This fact not only gives rise to a question of how the low-latitude entry become dominant during northward IMF periods, but also gives rise to renewed interest in the well-known plasma transport during southward IMF periods: How does the hot-tenuous plasma manage to almost exclude the cold-dense plasma from the plasma sheet? So far as the solar wind source is concerned, the near-Earth plasma sheet may well be envisaged as the mixture, or the mosaic of two different plasmas entrained into the magnetosphere by two different processes. Indeed, sudden switches from cold-dense to hot-tenuous state are occasionally observed [*Fujimoto et al.*, 1996; *Fujimoto et al.*, 1997], and they may be indicative mosaic-like structures of the plasma sheet. The statistical study reviewed in this paper has assumed that data at a point in the plasma sheet represents a global status. A revision in the future may be required to take this possible non-uniformity of the plasma sheet into account, which may become one of the targets of multi-satellite studies in the future.

Acknowledgments. M. F. acknowledges fruitful discussion with A. Nishida, T. Nagai, D. Farifield, T. Eastman, and D. Mitchell. M. F. and T. T. acknowledge stimulating discussion at the ISSI workshop "Source and loss processes of the magnetospheric plasma" (convener, B. Hultqvist) held in October, 1996. The key parameter data of WIND were provided by the NASA/GSFC data processing team.

REFERENCES

Baumjohann, W., G. Paschmann, and C. A. Cattell, Average plasma properties in the central plasma sheet, J. Geophys. Res., *94*, 6597, 1989.

Baumjohann, W., The Near-Earth Plasma Sheet: AMPTE CCE Perspective, *Space Sci. Rev.,64*, 141, 1993.

DeCoster, R. J., and L. A. Frank, Observations pertaining to the dynamics of the plasma sheet boundary layer, J. Geophys. Res., *84*, 5099, 1979.

Drakou, E, B. U. O. Sonnerup, and, W. Lotko, Self-consistent steady state model of the low-latitude boundary layer, J. Geophys. Res., *99*, 2351, 1994.

Dungey, J. W., Interplanetary magnetic field and the auroral zone, *Phys. Rev. Lett., 6,* 47, 1961.

Eastman, T. E., et al., The magnetopause boundary layer: Site of plasma, momentum and energy transfer from the magnetosheath into the magnetosphere, Geophys. Res. Lett., *3*, 685, 1976.

Eastman, T. E., L. A. Frank, and C. Y. Huang, The boundary layers as the primary transport region regions of the Earth's magnetotail, J. Geophys. Res., *90*, 9541, 1985.

Fairfield, D. H., et al., GEOTAIL observations of the K-H instability at the magnetotail boundary for parallel northward field, *Eos trans., AGU, 77,* Fall meeting, F615, 1996.

Fujimoto, M., et al., Structure of the low-latitude boundary layer: A case study with Geotail data, J. Geophys. Res., in press, 1998a.

Fujimoto, M., et al., Plasma Entry from the Flanks of the Near-Earth Magnetotail: GEOTAIL Observations, J. Geophys. Res., in press, 1998b.

Fujimoto, M., et al., The cold-dense plasma sheet: A GEOTAIL perspective, *Space Sci. Rev.,80*, 325, 1997.

Fujimoto, M., et al., Plasma Entry from the Flanks of the Near-Earth Magnetotail: GEOTAIL Observations in the Dawnside-LLBL and the Plasma Sheet, *J. Geoelectr. Geomag.,48*, 711, 1996.

Fujimoto, M., and T. Terasawa, Anomalous ion mixing within an MHD scale K-H vortex, J. Geophys. Res., *99*, 8601, 1994.

Fuselier, S. A., B. J. Anderson, and T. G. Onsager, Particle signatures of magnetic topology at the magnetopause: AMPTE/CCE observations, J. Geophys. Res., *100*, 11, 805, 1995.

Goslilng, J. T., et al., Accelerated plasma flows at the near-tail magnetopause, J. Geophys. Res., *91*, 3029, 1986.

Hall, D. S., et al., Electrons in the boundary layer near the dayside magnetopause, J. Geophys. Res., *96*, 7869, 1991.

Le, G., et al., ISEE observations of low-latitude boundary layer for northward interplanetary field: Implications for cusp reconnection, J. Geophys. Res., *101*, 27,239, 1996.

Lennartsson, W., A scenario for solar wind penetration of Earth's magnetic tail based on ion composition data from ISEE 1 spacecraft, J. Geophys. Res., *97*, 19,221, 1992.

McPherron, R. L., Physical Processes Producing Magnetospheric Substorms and Magnetic Storms, in *Geomagnetism,4*, Academic Press, 1991.

Mitchell, D. G., et al., An extended study of the low-latitude boundary layer on the dawn and dusk flanks of the magnetosphere, J. Geophys. Res., *92*, 7394, 1987.

Ogilvie, K. W., R. J. Fitzenreiter, and J. D. Scudder, Observations of electron beams in the low-latitude boundary layer, J. Geophys. Res., *89*, 10,723, 1984.

Ogino, T., R. J. Walker, and M. A. Abdalla, A global MHD simulation of the response of the magnetosphere to a northward turning of the interplanetary magnetic field, J. Geophys. Res., *99*, 11,027, 1994.

Phan, T. D., et al., The low-latitude dusk flank magnetosheath, magnetopause, and boundary layer foe low magnetic shear: WIND observations, J. Geophys. Res., *102*, 19,883, 1997.

Phan, T. D., B. U. O. Sonnerup, and, W. Lotko, Self-consistent model of the low-latitude boundary layer, J. Geophys. Res., *94*, 1281, 1989.

Popielawska, B., et al., An imprint of the quiet plasma sheet structure at the orbit of VIKING: Magnetosphere without substorms, in *Proc. of ICS-3*, ESA SP-389, 1996.

Raeder, J., R. J. Walker, and M. Ashour-Abdalla, The structure of the distant geomagnetic tail during long periods of northward IMF, Geophys. Res. Lett., *22*, 349, 1995.

Sckopke, N. G., et al., Structure of the low-latitude boundary layer, J. Geophys. Res., *86*, 2099, 1981.

Song, P., and C. T. Russell, Model of the formation of the low-latitude boundary layer for strongly northward IMF, J. Geophys. Res., it 97, 1411, 1992.

Spence, H. E., M. G. Kivelson, Contributions of the low-latitude boundary layer to the finite width magnetotail convection, J. Geophys. Res., *98*, 15,477, 1994.

Terasawa, T., et al., Solar wind control of density and temperature in the near-Earth plasma sheet: WIND and GEOTAIL collaboration, Geophys. Res. Lett., *24*, 935, 1997.

Terasawa, T., M. Fujimoto, H. Karimabadi, and N. Omidi, Anomalous ion mixing within a K-H vortex in a collisionless plasma, *Phys. Rev. Lett., 68,* 2778, 1992.

Thomas, V. A., and D. Winske, Kinetic simulation of the K-H instability at the magnetopause, J. Geophys. Res., *98*, 11,425, 1993.

Traver, D. P., et al., Two encounters with the flank low-latitude boundary layer: further evidence for closed field topology and investigation of the internal structure, J. Geophys. Res., *96*, 21,025, 1991.

Treumann, R. A., et al., Diffusion processes: An observational perspective, in *Physics of the magnetopause*, edited by P. Song et al., AGU monograph 90, AGU, Washington, D.C., 1995.

Williams, D. J. et al., Energetic particle observations in the low-latitude boundary layer, J. Geophys. Res., *90*, 5097, 1985.

Winske, D., et al., Diffusion at the magnetopause: A theoretical perspective, in *Physics of the magnetopause*, edited by P. Song et al., AGU monograph 90, AGU, Washington, D.C., 1995.

Zwolakowska, D., and B. Popielawska, Tail plasma domains and the auroral oval - Results of mapping based on the T89 magnetosphere model, *J. Geoelectr. Geomag.*, *44*, 1145, 1992.

M. Fujimoto, Department of Earth and Planetary Sciences, Tokyo Institute of Technology, Meguro 152-8551, Japan. (email: fujimoto@geo.titech.ac.jp)

T. Mukai, ISAS, Sagamihara 229-0022, Japan.

T. Tersawa, Department of Earth and Planetary Physics, Univ. of Tokyo, Bunkyo 113-0033, Japan.

Cold Dense Ion Flows in the Distant Magnetotail: The Geotail Results

Masafumi Hirahara,[1,2,3] Kanako Seki,[1] and Toshifumi Mukai[4]

The Geotail spacecraft frequently observed the cold dense ion flows (CDIFs) containing heavy ion species in the distant tail lobe/mantle region. The ion components always streamed tailward along magnetic field lines, and the flow velocities, parallel and perpendicular to the fields, were nearly the same for all ion species. The multicomposition ion flows are a consequence of the solar wind penetration into the cusp or the mantle and the mixing with the terrestrial ions outflowing from the ionosphere/plasmasphere, probably in the vicinity of the Earth. The ions flowing in the lobe/mantle were generally believed to be injected into the plasma sheet through the plasma sheet boundary layer (PSBL). The two proton flows with different bulk velocities were usually observed in the PSBL by Geotail: One is the high-speed flow of central plasma sheet (CPS) origin, and another is the component carried from the lobe/mantle. The parallel velocity of the high-speed flow was higher near the lobe/mantle than that near the CPS. On the other hand, the perpendicular velocities were almost equal for the two proton components and significantly increased as approaching the CPS. It is usual that the double components finally merged with each other in the CPS. The Geotail observations also suggest that a large amount of the ions flowing in the lobe/mantle would escape to the magnetosheath when the magnetosheath field lines were reconnected with those in the magnetotail. Then, the plasma at the magnetopause would consist of two components: A solar wind component injected into the magnetotail and a magnetospheric component outgoing from the lobe/mantle. The perpendicular velocities of the ions in the vicinity of the magnetopause generally increased as Geotail moved toward the magnetosheath. The velocity of the solar wind component changed at kinks of the open field lines. This type of plasma escape from the magnetotail into the magnetosheath could occur under not only northward but also southward interplanetary magnetic field conditions.

1. INTRODUCTION

It is prevalently known that the Earth's magnetosphere is formed by the interaction of the strong geomagnetic field with the solar wind carrying dense supersonic plasmas and weak interplanetary magnetic fields (IMF). According to the model proposed by *Dungey* [1963], the solar wind plasmas are injected efficiently into the magnetosphere, particularly in the cusp and mantle regions, along open geomagnetic field lines reconnected with the IMF at the dayside magnetopause under southward IMF B_z conditions or at the high-latitude magnetopause under northward B_z conditions. Both dayside and high-latitude reconnection processes were confirmed by many observational studies [e.g., *Paschmann et al.*, 1979,

[1]Department of Earth and Planetary Physics, Faculty of Science, University of Tokyo, Bunkyo-ku, Tokyo.

[2]On leave at University of Alabama and NASA Marshall Space Flight Center, Huntsville, Alabama.

[3]Now at Department of Physics, College of Science, Rikkyo University, Toshima-ku, Tokyo.

[4]Institute of Space and Astronautical Science, Sagamihara, Kanagawa, Japan.

New Perspectives on the Earth's Magnetotail
Geophysical Monograph 105

1985; *Sonnerup et al.*, 1981; *Gosling et al.*, 1982; *Fuselier et al.*, 1991; *Nakamura et al.*, 1996; *Maezawa*, 1976; *Gosling et al.*, 1991, 1996; *Kessel et al.*, 1996; *Le et al.*, 1996].

For an ideal dayside reconnection case, an amount of the solar wind plasma injected into the magnetosphere is penetrating into the cusp/mantle, reflected by magnetic mirror forces at low altitudes, and then carried into the magnetotail [*Smith and Lockwood*, 1996, and references therein]. Another important injection route of the solar wind is through open field lines crossing the nightside lobe/mantle magnetopause. In this process, the high-density solar wind component penetrates directly into the magnetotail without reflections at low altitudes and streams tailward in the lobe/mantle [*Gosling et al.*, 1984, 1985a; *Mozer et al.*, 1994]. When the solar wind exclusively contributes to the population in the magnetotail in this manner, the mass of the whole Earth's magnetotail would be dominated by the ions, namely with protons and α particles of solar wind origin in the lobe/mantle. *Lee and Roederer* [1982] theoretically estimated the total amount of the solar wind injected along the "open" lobe/mantle magnetopause. In this scenario, the plasma density of solar wind origin in the lobe/mantle and/or the volume occupied by the high-density plasma increase with increasing tailward distance from the Earth, as shown by *Zwickl et al.* [1984] and *Yamamoto et al.* [1994]. These features are consistent with an "expansion fan" model of the magnetotail, as theoretically studied by *Siscoe and Sanchez* [1987] and observationally diagnosed by *Siscoe et al.* [1994]. It should also be noted that the IMF components have significant effects on the spatial distributions of the solar wind plasma in the tail lobe/mantle, as demonstrated by *Gosling et al.* [1985b].

On the other hand, since late 1970's, ion mass spectrometers onboard high-altitude satellites began to provide us with a new perspective regarding source/origin of the magnetospheric plasma. *Balsiger et al.* [1980] studied the mass composition and population of energetic ions in the equatorial magnetosphere during storm times and showed that the majority of the ions observed under highly disturbed conditions is of terrestrial origin. The dependence of the heavy ion densities on the geomagnetic and solar activities was also investigated by *Young et al.* [1982]. The observations of ion streams and their composition in the magnetotail were crucial to study the acceleration and transport processes of the energetic ions of ionospheric origin [*Sharp et al.*, 1981; *Peterson et al.*, 1981; *Sharp et al.*, 1982; *Orsini et al.*, 1986]. Particularly in active periods, the behaviors of tailward flowing heavy ions are interesting from the viewpoint of the ionospheric plasma supply from the polar ionosphere into the plasma sheet [e.g., *Orsini et al.*, 1985; *Lennartsson et al.*, 1981, 1985; *Shelley et al.*, 1985]. On the basis of a statistical survey in low-altitude ion data, *Yau et al.* [1985] estimated outflowing ion fluxes originating from typical auroral/polar regions. Moreover, *Chappell et al.* [1987] qualitatively evaluated the total ionospheric ion fluxes supplied to the magnetosphere and proposed that the ionosphere could be a dominant contributor of the magnetospheric plasma. These works established the importance of the ionospheric plasma supply to the magnetosphere.

The works by *Frank et al.* [1977], *Orsini et al.* [1982], and *Mukai et al.* [1994a] suggested the existence of heavy ion (O^+) in the mid-distance magnetotail lobe. *Candidi et al.* [1982] reported that the slow and fast flowing O^+ ions tended to be detected near the plasma sheet boundary layer (PSBL) and the magnetopause, respectively. The results are consistent with the velocity filter effect in the magnetotail [*Lockwood et al.*, 1985; *Horwitz*, 1986; *Candidi et al.*, 1988].

Hirahara et al. [1996] adapted the $\mathbf{E}\times\mathbf{B}$ drift analysis for the Geotail energy-per-charge ion spectral data obtained with high-sensitivity, and successfully identified the heavy ion species in the distant lobe/mantle. *Seki et al.* [1996] also reported O^+ population coexisting with solar H^+ and He^{2+} components in the distant lobe/mantle during an active interval. These results of the heavy ions of ionospheric origin in the distant magnetotail raise an essential question concerning the magnetospheric plasma dynamics: How can the ionospheric ions be accelerated and transported to the distant magnetotail?

Recently, *Baker et al.* [1996] suggested that the cold ion beams originating from the near-Earth region could remain in the distant lobe/mantle when the plasma convection toward the plasma sheet becomes weaker than expected. On the other hand, the latest Geotail results by *Seki et al.* [1998] indicate that it is not decrease of the perpendicular velocity but increase of the parallel velocity that keeps the cold O^+ beams of ionospheric origin flowing in the distant lobe/mantle without being injected into the plasma sheet.

With respect to the works by *Seki et al.* [1998], it should also be mentioned that the cold O^+ beam events have a clear correlation with the geomagnetic activity and that the spatial distribution is strongly controlled by the IMF, which is similar to the spatial distribution of the mantle plasma of solar wind origin, as discussed by *Gosling et al.* [1985b]. Their statistical results also showed that the O^+ flows were observed particularly near the magnetopause. The dependence of the spatial distribution on the IMF B_y component also indicates that the dayside reconnection under the southward IMF conditions has a considerable influence on the outflowing ionospheric ions. These results regarding the features of the heavy ion flows are important in order to reveal the acceleration and transport of ionospheric ions into the distant magnetotail.

The heavy ion composition at the dayside and flankside magnetopause was studied by a number of works [e.g., *Peterson et al.*, 1982; *Fuselier et al.*, 1989a, b; *Gosling et al.*, 1990; *Eastman et al.*, 1990]. On the basis of the Geotail results, *Fujimoto et al.* [1997] presented signatures of O^+ ions of ionospheric origin in the duskside low latitude boundary layer (LLBL) and also showed that the structure of the LLBL was affected by strong IMF B_y components. Some of these works suggested the importance of dayside reconnection, and the results are useful in examining the dynamics of the heavy ions on the open field lines on the dayside and in the tail lobe/mantle, while the acceleration processes acting on the ionosphere-origin ions were not completely specified.

Here, we should also answer a fundamental question concerning the distant magnetotail plasma: Where will the distant lobe/mantle plasma of solar wind and/or ionospheric origins finally be transported? In the near-Earth regions, the low-energy lobe plasma is thought to be injected into the plasma sheet earthward of the neutral line and carried with closed geomagnetic field lines toward the Earth due to the global plasma circulation, as illustrated by *Freeman et al.* [1977], *Balsiger* [1981], and *Chappell et al.* [1987]. Also in the magnetotail beyond the distant neutral line, the lobe/mantle plasma could penetrate into the plasma sheet along the field lines across the PSBL.

Akinrimisi et al. [1990] suggested that characteristics of the mantle plasma gradually changed to those of the plasma sheet. *Orsini et al.* [1984, 1990] investigated flow patters of O^+ ions and structures of the electric fields on the basis of the ion measurements in the PSBL. *Eastman et al.* [1984, 1985] and *Nakamura et al.* [1992] exhibited properties of the PSBL and reported coexistence of the higher- and lower-speed ion flows. While these studies were important in terms of the plasma dynamics in the PSBL, the regions of their observations were limited relatively near-Earth. The detailed ion features in the distant PSBL remained unknown until Geotail realized the high-quality ion measurements in the farther magnetotail [*Hirahara et al.*, 1994].

The Geotail data are also useful to characterize micro-scale mechanisms occurring in the PSBL. The slow-mode shocks and their foreshocks between the dense plasma sheet and the tenuous lobe have been identified with the Geotail data by *Saito et al.* [1995, 1996]. It is generally believed that the shocks in the PSBL are responsible for the acceleration and heating of the plasma transported from the lobe into the plasma sheet. However, a number of the PSBL crossings of Geotail also show that the density in the lobe/mantle is higher than in the plasma sheet, and even under such a condition, the ions transported from the lobe/mantle to the plasma sheet are actually accelerated and heated in the PSBL. There are no definite answers for the question: What kinds of shock or discontinuity exist in the PSBL to energize the ions?

The reconnection actually occurs not only in the plasma sheet and at the high-latitude magnetopause but also at the tail-flank magnetopause adjacent to the plasma sheet and the distant lobe/mantle, as studied by *Gosling et al.* [1986] and *Hirahara et al.* [1997a]. These studies suggest that, after the injection into the Earth's magnetotail, the high density plasma of solar wind origin would finally be ejected from the farther magnetotail and flow away in the interplanetary space, sometimes with the unexpectedly dense heavy ions of terrestrial origin. The dynamics of the open field lines reconnected with the IMF should be subject to detailed studies because they have significant effects on the acceleration/deceleration and transport of the injected solar wind component and the magnetotail ions, especially in the vicinity of the magnetopause.

In this paper, we present the observational features of the cold dense ion flows (CDIFs) observed by the Geotail spacecraft in the distant magnetotail in order to discuss the behaviors, dynamics, and transport of plasmas in the magnetotail. First part of the next section describes general features of the single/multiple CDIFs observed during two periods when Geotail surveyed different magnetospheric regions. In the second part, we investigate the characteristics of the multicomposition ion flows of ionospheric origin and solar wind origin in the lobe/mantle. Third part is devoted to demonstrate the process of ion injection from the distant lobe/mantle into the plasma sheet through the PSBL. Properties of high-speed ion flows in the distant PSBL will also be exhibited. Through the magnetopause formed by open field lines, the solar wind ions penetrate into the lobe/mantle. The lobe/mantle plasma can be not only injected into the plasma sheet but also injected again into the magnetosheath through the magnetopause due to lobe/mantle reconnection. In the fourth part, these two types of ion transport through the magnetopause are examined on the basis of two observational results of magnetopause crossings. After then, we discuss possible sources and transport processes of the distant magnetotail plasma.

2. OBSERVATIONAL ASPECTS

We study three types of cold dense ion flows (CDIFs) observed by the Geotail spacecraft in the different regions of the distant magnetotail: The lobe/mantle, the PSBL, and the magnetopause. The data analyses presented in this paper are based on the measurements of the low-energy particle (LEP, *Mukai et al.* [1994b]) and the magnetic field (MGF, *Kokubun et al.* [1994]) experiments during two observational intervals: October 8, 1993 and November 17, 1993. These two cases contain many important aspects for investigating the properties of the magnetotail ion flows in detail and indicate fundamental physical processes occurring in the distant magnetotail.

2.1. General Features on the Regions and the Ion Flows

Figure 1 shows the general plasma and magnetic field features during the two intervals: the October 8 case (Figure 1a) and the November 17 case (Figure 1b). The observation sites were at (−142, 8, 8) R_E and (−209, −29, 3) R_E in the GSM coordinates, respectively. In both cases, Geotail moved from the distant magnetotail to the magnetosheath, where the ion flux was high and the magnetic field was fluctuating, as seen near the ends of the displayed intervals.

The October 8 data in Figure 1a indicate that Geotail stayed in the distant lobe/mantle region of the northern geomagnetic hemisphere, where both magnitude and direction of the magnetic field were pretty steady until the magnetopause crossing at ~1013 UT. On November 17, in the northern hemisphere Geotail frequently encountered the distant PSBL consisting of two-proton streams with different bulk velocities [*Eastman et al.*, 1984], as seen during the first 20-minute interval of Figure 1b (also see Figure 6). Geotail finally moved to the magnetosheath across the fairly thick magnetopause current layer during 1823-1828 UT when a clear energy-dispersed ion signature was observed. Also in this case, the region between the PSBL and the magnetosheath was the lobe/mantle, and the characteristics were basically similar to those of the October 8 event, although there was a rapid back and forth movement between the lobe/mantle and the magnetosheath at ~1807 UT.

The most noticeable feature common in the lobe/mantle and the magnetopause seen in Figure 1a and the PSBL in Fig-

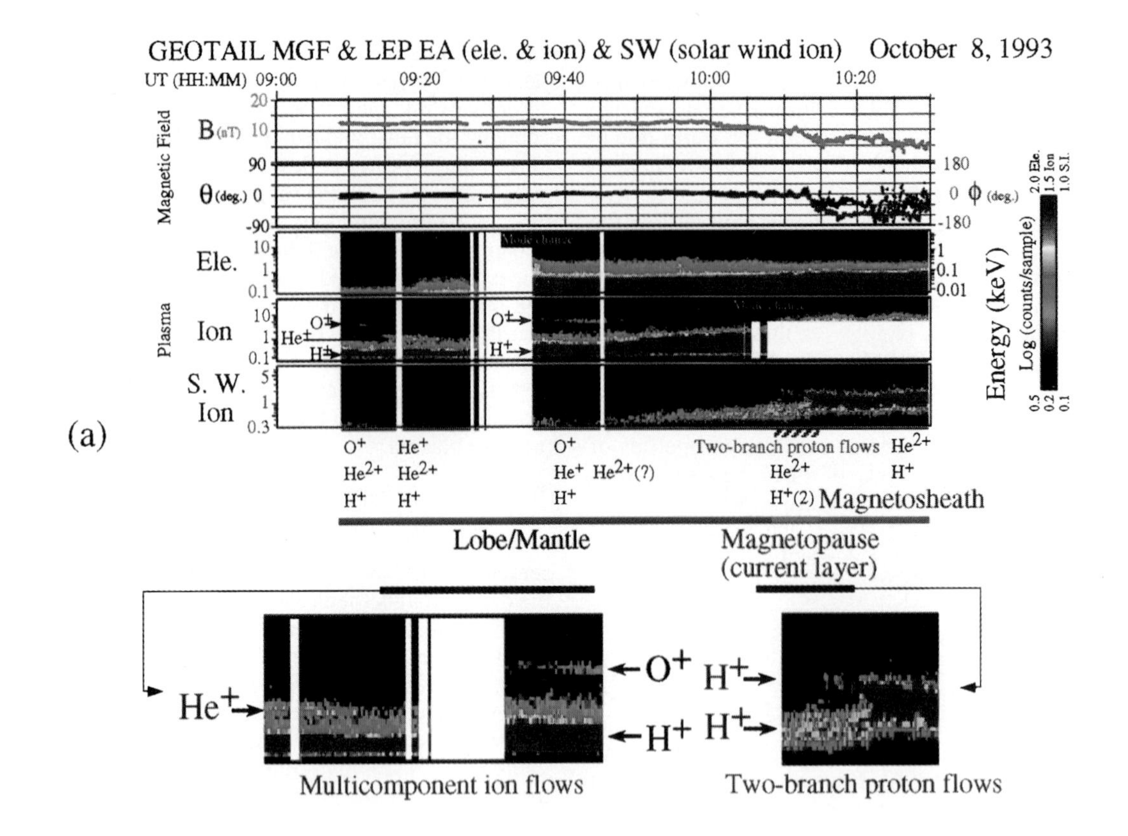
GEOTAIL MGF & LEP EA (ele. & ion) & SW (solar wind ion) October 8, 1993
UT (HH:MM) 09:00 09:20 09:40 10:00 10:20
Magnetic Field
B (nT)
θ (deg.)
φ (deg.)
Plasma
Ele.
Ion
S. W. Ion
Energy (keV)
Log (counts/sample)
Two-branch proton flows
Magnetosheath
Lobe/Mantle
Magnetopause (current layer)
He+
O+
H+
Multicomponent ion flows
Two-branch proton flows
(a)

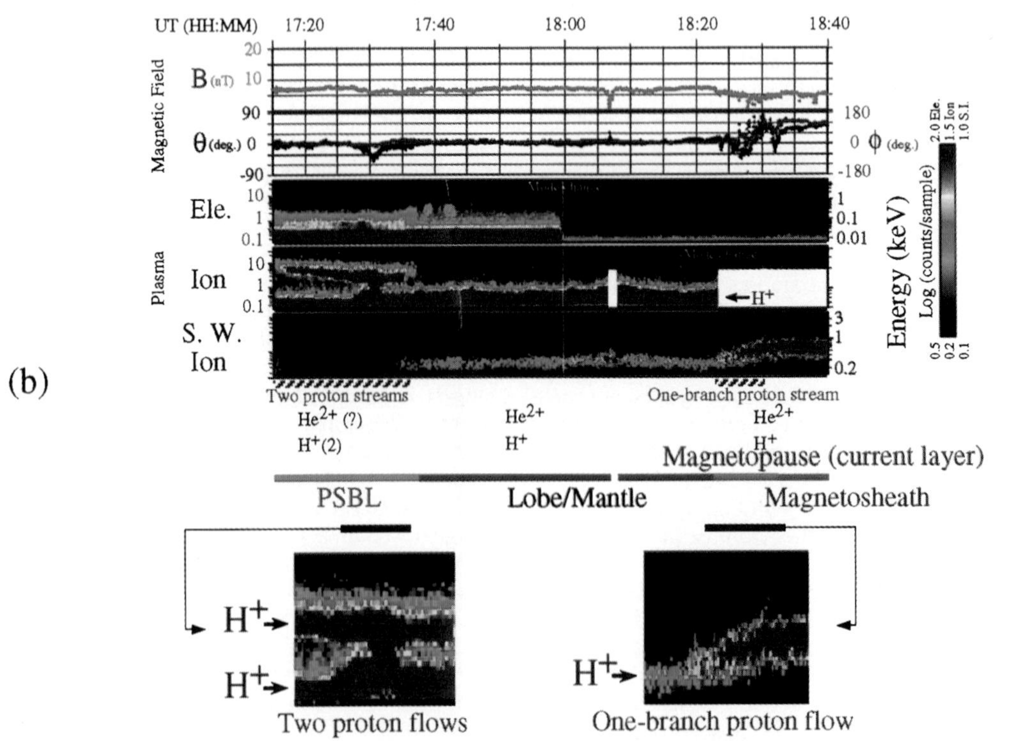
GEOTAIL MGF & LEP EA (ele. & ion) & SW (solar wind ion) November 17, 1993
UT (HH:MM) 17:20 17:40 18:00 18:20 18:40
Magnetic Field
B (nT)
θ (deg.)
φ (deg.)
Plasma
Ele.
Ion
S. W. Ion
Energy (keV)
Log (counts/sample)
Two proton streams
One-branch proton stream
Magnetopause (current layer)
PSBL
Lobe/Mantle
Magnetosheath
H+
Two proton flows
One-branch proton flow
(b)

Figure 1. Two examples of the cold dense ion flows observed in the lobe/mantle, the PSBL, and the magnetopause current layer by Geotail on (a) October 8, 1993 and (b) November 17, 1993, respectively. Each subset shows the magnetic field and low-energy plasma data obtained by MGF and LEP. The upper two panels show the magnitude (green dots) and directions (θ angle from the X-Y plane: blue dots, ϕ angle from the X axis in the X-Y plane: red dots) of the magnetic field. In the next two panels, omnidirectional energy-time (E-t) spectrograms of electrons and ions from the LEP-EA data are plotted. Third panel is an E-t spectrogram of tailward ($135° \leq \phi \leq 225°$) flowing ions observed by LEP-SW. The ordinate of the E-t spectrograms is energy-per-charge (keV/q), and the scales are logarithmic except for the range of LEP-EA below ~600 eV/q. Energy ranges of electron measurements changed once in each case (in Plate 1a from a wide energy mode shown on the left hand ordinate to a low energy mode on the right hand ordinate, and reversely in Plate 1b). The identifications of the regions are shown by color underlines below the third E-t spectrogram (red: magnetosheath, blue: lobe/mantle, green: magnetopause). The labels of ion species above these color lines of the region identifications show the ion composition specified in each region by assuming that all ion flows consisting of multiple composition have the same parallel and/or perpendicular velocities.

ure 1b is multicomponent ion flows at different energies (also see Figures 2 and 8 to know the flow directions). In Figure 1b, on the other hand, the distant lobe/mantle and the magnetopause were composed of one ion component. While the velocity-dispersed signature of the high-speed proton stream in the PSBL is well known [e.g., *DeCoster and Frank*, 1979; *Forbes et al.*, 1981; *Williams*, 1981; *Takahashi and Hones*, 1988; *Hirahara et al.*, 1994], as observed in the November 17 event, the Geotail results suggest that the energy dispersions are a common feature of the CDIFs not only in the PSBL but also in the lobe/mantle and the magnetopause current layer (see the bottom energy-time (E-t) spectrograms of Figure 1). It should also be noted that the densities of the lobe/mantle ion flows were unexpectedly high in both cases, ranged in 0.1-1 cm^{-3}, as discussed in the next subsection.

2.2. Multicomposition Ion Flows in the Distant Lobe/Mantle

At first, we focus on the lobe/mantle observations of the multiple component cold ion flows, as seen before ~1000 UT of Figure 1a. The geomagnetic activity was fairly high, and Kp was 5_- to 3_+ during 0600-1200 UT. The ground magnetogram data from the Canadian zone showed large negative bays for 3-6 hours at ~0900 UT. The provisional AE index shows large geomagnetic activities during 0600-1500 UT, and the peak value was about 1100 nT at ~0800 UT [*World data center C2 for geomagnetism*, 1995].

In the second E-t spectrograms labeled "Ion" on the left-hand of Figure 1a, we can see several ion components at different energy-per-charges. The multicomponent ion flows are identified as H^+, He^{2+}, He^+, and O^+ from the lowest-energy component, as indicated by arrows in the panel and by labels below the bottom ion spectrogram. The validity of the identification is shown later by using the parallel flow speed and $\mathbf{E}\times\mathbf{B}$ drift velocity analyses. It is worthy to note that the clear He^+ and O^+ components rarely overlapped with each other, while the CDIF consisting of H^+ was continuously observed.

Figures 2a and 2b present the velocity distribution functions of the multicomponent ion flows during the intervals without and with the O^+ component, respectively. The ion species expected from the velocity moment analyses are indicated by arrows at the corresponding peaks of the distributions. Although the distribution corresponding to He^+ was also seen in Figure 2b, the peak was almost by an order lower than that of Figure 2a in which no O^+ flux was detected. The He^+ count rates were accumulated for 156 s (52 spins) in Figure 2b, and the count rates in each of energy spectra (12-s snapshots) were too low to calculate the moments precisely. Although it is not evident because of finite energy resolution, the distributions of the multiple ion components could be fitted by shifted Maxwellians, and there is a small bump on the high-energy tail of the H^+ distribution, probably due to the solar He^{2+} population.

The density and velocity moment results for each ion species are shown in Figure 3. Although there are small discrepancies in the parallel and perpendicular velocity components because of ambiguities by finite energy resolution and low count rates, we can conclude that the ion mass-per-charges are successfully identified since the velocity components are nearly equal for all ion species.

The H^+ density ranged 0.4-0.9 cm^{-3}, and the He^{2+} population was several percents of the major H^+ ions, which is roughly consistent with the density ratio of the typical solar wind population. However, we should note that the He^{2+} density may be somewhat underestimated and the velocity would be overestimated since the distribution generally overlapped the high-energy tail of the dominant H^+ distribution and the whole counts of He^{2+} cannot be picked up correctly, as inferred from Figure 2a. It is interesting that the He^+ density and flux during 0915-0925 UT were larger than those of O^+ at ~0910 UT and during 0936-0948 UT on the average. This feature is important for the discussion of possible sources and acceleration processes producing the heavy ion flows in the distant magnetotail.

As seen in the bottom ion spectrogram, the CDIF density and velocity gradually increased as the spacecraft approached the magnetopause. Figure 4 depicts an example of density-velocity (n-V) relation for the lobe/mantle ions during 0940-1010 UT, which is similar to the previous Geotail results reported by *Siscoe et al.* [1994] and *Seki et al.* [1996]. The striking n-V correlation is one of evidence for the solar wind injection into the lobe/mantle region through the "open" magnetopause, besides the existence of the He^{2+} population (also see the later discussion on the magnetopause CDIFs using Figure 8). It is apparent that the magnetic field strength began to decrease at ~1000 UT, probably due to the high-density H^+ population near the magnetopause. Although

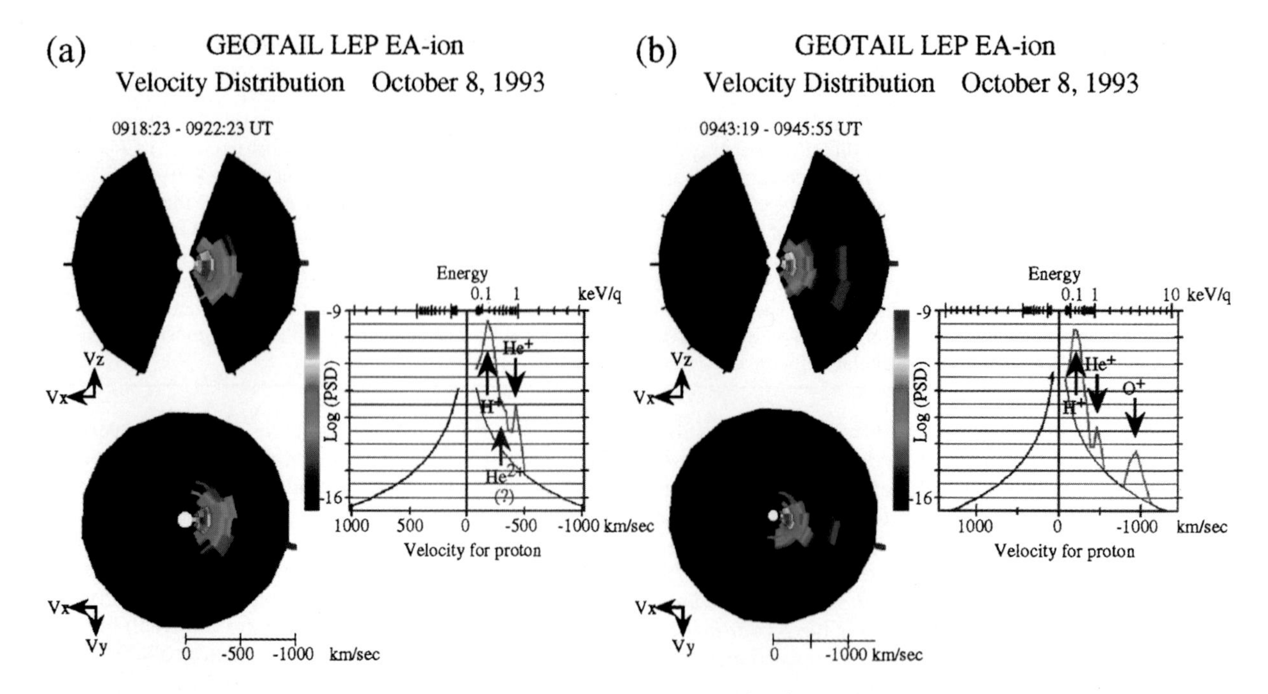

Figure 2. Two examples of velocity distribution functions of cold H^+, He^+, and O^+ flows observed by LEP-EA in the distant lobe/mantle. Two left-hand color contour plots show two-dimensional cross sections in the V_x-V_z and the V_x-V_y planes in the GSE coordinates from the top, respectively. Smooth thin curves in the cuts of the distributions plotted in the right-hand graphs indicate the one-count level. Because the speed is calculated on the assumption that all ions are protons, it is necessary to convert to the speeds of the actual ions by division of factor of square roots of ion mass-per-charges. (a) Two clear peaks and a small bump show the tailward cold ion flows consisting of H^+, He^+, and He^{2+}. (b) Three clear peaks correspond to the cold H^+, He^+, and O^+ flows.

a similar n-V correlation was also seen in the lobe/mantle (1735-1823 UT) of the November 17 event (not shown here), it is not as clear as the result shown in Figure 4. It is noteworthy that the time variation of the O^+ density was different from that for H^+ while the parallel velocities of both ion species varied in the same manner, as seen in Figures 3a and 3b during the interval 0935-0948 UT. This implies that the O^+ density anticorrelated or had no correlation to the parallel velocity component.

Taking into account that this lobe/mantle observation was made in the northern hemisphere, the negative $V_{\perp_z}$ components are reasonable, which suggests that the lobe/mantle plasma was convected toward the plasma sheet. The features are consistent with the tailward and southward flow vectors seen in the upper contour plots of Figure 2 (the cuts in the V_x-V_z plane).

Finally, we mention that electron anisotropic flow distributions during the lobe/mantle CDIF observations in Figure 1 were similar to polar rain components, as discussed by *Baker et al.* [1997] on the basis of the Geotail data. Also in their polar rain events, the multicomponent CDIFs were clearly observed. The upper cutoff energy of the polar rain components in the cases of Figure 1a was about 600 eV, and the components flowing tailward and earthward were slightly more intense than the dawn-dusk components.

2.3. *Double Proton Flows in the Distant PSBL*

In this subsection, we examine properties of the CDIFs observed in the distant PSBL during 1715-1735 UT of the November 17 event in Figure 1b. The characteristic flow energies (bulk velocities) of the higher-energy ions in the PSBL gradually decreased as approaching the central plasma sheet (CPS), while the lower-energy component showed the opposite tendency (see the energy variations at ~1731 UT where the θ angle of the magnetic field had a large V-shaped negative excursion).

We can determine the ion composition in the PSBL by using the $\mathbf{E}\times\mathbf{B}$ drift analysis, in the same way for the multicomposition cold ion flows in the lobe/mantle. Assuming that all components of the ion flows in the PSBL consisted of protons, the perpendicular components were almost equal for the two components, as shown in Figure 5d, although there were a few short-term deviations near the beginning and end of the plots and at ~1726 UT. These discrepancies took place when the density of the higher-energy component was low or suddenly dropped. Compared with those in the lobe/mantle, the perpendicular velocities of the double proton flows in the PSBL were significantly enhanced with increasing elevation angle of the magnetic field, while the variations of the parallel velocities were opposite. The difference of the parallel velocity components between the higher- and lower-energy proton flows, at times, was >600 km/s. The difference was significantly larger than the local Alfvén speed (<~250 km/s) in the PSBL, suggesting that plasma waves could be driven in the PSBL by two-beam ion instability.

The density of the higher-energy proton flow gradually increased up to ~0.1 cm^{-3} in the deep PSBL near the CPS, while that of the lower-energy protons gradually decreased from 1 cm^{-3} to 0.1 cm^{-3}. Although the spacecraft did not entirely enter the CPS in the present case, the density in the CPS might be less than 1 cm^{-3}. On the other hand, the ion density in the lobe/mantle was always more than 1 cm^{-3}. One important possibility expected from the PSBL observations presented here is that the density in the lobe/mantle adjacent to the PSBL was higher than those in the PSBL and the plasma sheet, which is discussed in the next section.

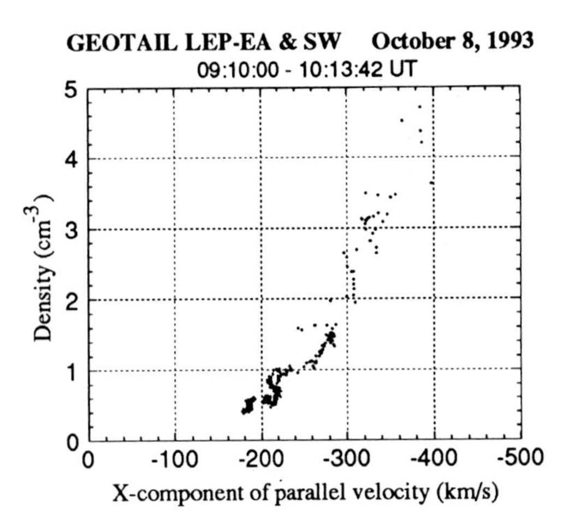

Figure 4. Density-velocity (x component of parallel velocity) scatter plot for the ions observed in the lobe/mantle on October 8.

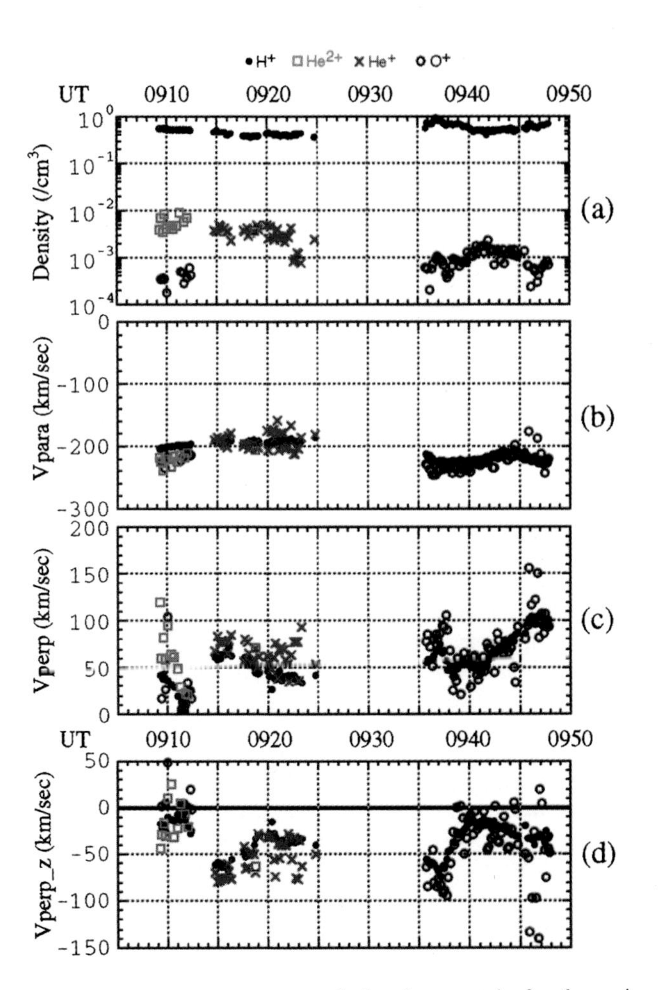

Figure 3. Flow parameters calculated separately for the major discrete ion components observed in the distant lobe/mantle (black dots: H^+, green open rectangles: He^{2+}, red crosses: He^+, blue open circles: O^+). From the top, (a) ion densities, (b, c) velocities parallel and perpendicular to the magnetic field, and (d) z component of the perpendicular velocity are plotted.

Figure 6 shows two examples of the distribution functions of the double proton flows measured in the middle of the PSBL and in the PSBL near the CPS, respectively. In both examples, each of the higher-energy proton components showed a "crescent" distribution, as surrounded with solid curves. Also in the cut of the distribution shown on the right-hand of each subset, the distribution function could not be fitted to a shifted Maxwellian, which is different from the cases for the multicomposition ion flows in the lobe/mantle. In contrast, the distribution of the lower-energy CDIFs can be expressed by a shifted Maxwellian.

The higher-energy proton flow directions were almost field-aligned, as illustrated in an expanded contour plot with the magnetic field direction in Figure 6b. On the other hand, the comparison of the lower-energy CDIF distributions in Figures 6a and 6b suggests that the velocity component increased mainly in the perpendicular direction. Because the perpendicular velocity components were nearly the same for both proton flows, the $\mathbf{E}\times\mathbf{B}$ drift velocity should be thought to increase, particularly in the deep PSBL. As seen in Figure 5c, because the magnetic field strength did not change considerably in the PSBL encounter, the increase of the electric field component would mainly be responsible for the $\mathbf{E}\times\mathbf{B}$ drift enhancement.

The evolution of the energy and the azimuthal flow angle for the lower-energy CDIFs of lobe/mantle origin is shown in Figure 7. The energy of the lower-energy component significantly increased with approaching the CPS, while that of the higher-energy flow slightly decreased. The azimuthal flow direction (ϕ) of the higher-energy component did not show as large of variations as seen in the lower-energy component of lobe/mantle origin.

It is logical to consider that the distant neutral line was located earthward of the observation site because all of the ions were flowing tailward and the θ angle of the magnetic field in the PSBL was negative. This is consistent with that the observation site was extremely downtail (~209 R_E in tailward).

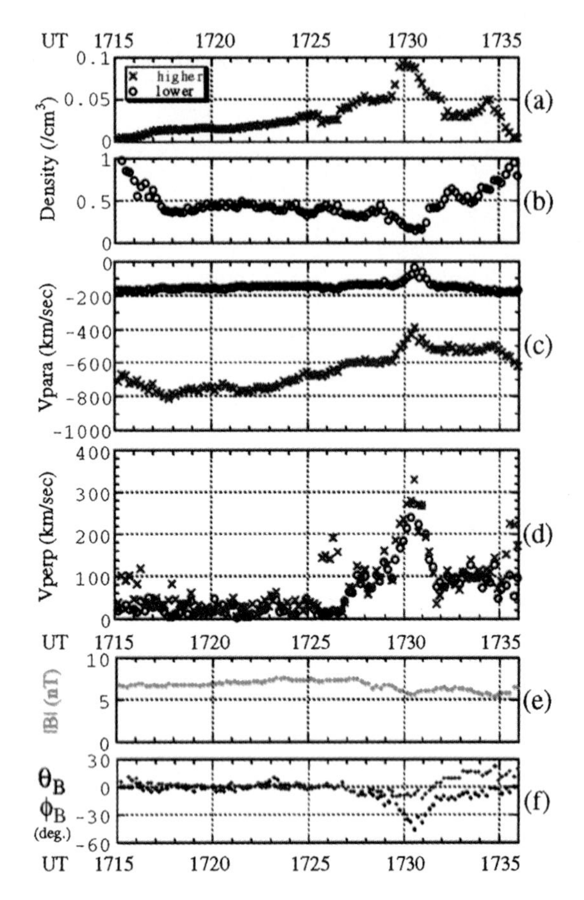

Figure 5. Flow parameters calculated separately for the double proton flows observed in the distant PSBL (red crosses: higher-energy component, blue circles: lower-energy component). The top two panels (a and b) show ion densities of the higher- and lower-energy components. The panels c and d show the flow velocities parallel and perpendicular to the magnetic field (**B**). The bottom two panels (e and f) are for the magnetic field (green, blue, red dots: magnitude, θ and ϕ angles of **B**).

2.4. Single/Double Proton Flows at the Distant Magnetopause

Geotail often crossed or encountered the distant lobe/mantle magnetopause. The examples shown in Figure 1 show two kinds of the plasma features during the magnetopause crossings, both of which are typical in the Geotail dataset. The most noticeable difference between these two magnetopause crossing events is a characteristic of the ion flows. Namely, the October 8 event showed double-component CDIFs, whereas the ion flow was single in the November 17 event. Flow properties of the single CDIF in Figure 1b changed continuously from the solar wind component in the magnetosheath. With regard to the "two-branch" CDIFs at the magnetopause of Figure 1a, the lower-energy component was smoothly connected with the proton flow in the lobe/mantle. It is also likely that a portion of the magnetosheath component existed as the higher-energy CDIF in the magnetopause current layer.

Figure 8 displays the contour plots in the V_x-V_y plane and the cuts in the directions containing the major peaks of the ion distribution functions during the two magnetopause crossings. *Candidi et al.* [1984] also reported a double proton population in the magnetotail, similar to our October 8 event (see Figure 8a). The ion mass identification for the double CDIFs on the basis of the $\mathbf{E}\times\mathbf{B}$ velocity analysis indicates that both higher- and lower-energy CDIFs consisted of protons, as evident from the equality of the perpendicular velocity component ($V_\perp$) calculation shown in Figure 9c. Although two velocity moment results at 1009:52 and 1012:25 UT show large discrepancies between the $V_\perp$ components of the higher- and lower-energy CDIFs, it should also be noticed that the count rates of the higher-energy component might not be enough to obtain precise results.

It is worthy to mention that the He^{2+} population of solar wind origin could also be measured as a small bump at the higher-energy tail of the dominant proton distribution in each case of Figures 8a and 8b, as indicated in the right-hand graphs. The He^{2+} component which should also be detected with the higher-energy proton flow of the October 8 event was unclear due to the coarse resolution at high-energies.

The large variations of the velocities for both components were seen mainly during the interval 1013-1014 UT when the magnetic field direction significantly changed, as plotted in Figure 9e. The parallel velocities ($V_{//}$) of the double CDIFs were different from each other and decreased as the spacecraft approached the magnetosheath. The velocity difference was usually larger than the local Alfvén speed, which is similar to the observations of the double proton flows in the PSBL. After the Geotail entered the magnetosheath, the two components were not distinguishable because the dominant velocity component in the magnetosheath was not $V_{//}$ but $V_\perp$.

During the magnetopause crossing (1013-1015 UT), the magnitude and direction of the magnetic field gradually varied, probably because of the finite thickness of the current layer. The magnetic field direction in the magnetosheath was basically south-dawnward, and the ϕ_B angle was $<-90°$, suggesting there was a small antiparallel component between the magnetic fields in the magnetosheath and the northern lobe/mantle.

The higher-energy CDIF was observed not only in the magnetopause current layer but also in the lobe/mantle adjacent to the magnetopause, as seen in Figures 1a and 9. The higher-energy signature in the lobe/mantle was intermittent, probably because the magnetic flux tube containing the higher-energy CDIF was thin and Geotail often missed the tube. This interpretation may indicate that the solar wind component was not always accelerated and injected efficiently along the open magnetic field lines. In the lobe/mantle before 1013 UT, the flow velocity of the higher-energy component showed no large variations except for the two excursions mentioned above, and the number density was small ($\sim$0.2 cm^{-3} on the average), almost by an order lower than the magnetosheath ion density.

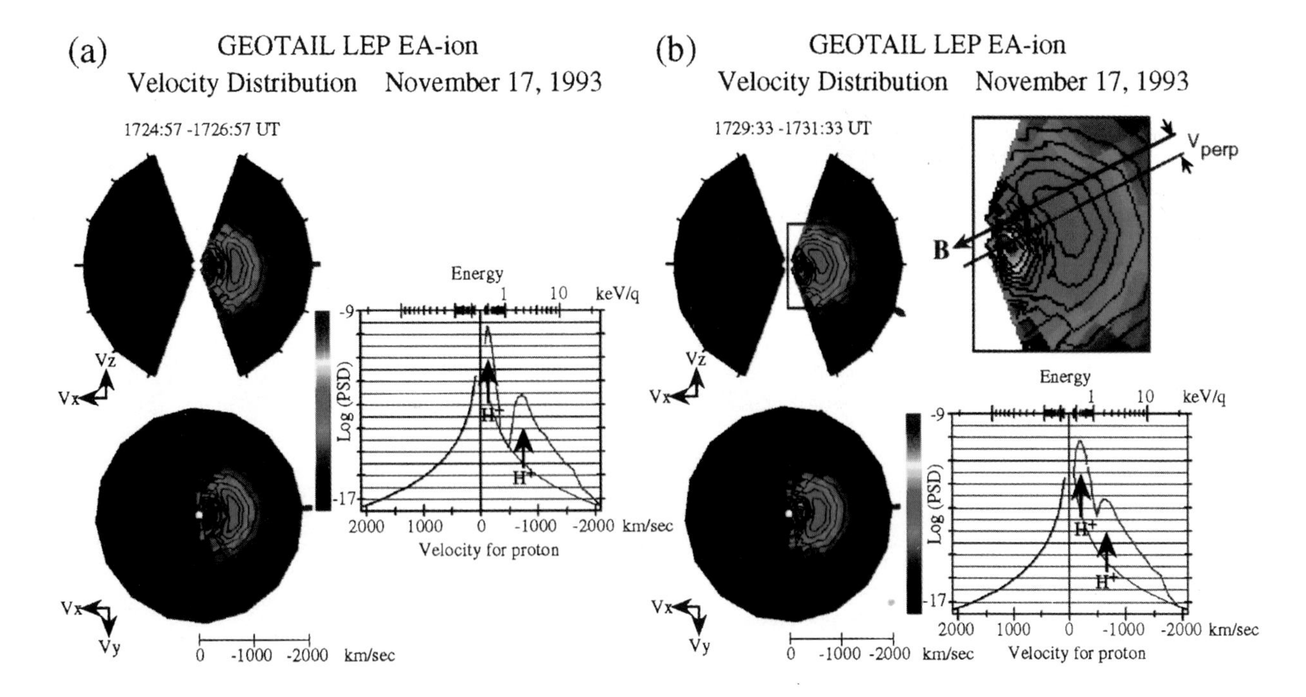

Figure 6. Two examples of the velocity distribution functions of the double proton flows observed by LEP-EA, (a) in the middle of the PSBL and (b) in the PSBL near the CPS, respectively. The contour plots show two-dimensional cross sections in the directions marked by red ticks. An expanded contour in b shows that the perpendicular velocities of the two proton flows due to the large **E**×**B** drift were almost the same with each other and that the convection was toward the plasma sheet.

3. DISCUSSION ON SOURCE AND TRANSPORT OF MAGNETOTAIL PLASMAS

3.1. *On Ion Flows in the Lobe/Mantle*

The largest difference between the observations in the lobe/mantle on October 8 and November 17 is regarding the existence of heavy ion flows. One of the causes is due to the difference of the geomagnetic activity. The multi-composition CDIFs were observed more frequently during active times than during quiet times, as reported by *Seki et al.* [1998], which is consistent with the results presented in this paper (Kp=5_- and the maximum provisional AE value was ~1100 nT for the October 8 event, whereas Kp=0_+ and the maximum AE was ~200 nT for the November 17 event [*World data center C2 for geomagnetism*, 1995]).

Hirahara et al. [1996] reported the existence of abundant He^+ ions in the lobe/mantle at $(-90, +17, +5)\ R_E$ of the GSM during the recovery phase (Kp=5_+). The geomagnetic condition was similar to the October 8 event. Also in both events, the peak flux of He^+ was larger than that of O^+. *Gosling et al.* [1990] presented velocity distributions of multicomposition ion flows consisting of H^+, He^+, and O^+ observed in the dayside LLBL, in which all ion species were flowing in nearly the same velocities. These features are similar to the distributions observed in the distant lobe/mantle, as shown in Figure 2b. It is likely that the multicomposition ion flows observed in the dayside LLBL could be transported to the distant tail lobe/mantle [*Fujimoto et al.*, 1997].

The ion composition at the dayside magnetopause may be of plasmaspheric origin, as discussed by *Fuselier et al.* [1989a] and *Gosling et al.* [1990]. The heavy ions could be accelerated due to reconnection process and/or motion of open field lines reconnected at the dayside magnetopause, and transported to the magnetotail owing to the global plasma circulation, as illustrated by *Freeman et al.* [1977]. *Elphic et al.* [1997] recently proposed that the high-density cold heavy ions observed in the distant lobe/mantle are a consequence of the acceleration and circulation process acting on the dayside plasmaspheric ions during active times. The density ratio between He^+ and O^+ observed in the distant lobe/mantle by Geotail seems consistent with this scenario. On the other hand, the ionosphere is also important to supply the heavy ions directly into the dayside magnetopause, as discussed by *Fuselier et al.* [1989b].

In the Geotail dataset of the distant lobe/mantle observations, the occurrence probability of the He^+ flows is much less than that of the O^+ flows (more detailed statistical surveys of the He^+ flows are being performed). If most of the heavy ions of the multicomposition flows had plasmaspheric origin, Geotail should have observed the He^+ signatures more frequently. On the other hand, the occurrence probability of the heavy ions in the distant tail seems consistent with the idea that the ultimate source is polar ion outflows or upward flowing ions (UFIs), while the acceleration process for the polar ion outflows and the flux balance between the ion outflows and the magnetotail CDIFs remain unclear.

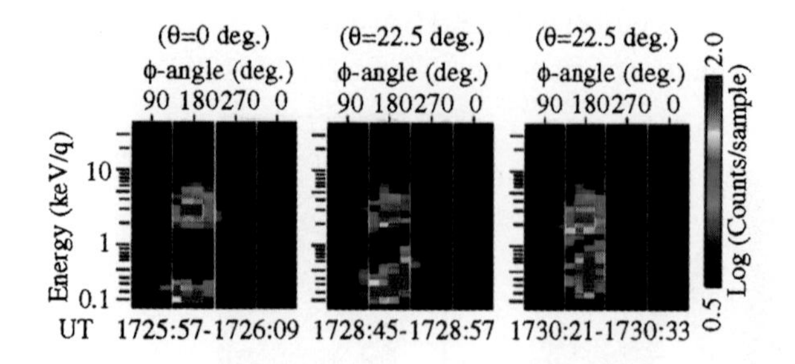

Figure 7. Energy-ϕ angle (E-ϕ) spectrograms of the double proton flows in the distant PSBL. Three spectrograms show the variations of energy and flow angle observed in the middle to the deep PSBL from the left. The first and last intervals of the E-ϕ spectrograms are included in the distribution plots in Plates 5a and 5b, respectively.

It is possible that the extremely high-density ($\sim$1 cm^{-3}) and/or high-temperature (>$\sim$50 eV) H^+ in the lobe/mantle may mask tenuous components whose energy-per-charge ratios are close to that of H^+, so that it might be difficult to identify the He^+ populations. However, detailed surveys of the ion *E-t* spectrograms guided by eyes indicate that the occurrence probability of He^+ flow is much lower that of O^+ even for the cases in which the H^+ density nor temperature was not high (the results will be presented in the near future).

Here, we focus on another possible model in order to explain the density (flux) ratio and non-coexistence of the heavy ion (He^+ and O^+) flows in the distant lobe/mantle in Figures 1-3. On the basis of the other Geotail observations in the distant tail, it can also be suggested that the maximum density of He^+ is frequently higher than that of O^+, whereas the occurrence probability of the He^+ flows is much lower than that of the O^+ flows. If the high-energy UFI beams were the source of the heavy ion flows in the distant tail, these features could consistently be explained. In fact, Geotail observed field-aligned counterstreaming ions in the equatorial region near the dayside magnetopause, and we interpret that the ion flows were of ionospheric origin and low-altitude parallel electrostatic potential drops were responsible for the acceleration [*Hirahara et al.*, manuscript in preparation].

Parallel electrostatic potential drops in the low-altitude auroral regions can accelerate the ionospheric ions to >10 keV/*q* [e.g., *Hultqvist et al.*, 1988]. The field lines with the high-energy UFIs are usually closed, and the UFI beams would be bouncing on the field lines. *Hirahara et al.* [1997b] reported the energy-dispersed ionospheric ion signatures bouncing in the dawnside sector on the basis of low-altitude satellite data, and concluded that these heavy ions with peak energies of $\sim$10 keV/*q* were of UFI beam origin.

It is generally believed that the global convection transports the dawnside and duskside closed field lines toward the dayside magnetopause. After the sunward convection, the field lines could be reconnected with the IMF at the dayside magnetopause to be open, preferentially under southward IMF conditions. Then, some amount of the high-energy ionospheric ions of UFI beam origin would be transported into the magnetotail with the solar wind component injected along the open field lines.

It is plausible that the velocity filter effect would discriminate the energy-per-charges of these multicomposition ion flows of ionospheric and solar wind origins, so that the parallel velocities are equal for all ion species in the distant lobe/mantle. It should be mentioned that parallel electrostatic potential drops accelerate ionospheric ions not to same velocities but to same energy-per-charges. The energy width of UFIs accelerated by a potential drop is sometimes narrow at a given position, namely, on a given field line, although the whole distribution of a potential drop usually shows a wide energy range from several tens of eV to $\sim$10 keV, as seen in "inverted-V" electron events. If the energy width is narrower than the mass-per-charge ratio (1:4) between the He^+ and O^+ ions, we cannot observe both ion flows at the same time (on the same field line), because He^+ and O^+ with the same parallel velocities do not originally coexist on the field line. Furthermore, the fluxes of ions accelerated by potential drops are usually lower at higher energies than at lower energies. If we adopt the UFI-source model that the acceleration

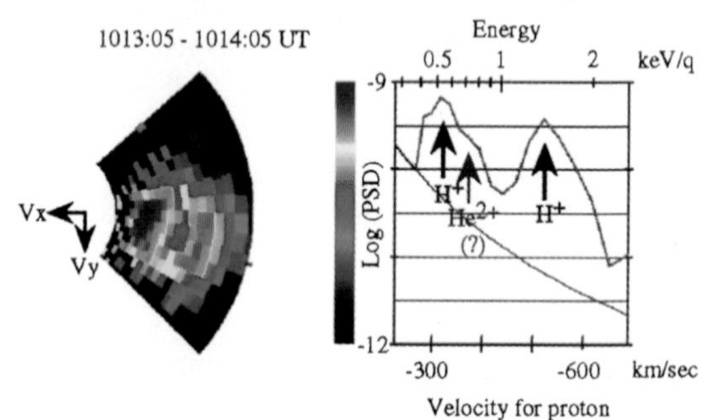

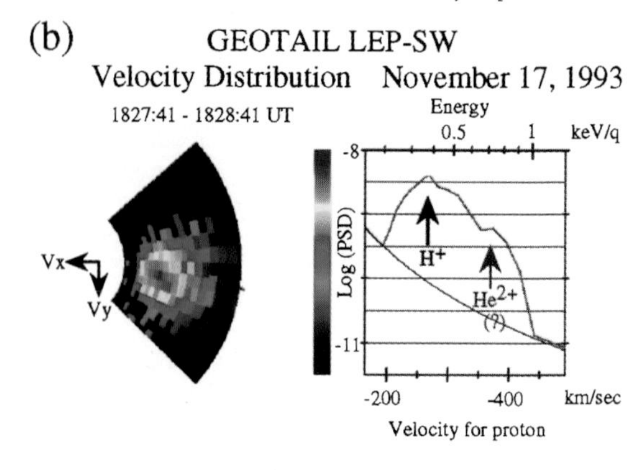

Figure 8. Two examples of the ion velocity distribution functions observed in the magnetopause current layer by LEP-SW, (a) for the October 8 event with double proton components and (b) for November 17 event with a single proton component, respectively. Color contour plots are in the V_x-V_y plane in the GSE coordinates. The right graphs show the distributions in the directions marked by red ticks in the contour plots.

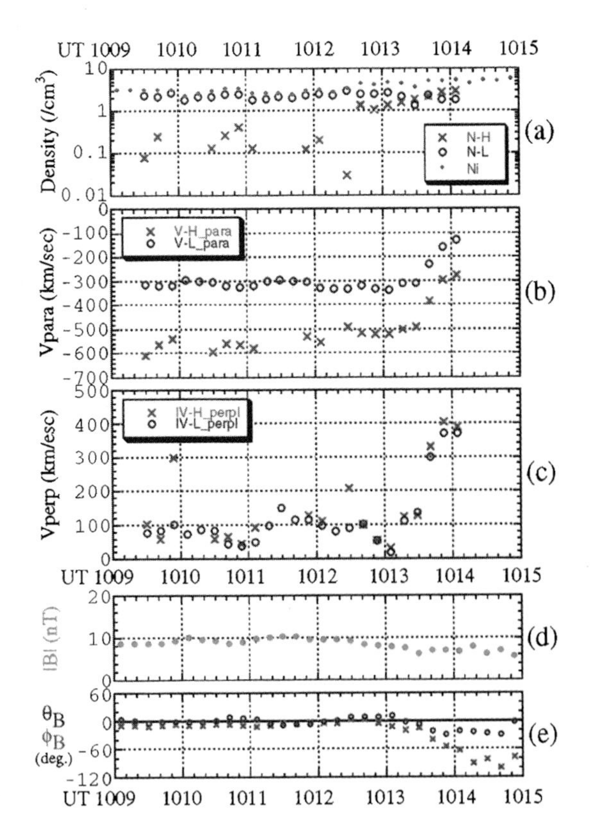

Figure 9. Flow parameters calculated separately for the double proton flows observed at the distant magnetopause of the October 8 event (red crosses: higher-energy component, blue large crosses: lower-energy component). From the top, (a) ion densities, (b, c) velocities parallel and perpendicular to the magnetic field (**B**) are shown, respectively. The bottom two panels (d, e) are for the magnetic field (green dots: magnitude of **B**, blue small circles: θ of **B**, red small crosses: ϕ of **B**).

process of the heavy ions is a parallel electrostatic potential drop at low altitudes and we assume that the ion velocities in the distant lobe/mantle are discriminated by the velocity filter effect during the transport, we could qualitatively interpret the observational results about the He^{+}-O^{+} density (flux) ratio and the non-coexistence of these ion flows. While *Seki et al.* [1998] discuss the several other mechanisms which could cause the multicomposition ion flows with the same bulk velocities in the distant lobe/mantle, we propose that the alternating (non-coexisting) occurrence of O^{+} and He^{+} ions would provide a crucial key in revealing the source, acceleration, and transport of the ionosphere-origin ion flows.

3.2. *On Ion Flows in the PSBL*

While it is generally believed that the slow-mode shocks and their foreshocks are often responsible for the acceleration and heating of the ions in the PSBL [e.g., *Saito et al.*, 1995, 1996], slow-mode shocks would not be identified in the PSBL if the CDIF density in the lobe/mantle was higher than that of the plasma sheet. The variations of the plasma and magnetic field features were gradual, and the slow-mode shock could not be identified in the PSBL observations presented here, although Geotail did not completely cross the PSBL unfortunately. On the other hand, not only the present case but a number of Geotail observations in the distant tail demonstrate that the plasma sheet density was lower than the lobe/mantle density and the magnetic field variation was gradual in the PSBL. The energization of the lower-energy component is obviously recognized even during such PSBL crossings. These results would require us to deduce a new paradigm with respect to the plasma transport and acceleration mechanism occurring in such a peculiar type of the PSBL structure.

The magnitude of the electric fields in the distant PSBL can be roughly estimated to be 0.2-2 mV/m for the November 17 event, in which the perpendicular velocities ranged 30-300 km/s. This estimate is smaller than the amplitudes (5-10 mV/m) of the electric fields measured in the near-Earth regions ($X_{GSE} > -22\, R_E$) [*Pederson et al.*, 1985]. While *Hirahara et al.* [1994] reported that the magnitude of the electric fields in the distant PSBL was 2-5 mV/m from the **E**×**B** analysis, the direction of the electric fields was not always stable in the dawn-to-dusk direction but often fluctuated in the north-southward direction.

3.3. *On Ion Flows at the Magnetopause*

Figure 10 depicts two types of the magnetic field configurations in and near the magnetopause current layer: One is the usual case of the lobe/mantle structure in which open field lines connect the Earth's ionosphere with the magnetosheath, and another is the case that re-reconnection occurs on originally open field lines. While the former case is a consequence of the dayside reconnection and the convection of the reconnected field lines toward the magnetotail, the latter would be caused by the reconnection at the near-Earth lobe magnetopause or at the distant lobe/mantle magnetopause [*Hirahara et al.*, 1997a]. The solar wind component injected into the lobe/mantle would be decelerated or accelerated in the direction parallel or antiparallel to the magnetic field by **J**×**B** force in the magnetopause current layer.

If the open magnetopause model on the basis of the dayside reconnection is adaptable to the November 17 case, the solar wind component should be able to injected everywhere in the magnetotail more or less, as discussed by *Gosling et al.* [1984, 1985a] and *Mozer et al.* [1994]. As a result, the injected solar wind component could be observed as the CDIFs in the lobe/mantle after being decelerated by the **J**×**B** force in the magnetopause current layer, as shown in Figure 10a.

In order to explain the two-branch CDIFs in and near the magnetopause in the October 8 event, it is realistic to consider that the magnetic field lines were being detached from the magnetotail due to the re-reconnection occurring somewhere at the lobe/mantle magnetopause between the near-Earth region and the observation site, as shown in Figure 10b. This means that the plasma on the field lines re-reconnected with the IMF would finally be swept into the interplanetary space

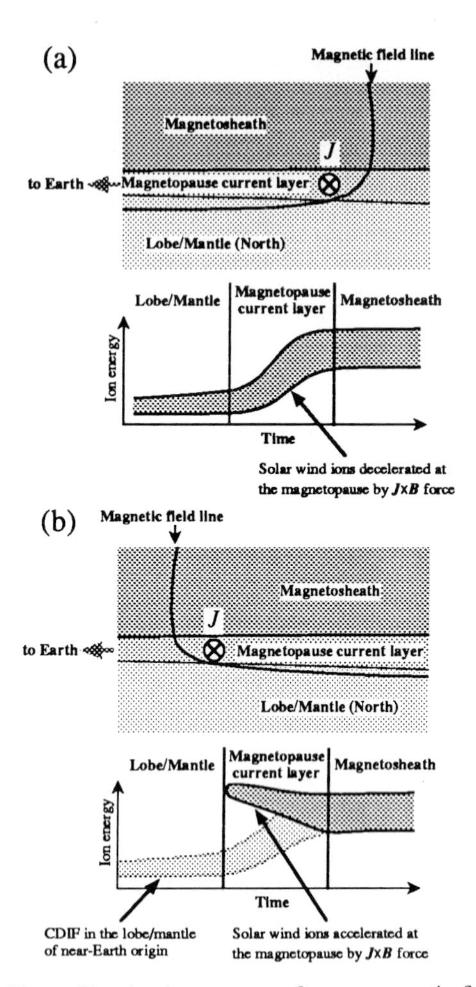

Figure 10. Sketch of two types of open magnetic field line configurations in and near the magnetopause current layer. Energy variations of tailward flows expected from the configurations are also illustrated.

without being injected into the plasma sheet. This transport would cause another fate of the ionospheric ions flowing in the magnetotail besides the process as schematically shown by *Chappell et al.* [1987]. In this scenario, the upper-branch component should be considered to be the solar wind component which was recently injected through the distant magnetopause after the latest reconnection, and accelerated by the $\mathbf{J}\times\mathbf{B}$ force in the current layer. We expect that the lower-branch CDIFs, which were continuously observed in both lobe/mantle and magnetopause current layer, were a mixture of the outflowing ionospheric/plasmaspheric ions (H^+, He^+, O^+) and the solar wind component (H^+, He^{2+}) injected in the near-Earth region along the open field lines before the latest reconnection.

In both cases, if the magnetopause was ideally a rotational discontinuity, we could estimate the deceleration and acceleration ratios of the solar wind component injected into the lobe/mantle. Assuming that the plasma pressure of the proton flows in the magnetosheath was isotropic, we apply the following equations;

$$\begin{aligned}\Delta\mathbf{V} &= \left(\frac{\rho_{\mathrm{MS}}}{4\pi}\right)^{\frac{1}{2}}\left(\frac{\mathbf{B}_{\mathrm{L/M}}}{\rho_{\mathrm{L/M}}}-\frac{\mathbf{B}_{\mathrm{MS}}}{\rho_{\mathrm{MS}}}\right)\\ &= \left(\frac{1}{4\pi\rho_{\mathrm{MP}}}\right)^{\frac{1}{2}}(\mathbf{B}_{\mathrm{L/M}}-\mathbf{B}_{\mathrm{MS}}),\end{aligned}$$

where $\mathbf{V}$, $\mathbf{B}$, and ρ denote velocity, magnetic field, and density, respectively. The symbols with three subscripts, MS, L/M, and MP mean the values in the magnetosheath, the lobe/mantle, and the magnetopause current layer, respectively. The ideal calculation would induce errors of $\sim$10 percent at most when the pressure anisotropy was ± 0.2, as discussed in detail by *Sonnerup et al.* [1981].

Tables 1 and 2 summarize the observed and estimated flow velocities of the solar wind components which were supposed to be injected into the lobe/mantle through the magnetopause for the October 8 and November 17 events, respectively. In each of events, two cases are evaluated: First is the combination of two densities in the magnetosheath and the lobe/mantle, and second case uses the typical density at the magnetopause. In both events, the two-density cases show that the estimates of the velocity variation (ΔV) are larger than observed. Particularly for the November 17 event, the relatively low-density feature in the lobe/mantle is mainly responsible for the discrepancy between the estimation and the observation. This result may imply that the solar wind component was not injected efficiently through the distant open magnetopause in this event. On the other hand, the estimates for the one-density cases show better agreements with the observations in both events. We, however, should mention that it is generally rare to find a perfect agreement between estimates and observational results because of restrictions in instrumental performance and assumptions on the ion velocity distributions. We conclude that the tangential stress balance test verifies that the open magnetopause model is valid for the Geotail observations presented here and that the field line configurations shown in Figure 10 are realistic at the magnetopause.

4. CONCLUDING REMARKS

In this paper, we have presented important properties of the CDIFs observed in the magnetotail by Geotail. They often showed the signatures consisting of multiple ion species or double proton populations. The ion composition of the ion flows in the distant lobe/mantle is an important clue to investigate the source/origin and the transport of the magnetotail plasma. From the behaviors of the proton flows in the PSBL and the magnetopause current layer, we can examine the injection and transport processes occurring in these magnetotail boundary layers.

It is most likely that solar wind ions are continuously injected into the lobe/mantle along the open field lines through the magnetopause. The solar wind penetration into the Earth's magnetotail has been discussed theoretically by *Lee and Roederer* [1982] and observationally by *Gosling et al.* [1984, 1985a], respectively. On the basis of the results obtained by the global surveys of Geotail in the magnetotail, we

Table 1. Velocity Components of Decelerated CDIF in Lobe/Mantle

Observation, km/s	Estimate from tangential stress balance, km/s	
	n_{MS} = 4.5/cm^3; n_{LLM} =1.3/cm^3	n_{MP} = 3.5/cm^3
(-191, +20, -5)	(-75, +39, -2)	(-240, +7, -21)

Table 2. Velocity Components of Accelerated Upper Branch CDIF

Observation, km/s	Estimate from tangential stress balance, km/s	
	n_{MS} = 5.5/cm^3; n_{LLM} = 3/cm^3	n_{MP} = 4/cm^3
(-550, +120, -70)	(-592, +50, -78)	(-539, +30, -83)

suggest that the plasma population in the whole magnetotail is frequently occupied mainly by the CDIFs in the distant lobe/mantle. Thus, the Geotail observations over a wide region of the magnetotail would shed a light on problems with respect to the income flux of the solar wind through the open magnetopause, and the population and dynamics of the magnetotail plasma.

It is also plausible that the majority of the solar wind component which is injected into the magnetotail through the open magnetopause and streaming in the distant lobe/mantle tailward of the flankside (high-latitude) re-reconnection site would finally escape into the interplanetary space without being injected into the plasma sheet. We propose that a significant amount of outflowing ionospheric ions transported into the distant magnetotail would also escape into the interplanetary space, together with the large amount of the solar wind component flowing in the magnetotail. While the geomagnetic field has a significant influence on the formation of the distant magnetotail and the unexpectedly high-density heavy ions of ionospheric origin have been detected in the distant lobe/mantle, the plasma population in the whole magnetotail is affected mostly by the solar wind contribution.

The transport and acceleration of the plasma consisting of multiple heavy ion species in and near the dayside magnetopause were observationally studied [e.g., *Fuselier et al.*, 1989a, b; *Gosling et al.*, 1990; *Fujimoto et al.*, 1997]. *Hirahara et al.* [1996] and *Seki et al.* [1996] presented the existence of the multicomposition CDIFs of solar wind and ionospheric origins in the distant lobe/mantle. Furthermore, from the viewpoint of the lobe/mangle magnetopause reconnection, *Gosling et al.* [1991, 1996] and *Hirahara et al.* [1997a] demonstrated the transport and dynamics of the multicomponent CDIFs in the vicinity of the near-Earth and distant magnetopause, respectively. The magnetotail observations [*Gosling et al.*, 1985b] and the recent Geotail results [*Seki et al.*, 1998] indicated asymmetries of the distributions of dense plasma and cold O^+ flows in the lobe/mantle and the dependence on the IMF and geomagnetic activity. The spatial distributions of the multicomposition flows in the magnetotail strongly suggest that the transport of ion flows is related with the dayside reconnection and the convection near the magnetopause. We expect that more detailed dynamics of the ionospheric ions in the magnetotail would be revealed in near future by using the knowledge from these works and the ISTP (International Solar-Terrestrial Program) results.

Acknowledgments. We thank T. Kamei and T. Iyemori of WDC-C2 for geomagnetism of Kyoto University for providing the provisional plots of the ground magnetogram data. We are also indebted to S. Kokubun and T. Yamamoto for the Geotail-MGF data distribution. The discussions with Y. Saito, M. Nakamura, M. Hoshino, and T. Terasawa were valuable in interpretations of the Geotail data. T. E. Moore and J. L. Horwitz gave us useful comments on the dynamics of the ionospheric heavy ions in the auroral and polar cap regions.

REFERENCES

Akinrimisi, J., S. Orsini, M. Candidi, and H. Balsiger, Ion dynamics in the plasma mantle, *Ann. Geophys.*, *8*, 739-754, 1990.

Baker, D. N., T. I. Pulkkinen, P. Toivanen, M. Hesse, and R. L. McPherron, A possible interpretation of cold ion beams in the Earth's tail lobe, *J. Geomagn. Geoelectr.*, *48*, 699-710, 1996.

Baker, D. N., A. Nishida, T. Mukai, T. Yamamoto, Y. Saito, Y. Matsuno, S. Kokubun, and T. I. Pulkkinen, Observations of bidirectional electrons in the distant tail lobes: GEOTAIL results, *Geophys. Res. Lett.*, *24*, 959-962, 1997.

Balsiger, H., Composition of hot ions (0.1-16 keV/e) as observed by the GEOS and ISEE mass spectrometers and inferences for the origin and circulation of magnetospheric plasmas, *Adv. Space Res.*, *1*, 289-303, 1981.

Balsiger, H., P. Eberhardt, J. Geiss, and D. T. Young, Magnetic storm injection of 0.9- to 16-keV/e solar and terrestrial ions into the high-altitude magnetosphere, *J. Geophys. Res.*, *85*, 1645-1662, 1980.

Candidi, M., S. Orsini, and V. Formisano, The properties of ionospheric O^+ ions as observed in the magnetotail boundary layer and northern plasma lobe, *J. Geophys. Res.*, *87*, 9097-9106, 1982.

Candidi, M., S. Orsini, and A. G. Ghielmetti, Observations of multiple ion beams in the magnetotail: Evidence for a double proton population, *J. Geophys. Res., 89*, 2180-2184, 1984.

Candidi, M., S. Orsini, and J. L. Horwitz, The tail lobe ion spectrometer: Theory and observations, *J. Geophys. Res., 93*, 14,401-14,409, 1988.

Chappell, C. R., T. E. Moore, and J. H. Waite Jr., The ionosphere as a fully adequate source of plasma for the Earth's magnetosphere, *J. Geophys. Res., 92*, 5896-5910, 1987.

DeCoster, R. J., and L. A. Frank, Observations pertaining to the dynamics of the plasma sheet, *J. Geophys. Res., 84*, 5099-5121, 1979.

Dungey, J. W., The structure of the ionosphere, or adventures in velocity space, in *Geophysics: The Earth's Environment*, edited by *C. Dewitt, J. Hiebolt, and A. Lebeau*, pp. 526-536, Gordon and Breach, New York, 1963.

Eastman, T. E., L. A. Frank, W. K. Peterson, and W. Lennartsson, The plasma sheet boundary layer, *J. Geophys. Res., 89*, 1553-1572, 1984.

Eastman, T. E., L. A. Frank, and C. Y. Huang, The boundary layers as the primary transport regions of the Earth's magnetotail, *J. Geophys. Res., 90*, 9541-9560, 1985.

Eastman, T. E., E. A. Greene, S. P. Christon, G. Gloeckler, D. C. Hamilton, F. M. Ipavich, G. Kremser, and B. Wilken, Ion composition in and near the frontside boundary layer, *Geophys. Res. Lett., 17*, 2031-2034, 1990.

Elphic, R. C., M. F. Thomsen, and J. Borovsky, The fate of the outer plasmasphere, *Geophys. Res. Lett., 24*, 365-368, 1997.

Forbes, T. G., E. W. Hones, S. J. Bame, J. R. Asbridge, G. Paschman, N. Sckopke, and C. T. Russell, Evidence for the tailward retreat of a magnetic neutral line in the magnetotail during substorm recovery, *Geophys. Res. Lett., 8*, 261-264, 1981.

Frank, L. A., K. L. Ackerson, and D. M. Yeager, Observations of atomic oxygen (O^+) in the Earth's magnetotail, *J. Geophys. Res., 82*, 129-134, 1977.

Freeman, J. W., H. K. Hills, T. W. Hill, P. H. Reiff, and D. A. Hardy, Heavy ion circulation in the Earth's magnetosphere, *Geophys. Res. Lett., 4*, 195-197, 1977.

Fujimoto, M., T. Mukai, A. Matsuoka, A. Nishida, T. Terasawa, K. Seki, H. Hayakawa, T. Yamamoto, S. Kokubun, and R. P. Lepping, Dayside reconnected field lines in the south-dusk near-tail flank during an IMF $B_y > 0$ dominated period, *Geophys. Res., Lett., 24*, 931-934, 1997.

Fuselier, S. A., W. K. Peterson, D. M. Klumpar, and E. G. Shelley, Entry and acceleration of He^+ in the low latitude boundary layer, *Geophys. Res. Lett., 16*, 751-754, 1989a.

Fuselier, S. A., D. M. Klumpar, W. K. Peterson, and E. G. Shelley, Direct injection of ionospheric O^+ into the dayside low latitude boundary layer, *Geophys. Res. Lett., 16*, 1121-1124, 1989b.

Fuselier, S. A., D. M. Klumpar, and E. G. Shelley, Ion reflection and transmission during reconnection at the Earth's subsolar magnetopause, *Geophys. Res. Lett., 18*, 139-142, 1991.

Gosling, J. T., J. R. Asbridge, S. J. Bame, W. C. Feldman, G. Paschmann, N. Sckopke, and C. T. Russell, Evidence for quasi-stationary reconnection at the dayside magnetopause, *J. Geophys. Res., 87*, 2147-2158, 1982.

Gosling, J. T., D. N. Baker, S. J. Bame, E. W. Hones Jr., D. J. McComas, R. D. Zwickl, J. A. Slavin, E. J. Smith, and B. T. Tsurutani, Plasma entry into the distant tail lobes: ISEE-3, *Geophys. Res. Lett., 11*, 1078-1081, 1984.

Gosling, J. T., M. F. Thomsen, D. W. Swift, and L. C. Lee, A note on the nature of the distant geomagnetic tail magnetopause and boundary layer, *Geophys. Res. Lett., 12*, 153-154, 1985a.

Gosling, J. T., D. N. Baker, S. J. Bame, W. C. Feldman, R. D. Zwickl, and E. J. Smith, North-south and dawn-dusk plasma asymmetries in the distant tail lobes: ISEE 3, *J. Geophys. Res., 90*, 6354-6360, 1985b.

Gosling, J. T., M. F. Thomsen, S. J. Bame, and C. T. Russell, Accelerated plasma flows at the near-tail magnetopause, *J. Geophys. Res., 91*, 3029-3041, 1986.

Gosling, J. T., M. F. Thomsen, S. J. Bame, R. C. Elphic, and C. T. Russell, Cold ion beams in the low latitude boundary layer during accelerated flow events, *Geophys. Res. Lett., 17*, 2245-2248, 1990.

Gosling, J. T., M. F. Thomsen, S. J. Bame, and R. C. Elphic, Observations of reconnection of interplanetary and lobe magnetic field lines at the high-latitude magnetopause, *J. Geophys. Res., 96*, 14,097-14,106, 1991.

Gosling, J. T., M. F. Thomsen, G. Le, and C. T. Russell, Observations of magnetic reconnection at the lobe magnetopause, *J. Geophys. Res., 101*, 24,765-24,773, 1996.

Hirahara, M., M. Nakamura, T. Terasawa, T. Mukai, Y. Saito, T. Yamamoto, A. Nishida, S. Machida, and S. Kokubun, Acceleration and heating of cold ion beams in the plasma sheet boundary layer observed with GEOTAIL, *Geophys. Res. Lett., 21*, 3003-3006, 1994.

Hirahara, M., T. Mukai, S. Machida, T. Terasawa, Y. Saito, T. Yamamoto, and S. Kokubun, Cold dense ion flows with multiple components observed in the distant tail lobe by Geotail, *J. Geophys. Res., 101*, 7769-7784, 1996.

Hirahara, M., T. Terasawa, T. Mukai, M. Hoshino, Y. Saito, S. Machida, T. Yamamoto, and S. Kokubun, Cold ion streams consisting of double proton populations and singly charged oxygen observed at the distant magnetopause by Geotail: A case study, *J. Geophys. Res., 102*, 2359-2372, 1997a.

Hirahara, M., T. Mukai, E. Sagawa, N. Kaya, and H. Hayakawa, Multiple energy-dispersed ion precipitations in low-latitude auroral oval: Evidence of $\mathbf{E}\times\mathbf{B}$ drift effect and upward flowing ion contribution, *J. Geophys. Res., 102*, 2513-2530, 1997b.

Horwitz, J. L., The tail lobe ion spectrometer, *J. Geophys. Res., 91*, 5689-5699, 1986.

Hultqvist, B., R. Lundin, K. Stasiewicz, L. Block, P.-A. Lindqvist, G. Gustafsson, H. Koskinen, A. Bahnsen, T. A. Potemra, and L. J. Zanetti, Simultaneous observation of upward moving field-aligned energetic electrons and ions on auroral zone field lines, *J. Geophys. Res., 93*, 9765-9776, 1988.

Kessel, R. L., S.-H. Chen, J. L. Green, S. F. Fung, S. A. Boardsen, L. C. Tan, T. E. Eastman, J. D. Craven, and L. A. Frank, Evidence of high-latitude reconnecting during northward IMF: Hawkeye observations, *Geophys. Res. Lett., 23*, 583-586, 1996.

Kokubun, S., T. Yamamoto, M. H. Acuña, K. Hayashi, K. Shiokawa, and H. Kawano, The GEOTAIL magnetic field experiment, *J. Geomagn. Geoelectr., 46*, 7-21, 1994.

Le, G., C. T. Russell, J. T. Gosling, and M. F. Thomsen, ISEE observations of low-latitude boundary layer for northward interplanetary magnetic field: Implications for cusp reconnection, *J. Geophys. Res., 101*, 27,239-27,249, 1996.

Lee, L. C., and J. G. Roederer, Solar wind energy transfer through the magnetopause of an open magnetosphere, *J. Geophys. Res., 87*, 1439-1444, 1982.

Lennartsson, W., R. D. Sharp, E. G. Shelley, R. G. Johnson, and H. Balsiger, Ion composition and energy distribution during 10 magnetic storms, *J. Geophys. Res., 86*, 4628-4638, 1981.

Lennartsson, W., R. D. Sharp, and R. D. Zwickl, Substorm effects on the plasma sheet ion composition on March 22, 1979 (CDAW 6), *J. Geophys. Res., 90*, 1243-1252, 1985.

Lockwood, M., T. E. Moore, J. H. Waite Jr., C. R. Chappell, J. L. Horwitz, and R. A. Heelis, The geomagnetic mass spectrometer - Mass and energy dispersions of ionospheric flows into the magnetosphere, *Nature, 316*, 612-613, 1985.

Maezawa, K., Magnetospheric convection inducing by the positive and negative Z components of the interplanetary magnetic field: Quantitative analysis using polar cap magnetic records, *J. Geophys. Res., 81*, 2289-2303, 1976.

Mozer, F. S., H. Hayakawa, S. Kokubun, M. Nakamura, T. Okada, T. Yamamoto, and T. Tsuruda, Direct entry of dense flowing plasmas into the distant tail lobes, *Geophys. Res. Lett., 21*, 2959-2962, 1994.

Mukai, T., M. Hirahara, S. Machida, Y. Saito, T. Terasawa, and A. Nishida, Geotail observation of cold ion streams in the medium distance magnetotail lobe in the course of a substorm, *Geophys. Res. Lett., 21*, 1023-1026, 1994a.

Mukai, T., S. Machida, Y. Saito, M. Hirahara, T. Terasawa, N. Kaya, T. Obara, M. Ejiri, and A. Nishida, The low energy particle (LEP) experiment onboard the GEOTAIL satellite, *J. Geomagn. Geoelectr., 46*, 669-692, 1994b.

Nakamura, M., G. Paschmann, W. Baumjohann, and N. Sckopke, Ion distributions and flows in and near the plasma sheet boundary layer, *J. Geophys. Res., 97*, 1449-1460, 1992.

Nakamura, M., T. Terasawa, H. Kawano, M. Fujimoto, M. Hirahara, T. Mukai, S. Machida, Y. Saito, S. Kokubun, T. Yamamoto, and K. Tsuruda, Leakage ions from the LLBL to MSBL: Confirmation of reconnection events at the dayside magnetopause, *J. Geomagn. Geoelectr., 48*, 65-70, 1996.

Orsini, S., M. Candidi, H. Balsiger, and A. Ghielmetti, Ionospheric ions in the near Earth geomagnetic tail plasma lobes, *Geophys. Res. Lett., 9*, 163-166, 1982.

Orsini, S., M. Candidi, V. Formisano, H. Balsiger, A. Ghielmetti, and K. W. Ogilvie, The structure of the plasma sheet-lobe boundary in the Earth's magnetotail, *J. Geophys. Res., 89*, 1573-1582, 1984.

Orsini, S., E. Amata, M. Candidi, H. Balsiger, M. Stokholm, C. Huang, W. Lennartsson, and P.-A. Lindqvist, Cold streams of ionospheric oxygen in the plasma sheet during the CDAW 6 event of March 22, 1979, *J. Geophys. Res., 90*, 4091-4098, 1985.

Orsini, S., K. Altwegg, and H. Balsiger, Composition and plasma properties of the plasma sheet in the Earth's magnetotail, *Ann. Geophys., 4*, 391-398, 1986.

Orsini, S., M. Candidi, M. Stokholm, and H. Balsiger, Injection of ionospheric ions into the plasma sheet, *J. Geophys. Res., 95*, 7915-7928, 1990.

Paschmann, G., B. U. Ö. Sonnerup, I. Papamastorakis, N. Sckopke, G. Haerendel, S. J. Bame, J. R. Asbridge, J. T. Gosling, C. T. Russell, and R. C. Elphic, Plasma acceleration at the Earth's magnetopause: Evidence for reconnection, *Nature, 282*, 243-246, 1979.

Paschmann, G., I. Papamastorakis, N. Sckopke, B. U. Ö. Sonnerup, S. J. Bame, and C. T. Russell, ISEE observations of the magnetopause: Reconnection and the energy balance, *J. Geophys. Res., 90*, 12,111-12,120, 1985.

Pederson, A., C. A. Cattell, C.-G. Fälthammar, K. Knott, P.-A. Lindqvist, R. H. Manka, and F. S. Mozer, Electric fields in the plasma sheet and plasma sheet boundary layer, *J. Geophys. Res., 90*, 1231-1241, 1985.

Peterson, W. K., R. D. Sharp, E. G. Shelley, and R. G. Johnson, Energetic ion composition of the plasma sheet, *J. Geophys. Res., 86*, 761-767, 1981.

Peterson, W. K., E. G. Shelley, G. Haerendel, and G. Paschmann, Energetic ion composition in the subsolar magnetopause and boundary layer, *J. Geophys. Res., 87*, 2139-2145, 1982.

Saito, Y., T. Mukai, T. Terasawa, A. Nishida, S. Machida, M. Hirahara, K. Maezawa, S. Kokubun, and T. Yamamoto, Slow-mode shocks in the magnetotail, *J. Geophys. Res., 100*, 23,567-23,581, 1995.

Saito, Y., T. Mukai, T. Terasawa, A. Nishida, S. Machida, S. Kokubun, and T. Yamamoto, Foreshock structure of the slow-mode shocks in the Earth's magnetotail, *J. Geophys. Res., 101*, 13,267-13,274, 1996.

Seki, K., M. Hirahara, T. Terasawa, I. Shinohara, T. Mukai, Y. Saito, S. Machida, T. Yamamoto, and S. Kokubun, The coexistence of the Earth-origin O^+ and the solar wind-origin H^+/He^{++} in the distant magnetotail, *Geophys. Res. Lett., 23*, 985-988, 1996.

Seki, K., M. Hirahara, T. Terasawa, T. Mukai, Y. Saito, S. Machida, T. Yamamoto, and S. Kokubun, Statistical properties and possible supply mechanisms of tailward cold O^+ beams in the lobe/mantle regions, *J. Geophys. Res., 103*, 4477-4493, 1998.

Sharp, R. D., D. L. Carr, W. K. Peterson, and E. G. Shelley, Ion streams in the magnetotail, *J. Geophys. Res., 86*, 4639-4648, 1981.

Sharp, R. D., W. Lennartsson, W. K. Peterson, and E. G. Shelley, The origins of the plasma in the distant plasma sheet, *J. Geophys. Res., 87*, 10,420-10,424, 1982.

Shelley, E. G., D. M. Klumpar, W. K. Peterson, and A. Ghielmetti, AMPTE/CCE observations of the plasma composition below 17 keV during the September 4, 1984 magnetic storm, *Geophys. Res. Lett., 12*, 321-324, 1985.

Siscoe, G. L., and E. Sanchez, An MHD model for the complete open magnetotail boundary, *J. Geophys. Res., 92*, 7405-7412, 1987.

Siscoe, G. L., L. A. Frank, K. L. Ackerson, and W. R. Paterson, Properties of mantle-like magnetotail boundary layer: Geotail data compared with a mantle model, *Geophys. Res. Lett., 21*, 2975-2978, 1994.

Smith, M. F., and M. Lockwood, Earth's magnetospheric cusps, *Rev. Geophys., 34*, 233-260, 1996.

Sonnerup, B. U. Ö., G. Paschmann, I. Papamastorakis, N. Sckopke, G. Haerendel, S. J. Bame, J. R. Asbridge, J. T. Gosling, and C. T. Russell, Evidence for magnetic field reconnection at the Earth's magnetopause, *J. Geophys. Res., 86*, 10,049-10,067, 1981.

Takahashi, K., and E. W. Hones Jr., ISEE 1 and 2 observations of ion distributions at the plasma sheet-tail lobe boundary, *J. Geophys. Res., 93*, 8558-8582, 1988.

Williams, D. J., Energetic ion beams at the edge of the plasma sheet: ISEE 1 observations plus a simple explanatory model, *J. Geophys. Res., 86*, 5507-5518, 1981.

World Data Center C2 for geomagnetism, Provisional Auroral Electrojet Indices (AE11) for January - December 1993, Data Analysis Center for Geomagnetism and Space Magnetism, Faculty of Science, Kyoto University, Kyoto, Japan, 1995.

Yamamoto, T., A. Matsuoka, K. Tsuruda, H. Hayakawa, A. Nishida, M. Nakamura, and S. Kokubun, Dense plasmas in the distant magnetotail as observed by GEOTAIL, *Geophys. Res. Lett., 21*, 2879-2882, 1994.

Yau, A. W., E. G. Shelley, W. K. Peterson, and L. Lenchyshyn, Energetic auroral and polar ion outflow at DE 1 altitudes: Magnitude, composition, magnetic activity dependence, and long-term variations, *J. Geophys. Res., 90*, 8417-8432, 1985.

Young, D. T., H. Balsiger, and J. Geiss, Correlations of magnetospheric ion composition with geomagnetic and solar activity, *J. Geophys. Res., 87*, 9077-9096, 1982.

Zwickl, R. D., D. N. Baker, S. J. Bame, W. C. Feldman, J. T. Gosling, E. W. Hones Jr., D. J. McComas, B. T. Tsurutani, and J. A. Slavin, Evolution of the Earth's distant magnetotail: ISEE 3 electron plasma results, *J. Geophys. Res., 89*, 11,007-11,012, 1984.

M. Hirahara, Department of Physics, College of Science, Rikkyo University, 3-34-1 Nishi-Ikebukuro, Toshima-ku, Tokyo 171-8501, Japan. (e-mail: hirahara@rikkyo.ac.jp)

T. Mukai, Institute of Space and Astronautical Science, 3-1-1 Yoshinodai, Sagamihara, Kanagawa 229-0022, Japan. (e-mail: mukai@stp.isas.ac.jp)

K. Seki, Department of Earth and Planetary Physics, Faculty of Science, University of Tokyo, 7-3-1 Hongo, Bunkyo-ku, Tokyo 113-0033, Japan. (e-mail: seki@grl.s.u-tokyo.ac.jp)

(Received March 24, 1997; revised February 2, 1998;
accepted February 2, 1998.)

Convection and Reconnection in the Earth's Magnetotail

A. Nishida

Institute of Space and Astronautical Science, Sagamihara, Kanagawa 229 , Japan

T. Ogino

Solar Terrestrial Environment Laboratory, Toyokawa, Aichi 442 , Japan

This paper presents an overview of the convection in the magnetotail that has been revealed by extensive observations by GEOTAIL as well as by recent advances in the three-dimensional simulation. Over a wide range of the IMF conditions the magnetic reconnection is the basic element of the convection in the magnetotail. In geomagnetically active times that correspond to the southward IMF, the observations confirm that there are two preferred sites for the reconnection; one is in the distant tail and at about 140 Re, while the other is in the near-earth region and is formed earthward of 50 Re. Larger flux of the open field lines is reconnected at the distant neutral line than at the near-earth neutral line. In geomagnetically quiet times when IMF $|B_y| \geq B_Z > 0$, the reconnection still occurs in the magnetotail but it generates an apparently different signature because of the twisting of the tail due to the IMF B_y; the field lines that are convected tailward beyond the neutral line have the northward B_Z polarity although they cross the twisted neutral sheet from the northern side to the southern side. Since the northward B_Z is not the unique signature of the closed field lines, the amount of the closed field lines in the tail is not easy to estimate and the conjecture that the magnetosphere is tadpole shaped when the IMF is almost due northward is yet to be proven.

INTRODUCTION

The magnetosphere is a dynamic entity. The plasma contained therein is almost always in motion, and the large-scale motion forms systematic convection patterns that circulate between the day and the night sides. Since the frozen-in theorem is applicable except in some specific regions, the plasma convection is accompanied by the convection of the magnetic field lines in the bulk of the magnetosphere. In the decades following the original suggestion by Axford and Hines [1961], the convection has been established as a causal agency that provides basis for various phenomena observed in the magnetosphere and the polar ionosphere.

New Perspectives on the Earth's Magnetotail
Geophysical Monograph 105

The convection is generated by the interaction with the solar wind. The supply of momentum and energy that drives the anti-sunward motion in the boundary region of the magnetosphere should also stretch the geomagnetic field lines. Hence Axford et al. [1965] noted that the generation of the convection and the formation of the magnetotail are expected from the same interaction process.

The most important driving mechanism that has stood the test of time is reconnection of interplanetary (IMF) and geomagnetic field lines that was proposed by Dungey [1961]. Although other interaction processes also seem to operate at the magnetopause and involve interesting physics [e.g., Fujimoto et al., 1996], they are less significant as the driving agent of the large-scale convection [see, e.g., Boyle et al., 1997]. The pattern of the convection depends upon the

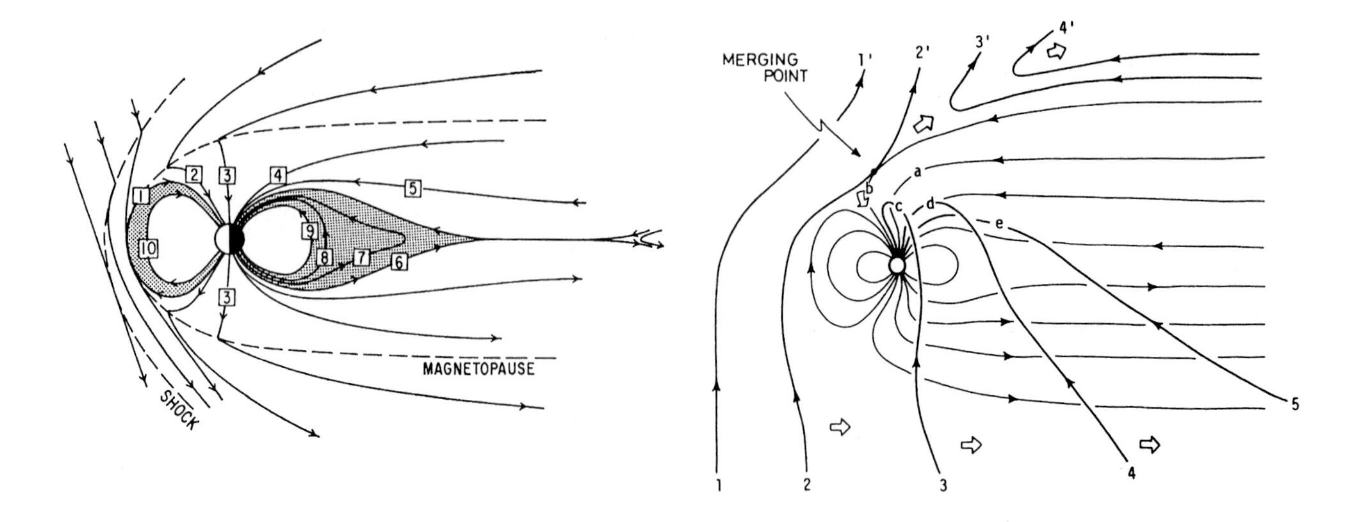

Figure 1. Generic patterns of the magnetospheric convection for the southward IMF case (left, after Levy et al. [1964]) and for the northward IMF case (right, after Maezawa [1976]).

direction of the IMF which governs the location of the reconnection line on the magnetopause.

Figure 1 shows two generic patterns of the convection. The left panel is for the southward IMF case when the reconnection occurs in the equatorial region of the dayside magnetopause [Levy et al., 1964]. The closed field lines (e.g., "10") that reach the dayside magnetopause are transformed into open field lines by the reconnection, and the open field lines flow tailward ("2" through "5"). The return flow of the closed field lines ("6" through "10") toward the dayside results from another reconnection process that occurs in the neutral sheet of the magnetotail. The characteristic feature of this convection pattern is the transition of the field line topology between the open and the closed, and there are reconnection lines both on the dayside magnetopause and in the nightside neutral sheet. A variety of this configuration pattern is that of Hones [1979] in which there are two preferred locations of the tail reconnection line, one being in the near tail and the other in the distant tail. The right panel is for the northward IMF case when the reconnection occurs on the high-latitude surface of the magnetosphere where the open field lines (e.g., "b") make contact with the IMF field line ("2'") in the solar wind [Maezawa, 1976]. In this case only the open field lines ("a" through "e") are involved in the convection and there is no reconnection line inside the tail. Several varieties of the field-line topology and the convection that can result from the reconnection of the open or closed geomagnetic field lines on the tail surface with the northward IMF are illustrated by Cowley [1982].

In this article we shall present an updated overview of our understanding of the convection in the magnetotail. Substantial advances have been made in recent years by virtue both of the advent of the dedicated GEOTAIL satellite mission and of progress in numerical simulations of the global magnetospheric dynamics.

CONVECTION IN ACTIVE TIMES

Since the primary direction of the geomagnetic field is northward and that of the magnetospheric convection is earthward/tailward, the basic signature of the convection can be seen by plotting the northward component B_Z of the magnetic field versus the x component $V_{perp,x}$ of the bulk velocity $\mathbf{V}_{perp}$ that is perpendicular to the magnetic field. Figure 2 shows such plots at two locations for the times of high magnetic activity (Kp > 3). Each data point represents an observation at 12 s intervals during contiguous 24 hours, and the coordinates used are the Solar Magnetospheric Coordinates. At x = - 169 Re in the distant tail (left panel), $V_{perp,x}$ is negative, namely, directed tailward most of the time, and B_Z is very often negative, namely, southward. This demonstrates that the reconnection occurs earthward of this distance. At x = - 44 Re (right panel) in the medium distance tail, earthward and tailward $V_{perp,x}$ occur just as often, and B_Z is divided between northward and southward when $V_{perp,x}$ is tailward, while it is northward most of the time when $V_{perp,x}$ is earthward. This means that sometimes the plasmoids/flux ropes are produced on the earthward side of this distance and are convected tailward, while in some other times the northward magnetic field lines are produced by reconnection beyond this distance and are convected earthward. In both cases the y component E_y of the electric field calculated by the $\mathbf{E} = - \mathbf{V} \times \mathbf{B}$ relation is positive and a few times 0.1 mV/m.

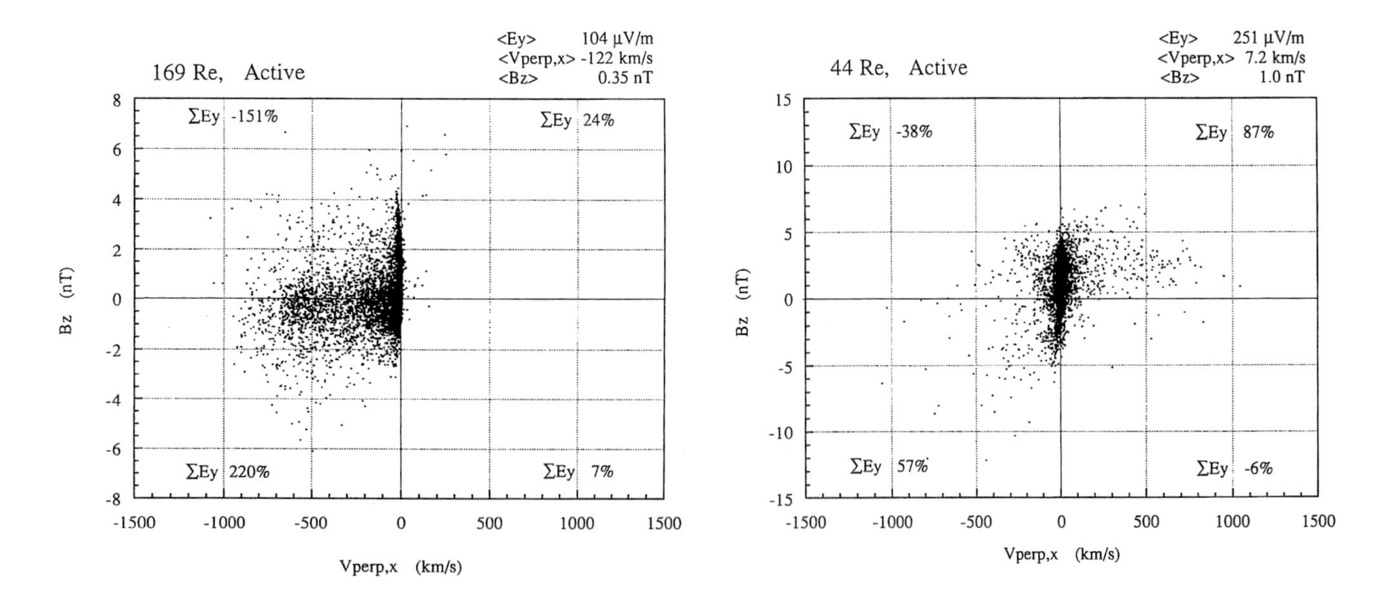

Figure 2. Plots of B_Z versus $V_{perp,x}$ for two 24-h intervals in active times (Kp > 3) at different distances indicated on the upper left of each diagram. Averages of E_y, $V_{perp,x}$ and B_Z are given on the upper right of each panel. ΣE_y given at the corners of the quadrants expresses the amount of the magnetic flux transport associated with the data points in each quadrant and is given in percentage relative to the total for all the points [Nishida et al., 1996a].

Nishida et al. [1996a] have produced B_Z versus $V_{perp,x}$ diagrams for several active intervals (Kp > 3) when GEOTAIL was x = -36 Re to -169 Re in the magnetotail, and estimated the position of the reconnection line by using E_y to measure the rate of the earthward/tailward transport of the magnetic field lines. It is assumed throughout that the state of the magnetotail is kept the same (when averaged for individual disturbance events) for the duration (as long as 24 hours) of each data set. Although $E_y = -V_Z B_X + V_X B_Z$ involves a term $-V_Z B_X$ which means the flux transport in the z direction, this transport would be converted into the transport in the x direction if there is symmetry with respect to the $B_X = 0$ plane as would statistically be the case over the durations of each data set. The same conclusions as below have been obtained when $V_{perp,x} B_Z$ in the plasma sheet where the ion beta is greater than 0.5 is used instead of E_y as the measure of the flux transport in the x direction [Nishida et al., 1996a].

The grand total of E_y for each 24-h interval expresses the fluxes of the open field lines and products therefrom (by reconnection) that are transported earthward/tailward across the satellite position over the duration of the data set. The products that count here are closed field lines on the earthward side of the neutral line and the IMF-type field lines on its tailward side. (The"IMF-type" field lines are defined to be those that have no feet on the earth and they are not necessarily in the solar wind.) The other closed field lines which do not originate from reconnection of open field lines would not contribute to the above grand total. This is because the same closed field line passes the satellite twice with opposite signs of E_y as it flows tailward and then returns earthward. The magnetic loops which result from reconnection of closed field lines or from the tearing mode instability are not included in this grand total either, since the contributions to ΣE_y from the front part and the rear part of the loop cancel each other as they have the same magnitude but are opposite in sign.

The sums of E_y are calculated also for each quadrant of Figure 2 separately, and they are written at the corners of the quadrants in % of the grand total. The use of the percentage means that these sums are normalized by the grand total of E_y for each data set. In the case of the right panel for 44 Re, 87 % of the flux transport is due to earthward convection of the northward field, and 57 % is due to tailward convection of the southward field, but 38 % of the latter is canceled by tailward transport of the northward field which contributes negatively to E_y. The tailward transport of the northward field lines includes those which occurred prior to the onset of reconnection since we are using the data obtained over extended periods of active times during which dynamical state of the tail can be considered to be kept the same on average. In the case of the left panel for 169 Re, the flux transport due to tailward convection of the southward field is as much as 220 % of the grand total, but much of this is canceled by tailward transport of the northward field since roughly 151/ 220 = 68 % of the southward field is tied to the northward field by forming magnetic loops. (ΣE_y for the fourth quadrant of the left panel is positive, while $V_{perp,x} B_Z$ for this quadrant

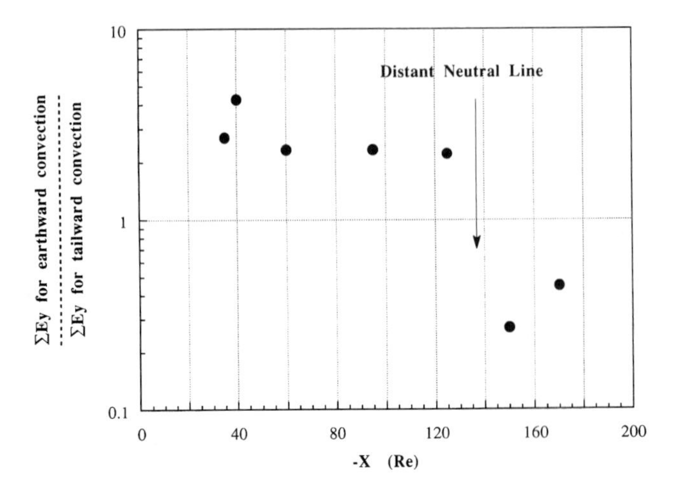

Figure 3. The ratio between the net earthward transport of the northward field lines and the net tailward transport of the southward field lines is plotted versus x. The inferred position of the distant neutral line is indicated [Nishida et al., 1996a].

should be negative, because of the contribution from $-V_{perp,z}B_x$ to E_y. This difference is due to limited size of the data set.)

The location where open field lines are reconnected can be obtained by observing how the ratio between the net earthward transport of the northward field and the net tailward transport of the southward field varies with the distance. The former is the difference between the earthward transports of the northward and the southward fields, and it is given by the sum ΣE_y in the first and the fourth quadrants of Figure 2. The latter is the difference between the tailward transports of the southward and the northward fields, and is given by the sum ΣE_y in the second and the third quadrants. The net transports do not involve contributions from plasmoids/flux ropes since the transports of northward and southward field lines in the same direction have been combined. In the case of Figure 2, the ratio between them is (24+7)/(220-151) = 0.45 and thus less than 1 for the left panel, while it is (87-6)/(57-38) = 4.2 and thus larger than 1 for the right panel. This ratio is calculated for all seven data sets and they are plotted in Figure 3 versus x coordinate of the observing site. It is seen that the ratio takes relatively fixed values of 2 to 4 for the five data sets obtained at distances closer than about 140 Re, but is 0.2 to 0.4 for the other two data sets which are obtained at further distances. Thus there is a sharp transition around x = -140 Re; inside this distance the net earthward convection of the northward field is dominant, while the net tailward convection of the southward field is dominant beyond it. This transition can be interpreted as the location of the distant neutral line.

The variation of the reconnection rate across the tail is reflected in the dependence of the average E_y on the y coordinate of the observing site. As shown in Figure 4 the reconnection rate of the open field lines is a few times higher near the expected position of the tail axis than at the flanks [Nishida et al., 1995a].

The location where the closed as well as the open field lines are reconnected can be obtained by taking ratios between ΣE_y of the first and the third quadrants, that is, between the earthward transport of the northward field and the tailward transport of the southward field. In contrast to the net transport used in Figure 3, ΣE_y for the third quadrant alone contains the contributions from the magnetic loops. When these ratios are plotted versus x in Figure 5, sharp decreases are seen at two locations. The more distant decrease corresponds to the distant neutral line, and another decrease that occurs inside 50 Re suggests that the near-earth neutral line is located inside that distance. Thus, it is confirmed that there are two preferred sites of reconnection in the magnetotail convection in magnetically active times that correspond to the southward IMF. When magnetic reconnection occurs simultaneously at both of these sites magnetic loops (i.e., quasi-stagnant plasmoid) are embedded between them [Hoshino et al., 1996; Kawano et al., 1996].

Since the ratio between the net earthward transport of the northward field and the net tailward transport of the southward field is greater than 1 inside 140 Re (Figure 3), a larger flux of open field lines is reconnected at the distant neutral line than at the near-earth neutral line. Furthermore, from the result

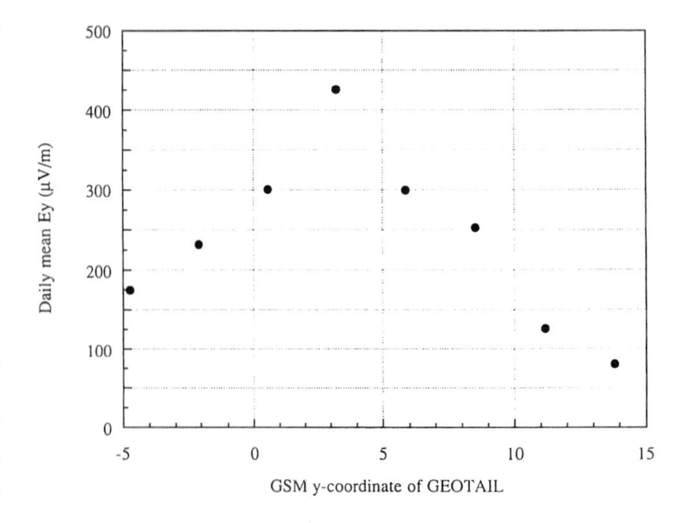

Figure 4. Variation of the daily averages of E_y with the y coordinate during an active interval of January 12 to 19, 1994 when the x coordinate was about -95 Re. The tail axis would be at y = 6.6 Re when the aberration angle of the tail is 4 ° [Nishida et al., 1995a].

that the decrease of the ratio inside 50 Re is seen in Figure 5 but is not clear in Figure 3, it could be suggested that mostly the closed field lines are reconnected at the near-earth neutral line. However, since ΣE_y from the second and the third quadrants of the right panel of Figure 2 do not fully cancel, the result may not be strong enough to argue that no open field lines are reconnected at this neutral line.

The above analyses were performed by selecting the active times when Kp was greater than 3 at least for 24 hours. Since such intervals were not very frequent the number of the data points in Figures 3 and 5 is not very large, and hence the possible variability in the location of the distant neutral line could not be addressed. The inferred distance of inside 50 Re for the near-earth neutral line is an upper limit since only the data beyond about 36 Re were used. Nagai et al. [1998] and Ieda et al. [1998] have studied the formation of the near-earth neutral line in relation to substorms by using bulk of the GEOTAIL observations in the 10 to 50 Re range, and have found that the reconnection starts in the premidnight sector at about x = - 25 Re.

In the case of the left panel of Figure 2, the average E_y is positive while average $V_{perp,x}$ and average B_z are negative and positive respectively; that is, the electric field is directed from dawn to dusk on average while the convection velocity is tailward and the magnetic field is directed northward on average. This is not because the contribution from another term $-V_{perp,z}B_x$ in E_y is dominant. Rather, it is because the magnetic field tends to be more often southward when the tailward convection is fast. This tendency was recognized earlier in relation to the near-earth reconnection [Hayakawa et al., 1982].

The y component $V_{perp,y}$ of the convective motion is governed by the y component of the magnetic field which is produced under the influence of the IMF B_y. In the region of the earthward convection, $V_{perp,y}$ represents the rotation of the closed field lines toward being more parallel to the dipole axis since the effect of B_y on their configuration becomes less pronounced as they move toward the earth. In the region of the tailward convection, $V_{perp,y}$ represents the field line rotation associated with the tailward propagation of flux ropes [Nishida et al., 1996b].

CONVECTION IN QUIET TIMES

The geomagnetic quiescence corresponds generally to the northward polarity of the IMF. Low-altitude observations of the convection in such times (made extensively by using the data collected by polar orbiting satellites, radars installed in the polar cap, and the magnetometers on the ground) have shown a characteristic dependence on the relative magnitudes of the IMF B_y and B_z [Weimer, 1995; Ruohoniemi and

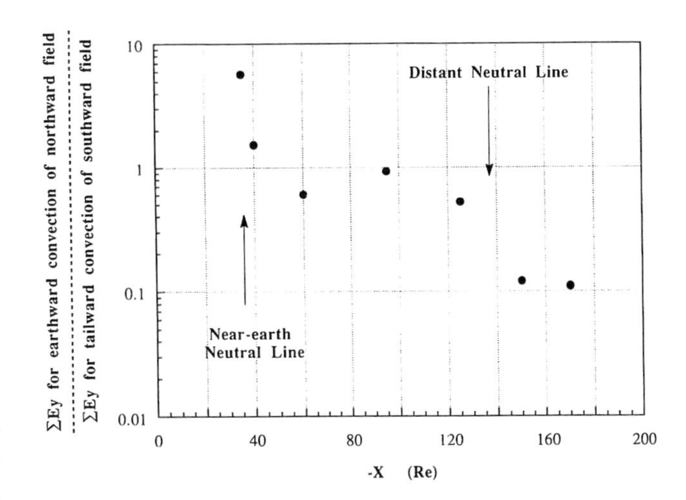

Figure 5. The ratio between the earthward transport of the northward field lines and the tailward transport of the southward field lines is plotted against x. The inferred positions of the distant and the near-earth neutral lines are indicated [Nishida et al., 1996a].

Greenwald, 1996; and references therein]. This section deals with the magnetotail convection that is generated in the magnetotail when the IMF B_z is northward but $|B_y|$ is comparable to or larger than B_z [after Nishida et al., 1995b and 1998].

Figure 6 shows the B_z versus $V_{perp,x}$ diagrams for the quiet times when such IMF condition lasted for 24 hours. At x = - 84~89 Re in the distant tail (left panel), $V_{perp,x}$ is negative, namely, directed tailward most of the time as was the case in the distant tail in active times (Figure 2, left panel). However, unlike in the active times B_z is northward most of the time and E_y is negative on average; the tailward convection of the apparently northward magnetic field is the characteristic feature in such times. At x = - 19 ~ 29 Re (right panel), in contrast, both $V_{perp,x}$ and B_z are positive for a great majority of the points, so that E_y is positive on average and the northward magnetic field is convected earthward. The variation of E_y with the distance is shown with a larger number of daily samples in Figure 7. All of the 11 data points displayed in this plot represent contiguous 24-h intervals when IMF $|B_y| \geq B_z > 0$ and Kp $\leq$ 1+ most of the time. The plot shows a distinct change in averages of E_y between 50 and 80 Re; earthward of this distance E_y is positive, namely directed from dawn to dusk, while beyond this distance E_y is very small and often negative. Note that E_y is expected to be almost uniform if the tail midplane runs parallel to the y axis since the tail structure would be two dimensional to the first order approximation,

In the distant tail where $E_y < 0$, the convection velocities projected to the yz plane are directed northward (southward)

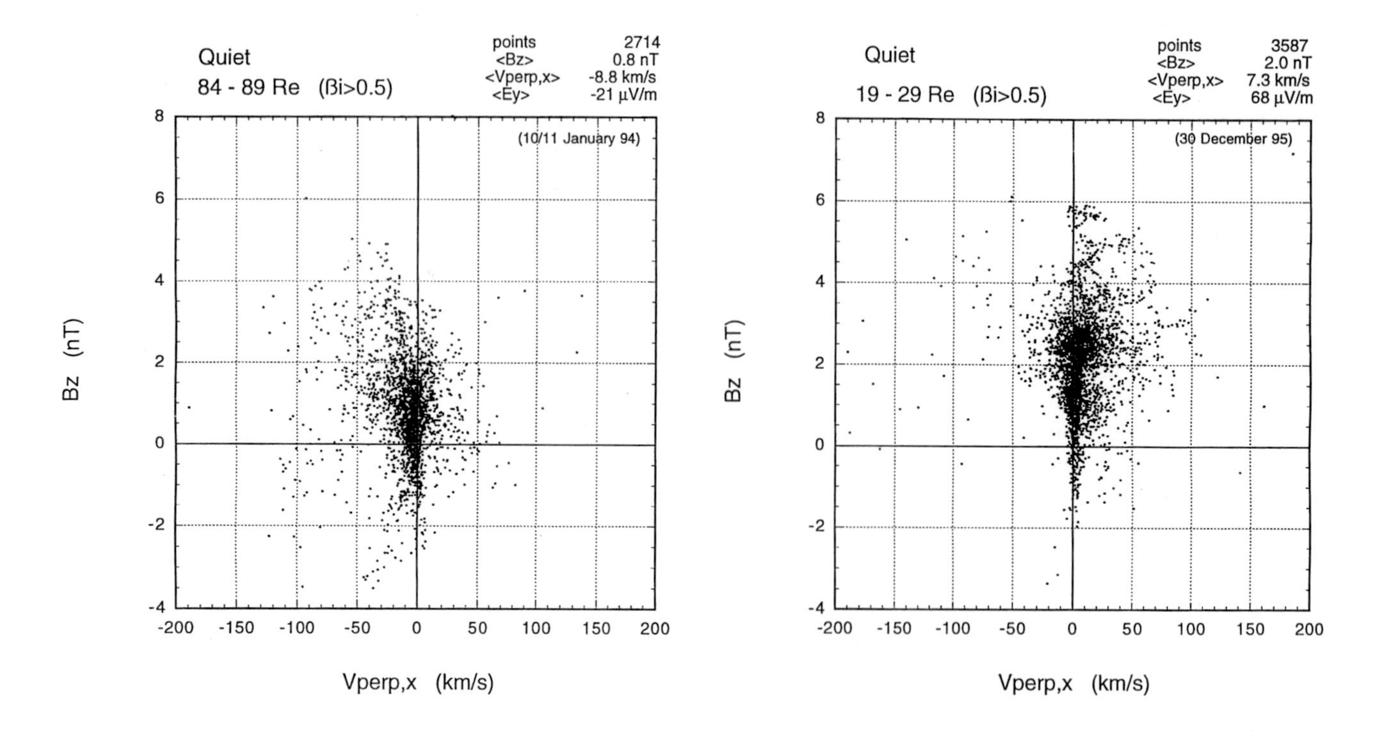

Figure 6. Plots of B_Z versus $V_{perp,x}$ for two quiet-time observations in the plasma sheet (where ion beta is larger than 0.5) at two different distances indicated on the upper left of each diagram [Nishida et al., 1998].

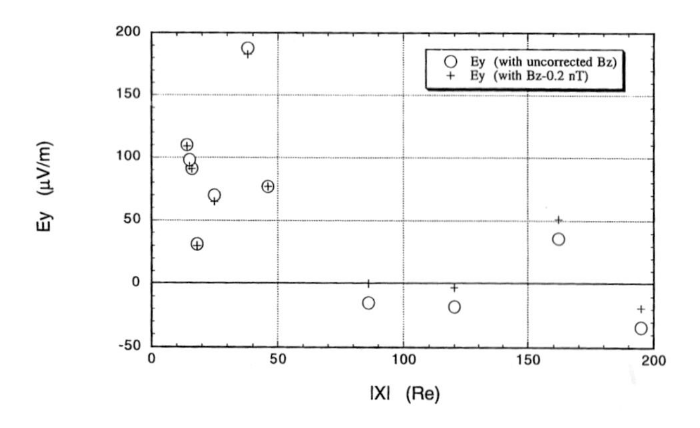

Figure 7. Average of E_y against x for eleven sets of 24-h intervals of the geomagnetic quiescence when Kp < 1+ most of the time [Nishida et al., 1998]. Circles and plus signs correspond to the use of the GEOTAIL data set values and the values corrected for a possible offset of 0.2 nT in the B_Z measurements, respectively.

in the sector which contains the field lines with the $B_X > 0$ ($B_X < 0$) polarity originating from the northern (southern) polar cap, and they are far from being meridional. This is shown in Figure 8 where y and z components $V_{perp,y}$ and $V_{perp,z}$ of the convection velocity are plotted versus B_X which is a measure of the distance from the neutral sheet. The above direction of $V_{perp,z}$ is opposite to the observations in active times where it is southward (northward) in the region of $B_X > 0$ ($B_X < 0$).

To understand this feature, it is essential to note that the neutral sheet is twisted under the influence of the IMF, and that the twisting angle is much larger and is toward the projection of the IMF onto the yz plane under the northward IMF [Owen et al., 1995; Nishida et al., 1995b; Maezawa et al., 1997]. The twisting of the tail in response to the IMF B_y was envisaged earlier by Cowley [1981]. Cross section of the distant tail is illustrated in Figure 9 for the $B_Z > 0$ and $B_y > 0$ case. Substantial part of the region of $B_X > 0$ ($B_X < 0$) is located on the southern (northern) side of the equatorial plane because of this twist, and the convection velocity has the northward (southward) component in the $B_X > 0$ ($B_X < 0$) region even if it is directed toward the neutral sheet. Since the incidence of the convection toward the neutral sheet is highly oblique, the yz projection of the electric field vector **E** makes a large angle with the neutral sheet, so that the y-component E_y can be negative even if the component parallel to the neutral sheet is still directed from dawn to dusk.

The neutral sheet is highly twisted under the northward IMF because the open field lines extend to the other side of the equatorial plane. As illustrated in Figure 10-a, the open field lines that are rooted in the northern (southern) polar cap extend to the south (north), and the neutral sheet is twisted as it reflects the direction of these field lines. The open field lines are

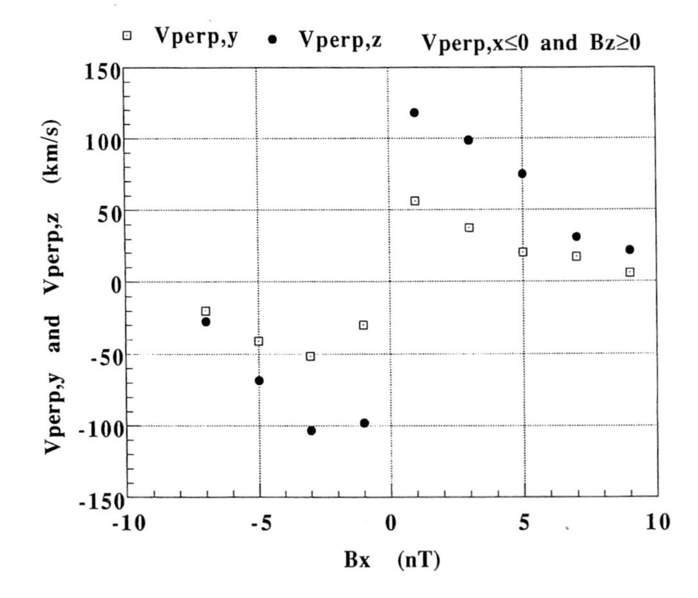

Figure 8. Components $V_{perp,y}$ and $V_{perp,z}$ of the convection velocity are plotted against B_x when the x component $V_{perp,x}$ is directed tailward and B_z is northward. The data are taken over a 72-h interval of the geomagnetic quiescence in April 20 to 23, 1994 when GEOTAIL was at x = -200 Re. B_y in this interval is positive on average [Nishida et al., 1995b].

convected toward the neutral sheet at the same time as they move across the tail, and they are reconnected when they reach the reconnection line on the neutral sheet. The other panels of Figure 10 show the configurations of the resultant field lines on the earthward side (Figure 10-b) and the tailward side (Figure 10-c) of the reconnection line. Field lines in Figure 10-c are of the IMF-type and move tailward, and they cross the neutral sheet from the northern side to the southern side although they are directed northward everywhere. In this manner the tailward convection of the apparently northward field lines is produced due to the twist of the neutral sheet. These field lines are loaded with the hot plasma that has resulted from the reconnection process. Field lines in Figure 10-b are closed and move earthward, so that the earthward injection of the hot plasma is observed at the earthward boundary of the plasma sheet even during extremely quiet geomagnetic times under the northward IMF [Nishida et al., 1997].

Under the northward IMF, not all the field lines are convected to the neutral sheet ; in Figure 9 the field lines on the streamlines in the shaded area reach the neutral sheet, but others merely traverse the tail without experiencing reconnection in the tail. Observations show that the principal constituent of $V_{perp,y}$ and $V_{perp,z}$ is the perpendicular projection of the predominant x-directed flow velocity V_x in the presence of B_y and B_z as well as B_x component of the magnetic field, and this is given by $V_{0,y} = -V_x B_x B_y/B^2$ and $V_{0,z} = -V_x B_x B_z/B^2$.

Our analysis has not been advanced far enough to see if the neutral line tends to be formed at two preferred locations, namely in the distant tail and the near-earth tail, in the quiet times as well. This relates to the question of whether there are sporadic enhancements in the reconnection rate in the tail under the condition of $|By| \geq Bz > 0$ as is the case under the southward IMF.

In an independent study, Fairfield et al. [1996] presented a case where the direction of the flow as projected to the yz plane remained the same for about 12 hours at x = -135 Re although the IMF B_y was variable while B_z was northward. However, when the component of these flow vectors perpendicular to the magnetic field is taken, the direction of the convection has been found to vary with the IMF B_y as well as with the local direction of B_x in a manner consistent with the present model.

CONVECTION IN THE THREE-DIMENSIONAL SIMULATION MODEL

Many of the basic features of the convection in the magnetotail have been reproduced in the three-dimensional simulation of the dynamical magnetosphere [Ogino and Walker, 1998]. In this simulation, MHD and Maxwell's equations are solved as an initial value problem in order to study the interaction between the solar wind and the

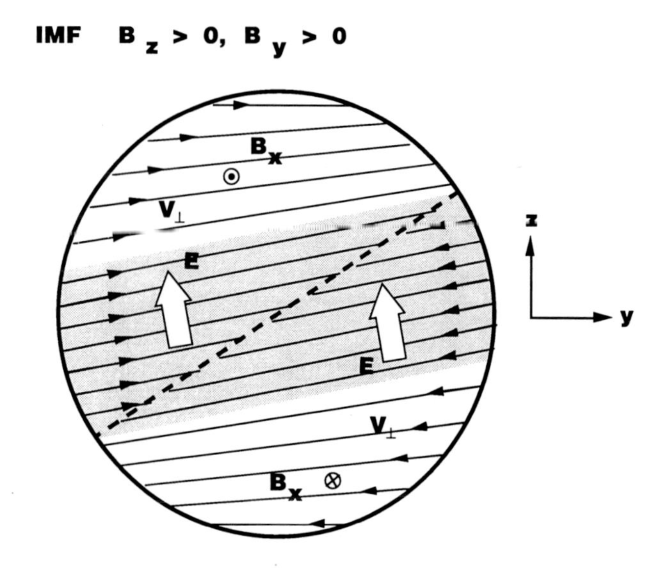

Figure 9 Schematic drawing of the projections of the electric field **E** and the convection velocity $\mathbf{V}_{perp}$ (labeled $\mathbf{V}_\perp$) to the yz plane. The dashed line is the neutral sheet. The convection streamlines enter the neutral sheet in the shaded region, while they merely traverse the tail from one side to other elsewhere.

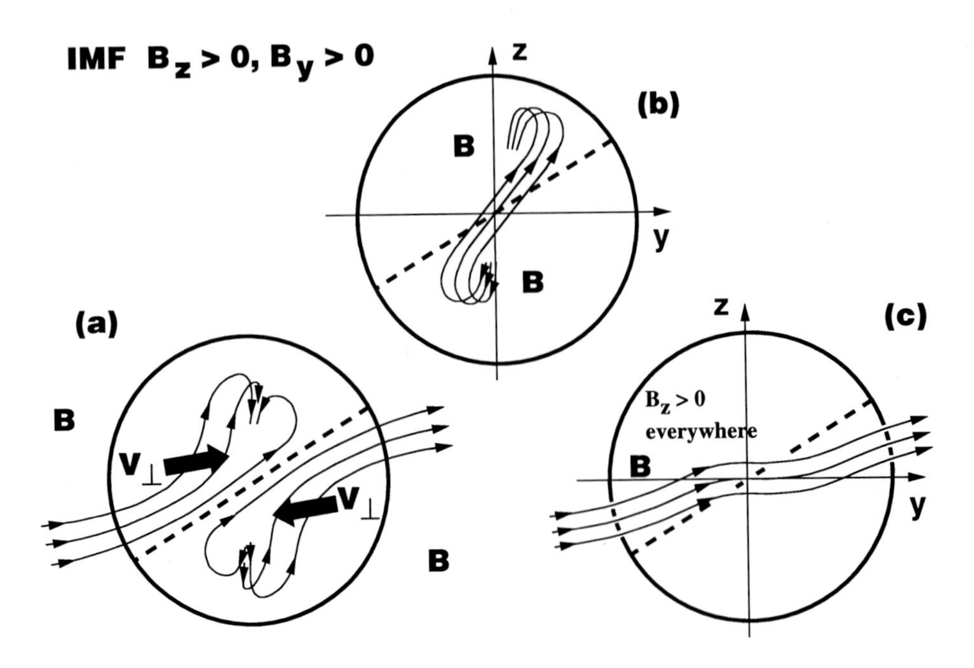

Figure 10. Magnetic field lines and their convection in the magnetotail under the northward IMF as projected to the yz plane. Dashed line represents the neutral sheet which is twisted. Panel (a) shows the open field lines, and panels (b) and (c) show the field lines produced by the reconnection on the earthward side and the tailward side of the reconnection line, respectively [Nishida et al., 1998].

magnetosphere. The MHD equations involve a resistivity term where the magnetic Reynold's number is larger than 200. For the numerical integration the modified leapfrog scheme is used where the leapfrog and the two-step Lax-Wendroff schemes are combined. The basic version of the global MHD model using this scheme was presented in detail by Ogino et al. [1992].

At the initial stage, the geomagnetic field is confined by a mirror dipole magnetic field. The inner boundary condition that is imposed near the earth is determined by requiring a static equilibrium. The solar wind with density of 5 /cm^3, velocity of 300 km/s, and temperature of 2 x 10^5 K flows into a large simulation box with dimensions of 30 Re > x > -220 Re, 50 Re > y > 0 Re and 50 Re > z > -50 Re in solar magnetospheric coordinates. The mesh size is 0.5 Re and the number of grids is (500, 100, 200). The time step is 0.937 s which is one fourth of the minimum Alfven transit time of the mesh. The y and z components of the IMF are expressed as $(B_y, B_z) = B_0(\cos\theta, \sin\theta)$ where B_0 = 5 nT and θ is measured from the dusk direction counterclockwise as viewed toward the earth.

The IMF is imposed after the simulation has been run without the magnetic field for 4 hours and the steady magnetospheric configuration has been formed. Subsequently the MHD code is run for 12 hours to produce the steady-state configuration of the magnetosphere under the fixed orientation of the IMF.

Figure 11 shows the field-line configurations obtained for three assumed directions of the IMF. The three panels are for the IMF directions (a) 30 ° south, (b) 15° north, and (c) 60° north of the dusk direction. The field lines are viewed from the dusk side, and the green, blue, and red colors are used for the closed, open, and IMF-type field lines, respectively.

The configuration in panel (a) for the southward B_Z shows the basic elements of the magnetosphere that was illustrated in the left panel of Figure 1. On the dayside the open field lines (blue) are peeled away from the equatorial region where they are formed by reconnection with the IMF, and in the nightside equatorial region there are IMF-type field lines (red) that are formed by reconnection of the open field lines of the tail lobes. (The only problem of this simulation is that the region of the closed field lines does not extend far enough on the nightside and the distant and the near-earth neutral lines are not resolved.) In contrast, the configuration in panel (c) for the highly northward B_Z shows no indications of the peeling of the field lines from the dayside low-latitude region nor of the formation of the IMF-type field lines (red) inside the tail. In the case (c) the reconnection with the IMF takes place on the surface of the tail lobe and involves only the open field lines as illustrated in in the right panel of Figure 1.

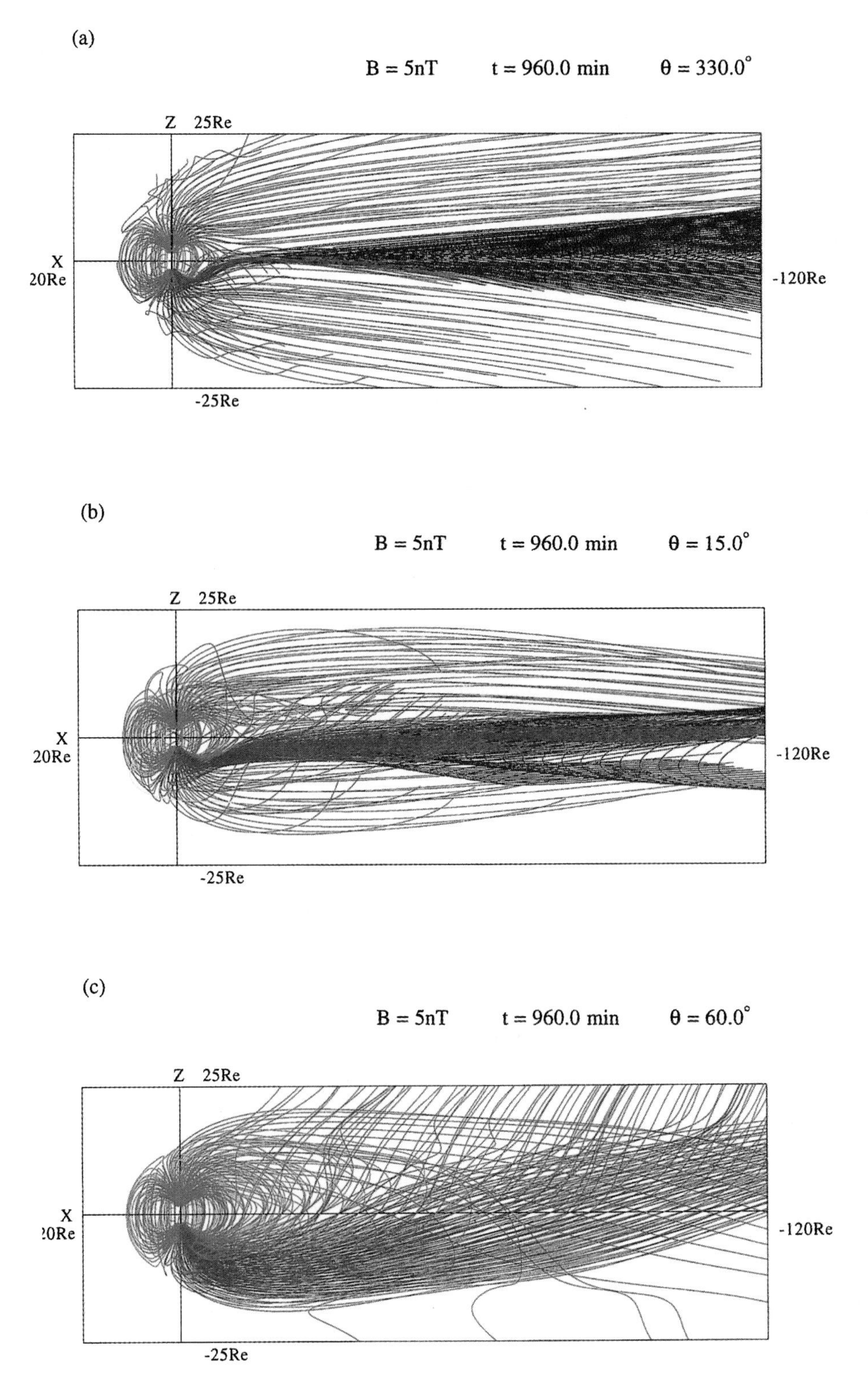

Figure 11. Views of the calculated magnetic field lines from the duskside. The simulations were performed for the IMF directions (a) 30 ° south, (b) 15° north, and (c) 60° north of the dusk direction (i.e., the y axis). The green, blue, and red colors correspond to the closed, open, and IMF-type field lines, respectively [Ogino and Walker, 1998].

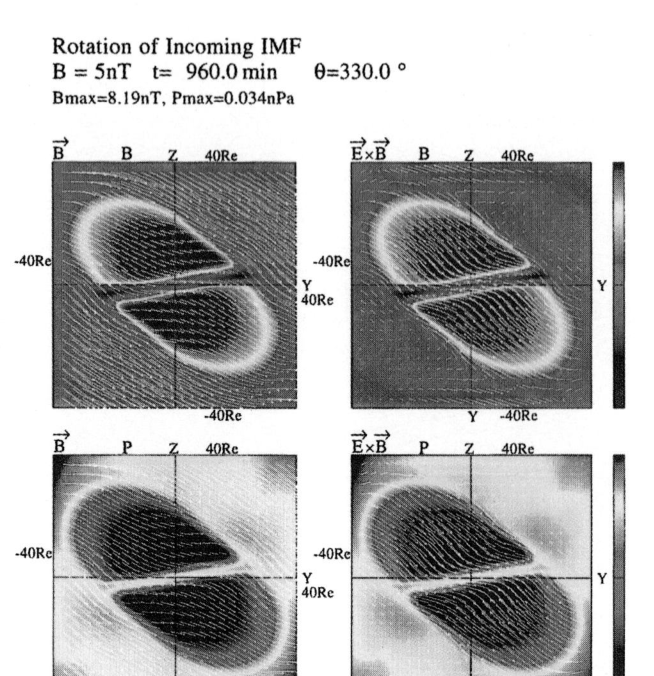

Figure 12a. Projections of the magnetic field vectors (left) and convection velocities (right) to the yz plane for the case (a) of the IMF directions assumed in Figure 11 [Ogino and Walker, 1998]. They are shown on the background of the color-coded contours of the magnetic field strength (top two diagrams in each panel) and of the pressure (bottom two diagrams).

In the intermediate case (b) of the slightly northward IMF, the reconnection with the IMF seems to occur both on lower-latitude and higher-latitude sides of the cusp, and the tail reconnection produces both IMF-type field lines (red) and closed field lines (green).

Figure 12 shows the simulation results of the magnetic field vectors (left column) and the convection velocities (right column) projected to the yz plane at x = -120 Re in the distant tail. The three panels (a) through (c) correspond to three cases of the IMF direction as in the preceding Figure 11. These vectors are shown on the background of the color-coded presentation of the magnetic field strength (top) and the pressure (bottom).

In the case of the southward IMF (panel a), the twisting of the neutral sheet is small. The magnetic field is directed from the northern tail to the southern tail across the neutral sheet as is expected for the IMF-type field lines that are formed beyond the tail reconnection line. The convection velocities are directed toward the neutral sheet, and most of the field lines are transported from the lobes to the neutral sheet by the convection. In the case of the slightly northward IMF (panel b), the twisting of the neutral sheet is more pronounced. The magnetic field in the region of the neutral sheet is nearly parallel to the neutral sheet, but it still has a weak component that is directed from the northern tail to the southern tail. The B_Z component of the magnetic field is northward everywhere including the neutral sheet because of the twist. The convection velocities also tend to be parallel to the neutral sheet but some field lines still flow into the neutral sheet. Most field lines in the tail, however, are convected from one side of the tail to the other and exit from the magnetopause without ever reaching the neutral sheet. These features of the case (b) simulation for the slightly northward IMF agree well with the observationally derived pictures (Figures 9 and 10c) for the times of IMF $|B_Y| \geq B_Z > 0$.

In the case of the highly northward IMF (panel c), the neutral sheet is highly inclined to the equatorial plane and the magnetic field vectors are parallel to the neutral sheet. The convection velocities are also parallel to the neutral sheet and almost all the field lines seem to be convected across the magnetotail from one side to the other. This means that there are no signatures of the magnetic reconnection occurring at the neutral sheet, so that the convection in this case involves only the open field lines. In this convection profile the open field

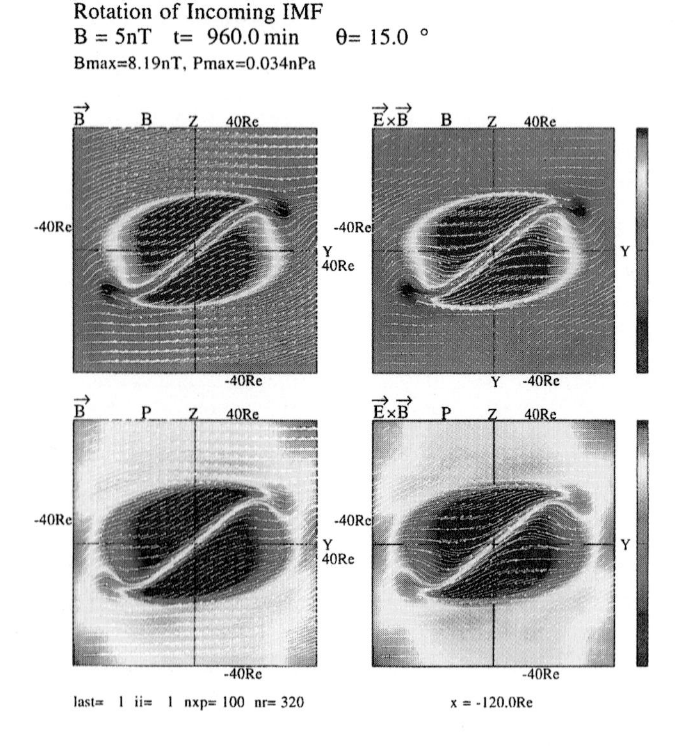

Figure 12b. For the case (b) of the IMF direction.

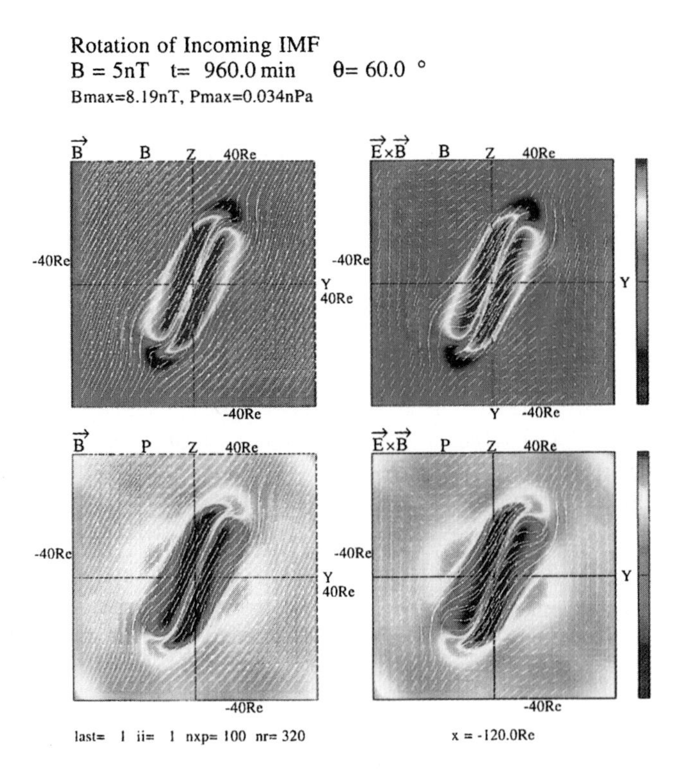

Figure 12c. For the case (c) of the IMF direction.

lines at the entry and the exit points of the tail surface are projected to the same point in the reconnection region on the dayside lobe surface, so that the convection streamlines are closed at low altitudes to form the lobe cell.

THE CASE OF THE ALMOST DUE NORTHWARD IMF

When the IMF is due northward, computer simulations suggest that all the magnetic field lines in the magnetosphere become closed and the length of the magnetotail becomes very short - less than 200 Re [Usadi et al., 1993; Ogino et al., 1994; Fedder and Lyon, 1995; Raeder et al., 1995]. Although the intervals of the due northward field are not expected to last for any finite length, there have been times when the solar magnetospheric latitude angle of the IMF was larger than 60° for several hours [Fairfield, 1993; Fairfield et al., 1996].

In such times, Fairfield [1993] found that the larger than normal B_Z was observed at 230 Re in the distant magnetotail, and interpreted it to belong to the closed magnetic field lines. He then pointed out that the flux of the closed field lines was so large that the entire tail could be easily closed earthward of 230 Re, leaving no room for the open field lines. However, it is now known that the magnetotail is severely twisted under the northward IMF, so that the northward B_Z is not the unique signature of the closed field lines. As illustrated in Figures 10, 11 and 12, the open field lines as well as the IMF-type field lines in the tail have the northward B_Z under the northward IMF. Hence the observed northward B_Z values cannot be used to estimate the closed magnetic flux that is rooted in the polar cap at both ends, and the shortness of the tail cannot be concluded on their basis.

The length of the tail may be estimated indirectly from the diameter of the tail. This however requires a large body of data on the crossing positions of the tail magnetopause in order that other effects such as the variations in the solar wind and the activity of the surface wave can be eliminated. The observed cases of the very northward IMF are so infrequent that a reliable statistics is still difficult to obtain. For the northward and southward B_Z intervals in general, Maezawa and Hori [1998] obtained the average diameter from the magnetopause crossing position at $x < -100$ Re and found that the diameter is smaller when the IMF is northward than when it is southward. The diameter for the northward IMF conditions, however, varies with the solar wind pressure in such a way as to suggest approximate conservation of the tail magnetic flux, and the shortness of the tail length could not immediately be concluded.

Under the almost due northward IMF the plasma sheet is expected to be very cold since the reconnection would not operate in the plasma sheet to heat the plasma. The only published case where this expectation could be tested by the observation was that of Fairfield et al. [1996] (their Figure 5) where GEOTAIL observed the plasma sheet at $x = -134$ Re when the θ angle of IMF was larger than 60°. Unfortunately, however, a plasmoid/flux rope was present in the plasma sheet on that occasion, and the comparison with the model could not be made. The presence of the plasmoid/flux rope in this single case is not to be construed against the model we have presented. Our model of the magnetotail under the northward IMF is to apply in a great majority of the times, but in almost all the northward IMF intervals we have studied there are other, though much less frequent, times when plasmoids were seen in the magnetotail [e.g., Nishida et al., 1994]. It seems that the solar wind magnetosphere interaction processes weaker than the global reconnection (such as the random reconnection [Nishida, 1989] and the plasma mixing at the boundary layer [Fujimoto et al., 1996]) are still operative under the almost due northward IMF condition and cause the stretching and tipping off of the closed field lines.

When its perigee of 10 Re is on the dayside, GEOTAIL skims along the dayside magnetopause. In a few of such occasions Katamura [1997] has found the anti-sunward convection in the dayside nose region of the magnetosphere during brief intervals of very northward IMF. Such a flow could be produced when the northward IMF field lines are reconnected with the geomagnetic field in both northern and southern cusp regions and engulfed into the magnetosphere [Song and Russell, 1992], but this interpretation was not adopted since the anti-sunward flow was not accompanied by the magnetosheath-type plasma. Katamura [1997] chose to consider these flows as a transient feature; when IMF turns from southward to northward, the geomagnetic field lines which used to flow toward the dayside magnetopause can no longer merge with the IMF so that they are forced to return toward the earth.

From the particle observations at the ionospheric end of the field lines, Newell et al. [1997] recently suggested that the polar cap completely closes when IMF $B_Z > |B_y| > 0$ for about 4 hours, but that even brief (only 6-7 min long) southward excursions of the IMF can fill the polar cap with open field lines. This provides an important measure for the IMF signature to look for in future in order to isolate the effect of the $B_Z > |B_y| > 0$ condition on the magnetotail configuration and dynamics.

The simulation of Ogino and Walker [1998] suggests that the reconnection does not occur in the magnetotail when IMF is very northward. Nevertheless the plasma sheet is present in their simulation result, and the question remains as to how the plasma sheet is formed and maintained under the very northward IMF.

DISCUSSION

It was considered earlier that ISEE 3 observations were not compatible with a simple model of the neutral line since the average B_Z of the magnetic field does not turn southward until x = -210 Re which is about 100 Re further downstream than where the flow becomes tailward and super-Alfvenic [Zwickl et al., 1984: Daly et al., 1984: Slavin et al., 1985]. Slavin et al. [1985] attempted to resolve this apparent difficulty in terms of the dependence of the neutral line position on the y coordinate across the tail. By tracing the position where the average B_Z turns southward, they suggested that the distant neutral line is curved and is the nearest to the Earth at midnight. They also suggested that closed field lines are being swept tailward along the flanks even at x = -210 Re.

However, the primary reason for the ISEE 3 observations of the northward field in the region of the tailward flow should have been the contribution from the data of geomagnetically quiet times, that is, under the northward IMF conditions. We have shown that the tailward-convecting IMF-type field lines beyond the reconnection line have the northward polarity under such conditions.

The present model of the convection under northward IMF conditions consists of the merging cell as well as the lobe cell in terms of the nomenclature used by Reiff and Burch [1985]. We compare our model with the earlier conjectures.

Heikkila [1988] noted that B_Z in the neutral sheet is northward on average when the speed of the tailward flowing plasma is moderate (< 400 km/s) and advocated that the closed field lines are convected tailward in quiet times. This interpretation has a serious difficulty as it implies that unrealistically large amounts of the magnetic flux are transported tailward when the quiet-time condition lasts for a day or more [Nishida et al., 1994]. With the tail radius of 20 Re, the plasma sheet half-thickness of 5 Re, and the lobe field strength of 8 nT the magnetic flux content of the lobe is estimated to be 1.4×10^8 Wb. The electric field of only - 0.1 mV/m, on the other hand, corresponds to the tailward transport of 2.6×10^4 Wb/s through the width of 40 Re of the plasma sheet, and it takes only 5.4×10^3 s to completely replenish the tail.

As a variation of the above it may be conjectured that the closed field lines are stretched tailward not continually but sporadically or quasi-periodically, and their tips are chopped off to form the tailward propagating magnetic loops while the remaining closed field lines retreat temporarily toward to Earth. While such features are sometimes observed in quiet times as described in the last section, this cannot explain the bulk of the quiet-time observations of the net tailward transport of the northward magnetic field lines; the tailward propagation of the magnetic loops does not represent the net transport since the magnetic fields in the forward and the rear parts of a loop are oppositely directed. If alternatively it is supposed that the tips are cut off from the closed field lines that are stretched continually tailward, the difficulty of the unrealistically large amount of the transport of the closed magnetic flux arises again.

Richardson et al. [1989] noted that in quiet-time cases of CDAW 8 that B_Z in the plasma sheet was northward on average when the deflection to the south was expected as the observations were made tailward of the neutral line. They suggested that the tail twisting or tilting may have masked the southward component, and used in particular the tilting of the tail axis to interpret their interval D. In our model the systematic twisting of the tail under the influence of B_y plays a more fundamental role, so that B_Z need not be southward in the plasma sheet tailward of the neutral line.

Reiff and Burch [1985] have suggested that under the northward IMF the reconnection between the open field lines, which characteristically traverse the equatorial plane, produces a finger of closed field lines near midnight which move sunward just poleward of the polar cap boundary. They called this convection as the reclosure cell. While our model asserts that the reconnection occurs in the magnetotail under the slightly northward IMF condition, the resulting closed field lines are not considered to flow sunward across the polar cap but on the equatorward side of the polar cap boundary. This agrees [Nishida et al., 1998] with the properties of the convection in low altitudes observed by the DE 2 satellite [Taguchi and Hoffman, 1996].

In the present model for the IMF $|B_Y| \geq B_Z > 0$ condition, the dayside reconnection occurs both with the open field lines poleward of the cusp and with the closed field lines equatorward of the cusp. In the Cowley [1982]'s diagram the former element corresponds to the panel (e) but the latter element was not considered. The inferred geometry of the reconnection lines that are supposed to be simultaneously present both poleward and the equatorward of the cusp is illustrated in Nishida et al. [1998].

A characteristic feature of the magnetotail under the northward IMF condition is the prevalence of the cold and dense plasmas whose temperature and density are intermediate between the magnetosheath and the plasma-sheet values [Fairfield, 1993; Fairfield et al., 1996; Fujimoto et al., 1996; Terasawa et al., 1997]. In our model, these plasmas are on the open or IMF-type field lines and are supplied from the magnetosheath by flowing along these field lines. The entry in the form of the slow expansion fan is a possible cause of the transition in density [Siscoe et al., 1994]. The cold and dense plasmas, however, have sometimes been found with the signatures of the closed character of the magnetic field lines such as the isotropic energetic electron flux [Richardson et al., 1989] and the bistreaming electrons [Fujimoto et al., 1996], and this may suggest that other mechanisms are also operative. Richardson et al. [1989] have suggested that some of the closed field lines that are produced earthward of the neutral line are drawn tailward along the magnetotail boundary and form a slow plasma sheet adjacent to the magnetopause.

The structure of the tail magnetopause and the boundary region under the northward IMF condition warrants further investigation in order that the understanding of the convection be made more complete. The dominant process may vary with the downstream distance x. Examination of the kinetic properties of the boundary layer plasma is particularly important for clarifying the plasma transport and relaxation processes. The interaction through the reconnection between IMF and the tail field lines should be accompanied by the exchange of the magnetosheath and the tail plasma via field-aligned flows. Such flows have been identified with GEOTAIL observations [Hirahara et al., 1998: Shirai et al., 1998] and future studies of their dependence on the IMF condition will provide very important clues on the nature of the interaction process.

CONCLUSION

Magnetotail observations over a wide range of the IMF conditions including the southward and slightly northward cases have shown that the magnetic reconnection is the basic element of the convection in the magnetotail. In geomagnetically active times that correspond to the southward IMF on average, there are two preferred sites for the reconnection. One is in the distant tail and at about 140 Re, while the other is in the near-earth tail and is formed inside about 50 Re. More open field lines are reconnected at the distant neutral line than at the near-earth neutral line. In geomagnetically quiet times that correspond to IMF $|B_Y| \geq B_Z > 0$, the reconnection still occurs in the magnetotail but it generates an apparently different signature because of the twisting of the tail due to the IMF B_Y; the field lines that are convected tailward beyond the neutral line have the northward B_Z polarity although they cross the neutral sheet from the northern side to the southern side. In the presence of the tail twist, the northward B_Z is not the unique signature of the closed field lines.

The numerical simulation of the magnetospheric structure and dynamics under the influence of the IMF has clearly reproduced the twisting of the tail as well as the occurrence of reconnection in the tail for the IMF $|B_Y| \geq B_Z > 0$ condition. It has also suggested that the reconnection would not occur in the tail when the IMF direction is almost due northward. Correspondingly, the simulation when IMF $|B_Y| \geq B_Z > 0$ shows that some of the streamlines of the convection flow from the lobe into the neutral sheet, while all the streamlines traverse the tail without reaching the neutral sheet when the IMF is almost due northward. Thus the ionospheric convection consists of both the lobe and the merging cells when the IMF $|B_Y| \geq B_Z > 0$ but it has only the lobe cell when the IMF is almost due northward.

Although numerical simulations suggest that the length of the magnetotail becomes very small and the magnetosphere becomes tadpole shaped when the IMF is almost due northward, a convincing observational proof has not yet been given. This is because the northward polarity of B_Z is not a unique signature of the closed character of the field lines in the tail which is severely twisted under such IMF conditions. Statistical studies of the tail diameter using a substantial body of data under the IMF $B_Z > |B_Y|$ condition are needed to test this conjecture.

REFERENCES

Axford, W.I., and C.O. Hines, A unifying theory of high-latitude geophysical phenomena and geomagnetic storms, Can. J. Phys., 39, 1433-1464, 1961.

Axford, W.I., H.E. Petschek, and G.L. Siscoe, Tail of the magnetosphere, J. Geophys. Res., 70, 1231-1236, 1965.

Boyle, C.B., P.H. Reiff, and M.R. Hairston, Empirical polar cap potentials, J. Geophys. Res., 102, 111-125, 1997.

Cowley, S.W.H., Magnetospheric asymmetries associated with the

Y-component of the IMF, Planet. Space Sci., 29, 77-96, 1981.

Cowley, S.W. H., Magnetospheric and ionospheric flow and the interplanetary magnetic field, in The Physical Basis of the Ionosphere in the Solar-Terrestrial System, pp. 4-4—4-14, AGARD, Neuilly sur Seine, France, 1982.

Daly, P.W., T.R. Sanderson, and K.-P. Wenzel, Survey of energetic (E > 3 5 keV) ion anisotropies in the deep geomagnetic tail, J. Geophys. Res., 89, 10733-10739, 1984.

Dungey, J.W., Interplanetary magnetic field and the auroral zones, Phys. Rev. Lett., 6, 47-48, 1961.

Fairfield, D. H., Solar wind control of the distant magnetotail: ISEE 3, J. Geophys. Res., 98, 21265-21276, 1993

Fairfield, D.H., R.P. Lepping, L.A. Frank, K.L. Ackerson, W.R. Paterson, S. Kokubun, T. Yamamoto, K. Tsuruda, and M. Nakamura, Geotail observations of an unusual magnetotail under very northward IMF conditions, J. Geomag. Geoelectr., 48, 473-487, 1996.

Fedder, J.A., and J. G. Lyon, The Earth's magnetosphere is165 Re long: Self-consistent currents, convection, magnetospheric structure, and processes for northward interplanetary magnetic field, J. Geophys. Res., 100, 3623-3635, 1995.

Fujimoto, M., A. Nishida, T. Mukai, Y. Saito, T. Yamamoto, and S. Kokubun, Plasma entry from the flanks of the near-earth magnetotail: GEOTAIL observations in the dawnside LLBL and the plasma sheet, J. Geomag. Geoelectr., 48, 711-727, 1996.

Hayakawa, H., A. Nishida, and E. W. Hones, Jr., Statistical characteristics of plasma flow in the magnetotail, J. Geophys. Res., 87, 277-285, 1982.

Heikkila, W.J., Current sheet crossings in the distant magnetotail, Geophys. Res., Lett., 15, 299-302, 1988.

Hirahara, M., T. Terasawa, T. Mukai, M. Hoshino, Y. Saito, S. Machida, T. Yamamoto, and S. Kokubun, Cold ion streams consisting of double proton populations and singly charged oxygen observed at the distant magnetopause by Geotail: A case study, J. Geophys. Res., 102, 2359-2372, 1997.

Hones, E.W., Jr., Plasma flow in the magnetotail and its implications for substorm theories, in Dynamics of the Magnetosphere, S.I. Akasofu, ed., p. 545-562, D. Reidel Publ. Co., 1979.

Hoshino, M., T. Mukai, A. Nishida, Y. Saito, T. Yamamoto, and S. Kokubun, Evidence of two active reconnection sites in the distant magnetotail, J. Geomag. Geoelectr., 48, 515-523, 1996.

Ieda, A., S. Machida, T. Mukai, Y. Saito, T. Yamamoto, A. Nishida, T. Terasawa, and S. Kokubun, Statistical analysis of the plasmoid evolution with GEOTAIL observations, J. Geophys. Res., 102, 4453, 1997.

Katamura, C., Near-earth magnetospheric convection, M.Sc. thesis, University of Tokyo, 1997.

Kawano, H., A. Nishida, M. Fujimoto, T. Mukai, S. Kokubun, T. Yamamoto, T. Terasawa, M Hirahara, Y. Saito, S. Machida, K. Yumoto, H. Matsumoto, and T. Murata, A quasi-stagnant plasmoid observed with Geotail on October 15, 1993, J. Geomag. Geoelectr., 48, 525-539, 1996.

Levy, R.H., H.E. Petschek, and G.L. Siscoe, Aerodynamic aspects of the magnetospheric flow, AIAA J., 2, 2065-2076, 1964.

Maezawa, K., Magnetic convection induced by the positive and negative z-component of the interplanetary magnetic field: Quantitative analysis using polar cap magnetic records, J. Geophys. Res., 81, 2289-2303, 1976.

Maezawa, K., T. Hori, T. Mukai, Y. Saito, T. Yamamoto, S. Kokubun, and A. Nishida, Structure of the distant magnetotail and its dependence on the IMF By component: GEOTAIL observations, Adv. Space Res., 20, 949-959, 1997.

Maezawa, K., and T. Hori, Distant magnetotail: its structure, IMF dependence and thermal properties, this volume, 1998.

Nagai, T., M. Fujimoto, Y. Saito, S. Machida, T. Terasawa, R. Nakamura, T. Yamamoto, T. Mukai, A. Nishida, and S. Kokubun, Structure and dynamics of magnetic reconnection for substorms onsets with GEOTAIL observations, J. Geophys. Res., 103, 4419, 1998.

Newell, P.T., D. Xu, C.-I. Meng, and M. Kivelson, Dynamic polar cap: a unifying approach, J. Geophys. Res. 102, 127-139, 1997.

Nishida, A., Can random reconnection on the magnetopause produce the low latitude boundary layer? Geophys. Res. Lett., 16, 227-230, 1989.

Nishida, A., T. Yamamoto, K. Tsuruda, H. Hayakawa, A. Matsuoka, S. Kokubun, M. Nakamura, and K. Maezawa, Structure of the neutral sheet in the distant tail (x= -210 Re) in geomagnetically quiet times, Geophys. Res. Lett., 21, 2951-2954, 1994.

Nishida, A., T. Mukai, T. Yamamoto, Y. Saito, and S. Kokubun, GEOTAIL observations on the reconnection process in the distant tail in geomagnetically active times, Geophys. Res. Lett., 22, 2453-2456, 1995a.

Nishida, A., T. Mukai, T. Yamamoto, Y. Saito, S. Kokubun, and K. Maezawa, GEOTAIL observation of magnetospheric convection in the distant tail at 200 Re in quiet times, J. Geophys. Res., 100, 23663-23675, 1995b.

Nishida, A., T. Mukai, T. Yamamoto, Y. Saito, and S. Kokubun, Magnetotail convection in geomagnetically active times, 1. Distance to neutral lines, J. Geomag. Geoelectr., 48, 489-501, 1996a.

Nishida, A., T. Mukai, T. Yamamoto, Y. Saito, and S. Kokubun, Magnetotail convection in geomagnetically active times, 2. Dawn-dusk motion in the plasma sheet, J. Geomag. Geoelectr., 48, 503-513, 1996b.

Nishida, A., T. Mukai, T. Yamamoto, Y. Saito, S. Kokubun, and R.P. Lepping, Traversal of the nightside magnetosphere at 10 to 15 Re during northward IMF, Geophys. Res. Lett., 24, 939-942, 1997.

Nishida, A., T. Mukai, T. Yamamoto, S. Kokubun, and K. Maezawa, A unified model of the magnetotail convection in geomagnetically quiet and active times, J. Geophys. Res., 103, 4409, 1998.

Ogino, T., and R.J. Walker, A global MHD simulation of magnetotail configuration depending on the IMF orientation, in preparation, 1998.

Ogino, T., R.J. Walker, and M. Ashour-Abdalla, A global magnetohydrodynamic simulation of the magnetosheath and magnetosphere when the interplanetary magnetic field is northward, IEEE Trans. Plasma Sci., 20(6), 817-828, 1992.

Ogino, T., R.J. Walker, and M. Ashour-Abdalla, A global magnetohydrodynamic simulation of the response of the magnetosphere to a northward turning of the interplanetary magnetic field, J. Geophys. Res., 99, 11027-11042, 1994.

Owen, C.J., J.A. Slavin, I.G. Richardson, N. Murphy, and R.J. Hynds, Average motion, structure and orientation of the distant magnetotail determined from remote sensing of the edge of the plasma sheet boundary layer with E > 35 keV ions, J. Geophys. Res., 100, 185-204, 1995.

Raeder, J., R.J. Walker and M. Ashour-Abdalla, The structure of the distant geomagnetic tail during long periods of northward IMF, Geophys. Res. Lett., 22, 349-352, 1995.

Reiff, P.H., and J.L. Burch, IMF By-dependent plasma flow and Birkeland currents in the dayside magnetosphere, 2. A global model for northward and southward IMF, J. Geophys. Res., 90, 1595-1609, 1985.

Richardson, I.G., C.J. Owen, S.W.H. Cowley, A.B. Galvin, T.R. Sanderson, M. Scholer, J.A. Slavin, and R.D. Zwickl, ISEE 3 observations during the CDAW 8 intervals: Case studies of the distant geomagnetic tail covering a wide range of geomagnetic activity, J. Geophys. Res., 94, 15189-15220, 1989.

Ruohoniemi, J.M., and R.A. Greenwald, Statistical patterns of high-latitude convection obtained from Goose Bay HF radar observations, J. Geophys. Res., 101, 21743-21763, 1996.

Shirai, H., K. Maezawa, T. Mukai, T. Yamamoto, Y. Saito, M. Fujimoto, and S. Kokubun, Entry process of low-energy electrons into the magnetosphere along open field lines: Polar rain electrons as field line tracers, J. Geophys. Res., 103, 4379, 1998.

Siscoe, G.L., L.A. Frank, K.L. Ackerson, and W.R. Paterson, Properties of the mantle-like magnetotail boundary layer: GEOTAIL data compared with a mantle model, Geophys. Res. Lett., 25, 2975-2978, 1994.

Slavin, J.A., E.J. Smith, D.G. Sibeck, D.N. Baker, R.D. Zwickl, and S.-I. Akasofu, An ISEE 3 study of average and substorm conditions in the distant magnetotail, J. Geophys. Res., 90, 10875-10895, 1985.

Song, P., and C.T. Russell, Model of the formation of the low-latitude boundary layer for strongly northward interplanetary magnetic field, J. Geophys. Res., 97, 1411-1420, 1992.

Taguchi, S., and R.A. Hoffman, Ionospheric plasma convection in the midnight sector for northward interplanetary magnetic field, J. Geomag. Geoelectr., 48, 925-933, 1996.

Terasawa, T., M. Fujimoto, T. Mukai, I. Shinohara, Y. Saito, T. Yamamoto, S. Machida, S. Kokubun, A.J. Lazarus, J.T. Steinberg, and R.P. Lepping, Solar wind control of density and temperature in the near-earth plasma sheet: WIND-GEOTAIL collaboration, Geophys. Res. Lett., 24, 935-938, 1997.

Usadi, A., A. Kageyama, K. Watanabe, and T. Sato, A global simulation of the magnetosphere with a long tail: Southward and northward interplanetary magnetic field, J. Geophys. Res., 98, 7503-7517, 1993.

Weimer, D.R., Models of high-latitude electric potentials derived with a least error fit of spherical harmonic coefficients, J. Geophys. Res., 100, 19595-19607, 1995.

Zwickl, R.D., D.N. Baker, S.J. Bame, W.C. Feldman, J.T.Gosling, E.W. Hones, Jr., D.J. McComas, B.T. Tsurutani, and J.A. Slavin, Evolution of the earth's magnetotail: ISEE 3 electron plasma results, J. Geophys. Res., 89, 11007-11012, 1984.

A. Nishida, Institute of Space and Astronautical Science, 3-1-1 Yoshinodai, Sagamihara, Kanagawa 229, Japan

T. Ogino, Solar Terrestrial Environment Laboratory, Nagoya University, 3-13 Honohara, Toyokawa, Aichi 442, Japan

Magnetotail Structure and its Internal Particle Dynamics During Northward IMF

M. Ashour-Abdalla, J. Raeder, M. El-Alaoui, V. Peroomian

Institute of Geophysics and Planetary Physics, University of California, Los Angeles, CA 90095-1567

This study uses Global magnetohydrodynamic (MHD) simulations driven by solar wind data along with Geotail observations of the magnetotail to investigate the magnetotail's response to changes in the interplanetary magnetic field (IMF); observed events used in the study occurred on March 29, 1993 and February 9, 1995. For events from February 9, 1995, we also use the time-dependent MHD magnetic and electric fields and the large-scale kinetic (LSK) technique to examine changes in the Geotail ion velocity distributions. Our MHD simulation shows that on March 29, 1993, during a long period of steady northward IMF, the tail was strongly squeezed and twisted around the Sun-Earth axis in response to variations in the IMF B_y component. The mixed (magnetotail and magnetosheath) plasma observed by Geotail results from the spacecraft's close proximity to the magnetopause and its frequent crossings of this boundary. In our second example (February 9, 1995) the IMF was also steady and northward, and in addition had a significant B_y component. Again the magnetotail was twisted, but not as strongly as on March 29, 1993. The Geotail spacecraft, located ~30 R_E downtail, observed highly structured ion distribution functions. Using the time-dependent LSK technique, we investigate the ion sources and acceleration mechanisms affecting the Geotail distribution functions during this interval. At 1325 UT most ions are found to enter the magnetosphere on the dusk side earthward of Geotail with a secondary source on the dawn side in the low latitude boundary layer (LLBL). A small percentage come from the ionosphere. By 1347 UT the majority of the ions come from the dawn side LLBL. The distribution functions measured during the later time interval are much warmer, mainly because particles reaching the spacecraft from the dawnside are affected by nonadiabatic scattering and acceleration in the neutral sheet.

1. INTRODUCTION

Over the past two decades, researchers have developed a number of global magnetohydrodynamic (MHD) models of the interaction of the solar wind with Earth's magnetosphere. These models have used as input various idealized solar wind conditions (e.g., where the IMF B_z is constant and northward or southward) to study the principal features of its coupling with the magnetosphere [*Leboeuf et al.*, 1978, 1981; *Lyon et al.*, 1981; *Brecht et al.*, 1982; *Ogino et al.*, 1984, 1985]. The actual state of the magnetosphere is far from being a simple superposition of idealized states [*Frank et al.*, 1995; *Raeder et al.*, 1995, 1997b] because

New Perspectives on the Earth's Magnetotail
Geophysical Monograph 105

the magnetosphere's internal evolution prevents it from responding in a simple, linear way to changes in the solar wind. Instead it is influenced by the history of the solar wind. While the solar wind varies on a time scale of minutes, the magnetosphere's internal evolution can take many hours to relax to a steady state. Consequently, in order to produce realistic configurations, it is necessary to use actual data from spacecraft located upstream in the solar wind as driving input to our models. Indeed recent MHD simulations with real data input can reproduce many of the changes in the global configuration of the magnetosphere [*Frank et al.*, 1995; *Raeder et al.*, 1997a, b; *Berchem et al.*, 1997a, b].

One of the goals of magnetospheric physics has been to understand the transport of plasmas through the solar wind-magnetosphere-ionosphere system. To attain such an understanding, it is necessary to determine the sources of plasmas, their trajectories through the magnetospheric electric and magnetic fields to the points of observation and the acceleration processes they undergo en route. There are too few spacecraft in the magnetosphere at any given time to make using observations alone an effective method of studying transport through the entire system. However, theory and modeling can be used to augment the available observations.

In this paper we detail an approach in which theory and numerical simulations are used in concert with observations to study plasma transport through the magnetospheric system. This approach is based on a global magnetohydrodynamic (MHD) simulation of the interaction between the solar wind and the magnetosphere which employs observed solar wind and interplanetary magnetic field (IMF) parameters to determine the solar wind boundary conditions and models the time-dependent response of the magnetosphere to the changing solar wind. We carefully calibrate the MHD model by comparing the fields and plasma moments from the model with observations within the magnetosphere, and when we are convinced a model is sound we use it to delineate plasma transport.

The MHD models yield a picture of the overall configuration of the magnetosphere and bulk transport, but the MHD paradigm neglects important physics such as particle drift motion. Magnetospheric observations show that plasma distribution functions are not simple thermal distributions. Instead they are frequently highly structured and complex [*Frank et al.*, 1995; *Ashour-Abdalla et al.*, 1996], and encoded in this complex structure is information about the transport and acceleration of the plasmas. We have developed a version of the large-scale kinetic (LSK) code [*Ashour-Abdalla et al.*, 1993] to extend the capability of MHD models by determining the history of the particles in the measured distribution functions. Starting from an observed distribution function we follow the trajectories of thousands of ions backwards in time through the electric and magnetic fields as determined from the corresponding MHD calculation [*El-Alaoui et al.*, 1995; *Ashour-Abdalla et al.*, 1997].

We apply this approach first to study the "quiet" magnetosphere when the IMF is northward for an extended period of time. Although the magnetosphere is less dynamic during intervals of northward IMF the magnetic configuration is still very complex. Observational, theoretical and simulation studies all indicate that when the IMF is northward for an extended period of time the magnetotail configuration is very different from the general configuration [*Frank et al.*, 1995; *Fairfield et al.*, 1996; *Raeder et al.*, 1995, 1997a, b; *Berchem et al.*, 1997a, b]. The simulation results support the observationally based evidence that high latitude reconnection between northward IMF field lines and tail field lines poleward of the polar cusps strongly influences the magnetospheric configuration [*Song and Russell,* 1992 and references therein]. For extremely prolonged periods of northward IMF, both observations [*Fairfield,* 1993] and simulations [*Ogino et al.*, 1992, 1994; *Usadi et al.*, 1993; *Raeder et al.*, 1995, 1997a; *Fedder and Lyon,* 1995] indicate that the magnetospheric configuration either is closed or has significantly reduced open field lines, resulting in the tail lobe having a smaller than average cross section. When the IMF has a B_y component the magnetotail can become twisted [*Cowley* 1981; *Brecht et al.*, 1981; *Ogino et al.*, 1985; *Frank et al.*, 1995; *Berchem et al.*, 1997a, b]. Field lines connecting the northern ionosphere twist to the southern hemisphere in the tail, and field lines from the southern ionosphere twist to the northern ionosphere, thereby giving the tail field lines a braided appearance [*Berchem et al.*, 1997a, b]. A changing B_y can cause the tail boundary to shift, resulting in a spacecraft near the magnetopause alternating between being in the magnetosheath and being in the magnetosphere [*Frank et al.*, 1995].

We briefly introduce the MHD model used throughout this study in section 2. Our first case study considers an interval from Geotail observations first noted by *Fairfield et al.* [1996]. On March 29, 1993, while the IMF remained steadily northward with constant sunward B_x, the B_y component changed from duskward to dawnward. We use this event to calibrate our MHD models by studying the response of the magnetosphere to the B_y component of the IMF during prolonged intervals of northward IMF. Our analysis of the configuration of the magnetosphere for this case is presented in section 3.

In the second case study presented in this paper, we address the question of ion transport through the magnetosphere during a period on February 9, 1995, when the IMF was northward and had a B_y component. In this study our goal is twofold: first we wish to ascertain the source of the ions observed by Geotail in the near-Earth tail, and second, we wish to determine how these ions were accelerated after entering the magnetosphere.

Three main sources of magnetospheric particles have been identified: the low latitude boundary layer (LLBL), the plasma mantle and the ionosphere. Observations in the magnetotail near the equator identify the LLBL as a transition region through which plasmas can enter or leave the magnetosphere [*Hones et al.*, 1972a; *Eastman et al.*, 1976]. At higher latitudes in the tail, solar wind particles can enter the magnetosphere through the plasma mantle on open magnetospheric field lines [*Hones et al.*, 1972b; *Akasofu et al.*, 1973; *Rosenbauer et al.*, 1975; *Hardy et al.*, 1975; *Haerendel et al.*, 1978; *Pilipp and Morfill*, 1978]. The ionosphere, too, contributes plasma to the magnetosphere [*Shelley et al.*, 1972; *Geiss et al.*, 1978; *Lennartsson et al.*, 1979, 1981; *Lundin et al.*, 1980; *Balsiger et al.*, 1980; *Peterson et al.*, 1981; *Sharp et al.*, 1982; *Lennartsson and Shelley*, 1986]. *Shelley et al.* [1982] argue that the solar wind is the dominant source of ions in the magnetosphere at quiet times and that the ionosphere is dominant during more active times. *Chappell et al.* [1987] argue the ionosphere supplies enough ions to make up the entire ion component of the magnetosphere, an assertion which *Lennartsson* [1992] challenges.

Once inside the magnetosphere ions move through the magnetospheric electric and magnetic fields to the points of observation. Of special interest is the motion of ions near current sheets and neutral lines, where their behavior is sometimes nonadiabatic [*Speiser*, 1965; *Lyons and Speiser*, 1982; *Martin*, 1986; *Büchner and Zelenyi*, 1986, 1989; *Chen and Palmadesso*, 1986]. *Ashour-Abdalla* [1993, 1994, 1995] use the LSK approach with empirical magnetic and electric field models and demonstrate that the distant plasma sheet could be populated by the plasma mantle source. *Peroomian and Ashour-Abdalla* [1995, 1996] show that nearer Earth all three sources – the LLBL, the plasma mantle and the ionosphere – contribute to the plasma population and that the relative contribution of each source to the local distribution depends on the local time and radial distance from the Earth. *Richard et al.* [1994] and *Walker et al.* [1995, 1996] study the entry of particles into the magnetosphere by launching a distribution of solar wind ions into the electric and magnetic fields from a global MHD model. Their results show that for purely northward IMF is LLBL was the dominant source for solar wind ions, while for purely southward IMF the plasma mantle dominates, though the LLBL also contributes ions.

In section 4 we use the LSK method to investigate the source of ions observed by Geotail in the near-Earth plasma sheet during an interval on February 9, 1995, with steady northward, sunward and duskward IMF. We show how the sources change with time, leading to a complex distribution function even in the near-Earth plasma sheet during quiet times, and how nonadiabatic motion can influence the observed distribution function. We summarize our findings in section 5.

2. THE MHD MODEL

For this study we use a global MHD code that includes an ionospheric model to provide for the closure of field-aligned currents. In order to accommodate the large simulation volume with a 400 R_E tail and long simulation times, the simulation code has been parallelized for running on MIMD (Multiple Instruction - Multiple Data) machines by using a domain decomposition technique. The model essentially solves the ideal MHD equations, modified to include an anomalous resistivity term, for the magnetosphere and a potential equation for the ionosphere. Numerical effects, such as diffusion, viscosity, and resistivity, are necessarily introduced by the numerical methods. These permit viscous interactions and also, to a limited extent, magnetic field reconnection. However, the numerical scheme is optimized to minimize numerical effects. In particular, numerical resistivity is so low that it is necessary to introduce an anomalous resistivity term in order to model substorms correctly [see *Raeder et al.*, 1996]. The ionospheric part of the model takes into account three sources of ionospheric conductance: solar EUV ionization is modeled using the empirical model of *Moen and Brekke* [1993], diffuse auroral precipitation is modeled by assuming full pitch angle scattering at the inner boundary of the MHD simulation (at 3.7 R_E), and accelerated electron precipitation associated with upward field-aligned currents is modeled in accordance with the approach of *Knight* [1972] and *Lyons et al.* [1979]. The empirical formulas of *Robinson et al.* [1987] are used to calculate the ionospheric conductances from the electron mean energies and the energy fluxes. A detailed description of the MHD model, including initial and boundary conditions, can be found in *Raeder et al.* [1996, 1997b].

3. CASE I: MARCH 29, 1993

3.1. *Solar Wind Observations*

On March 29, 1993, from about 1200 UT to 2200 UT, the IMP 8 spacecraft was near (35, –16, –9) R_E in GSE coordinates, upstream of the bow shock in the dawn sector. Figure 1 shows the IMP 8 solar wind plasma and field measurements during this time period (black diamonds). The total magnetic field held fairly steady during the entire interval, rising slowly from about 12 nT at 1000 UT to about 15 nT at 2200 UT, which is about twice as strong as is usual for the solar wind. At about 1200 UT, the field

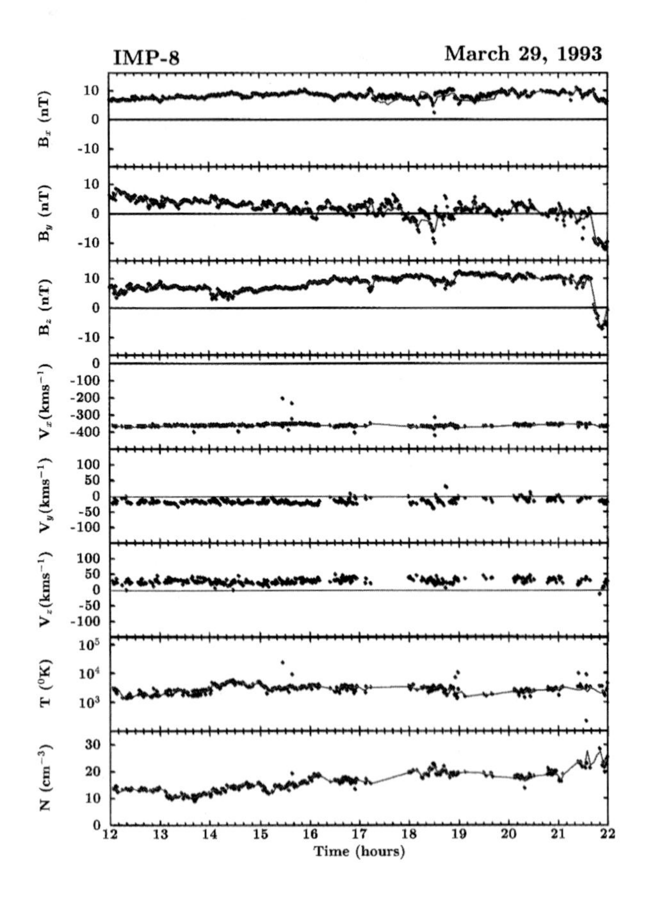

Figure 1. Solar wind parameters for March 23, 1993. The figure shows, from top to bottom: magnetic field components, velocity components, proton temperature, and proton density. The black dots indicate the IMP-8 measurements, and the red line shows the cleaned and averaged data that were used as input for the MHD simulation.

turned strongly northward and retained this northward orientation until 2130 UT. The IMF B_y component was fairly strong and positive at the beginning of the time interval, and until the end of the time interval it drifted slowly, almost linearly, towards zero. However, while B_z held fairly steady for most of the time interval of interest, B_y exhibited large fluctuations and sometimes changed sign. The IMF B_x component was quite large and steady with a value of about 8 nT throughout the time interval. The solar wind plasma measurements are less complete than the magnetic field measurements; however, inspection of Figure 1 shows that the solar wind density, velocity, and temperature were very steady, which allows us to interpolate and to fill in the missing values. The data also include some unrealistic and seemingly erroneous values in the plasma time series which we removed. Although the V_y and V_z solar wind velocity components were available, we did not use these data, but instead set V_y and V_z to zero, because the data not only had the same gaps that the other plasma parameters had but also fluctuated strongly on a short time scale (1 min), with amplitudes reaching a few degrees in direction. Because a one-degree deflection of the solar wind direction causes a 2.35 R_E deflection of the tail at the Geotail position (134 R_E downtail) we deemed these measurements not accurate enough for use in this study.

The red curves in Figure 1 show the processed solar wind data used as input to the model. Data gaps, primarily in the plasma data, were filled by interpolation or extrapolation. The input data are 3 minute averages and are produced by the solar wind model described in *Raeder et al.* [1996, 1997b]. The IMF **B** vector that is used as input is nearly identical to the observed IMF. However, around 1800 UT some strong rotations of the IMF vector occur that are not completely resolved by the solar wind model, and slight deviations of the input field from the measured field become visible. Although this is not entirely satisfactory, this solar wind model is the best that can be constructed with single point IMF measurements, because the alternative is to keep B_x constant and thereby miss some major rotations of the IMF vector.

3.2. Geotail Observations and Comparison with the MHD Model

During the latter half of March 29, 1993, the Geotail spacecraft was located in the distant tail, where it moved from its position at (–134.4, 14.4, –0.8) R_E (in GSE coordinates) at 1200 UT to (–132.4, 13.2, –0.9) R_E at 2200 UT. For a 370 km/s solar wind the aberration resulting from Earth's motion would cause the tail to rotate by 4.5° toward dusk, thus reducing the effective Geotail y position by 10.5 R_E. This rotation would put Geotail very close to the tail center, i.e., in aberrated coordinates the Geotail position would be near (–133, 3.5, –0.8) R_E. Additional tail aberrations, caused by a non-radial solar wind flow vector, may also be found in the data. *Fairfield et al.* [1996] estimate from the IMP 8 observations that the solar wind flow vector points on average 2.5° toward dawn and 2.6° towards north. Using this estimate, the Geotail position in aberrated coordinates would be near (–133, 9.3, –6.8) R_E, still relatively very close (11.5 R_E) to the tail center. We note, however, that the IMP 8 solar wind directional measurements fluctuate substantially during this time interval. By all estimates, Geotail was very close to the tail center at this time, and it is thus expected that it observed only plasmas and fields that are typical for the tail.

Figure 2 shows the Geotail plasma and magnetic field observations at 48 s time resolution from 1200 UT to 2200 UT on March 29, 1993 (black curves), results from the

MHD model with no aberration (red curves), with nominal (only Earth motion) aberration (green curves), and with the aberration vector estimated by *Fairfield et al.* [1996] (blue curves).

The Geotail observations can be ordered most easily by the x component of the velocity (V_x). During most of the interval V_x is between –400 km/s and –350 km/s, which is close to the observed solar wind velocity. At these times, the ion density is fairly high, near 10 cm^{-3}, and the plasma is cool, about 2×10^5 °K. Apparently, Geotail is in the magnetosheath during these times or possibly in a boundary layer with plasmas and fields that are very like the magnetosheath. Between these time intervals of magnetosheath-like plasmas and fields, the tailward velocity drops notably, the plasma is at least one order of magnitude hotter, the density is at least one order of magnitude lower, and the magnetic field assumes a tail-like orientation. Before 1900 UT this tail-like orientation corresponds to the southern lobe. Later, the direction is that expected for the northern lobe.

The colored curves show the MHD results for three aberration angles that lie within a cone of about 5°. None matches the observations very well. However, taken together, they essentially bracket B_x and V_x. B_x and V_x have sharp spatial gradients and therefore are the most revealing quantities of the tail topology. It is reasonable to assume that the tail boundary corresponding to these gradients is also well bracketed by the simulation results. As we show later, this boundary is the distant tail magnetopause. Small changes in the aberration angle, which are about of the same order as the accuracy of the solar wind directional measurements, cause drastic changes in the model output. Thus, even if we had a perfect model, it would be necessary to know the solar wind direction to within a few tenths of a degree in order to predict the Geotail observations accurately. Given the limitations of the solar wind measurements and the inaccuracies that are inherent in our model, in particular the coarse resolution in the distant tail (about 1.5 R_E in y and z, and about 5 R_E in x), the comparison is reasonably good. This assessment also holds for the B_y and B_z magnetic field components, except for small scale fluctuations that are beyond the realm of the model. In particular, the model predicts, in accordance with the observations, a B_y component that slowly drifts from positive to negative values, and a strongly positive B_z. The V_y and V_z components are also basically in accordance with the observations, considering that they are also affected by the solar wind aberration. In particular, the model predicts slightly duskward flows, i.e., the flow vector is oriented towards the tail center. Because of the low resolution, the model fails to predict the sharp gradients in the plasma density and temperature, i.e., these quantities are somewhat smeared out over the grid and the predictions generally fall between the extremes of the observed values.

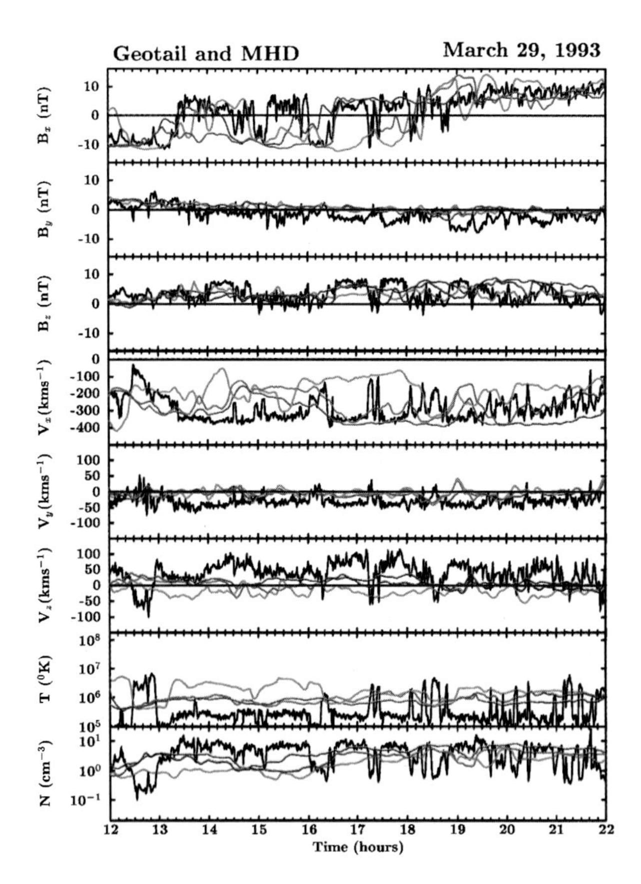

Figure 2. A comparison of time series from Geotail (black curves) and the MHD simulation (colored curves based on different assumptions of aberration; see text). The figure shows, from top to bottom: magnetic field components, velocity components, proton temperature, and proton density.

3.3. The Tail Cross Section at Geotail

Figures 3a and 3b show y_{GSE} - z_{GSE} cross sections at the Geotail position, i.e., at x = –134 R_E, at different times. The left panels show the color coded magnetic field's B_x component, and lines that are tangential to the magnetic field projection in the y_{GSE} - z_{GSE} plane. The right panels show the color coded V_x component of the flow and the (V_y, V_z) vectors. The dots indicate the Geotail position, according to the different tail aberration assumptions. The color of the dots corresponds to the colors in Figure 2: red is the nonaberrated position, green the position with nominal (only Earth motion) aberration, and blue the *Fairfield et al.* [1996] estimate. The six snapshots correspond to 1200 UT, 1400 UT, 1500 UT, 1700 UT, 1900 UT, and 2100 UT, respectively.

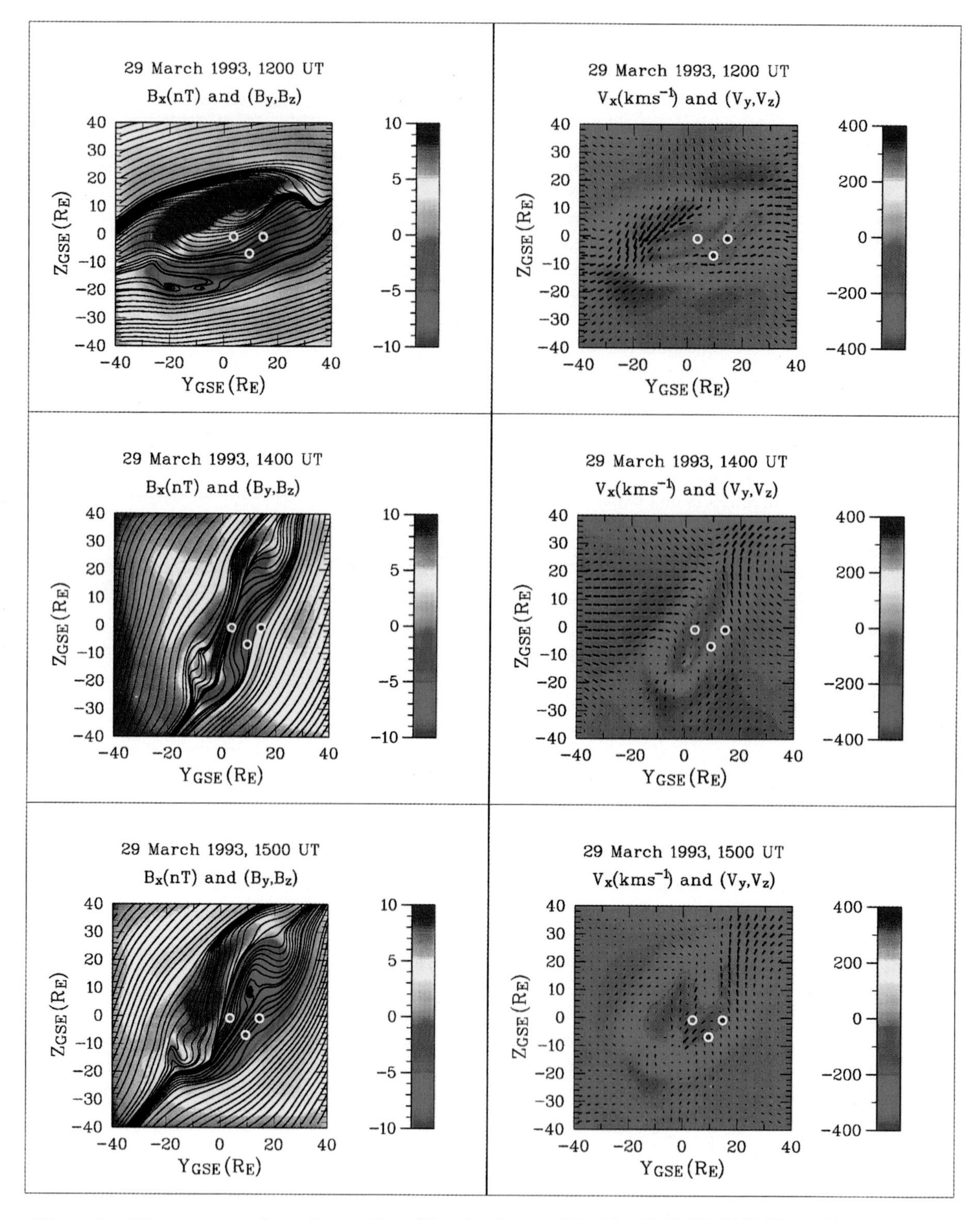

Figure 3. (a) Cross sections of the tail at the Geotail location ($x_{GSE} = -134\ R_E$) at 1200 UT, 1400 UT, and 1500 UT. The left panels show the color-coded magnetic field B_x component and lines that are tangential to the field. The right panels show the color coded V_x velocity component along with the (V_y, V_z) velocity vectors. The colored dots indicate the Geotail location in the y_{GSE} -z_{GSE} plane according to different assumptions of aberration (see text for details); (b) Similar to (a), but at 1700 UT, 1900 UT, and 2100 UT.

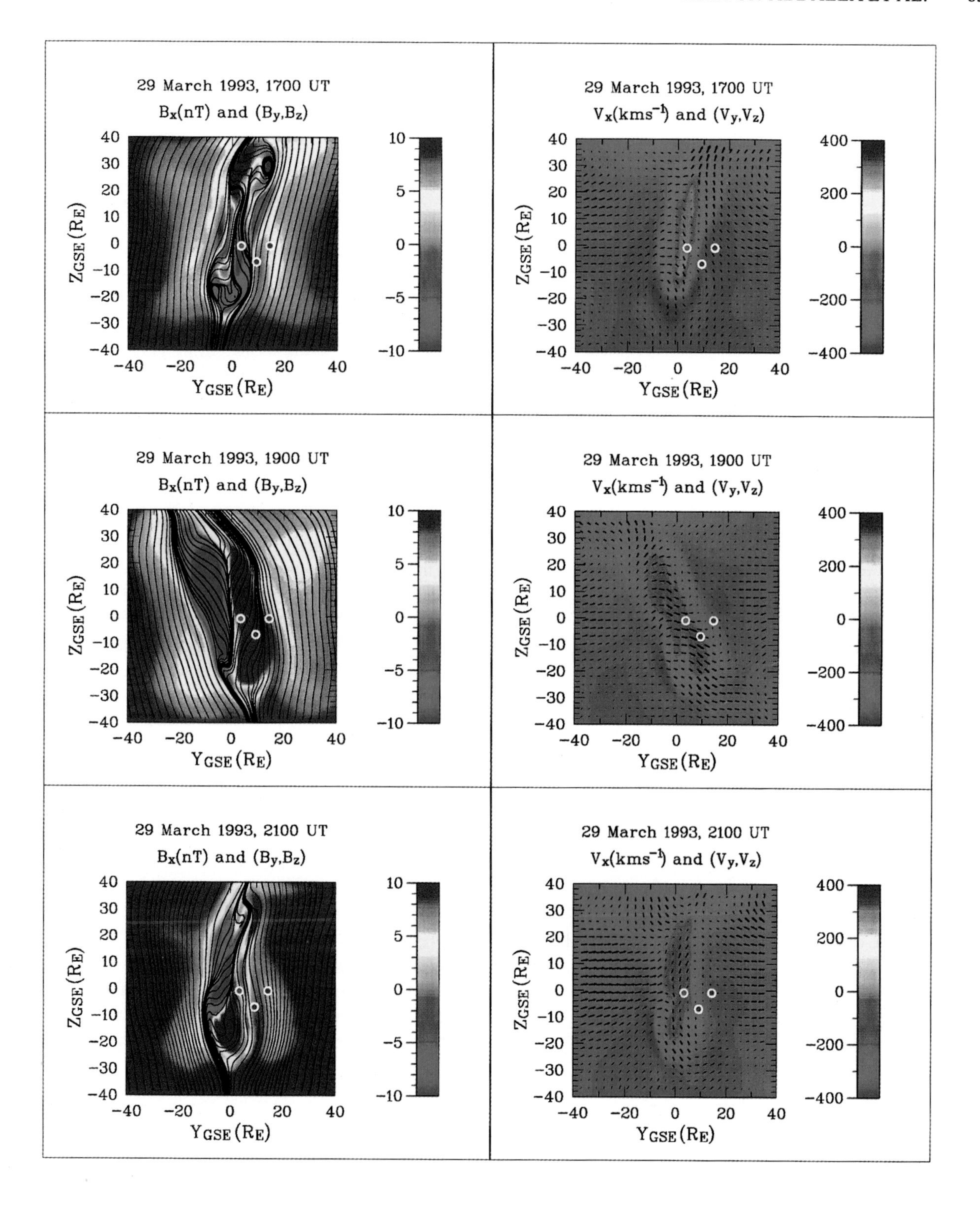

Figure 3. Continued from previous page.

At 1200 UT, Geotail is located in the southern lobe (compare with Figure 2). The lobe at this time is not circular, as many models predict (see, for example, *Sibeck et al.*, 1986), but instead is flattened and twisted. The overall shape of the tail is roughly elliptical, with the semimajor axis about twice as large as the semiminor axis. The flow velocity is lowest in the lobes. In the magnetosheath and the tail plasma sheet, it is comparable to the solar wind velocity.

At 1400 UT, the lobe has become even more flattened and twisted. Along the minor axis, the tail dimension is reduced to about 20 R_E, while the major dimension is about 80 R_E. At the same time, the lobe is twisted by almost 90 degrees counter-clockwise as viewed from the sun, such that the northern lobe is in the dawn sector and the southern lobe is in the dusk sector. Except for the case of nominal aberration, the Geotail position is very close to the magnetopause. The Geotail observations (Figure 2) show that Geotail is at this time in a region of magnetosheath like plasma and fields.

At 1500 UT the tail has somewhat relaxed and now looks more like that at 1200 UT. This rotation and the increase of the minor dimension, given our assumptions regarding tail aberration, put Geotail back into the southern lobe at this time. Unlike the 1200 UT and 1400 UT intervals, the plasma sheet at 1500 UT is inclined by about 20 degrees with respect to the major axis.

At 1700 UT, the tail has narrowed again and is twisted to an almost upright orientation. As was the case around 1400 UT, the minor dimension is reduced to about 20 R_E, leaving Geotail in magnetosheath-like plasmas and fields. However, this case is different from that for 1400 UT, in that the northern lobe lies mainly in the northern hemisphere and the southern lobe lies mainly in the southern hemisphere. However, there are some structures of opposite sign in both hemispheres.

A dramatic change has occurred by 1900 UT. While at earlier times the tail was twisted such that the northern lobe tended to be on the dawn side and the southern lobe on the dusk side, the opposite is now true. The fact that the tail has also expanded somewhat results in Geotail's now being in the northern lobe for the first time. Inspection of Figure 2 reveals that after this time Geotail does not enter the southern lobe again for the subsequent 3 hours. However, the observations indicate that from this point on, Geotail repeatedly enters the northern lobe because B_x stays positive during intervals of low density and high temperature. The simulation shows exactly the same picture.

At 2100 UT, the tail has essentially retained its previous structure. However, the twisting angle has become considerably larger than 90 degrees. As a consequence, the northern lobe is at this point entirely in the southern hemisphere and vice versa.

A number of features are common to the tail cross sections at all times. First, the major axis of the elliptically distorted tail is always aligned with the magnetosheath field direction in the y - z plane. This behavior of the tail is similar to that found in ISEE-3 data by *Sibeck et al.* [1985, 1986], who deduced a tail cross section of similar shape, but with the major axis in the east-west direction. Second, the flow in the magnetosheath is primarily directed towards the tail center where the tail dimension is smallest and away from the tail center at the "long ends." This is consistent with a flaring of the tail in the larger dimension, and a negative flaring, i.e., convergence, in the smaller dimension. Third, the flow velocities in the lobes are not uniform. Towards the magnetopause, the velocities are about 100 - 200 km/s smaller than the magnetosheath velocity. Towards the tail center, the velocity becomes very small, of the order of 100 km/s or less. This indicates that the topology of the inner part of the tail is different from that of the outer part. Because of the strong northward IMF, the outer part may well be a boundary layer just inside the magnetopause. Such a broad boundary layer has been found in earlier studies of the middle tail [*Raeder et al.*, 1997*b*], but little is known about its extension into the distant tail. Fourth, the simulation results show that Geotail traverses the distant tail magnetopause when it observes steep gradients in the plasma and field parameters. Such an interpretation is not obvious from the Geotail observations alone, and *Fairfield et al.* [1996] suggest that the magnetosheath-like plasmas and fields may be a signature of a closed tail.

In summary, our global simulation for March 29, 1993, shows that the structure of the distant tail under strongly northward IMF conditions is quite different from that predicted by simple models. The simulation results are in basic accordance with the Geotail observations and provide a framework for interpreting the data. However, additional analysis will be required to determine why the tail reacts to northward IMF in this way. This issue will be the topic of a forthcoming study.

4. CASE II: FEBRUARY 9, 1995

4.1. Observations

On February 9, 1995, between 0800 UT and 1400 UT the Wind spacecraft was located at (190, 40, 2) R_E in Earth-centered solar-ecliptic (GSE) coordinates. The solar wind magnetic field observations in Figure 4 were provided by the GSFC magnetometer on Wind [*Lepping et al.*,

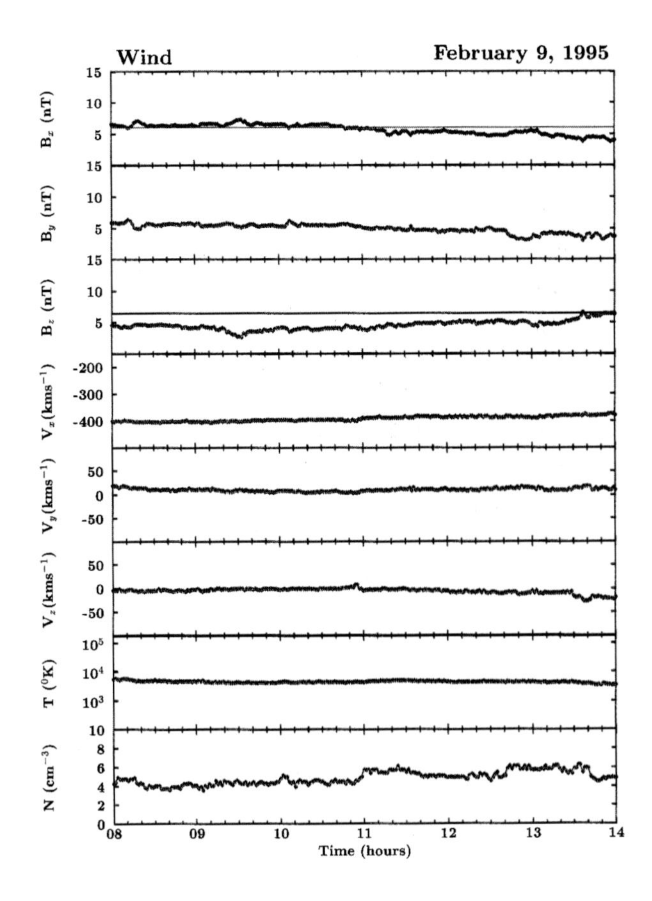

Figure 4. Solar wind parameters on February 9, 1993. The figure shows, from top to bottom: magnetic field components, velocity components, proton temperature, and proton density. The black dots indicate Wind measurements, and the red line shows the cleaned and averaged data that were used as input for the MHD simulation.

1995], while the plasma parameters were obtained from the Solar Wind Experiment (SWE) [*Ogilvie et al.*, 1995]. We use Wind data as input to the MHD model, even though the spacecraft was far from Earth, because it is the only solar wind monitor available. (Although IMP-8 was closer to Earth it was not in the solar wind at this time.) The IMF and the solar wind parameters were very stable for the entire 6 hours. During this time interval of steady northward IMF (B_z = 5 nT), there was also a significant positive B_y component (B_y = 5 nT) to the IMF which was comparable to its B_z component. The values of the solar wind speed, density, and thermal pressure were about 400 km/s, 5 cm^{-3}, and 3.5 pPa, respectively, consistent with the average solar wind.

During this time interval the Geotail satellite was located in the tail, and moved from (–31.0, 3.0, –3.0) R_E to (–27.8, –0.6, –2.3) R_E (GSE). Three-dimensional distribution functions were obtained from the Hot Plasma Analyzer (HP) of the Comprehensive Plasma Instrumentation (CPI) [*Frank et al.,* 1994]. For the observations presented here, the HP analyzer was operated in a mode in which the ion velocity distributions were determined, with more than 3000 samples falling within the energy-per-unit charge (E/Q) range of 22 V to 48 kV with a repetition rate of 22 s. Figure 5 shows V_y - V_z, V_x - V_z, and V_x - V_y cuts of the 3-D Geotail ion distribution functions in GSE coordinates. The two rows in this figure correspond to cuts of the Geotail velocity distribution functions for the 1325 UT and 1347 UT time intervals. The arrows show the direction of the magnetic field projected onto the x - z plane. At 1325 UT, the ion distribution exhibits counterstreaming structures in the direction parallel to **B** and relatively cold structures in the direction perpendicular to **B**. Geotail was in the southern plasma sheet boundary layer (PSBL) at this time since $B_x < 0$ and is the largest component (Figure 6). By 1347 UT, Geotail observed earthward flow and a stronger B_z, suggesting that the spacecraft was in the outer central plasma sheet (CPS). The perpendicular temperature of the ion distribution at 1347 UT was higher than in the earlier time period.

4.2. Comparison of Geotail Observations with the MHD Model

Our previous example (March 29, 1995) demonstrated that the magnetosphere displays a great deal of complexity that is a consequence of the interaction between external driving, i.e. the IMF and the solar wind, and the magnetosphere's internal evolution. This effect is also seen in our second example, February 9, 1995, when the magnetosphere was not in a steady state despite an IMF and a solar wind that were steady for more than 6 hours.

IMF and plasma data from the Wind spacecraft on February 9, 1995 (Figure 4), were used as input to the global MHD simulation outlined in section 2. The red curves in Figure 4 indicate the values used as input in the MHD simulation. Note that a constant B_x was used in the simulation. Late in the simulation run this caused us to overestimate the value of B_x.

Figure 6 shows as black curves the magnetic field (B_x, B_y, B_z), the three components of the velocity (V_x, V_y, V_z), the ion temperature (T), and the number density (N) observed by the Geotail magnetometer [*Kokubun et al.*, 1994] and CPI versus time from 1200 UT to 1400 UT. The magnetic field data are 3-second averages while the resolution of the plasma moments is 64 seconds. The time series from the MHD simulation at the location of Geotail are

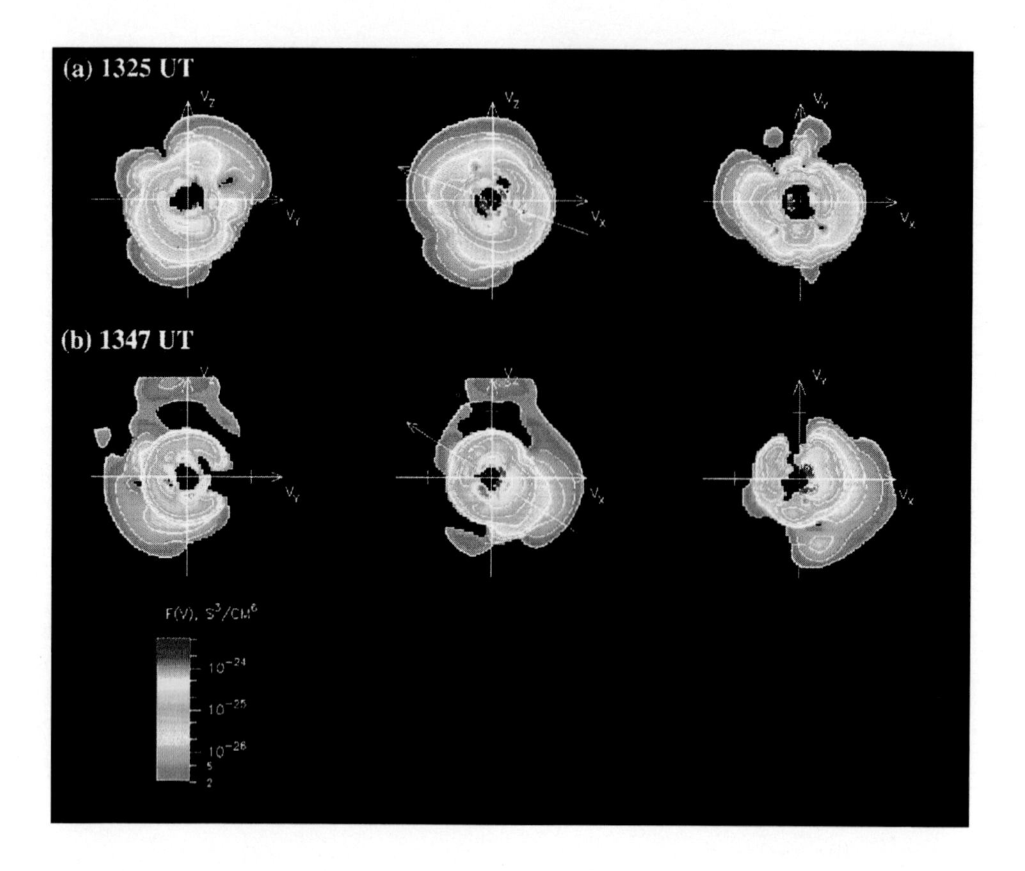

Figure 5. Cuts of the three-dimensional Geotail velocity distribution functions from the CPI instrument for 1325 UT (upper row) and 1347 UT (lower row). The three panels in each row correspond to V_y - V_z, V_x - V_z, and V_x - V_y cuts, and the arrows indicate the projection of the average magnetic field onto the plane of the cut.

plotted in green. The simulation does a poor job of reproducing the observations. The reason for this can be seen in Figure 7 where we have plotted B_x and V_x in the y - z plane and V_x in the x - z plane at 1325 UT and 1347 UT. The location of Geotail is shown by a blue dot. Throughout this interval in the simulation Geotail is in the southern tail lobes, just outside of the PSBL. This is a region of steep gradients in the field and plasma parameters. The MHD model cannot resolve phenomena on scales smaller than a few grid points. During this interval the grid spacing is $\Delta x = 0.95\ R_E$, $\Delta y = 0.50\ R_E$, and $\Delta z = 0.50\ R_E$. In addition to uncertainties caused by the grid spacing, errors in the location of the spacecraft with respect to magnetospheric boundaries can result from the use of constant IMF B_x in the simulations and because we didn't update Earth's dipole tilt during the simulation. A displacement of a few grid spaces would greatly change the location of Geotail with respect to the physical boundaries in the magnetotail.

As can be seen in Figure 6, the simulation does an especially poor job of reproducing the observations in the interval between 1200 UT and 1300 UT. In particular the simulation does not reproduce the observed strong Earthward flows ($V_x \approx 200$ km/s). The reason for this can be seen in Figure 7, which shows that in the simulation the flow diverges just earthward of the spacecraft position with earthward flows about 4 R_E or 5 R_E earthward of Geotail. Displacing Geotail 5 R_E in x, 2 R_E in y, and 1 R_E in z yields the blue curve in Figure 6. This displacement moves the spacecraft into the earthward flow region and more closely reproduces B_x as well. However, now the model gives strong earthward flows between 1300 UT and 1400 UT, flows that are not observed.

After trying many different displacements we concluded that it is not possible to find any single displacement which can correct the model for the entire interval. The variability of the results in Figure 6 strongly suggests that

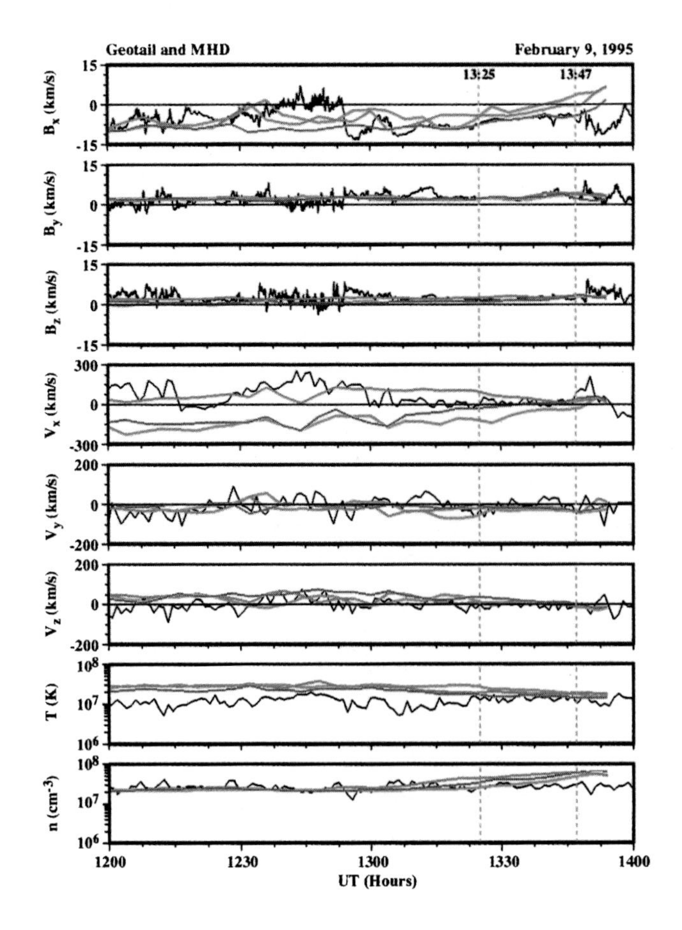

Figure 6. A comparison of time series from Geotail (black curves) and the MHD simulation (colored curves according to different spacecraft displacement, see text). The figure shows, from top to bottom: magnetic field components, velocity components, proton temperature, and proton density. The vertical yellow lines indicate the times at which particles from measured distribution functions where launched in the model.

the magnetosphere remains dynamic even when the IMF is northward for a long period of time. As the magnetosphere changes the gradients change, and this causes the error in the model to change. Therefore we decided to try to optimize the MHD model only for this study's interval of prime interest. Accordingly, we looked for displacements giving the best fit for the Geotail observations between 1300 UT and 1400 UT. During this interval the ions observed at 1325 UT and 1347 UT enter the magnetosphere and travel to Geotail. The red curve in Figure 6 gives the desired best fit to the observed values between 1300 UT and 1400 UT. It was obtained by displacing Geotail by 0 in x, 1.5 R_E in y, and 0.5 R_E in z. Except for giving too large a number density the fit is very good. Although this good fit to observations from one spacecraft is no guarantee that the MHD model is correct, it is the best indicator that we presently have.

The magnetospheric configuration at this time is very complex. Figure 8 shows a perspective view of the magnetosphere at 1312 UT. The three color contours shown in this figure correspond to the total current density from the MHD simulation in the $x = -30\ R_E$, $x = -90\ R_E$, and $x = -150\ R_E$ planes. The overall topology of the magnetosphere remains relatively unchanged during the simulation, and these planes are representative of the entire interval. Geotail's location, which is around $x = -28.2\ R_E$, $y = -0.2\ R_E$, and $z = -2.4\ R_E$, is indicated by the white dot. Characterized by large current densities, the current sheet and the magnetopause are clearly discernible. The structure of the plasma sheet has undergone significant changes as a result of the more than 5 hours of steady northward IMF with a significant B_y component. Because of the IMF B_y, the current sheet is twisted around the Sun-Earth axis toward the duskside at $x = -30\ R_E$, and this effect gets more pronounced further downtail. The field lines plotted in Figure 8 show that because of the twisting of the geomagnetic tail, portions of the northern (southern) lobe fall below (above) the equatorial plane, and field lines emanating from the northern (southern) hemisphere cross the equatorial plane and are found downtail in the opposite hemisphere, just as occurred on March 29, 1995 (previous section).

Figure 9 shows plots of field lines within 1 R_E of Geotail at the two times chosen for this study. At 1325 UT, the spacecraft is in a mixed region of open (red) and closed (blue) field lines. At 1347 UT, Geotail is mostly in a region of closed field lines. In order to gain a better understanding of the changes in the Geotail velocity distribution functions as the spacecraft traverses from a region of open to closed field lines, we integrate ion orbits backwards in time in a time-dependent large-scale kinetic (LSK) study, using the Geotail velocity distribution functions as input; the results are discussed below.

4.3. Sources of Ions for Geotail Distributions

As already noted, despite the quiet solar wind conditions observed on February 9, 1995, significant changes in the Geotail distributions occurred during this interval (Figure 5). In this section, we investigate the sources of the ions comprising the Geotail distributions in order to determine the origin and delineate the transport of magnetotail plasma during this seemingly quiet period.

Because of the time-dependent nature of the problem, we have used the time-dependent magnetic and electric fields

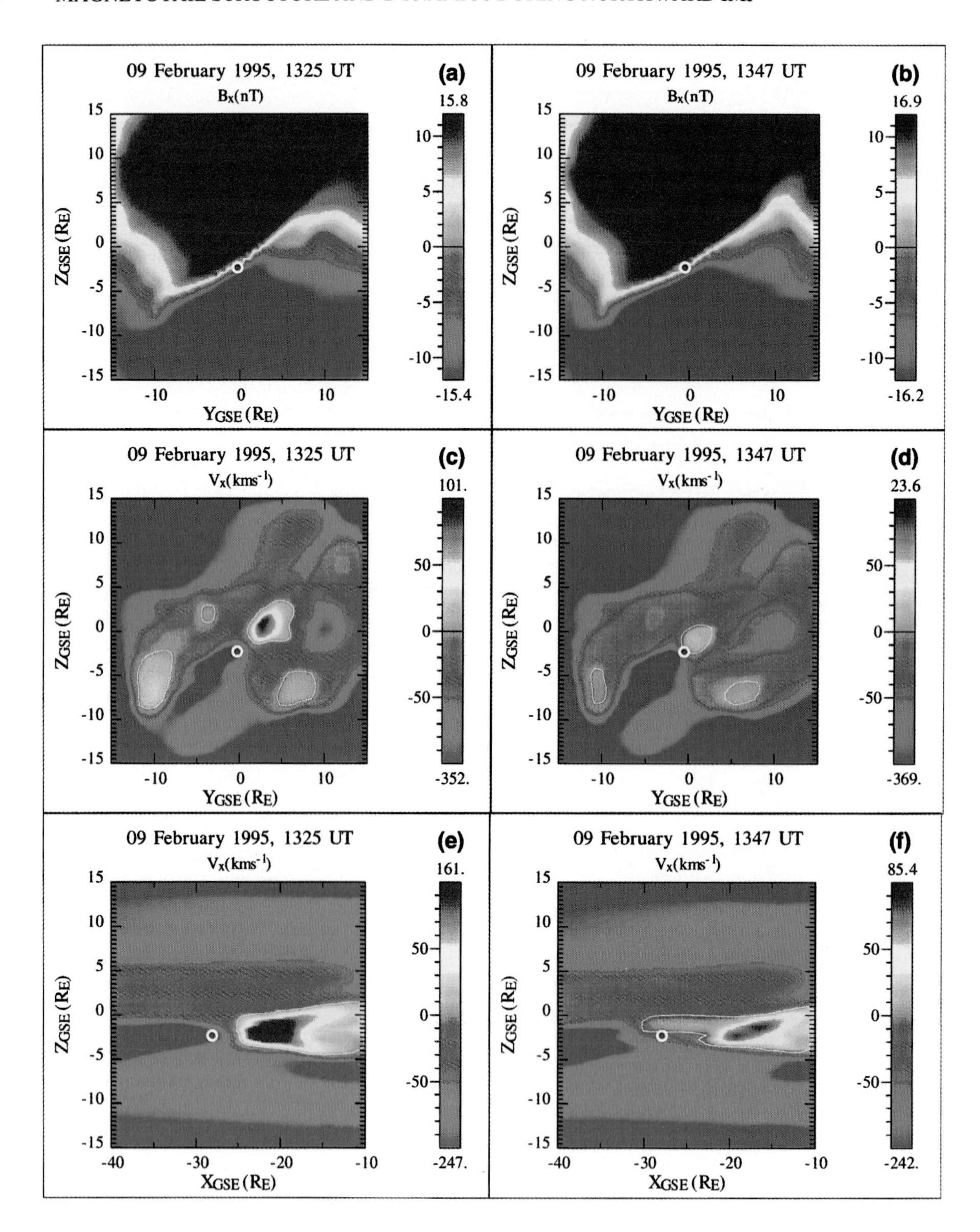

Figure 7. Top panel: Cross sections of the tail at the Geotail location ($x = -29\ R_E$) at 1325 UT and 1347 UT. The magnetic field component B_x is color coded. Middle panels: Same as top panel, but the V_x component of the velocity is color coded. Lower panel: Like the middle panel, but a cut of the x_{GSE} - z_{GSE} plane through the Geotail location. The dots indicate the nominal Geotail location.

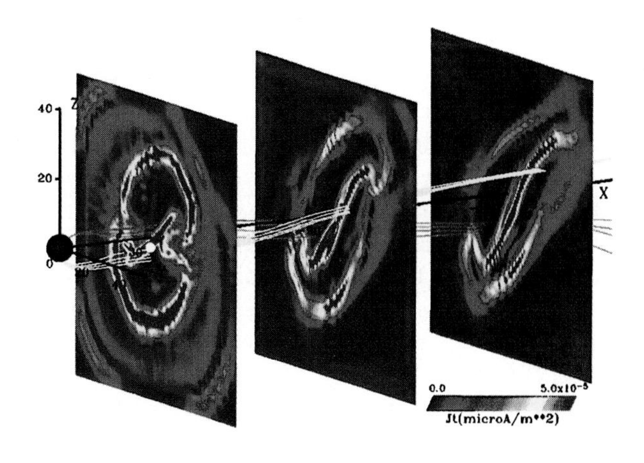

Figure 8. y - z cuts of the MHD current at 1312 UT on February 9, 1995, at $x = -30\ R_E$, $x = -90\ R_E$, and $x = -150\ R_E$. The white dot indicates the location of Geotail.

from the MHD simulation in our LSK calculation. The MHD electric field used in this calculation is given by:

$$\mathbf{E} = -\mathbf{v} \times \mathbf{B} + \eta \mathbf{J} \quad (1)$$

where **v** is the bulk velocity, **B** is the local magnetic field, η is the resistivity which is a function of the local magnetic field gradients, and **J** is the local current density. The electric field has both a convective ($-\mathbf{v} \times \mathbf{B}$) and a resistive ($\eta \mathbf{J}$) term. The resistive term becomes important near the magnetopause and near x-lines but is negligible elsewhere. Starting with the Geotail distributions at 1325 UT and 1347 UT in Figure 5, we construct three-dimensional distributions in velocity space by placing ~90,000 ions in V_x - V_y - V_z bins (100 km/s × 100 km/s × 100 km/s) such that the number in each bin is proportional to the observed phase space density of that bin. One particle in the computational distribution function corresponds to 5×10^{-27} $s^3 cm^{-6}$ in the Geotail distribution function. Since the scale in Figure 5 ranges from 2×10^{-27} to 5×10^{-24}, the fine-scale structure of the observed distribution is preserved in the computational distribution function. For each particle we integrate the equation of motion ($d\mathbf{v}/dt = q\mathbf{v} \times \mathbf{B} + q\mathbf{E}$) backward in time until the particle encounters the magnetopause (defined by using the MHD currents) or the ionosphere (taken as the inner boundary of the MHD simulation, at $r = 3.7\ R_E$). Since the minimum grid spacing in the global MHD simulation is relatively large (~0.5 R_E), and the simulation data are saved at four minute time intervals, we use linear interpolation (over space and time) to determine the instantaneous values of the MHD fields on scales smaller than the grid spacing. We calculate the ion trajectories in the evolving magnetic and electric fields by using a fourth-order Runge-Kutta method. The time step in the particle trajectory calculation is nominally set to 0.002 times the local ion gyro-period, with an upper limit imposed to ensure that the time step does not get too large in weak field regions.

The two panels in Figure 10 show the entry points of the ions in the measured distribution functions shown in Figure 5. Each color-coded dot in the figure represents the number of ions originating from a 1 $R_E \times 1\ R_E \times 1\ R_E$ region centered at the point. The green sphere shows the location of Geotail, and the contour plot in each panel represents a cut of the MHD current in the $x = -150\ R_E$ plane at the time of the measurement and is shown to help locate the magnetopause.

Between 1325 UT and 1347 UT, Geotail slowly moves from the PSBL into the outer CPS. Figure 10*a* shows the entry points of particles seen by Geotail at 1325 UT, when the spacecraft has just entered the closed field line region. The total current in the y - z plane at x = -150 R_E is also

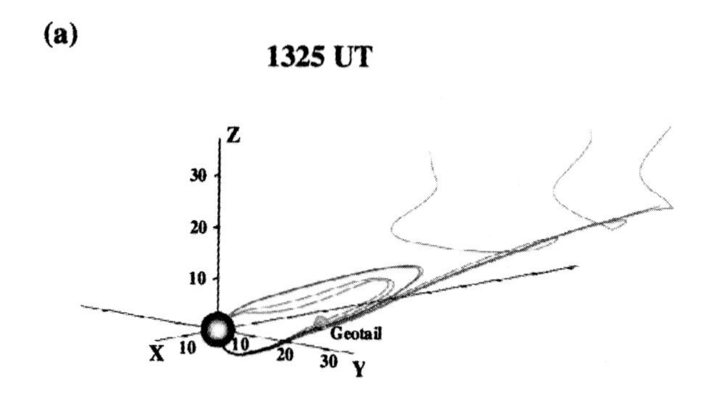

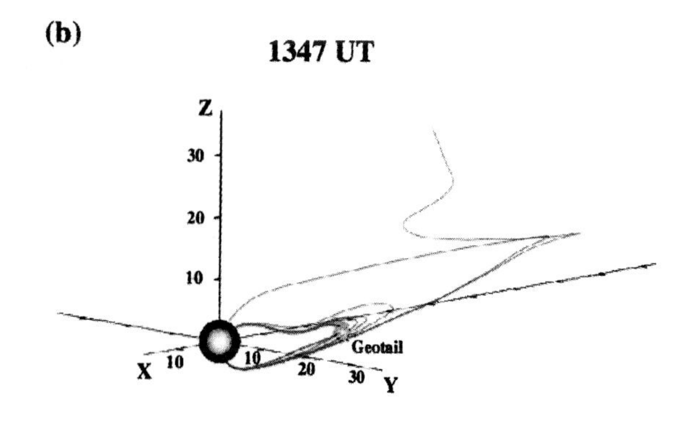

Figure 9. Plot of field lines for the two time intervals examined in the paper. Field lines within 1 R_E of the Geotail location are shown for each interval. Red field lines are open, and blue field lines are closed.

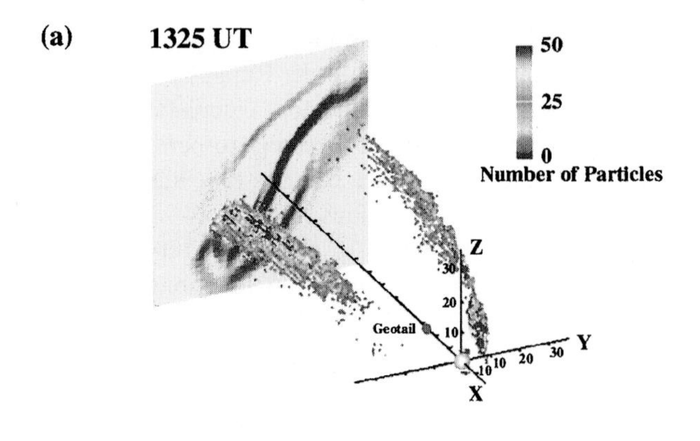

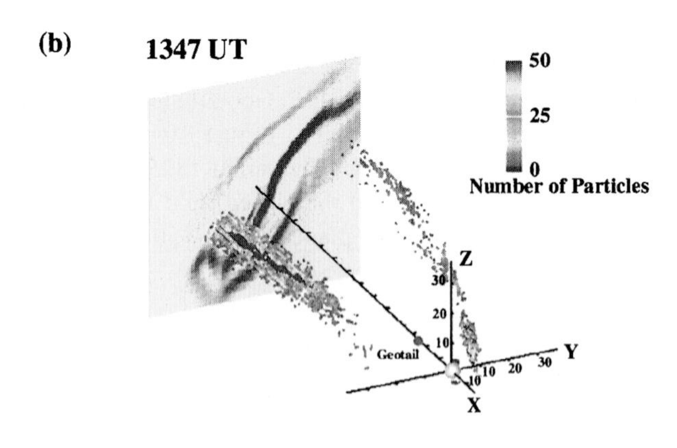

Figure 10. Magnetospheric entry plots for the two time intervals examined in the paper. The dots are color coded according to the number of particles originating from a 1 $R_E \times 1$ $R_E \times 1$ R_E bin centered on the point. The total current in the $x = -150$ R_E plane is also shown in each panel.

plotted. This gives an indication of the relationship between the ion entry points and magnetospheric boundaries. Ions from both the dawn and the dusk side LLBL have access in nearly equal numbers to the Geotail location. Dawnside LLBL ions (50% of ions measured at Geotail) originate in a wide region in z centered at the equatorial plane extending between -10 $R_E > x > -150$ R_E. However, the bulk of the dawnside entry occurs tailward of $x = -100$ R_E. As we shall see below, the dawnside ions are on primarily nonadiabatic orbits. These ions are responsible for the bulk plasma seen in the direction perpendicular to **B** in the Geotail velocity distribution functions at this time. The duskside LLBL is another major source of ions for this time period (48% of ions seen by Geotail), with the bulk of the ions from this source originating closer to Earth, in the region -10 $R_E > x > -60$ R_E. The duskside entry of ions for this interval occurs along closed LLBL field lines. Ionospheric particles comprise about 2% of the particles measured by Geotail at this time.

At 1347 UT, Geotail is embedded in a region of closed field lines. Figure 10*b* shows that the source of the ions seen by Geotail is no longer split equally between the dawn and dusk flanks, and the dawnside LLBL is the primary supplier of particles to the Geotail distribution function (88% of ions measured by Geotail). To provide insight into the particle dynamics and access to Geotail during the 1325 UT and 1347 UT time intervals, we plot in Figure 11 the trajectory of a typical particle originating from the duskside LLBL and measured by Geotail at 1347 UT. The upper panel of this figure shows a three-dimensional view of the particle's orbit in GSE coordinates. The black curve in the lower panel shows the particle's kinetic energy (from 0 to 10 keV, scale shown on the left), and the red dots in the lower panel show the particle's parameter of adiabaticity κ (from 0 to 10, scale on the right of the panel) as a function of time [*Büchner and Zelenyi*, 1986]. κ is defined as

$$\kappa = \sqrt{R_c / \rho_L} \tag{2}$$

where R_c is the local magnetic field radius of curvature and ρ_L is the ion Larmor radius. When $\kappa > 1$, particles follow guiding center orbits and are adiabatic. When $\kappa \leq 1$, however, particles no longer conserve their first adiabatic invariant and follow stochastic, or quasi-adiabatic, trajectories. The upper panel of this figure shows that the particle enters

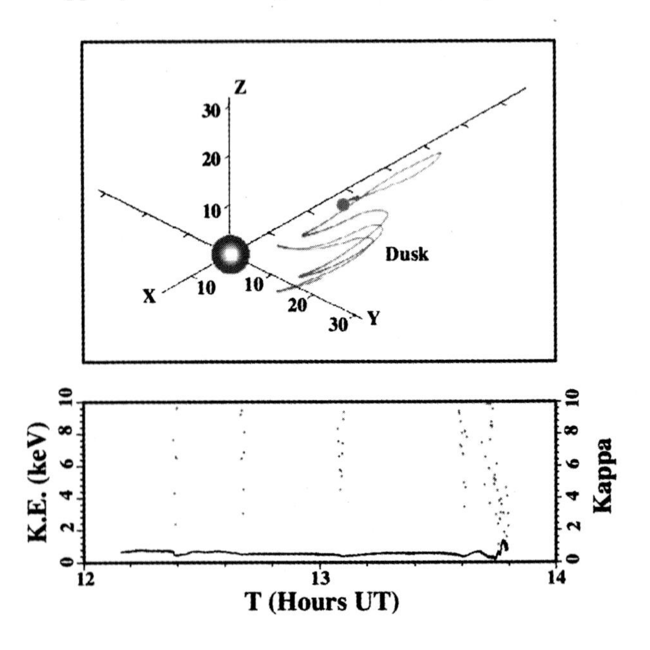

Figure 11. Trajectory of a particle originating in the duskside magnetopause and arriving at Geotail at 1347 UT. The upper panel shows a three-dimensional projection of the particle's orbit, and the lower panel depicts the particle's kinetic energy (black curve, scale on left), and κ (red dots, scale on right).

the magnetosphere at $x \sim -20\ R_E$ on the dusk side of the magnetopause in the region with the largest number of ions entering via the dusk side LLBL. After entering the magnetosphere, the ion is immediately trapped in the near-Earth region. It executes several adiabatic ($\kappa >> 1$) bounces and convects dawnward until it encounters Geotail (shown by a green dot). This orbit is representative of that of the ions entering the magnetosphere on the dusk flank between $-10\ R_E > x > -30\ R_E$ and shows that the bulk of the ions from the duskside source reaches Geotail on trapped orbits. At the same time, the particle's energy does not change significantly and the particle arrives at Geotail with an energy of ~ 2 keV.

Figure 12 shows a particle originating in the dawnside LLBL in a format similar to Figure 11. The three-dimensional rendering of the particle's orbit (upper panel) shows the particle entering the magnetosphere on the dawnside at $x \sim -90\ R_E$ with a kinetic energy of ~ 2 keV. The particle then crosses the magnetotail from dawn to dusk, executes several nonadiabatic ($\kappa \leq 1$) interactions with the current sheet (red dots in lower panel) in which it gains energy, and finally arrives at Geotail (green dot) with an energy of ~ 4 keV. This type of particle accounts for the bulk of the distribution measured in the direction perpendicular to **B** as well as for the warmer perpendicular temperature of the measured distribution compared to that at 1325 UT.

5. SUMMARY AND DISCUSSION

In this paper, we have examined two separate events that are characterized by steady, northward IMF. Using a global MHD calculation driven by IMP-8 and Wind spacecraft solar wind and IMF data we have simulated these events and compared the results with Geotail observations. In the second case (Feb. 9, 1995), we have gone one step further and applied the LSK technique using time-dependent MHD fields and observed distribution functions in order to examine plasma transport and energization.

1. Both examples shown in this paper demonstrate that the magnetosphere responds in a complex, nonlinear way to changes in the IMF and solar wind, and is significantly influenced by the past history of solar wind changes. This can be understood as being a consequence of the interaction between external driving by the solar wind and the IMF and the magnetosphere's internal dynamics.

2. On March 29, 1993, the presence of a changing IMF B_y component causes the magnetotail to become severely twisted and flattened. The longer dimension of the tail cross section follows the changes in the IMF clock angle rapidly, i.e. within a few minutes, such that the semimajor tail axis remains largely aligned with the magnetosheath field direction in the y - z plane. At times, the twisting is so strong that the northern (southern) lobe is displaced to the southern (northern) hemisphere in the distant tail.

3. On February 9, 1995, the IMF B_y component also causes a twisting of the magnetotail current sheet. However, because the IMF B_y is relatively steady during the time interval considered, the effect on the magnetotail is not as drastic as that of March 29, 1993, and the magnetotail cross-section shows very little change as a function of time. Although the IMF shows little variation over several hours, subtle changes in the interior structure of the tail occur owing to its internal dynamics and convection. These changes have a substantial effect on the ion distribution functions, as is evident from the Geotail observations and the LSK calculations.

4. By applying the LSK technique in the February 9, 1995 case, we have been able to examine these changes of the distribution functions in detail. Specifically, we found that

a) At 1325 UT, Geotail was entering a region of closed field lines, allowing ions from both the dawn and dusk side LLBL to gain access to the spacecraft location. Ions from both the duskside and the dawnside sources originated from broad regions in z centered on the equatorial plane which had rotated out of the $z = 0$ plane because IMF $B_y \neq 0$. The ionosphere also made a contribution to the Geotail distribution, supplying ~2% of the measured ions. Ions from the duskside source followed guiding center orbits. However, ions from the dawnside source were characterized by multiple nonadiabatic interactions with the current sheet and accounted for the component of the distribution measured by Geotail in the direction perpendicular to **B**.

b) At 1347 UT, Geotail was embedded in a region of closed field lines. Because the spacecraft entered the outer CPS from the PSBL, it measured more particles from the dawn LLBL so that the dawnside ion source became dominant. It is interesting to note that because of their stochastically trapped orbits, dawn side ions are seen in numbers only when the satellite enters into a region of closed field lines. The ionospheric contribution to the Geotail distribution increased slightly, to ~3% of the measured particles. Once again, dawnside ions moved on principally nonadiabatic trajectories which scattered them in pitch angle and energized them before they arrived at Geotail. The density of particles entering the magnetosphere from the dawnside ion source increased steadily downtail. In contrast, the peak in duskside entry occurred earthward of Geotail. This caused the duskside ions to be trapped in adiabatic trajectories as soon as they entered the magnetosphere and resulted in their arriving at Geotail while moving in the tailward direction.

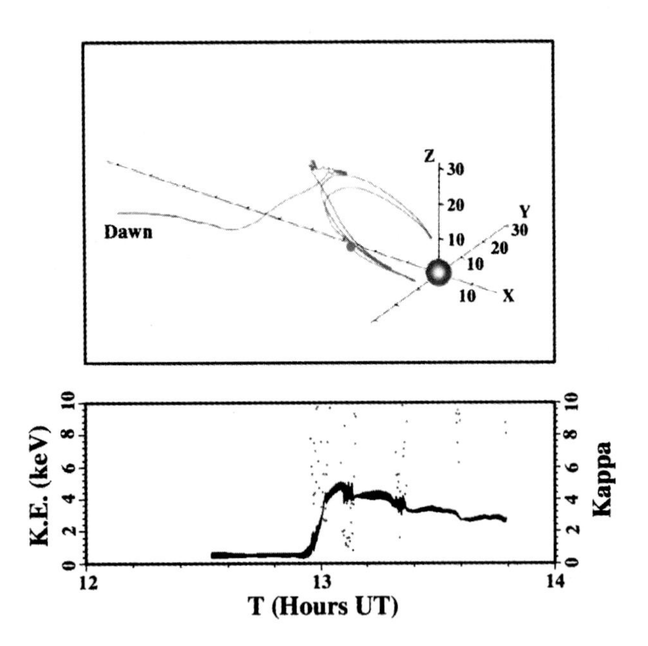

Figure 12. Trajectory of a particle originating on the dawnside and arriving at Geotail at 1347 UT. The format of the figure is similar to that of Fig. 11.

The MHD simulations presented in this paper have shown that the structure of the magnetotail during periods of northward IMF with a significant IMF B_y component is quite different from the picture predicted by simple models and that the overall picture cannot be obtained from a simple superposition of idealized configurations. An analysis of why the magnetotail responds to northward IMF in the manner it does will be provided in a forthcoming paper.

The MHD simulations used in this paper are far from ideal. Using a single point measurement of the solar wind ~200 R_E upstream of Earth does not allow for compensating for structures in the solar wind. For example, structures measured by the Wind spacecraft and used as input for our simulations may never encounter Earth and result in discrepancies between the model and observations. At the same time, the spatial grid used in the simulation is often too coarse to resolve sharp gradients in plasma and magnetic field parameters. Also, we use a simple model for the IMF B_x in our simulations and hold Earth's dipole tilt constant throughout the simulation. Both of these approximations can lead to errors, especially late in the simulation run. Despite these shortcomings we have used MHD simulations simply because they are the best global models available for our purposes.

We have outlined our initial efforts to use the LSK technique along with time-dependent MHD electric and magnetic fields in order to assess the transport and acceleration processes affecting particle behavior in the magnetotail. A natural extension of these efforts would be the application of this technique to more dynamic magnetospheric intervals and to times when there is more than one satellite in the magnetotail. So far, our studies have been limited to periods of northward IMF, but as the magnetosphere is much more active during southward IMF intervals, it represents an interesting challenge to our modeling effort. Similarly, a study using Geotail data together with data from either Polar or Interball would allow us to investigate not only the transport and acceleration of plasma, but would also allow us to gain insight into the subject of common particle sources supplying the magnetospheric regions covered by these spacecraft.

Acknowledgments. We thank J. M. Bosqued, L. A. Frank, W. R. Paterson, R. L. Richard, R. J. Walker, and L. M. Zelenyi for many useful discussions. We also thank L. A. Frank and W. R. Paterson for providing the Geotail distribution functions, K. Ogilvie for providing the Wind plasma data and R. P. Lepping for providing the Wind IMF data. This work was supported by NASA grants NAG5-1100 and NAGW-4553. Computing support was provided by the Cornell Theory Center, the Office of Academic Computing at UCLA, the San Diego Supercomputer Center, and the Maui High Performance Computing Center. UCLA/IGPP publication number 4934.

REFERENCES

Akasofu, S.-I., E. W. Hones, Jr., S. J. Bame, J. R. Asbridge, and A. T. Y. Lui, Magnetotail and boundary layer plasmas at a geocentric distance of ~18 R_E: VELA 5 and 6 observations, *J. Geophys. Res.*, *78*, 7257, 1973.

Ashour-Abdalla, M., J. Berchem, J. Büchner, and L. M. Zelenyi, Shaping of the magnetotail from the mantle: Global and local structuring, *J. Geophys. Res.*, *98*, 5651, 1993.

Ashour-Abdalla, M., L. M. Zelenyi, V. Peroomian, and R. L. Richard, Consequences of magnetotail ion dynamics, *J. Geophys. Res.*, *99*, 14,891, 1994.

Ashour-Abdalla, M., L. M. Zelenyi, V. Peroomian, R. L. Richard, and J. M. Bosqued, The mosaic structure of plasma bulk flows in the Earth's magnetotail, *J. Geophys. Res.*, *100*, 19,191, 1995.

Ashour-Abdalla, M., L. A. Frank, W. R. Paterson, V. Peroomian, and L. M. Zelenyi, Proton velocity distributions in the magnetotail: Theory and observations, *J. Geophys. Res.*, *101*, 2587, 1996.

Ashour-Abdalla, M., M. El Alaoui, V. Peroomian, J. Raeder, R. J. Walker, R. L. Richard, L. M. Zelenyi, L. A. Frank, W. R. Paterson, J. M. Bosqued, R. P. Lepping, K. Ogilvie, S. Kokubun, and T. Yamamoto, Ion sources and acceleration mechanisms inferred from local distribution functions, *Geophys. Res. Lett.*, *24*, 955, 1997.

Balsiger, H., P. Eberhardt, J. Geiss, and D. T. Young, Magnetic storm injection of 0.9 - 16 keV/e solar and terrestrial ions into the high-altitude magnetosphere, *J. Geophys. Res.*, *85*, 1645, 1980.

Berchem, J., J. Raeder, M. Ashour-Abdalla, L. A. Frank, W. R. Paterson, K. L. Ackerson, S. Kokubun, T. Yamamoto, R. P. Lepping, and K. Ogilvie, The distant tail at 200 R_E: Comparisons between Geotail observations and the result of global MHD simulations, *J. Geophys. Res.*, submitted 1997*a*.

Berchem, J., J. Raeder, M. Ashour-Abdalla, L. A. Frank, W. R. Paterson, K. L. Ackerson, S. Kokubun, T. Yamamoto, and R. P. Lepping, Large-scale dynamics of the magnetospheric boundary: Comparisons between global MHD simulation results and ISTP observations, in *Encounter Between Global Observations and Models in the ISTP Era, Geophysical Monograph Series*, J. Horwitz, ed., submitted, 1997*b*.

Büchner, J., and L. M. Zelenyi, Deterministic chaos in the dynamics of charged particles near a magnetic field reversal, *Phys. Lett. A*, *118*, 395, 1986.

Büchner, J., and L. M. Zelenyi, Regular and chaotic charged particle motion in magnetotail-like field reversal, 1, Basic theory of trapped motion, *J. Geophys. Res.*, *94*, 11,821, 1989.

Brecht, S. H., J. G. Lyon, J. A. Fedder, and K. Hain, A simulation study of east-west IMF effects on the magnetosphere, *Geophys. Res. Lett.*, *8*, 397, 1981.

Brecht, S. H., J. G. Lyon, J. A. Fedder, and K. Hain, A time dependent three dimensional simulation of the Earth's magnetosphere: Reconnection events, *J. Geophys. Res.*, *87*, 6098, 1982.

Chappell, C. R., T. E. Moore, and J. H. Waite, Jr., The ionosphere as a fully adequate source of plasma for the Earth's magnetosphere, *J. Geophys. Res.*, *92*, 5896, 1987.

Chen, J., and P. J. Palmadesso, Chaos and nonlinear dynamics of single particle orbits in a magnetotail-like field, *J. Geophys. Res.*, *91*, 1499, 1986.

Cowley, S. W. H., Magnetospheric asymmetries associated with the Y-component of the IMF, *Planet. Space. Sci.*, *29*, 79, 1981.

Eastman, T. E., E. W. Hones, Jr., S. J. Bame, and J. R. Asbridge, The magnetospheric boundary layer: Site of plasma, momentum and energy transfer from the magnetosheath into the magnetosphere, *Geophys. Res. Lett.*, *3*, 685, 1976.

El-Alaoui, M., M. Ashour-Abdalla, J. Reader, J. M. Bosqued, Simulation of ion trajectories in the magnetotail using time-dependent electromagnetic fields, AGU Spring Meeting, Baltimore (EOS, vol. 76, no. 16), 1995.

Fairfield, D. H., Solar wind control of the distant magnetotail: ISEE 3, *J. Geophys. Res.*, 98, 21,625, 1993.

Fairfield, D. H., R. P. Lepping, L. A. Frank, K. L. Ackerson, W. R. Paterson, S. Kokubun, T. Yamamoto, K. Tsuruda, and M. Nakamura, Geotail observations of an unusual magnetotail under very northward IMF conditions, *J. Geomag. Geoelec.*, *48*, 473, 1996.

Fedder, J. A., and J. G. Lyon, The Earth's magnetosphere is 165 R_E long: Self consistent currents, convection, magnetospheric structure and processes for northward interplanetary magnetic field, *J. Geophys. Res.*, *100*, 3623, 1995.

Frank, L. A., K. L. Ackerson, W. R. Paterson, J. A. Lee, M. R. English, and G. L. Pickett, The Comprehensive Plasma Instrumentation (CPI) for the Geotail spacecraft, *J. Geomag. Geoelec.*, *46*, 23, 1994.

Frank, L. A., M. Ashour-Abdalla, J. Berchem, J. Raeder, W. R. Paterson, S. Kokubun, T. Yamamoto, R. P. Lepping, F. V. Coroniti, D. H. Fairfield, and K. L. Ackerson, Observations of plasmas and magnetic field in Earth's distant magnetotail: Comparison with a global MHD model, *J. Geophys. Res.*, *100*, 19,177, 1995.

Geiss, J., H. Balsiger, P. Eberhardt, H. P. Walker, L. Weber, D. T. Young, and H. Rosenbauer, Dynamics of magnetospheric ion composition as observed by the Geos mass spectrometer, *Space Sci. Rev.*, *22*, 537, 1978.

Haerendel, G., G. Paschmann, N. Sckopke, H. Rosenbauer, and P. C. Hedgecock, The frontside boundary layer of the magnetosphere and the problem of reconnection, *J. Geophys. Res.*, *83*, 3195, 1978.

Hardy, D. A., H. K. Hills, and J. W. Freeman, A new plasma regime in the distant geomagnetic tail, *Geophys. Res. Lett.*, *2*, 169, 1975.

Hones, Jr., E. W., J. R. Asbridge, S. J. Bame, M. D. Montgomery, S. Singer, and S.-I. Akasofu, Measurements of magnetotail plasma flow made with Vela 4B, *J. Geophys. Res.*, *77*, 5503, 1972*a*.

Hones, Jr., E. W., S.-I. Akasofu, S. J. Bame, and S. Singer, Outflow of plasma from the magnetotail into the magnetosheath, *J. Geophys. Res.*, *77*, 6688, 1972*b*.

Knight, S., Parallel electric fields, *Planet. Space Sci.*, *21*, 741, 1972.

Kokubun, S., T. Yamamoto, M. H. Acuña, K. Hayashi, K. Shiokawa, and H. Kawano, The Geotail magnetic field experiment, *J. Geomag. Geoelec.*, *46*, 7, 1994.

Leboeuf, J. N., T. Tajima, C. F. Kennel, and J. M. Dawson, Global simulation of the time-dependent magnetosphere, *Geophys. Res. Lett.*, *5*, 609, 1978.

Leboeuf, J. N., T. Tajima, C. F. Kennel, and J. M. Dawson, Global simulations of the three-dimensional magnetosphere, *Geophys. Res. Lett.*, *8*, 257, 1981.

Lennartsson, W., A scenario for solar wind penetration of Earth's magnetic tail based on ion composition data from the ISEE 1 spacecraft, *J. Geophys. Res.*, *97*, 19,221, 1992.

Lennartsson, W., and E. G. Shelley, Survey of 0.1- to 16-keV/e plasma sheet ion composition, *J. Geophys. Res.*, *91*, 3061, 1986.

Lennartsson, W., E. G. Shelley, R. D. Sharp, R. G. Johnson, and H. Balsiger, Some initial ISEE-1 results on the ring current composition and dynamics during the magnetic storm of December 11, 1977, *Geophys. Res. Lett.*, *6*, 483, 1979.

Lennartsson, W., R. D. Sharp, E. G. Shelley, R. G. Johnson, and H. Balsiger, Ion composition and energy distribution during 10 magnetic storms, *J. Geophys. Res.*, *86*, 4628, 1981.

Lepping, R. P., et al., The Wind magnetic field investigation, *Space Sci. Rev.*, *71*, 207, 1995.

Lundin, R., L. R. Lyons, and N. Pissarenko, Observations of the ring current composition at L < 4, *Geophys. Res. Lett.*, *7*, 425, 1980.

Lyon, J, G., S. H. Brecht, J. D. Huba, J. A. Fedder, J. A. Fedder, and P. J. Palmadesso, Computer simulation of a geomagnetic substorm, *Phys. Rev. Lett.*, *46*,1038, 1981.

Lyons, L. R., and T. W. Speiser, Evidence for current sheet acceleration in the geomagnetic tail, *J. Geophys. Res.*, *87*, 2276, 1982.

Lyons, L. R., D. Evans, and R. Lundin, An observed relation between magnetic field aligned electric fields and downward electron energy fluxes in the vicinity of auroral forms, *J. Geophys. Res.*, *84*, 457, 1979.

Martin, R. F., Jr., Chaotic particle dynamics near a two-dimensional neutral point, with application to the Earth's magnetotail, *J. Geophys. Res.*, *91*, 11,985, 1986.

Moen, J., and A. Brekke, The solar flux influence on quiet time conductances in the auroral ionosphere, *Geophys. Res. Lett.*, *20*, 971, 1993.

Ogilvie, K. W., et al., SWE, A comprehensive plasma instrument for the Wind spacecraft, *Space Sci. Rev.*, *71*, 55, 1995.

Ogino, T., and R. J. Walker, A magnetohydrodynamic simulation of the bifurcation of the tail lobes during intervals with a northward interplanetary magnetic field, *Geophys. Res. Lett.*, *11*, 1018, 1984.

Ogino, T., R. J. Walker, M. Ashour-Abdalla, and J. M. Dawson, An MHD simulation of B_y dependent magnetospheric convection and field-aligned currents during northward IMF, *J. Geophys. Res.*, *90*, 10,835, 1985.

Ogino, T., R. J. Walker, and M. Ashour-Abdalla, A global magnetohydrodynamic simulation of the magnetosheath and magnetosphere when interplanetary magnetic field is northward, *IEEE Trans. Plasma Sci.*, *20*, 817, 1992.

Ogino, T., R. J. Walker, and M. Ashour-Abdalla, A global magnetohydrodynamic simulation of the response of the magnetosphere to a northward turning of the interplanetary magnetic field, *J. Geophys. Res.*, *99*, 11,027, 1994.

Peroomian, V., and M. Ashour-Abdalla, Relative contribution of the solar wind and the auroral zone to near-Earth plasmas, in *Cross-Scale Coupling in Space Plasmas, Geophysical Monograph Series*, vol. 93, edited by J. Horwitz et al., pp. 213-217, AGU, Washington, D. C., 1995.

Peroomian, V., and M. Ashour-Abdalla, Population of the near-Earth magnetotail from the auroral zone, *J. Geophys. Res.*, *101*, 15,387, 1996.

Peterson, W. K., R. D. Sharp, E. G. Shelley, R. G. Johnson, and H. Balsiger, Energetic ion composition in the plasma sheet, *J. Geophys. Res.*, *86*, 761, 1981.

Pilipp, W. G., and G. Morfill, The formation of the plasma sheet resulting from plasma mantle dynamics, *J. Geophys. Res.*, *83*, 5670, 1978.

Raeder, J., R. J. Walker, and M. Ashour-Abdalla, The structure of the distant geomagnetic tail during long periods of northward IMF, *Geophys. Res. Lett.*, *22*, 349, 1995.

Raeder, J., J. Berchem, and M. Ashour-Abdalla, The importance of small-scale processes in global MHD simulations: Some numerical experiments, in *The Physics of Space Plasmas*, vol. 14, T. Chang and J. R. Jasperse, eds., p. 403, MIT Center for Theoretical Geo/Cosmo Plasma Physics, Cambridge, MA, 1996.

Raeder, J., J. Berchem, M. Ashour-Abdalla, L. A. Frank, W. R. Paterson, K. L. Ackerson, R. P. Lepping, K. Ogilvie, S. Kokubun, T. Yamamoto, and D. H. Fairfield, The distant tail under strong northward IMF conditions: Global MHD results for the Geotail March 29, 1993 observations, *J. Geophys. Res.*, in press, 1997*a*.

Raeder, J., J. Berchem, M. Ashour-Abdalla, L. A. Frank, W. R. Paterson, K. L. Ackerson, J. M. Bosqued, R. P. Lepping, S. Kokubun, T. Yamamoto, S. A. Slavin, Boundary layer formation in the magnetotail: Geotail observations and comparisons with a global MHD model, *Geophys. Res. Lett.*, in press, 1997*b*.

Richard, R. L., R. J. Walker, and M. Ashour-Abdalla, The population of the magnetosphere by solar wind ions when the interplanetary magnetic field is northward, *Geophys. Res. Lett.*, *21*, 2455, 1994.

Robinson, R. M., R. R. Vondrak, K. Miller, T. Dabbs, and D. Hardy, On calculating ionospheric conductances from the flux and energy of precipitating electrons, *J. Geophys. Res.*, *92*, 2565, 1987.

Rosenbauer, H., H. Grünwaldt, M. D. Montgomery, G. Paschmann, and N. Sckopke, Helios 2 plasma observations in the distant polar magnetosphere: The plasma mantle, *J. Geophys. Res.*, *80*, 2723, 1975.

Sharp, R. D., W. Lennartsson, W. K. Peterson, and E. G. Shelley, The origins of the plasma in the distant plasma sheet, *J. Geophys. Res.*, *87*, 10,420, 1982.

Shelley, E. G., R. G. Johnson, and R. D. Sharp, Satellite observations of energetic heavy ions during a geomagnetic storm, *J. Geophys. Res.*, *77*, 6104, 1972.

Shelley, E. G., W. K. Peterson, A. G. Ghielmetti, and J. Geiss, The polar ionosphere as a source of energetic magnetospheric plasma, *Geophys. Res. Lett.*, *9*, 941, 1982.

Sibeck, D. G., G. L., Siscoe, J. A. Slavin, E. J. Smith, and B. T. Tsurutani, and R. P. Lepping, The distant magnetotail's response to a strong interplanetary magnetic field B_y: Twisting, flattening, field line bending, *J. Geophys. Res.*, *90*, 4011, 1985.

Sibeck, D. G., J. A. Slavin, E. J. Smith, and B. T. Tsurutani, Twisting of the geomagnetic tail, in *Solar Wind Magnetosphere Coupling*, edited by Y. Kamide and J. A. Slavin, p. 731, Terra Scientific, Tokyo, 1986.

Song, P. and C. T. Russell, Model of the formation of the low-latitude boundary layer for strongly northward interplanetary magnetic field, *J. Geophys. Res.*, *97*, 1411, 1992.

Speiser, T. W., Particle trajectories in model current sheets, *J. Geophys. Res.*, *70*, 4219, 1965.

Usadi, A., A. Kageyama, K. Watanabe, and T. Sato, A global simulation of the magnetosphere with a long tail: Southward and northward interplanetary magnetic field, *J. Geophys. Res.*, *98*, 7503, 1993.

Walker, R. J., R. L. Richard, and M. Ashour-Abdalla, The entry of solar wind ions into the magnetosphere, in *Physics of the Magnetopause*, *Geophysical Monograph Series*, vol. 90, edited by P. Song et al., pp. 311-319, AGU, Washington, D. C., 1995.

Walker, R. J., R. L. Richard, T. Ogino, and M. Ashour-Abdalla, Solar wind entry into the magnetosphere when the interplanetary magnetic field is southward, in *Physics of Space Plasmas (1995)*, *SPI Conference Proceedings and Reprint Series*, edited by T. Chang, p. 561, MIT, Cambridge, MA, 1996.

Maha Ashour-Abdalla, Joachim Raeder, Mostafa El-Alaoui, Vahé Peroomian, Institute of Geophysics and Planetary Physics, University of California, Los Angeles, CA 90095-1567 (e-mail mabdalla@igpp.ucla.edu).

Ion and Electron Heating in the Near-Earth Tail

Wolfgang Baumjohann

Max-Planck-Institut für extraterrestrische Physik, D-85740 Garching, Germany

The remarkable anticorrelation between ion temperature and density often seen in the plasma sheet is not by itself indicative of a non-adiabatic heating process, but rather a result of non-isentropy and the need for pressure equilibrium. The plasma consists of high-density low-temperature blobs containing low-entropy plasma and low-density bubbles containing accelerated hot plasma of higher entropy. Typically the heated plasma bubbles are associated with substorm expansion phases and bursty bulk flow events. During these periods heating occurs in a non-adiabatic fashion with a strong increase in specific entropy. In addition to non-adiabatic heating, the tail plasma is heated adiabatically when being convected earthward. Moreover, ions and electrons behave like a single fluid. The ratio of their temperatures is about constant, $T_i/T_e \approx 7$, and remains at that value even when the plasma is heated. The particular temperature ratio and its constancy gives some hints on possible heating and cooling mechanisms.

1. INTRODUCTION

The question whether convection in the Earth's magnetotail is adiabatic or not had been a matter of considerable debate [*Goertz and Baumjohann*, 1991, and references therein]. As often in our field of research, it turned out that there is no unique answer to this question. The Earth's plasma sheet behaves on average adiabatic, but also non-adiabatic heating of ions and electrons is observed.

Moreover, ions and electrons are observed to behave like a single fluid. The ratio of their temperatures (defined via the second-order moment of the distribution function) is about constant, $T_i/T_e \approx 7$, and does not change even when the plasma is heated. There are indications that the particular value of this ratio is associated with heating at slow-mode shocks. The constancy of this ratio throughout all levels of temperatures and geomagnetic activity hints to adiabatic cooling as the most likely loss mechanism.

New Perspectives on the Earth's Magnetotail
Geophysical Monograph 105

2. ADIABATICITY

If one follows particle trajectories, the prognostic equation for scalar pressure in space plasmas can be written as $P = \alpha N^{\gamma}$, where P denotes thermal plasma pressure, N number density, γ the polytropic index, and α is a constant which depends on the specific entropy of the plasma. For adiabatic behavior the polytropic index takes the value $\gamma = (f+2)/f$, with f denoting the degrees of freedom. Since the plasma sheet ions are fairly isotropic [*Baumjohann et al.*, 1988] we have $\gamma = 5/3$.

The plasma sheet electron pressure, P_e, is much smaller than the ion pressure, P_i, and, moreover, $P_e \propto P_i$ (see below). Hence, one can determine γ by performing a linear regression of samples of log P_i and log N_i, where N_i denotes the ion number density. Ideally, all the sample pairs should come from the same plasma element on its pass further in or outward. Using data from a single satellite, it is nearly impossible to get such data. Hence, determining the polytropic index by a linear regression is only possible, if the plasma elements have the same α or entropy, i.e., if the flow is isentropic or if at least the variation in entropy is much smaller than the variation in pressure and density.

Figure 1 shows a scatter diagram of all ion pressure-density pairs measured in the plasma sheet during the lifetime of the

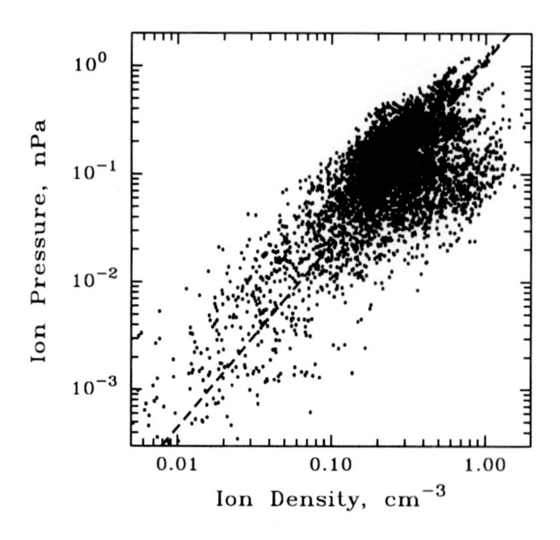

Figure 1. Ion density and thermal pressure in the plasma sheet [after *Baumjohann and Paschmann*, 1989]; the dashed line gives the regression line.

AMPTE/IRM spacecraft. Most of the high pressure-high density pairs were obtained in the central plasma sheet while most other pairs originate from plasma sheet boundary layer encounters. It is obvious that there is a good linear relation between the logarithmic values of ion pressure and density. In fact, a regression analysis yields a correlation coefficient of $r = 0.74$. There is scatter perpendicular to the regression line, indicating varying α, but the overall changes in entropy are much smaller than the changes in pressure and density.

The regression coefficient, i.e., the magnitude of the polytropic index, is $\gamma = 1.69$, very close to 5/3. Hence, a plasma element move from the outer plasma sheet boundary layer to the neutral sheet (or a flux tube moving from the distant tail to the near-Earth tail) more or less adiabatically. There may be changes in entropy, but on average heating and dissipation in the tail are in rough equilibrium.

3. HEATING AND COOLING

Figure 2 shows ion pressure versus density scattergrams for the quiet ($AE < 100$ nT) and the disturbed plasma sheet boundary layer ($AE > 300$ nT). In both cases the two quantities are highly correlated, with $r = 0.76$ for quiet times and $r = 0.82$ for disturbed periods. Again, the scatter indicates that the flow is not totally isentropic, but near-adiabatic changes prevail.

For the $AE < 100$ nT samples the polytropic index is $\gamma = 1.36$, somewhat lower than the adiabatic index of $\gamma = 5/3$. Hence, the quiet plasma sheet boundary layer is undergoing some cooling. Possible non-adiabatic cooling mechanisms include particle losses to the Earth's ionosphere via precipitation or to the solar wind (through the dayside and/or flank magnetopause).

During disturbed times, on the other hand, the polytropic index is $\gamma = 1.90$, indicating that in the active plasma sheet boundary layer heating dominates dissipation. Possible non-adiabatic heating mechanisms are current sheet heating and reconnection. Hence, the plasma sheet boundary layer behaves nearly adiabatic, but depending on the level of geomagnetic activity, either heating or dissipation are prevalent.

4. NON-ISENTROPY

If one focuses on the central plasma sheet part of Figure 1, i.e., those samples with $P_i > 0.1$ nPa and $N_i > 0.1\,cm^{-3}$, it becomes clear that this near-circular cloud of data points cannot give any meaningful regression result. Indeed, the correlation coefficient for these data pairs is very low, $r = 0.19$ [*Baumjohann and Paschmann*, 1989]. However, this does not mean that the central plasma sheet does not behave adiabatically, it rather is a result of the restricted range of possible pressure and density values and of the flow being non-isentropic over longer time intervals and larger regions [*Baumjohann and Paschmann*, 1989; *Goertz and Baumjohann*, 1991].

As stated above, non-isentropy means varying α, which naturally leads to scatter perpendicular to the $\gamma \approx 5/3$ line. Apparently, the variation in entropy is rather large if one takes data from many different intervals. Yet, comparing traces of ion density and pressure during single traversals of a spacecraft through the central plasma sheet like in the example shown in Figure 3, one clearly notices a good correlation between those two quantities.

5. BUBBLES AND BLOBS

One can check if the density-pressure changes within a restricted time interval follows an isentropic and adiabatic path. *Baumjohann and Paschmann* [1989] took about 500 intervals of 30 min of contiguous central plasma sheet data and performed a linear regression on each of them. In about half the cases the correlation was pretty poor and thus the flow non-isentropic, but in 42% of the intervals they obtained correlation coefficients greater than 0.7.

Form the probability distribution of the polytropic index of the latter cases in Figure 4, one notices that adiabaticity is the most common behaviour. In fact, more than half of the intervals with $r > 0.7$ have polytropic index values between 4/3 and 6/3. The average regression coefficient is with 1.59 rather close to $\gamma = 5/3$. The values of α, however, are spread over a wide range.

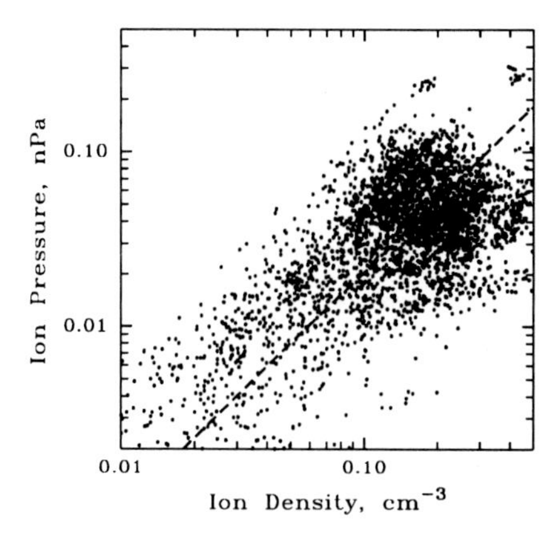

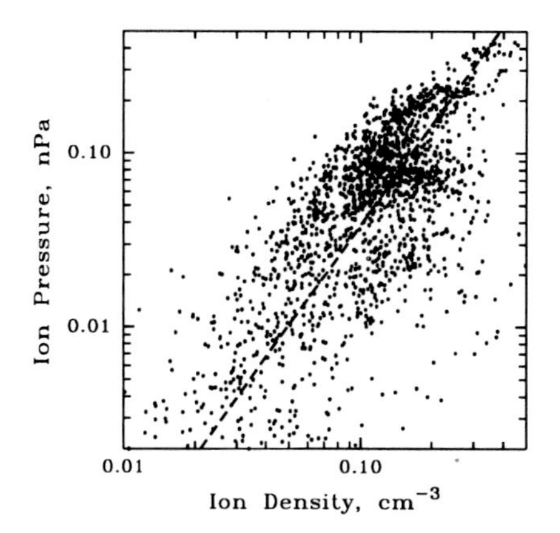

Figure 2. Ion density and thermal pressure in the quiet (left-hand panel; $AE < 100$ nT) and the disturbed (right-hand panel; $AE > 300$ nT) plasma sheet boundary layer [after *Goertz and Baumjohann*, 1991]; the dashed lines represent regression lines.

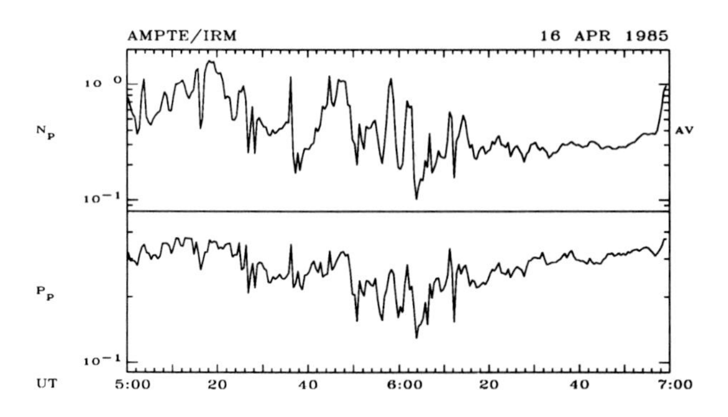

Figure 3. Ion density and ion pressure measured by the IRM spacecraft on 16 April 1985 during a traversal of the premidnight (22 MLT) central plasma sheet at a radial distance of about 13 R_E.

Based on these results, *Goertz and Baumjohann* [1991] concluded that the Earthward convecting plasma consists of fairly large regions of isentropic plasma with relatively sharp boundaries between regions of different specific entropy. Using the terminology of *Pontius and Wolf* [1990], the central plasma sheet contains high-entropy "bubbles" of low-density high-temperature plasma and low-entropy "blobs" with dense but cool plasma. For each component the entropy is constant and density and temperature are positively correlated.

6. SUBSTORM HEATING

Baumjohann et al. [1991], found that during the expansion phase the central plasma sheet is heated to about twice its pre-substorm temperature while the density decreases by about 25%. This corroborates a similar study by *Huang et al.* [1992] and clearly indicates that the substorm heating cannot occur in an adiabatic fashion, where temperature and density changes must be positively correlated.

Figure 5 shows the results of a superposed epoch analysis of the average specific entropy, $\alpha = kT_i/N_i^{2/3}$. The trace was constructed by binning central plasma sheet measurements taken between 45 min before and 90 min after substorm onset with respect to onset time into 9-min-wide bins and averaging over all samples in a particular bin.

The specific entropy trace is not constant, but rather rapidly changing during the expansion phase. Actually, the specific entropy increase is stronger than the enhancement of the ion

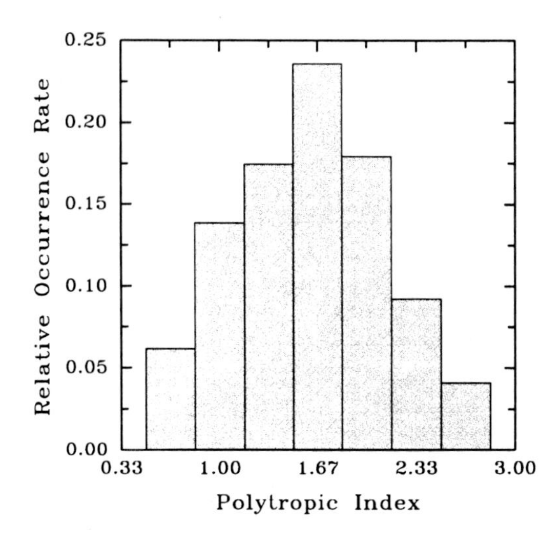

Figure 4. Probability distribution of polytropic index values for 30-min intervals of contiguous central plasma sheet data [from *Baumjohann and Paschmann*, 1989].

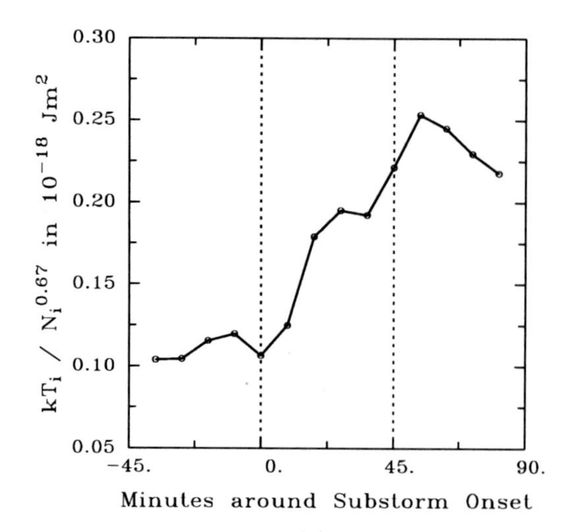

Figure 5. Change of specific entropy during substorm onset [from *Baumjohann*, 1991]; the dashed vertical lines mark substorm onset and the start of the recovery phase.

temperature during the expansion phase and, as a result, pre- and post-substorm plasma are at totally different specific entropy levels. Thus substorm-associated heating is clearly non-adiabatic, most likely current sheet heating or heating by reconnection.

7. HIGH-ENTROPY FLOWS

Another example of a high-entropy plasma is shown in Figure 6, where the pressure-density relationship is plotted for all plasma elements travelling in the central plasma sheet with bulk velocities of more than 400 km/s. As can be seen, these value pairs are reasonably well correlated with $r = 0.61$, indicating that all high-speed plasma elements have a similar entropy and originate from a similar source. A regression coefficient of 1.65 indicates that the high-speed plasma behaves adiabatic on its way further in.

On the other hand, the regression constant, which is related to the specific entropy of the plasma, has about twice the value obtained for all plasma sheet samples, indicating that the plasma in the high-speed flows has a high entropy. The entropy level is, in fact, similar to that found during a substorm expansion phase. Hence, it is thus likely that high-speed and expansion-phase plasma have undergone the same kind of heating mechanisms.

8. ELECTRON TEMPERATURE

It is not only the ion temperature that increases strongly with geomagnetic activity, the electron temperature varies the same way, too. The lower panel of Figure 7 presents traces of ion (upper curve) and electron temperature (lower curve) during a traversal of the central plasma sheet. While both temperatures vary by nearly a decade between the northern and southern boundary of the central plasma sheet and the neutral sheet (cf. magnetic field traces in the upper three panels), both traces run nearly parallel all the way.

9. SINGLE FLUID BEHAVIOR

Figure 8 presents the relation between electron and ion pressure in the central plasma sheet and in the plasma sheet boundary layer [see also *Baumjohann et al.*, 1989*b* and *Baumjohann*, 1991]. (Here pressure is plotted instead of temperature, since the cold photo electrons in the plasma sheet boundary layer [cf. *Baumjohann et al.*, 1988] hardly distort the electron pressure, but strongly affect the electron density, which is needed to calculate the temperature from the measured pressure tensor.)

The two pressures and thus temperatures show a high degree of correlation, with a linear correlation coefficient of $r = 0.94$ and $T_i/T_e = 7.05$ for the central plasma sheet and $r = 0.95$ and $T_i/T_e = 6.34$ for the plasma sheet boundary layer. More than 80% of the central plasma sheet data points have an ion-to-electron temperature ratio between 5 and 10. If it were not for the samples with $P_i < 10^{-2}$ nPa in the plasma sheet boundary layer panel, where the electron pressure is somewhat corrupted by the cold yet abundant photo electrons, $T_i/T_e \approx 7$ might hold for the plasma sheet boundary layer, too.

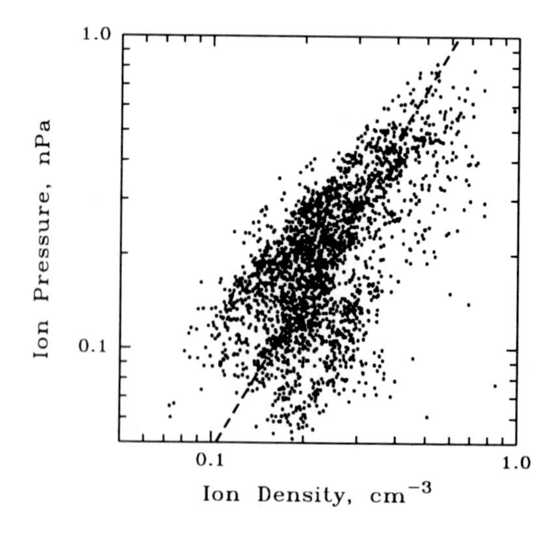

Figure 6. Ion density and thermal pressure during high-speed flow (>400 km/s) [from *Baumjohann*, 1993].

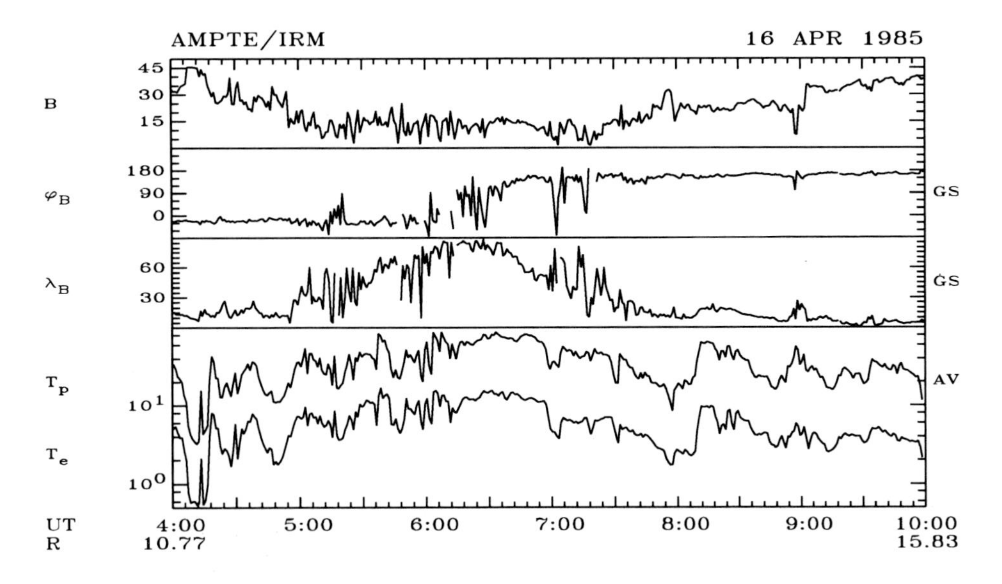

Figure 7. Magnetic field (magnitude, azimuth, and elevation) and ion and electron temperature measured by the IRM spacecraft on 16 April 1985 during a crossing of the premidnight (22–23 MLT) central plasma sheet at radial distances of 11–16 R_E.

The linear relation between T_i and T_e holds for temperature variations over nearly two decades, i.e., both in a hot and cold plasma sheet and is independent of disturbance level or radial distance. It corroborates the conclusion of *Christon et al.* [1988], that plasma as well as energetic particles respond collectively as a single unified particle population during plasma sheet temperature transitions.

10. HEATING MECHANISM

For quite some years it has been unclear what causes this particular temperature ratio. An answer to this question is even more important, because a process which causes this particular ratio is also the most likely heating mechanism for the tail plasma sheet.

Even today this question cannot be answered. However, there are some hints from recent Geotail data. *Saito et al.* [1995, 1996] looked at the behavior of ion and electron temperatures across the slow mode shock associated with magnetotail reconnection. Upstream of the shock, in the tail lobe, the electrons are typically hotter than the ions, but when crossing the shock front and thus entering the plasma sheet, the ions gain considerably more energy than the electrons and the temperature ratio becomes greater than unity (as is the case for the fast-mode bow shock).

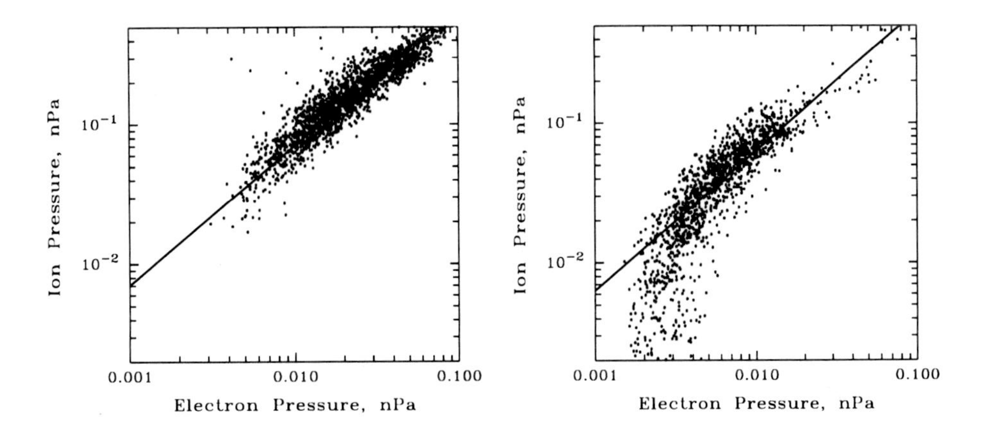

Figure 8. Ion and electron pressure in the central plasma sheet (left panel) and plasma sheet boundary layer (right panel); the solid line gives the regression line [from *Baumjohann*, 1993].

In three out of the five cases presented in two papers by *Saito et al.*, the temperature ratio is close to seven in the downstream plasma sheet and the average over all five cases is $T_i/T_e = 6.6$, rather close to to the average value found in the near-Earth plasma sheet. Hence, it seems quite possible that the particular temperature ratio typical for the near-Earth plasma sheet is attained at a slow mode shock front associated with tail reconnection. However, more cases have to be studied and more theoretical work on particle acceleration at slow mode shocks has to be done before this suggestion can be based on firm grounds.

11. LOSS MECHANISM

Looking at the ion-to-electron temperature ratio does not only allow to draw conclusions about possible heating mechanisms, the constancy of this ratio throughout all levels of temperatures and geomagnetic activity also tells us something about possible loss mechanisms.

If radiation were to cool the tail plasma after a substorm, it is very unlikely that the temperature ratio would remain constant, since electrons and ions are receptive to a totally different set of plasma waves. Loss of energetic particles to the ionosphere may be one way to get the average plasma temperature down. However, refilling of the loss cone, necessary for any effective loss, is again done by plasma waves.

The one mechanism which clearly retains a given temperature ratio is adiabatic cooling, i.e., an expansion of the heated plasma refilling the plasma sheet during the recovery phase to substitute any losses due to plasmoid ejection downtail into the solar wind. Hence, it may be the most likely candidate.

Acknowledgment. I would like to thank Rudolf Treumann for helpful discussions about heating and loss mechanisms.

REFERENCES

Baumjohann, W., Heating and fast flows in the near-Earth tail, in *Magnetospheric Substorms*, ed. by J. R. Kan et al., pp. 141–145, American Geophysical Union, Washington, 1991.

Baumjohann, W., The near-Earth plasma sheet: An AMPTE/IRM perspective, *Space Sci. Rev.*, *64*, 141–163, 1993.

Baumjohann, W., and G. Paschmann, Determination of the polytropic index in the plasma sheet, *Geophys. Res. Lett.*, *16*, 295–298, 1989.

Baumjohann, W., G. Paschmann, N. Sckopke, C. A. Cattell, and C. W. Carlson, Average ion moments in the plasma sheet boundary layer, *J. Geophys. Res.*, *93*, 11,507–11,520, 1988.

Baumjohann, W., G. Paschmann, and C. A. Cattell, Average plasma properties in the central plasma sheet, *J. Geophys. Res.*, *94*, 6597–6606, 1989b.

Baumjohann, W., G. Paschmann, T. Nagai, and H. Lühr, Superposed epoch analysis of the substorm plasma sheet, *J. Geophys. Res.*, *96*, 11,605–11,608, 1991.

Christon, S. P., D. G. Mitchell, D. J. Williams, L. A. Frank, C. Y. Huang, and T. E. Eastman, Energy spectra of plasma sheet ions and electrons from $\sim$50 eV/e to $\sim$1 MeV during plasma temperature transitions, *J. Geophys. Res.*, *93*, 2562–2572, 1988.

Goertz, C. K., and W. Baumjohann, Thermodynamics of the plasma sheet, *J. Geophys. Res.*, *96*, 20,991–20,998, 1991.

Huang, C. Y., L. A. Frank, G. Rostoker, J. Fennel, and D. G. Mitchell, Nonadiabatic heating of the central plasma sheet at substorm onset, *J. Geophys. Res.*, *97*, 1481–1495, 1992.

Pontius, D. H., and R. A. Wolf, Transient flux tubes in the terrestrial magnetosphere, *Geophys. Res. Lett.*, *17*, 49–53, 1990.

Saito, Y., T. Mukai, T. Terasawa, A. Nishida, S. Machida, M. Hirahara, K. Maezawa, S. Kokubun, and T. Yamamoto, Slow-mode shocks in the magnetotail, *J. Geophys. Res.*, *100*, 23,567–23,581, 1995.

Saito, Y., T. Mukai, T. Terasawa, A. Nishida, S. Machida, S. Kokubun, and T. Yamamoto, Fore-shock structure of the slow-mode shocks in the Earth's magnetotail, *J. Geophys. Res.*, *101*, 13,267–13,274, 1996.

W. Baumjohann, MPI für extraterr. Physik, D-85740 Garching, Germany (e-mail: bj@mpe.mpg.de)

Kinetic Structure of the Slow-mode Shocks in the Earth's Magnetotail

Yoshifumi Saito, Toshifumi Mukai

Institute of Space and Astronautical Science, Sagamihara, Kanagawa, Japan

Toshio Terasawa

Department of Earth and Planetary Physics, University of Tokyo, Hongo, Tokyo, Japan

We have identified slow-mode shocks between plasmasheet and lobe in the mid-tail to distant-tail regions by using three-dimensional magnetic field data and three-dimensional plasma data observed by the GEOTAIL satellite. Analyzing the data obtained between 14 September, 1993 and 16 February, 1994, we have found 303 plasmasheet-lobe boundary crossings at distances between $X_{GSE} \sim -30$Re and $X_{GSE} \sim -210$Re. Thirty-two out of these 303 boundaries are identified as slow-mode shocks. Using the identified slow-shock boundary data, we have investigated the ion distribution function in the upstream region of the slow-mode shocks in detail. We have found the existence of a region called the "fore-shock region" in the upstream region that can be clearly distinguished from the main dissipation region of the slow-mode shocks. This "fore-shock region" is characterized by counterstreaming ions: backstreaming ions that flow from the slow-shock boundary toward the upstream region along the magnetic field and the lobe cold ions that flow from the upstream region into the slow-shock boundary. Backstreaming ions are hot plasmasheet ions escaping from the downstream region toward the upstream region. The pitch angle distributions of these backstreaming ions show that only 12% to 43% of the ions that try to escape from the downstream region toward the upstream region can successfully escape to the upstream region and the rest of these ions are reflected back toward the downstream region by a magnetic mirror. The incident cold ions are heated about 3% - 20% of the total ion heating throughout the slow-shock transition in the "fore-shock region". Wave-particle interaction with the electromagnetic ion cyclotron waves generated by the counterstreaming ions is a candidate of the cold ion heating mechanism in the "fore-shock region". The width of the main dissipation region of the slow-mode shocks is estimated to be several times the upstream ion inertial length, while the total width of the slow-mode shocks including the "fore-shock region" is as wide as 20 times the upstream ion inertial length.

New Perspectives on the Earth's Magnetotail
Geophysical Monograph 105

1. INTRODUCTION

The existence of an X-type neutral line in the magnetotail, which is expected from the reconnection model of the Earth's magnetosphere [*Dungey*, 1961], has long been a controversial problem. If magnetic field reconnection occurs in the magnetotail, slow-mode shocks are predicted to be formed at the boundaries between plasmasheet and lobe bounding the X-type neutral line [*Vasyliunas*, 1975, and references therein].

The existence of the slow-mode shocks in the distant-tail regions was for the first time reported by the ISEE 3 deep-tail mission [*Feldman et al.*, 1984a, b, 1985; *Scarf et al.*, 1984; *Smith et al.*, 1984]. Analyzing the three-dimensional magnetic field data and two-dimensional electron data, *Feldman et al.* [1985] identified the slow-mode shocks and concluded that the slow-mode shocks were observed under all states of geomagnetic activity. *Ho et al.* [1996] found 86 slow-shock boundaries from 439 plasma sheet/lobe crossings using two distant ISEE 3 geomagnetic tail passes.

Recent observations in the distant magnetotail by the GEOTAIL satellite have proved the existence of the slow-mode shocks using three-dimensional magnetic field data and three-dimensional plasma velocity moments [*Saito et al.*, 1995; *Seon et al.*, 1995, 1996a]. The greatest difference between the results of the ISEE 3 and the GEOTAIL observations exists in the occurrence frequency of the observed slow-shock boundaries. *Saito et al.* [1995] reported that only about 10% of the plasmasheet-lobe boundaries were identified as slow-mode shocks. Using the data set obtained during the almost different period, *Seon et al.* [1996a] reported that only three cases out of 300 plasmasheet-lobe boundaries were identified as slow-mode shocks.

The observational identification of slow-mode shocks was conducted under the assumption of one-fluid MHD in the upstream and downstream plasma. However, the real plasma is composed of both electrons and ions, which show kinetic structure of slow-mode shocks.

Using the ISEE 3 data, *Cowley et al.* [1984] and *Richardson et al.* [1985] reported the existence of the high energy ion beams (> 35keV) around the plasmasheet-lobe boundaries in the distant-tail region. *Scarf et al.* [1984] and *Smith et al.* [1984] suggested the existence of foreshock region in the front of slow-shocks that is similar to the foreshock structure observed near bow shock.

The characteristic electron distributions around slow-mode shocks were investigated in detail using ISEE 3 data [*Feldman et al.*, 1984a, b, 1985; *Schwartz et al.*, 1987]. Electron heat flux leakage was observed in the upstream region, while flat-top electron distributions were observed in the downstream region. These electron distributions were also observed by the GEOTAIL satellite [*Saito et al.*, 1995; *Seon et al.*, 1996b].

The low-energy ion distributions around the slow-mode shocks in the distant magnetotail were for the first time observed by the GEOTAIL satellite. According to the GEOTAIL observations, ion distributions around slow-mode shocks are characterized by accelerated cold ions and backstreaming ions [*Saito et al.*, 1995; *Seon et al.*, 1996b].

The existence of the slow shock structure in the near-Earth magnetotail at X ~ –20Re was also reported by *Feldman et al.* [1987], based on three-dimensional magnetic field data and two-dimensional electron and ion data obtained by the ISEE 2 satellite. They reported that the lobeward edge of the plasmasheet-lobe boundary was identified as a slow-mode shock, which was different from the plasmasheet-lobe boundaries in the distant magnetotail where the entire boundary region had a slow-shock structure. Using the Active Magnetospheric Particle Tracer Explorers / Ion Release Module (AMPTE/IRM) data, *Cattel et al.* [1992] investigated ~80 plasmasheet-lobe crossings at downtail distances of 10Re to 18Re. They concluded that the plasmasheet-lobe boundaries in the near-tail region were not well modeled by a planar MHD discontinuity with a normal component of the magnetic field and, in particular, did not contain a slow-mode shock. They suggested that some factor operating Earthward of the neutral line prevents the formation of slow-mode shocks in the near-tail region. They also suggested that the plasmasheet boundary in the near-Earth region is usually a tangential discontinuity or there are effects such as time variations, single particle effects, or three-dimensional structures that invalidate the assumptions of planar time-stationary magnetohydro-dynamics.

The large-scale structure and dynamics of the magnetotail including the plasmasheet boundary layer (PSBL) have been numerically studied using MHD simulations. The existence of the slow-mode shocks in the PSBL is shown by *Sato* [1979], *Scholer and Roth* [1987], and *Ugai* [1993]. *Birn et al.* [1986] and *Hesse and Birn* [1991], on the other hand, reported that slow-mode shocks were not formed on the Earthward side of the X-type neutral line, where the magnetic field is connected to the ionosphere.

The first theoretical work on the structure of slow shocks was done by *Coroniti* [1971], who showed that the slow shocks are associated with a dispersive wave train in the downstream region. He suggested that the energy dissipation of the slow shock is provided by dumping of the dispersive wave train.

Several numerical simulations concerning the kinetic structure of slow-mode shocks have been reported as well [*Swift*, 1983; *Winske*, 1985; *Omidi and Winske*, 1992;

Fujimoto and Nakamura, 1994; *Omidi*, 1995]. These numerical simulations are hybrid code simulations, where electrons are treated as a fluid and ions are treated as particles. *Swift* [1983] reported the existence of the backstreaming particles in the upstream region for the first time. *Omidi and Winske* [1992] noted the importance of the electromagnetic ion/ion cyclotron instability that may be generated by the backstreaming ions and the incident lobe ions in the upstream region.

The purpose of this paper is to summarize the observed characteristics of the ion and electron distribution functions around slow-mode shocks identified by the GEOTAIL satellite in the Earth's magnetotail .

2. OBSERVATION

We have analyzed the data obtained by the Low Energy Particle - Energy Analyzer (LEP-EA) and Magnetic Field (MGF) experiment on the GEOTAIL satellite. LEP-EA provides three-dimensional distribution functions of electrons (8eV – 38keV) and ions (32eV/e – 43keV/e) every four spin periods (about 12 seconds). MGF provides vector magnetic field data with a time resolution of 1/16 seconds. In our present analysis, we have used spin-time averaged vector magnetic field data. A detailed description of the LEP-EA and MGF is given in *Mukai et al.* [1994] and *Kokubun et al.* [1994], respectively.

Figure 1 shows an example of a slow-shock boundary observed on 13 February, 1994, when the GEOTAIL satellite was at $X_{GSE} \sim -63.0Re$, $Y_{GSE} \sim 6.8Re$, and $Z_{GSE} \sim -3.8Re$. We have observed a slow-mode shock in the transition from lobe to plasmasheet around 1936 UT. Before this time, the spacecraft was in the lobe where the magnetic field was about 10 nT and plasma density was about 0.028 /cm^3. The azimuthal angle ϕ_{MAG} of the magnetic field was ±180°, which means that the satellite was in the southern lobe. The plasma bulk velocity was about 150 km/s and its direction was tailward. At 1936 UT the satellite entered the plasmasheet, where the magnetic field decreased to be about 6 nT. The plasma density increased to be about 1.6 times the density in the lobe, and the ion and electron temperature increased simultaneously. The plasma bulk flow velocity increased to be about 600 km/s and the direction changed to be earthward.

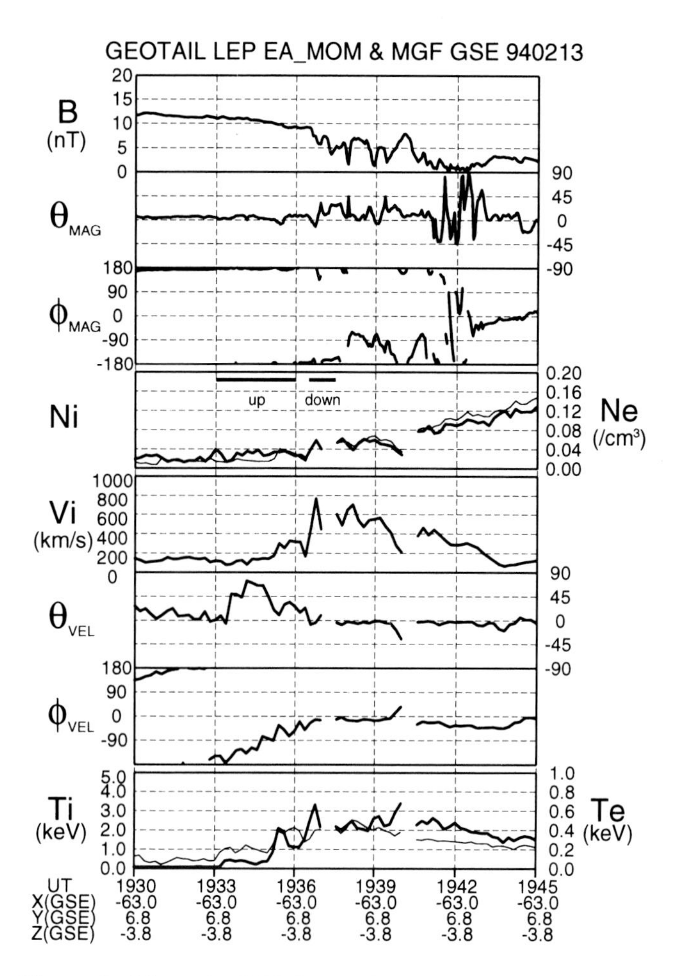

Figure 1. Magnetic field data and plasma velocity moments obtained on 13 February, 1994. From top to bottom, total magnetic field, polar and azimuthal angles of the magnetic field direction in the GSE polar coordinate system, ion (thick line) and electron (thin line) density, magnitude of ion bulk velocity, polar and azimuthal angles of the ion bulk flow velocity vector in the GSE coordinate system, and ion (thick line) and electron (thin line) temperature are shown. The upstream region and downstream region are indicated in the figure.

We have checked if the upstream and downstream magnetic field and plasma data satisfy the one-dimensional Rankine-Hugoniot relation. The detailed method of identifying slow-mode shocks is given in *Saito et al.* [1995]. Figure 2 shows the summary of the upstream and downstream parameters of the slow-mode shock in the normal incidence frame. We have calculated the downstream parameters by giving the upstream parameters and the downstream temperature. The calculated downstream parameters (inside the parentheses) and the observed downstream parameters show reasonable agreement. We have checked the energy flux conservation by calculating the total energy flux E_{tot}. Figure 2 also shows the upstream Alfvén Mach number $M_A = V_n / V_A$, and the intermediate and slow Mach numbers $M_i = V_n / V_{imd}$, $M_s = V_n / V_{sl}$ in both the upstream and downstream regions of the shock, where V_A, V_{imd}, V_{sl}, and V_n are Alfvén velocity, intermediate-mode velocity, slow-mode velocity and ion velocity along the shock normal direction, respectively. The

	1994/2/13 UP 1933:00-1936:00 DN 1936:30-1937:30	
	UP	DOWN
ρ (10^{-2} cm^{-3})	2.8	4.6 (4.2) 9%
B (nT)	10.9	6.2 (8.1) 23%
θ_b (deg.)	77.3	67.4 (72.8) 5.4
V (km/s)	227	647(544) 20%
θ_V (deg.)		77.8(74.0) 3.8
T (keV)	0.51	3.3
Te (keV)	0.22	0.39
Ti (keV)	0.29	2.9
E_{tot} (10^{-5} J/m^2)	1.9	1.6 | 16%
M_A	0.16	
M_i	0.72	0.59
M_s	3.68	0.88
$\Delta\phi_{bv}$ (deg.)	9.1	
n (GSE)	(0.25,0.91,-0.33)	
Vs (km/s)	144	
position (Re) (GSE)	(-63.0,6.8,-3.8) north lobe Earthward tailward south lobe	

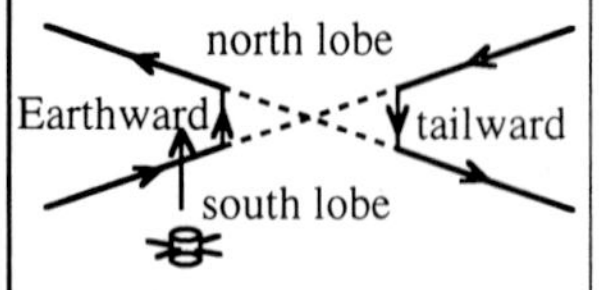

Figure 2. Upstream and downstream shock parameters of a slow-mode shock observed on February 13, 1994. From top to bottom, observed date, time, and position in the GSE coordinate system, upstream and downstream shock parameters, upstream Alfvén Mach number, upstream and downstream intermediate and slow Mach number, shock normal information and schematic figure of satellite position with respect to the X-type neutral line are shown. The downstream parameters of the shock are the observed value (left side value), the calculated value (in the parenthesis), and the difference between calculated and observed values (right side value).

parameters V_n, B_n, and V_{sl} can be calculated from the other parameters in the table as V_n = 227 km/s (upstream), 137 km/s (downstream), B_n = 2.4 nT, and V_{sl} = 62 km/s (upstream), 187 km/s (downstream), where B_n is the magnetic field intensity along the shock normal. The result of the coplanarity check, the shock normal direction, and the shock velocity are also shown in Figure 2. The satellite position in the GSE coordinate system and a schematic configuration of the satellite position with respect to the expected X-type neutral line are shown at the bottom. Since the flow direction of the downstream plasma was earthward, the shock was expected to be on the earthward side of the X-type neutral line.

We have checked if the slow-mode shock conditions are satisfied for all the plasmasheet-lobe boundary crossings when three-dimensional plasma (LEP) data and magnetic field (MGF) data are both available between 14 September, 1993 and 16 February, 1994. Figure 3 shows the orbit of the GEOTAIL satellite in the GSE coordinate system during this period. The dots in the figure show the position of the satellite at 0000 UT every day. The large dots show the first day in each month. The periods when plasmasheet-lobe crossings were observed are shown by thick lines in the orbit projected on the GSE X-Y plane. GEOTAIL satellite crossed the plasmasheet-lobe boundaries in the magnetotail of X_{GSE} ~ −30Re to ~ −210Re during this period. Some of the plasmasheet-lobe crossings were too rapid to be identified as slow-mode shocks. We have selected the crossings that have at least five minutes continuous upstream (lobe) and downstream (plasmasheet) regions, which totally amount 303 crossings. Thirty-two out of these 303 plasmasheet-lobe boundaries were identified as slow-mode shocks. In this paper we will investigate the kinetic properties of the slow-mode shocks in detail using the identified slow-shock boundaries.

Figure 4 shows the omni-directional Energy-time (E-t) diagram of ions observed on February 13, 1994. The spacecraft was in the lobe region where only cold ions below 1 keV/q were observed between 1930 UT and 1933 UT. At 1933 UT, two components of ions began to be observed. One was high energy ions that gradually lost energy, and the other was cold ions that gradually gained energy. About 1936:30 UT, these two components merged, such that only one-component (hot ions) was observed. This region with one-component hot ions was plasmasheet.

Figure 5a shows upstream electron distribution function parallel to the magnetic field in the de-Hoffmann Teller frame corresponding to the slow-mode shock shown in Figure 1. The direction of the positive velocity is the magnetic field direction. The large dot shows the distribution function and the small dots are also distribution function plotted inverting the horizontal axis to show the asymmetry of the distribution function with respect to the center (velocity=0) line. Electron heat flux toward the upstream direction (opposite to the magnetic field direction) can be clearly seen in the high energy part of the distribution function in the upstream region (see hatched region in Figure 5a). Since the plasmas are flowing into the shock, the low energy part of the electron distribution function is slightly deviated to the downstream direction (magnetic field direction). Figure 5b shows downstream electron distribution function parallel to the magnetic field in the de-Hoffmann Teller frame. As the spacecraft moves into the plasmasheet, the electron distribution is thermalized, and then a flat-top distribution can be seen in the lowest energy steps of the distribution function in the downstream region.

We have investigated the distribution function of ions between the upstream and downstream regions of this slow-shock boundary in the de-Hoffmann Teller frame. We can transform the distribution function from the spacecraft coordinate system to the de-Hoffmann Teller frame using the slow-shock parameters. In this frame, plasma flows along the magnetic field in both the upstream and the downstream regions. Figure 6a shows the total magnetic field observed on 13 February, 1994. Figure 6b-6g show the pitch angle distribution of ions in the de-Hoffmann Teller frame. The vertical and horizontal axes on the right hand side of each panel show ion velocity parallel and

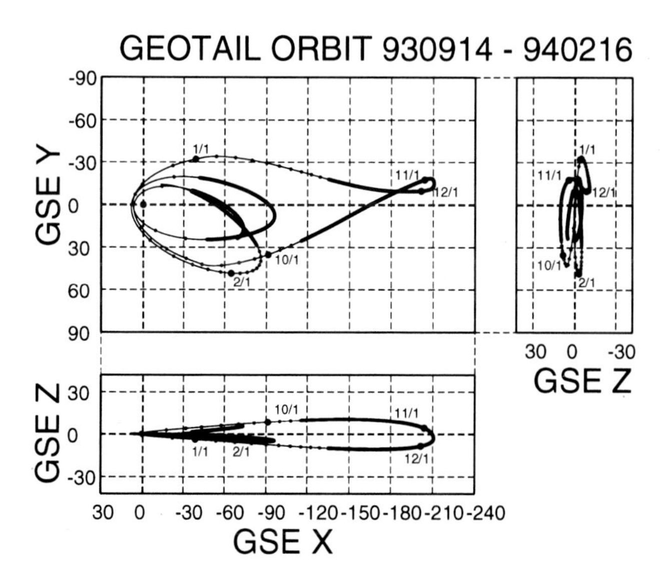

Figure 3. Orbit of GEOTAIL satellite in the GSE coordinate system between 14 September, 1993 and 16 February, 1994. Small dots show the position of the satellite at 0000 UT every day. Large dots show the first day in each month. The periods when plasmasheet-lobe crossings are observed are indicated by thick line in the orbit projected on GSE X-Y plane.

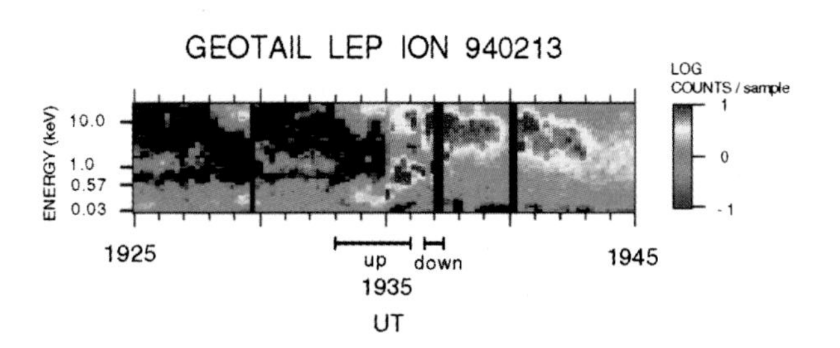

Figure 4. Omni-directional Energy-time (E-t) diagram of ions around a slow-mode shock observed on February 13, 1994

perpendicular to the magnetic field. The vertical axis on the left-hand side of each panel shows the off-plane angle, which is the angle between the ion velocity vector and the plane in which both the upstream and the downstream magnetic field and ion bulk flow vector exist. The horizontal axis on the left-hand side of each panel shows the ion velocity. In the upstream region (Figure 6b), only cold ions with positive parallel energy were observed. These cold ions were flowing into the shock along the magnetic field with the velocity of about 1500 km/s in the de-Hoffmann Teller frame. Since cold ions were flowing along the magnetic field, the off-plane angle of these cold ions is distributed between 0° and 360°. When the spacecraft approached the shock surface, ions with negative parallel velocity began to be observed (Figure 6c, 6d). Since these ions were flowing from the shock surface toward the upstream region, these ions were backstreaming ions. As the spacecraft came nearer to the shock surface, the density of the backstreaming ions increased and, at the same time, cold ions began to be heated (Figure 6e). The backstreaming ions and cold ions that flow from the upstream to the downstream began to merge at 1936:23 UT (Figure 6f). In the downstream region, only one-component hot ions flowing from the shock surface toward the downstream plasmasheet were observed (Figure 6g). Comparing the upstream ion pitch angle distributions (Figure 6c) with the downstream ion pitch angle distributions (Figure 6g), we recognize that the backstreaming ions correspond to the downstream hot ion distributions with negative velocity. This suggests that the backstreaming ions are the ones leaking from the shock surface toward the upstream region. Cold ions were continuously heated while the backstreaming ions are observed. The parallel velocity of the cold ions was slightly decreased in the de-Hoffmann Teller frame.

In order to show the variations of the ion distributions more quantitatively, we have calculated the ion velocity moments of the cold ions and the backstreaming ions separately. Figure 7 shows total magnetic field (a), cold ion temperature (b), cold ion velocity (c), backstreaming ion

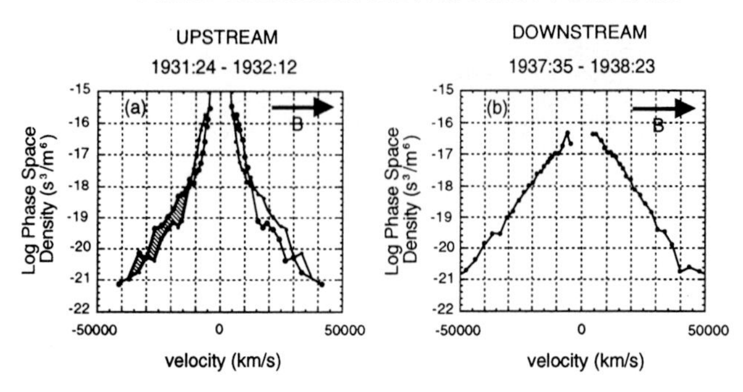

Figure 5. (a)Upstream electron distribution function parallel to the magnetic field in the de-Hoffmann Teller frame corresponding to the slow-mode shock shown in Figure 1. The large dot indicates the distribution function and the small dots are plotted inverting the horizontal axis in order to show the asymmetry of the distribution function with respect to the V = 0 line. (b) Downstream electron distribution function parallel to the magnetic field in the de-Hoffmann Teller frame.

density (d), magnetic field fluctuation (e), and bulk velocity difference between the cold ions and the backstreaming ions in units of local Alfvén velocity (f) around the same slow-mode shock shown in Figure 1. At 1933 UT (the first dashed line), the total magnetic field began to decrease more rapidly than the preceding time period. The cold ion temperature began to increase and the backstreaming ion density also began to increase. At 1936:20 UT (the second dashed line), total magnetic field abruptly decreased. Around this time, cold ions and backstreaming ions merged and one component hot ions were observed (see Figure 4). The main dissipation region of the slow-mode shock was observed after 1936:20 UT. However, cold ions had already begun to be heated after 1933 UT. Since this region between 1933 UT and 1936:20 UT is a region before the main dissipation region of the slow-mode shock, we will call it a "fore-shock region". This fore-shock region is characterized by the existence of two component ions, the cold ions flowing into the shock and the backstreaming ions flowing out from the shock toward the upstream region. In the fore-shock region of the slow-mode shock shown in Figure 1, cold ions are heated to about 10% of the total ion heating throughout the shock. The cold ion velocity slightly decreased in this region (see Figure 7c). The magnetic field fluctuation is the root-mean-square of the magnetic field fluctuations on the 3-second average. It began to be observed at 1933 UT when cold ion temperature and backstreaming ion density began to increase. The bulk flow velocity difference between cold ions and backstreaming ions in the fore-shock region was between $1.6V_A$ and $2.2V_A$ for this slow-mode shock, where V_A is the local Alfvén velocity.

Figure 8 shows the cold ion temperature (a), backstreaming ion temperature (b) in the fore-shock region and total ion temperature (c) in both the upstream and downstream regions. $T_\parallel$, $T_{\perp 1}$, and $T_{\perp 2}$ are components of the temperature tensor parallel to the magnetic field and two components of the temperature tensor perpendicular to the magnetic field. These three components are diagonal elements of the temperature tensor in a coordinate system in which the three orthogonal axes are the magnetic field direction and two directions perpendicular to the magnetic field. The three components T_{off1}, T_{off2}, and T_{off3} in Figure 8c are off-diagonal elements of the temperature tensor. The off-diagonal elements in this coordinate system were much smaller than the diagonal elements. Cold ion temperature was nearly isotropic in the upstream region far from the shock surface (before 1935 UT). The perpendicular temperature began to increase at 1935 UT more rapidly than parallel temperature. However, the perpendicular temperature of the backstreaming ions decreased as the spacecraft approached the shock surface. The total ion temperature varied differently. In the upstream region (before 1933 UT) the temperature was nearly isotropic. In the fore-shock region, the parallel temperature increased first when two component ions (cold ions and backstreaming ions) began to be observed. As the satellite approached the main dissipation region of the slow-shock, both $T_\parallel$ and $T_\perp$ increased gradually, and in the main dissipation region, $T_\perp$ increased to be nearly the same as $T_\parallel$. In Figure 8c, the temperature in the downstream region about 1936:30 UT was nearly isotropic. At 1937:00 UT, the $T_\parallel$ increased and $T_\perp$ decreased. The satellite may have temporally moved back to the main dissipation region at this time.

The variation of the pitch angle distributions of the backstreaming ions also reflects the structure of the slow-mode shocks. Pitch angle distribution of ions between the upstream and the downstream regions are shown in Figure 6b-6g. The backstreaming ions began to be observed at 1934:11 UT (Figure 6c) . As the spacecraft approached the slow-mode shock, the edge of the pitch angle distribution of the backstreaming ions that is shown by white lines, gradually extended to lower pitch angles (Figure 6c-6e). The backstreaming ions and cold ions that flow from the upstream to the downstream began to merge at 1936:23 UT (Figure 6f), and only one component downstream ions were observed after 1936:47 UT (Figure 6g). This variation of the edge angle can be explained by the magnetic mirror reflection of the ions escaping from the downstream region toward the upstream region as shown schematically in Figure 9. Since the downstream magnetic field is weaker than the upstream magnetic field, the hot ions that try to escape from the downstream region toward the upstream region are magnetically mirror reflected and turned back

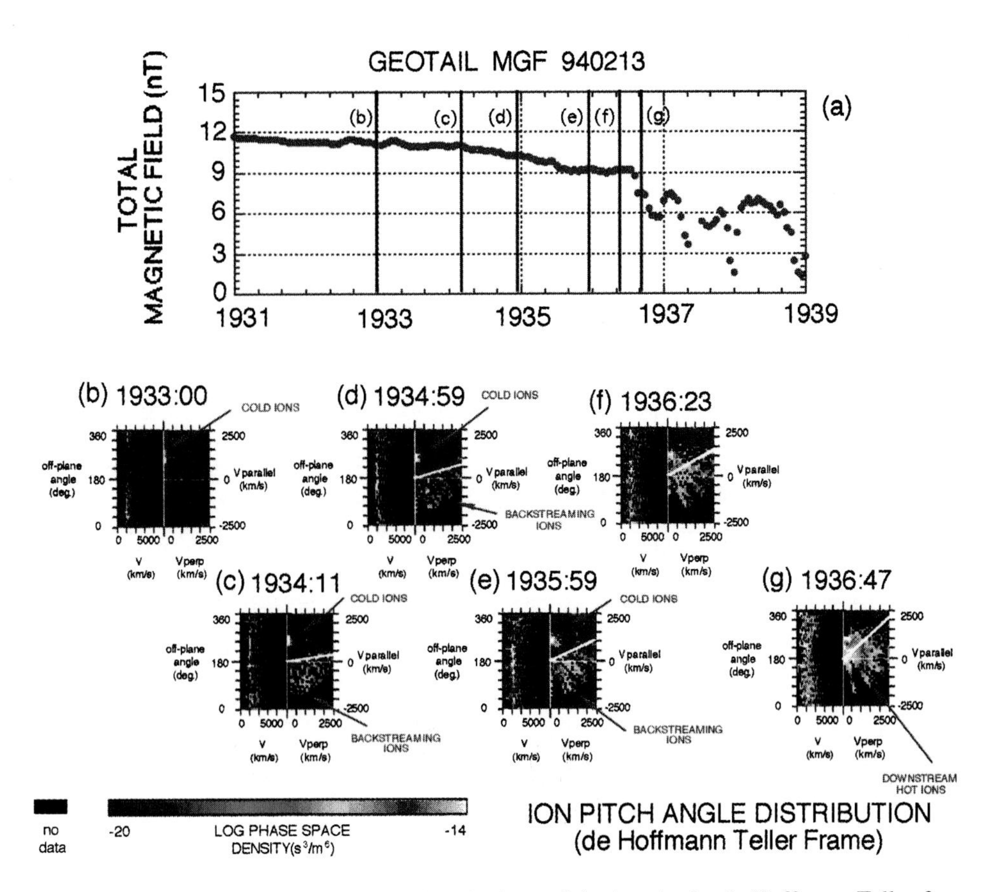

Figure 6. (a)Total magnetic field and (b)-(g)pitch angle distributions of the ions in the de-Hoffmann Teller frame around a slow-mode shock observed on 13 February, 1994. White lines in Panel(b)-(g) show calculated edge angle of the pitch angle distribution function of the backstreaming ions.

toward the downstream region. The resultant pitch angle distribution of ions has the observed characteristics that the edge angle extended to lower pitch angles as the spacecraft approached the downstream regions from the upstream region.

The white lines in Figure 6b-6g showing the edge angle of the pitch angle distribution of the backstreaming ions show calculated pitch angles assuming the conservation of the magnetic moment. The edge angle θ of the pitch angle distributions where magnetic field is B can be expressed using the magnetic field intensity B_0 at the upper most boundary of the magnetic mirror reflection region where $\theta = 90°$. Using B and B_0, we can express the edge angle θ as follows.

$$\theta = \sin^{-1}\sqrt{\frac{B}{B_0}}$$

In the case of the slow-shock boundary shown in Figure 6, $B_0 \sim 11$ nT at 1934:00 UT. Between 1933:00 UT and 1935:59 UT, the calculated white line corresponded with the observed edge angle. However at 1936:23 UT, the backstreaming ion distributions spread over the calculated white line and began to merge with the cold ion distribution. This shows that between 1933:00 UT and 1935:59 UT the ions behaved mostly adiabatically. This adiabatic motion broke down around 1936:23 UT. The magnetic field intensity also began to decrease abruptly at this time. This result shows that the main dissipation region of the slow-mode shock observed on 13 February, 1994 was between 1936:23 UT and 1936:47 UT.

Figure 10 shows the pitch angle distribution of incident cold ions in the main dissipation region. The cold ions are pitch angle scattered on a shell (indicated by a white circle) centered on a parallel velocity (indicated by a white dashed line) between the upstream and the downstream plasma bulk velocity.

So far, we have shown the detailed variation of the ion distribution function using a slow-shock boundary observed on 13 February, 1994. We have also calculated several parameters of identified slow-shock boundaries, and

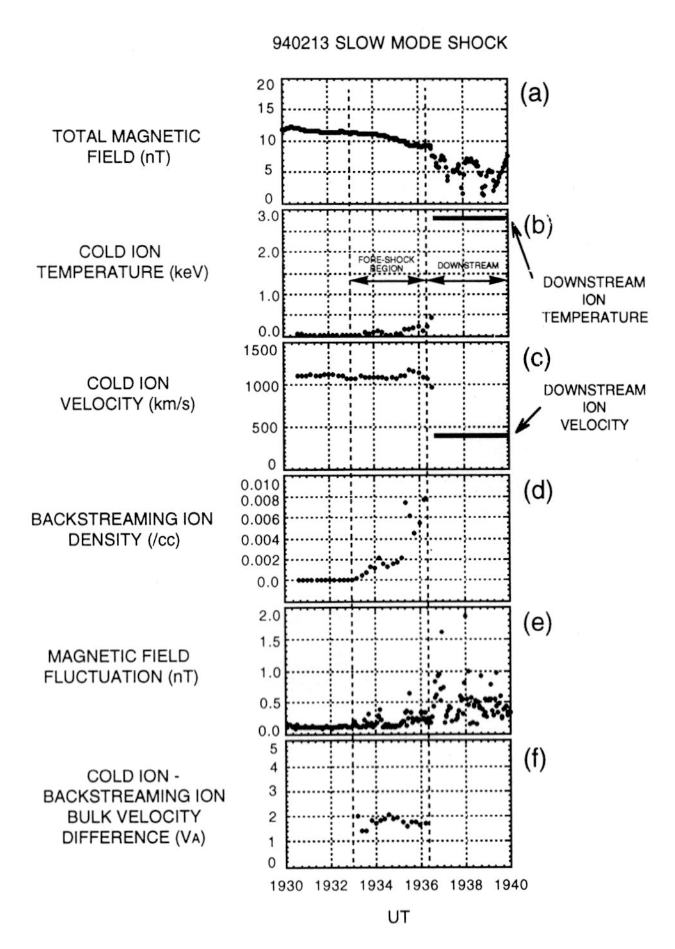

Figure 7. Total magnetic field (a), cold ion temperature (b), cold ion velocity (c), backstreaming ion density (d), magnetic field fluctuation (e), and bulk velocity difference between cold ions and backstreaming ions in units of local Alfvén velocity (f) around the same slow-mode shock in Figure 1.

investigated their characteristics statistically. We have calculated the density of the downstream hot ions, cold ions that flow from the upstream to the downstream region, backstreaming ions that successfully escaped from the plasmasheet to the upstream lobe region, and the ions that try to escape from the downstream plasmasheet to the upstream lobe region but magnetically mirror reflected back toward the plasmasheet, separately. The result is shown in Figure 11. Between 15% and 43% of the downstream ions try to escape from the downstream region. However, only 12% to 43% of these ions can successfully escape to the upstream region and the rest of these ions are magnetically mirror reflected back to the downstream region. As a result, the ratio of the backstreaming ion density to the downstream ion density is between 2% and 13%.

We can calculate the width of the slow-mode shocks using the shock velocity deduced from the mass conservation between the upstream and downstream regions, and the duration of the data. The duration of the main dissipation region is defined by the time interval when characteristic ion pitch angle distribution shown in Figure 6f is observed while the duration of the fore-shock region is defined by the time interval between Figure 6c and Figure 6e. Table 1 shows the calculated width of the slow-mode shocks. The width of the main dissipation region of the slow-mode shocks normalized by the upstream ion inertial length is between ~ 1.2 and ~ 7.2. While the total width of the slow-mode shocks including the width of the fore-shock region normalized by the upstream ion inertial length is ~ 20.

We have also calculated the ion velocity moments of cold ions and backstreaming ions in the fore-shock region

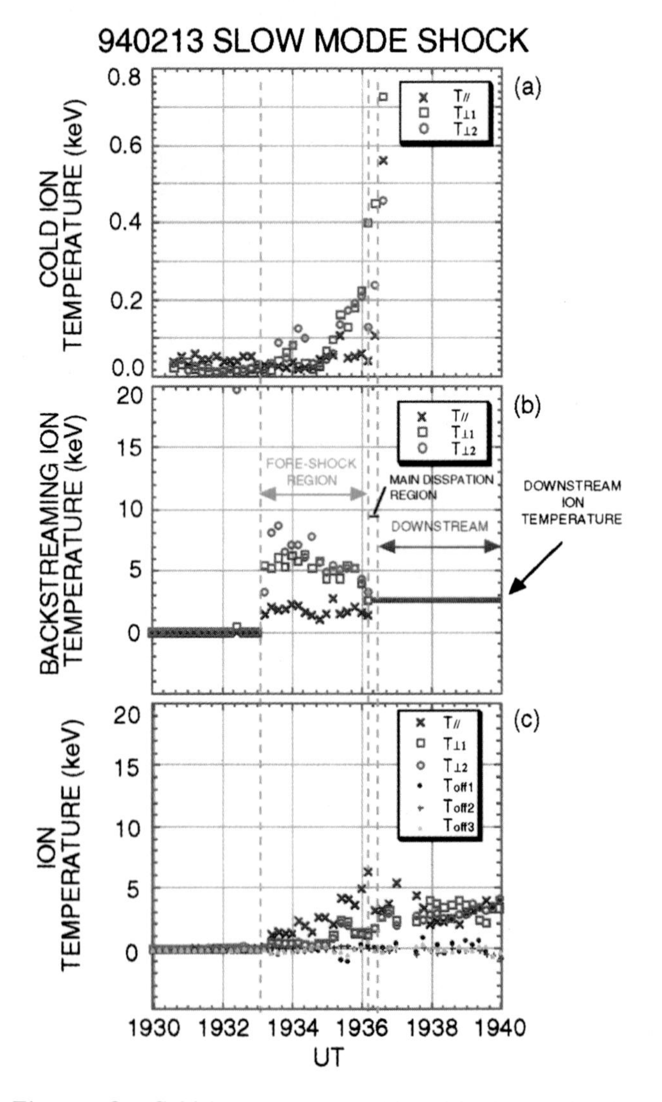

Figure 8. Cold ion temperature (a) and backstreaming ion temperature (b) in the fore-shock region. (c) Ion temperature in the upstream and downstream regions.

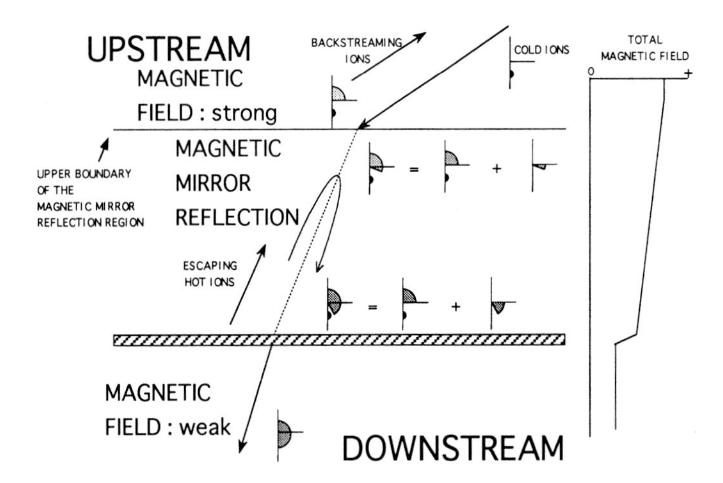

Figure 9. Schematic figure showing the variation of the ion distribution function in the foreshock region.

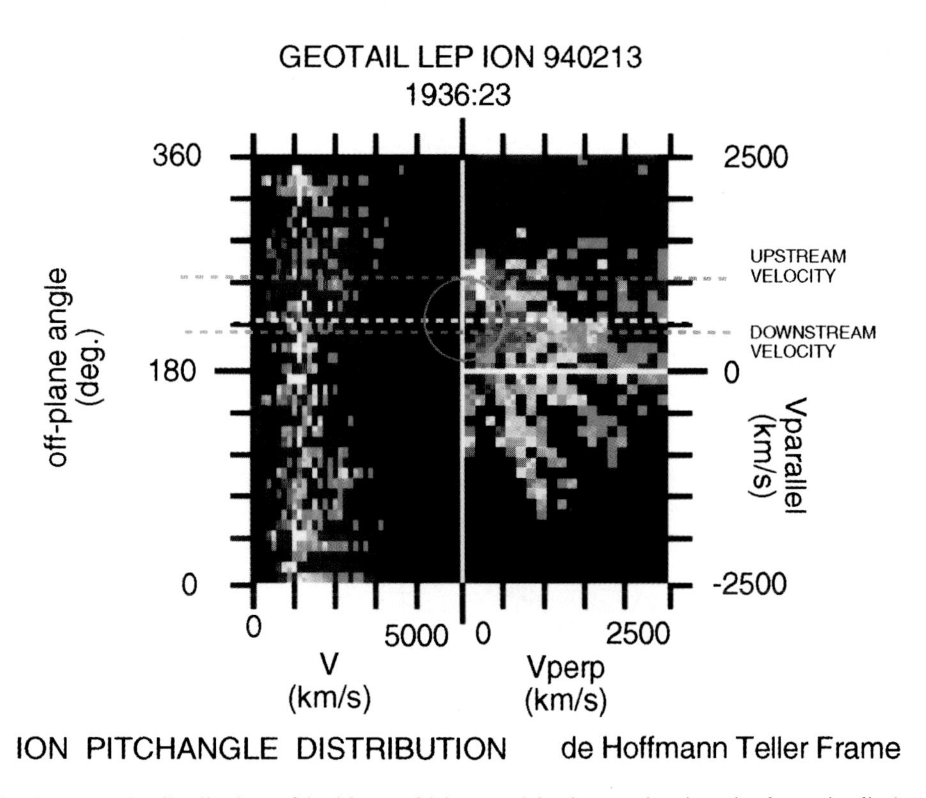

Figure 10. Pitch angle distribution of incident cold ions and backstreaming ions in the main dissipation region.

separately for the ten slow-shock boundaries. Table 2 shows the result. In the fore-shock region, cold ions are heated about 3% – 20% of the total ion heating throughout the slow-mode shocks.

Figure 12 shows the relation between the upstream shock normal-magnetic field angle θ_{Bn} and the backstreaming ion density normalized by the downstream ion density N_{back} / N_d. For most slow-mode shocks, the angle θ_{Bn} is near 90°, that indicates that most slow-mode shocks are near perpendicular shocks. When θ_{Bn} is small, N_{back} / N_d has large value. This result supports the idea that backstreaming ions are the ions escaping from the downstream toward the upstream region.

ION DENSITY

DATE		940213	940114	940112	940112	940212
TIME		1936	1540	1042	1541	1358
UPSTREAM COLD IONDENSITY (/cc)	Nc	0.028	0.019	0.0091	0.013	0.02
DOWNSTREAM ION DENSITY (/cc)	Nd	0.046	0.054	0.016	0.030	0.030
MIRROR ION DENSITY (/cc)	Nmirror	0.006	0.004	0.005	0.0053	0.004
BACKSTREAMING ION DENSITY (/cc)	Nback	0.002	0.001	0.002	0.0015	0.003
Nback/Nd		4.3%	2%	12.5%	5%	10%
(Nmirror+Nback)/Nd		15%	15%	43%	23%	23%
Nback/(Nmirror+Nback)		29%	12%	29%	22%	43%

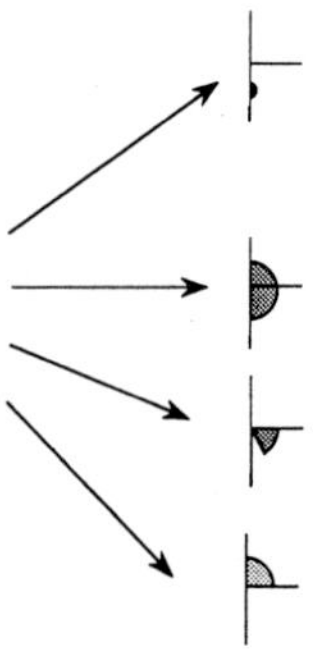

Figure 11. Ion density of incident cold ions (N_c), downstream hot ions (N_d), magnetic mirror reflected ions (N_{mirror}), and backstreaming ions (N_{back}). The ratio N_{back} / N_{mirror}, (N_{mirror}+N_{back}) / N_d, and N_{back} / (N_{mirror}+N_{back}) are also shown. The right-hand side schematic figures show the pitch angle distributions of each component.

3. DISCUSSION

We have identified slow-mode shocks at the mid-tail to distant-tail regions using three-dimensional magnetic field and plasma data obtained by the GEOTAIL satellite. We have checked the one-dimensional Rankine-Hugoniot relation between the upstream and downstream parameters in NIF. The detailed method of identifying slow-shock boundaries is given in *Saito et al.* [1995]. Recently *Seon et al.* [1995, 1996a] also reported the existence of slow-shock boundaries in the distant magnetotail using almost different data set obtained by the GEOTAIL satellite. *Seon et al.* [1996b] investigated the effect of the pressure anisotropy of the downstream plasma. Since the downstream plasma of the slow-shock boundary we observed on 13 February, 1994 was nearly isotropic as shown in Figure 8c, we did not consider the anisotropy of the plasma pressure in this particular case.

The determined shock normal direction is not always restricted to in the GSE X-Z plane. Especially in the case shown in Figure 2, the shock normal has a large GSE-Y component. Magnetotail may be twisted in such a case.

Backstreaming ions are observed for most of the slow-mode shocks that we have identified. These backstreaming ions have been predicted by several numerical simulations. *Swift* [1983] noticed the existence of the upstream escape of hot shocked plasma. According to *Winske et al.* [1985], the density of the backstreaming ions is roughly 50% of the incident ion density. The relative velocity of the backstreaming ions with respect to the incident ions along the magnetic field is $1.8V_A$. They concluded that most of the backstreaming ions are hot ions that have leaked from downstream. Our observational results are also consistent with the idea that the backstreaming ions are those that escape from the hot plasmasheet toward the upstream region. The backstreaming ion density ranged from 1% to 15% of the downstream ion density, which is comparable to a backstreaming ion density / incident cold ion density ratio of between 5% and 30%. This value of backstreaming ion density is a little smaller than the result of *Winske et al.* [1985]. However, *Omidi and Winske* [1992] used a different method of slow-shock formation and reported that the

Table 1. Width of the slow-mode shocks.

DATE	940213	940114	940112	940112	940212
TIME	1936	1540	1042	1541	1358
SHOCK VELOCITY (km/s)	143.6	158.8	714.7	97.8	157.1
MAIN DISSIPATION REGION					
DURATION (sec)	~ 20	~ 36	~ 24	~ 24	~ 24
WIDTH (km)	~ 2900	~ 5720	~ 17152	~ 1960	~ 3770
WIDTH (/upstream ion inertial length)	~ 2.1	~ 3.5	~ 7.2	~ 1.2	~ 2.4
FORE-SHOCK REGION					
DURATION (sec)	~ 180	~ 168	~ 168	~ 48	~ 192
WIDTH (km)	~ 25800	~ 26700	~ 120000	~ 4700	~ 30200
WIDTH (/upstream ion inertial length)	~ 18	~ 16	~ 19	~ 2.3	~ 19

Width of the main dissipation region of the slow-mode shocks and the width of the fore-shock region.

Table 2. Characteristic parameters of the fore-shock region of slow-mode shocks.

DATE	TIME	$\Delta T_{fore,cold}$ (eV)	$\Delta T_{total,cold}$ (eV)	$R_{\Delta T,fore}$ (%)	N_{down} ($/cm^3$)	N_{back} ($/cm^3$)	$R_{n,fore}$ (%)	θ_{Bn} (deg.)	β_{up}
940213	1936	300	2800	10	0.046	0.002	4.3	77.3	0.048
940113	1310	200	750	13	0.133	0.0008	0.6	83.1	0.069
930918	1035	100	2000	5	0.095	0.012	12.6	55.8	0.012
940114	1540	500	3200	16	0.054	0.001	1.9	83.7	0.012
940112	1042	800	4300	19	0.016	0.002	12.5	56.0	0.049
930918	1055	100	1000	10	0.158	0.008	5.1	67.2	0.025
940117	1535	180	3600	5	0.029	0.0006	2.1	84.7	0.003
940112	1541	350	2000	17	0.030	0.0015	5.0	80.7	0.010
940212	1358	100	4000	12	0.030	0.003	10.0	81.2	0.071
941208	1629	20	670	3	0.180	0.005	2.8	85.7	0.055

Velocity moments of cold ions and backstreaming ions in the fore-shock region are separately calculated for the ten slow-shock boundaries observed between 14 September, 1993 and 16 February, 1994. From left to right, date, time, fore-shock cold ion heating, total cold ion heating, fore-shock cold ion heating/total cold ion heating ratio, downstream ion density, backstreaming ion density, backstreaming ion density/downstream ion density ratio, the angle between the upstream magnetic field direction and the shock normal direction, and upstream plasma beta are shown.

backstreaming ion density varies from case to case, ranging from ~5% to ~10%, that is consistent with our result. The observed relative velocity of the backstreaming ions with respect to the incident cold ions ranged from $1.6V_A$ and $2.2V_A$. This result is also consistent with the result of numerical simulations [*Swift*, 1983; *Winske*, 1985; *Omidi and Winske*, 1992; *Fujimoto and Nakamura*, 1994; *Omidi*, 1995]. *Winske et al.*[1985] report that the characteristics of the backstreaming population are not strongly dependent on the upstream ion beta. Our result in the last column of Table 2 shows no clear relation between upstream beta and backstreaming ion density.

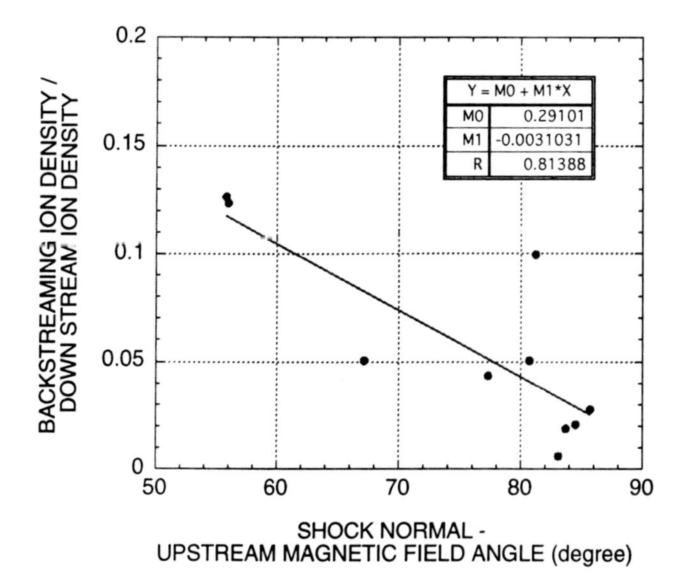

Figure 12. Scatter plot showing the relation between the upstream shock normal - magnetic field angle θ_{Bn} and backstreaming ion density normalized by the downstream ion density N_{back} / N_d.

Lobe cold ions are convected into the plasmasheet. These ions are heated at the slow-mode shock that exist between plasmasheet and lobe. Incident cold ions begin to be heated in the fore-shock region, that can be clearly distinguished from the main dissipation region of a slow-mode shock. Comparing the calculated and observed edge angle of the backstreaming ions, we have succeeded in identifying the fore-shock region and the main dissipation region of a slow-shock boundary. One of the candidates of the cold ion heating in the fore-shock region is the wave particle interaction with electromagnetic ion cyclotron wave generated by electromagnetic ion/ion instability. According to the theoretical work of *Gary et al.* [1986], ion/ion resonant mode has a non-negligible growth rate when counterstreaming ions have a relative velocity above the threshold of ~$2V_A$. Using the ISEE 3 data, *Tsurutani et al.* [1985] observed the right-hand resonant ion beam instability in the distant plasma sheet boundary layer. *Winske and Omidi* [1990, 1992] discussed the linear and non-linear properties of the electromagnetic ion/ion cyclotron instability, which has the lower threshold of the relative velocity of the counter-streaming ions than ion/ion resonant mode. The fore-shock region is characterized by the counter streaming ions consisting of backstreaming ions and cold ions. Our observation shows that the relative velocity between the incident cold ions and backstreaming ions is between $1.6V_A$ and $2.2V_A$. The coincidence of the occurrence of the counterstreaming ions and the magnetic field fluctuations shown in Figure 7 suggests that some electromagnetic waves are generated by counterstreaming ions. The variation of ion temperature shown in Figure 8 is also consistent with the cold ion heating by the interaction with cyclotron waves which are excited in the fore-shock region near the shock surface. The parallel temperature of the backstreaming ions is a little lower than the perpendicular temperature in the shock surface because the backstreaming ions are part of the downstream hot ions with velocity toward the upstream region along the magnetic field. Backstreaming ions are heated perpendicularly to the magnetic field in the region close to the shock surface. With the effect of the conservation of the magnetic moment, the perpendicular temperature becomes much higher than the parallel temperature in the far upstream region. On the contrary, incident cold ions have isotropic temperature in the far upstream region. They are heated perpendicularly to the magnetic field in the region near to the shock surface, and the perpendicular temperature becomes much higher than the parallel temperature at the shock surface.

We have estimated the width of the slow-shock boundaries. *Seon et al.* [1996a] also estimated the width of the slow-mode shocks. According to their results, the width is as wide as 70 times the upstream ion inertial length. Our results show that the width of the total slow-shock boundary including the fore-shock region is very wide. However, the width of only the main dissipation region of slow-shock boundary is several times the upstream inertial length.

Investigating the heating of the incident cold ions and the pitch angle distribution function of the backstreaming ions, we have succeeded in identifying several inside structures of slow mode shocks. Figure 13 shows a schematic view of the structure of a slow-mode shock as we envision. We have found the existence of a fore-shock region that is characterized by cold ion heating and counterstreaming ions, which are backstreaming ions and incident cold ions. In this fore-shock region, ions are heated to about 3% – 20% of the total ion heating throughout the slow-mode shock. In the fore-shock region, there is also a region where ions that try to escape from the downstream region are magnetically mirror reflected. In the main dissipation region whose width is several times the upstream ion inertial length, magnetic field abruptly decreased, and incident cold ions are pitch angle scattered and they are heated to the downstream ion temperature.

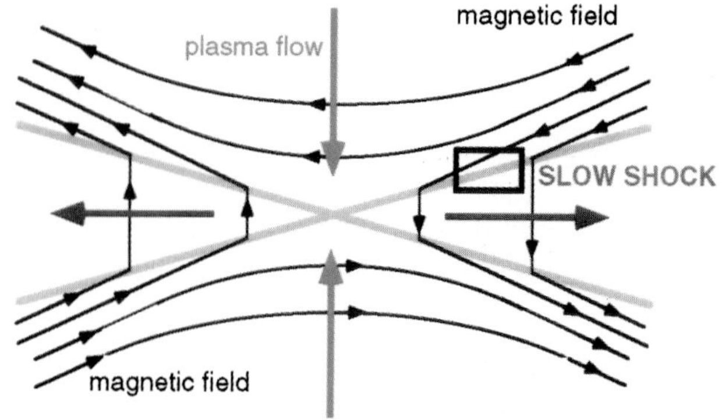

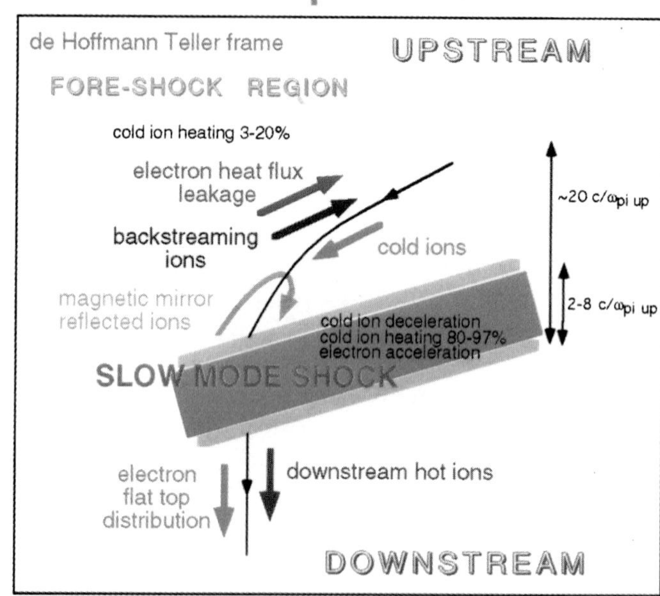

Figure 13. A schematic view of the structure of a slow-mode shock in the Earth's magnetotail.

Acknowledgments. We are greatly indebted to all members of GEOTAIL project team. We express special thanks to M. Hoshino of ISAS and M. Fujimoto of Tokyo Institute of Technology for valuable comments and discussions.

REFERENCES

Cattell, C. A., C. W. Carlson, W. Baumjohann, and H. Lühr, The MHD structure of the plasmasheet boundary, 1, Tangential momentum balance and consistency with slow mode shocks, *Geophys. Res. Lett., 19*, 2083, 1992.

Coroniti, F. V., Laminar wave-train structure of collisionless magnetic slow mode shocks, *Nucl. Fusion, 11*, 261, 1971.

Cowley, S. W. H., R. J. Hynds, J. G. Richardson, P. W. Daly, T. R. Sanderson, K. P. Wenzel, J. A. Slavin, and B. T. Tsurutani, Energetic ion regimes in the deep geomagnetic tail: ISEE-3, *Geophys. Res. Lett., 11*, 275, 1984.

Dungey, J. W., Interplanetary magnetic field and the auroral zones, *Phys. Rev. Lett., 6*, 47, 1961.

Feldman, W. C., et al., Evidence for slow-mode shocks in the deep geomagnetic tail, *Geophys. Res. Lett., 11*, 599, 1984a.

Feldman, W. C., D. N. Baker, S. J. Bame, J. Birn, E. W. Hones Jr., S. J. Schwaltz, and R. L. Tokar, Power dissipation at slow-mode shocks in the distant geomagnetic tail, *Geophys. Res. Lett., 11*, 1058, 1984b.

Feldman, W. C., D. N. Baker, S. J. Bame, J. Birn, J. T. Gosling, E. W. Hones Jr., and S. J. Schwartz, Slow-mode shocks: A semipermanent feature of the distant geomagnetic tail, *J. Geophys. Res., 90*, 233, 1985.

Feldman, W. C., R. L. Tokar, J. Birn, E. W. Hones Jr., S. J. Bame, and C. T. Russel, Structure of a slow mode shock observed in the plasma sheet boundary layer, *J. Geophys. Res., 92*, 83, 1987.

Fujimoto, M., and M. Nakamura, Acceleration of heavy ions in the magnetotail reconnection layer, *Geophys. Res. Lett., 15*, 2955, 1994.

Gary, S. P., C. D. Madland, D. Schriver, and D. Winske, Computer simulations of electromagnetic cool ion beam instabilities, *J. Geophys. Res., 91*, 4188, 1986.

Ho, C. M., B. T. Tsurutani, E. J. Smith, and W. C. Feldman, Properties of slow-mode shocks in the distant (> 200 R_E) geomagnetic tail, *J. Geophys. Res., 101*, 15,277, 1996.

Kokubun, S., T. Yamamoto, M. H. Acuna, K. Hayashi, K. Shiokawa, and H. Kawano, The GEOTAIL magnetic field experiment, *J. Geomag. Geoelectr., 46*, 7, 1994.

Mukai, T., S. Machida, Y. Saito, M. Hirahara, T. Terasawa, N. Kaya, T. Obara, M. Ejiri, and A. Nishida, The Low Energy Particle (LEP) experiment onboard the GEOTAIL satellite, *J. Geomagn. Geoelectr., 46*, 669, 1994b.

Omidi, N., Magnetic structure of slow shocks and the associated ion dissipation, *Adv. Space Res., 15*, 489, 1995.

Omidi, N., and D. Winske, Kinetic structure of slow shocks: Effects of the electromagnetic ion/ion cyclotron instability, *J. Geophys. Res., 97*, 14,801, 1992.

Richardson, I. G., and S. W. H. Cowley, Plasmoid-associated energetic ion bursts in the deep geomagnetic tail: properties of the boundary layer, *J. Geophys. Res., 90*, 12,133, 1985.

Saito, Y., T. Mukai, T. Terasawa, A. Nishida, S. Machida, M. Hirahara, K. Maezawa, S. Kokubun, T. Yamamoto, Slow-mode shocks in the magnetotail, *J. Geophys. Res., 100*, 23,567, 1995.

Saito, Y., T. Mukai, T. Terasawa, A. Nishida, S. Machida, S. Kokubun, T. Yamamoto, Foreshock structure of the slow-mode shocks in the Earth's magnetotail, *J. Geophys. Res., 101*, 13,267, 1996

Sato, T., Strong plasma acceleration by slow shocks resulting from magnetic reconnection, *J. Geophys. Res., 84*, 7177, 1979.

Scarf, F. L., F. V. Coroniti, C. F. Kennel, E. J. Smith, J. A. Slavin, B. T. Tsurutani, S. J. Bame, and W. C. Feldman, Plasma wave spectra near slow mode shocks in the distant magnetotail, *Geophys. Res. Lett., 11*, 1050, 1984.

Schwartz, S. J., M. F. Thomsen, W. C. Feldman, and F. T. Douglas, Electron dynamics and potential jump across slow mode shocks, *J. Geophys. Res., 92*, 3165, 1987.

Scholer, M., and D. Roth, A simulation study on reconnection and small-scale plasmoid formation, *J. Geophys. Res., 92*, 3223, 1987.

Seon, J., L. A. Frank, W. R. Paterson, J. D. Scudder, F. V. Coroniti, S. Kokubun, and T. Yamamoto, Observations of a slow-mode shock at the lobe-plasma sheet boundary in Earth's distant magnetotail, *Geophys. Res. Lett., 22*, 2981, 1995.

Seon, J., L. A. Frank, W. R. Paterson, J. D. Scudder, F. V. Coroniti, S. Kokubun, and T. Yamamoto, Observations of slow-mode shocks in Earth's distant magnetotail with the Geotail spacecraft, *J. Geophys. Res., 101*, 27383, 1996a.

Seon, J., L. A. Frank, W. R. Paterson, J. D. Scudder, F. V. Coroniti, S. Kokubun, and T. Yamamoto, Observations of ion and electron velocity distributions associated with slow-mode shocks in Earth's distant magnetotail, *J. Geophys. Res., 101*, 27399, 1996b.

Smith, E. J., J. A. Slavin, B. T. Tsurutani, W. C. Feldman, and S. J. Bame, Slow mode shocks in the Earth's magneto tail: ISEE-3, *Geophys. Res. Lett., 11*, 1054, 1984.

Swift, D. W., On the structure of the magnetic slow switch-off shock, *J. Geophys. Res., 88*, 5685, 1983.

Ugai, M., Computer studies on the fast reconnection mechanism in a sheared field geometry, *Phys. Fluids B, 5*, 3021, 1993.

Vasyliunas, V. M., Theoretical Models of Magnetic Field Line Merging, 1, *Rev. Geophys., 13*, 303, 1975.

Winske, D., and N. Omidi, Electromagnetic ion/ion cyclotron instability at slow shocks, *Geophys. Res. Lett., 17*, 2297, 1990.

Winske, D., and N. Omidi, Electromagnetic ion/ion cyclotron instability: Theory and simulations, *J. Geophys. Res., 97*, 14,779, 1992.

Winske, D., E. K. Stover, and S. P. Gary, The structure and evolution of slow-mode shocks, *Geophys. Res. Lett., 12*, 295, 1985.

T. Mukai, Y. Saito, Institute of Space and Astronautical Science, 3-1-1, Yoshinodai, Sagamihara, Kanagawa, 229-0022, Japan. (e-mail: mukai@stp.isas.ac.jp; saito@stp.isas.ac.jp)

T. Terasawa, Department of Earth and Planetary Physics, University of Tokyo, Hongo, Tokyo 113-0033, Japan. (e-mail: terasawa@grl.s.u-tokyo.ac.jp)

Dynamics and Kinetic Properties of Plasmoids and Flux Ropes: GEOTAIL Observations

T. Mukai and T. Yamamoto

The Institute of Space and Astronautical Science, Sagamihara, Kanagawa, Japan

S. Machida

Geophysical Institute, Kyoto University, Kyoto, Japan

GEOTAIL plasma and magnetic field observations have revealed several new features pertaining to the dynamics and kinetic properties of plasmoids/flux ropes in the distant magnetotail. In particular, three-dimensional ion distribution functions and their moments have provided a key clue for this study. In the lobe, as theoretically expected for the formation of a plasmoid by magnetic reconnection, cold plasmas are pushed away from the plasma sheet before the arrival of plasmoids, while after the plasmoid passage the convection is enhanced toward the normal direction to the plasma sheet. Cold ions of the lobe origin flow into the plasmoid along magnetic field lines, are heated and accelerated perpendicularly at the boundary, and finally merge with hot plasmas deeper inside the plasmoid. Deep inside the plasmoid (even near the neutral sheet), however, their remnants are still discernible and the ion distribution functions often show the existence of counterstreaming ion beams, which suggests that those field lines are connected through the northern and southern lobes to the interplanetary magnetic field (IMF). The counterstreaming ion beams are observed predominantly in the latter part of the plasmoid shortly after southward turning of the magnetic field, especially after the plasma bulk speed has increased stepwise and the magnitude of the By or Bz field has attained the peak value. The compression due to the draping of these field lines may lead to the core field enhancement. The ion distribution functions in a long-duration post-plasmoid plasma sheet indicate the signature of continual but highly variable reconnections on the earthward side of the spacecraft location. In addition to the counterstreaming ion beams, non-gyrotropic ion distribution functions are at times observed, which suggests that the plasma sheet becomes as thin as comparable to the ion gyroradius. The high variability would represent a consequence of multiple and bursty reconnections, and coalescence of a

New Perspectives on the Earth's Magnetotail
Geophysical Monograph 105

number of vortices may constitute the structure of the post-plasmoid plasma sheet. Statistical results on evolution of the kinetic and thermal properties of the core part during the tailward propagation show that plasmoids are accelerated tailward until X ~ -100 Re, whereas the tailward speed is reduced beyond this distance probably owing to the interaction with pre-existing plasmas. It is also found that plasmoids expand toward the flank side until the mid-tail region (~70 Re downtail), and they have a full width of the magnetotail beyond this distance down the tail. The thermal pressure decreases steeply during the expansion and thereafter gradually, while the magnetic pressure also decreases until X ~ -200 Re, though the slope is more gradual than that of the thermal pressure decrease. Finally, outlook for future studies is addressed.

1. INTRODUCTION

It is widely believed that a neutral line is formed at a substorm onset in the near-Earth region of the magnetotail, and the associated earthward and tailward fast flows are produced in the plasma sheet on the earthward and tailward sides of the reconnection region, respectively [*McPherron et al.*, 1973; *Nishida et al.*, 1981]. As the reconnection proceeds, the open field lines in the lobe begin to be reconnected so that the plasma on the tailward side of the reconnection region is no longer magnetically connected to the Earth and is free to propagate downtail, forming a plasmoid [*Hones*, 1979]. The expected signature of the plasmoid is a north-to-south bipolar variation of the magnetic field that contains hot, fast tailward (antisunward) flowing plasmas.

Much effort has been spent trying to reveal the structure and kinetic properties of plasmoids by satellite observations of the plasma and magnetic field in the equatorial region of the magnetotail. The ISEE-3 electron plasma, energetic particle and magnetic field data on the geotail orbit were extensively used to develop a description of plasmoids [e.g., *Hones, et al.*, 1984; *Scholer et al.*, 1984; *Richardson and Cowley*, 1985; *Baker et al.*, 1987; *Richardson et al.*, 1987; *Moldwin and Hughes*, 1992; *Moldwin and Hughes*, 1993]. These previous studies have revealed that plasmoids are generally consistent with the formation due to the near-Earth neutral line in association with a substorm onset. Figure 1 illustrate a schematic picture of the magnetotail configuration when magnetic reconnection is in progress, based on the ISEE-3 observations [*Richardson et al.*, 1987]. At the stage (a), field lines being reconnected are the closed ones in the plasma sheet, and the magnetic loop produced by the near-Earth neutral line is still embedded in the pre-existing plasma sheet. As the reconnection proceeds, open field lines in the lobe begin to be reconnected, and at the stage (b), plasmas on the reconnected field lines that constitute a region designated as post-plasmoid plasma sheet (PPPS) are not taken from the original plasma sheet population but originate from the lobe population which is accelerated by the reconnection process. At the stage (c) the region filled by the newly accelerated population is enhanced. However, this scenario is drawn based mostly on theoretical considerations, and was not necessarily evidenced by the ISEE-3 observations.

Though the Hones original model was two-dimensional, as also shown in Fig. 1, and attention first focused on the north-south (Bz) component of the magnetic field, a strong core field with a flux-rope-like structure is often seen [*Slavin et al.*, 1989; *Moldwin and Hughes*, 1991; *Machida et al.*, 1994, *Frank et al.*, 1995]. The core field appears in the By component of the magnetic field, which peaks when Bz reverses the sign from north to south, and its intensity is at times as large as, or even greater than the lobe field magnitude. Recent attempts have succeeded in modeling the magnetic field variations as force-free flux ropes for some selected data sets [*Lepping et al.*, 1995; *Slavin et al.*, 1995; *Kivelson and Khurana*, 1995].

The formation process(es) of the tailward flowing rope-like structure and the temporal/spatial evolution have attracted much attention, since its formation due to reconnection of the plasma-sheet field lines under the pre-existing magnetic By field cannot simply lead to disconnection from the Earth. *Hughes and Sibeck* [1987] proposed a model of the formation of flux ropes that could be disconnected from the Earth, propagating downtail. *Hesse and Bim* [1991] showed evolution of a self-consistent topology of the rope-like field lines with the MHD simulation. *Hesse et al.* [1996] suggested a scenario for enhancement of the core field due to loss of plasmas through field lines to the low-latitude boundary layer (LLBL), or the flank region in the magnetotail. *Moldwin and Hughes* [1992] also suggested the connection between the

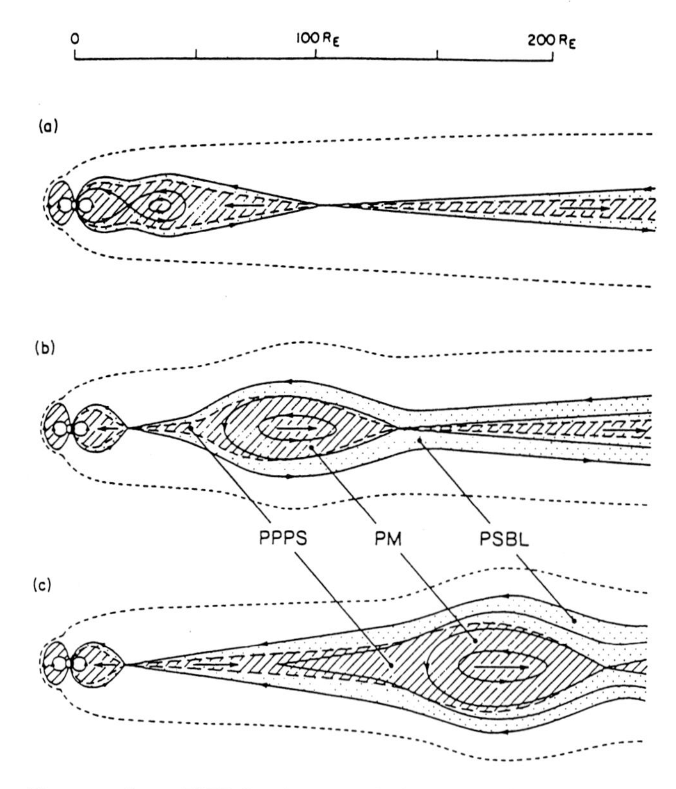

Figure 1. ISEE-3 picture of dynamic change in the magnetotail configuration during three stages of the substorm expansion [*Richardson et al.*, 1987]. The hatched area bounded by dashed lines is the plasma sheet which is filled with hot plasmas, while the dotted area is the plasma sheet boundary layer in which energetic particles are streaming along magnetic field lines.

LLBL and the rope-like plasmoid field lines, based on the ISEE-3 observations of the dependence of density and temperature in plasmoids on the distance downtail from the Earth. On the other hand, the magnetopause boundary region between the magnetosheath and the distant magnetotail seems to have only the mantle nature, though it can be classified into two distinct regions of the LLBL and the polar mantle at high latitudes in the near-Earth region. The mantle region in the distant tail is not restricted near the magnetopause, but extends well inside the magnetotail. Based on the ISEE-3 electron measurements, *Gosling et al.* [1985] suggested that the distant-tail lobe can be classified into two distinct regions of the mantle-like region and the empty lobe. Therefore the connection of field lines between the flux rope and the flank region is not necessarily evident observationally.

The actual appearance of the By field varies from case to case and cannot necessarily be expressed by a simple flux-rope model. In principle, force-free models of the flux ropes can be used in a high-β plasma while balance between the pressure gradient and the Lorenz force has to be considered for the high-β plasma [*Kivelson and Khurana*, 1995]. It would also be necessary to consider the inertia term in the developing stage of flux ropes when the velocity is varying with time. Actually a majority of the plasmoid/flux ropes are filled with hot plasmas of high-β. Whatever it is, the presence of the By field indicates the importance of three-dimensional reconnections in the formation and evolution of the structure. In this paper, hereafter, we use the term of plasmoid in a wider meaning that contains a rope-like field line topology as well.

As noted above, the ISEE-3 observations made significant progress in understanding of the structure and kinetic properties of plasmoids/flux ropes, but much has remained as speculations because of lack of the thermal ion measurement. GEOTAIL spacecraft has made it possible to advance beyond the ISEE-3 results, since it carries a comprehensive set of field and plasma instruments. As expected, GEOTAIL observations have confirmed that the ISEE-3 picture is generally valid, and in addition, have revealed several new results pertaining to the structure and kinetic properties of plasmoids/flux ropes and their boundary regions [*Machida et al.*, 1994; *Lui et al.*, 1994; *Frank et al.*, 1995; *Mukai et al.*, 1996]. In particular, thermal ion data have provided essential clues to study heating and acceleration of plasmas in association with the plasmoid formation due to the reconnection process [*Mukai et al.*, 1996; *Fujimoto et al.*, 1996; *Hoshino et al.*, 1998].

The purpose of this paper is to review new GEOTAIL results with the focus on the dynamics and kinetic properties of plasmoids/flux ropes. Section 2 provides a brief description of the GEOTAIL instrumentation on plasma and magnetic field measurements. The GEOTAIL orbit was designed to survey extensively the magnetotail at distances up to 210 Re. Plasmoids/flux ropes were encountered when the spacecraft was either in the lobe/mantle region or in the plasma sheet. In Section 3 we review GEOTAIL results on the interaction of tailward propagating plasmoids with the lobe/mantle cold plasmas. In Section 4, detailed distribution functions of electrons and ions in the pre-plasmoid plasma sheet, the main part of a plasmoid and the post-plasmoid plasma sheet are presented and discussed in terms of the structure and dynamics of the plasmoid. Section 5 presents statistical results on the three-dimensional evolution of the plasmoid during the tailward propagation, which has recently been obtained by *Ieda et al.* [1998]. Finally Section 6 gives conclusion and some remarks on unresolved problems.

2. GEOTAIL SPACECRAFT AND INSTRUMENTATION

The GEOTAIL spacecraft was launched on July 24, 1992 to study the structure and dynamics of magnetotail plasmas [*Nishida*, 1994]. As shown in Fig. 2, the GEOTAIL orbit is divided into distant-tail and near-Earth orbits. The change from the former to the latter took place in mid November 1994. GEOTAIL carries a comprehensive set of magnetic field and plasma instruments where measurements of plasma and energetic particles are made by two independent sets of instruments, LEP/CPI and HEP/EPIC, respectively [*Nishida*, 1994]. In this paper we use the plasma and magnetic field data obtained in the distant-tail orbit by the Low Energy Particle-Energy Analyzer (LEP-EA) and the Magnetic Field (MGF) experiments. LEP-EA consists of two nested sets of quadrispherical electrostatic analyzers to measure three-dimensional energy-per-charge distributions of electrons (with EA-e) and ions (with EA-i) simultaneously and separately. In the present observations, EA-i covers the energy range of 32 eV/q to 39 keV/q divided into 32 bins, in which 24 bins are equally spaced on logarithmic scale in energies higher than 630 eV/q and have width ±9.4 % of the center energy, while the lower-energy 8 bins are spaced linearly with width of ±40 eV/q (±20 eV/q for the lowest energy bin). The EA-e covers the energy range of 60 eV to 38 keV, which is also divided into 32 bins with logarithmic spacing for higher 24 bins (> 650 eV) and linear spacing for lower 8 bins in a similar way to EA-i. The full energy range is swept in a time which is 1/16 of a spin period (synchronized with the spacecraft spin motion). The field of view is fan-shaped with ~10°x145°, in which the longer dimension is perpendicular to the spin plane and divided into seven directions centered at elevation angles of 0°, ±22.5°, ±45° and ±67.5° with each width of 6-10° (wider for higher elevation angles). While the velocity moments are calculated onboard every spin period, the complete three-dimensional distributions can only be obtained in a period of four spins (about 12 seconds) owing to the telemetry constraints; the count data are accumulated during the four-spin period. A more detailed description of LEP instrumentation is given in *Mukai et al.* [1994a]. MGF provides vector magnetic field data with a time resolution of 1/16 second [*Kokubun et al.*, 1994]. The magnetic field data used in this paper are averaged over one spin period, or four spin periods when showing distribution functions with reference to the magnetic field direction. The coordinate system used in this paper is the Geocentric Solar Magnetospheric (GSM) coordinates for presentation of the magnetic field and the plasma (ion) bulk velocity. The spacecraft position is also expressed in GSM but modified with the solar wind aberration under the assumption of 400 km/s radial flow speed.

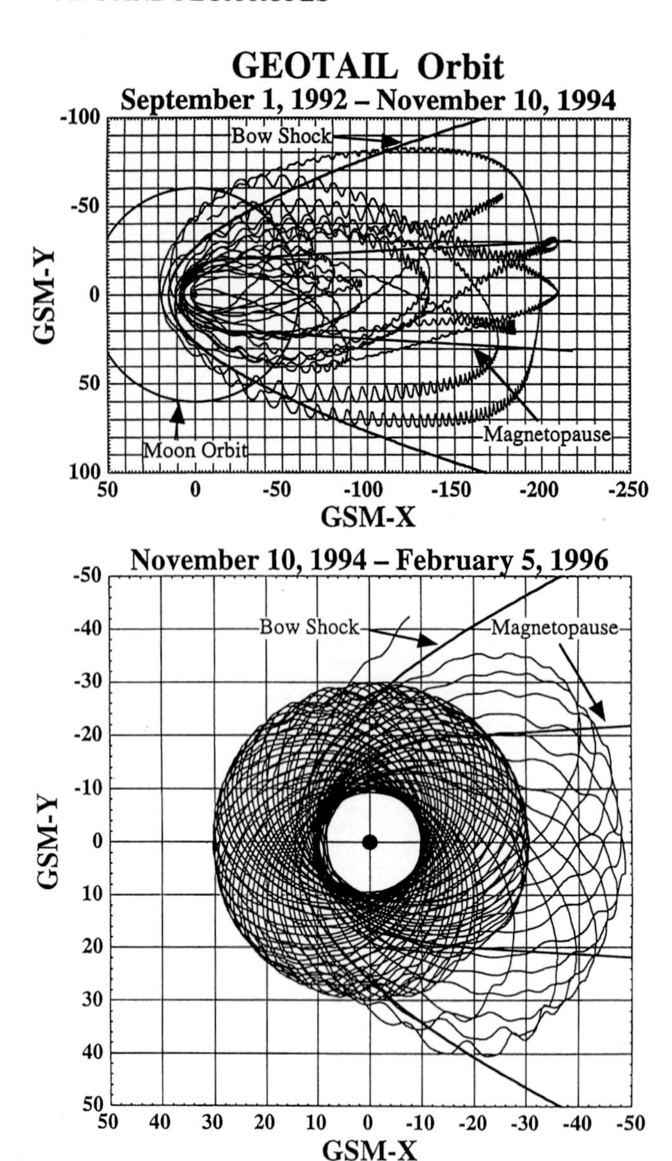

Figure 2. The GEOTAIL orbit in the modified GSM coordinates in which the solar wind aberration is taken under the assumption of 400 km/s radial flow speed. The upper and lower panels show orbits in the distant-tail and the near-Earth phases, respectively.

3. INTERACTION OF PLASMOIDS WITH THE SURROUNDING LOBE/MANTLE PLASMAS

Figure 3 shows an example of plasmoid observed in the distant tail at (-128, +21, +6) Re in the GSM coordinates on

October 5, 1993. The corresponding energy-time spectrograms of electrons and ions are shown in Fig. 4. The geomagnetic activity was weak (Kp= 2+), but the Pi-2 onset was detected at 2111 UT in ground magnetograms. The plasmoid is evident for a time interval of about 20 minutes starting from ~2120 UT (9 minutes after the Pi-2 onset), judging from a north-to-south bipolar signature of the magnetic field and the associated tailward flow of hot plasmas with 500-800 km/s. This example shows several important features of the interaction of a plasmoid with the surrounding lobe/mantle plasmas, though it also shows some unusual features, such as the magnetic By field variations. While it was generally negative, the magnetic By field turned

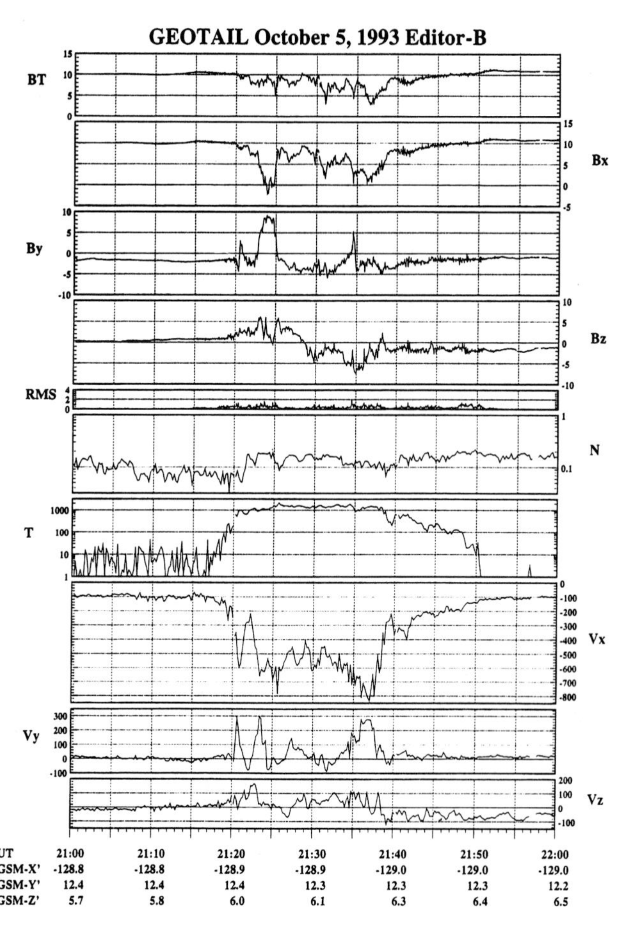

Figure 3. Magnetic field data and ion velocity moments during the time interval of 2100 to 2200 UT on October 5, 1993. From the top, time series of the total magnitude and three components of the magnetic field (nT), the root-mean squares of the magnetic field fluctuations (nT), density (cm^{-3}), temperature (eV), and three components of bulk velocity (km/s) are shown. Bottom three rows indicate the spacecraft location.

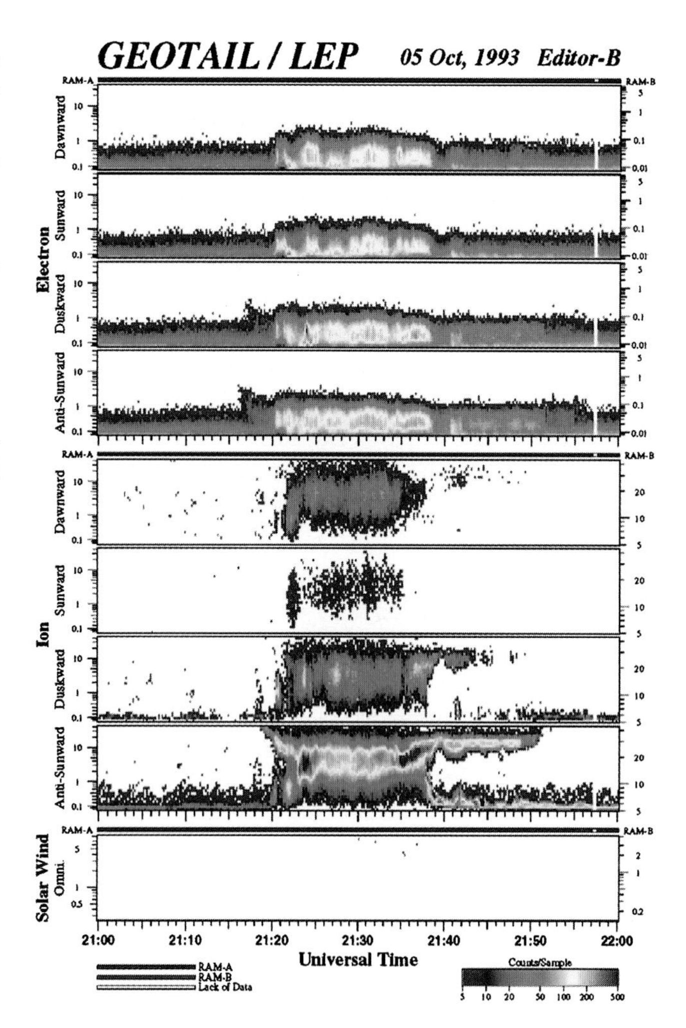

Figure 4. Electron and ion energy-time spectrograms in four azimuthal directions in the same time intervals as shown in Fig. 3. The ordinate scale is energy (keV/e), which is different between the right-hand and left-hand scales depending on the energy-scanning modes (RAM-A and RAM-B), but refer to the left-hand one in the present case. Data from the solar wind analyzer show almost nothing, indicating that the region is inside the magnetotail.

sporadically to strongly positive during a period of northward Bz, in which both the Bx and Bz values of magnetic field became small. Impulsive positive enhancements of the By field were also seen at the beginning and near the end of the plasmoid. Therefore, this structure cannot be expressed by a simple rope-like structure.

The region outside of the plasmoid is regarded as lobe in terms of the ion temperature. As already revealed by the ISEE-3 electron measurements [*Zwickl et al.*, 1984], the distant-tail lobe is not necessarily empty but the density becomes at times as high as 1 cm^{-3}, which constitute cold, dense plasmas. *Gosling et al.* [1985] reported that the distribution of cold, dense plasmas has dawn/dusk asymmetry such that the region is located preferentially in the dawn (dusk) sector in northern (southern) region under IMF By positive (and *vice versa* under IMF By negative). They suggested that the cold, dense plasma regime consists of field lines recently reconnected on the dayside, that is, the region may be the far-tail extension of the plasma mantle. The GEOTAIL initial measurements have also confirmed the presence of the cold, dense plasmas [*Yamamoto et al.*, 1994; *Mozer et al.*, 1994], and in addition, direct ion measurements have revealed new features in the dynamics of the distant-tail lobe/mantle region [*Siscoe et al.*, 1994; *Hirahara et al.*, 1996; *Seki et al.*, 1996]. In Figs. 3 and 4, it is evident that the lobe is not empty but filled with cold ion beams (and low-energy electrons) flowing tailward at a speed of ~100 km/s nearly along the magnetic field lines but with a small perpendicular component which represents the convection. The density in the lobe is comparable to that inside the plasmoid, so that the region cannot be discriminated by density, while the ion temperature in the lobe is by a few orders of magnitude lower than that inside the plasmoid.

Characteristic magnetic and plasma signatures in the lobe region are seen in association with passage of the plasmoid. One is the southward Bz value of the magnetic field and the associated negative Vz component of the bulk velocity behind the plasmoid. The Z component of $V_\perp$ (perpendicular component of the bulk velocity to the magnetic field) is also negative. Since the region is the northern lobe (Bx > 0), this signature is quite reasonably interpreted as the enhanced convection toward the plasma sheet as expected from the reconnection theory. The feature prior to arrival of the plasmoid is opposite; that is, northward excursion of the magnetic Bz field and the positive Vz component of the plasma bulk velocity. In addition, the total magnitude of the magnetic field began to increase from several minutes before the arrival of the main part of the plasmoid, which was somewhat earlier than the first appearance of field-aligned electron beams (mentioned later). This signature is interpreted as representing the field compression due to an effect of fast magnetosonic waves launched from the near-Earth reconnection region. It may also be related with a well-known signature of the traveling compression region (TCR) [e.g., *Slavin et al.*, 1984].

The most unexpected feature in the lobe is that the magnetic Bz component became gradually northward and the Vz component of the bulk velocity turned positive, or less negative from ~2106 UT, several minutes before the Pi-2 onset at 2111 UT. This feature indicates that the convection was weakened, or even reversed just before the substorm onset in the distant tail. The example shown here is not an exceptional case, but often seen in the distant tail-lobe. Figure 5 shows the average temporal profile of $V_{\perp Z}$ before and after the plasmoid, in which $V_{\perp Z}$ is the Z-component of the bulk velocity perpendicular to the magnetic field and its sign has been reversed for observations in the southern lobe. In addition to a clear enhancement signature after plasmoids as noted above, the convection is obviously weakened well before arrival of plasmoids. Though the standard deviation is large, it seems that the weakening of the convection takes place twice, from -60 to -35 min., and gradually from -20 min. The weakening of the convection from ~1 hour before arrival of plasmoids cannot simply be interpreted in terms of the near-Earth reconnection activity, but the latter weakening might be interpreted as representing the effect of fast waves that push the lobe plasma away from the plasma sheet. The similar but more pronounced signature is also reported in the GEOTAIL initial observation in the mid-tail region [*Mukai et al.*, 1994b].

As a signature of the plasmoid proper, field-aligned electron beams are first detected prior to the arrival of the plasmoid; see a blue horn starting from 2116 UT in the antisunward electron panel of Fig. 4. Then, energetic ions with velocity dispersion are observed starting from 2118 UT. The reverse-in-time/space ion signature is also evident behind the plasmoid. The velocity-dispersed electrons and ions are always unidirectional and tailward along the magnetic field lines, which suggests that the boundary layer is on open (IMF) field lines reconnected in a near-Earth region of the observation site. The outermost field line in which energetic electrons first appear corresponds to the separatrix layer of the reconnection region. The inverse velocity-versus-time regression for the ion dispersion ahead of the plasmoid is well fitted by a linear relationship (not shown here), and its extension toward infinite speed coincides roughly with the time of the first appearance of energetic electrons. It should be noted, however, that the regression coefficient does not simply give the source distance, since the velocity dispersion is due to the combined velocity-filter effect of the time-of-flight and the **E** x **B** drift through the boundary layer. This is the signature of the plasma sheet boundary layer that was well documented from the ISEE-3 energetic particle measurements [*Scholer et al.*, 1984; *Richardson and Cowley*, 1985], but the

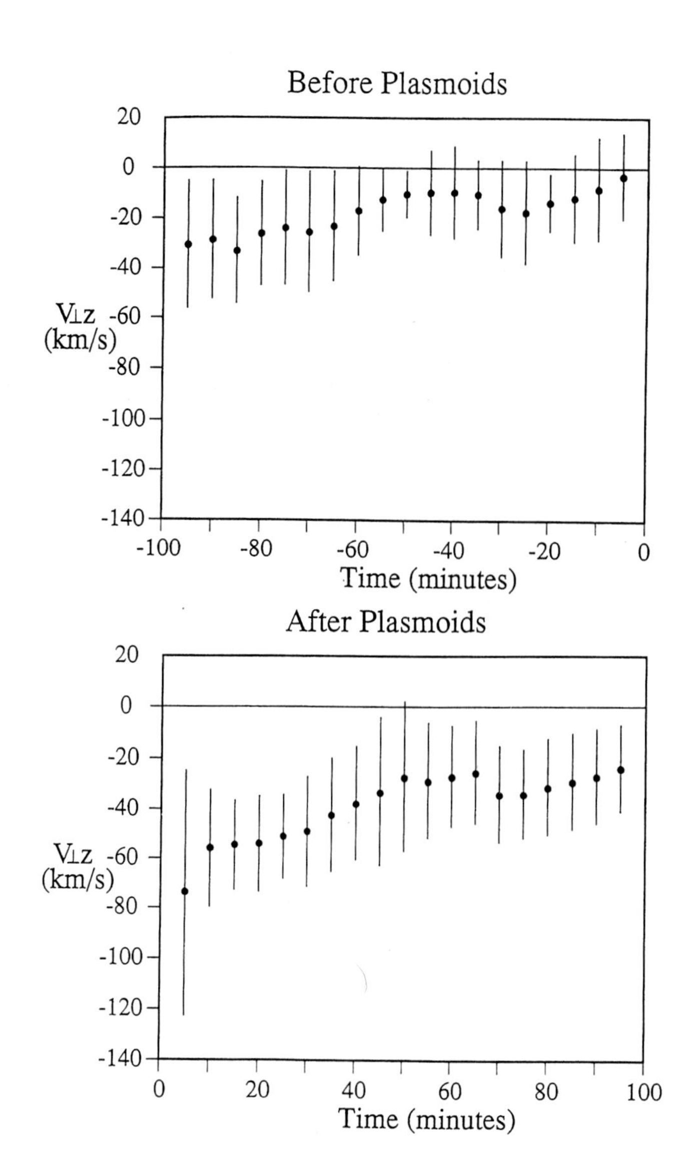

Figure 5. Average temporal profile of the Z-component of the lobe convection ($V_{\perp Z}$) in association with passage of plasmoids. The upper (lower) panel shows the profiles before arrival (after passage) of plasmoids. Here, the time is measured from the leading edge of plasmoids in the upper panel, while it is from the trailing edge in the lower panel. Taking the average value in each time bin of 5 min., the sign of $V_{\perp Z}$ in the southern lobe (Bx < 0) is changed, so that the negative sign indicates the convection toward the plasma sheet. The error bar indicate the standard deviation. The convection speed in the ordinary lobe (with no relation to passage of plasmoids) is 30-40 km/s; see the $V_{\perp Z}$ values apart from the plasmoid.

GEOTAIL observations clearly reveal that the velocity-dispersed signature extends down to thermal energies. A more important feature in the boundary layer is that cold ions of the lobe/mantle origin coexist with the energetic beams, as is clearly seen by two bright colors in low and high energies at ~2120 and ~2122 UT as well as from 2138 to ~2151 UT in the antisunward ion panel of Fig. 4.

Now we discuss plasma properties in the plasmoid in terms of its interaction with the lobe/mantle plasmas. *Richardson and Cowley* [1985] considered that the acceleration occurs only at the neutral line and the accelerated particles spread along magnetic field lines. In Fig. 1 taken from their paper the field lines do not bend at the boundary of the plasma sheet. *Sergeev et al.* [1987] noted, however, that the magnetic field lines are kinked at the edge of the plasma sheet. Although the direct comparison with the Rankine-Hugoniot relation was hampered by lack of thermal ion data in the ISEE-3 observations, *Feldman et al.* [1985] suggested that the plasma sheet boundary layer had the nature of the slow-mode shock. If it is the case, the slow-mode shock in the plasma sheet boundary layer should also be responsible for heating and acceleration of hot plasmas in plasmoids. Actually Fig. 4 shows clear evidence of heating and acceleration of ions in the plasma sheet boundary layer; see increasing energies of the low-energy ions at ~2120 and ~2122 UT as well as at ~2138-2139 UT in the antisunward ion panel. It is noteworthy that the energies of the accelerated ions seen at 2138-2139 UT in the post-plasmoid plasma sheet boundary layer are smoothly connected to those of cold ions in the lobe, which suggests that the lobe plasma is transported into the plasmoid, being accelerated and heated at the boundary. The acceleration of the cold ions are observed on the front side of the plasmoid as well, although they do not show smooth connection with the cold ions in the lobe. (Noted that a dip of the bulk velocity at ~2122 UT following the initial increase on the front side in Fig. 3 is due to the presence of significant cold ions coexisting with the hot component.) There are also other cases in which the smooth connection of the accelerated ions with the lobe ions is seen on the front side as well as in the post-plasmoid plasma sheet boundary layer. The similar feature of the cold ion acceleration is frequently observed in the plasma sheet boundary layer in general [*Hirahara et al.*, 1994]. Based on the GEOTAIL measurements of three-dimensional distribution functions of electrons and ions and the corresponding magnetic field, *Saito et al.* [1995] have found definite evidence that some of the plasma sheet-lobe boundary can be identified as the slow-mode shock, in which cold ions of the lobe origin are heated and accelerated. *Saito et al.* [1996] have also reported that significant ion heating takes place in the fore-shock region as well, in which cold ions of the lobe origin are streaming into the plasma sheet and hot ions are backstreaming from the plasma sheet, both along the

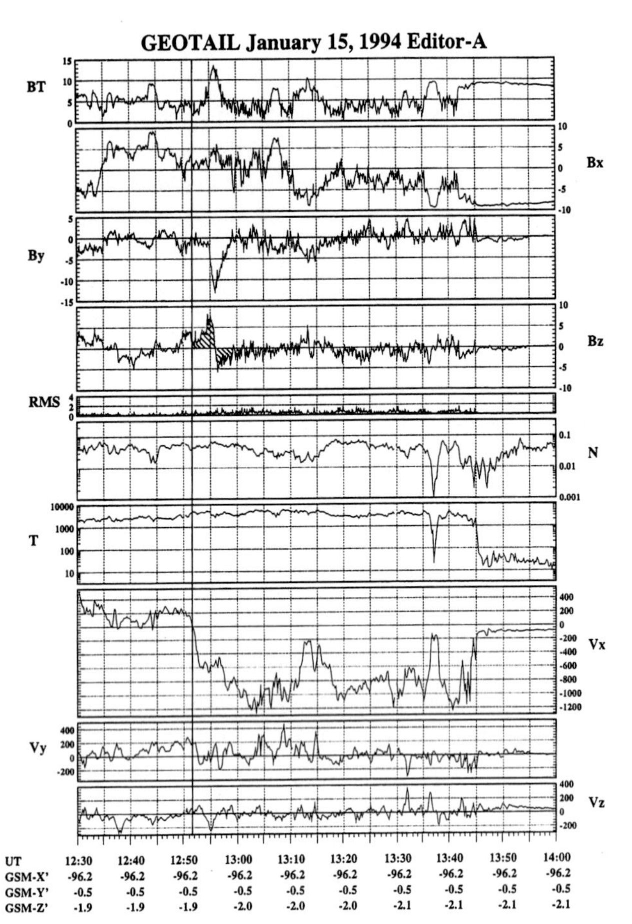

Figure 6. Same as Fig. 3, but the time period is from 1230 to 1400 UT on January 15, 1994.

magnetic field line in the De-Hoffman Teller frame. The coexistence of the two field-aligned beams may excite the ion-ion electromagnetic cyclotron instability, generating magnetic field fluctuations [*Tsurutani et al.*, 1985; *Kawano et al.*, 1994]. *Saito et al.* suggested that the magnetic turbulence is a cause for heating of cold ions streaming toward the slow shock and the plasma sheet.

The above signatures are commonly observed in the interaction of plasmoids with the surrounding lobe/mantle plasmas. However, one problem with the slow-shock acceleration is that the density in the lobe/mantle is sometimes comparable to, or even greater than that in plasmoids, and the condition for the slow-mode shock cannot be satisfied, but nonetheless the similar cold ion heating is observed near the plasma sheet-lobe boundary, as shown in Figs. 3 and 4 (and also the event on December 20, 1993, of *Mukai et al.* [1996]). The study on the heating and acceleration process(es) in the plasma sheet - lobe boundary is to be pursued in the future.

4. PLASMA DISTRIBUTION FUNCTIONS IN THE PRE-PLASMOID PLASMA SHEET, THE MAIN PART OF PLASMOID, AND THE POST-PLASMOID PLASMA SHEET

Figure 6 shows an example of plasmoid observed in the plasma sheet during the time period of 1230 to 1400 UT on January 15, 1994, when the spacecraft was located at (-96.2, -0.5, -1.9) Re. The corresponding energy-time spectrograms of electrons and ions are displayed in Fig. 7. A Pi-2 onset occurred at 1239 UT (at Kakioka) in association with this event. This example is selected here, since it reveals many common features observed in the plasma sheet regions of the pre-plasmoid, the main part of plasmoid, and the post-plasmoid. At the beginning (1230 UT), the spacecraft was in the southern ($Bx < 0$) plasma sheet, then crossed the neutral

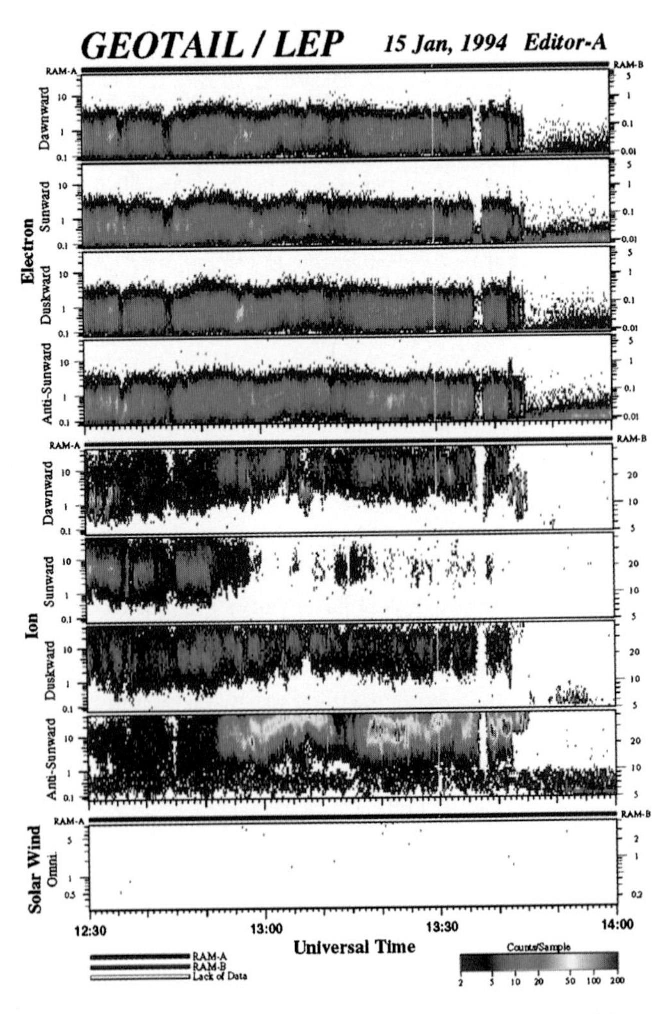

Figure 7. Same as Fig. 4, but the time period is from 1230 to 1400 UT on January 15, 1994.

sheet (Bx = 0) at 1235 UT, and remained in the northern plasma sheet with slow earthward flow until the encounter with the plasmoid. A hatched period of time is identified as the plasmoid. The magnetic Bx field magnitude remained less than a few nT throughout the plasmoid, so that the spacecraft may have been near the central portion (current sheet) of the plasmoid. A flow reversal from earthward to tailward occurred at 1251:30 UT, which we identify as the signature of the plasmoid encounter. The tailward flow speed increased rapidly up to 600 km/s, maintained this value for about four minutes, and then increased stepwise to 900 km. The tailward flow is accompanied by a clear bipolar signature in the magnetic Bz component. The magnitude of the magnetic By field increased toward dawnward (By < 0) during the period of 1255 to 1259 UT. Because of the strong By field, the total pressure during this period was significantly higher than in the ambient region. The ion temperature was 2-3 keV in the plasma sheet with slow earthward flow, and increased by a factor of 1.5-2 associated with the fast tailward flow (plasmoid). The ending of the plasmoid is ambiguous and somewhat arbitrary, followed by a long-duration post-plasmoid plasma sheet flowing tailward with high speeds. In what follows, we discuss plasma properties in three regions of the pre-plasmoid plasma sheet, the main part of the plasmoid and the post-plasmoid plasma sheet in terms of distribution functions of ions and electrons.

Pre-Plasmoid Plasma Sheet

The pre-plasmoid plasma sheet is characterized by slow earthward flow, which is interpreted as due to the distant neutral line located downtail from the spacecraft (at X = -96 Re). The flow speed is smaller than the tailward flow speed after the encounter with the plasmoid, but it is considerably higher than the usual earthward flow in this tail location. In particular, the earthward flow reached 500 km/s at 1230:30 UT and 300-400 km/s for a few minutes before the neutral sheet crossing at 1235 UT, which is also recognized in the ion energy-time spectrogram in the sunward sector. This enhancement of the earthward flow may indicate that the distant neutral line becomes active prior to the substorm onset. Similar features are at times observed in other substorm events as well. However, it is not clear whether it represents a growth phase signature in the distant tail, or whether the activation of the distant neutral line has some relation to triggering substorms. The ion energy-time spectrogram in the sunward sector shows that the activity of the earthward flow gradually decreased with intermittent quiescence. Judging from the Z-coordinate, the spacecraft is considered to be located close to the neutral sheet (Bx = 0), and actually crossed it at 1235 UT. Except for this neutral sheet crossing, however, the magnetic Bx field was rather stable and large, indicating that the region was the outer plasma sheet or the plasma sheet boundary layer. The plasma sheet might have become considerably thin before the arrival of the plasmoid.

Some selected distribution functions of electrons and ions in the pre-plasmoid plasma sheet are shown in Fig. 8. Here (and also in later panels for display of distribution functions), the distribution functions are displayed in a plane (hereafter, B-C plane), where the B axis is the magnetic field direction and the C axis is the direction of the convection (as defined by the perpendicular velocity to the magnetic field). The red arrow in each panel shows the bulk velocity obtained from the moment calculation. In the top panel (1231:08 - 1231:20 UT), the electron distribution function is somewhat elongated along the magnetic field line. The corresponding ion distribution function consists of a cold beam and energetic ions. Both components have nearly equal perpendicular velocity that is the convection velocity, while the parallel velocity of the cold component is small and directed toward the magnetic field direction (tailward), and that of the energetic component is large and earthward (toward the antiparallel direction to the magnetic field). The cold ion component, which is likely of the lobe origin, is heated and accelerated perpendicularly to the magnetic field; see the velocity shift of their distribution as high as ~300 km/s with similar velocity spread in the C-direction, which is significantly greater than the normal convection velocity and the thermal velocity of cold ions, both of order of 10 km/s, in the lobe. (It is also noted that the signature of the cold component is seen as an intense count rate at energies of 1-2 keV/e in the dawnward (!) sector in Fig. 7.) From these signatures, the region is regarded as the plasma sheet boundary layer. In other three panels, the electron and ion distribution functions consist essentially of a single, nearly isotropic component convecting earthward. In the third panel, the electron and ion distribution functions show simply lower density and lower temperature than those of the second and fourth panels. This is somewhat unexpected, since the region might be the plasma sheet boundary layer judging from enhancement of the total magnitude as well as the Bx component of the magnetic field and hence the coexistence of energetic ion beams and cold ions was expected to be observed somewhere in and near this region. Probably the distant neutral line became quiescent at this time (as also noted in the above paragraph).

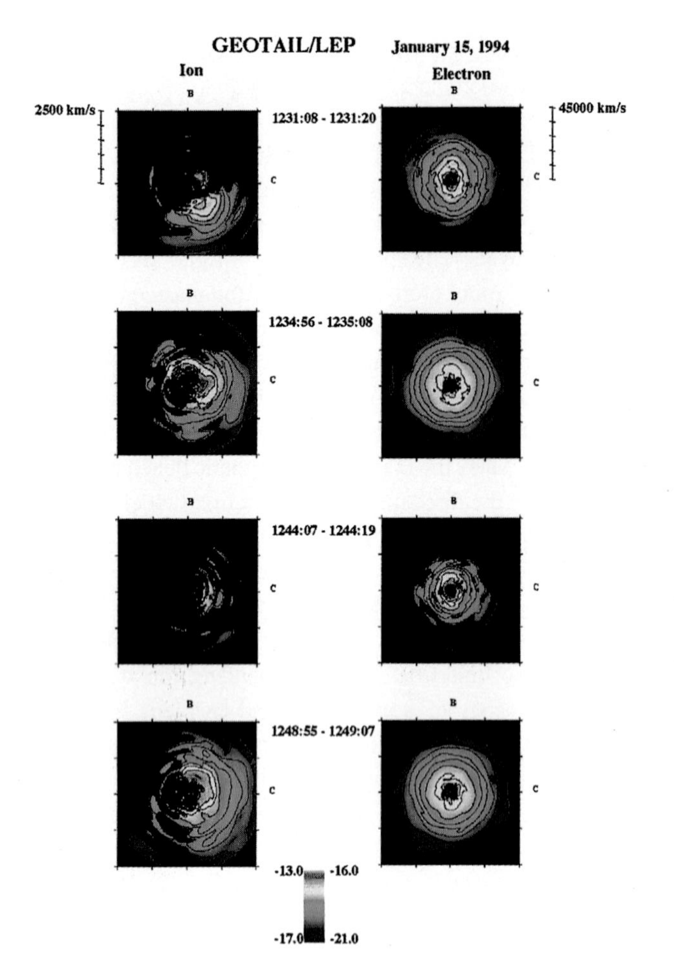

Figure 8. Examples of ion (left) and electron (right) distribution functions in the pre-plasmoid plasma sheet on January 15, 1994. Here, 'B' is the magnetic field direction, and 'C' is the direction of the convection velocity relative to the spacecraft frame. The phase space density is scaled by color coding on the logarithmic scale from -13.0 to -17.0 s^3/m^6 for ions (with assumption that all ion species are protons), and from -16.0 to -21.0 s^3/m^6 for electrons, as shown at the bottom. The contour lines are spaced every one tenth of the specified range on logarithmic scale.

Plasmoid

Distribution functions of electrons and ions in the main part of this plasmoid have been studied in detail by *Mukai et al.* [1996]. As shown in Fig. 9, the most remarkable feature is the existence of two beams counterstreaming along the magnetic field lines (relative to the convecting plasma frame), although the region is considered to be close to the neutral sheet; the magnetic Bx values at these times are very small (0.25 nT for E, and 2.15 nT for F). At time A, shortly after the northward enhancement of the Bz field (and just before start of the By enhancement), the ion distribution functions show the presence of two components (yellow blobs) counterstreaming along the magnetic field line relative to the convecting plasma frame (red arrow). The velocity difference between the two beams is ~1000 km/s along the magnetic field line, while the bulk flow speed as a whole is 700 km/s. At times B (~30 seconds before the peak of the By field) and C (around the peak of the By field), the ion distribution functions consist of a single hot component with anisotropy of $T_\perp > T_{//}$. At times D through F after the southward turning of the magnetic Bz field, the ion distribution functions consist essentially of the two components (red and yellow blobs) counterstreaming along the magnetic field lines relative to the convecting plasma frame. The velocity difference between the two beams along the magnetic field line is 910 km/s (500 km/s), 1210 km/s (640 km/s), and 1140 km/s (945 km/s) for times of D through F, respectively, while the values in the parentheses are the bulk flow velocities for reference. It is noted that the time E corresponds to the stepwise increase of the tailward velocity. In summary, the counterstreaming ion beams have the following properties:

a) At first, the "counterstreaming" is a fine structure inside the whole distribution function convecting tailward with high speeds. While both beams are convecting tailward in the inertia (spacecraft) frame, the direction of the counterstreaming in the plasma frame is dawnward/duskward and/or northward/southward because of the significant By and/or Bz components of the magnetic field.
b) These events are observed predominantly in the latter part of plasmoids after the southward turning of magnetic fields, but at times also in the leading part of the plasmoid. Their appearance is often associated with a stepwise increase of the tailward bulk velocity and intensification of the By/Bz field magnitude. However, they are observed even near the neutral sheet (Bx ~ 0).
c) Each beam has anisotropic distributions with the perpendicular temperature being several times higher than the parallel temperature.
d) The two beams are not necessarily similar in terms of the shape of the distribution functions and the phase space density, though there are some cases in which they are similar.
e) The velocity difference between the two components along the magnetic field line is as high as 1000-1500 km/s, which is much higher than the bulk flow speed of the whole

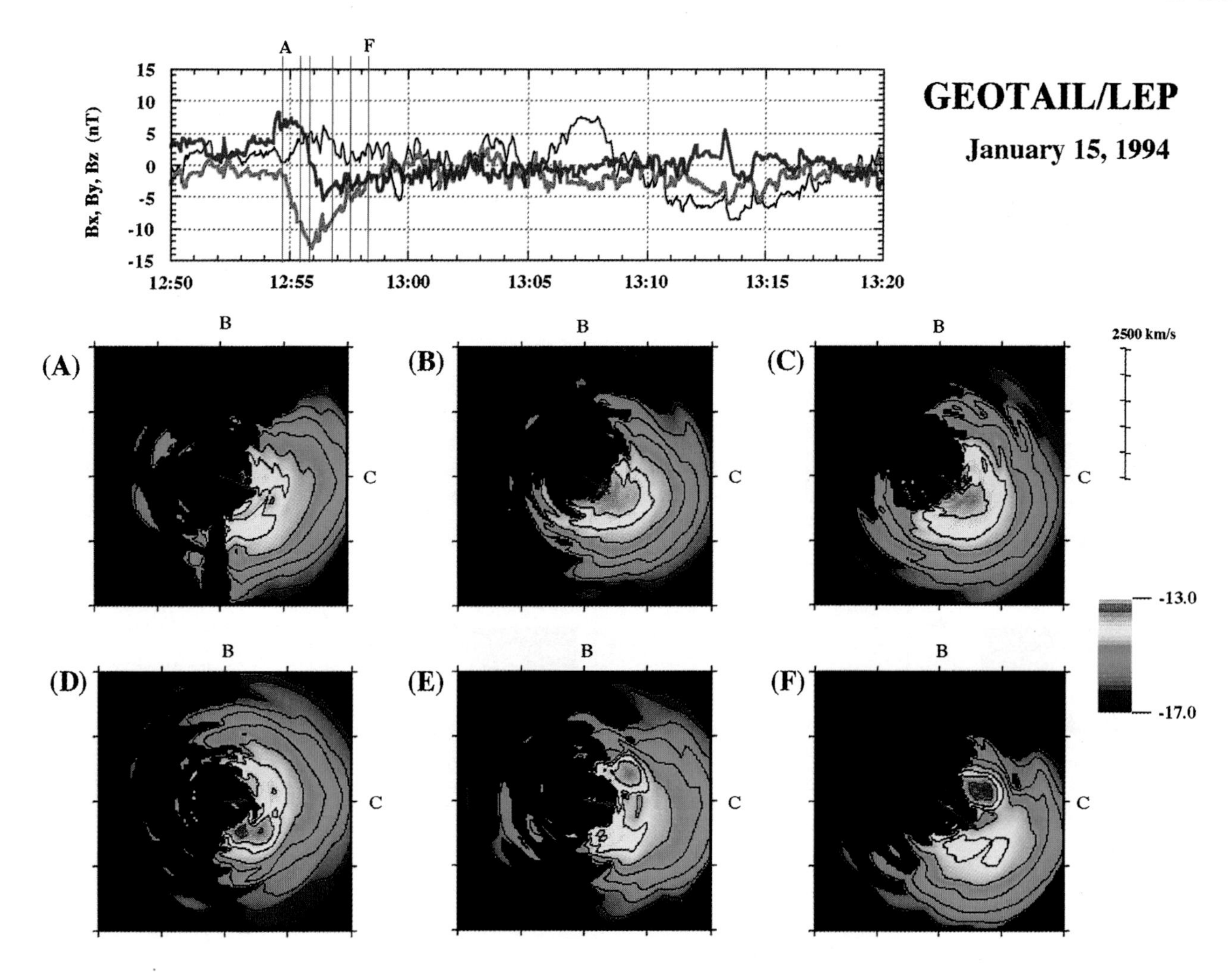

Figure 9. Three components (Bx: black; By: green; Bz: red) of the magnetic field data and six examples of the ion distribution functions during the time period of 1250 to 1320 UT on January 15, 1994. The times for the ion distribution functions are shown by vertical lines in the upper two panels from left (A) to right (F) as A (1254:42-54), B (1255:18-30), C (1255:42-54), D (1256:42-54), E (1257:18-30), and F (1258:06-18). The range of the phase space density is the same as that in Fig. 8.

distribution. It is also much higher than the local Alfven speed, and at times, exceeds twice the Alfven speed. In the latter case, the counterstreaming ion beams could be pitch-angle scattered by ion cyclotron waves generated by resonant electromagnetic ion-ion instability [*Tsurutani et al.*, 1985; *Kawano et al.*, 1994], forming a shell-like distribution [*Mukai et al.*, 1996].

The corresponding electron distribution functions are shown in Fig. 10, reproduced from *Mukai et al.* [1996]. In association with the appearance of the counterstreaming ion beams, the electron distribution functions consist of two components; low-energy electrons bi-streaming along the magnetic field lines superimposed on a higher-energy isotropic component. The bi-streaming electrons show a flat-top distribution, as shown by red curves on the right hand in each panel. The good correlation of the flat-top electron distribution with the counterstreaming ion beams suggests that both may be produced by the same process(es).

Mukai et al. [1996] suggested two possible models to explain the counterstreaming ion beams. One is a two-point reconnection model on the same field line. As noted previously, the distant neutral line may have been activated in the growth phase, and then a near-Earth neutral line may be activated when the distant neutral line remains still active.

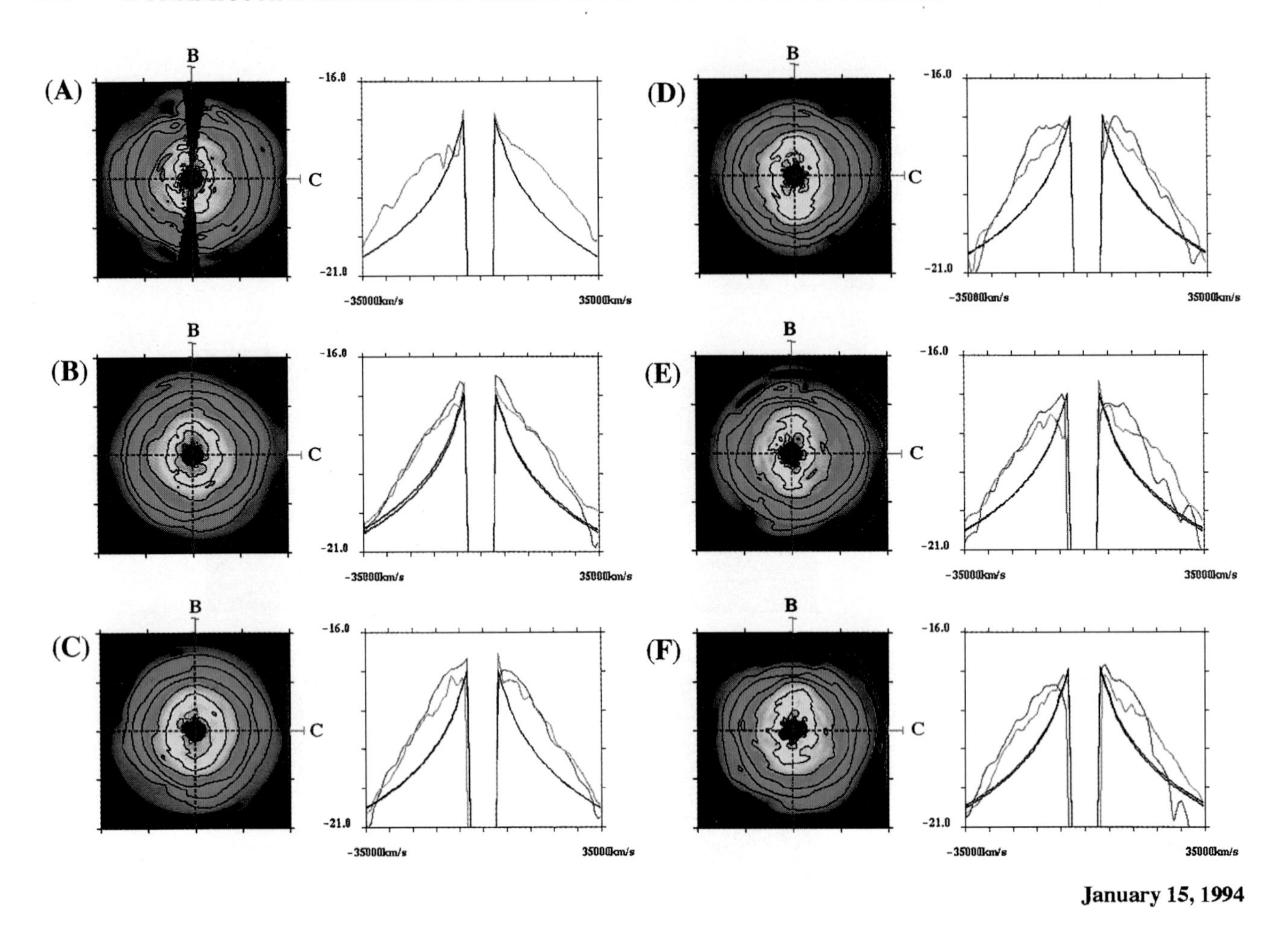

Figure 10. Electron distribution functions at the same times as shown in lower six panels of Fig. 9. In each panel, distribution functions in the parallel (red line along the B-axis) and perpendicular (green line along the C-axis) directions to the magnetic field are shown on the right hand side, in which a smooth curve shows the one-count level. Reproduced from Figure 8 of *Mukai et al.* [1996].

They suggested that the signature in Panel A of Fig. 9 might be interpreted in terms of this model. This model had also been proposed to interpret a stagnant plasmoid [*Nishida et al.*, 1986; *Kawano et al.*, 1996]. *Hoshino et al.* [1996] have also reported evidence for this model with the GEOTAIL observations. However, these cases are rare even though they may occur, and *Mukai et al.* suggested that the counterstreaming ion beams (observed after the southward turning of the magnetic Bz field) represent a signature of IMF field lines with both ends being connected, through the northern and southern lobes, to the interplanetary magnetic field. The two ion beams are most likely of the lobe origin, and they have not had time enough for thermalization. They might have entered the plasmoid from the northern and southern lobes, being heated and accelerated through slow-mode shocks at the boundaries. *Fujimoto et al.* [1997] have also reported a result of the hybrid-code computer simulation which supports this model. Each of the counterstreaming ion beams has strong anisotropy with the perpendicular velocity spread being a few times greater than the spread in the parallel direction. The heating predominantly in the perpendicular direction is consistent with the slow-shock signature [*Omidi*, 1995; *Saito et al.*, 1996]. The flat-top distribution of electrons is also consistent with the slow-shock signature [*Schwartz et al.*, 1987]. It is noted, however, that one of the counterstreaming beams is not necessarily similar to the other in terms of the shape and intensity. At times E and F, one component streaming toward the magnetic field direction is obviously colder and more intense than the other. The cold component might be fresh ions entering from the lobe, while the hotter component would have been scattered in the velocity space, or heated in the current sheet in addition to the heating through the slow-mode shock.

Let us consider the situation of near-Earth reconnection for plasmoid formation. Initially, the field lines in the plasma sheet begin to be reconnected and form flux ropes (or, closed loops). The structure would generally be the flux ropes, since non-negligible magnitude of the By field is usually present in the plasma sheet. In this period, however, both ends of the flux ropes are connected to the Earth, and hence the flux ropes can not simply move tailward freely. The flux ropes somewhere have to be reconnected to the lobe field line with the opposite direction so as to be released from the Earth [*Hughes and Sibeck*, 1987]. Then, the open field lines in the lobe begin to be reconnected with each other, and obviously these field lines do not form flux ropes, though the magnetic By field may exist in association with the convection in the dawn/dusk direction as an effect of IMF By. When the lobe field lines begin to be reconnected, the tailward flow speed increases from the initial phase in which the plasma-sheet field lines were reconnected, because it is generally determined by the Alfven velocity in the region of field lines to be reconnected. As time proceeds, the reconnection rate may further increase, and the reconnected field lines are draped around the trailing part of the plasmoid. The compression due to the draping may lead to the core field enhancement. The ion distribution functions in the core region (see Panels B and C in Fig. 9) show anisotropy of $T_{\perp} > T_{//}$ that may be regarded as the compression signature. It is also noted that the slope from north to south in the bipolar Bz variation is generally steep, which can also be explained by the compression of the draped field lines.

Post-Plasmoid Plasma Sheet

If the existence of counterstreaming ions is a signature of the IMF field lines, one may expect that similar distribution functions must be observed throughout the post-plasmoid plasma sheet. Actually, however, the observed distribution functions are not so simple but highly variable, as shown in Fig. 11. The high variability may be associated with (thin) current sheet dynamics, such as formation of tearing vortices and their coalescence [*Hoshino et al.*, 1994]. The shape of the distribution functions seems to have some correlation with magnetic Bx values and magnetic field fluctuations (fifth panel from the top in Fig. 6).

When the magnetic Bx value is high, the magnetic fluctuations become less and the distribution functions show the existence of two components, cold and hot ion beams counterstreaming along the magnetic field line relative to the convecting plasma frame; see the first and last three panels in Fig. 11(a). This is characteristic of the plasma sheet boundary layer, in which cold ions of the lobe origin are heated and accelerated perpendicularly to the magnetic field and coexist with hot ion beams. The signature is also seen in the ion energy-time spectrograms, in which cold ion beams are clearly seen in the dawnward sector simultaneously with the tailward flowing hot component. (This feature is similar to those observed in the pre-plasmoid plasma sheet boundary layer, but both the cold and hot components are flowing tailward in the post-plasmoid region.) Similar distribution functions are seen several times during the time period of this post-plasmoid plasma sheet as well as at the end (1342-1345 UT).

When the neutral sheet (Bx = 0) is approached, the magnetic fluctuations become enhanced, but nonetheless thermalization is not complete. Any distribution function is not represented by a simple, convecting Maxwellian, but most of the distribution functions show the existence of at least one cold beam, which can be regarded as fresh ions entering from the lobe. Some of panels in the middle row of Fig. 11 (a) clearly show the existence of the counterstreaming ion beams superimposed on a diffuse hot distribution.

Closer to the neutral sheet, the distribution functions show the existence of non-gyrotropic ion beams. Very pronounced examples of the non-gyrotropic distributions are shown in Fig. 12. Note that, if the distributions were gyrotropic in the convecting plasma frame, the two red blobs perpendicular to the magnetic field in the B-C plane (left-hand panel) must represent the cross-sectional points of a ring-shaped distribution in the E-C plane (right-hand panel), whereas each beam in the E-C plane is actually distributed in a limited range of gyro-phase angles around the magnetic field direction (perpendicular to the E-C plane) in the convecting plasma frame; that is, each beam is independent and non-gyrotropic. Temporal variations in Fig. 11 (b) seem to show that evolution of the non-gyrotropic distribution is closely related with the counterstreaming beams, which are observed in regions of a higher magnetic field intensity. Similar non-gyrotropic distributions are at times found in the post-plasmoid plasma sheet very close to the neutral sheet (Bx = 0) with very fast tailward flows [*Tu et al.*, 1997; *Hoshino et al.*, 1998]. This is a signature of a thin current sheet with high speed flows.

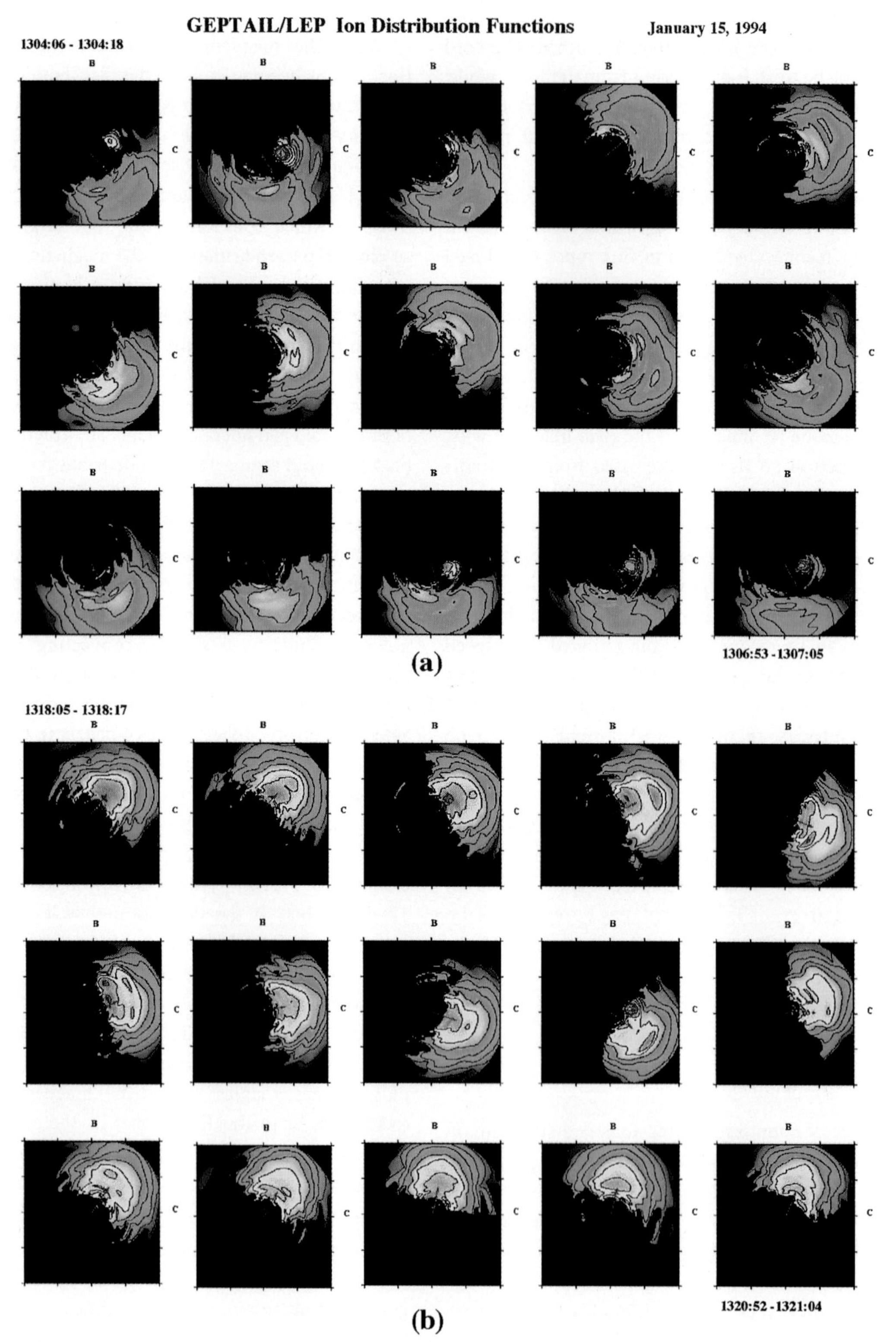

Figure 11. Temporal variation of ion distribution functions in the post-plasmoid plasma sheet during two selected time intervals (a) 1304:06 - 1307:05 UT and (b) 1318:05 - 1321:04 UT. In each time interval, the time shifts from the upperleft to the right-hand, then to the leftmost in the second row, and finally to the lower-right corner. Each panel represents a four-spin (~12 sec) averaged distribution measured consecutively.

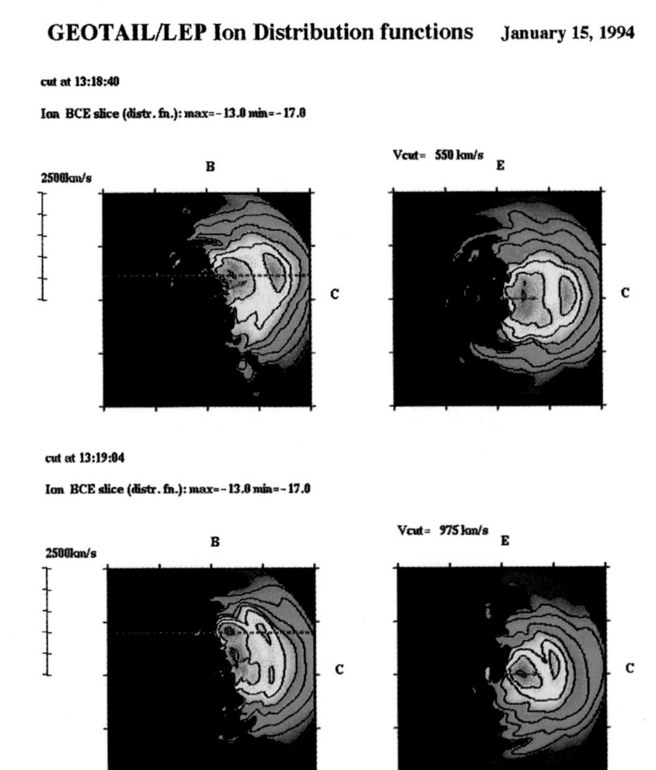

Figure 12. Examples of non-gyrotropic ion distribution functions in the post-plasmoid plasma sheet. The time for the upper panel corresponds to that for the fourth panel from the left in the top panel of Fig. 11 (b), while the lower panel to the leftmost one of the middle row. In each panel, the left-hand distribution is the same as shown in Fig. 11, and the right-hand one is the distribution in the E-C plane at the parallel velocity indicated by red line in the left-hand panel, where the E-C plane is perpendicular to the B-C plane, and the direction of the E-axis is calculated by $\mathbf{E} = -\mathbf{V} \times \mathbf{B}$.

One problem in the post-plasmoid plasma sheet is why it is observed for such a long time as ~45 minutes in spite that the signature of a thin current sheet is at times observed and the tail flapping might occur. While some signatures of the existence of cold ions are commonly observed in most times, the magnetic field and the ion distribution functions are highly variable. The region varies from the neutral sheet to the boundary layer and *vice versa*, but the spacecraft stays unexpectedly long in the plasma sheet, which suggests that the plasma sheet may be thick. The thickness of the plasma sheet might be variable depending on the reconnection rate in a region on the earthward side of the spacecraft. When the reconnection rate is enhanced (decreased), the angle of the separatrix becomes larger (smaller) and hence the plasma sheet becomes thicker (thinner). For example, see the data during a period of ~1307 to 1321 UT: The spacecraft at the beginning of this period was in the (northern) plasma sheet boundary layer, and then crossed the neutral sheet from north to south, in which the non-gyrotropic ion distribution were seen (though not shown here for this time period). The plasma sheet might be very thin during this period, and then the plasma flow became as slow as 200 km/s (tailward) with northward Bz. The energy-time spectrograms (and the distribution function, though not shown here) do not indicate any reconnection signature in this quasi-stagnant plasma region, and reconnections might be quenched. Then, the magnetic Bz field show a north-to-south bipolar variation, followed by fast flows. However, the quasi-stagnant bipolar signature might not represent the presence of closed loops or flux ropes, but rather simply show that the stagnant plasma was being pushed with compression by the following fast flows. The compression might also enhance the associated By field as seen in this period. The following fast flow was obviously caused by the increasing reconnection rate, followed by the thin current sheet as noted in the last paragraph. The nearly complete quenching of reconnections may be rare in the post-plasmoid plasma sheet, but in general the magnetic Bz field shows wavy variations with more or less north-to-south signatures which, however, cannot be regarded as plasmoids.

In summary the above features of the long-duration post-plasmoid plasma sheet indicate the prolonged presence of enhanced reconnections on the earthward side of the spacecraft. However, the reconnection region may not be a single point, but variable, bursty and multiple (but not necessarily simultaneous) reconnections are more likely to occur. Figure 13 shows schematic of a possible structure of a plasmoid and the post-plasmoid plasma sheet.

5. EVOLUTION OF PLASMOIDS DURING THE TAILWARD PROPAGATION

It is well established that plasmoids are detected in the distant tail at any Y coordinate in association with substorm onset (but of course, with time delay depending on distances down the tail) [*Moldwin and Hughes*, 1993; *Nagai et al.*, 1994], which suggests that plasmoids are widely spread in the dawn-dusk direction. Since the substorm onset occurs in a very localized longitude around midnight, and hence the region of the associated near-Earth neutral line must be very narrow in the Y direction, it is of interest to know three-dimensional evolution of plasmoids during the tailward propagation. *Ieda et al.* [1998] have made extensive statistical analysis of the kinetic and thermal properties of plasmoids based on GEOTAIL data. It should first be noted that their selection criterion of plasmoids is rigorous, and can exclude a wavy structure by taking account of the pressure enhancement such that the total pressure (sum of the ion pressure and the magnetic pressure) in the plasmoid region is more than 10 % higher than the ambient pressure. (The justification of this criterion is discussed in *Ieda et al.* [1998].) Because of this criterion, the post-plasmoid plasma sheet is not regarded as the plasmoid region in their analysis, and some different results from previous ones have been obtained. A brief summary of their results is given in this section.

Figure 14 shows that plasmoids are being accelerated toward tailward until X ~ -100 Re, while the tailward speed is reduced beyond this distance down the tail. This trend disagrees with previous ISEE-3 results [e.g., *Moldwin and Hughes*, 1992], probably due to the difference in the selection criterion of plasmoids. The acceleration of plasmoids during the tailward propagation is considered to be due to the field line draping and dynamic pressure of faster plasmas behind the plasmoids, but the field tension force will gradually decrease with distances away from the reconnected location, and the interaction of plasmoids with pre-existing plasmas may exceed the acceleration force beyond somewhere down the tail. Figure 14 shows that the acceleration is most effective until ~100 Re down the tail, and thereafter the deceleration seems to occur. The deceleration is greater in the flank region, which may be interpreted as due to the interaction with cold, dense plasmas. As noted in Section 3, the region of the mantle-like plasma appears to increase in the distant tail, and finally it will probably merge into the solar wind. Plasmoids will also

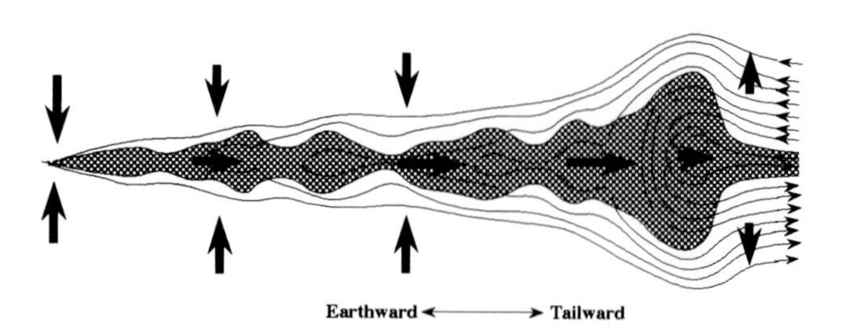

Figure 13. Schematic structure of a plasmoid and the post-plasmoid plasma sheet. The shaded area depicts the plasma sheet, in which the largest bulge on the right hand is the main part of the plasmoid propagating tailward. Thin arrow lines are examples of the magnetic field lines, while thick arrows show the plasma flows (not necessarily scaled).

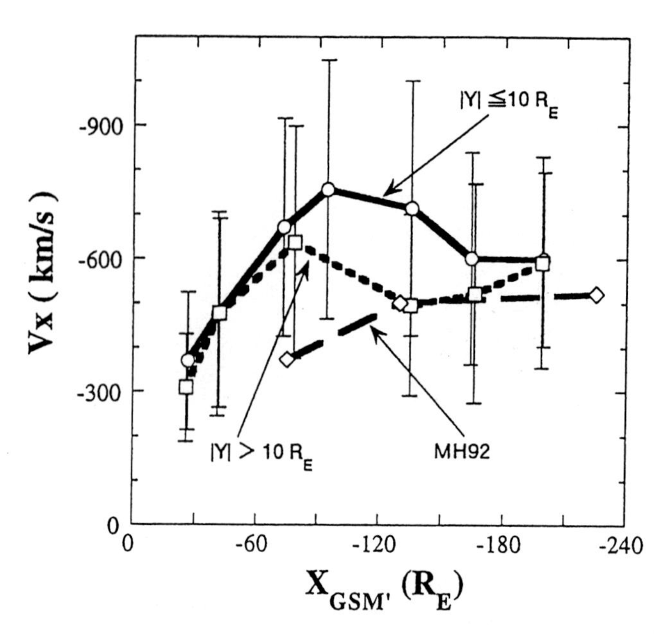

Figure 14. Dependence of tailward velocities (Vx) of plasmoids on the downtail distance (X_{GSM}). The solid line represents the averaged values in the central region ($|Y_{GSM}| \leq 10$ Re), while the dotted line corresponds to that in the flank region ($|Y_{GSM}| \geq 10$ Re). Bars show the standard deviations. The result of *Moldwin and Hughes* [1992] is also shown by a broken line for comparison. Reproduced from *Ieda et al.* [1998].

merge into the solar wind finally. Because of the difference of the tailward velocity between the central portion of the magnetotail and the flank region, plasmoids may be deformed in the distant tail.

Figure 15 shows an interesting result that plasmoids expand toward the flank side, duskward on the dusk side and dawnward on the dawn side, until X ~ -50 Re. Consequently, plasmoids would have a full width of the magnetotail and could be detected almost independent of the Y coordinate beyond the mid-tail region (~70 Re downtail). Figure 16 also shows expansion of plasmoids in the X direction during the tailward propagation. The average duration of plasmoids is ~1.2 minutes in the near-Earth region, and increases to ~1.7 minutes in the middle to distant tail, which leads to estimation of the scale length as ~4 Re in the near-Earth region and ~10 Re beyond the mid-tail region. The estimated scale length is reasonable if the shape of plasmoids is roughly circular, but contrast with previous results in which the scale length has a few tens of Re, and the shape is highly elongated along the X direction. The difference may also be due to the difference of the selection criterion.

Figure 17 shows that thermal pressure as well as magnetic pressure decrease with increasing distances down the tail. According to *Ieda* [private communication, 1997], the steep decrease of thermal pressure in the near-Earth region is mainly due to the density decrease during the expansion (see Figs. 15 and 16). On the other hand, *Moldwin and Hughes* [1992] reported that the density increases and the (electron) temperature decreases with increasing distances down the tail, and the resulting pressure are nearly constant beyond the mid-tail. *Ieda et al.* [1997] also reported the increasing density beyond the mid-tail region, which occurs mainly in the flank region but not so in the central portion of the magnetotail, while the ion temperature decreases regardless of the Y coordinate. The decrease of ion temperature is much steeper than that inferred from the tendency of the ISEE-3 electron results, and hence the thermal pressure continues to decrease gradually in the mid to distant tail. The magnetic pressure also decreases with increasing distances down the tail, but much more gradually than the slope of the thermal pressure. Consequently, plasma beta becomes close to unity in the distant tail.

Finally Fig. 18 summarizes schematically the plasmoid evolution in the X-Y plane, in which the inferred location of a near-Earth neutral line is also shown by a cross at X ~ -21 Re [*Ieda et al.*, 1997].

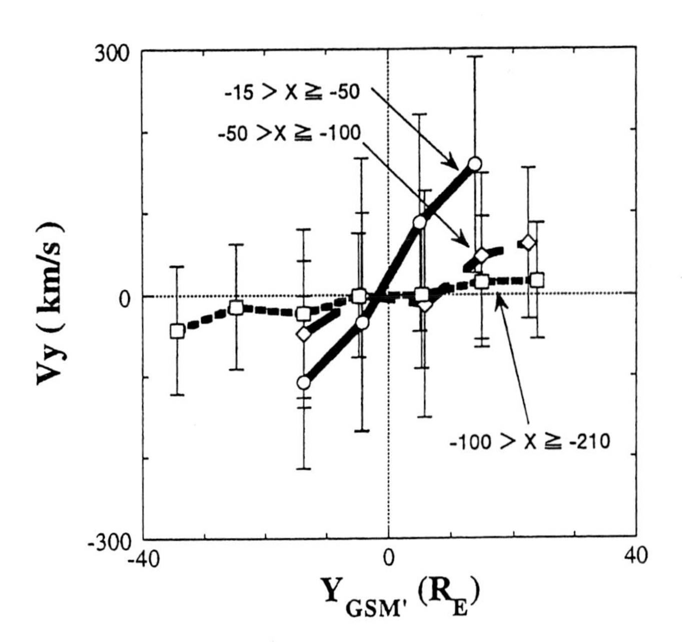

Figure 15. Average velocities of plasmoids in the Y_{GSM} direction (Vy) in 10-Re bins of the Y_{GSM} coordinate. The data sets were also divided into three distance ranges (X_{GSM}) down the tail. Reproduced from *Ieda et al.* [1998].

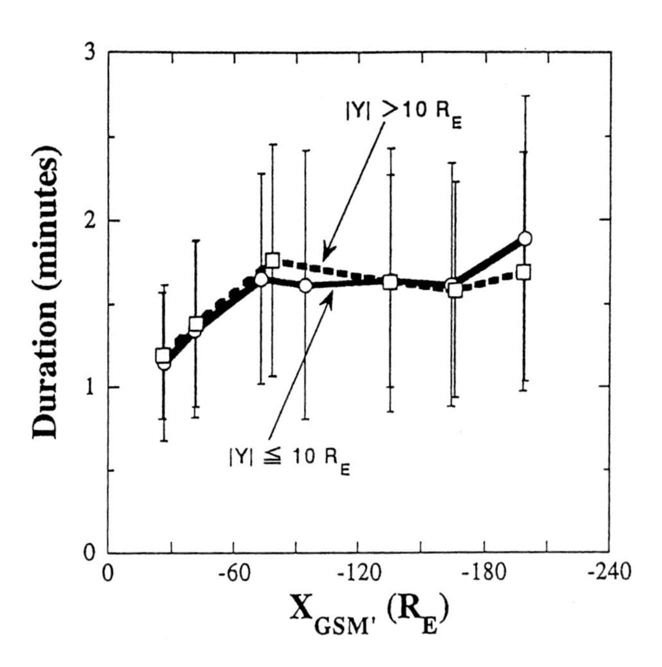

Figure 16. Duration of plasmoid events as a function of the X_{GSM} coordinate in the same format as that of Fig. 14. Reproduced from *Ieda et al.* [1998].

6. CONCLUDING REMARKS

We have reviewed dynamics and kinetic properties of plasmoids and their evolution during the tailward propagation, which have been newly obtained by the GEOTAIL plasma observations in the distant magnetotail. First of all, it is noted that most of the new results have been obtained by thermal ion measurements with high sensitivity and high time resolution. The observations have revealed clear signatures of the interaction of tailward propagating plasmoids with the surrounding lobe/mantle plasmas, which can be reasonably interpreted in terms of the formation of plasmoids due to the near-Earth magnetic reconnection. The most remarkable are the ion distribution functions in plasmoids and the post-plasmoid plasma sheet. They are not expressed by a simple convecting Maxwellian, but show the presence of beams which represent the memory of their source and route(s). Cold ions of the lobe origin flow into the plasmoid along magnetic field lines, are accelerated and heated at the boundary, and finally merge with hot components deeper inside the plasmoid. Deep inside plasmoids, however, their remnant(s) are still discernible and the ion distribution functions often show the existence of counterstreaming ion beams, which suggests that both ends of the field lines in this region are connected to IMF through the northern and southern lobes. The

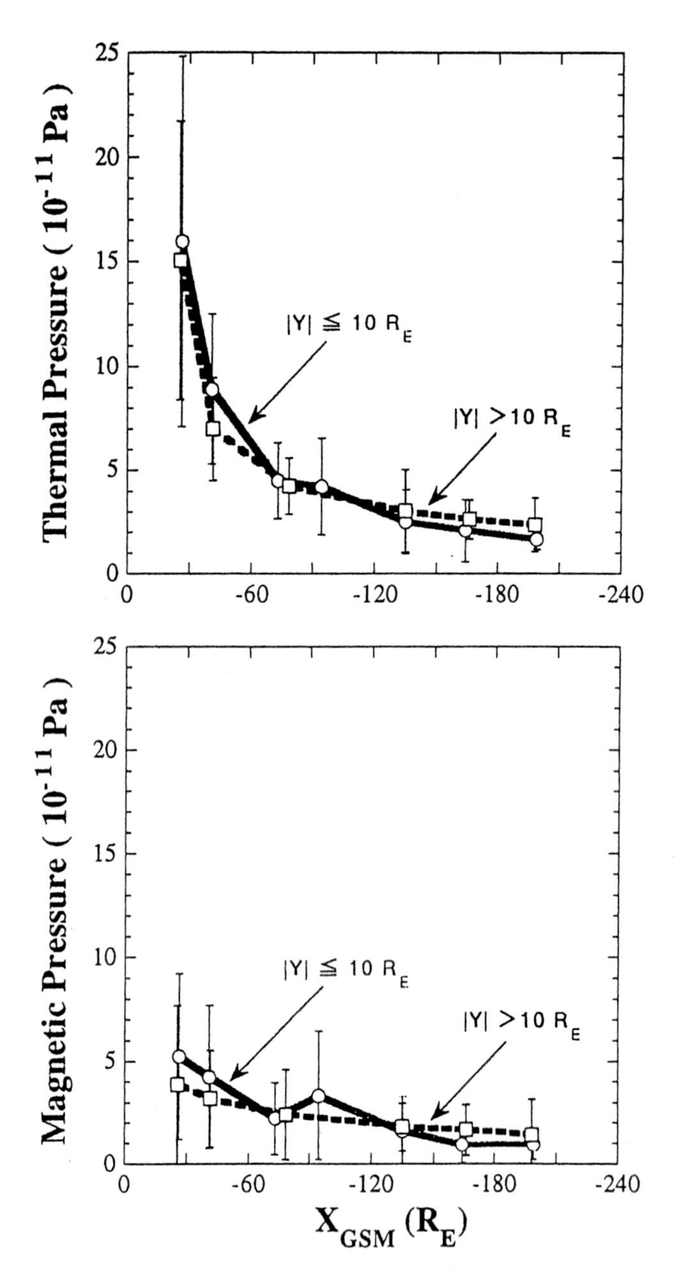

Figure 17. Thermal and magnetic pressures as a function of the X_{GSM} coordinate in the same format as that of Fig. 14.

counterstreaming ion beams are observed predominantly in the latter part of the plasmoid shortly after southward turning of the magnetic field, especially after the plasma bulk speed has increased stepwise and the magnitude of the By or Bz field has attained the peak value. That is to say, the field lines are connected to the lobe with a short distance (or, shorter travel time compared to the thermalization time) even in the regions in which the field lines have hitherto been considered as closed loops and/or flux ropes.

Based on the above signatures, we have suggested that the intensification of the By field, that is called the core field, could be due to compression caused by the field line draping behind the core region. On the other hand, based on the 2-1/2 dimensional MHD simulation, *Hesse et al.* [1996] suggested that the core field enhancement could result from a plasma pressure reduction caused by magnetic connection between the core region and the colder plasma in the flank region near the magnetopause. The pressure loss could result in the collapse of the plasmoid core and the shrinking cross section, and hence the field enhancement. The statistical results by *Ieda et al.* [1997] show the decrease of thermal pressure during the tailward propagation (see Fig. 17), but it can be interpreted as expansion rather than shrinking of the plasmoid region. On the other hand, their statistical results show also that the expansion in the dawn-dusk direction is completed at downtail distances of 50-100 Re (see Figs. 15 and 18). Then, the field lines in the core region might be connected to those in the magnetopause boundary region, which is filled with cold, dense plasmas. However, neither the signature of the field line connection to the LLBL region nor the shrinking of the plasmoid size has, up to now, been identified in the core region of flux ropes observed in the central region of the magnetotail. Even though it occurs in the flank region, the signature might have been lost by thermalization due to the long travel time of plasmas along long field lines to the center of the magnetotail, or the whole structure may consist of a number of rope-like substructures, each of which has a limited

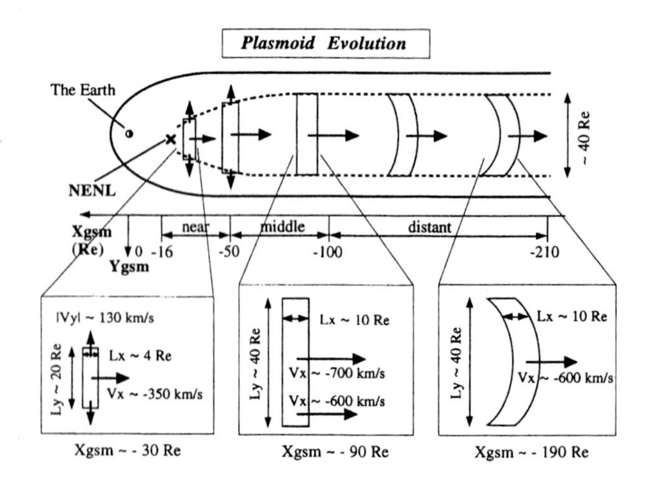

Figure 18. Schematic illustration of plasmoid evolution in the X_{GSM}-Y_{GSM} plane. The inferred location of a near-Earth neutral line is shown by a cross at X_{GSM} ~ -21 Re. Typical dimensions and shapes of plasmoids are displayed in the lower panel. Reproduced from *Ieda et al.* [1998].

extent in the dawn-dusk direction. The statistical study on the evolution of plasmoids has also shown that the shape of plasmoid might be deformed in the dawn-dusk direction. The plasma distribution functions of plasmoids in the flank region and in the adjacent magnetopause boundary region should be studied in the future to examine validity of the scenario by *Hesse et al.*

The presence of a long-duration post-plasmoid plasma sheet shows the continual reconnections on the earthward side of the spacecraft location, though the reconnection rate might be highly variable. In addition to the counterstreaming ion beams, non-gyrotropic ion distribution functions are at times observed in the post-plasmoid plasma sheet, which suggests that the plasma sheet becomes as thin as comparable to the ion gyroradius. High variability in the structure and kinetic properties of the post-plasmoid plasma sheet may be attributed to the flapping motion of the magnetotail, and in addition, to the variable reconnection rate. Earthward flows in the near-Earth region have a bursty nature, each of which is limited to a narrow region [*Angelopoulos et al.*, 1996]. The variable reconnection rate as seen in the post-plasmoid plasma sheet may represent a consequence of multiple and bursty reconnections, and coalescence of a number of vortices may constitute the structure of the post-plasmoid plasma sheet.

As noted above, we have suggested that the field lines are connected to the lobe (IMF field lines) with short distances even in the region which have previously been considered as closed loops and/or flux ropes with the significant By field, because of the presence of cold ion beams in the distribution functions. However, of course, there exists a region in which the ion distribution functions consist of a single hot component, and the field lines may constitute closed loops and/or flux ropes. The structure of this region has not been discussed in the present paper, but it is of interest to examine how well the field line structure in this region can be fitted by a force-free or more sophisticated model, as well as whether the field-aligned current in such a model is consistent with in-situ plasma measurements (difference between electron and ion bulk velocities). With this regard, *Frank et al.* [1995] reported an example in which the field-aligned current is present to form the rope-like structure, based on the GEOTAIL/CPI observation of electron and ion distribution functions. More extensive study on the structure of the core part of the plasmoid should be pursued in the future.

Finally, though the cold ion beams manifests their source(s) and transportation route(s), it is noted that a diffuse, hot component is simultaneously observed in the plasma sheet, which is different from the feature in the plasma sheet boundary layer in which the diffuse component is negligible. The diffuse component may represent thermalization of the ion beams due to pitch-angle scattering and energy diffusion during the transport of fresh ion beams into the deeper plasma sheet. The tracing of the fresh ion beams and the thermalization process(es) provide a key clue to heating mechanisms in the plasma sheet, which should also be further studied in the future.

Acknowledgments. We are indebted to T.K. Uesugi, GEOTAIL Project Manager, and all the GEOTAIL team members for the successful operations of the GEOTAIL spacecraft. We thank A. Nishida, M. Hoshino, M. Fujimoto, Y. Saito, T. Nagai, and A. Ieda for useful discussions and comments. The analysis of the lobe convection (Fig. 5) was conducted by Y. Matsuno. This work was supported in part by a grant-in-aid for scientific research project 07304038, Ministry of Education, Science and Culture, Japan.

REFERENCES

Angelopoulos, V., et al.; Multipoint analysis of a bursty bulk flow event on April 11, 1985, *J. Geophys. Res., 101*, 4,967, 1996.

Baker, D.N., R.C. Anderson, R.D. Zwickl, and J.A. Slavin; Average plasma and magnetic field variations in the distant magnetotail associated with near-Earth substorm effects, *J. Geophys. Res., 92*, 71, 1987.

Feldman, W.C., D.N. Baker, S.J. Bame, J. Birn, J.T. Gosling, E.W. Hones, Jr., and S.J. Schwartz, Slow shocks: a semipermanent feature of the distant geomagnetic tail, *J. Geophys. Res., 90*, 233, 1985.

Frank, L.A., W.R. Paterson, S. Kokubun, T. Yamamoto, and others; Direct detection of the current in a magnetotail flux rope, *Geophys. Res. Letters, 22*, 2,697, 1995.

Fujimoto, M., M.S. Nakamura, T. Nagai, T. Mukai, T. Yamamoto, and S. Kokubun; New kinetic evidence for the near-Earth reconnection, *Geophys. Res. Letters*, 23, 2,533, 1996.

Gosling, J.T., D.N. Baker, S.J. Bame, W.C. Feldman, R.D. Zwickl, and E.J. Smith; North-south and dawn-dusk plasma asymmetries in the distant tail lobes: ISEE-3, *J. Geophys. Res., 90*, 6,354, 1985.

Hesse, M., and J. Birn; Plasmoid evolution in an extended magnetotail, *J. Geophys. Res., 96*, 5,683, 1991.

Hesse, M., J. Birn, M. Kuznetsova, and J. Dreher; A simple model for core field generation during plasmoid evolution, *J. Geophys. Res., 101*, 10797, 1996.

Hirahara, M., M. Nakamura, T. Terasawa, T. Mukai, Y. Saito, T. Yamamoto, A. Nishida, S. Machida, and S. Kokubun; Acceleration and heating of cold ion beams in the plasma sheet boundary layer observed with Geotail, *Geophys. Res. Letters, 21*, 3,002, 1994.

Hirahara, M., T. Mukai, T. Terasawa, S. Machida, Y. Saito, T. Yamamoto, and S. Kokubun; Cold dense ion flows with multiple components observed in the distant tail lobe by GEOTAIL, *J. Geophys. Res., 101*, 7,769, 1996.

Hones, E.W., Jr.; Transient phenomena in the magnetotail and their relation to substorms, *Space Sci. Rev., 23*, 393, 1979.

Hones, E.W., Jr., J. Birn, D.N. Baker, S.J. Bame, W.C. Feldman, D.J. McComas, and R.D. Zwickl; Detailed examination of a plasmoid in the distant magnetotail, *Geophys. Res. Letters, 11*, 1,046, 1984.

Hoshino, M., A. Nishida, T. Yamamoto, and S. Kokubun; Turbulent magnetic field in the distant magnetotail: Bottom-up process of plasmoid formation ?, *Geophys. Res. Letters, 21*, 2,935, 1994.

Hoshino, M., T. Mukai, A. Nishida, T. Yamamoto, and S. Kokubun; Evidence of a pair of active reconnections in distant magnetotail, *J. Geomag. Geoelectr., 48*, 515, 1996.

Hoshino, M., T. Mukai, T. Yamamoto, and S. Kokubun; Ion dynamics in magnetic reconnection: Comparison between numerical simulation and GEOTAIL observations, *J. Geophys. Res., 103*, 4509, 1998.

Hughes, W.J., and D.G. Sibeck; On the 3-dimensional structure of plasmoids, *Geophys. Res. Letters, 14*, 636, 1987.

Ieda, A., S. Machida, T. Mukai, Y. Saito, T. Yamamoto, A. Nishida, T. Terasawa, and S. Kokubun; Statistical analysis of the plasmoid evolution with GEOTAIL observations, *J. Geophys. Res., 103*, 4453, 1998.

Kawano, Y., M. Fujimoto, T. Mukai, T. Yamamoto, T. Terasawa, Y. Saito, S. Machida, S. Kokubun, and A. Nishida; Right-handed ion-ion resonant instability in the plasma sheet boundary layer: Geotail observations, *Geophys. Res. Letters, 21*, 2,887, 1994.

Kawano, H., A. Nishida, M. Fujimoto, T. Mukai, S. Kokubun, T. Yamamoto, T. Terasawa, M. Hirahara, Y. Saito, S. Machida, K. Yumoto, H. Matsumoto, and T. Murata; A quasi-stagnant plasmoid observed with Geotail on October 15, 1993, *J. Geomag. Geoelectr., 48*, 525, 1996.

Kivelson, M.G., and K.K. Khurana; Models of flux ropes embedded in a Harris neutral sheet: force-free solutions in low and high beta plasmas, *J. Geophys. Res., 100*, 23,637, 1995.

Kokubun, S., T. Yamamoto, M.H. Acuna, K. Hayashi, K. Shiokawa, and H. Kawano; The GEOTAIL magnetic field experiment, *J. Geomag. Geoelectr., 46*, 7, 1994.

Lepping, R.P., D.H. Fairfield, J. Jones, L.A. Frank, W.R. Paterson, S. Kokubun, and T. Yamamoto; Cross-tail magnetic flux ropes as observed by the GEOTAIL spacecraft, *Geophys. Res. Letters, 22*, 1,193, 1995.

Lui, A.T.Y., D.J. Williams, S.P. Christon, R.W. McEntire, V. Angelopoulos, C. Jacquey, T. Yamamoto, and S. Kokubun; A preliminary assessment of energetic ions species in flux ropes/plasmoids in the distant tail, *Geophys. Res. Letters, 21*, 3,019, 1994.

Machida, S., T. Mukai, Y. Saito, T. Obara, T. Yamamoto, A. Nishida, M. Hirahara, T. Terasawa, and S. Kokubun; Geotail low energy particle and magnetic field observations of a plasmoid at $X = -142\ R_E$, *Geophys. Res. Letters, 21*, 2,995, 1994.

McPherron, R.L., C.T. Russell, and M.P. Aubry; Satellite studies of magnetospheric substorms, on August 15, 1968, 9, Phenomenological model for substorms, *J. Geophys. Res., 78*, 3,131, 1973.

Moldwin, M.B., and W.J. Hughes; Plasmoids as magnetic flux ropes, *J. Geophys. Res., 96*, 14,051, 1991.

Moldwin, M.B., and W.J. Hughes; On the formation and evolution of plasmoids: a survey of ISEE-3 geotail data, *J. Geophys. Res., 97*, 19,259, 1992.

Moldwin, M.B., and W.J. Hughes; Geomagnetic substorm association of plasmoids, *J. Geophys. Res., 98*, 81, 1993.

Mozer, F.S., H. Hayakawa, S. Kokubun, M. Nakamura, T. Okada, T. Yamamoto, and K. Tsuruda, Direct entry of dense flowing plasmas into the distant tail lobes, *Geophys. Res. Letters, 21*, 2,959, 1994.

Mukai, T., S. Machida, Y. Saito, M. Hirahara, T. Terasawa, N. Kaya, T. Obara, M. Ejiri, and A. Nishida; The Low Energy Particle (LEP) experiment onboard the GEOTAIL Satellite, *J. Geomag., Geoelectr., 46*, 669, 1994a.

Mukai, T., M. Hirahara, S. Machida, Y. Saito, T. Terasawa, and A. Nishida; Geotail observation of cold ion streams in the medium distance magnetotail lobe in the course of a substorm, *Geophys. Res. Letters, 21*, 1,023, 1994b.

Mukai, T., M. Fujimoto, M. Hoshino, S. Kokubun, S. Machida, K. Maezawa, A. Nishida, Y. Saito, T. Terasawa, and T. Yamamoto; Structure and kinetic properties of plasmoids and their boundary regions, *J. Geomag. Geoelectr., 48*, 541, 1996.

Nagai, T., K. Takahashi, H. Kawano, T. Yamamoto, S. Kokubun, and A. Nishida; Initial GEOTAIL survey of magnetic substorm signatures in the magnetotail, *Geophys. Res. Letters, 21*, 2,991, 1994.

Nishida, A., H. Hayakawa, and E.W. Hones, Jr.; Observed signatures of reconnection in the magnetotail, *J. Geophys. Res., 86*, 1,422, 1981.

Nishida, A., M. Scholer, T. Terasawa, S.J. Bame, G. Gloeckler, E.J. Smith, and R.D. Zwickl; Quasi-stagnant plasmoid in the middle tail: new preexpansion phase phenomenon, *J. Geophys. Res., 91*, 4,245, 1986.

Nishida, A.; The Geotail mission, *Geophys. Res. Letters, 21*, 2,871, 1994.

Omidi, N.; Magnetic structure of slow shocks and the associated ion dissipation, *Adv. Space Res., 15*, 489, 1995.

Richardson, I.G., and S.W.H. Cowley; Plasmoid-associated energetic ion bursts in the deep geomagnetic tail: properties of the boundary layer, *J. Geophys. Res., 90*, 12,133, 1985.

Richardson, I.G., S.W.H. Cowley, E.W. Hones, Jr., and S.J. Bame; Plasmoid-associated energetic ion bursts in the deep geomagnetic tail: Properties of plasmoids and the postplasmoid plasma sheet, *J. Geophys. Res., 92*, 9,997, 1987.

Saito, Y., T. Mukai, T. Terasawa, A. Nishida, S. Machida, M. Hirahara, K. Maezawa, S. Kokubun, and T. Yamamoto, Slow-mode shocks in the magnetotail, *J. Geophys. Res., 100*, 23,567, 1995.

Saito, Y., T. Mukai, T. Terasawa, A. Nishida, S. Machida, M. Hirahara, K. Maezawa, S. Kokubun, and T. Yamamoto; Foreshock structure of the slow-mode shocks in the Earth's magnetotail, *J. Geophys. Res., 101*, 13,267, 1996.

Scholer, M., G. Gloeckler, B. Klecker, F.M. Ipavich, D. Hovestadt, and E.J. Smith; Fast moving plasma structures in the distant magnetotail, *J. Geophys. Res., 89*, 6,717, 1984.

Schwartz, S.J., M.F. Thomsen, W.C. Feldman, and F.T. Douglas, Electron dynamics potential jump across slow mode shocks, *J. Geophys. Res., 92*, 3,165, 1987.

Seki, K., M. Hirahara, T. Terasawa, I. Shinohara, T. Mukai, Y. Saito, S. Machida, T. Yamamoto, and S. Kokubun; Coexistence of Earth-origin O^+ and solar wind-origin H^+/He^{++} in the distant magnetotail, *Geophys. Res. Letters, 23*, 985, 1996.

Sergeev, V.A., V.S. Semenov, and M.V. Sidneva; Impulsive reconnection in the magnetotail during substorm expansions, *Planet. Space Sci., 35*, 1,199, 1987.

Siscoe, G.L., L.A. Frank, K.L. Ackerson, and W.R. Paterson; Properties of mantle-like magnetotail boundary layer, *Geophys. Res. Letters, 21*, 2,975, 1994.

Slavin, J.A., E.J. Smith, B.T. Tsurutani, D.G. Sibeck, H.G. Singer, D.N. Baker, J.T. Gosling, E.W. Hones, Jr., and F.L. Scarf; Substorm associated traveling compression regions in the distant tail: ISEE 3 geotail observations during an extended interval of substorm activity, *Geophys. Res. Letters, 11*, 651, 1984.

Slavin, J.A., et al.; CDAW-8 observations of plasmoid signatures in the geomagnetic tail: an assessment, *J. Geophys. Res., 94*, 15,153, 1989.

Slavin, J.A., C.J. Owen, M.M. Kuznetsova, and M. Hesse; ISEE 3 observations of plasmoids with flux rope magnetic topologies, *Geophys. Res. Letters, 22*, 2,061, 1995.

Tsurutani, B.T., I.G. Richardson, R.M. Thorne, W. Butler, E.J. Smith, S.W.H. Cowley, S.P. Gray, S.-I. Akasofu, and R.D. Zwickl; Observations of the right-handed resonant ion beam instability in the distant plasma sheet boundary layer, *J. Geophys. Res., 90*, 12,159, 1985.

Tu, J.-N., T. Mukai, M. Hoshino, Y. Saito, Y. Matsuno, T. Yamamoto, and S. Kokubun; Geotail observations of ion velocity distributions with multi-beam structures in the post-plasmoid current sheet, *Geophys. Res. Letters, 24*, 2,247, 1997.

Yamamoto, T., A. Matsuoka, K. Tsuruda, H. Hayakawa, A. Nishida, M. Nakamura, and S. Kokubun; Dense plasmas in the distant magnetotail as observed by Geotail, *Geophys. Res. Letters, 21*, 2,879, 1994.

Zwickl, R.D., D.N. Baker, S.J. Bame, W.C. Feldman, J.T. Gosling, E.W. Hones, Jr., D.J. McComas, B.T. Tsurutani, and J.A. Slavin; Evolution of Earth's distant magnetotail: ISEE 3 electron plasma results, *J. Geophys. Res., 89*, 11,007, 1984.

T. Mukai, T. Yamamoto, The Institute of Space and Astronautical Science, 3-1-1 Yoshinodai, Sagamihara, Kanagawa 229. Japan.

S. Machida, Department of Geophysics, Kyoto University, Kyoto 606-01, Japan.

The Formation and Structure of Flux Ropes in the Magnetotail

Michael Hesse
Laboratory for Extraterrestrial Physics NASA/Goddard Space Flight Center, Greenbelt, MD

Margaret G. Kivelson
Institute of Geophysics and Planetary Physics and Department of Earth and Space Sciences
University of California, Los Angeles, CA

Flux ropes are a common signature in spacecraft observations in the plasma sheet of the nightside magnetosphere. The probability of observing the signature of a helical magnetic structure appears to increase with distance down the magnetotail, suggesting that there is a gradual transition with downtail distance from ordinary "bubble-like" plasmoids to flux ropes with significant magnetic fields at the core. On the other hand, a few very small flux ropes have been observed relatively close to the Earth. Thus observations of the distribution of flux ropes with distance may be biased by increases of the scale size as these structures propagate down the tail. Despite their obvious importance the formation process is not fully understood. A review of the present understanding of the formation, structure and evolution of magnetospheric flux ropes in the magnetotail is presented. The review includes a summary of flux rope observations by different spacecraft. Next, possible scenarios of flux rope formation from more "bubble-like" plasmoids are discussed. Analytical theory of flux rope structure is used to compare to observations. The review includes an outlook and analysis of flux rope importance in different plasma systems.

1. INTRODUCTION

Following theoretical investigations [e.g., Schindler, 1974] which predicted plasmoid-like structures as a result of a global magnetotail instability, Hones [1977] introduced the concept of a plasmoid to interpret the magnetotail response to a substorm. Hones envisioned that a plasmoid develops when a plasma sheet region somewhere within ~30 R_E of the Earth becomes disconnected from the geomagnetic field during the course of a substorm. The newly-detached flux bubble continues to move down the tail aided by trailing plasma which is accelerated by draped field lines linked to the interplanetary magnetic field. Thus Hones' plasmoids consist of looplike magnetic field structures extending from somewhere close to the Earth to the downtail neutral line with hot plasma sheet plasma trapped inside. The original plasmoid observations were made within the "trailing region," i.e., the part of the plasmoid volume closest to the Earth because the IMP orbits remained within ~40 R_E of Earth.

Recent evidence has called this picture of a plasmoid into question. Observations primarily in the far magnetotail indicate that many plasmoids exhibit a strong "core field," i.e., an enhancement of an axial magnetic field component, which produces a magnetic field in the plasmoid center, often approximately aligned with the magnetospheric *y* direction. These magnetic structures are referred to as "flux ropes." In some cases, the core magnetic field magnitude can be as large or larger than the ambient lobe magnetic field strength. If flows are not important, magnetic tension forces must balance the sum of the magnetic and thermal pressures both of which may have gradients directed towards the axis of the flux rope.

Several attempts have been made to model these flux rope-like structures. Moldwin and Hughes [1990] used an analytical model to fit the observed magnetic field

New Perspectives on the Earth's Magnetotail
Geophysical Monograph 105

variation, Birn [1991, 1992] developed a theoretical framework based on steady state magnetohydrodynamic (MHD) equilibrium for the self-consistent modeling of bubble-like plasmoid and flux rope structures, Lepping et al. [1995] employed a force-free magnetic field model (i.e., with $j \times B = 0$) adapted from the modeling of magnetic clouds in the solar wind, Slavin et al. [1995] combined a force-free magnetic field model with a kinematic model of the outer, not necessarily force-free, regions of the plasmoid, and recently, Kivelson and Khurana [1995] employed a self-consistent equilibrium theory to model flux rope type magnetic signatures. While all of these models have advanced our understanding of these structures, the formation process of flux ropes, and the relation between them and bubble-like plasmoids, which traditionally have been envisioned as closed magnetic loops with small cross-tail magnetic field components, remains poorly understood.

In this paper we review recent results pertaining to the formation and evolution of flux rope-like plasmoids. We begin with a brief overview of recent observations of flux ropes. Section 3 presents a synopsis of past MHD simulations of the formation of helical magnetic field lines. Section 4 discusses a simple model of flux rope formation from a bubble, based on the magnetic interconnection of the plasmoid magnetic field with colder plasma of the low-latitude boundary layer (LLBL) or the magnetosheath. Section 5 contains a description of past work on analytical modeling of flux rope structure based on equilibrium theory. Section 6 discusses the impact of the magnetic connection between a flux rope and the ionosphere during the earlier phases of its evolution. Finally, Section 7 consists of a summary and outlook into future developments.

2. OBSERVATIONS OF FLUX ROPES

A characteristic signature of a substorm in the magnetotail is a transient southward turning of the magnetic field. Thus, from the earliest analyses, attention focused on models that would explain the southward turning of the magnetic field in the magnetotail. The Hones [1997] model of a plasmoid or a disconnected magnetic bubble traveling down the tail accounts for typical substorm signatures within ~40 R_E in which the magnetic field rotates southward for several minutes near the beginning of the expansion phase. It also explains observations made farther down the tail, where the magnetic field rotates first northward and then southward. Such signatures have been identified in many far-tail observations, at distances from ~80 R_E to more than 200 R_E downtail [e.g., Hones et al., 1984; Baker et al., 1987; Richardson et al., 1987; Slavin et al., 1989, 1992; Moldwin and Hughes, 1991, 1992, 1993]. The model of the plasmoid as a bubble of magnetic flux propagating down the center of the tail also explains ISEE-3 observations of north-then-south rotations of the field associated with a field compression observed in the lobes of the magnetotail. Signatures of this sort are called "traveling compression regions" (TCRs) [Slavin et al., 1984] and are used to monitor remotely the passage of plasmoids. Thus, based on ISEE-3 observations, it was shown that each substorm is typically associated with the formation and propagation of one or more plasmoids [Slavin et al., 1992, 1993; Moldwin and Hughes, 1993].

Although attention first focused on the north-south component of the magnetic field (B_z in GSE), the observed signatures often show an additional magnetic perturbation that peaks when B_z reverses direction. Such signatures have been reported from both ISEE 3 and GEOTAIL data [Slavin et al., 1989; Fairfield et al., 1989; Moldwin and Hughes, 1991; Nagai et al., 1994; Machida et al., 1994; Frank et al., 1995; Lepping et al., 1995; Slavin et al., 1995]. Inside of lunar orbit, this component, referred to as the core field, is commonly oriented dawn-dusk (i.e., along the $\pm y$-direction in GSE) with B_y parallel to the y-component of the interplanetary magnetic field (IMF), consistent with rope-like form that we refer to as a flux rope. The core field is often as large as or larger than the lobe field [e.g., Slavin et al., 1995]. In the following we will denote such structures as "flux ropes," while plasmoids with no or small core B_y field enhancement will be referred to as "bubbles." Flux ropes are often encountered near neutral sheet crossings when the sign of the B_x GSE component changes.

Although our knowledge of most of the properties of flux ropes is based on magnetic field measurements, recent work based on data from Galileo and Geotail has used plasma measurements to augment our understanding of the structures. Spatial scales have in the past been estimated indirectly, but plasma flow velocities confirm the previously-inferred spatial scales. For example, near lunar orbit at 60 R_E downtail, Galileo observed two flux ropes, a small one of ~1000 km scale size and a larger one of a few R_E scale [Kivelson et al., 1993; Kivelson and Khurana, 1995]. Current has been measured only in smaller flux ropes [Khurana et al., 1995, 1996; Frank et al., 1995] because although the total current is proportion to the scale length, L, the current density varies like L^{-1}. Thus direct measurement was made in a flux rope moving earthward at ~6 km/s in a substorm recovery phase. It had an x-extent of ~1300 km and carried a current of 0.6×10^5 A. A second flux rope passed Galileo moving tailward at 150 km/s in a substorm expansion phase. It had an x-extent of 8 R_E. In this case, the current density was not directly measured but the total current was inferred as 2.4×10^6 A. This latter current is of the order of currents associated with other substorm-associated current systems. In both cases, the IMF orientation was towards positive y, and the core fields were both along the y-direction and positive.

Flux ropes have rarely been observed within the apogee of the ISEE 1 and 2 spacecraft [but see Elphic et al., 1986]. This is a bit puzzling as three-dimensional MHD simulations [Ogino et al, 1990; Walker and Ogino, 1996] show that flux ropes can initially form within 20 R_E. As in the observations, the B_y component of the core field is parallel to the IMF B_y but the core field may twist considerably away from the y-axis. This means that identifying flux ropes near Earth may not be easy. In particular, the bipolar perturbation in B_z produced by the toroidal field, even if comparable in magnitude with the lobe field may not produce a negative B_z. In Figure 1A we show an example of a flux rope observed in the ISEE data on March 5, 1979 that illustrates some of the points that are emphasized here. The event occurred during a quiet period more than 1 hour after the recovery phase of an earlier substorm. ISEE 2 at (-16.86,-12.04, -0.61, all in R_E) measured a bipolar signature in B_z but in a GSM coordinate system B_z remained predominantly positive during the ~2 minute perturbation. There is little doubt that the structure is a flux rope as may be seen in Figure 1B where the lowest panel shows the plasma (ion) pressure, the magnetic pressure, and the total pressure, revealing that the total pressure in the center of the structure exceeds the lobe pressure at 06:25 UT by ~35%. The curvature force implied by the bipolar perturbation of ±11 nT is roughly what is required to provide static force balance. The thermal pressure near the center of the flux rope is somewhat lower than the ambient plasma sheet pressure but still well above the pressure in the lobe, a point to which we return in a later section. The x-component of the flow velocity is seen to be ~100 km/s earthward. This is consistent with the fact that the bipolar signature was observed roughly 25 s earlier at ISEE 1 at (-17.47, -12.03, -0.33), implying an earthward motion at ~110 km/s.

ISEE 3 data, corrected for the delay in the solar wind between its position and Earth show that the IMF was southward with substantial positive B_x and negative B_y components. This may mean that the GSM reference frame is not optimum for the analysis of this event as the magnetotail was probably twisted about the tail axis in the manner sketched in Figure 1C. The data rotated by 20° about the tail axis resemble the classical flux rope signature with a significant negative contribution to the bipolar B_z signature as seen in Figure 1A. As flux ropes appear to be associated with finite B_y in the IMF, it seems reasonable that twisting of the tail is often associated with flux ropes and that the signatures depart from expectations because of the twisting of the tail.

The scale sizes of flux ropes observed within ~25 Re of the Earth are < 1 Re, small compared with the scale sizes of many Re typical of flux ropes observed at greater distance down the tail. It seems likely that the scale size grows with distance along the tail, thereby increasing the probability of observing flux ropes at large distance. Another alternative is that flux ropes observed near the Earth result from a different formation mechanism than larger flux ropes observed in the distant tail. In addition, the flux rope structure may become clearer as the structure propagates down the tail. Further observations will be important in establishing the importance of these two effects.

3. MHD MODELING OF THE FORMATION AND TOPOLOGY OF HELICAL MAGNETIC FIELD STRUCTURES

Initially, a plasmoid was defined as a volume of loop-like field lines closing on themselves [e.g., Hones, 1977]. The boundary of this volume is a magnetic surface, called a separatrix, made of field lines that connect to a separator line or X-line. The situation becomes more complicated if the symmetry of the system is relaxed by including a net cross-tail magnetic field component and a dependence on the cross-tail coordinate in the initial state of the simulation as was originally proposed by Hughes and Sibeck [1987]. It has been shown in an analytic kinematical model [Birn et al., 1989] that in such a general system the topology of the plasmoid changes from closed to open, i.e., to a structure that consists of helical field lines rather than closed loops. In fact, it was recently shown that the underlying three-dimensional reconnection process quite generically produces helical magnetic field lines [Hesse, 1995]. The investigation reviewed here is concerned with the self-consistent topological evolution of the magnetic field associated with the plasmoid. In order to study the topology in a self-consistent model, Hesse and Birn [1991] ran an MHD simulation starting from a self-consistent magnetotail equilibrium. The equilibrium shown is a self-consistent solution of the magnetohydrostatic force balance [Birn, 1987]. The initial state includes a net cross-tail magnetic field component of about 3% of the lobe field strength, consistent with typical observations [Fairfield, 1979]. Furthermore, the tail model includes a far separator line located at $x = -100R_E$.

The MHD equations were integrated within a tail region extending from $x = -20\ R_E$ to $x = -260\ R_E$. In this simulation, Hesse and Birn [1991] found that the plasmoid formed at about $t = 95$ (times are normalized to a characteristic Alfvén transit time over the distance of the current sheet half-width). Initially, shortly after formation

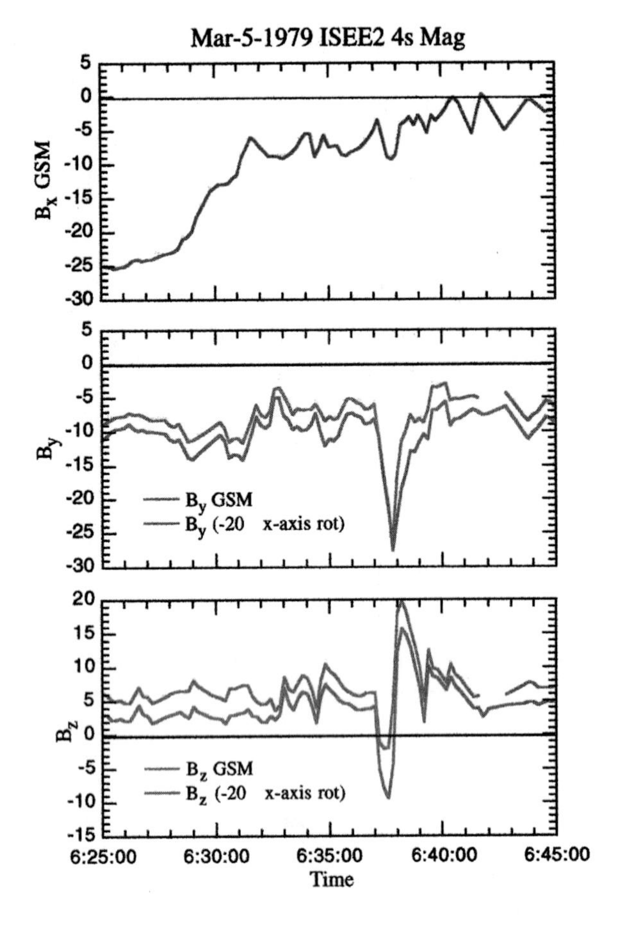

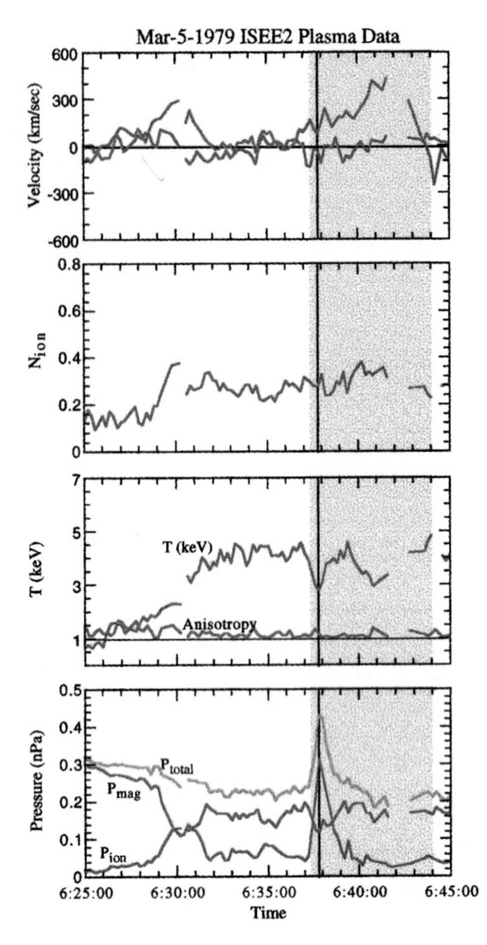

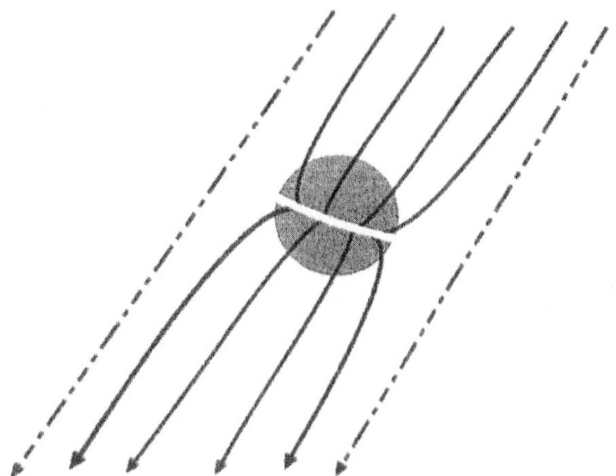

Figure 1. Data from the ISEE 2 spacecraft at a GSE position (-16.86,-12.04, -0.61) between 06:25 and 06:45 UT on March 5, 1979. Here distances are in R_E. In Figure 1a, the three components of the magnetic field (nT) in GSM are plotted in blue. A 20° rotation around the x-axis rotates the y and z components to the values plotted in red. Plasma parameters measured on ISEE 2 are presented in Figure 1b. The upper panels show the x and y components of the flow velocity, the ion density, the ion temperature (blue trace) and anisotropy (red trace). The lowest panel shows the ion pressure (red), the magnetic pressure (blue) and the total pressure (green). A vertical line marks the center of the flux rope. Pale blue shading marks an interval defined by Angelopoulos as a bursty bulk flow event which is closely linked to the flux rope. Figure 1c is a schematic diagram of a cross-section of the magnetotail which has rotated in response to stresses imposed by the finite cross tail component (y-component) of the IMF.

time, identified by the occurrence of negative B_z, the plasmoid consisted solely of closed field lines connected to the Earth at either "end." Figure 2 shows a set of typical field lines from the MHD results at $t = 170$. It is seen that there are six different types of field lines in the simulation: Helical field lines, which connect the Earth with itself (red), helical field lines which connect the Earth to the LLBL, or IMF (green), helical field lines which connect the LLBL to the LLBL (blue), and simple field lines connecting Earth to Earth (cyan), Earth to IMF (yellow), and IMF to IMF (purple). Note that the end planes of the simulation box are considered as magnetically connected to the Earth, or to the IMF. At this point it is evident that a connection of the plasmoid (helical field lines) to the LLBL appears to have been established (green and blue field lines).

The evolution of the magnetic field is revealed by considering the topology of field lines that cross the equatorial plane which then can be color-coded to reveal that topology as illustrated in Figure 3. The plasmoid region, defined by the region of helical field lines, is characterized by three possible colors: red, green, and blue. The panels show that the plasmoid initially consists entirely of closed field lines, i.e., field lines that connect to the Earth at both ends. The process of severance is found to be similar to that presented by Birn and Hesse [1990], and very different from two-dimensional models. Instead of an instantaneous severance of the entire plasmoid, the complicated change of topology of the plasmoid field lines in a three dimensional model occurs over a finite amount of time. In the intermediate stages of the evolution, the plasmoid exhibits a tangled structure of flux tubes of different magnetic topologies, qualitatively as predicted by the kinematical model of Birn et al. [1989]. For the purpose of the present review, it is noteworthy that a connection of the plasmoid to the LLBL becomes evident at $t = 150$. Following this time, an increasing number of helical flux tubes becomes disconnected from the Earth and newly connected to the LLBL. Similar results based on 3-D MHD simulations have been reported by Ogino et al. [1990] and by Walker and Ogino [1996], and complicated topologies at the magnetopause have also been found by Lee et al. [1993].

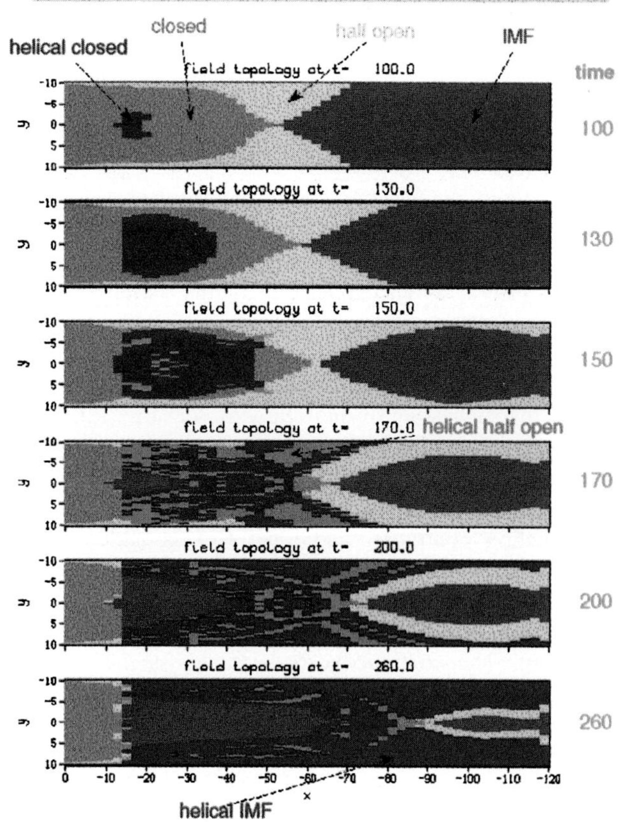

Figure 3. Evolution of the plasmoid. actual three-dimensional magnetic field topology in the MHD simulation of Hesse and Birn [1991]. Intersections of field lines with the equatorial plane of the simulation box are assigned a color depending on the topological type of the field line that intersects at this very point. The colors denote: "light blue": closed field lines, connected with the Earth at both ends and intersecting the equatorial plane once; "red": closed field lines that intersect the equatorial plane more than once ("closed plasmoid field lines"); "yellow": half open field lines that intersect the equatorial plane once, connecting to the Earth and the IMF through the sides of the simulation box; "green": half open field lines that intersect the equatorial plane more than once, ("lobelike plasmoid field lines") connecting to the Earth on one side and to the IMF on the other through either the tailward end plane or the sides of the simulation box; "magenta": open field lines (IMF type) that intersect the equatorial plane once; "blue": "open plasmoid field lines," intersecting the equatorial plane more than once, without connections to the Earth.

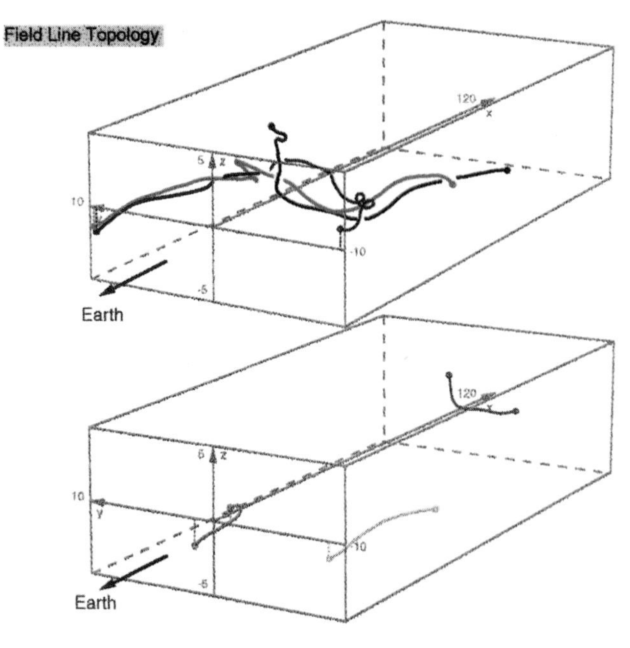

Figure 2. Topological structure of a plasmoid consisting of helical magnetic field lines. The field lines were integrated from the results of an MHD simulation at t=170 [Hesse and Birn, 1991]. The upper panel displays three characteristic field lines. The red field line connects to the Earth with both ends, the green field line connects to the Earth at one end and to the IMF at the other, and the blue field line is entirely open, i.e., connected to the IMF at either end. The lower panel displays an unperturbed plasma sheet field line (cyan), a lobe field line (yellow), and an IMF field line threading the plasma sheet (purple).

4. FLUX ROPE FORMATION BY ENERGY LOSS TO LLBL

A possible scenario for the formation of flux ropes was suggested by Hesse et al. [1996] based on modeling results similar to the one presented in the previous section [e.g., Hughes and Sibeck, 1987; Ogino et al., 1990; Birn and Hesse, 1990; Hesse and Birn, 1991], and observational evidence. While these models and simulations typically did not demonstrate B_y enhancements, they did establish that magnetic connections between the flank regions and the interior region of the plasmoid develop when helical field lines undergo reconnection during the disconnection of the plasmoid from the closed field line region. Similar connections have also been suggested by Moldwin and Hughes [1992] as an explanation of the variation of the average electron pressure and density with downtail distance.

The combination of observational and modeling results lead to the suggestion of the following scenario, illustrated in Figure 4. Initiating in a magnetotail with a net cross-tail magnetic field, three-dimensional magnetic reconnection generates helical field lines (panel a). As magnetic reconnection continues, additional flux is added to the plasmoid at outer, helical layers, while the inner helical field lines are extended to the flank regions by the progression of magnetic reconnection in the same direction (panel b). The continued expansion across the tail will ultimately lead to the establishment of a new magnetic connection of some plasmoid magnetic flux to the cold plasma of the magnetosheath or the LLBL, indicated by the shaded region in panel c. Indeed, Walker and Ogino [1996] find that the flux ropes in their simulations start to move down the tail only after they have extended fully across the tail to the magnetopause boundary. Once extended to the boundaries of the magnetotail, the pressure and/or temperature gradient along the magnetic field will cause mass flow and/or heat flux directed toward the flanks along the magnetic field, indicated by the black arrows. The ensuing pressure loss in the center of the tail causes a reduction of the diameter of the inner region of the plasmoid helix, until the remaining plasma pressure, enhanced by the volume reduction, is again sufficient to balance the ambient lobe pressure (panel d). Any pressure enhancement, however, will lead to an immediate additional drain to the flank regions. Therefore the only mechanism limiting the collapse is the buildup of the y component of the magnetic field and the magnetic pressure $B_y^2/2\mu_0$, which assumes the role of the plasma pressure (see below). Thus an increasingly force-free flux rope begins to form in the center of the original plasmoid structure. Finally, continuing reconnection will connect outer layers of the original plasmoid to the flanks of the tail as well, causing plasma pressure reductions in those regions (panel e). It should be noted that the question whether the final structure appears to be a force-free flux rope or not depends on the relative magnitudes of the dynamical evolution time scale of the plasmoid, and the energy loss time scale established by magnetic connections to the flanks of the magnetotail. For example, in the data of Figure 1B, the decrease of pressure within the flux rope relative to the plasma sheet pressure may result from the loss mechanism discussed here, but the evolution to an electromagnetically force-free configuration is not yet complete. In a different context, Ma et al. [1994] investigated the process of core field enhancement in flux ropes in the context of flux transfer events at the magnetopause, using two- and three-dimensional MHD simulations. They found that expulsion of plasma through forces along the axis of the flux rope could contribute to the increase in the core field.

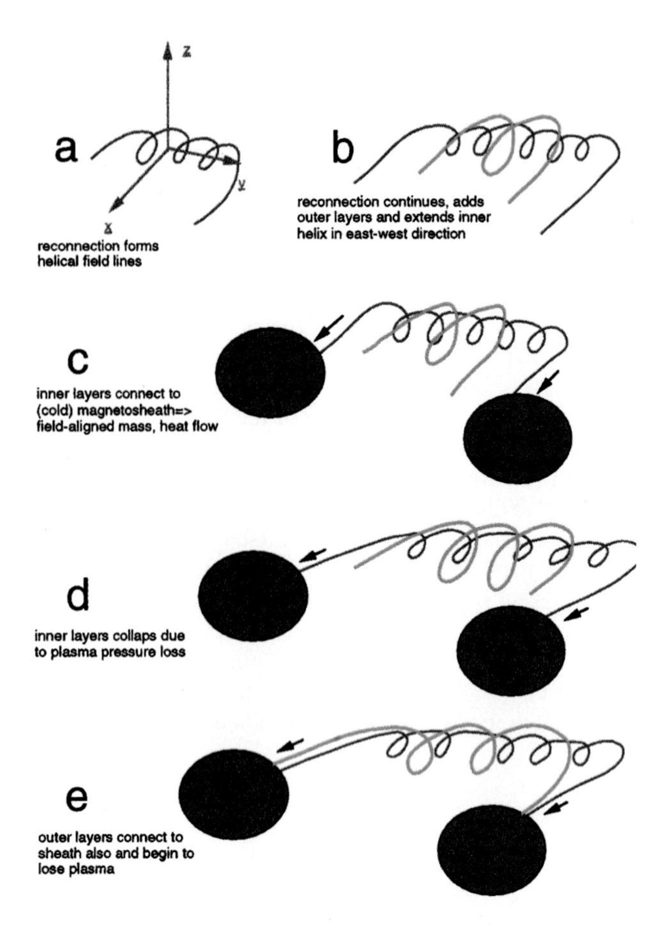

Figure 4. Sketch of the proposed transition scenario from plasmoid to flux rope. The meaning of the different panels is explained in the text.

In order to support their qualitative scenario, Hesse et al. [1996] also presented results of a simple MHD simulation of plasmoid formation, in which the evolution toward a force-free flux rope can be demonstrated. Owing to their simplicity, these MHD simulations are to be taken not as a final step of the investigation but rather as a simple model suggesting the viability of the above qualitative scenario.

For the purpose of self-consistent modeling, a 2.5 dimensional MHD model was modified to include energy drain to the LLBL along helical plasmoid field lines. Hesse et al. introduced a loss term of the form

$$\left.\frac{\partial p}{\partial t}\right|_{LOSS} = \frac{1}{\tau}(p_0 - p)$$

into the energy equation. Here p_0=0.03 (pressure normalized to half the square of a typical lobe magnetic field B_0=40*nT*) denotes a LLBL pressure, and τ a typical ion heat (or mass-) flux propagation time scale. The exact value of p_0 appears not to make a significant difference as long as p_0 is substantially smaller than the plasma sheet pressure, i.e., $p_0 << 1$. To model potential mass loss by bulk flow along field lines as well, Hesse et al. also included an adiabatic loss term in the continuity equation. This form of the loss is assumed for simplicity. (A loss based on heat flux leads to similar results, with the possibility of a density enhancement and low temperatures in the center of the flux rope structure). Their continuity equation takes the form

$$\frac{\partial \rho}{\partial t} = -\nabla\cdot(\rho \mathrm{v}) + \frac{3}{5}\frac{\rho}{p}\left.\frac{\partial p}{\partial t}\right|_{LOSS}$$

In order to model the pressure loss it is assumed that the loss time scales depend on the geometry as well as on the magnetic connectivity, which is modeled in an ad hoc fashion. Hesse et al. assumed a tail radius of about 20 R_E (corresponding to 10 units of a plasma sheet half-thickness of 2R_E in this model) and an equilibration time scale between plasma sheet and sheath/LLBL pressures determined by the sound speed in the plasma sheet, which is approximately equal to unity in their normalization (velocities are normalized to a typical Alfvén speed v_A=1000*km/s*, and time to the crossing time of Alfvén waves across a typical length). Further, the geometry of a helical field line will influence pressure reduction times. In the case of a simple helix with circular cross section this geometry is determined by the angle between the magnetic field and the y direction. In a simple approximation this scaling is adopted locally in the simulation. In order to model the time evolution of a three-dimensional plasmoid qualitatively, pressure relaxation is not initiated before the plasmoid has advanced to $x > 22.5$ (in units of 2R_E with x increasing down the tail) and B_y exceeds $B_y = 0.04$ locally. These parameters are arbitrary, modeling the fact that magnetic connection to the IMF appears to be established at some later time of plasmoid evolution and that some enhancement of B_y above the initial magnitude can be expected when a connection to the IMF is established. Changing these parameters influences only the quantitative evolution, not its qualitative properties. Thus the time scale τ is chosen as:

$$\tau = \begin{cases} 10\dfrac{B}{B_y} & \text{if } |B_y| > 0.04 \\ \infty & \text{else} \end{cases}$$

The initial configuration adopted for the simulation is a tail-like equilibrium with a constant y-component of the magnetic field, chosen to be B_y=0.03. The evolution of the magnetic field and the plasma flow for two runs, one with plasma contact (run 1) and the other without (run 2) are shown in Figure 5. The figure exhibits the common transition to instability growth, accompanied by plasmoid formation, and fast earthward and tailward flows.

The evolution of the B_y and B_z components of the magnetic field on the tail axis is displayed for both runs in Figures 6 and 7, respectively. The solid lines denote quantities from run 1, whereas the dash-dotted lines indicate the magnetic fields taken from the run without plasma contact (run 2). The figures show that the magnetic field evolution appears to be unchanged by the pressure loss until $t \approx 120$, when pressure loss has led to a B_y enhancement at the center of the bipolar B_z signature associated with a plasmoid. The upper panels exhibit the same features, albeit more pronounced. There the magnitude of the central B_y enhancement has reached the ambient lobe magnetic field strength. The maximum value of the B_y enhancement is found right at the B_z reversal, as found in the observations [e.g., Slavin et al., 1995]. The evident reduction of the scale size in the x direction of the plasmoid dimension is associated with the collapse of the plasmoid. Investigations of the scale size in the z direction show similar behavior.

5. ANALYTICAL MODELING OF FLUX ROPE STRUCTURE

Numerical modeling has led to new insight into the evolution of flux ropes in the magnetotail. However, analytical models in particular help to overcome the difficulties associated with the interpretation of single point measurements of flux rope magnetic fields and plasma pressure distribution. Thus, analytic models are useful in providing estimates of the net magnetic flux, mass, and internal energy that they transport and the currents that flow through them. In this section we discuss the use of such models for the interpretation of observations.

Models developed to describe solar coronal flux tubes or magnetic clouds in the solar wind are sometimes used to analyze data in the magnetotail. Such models assume that the structures are isolated, cylindrically symmetric about the flux rope axis and electromagnetically force-free [for

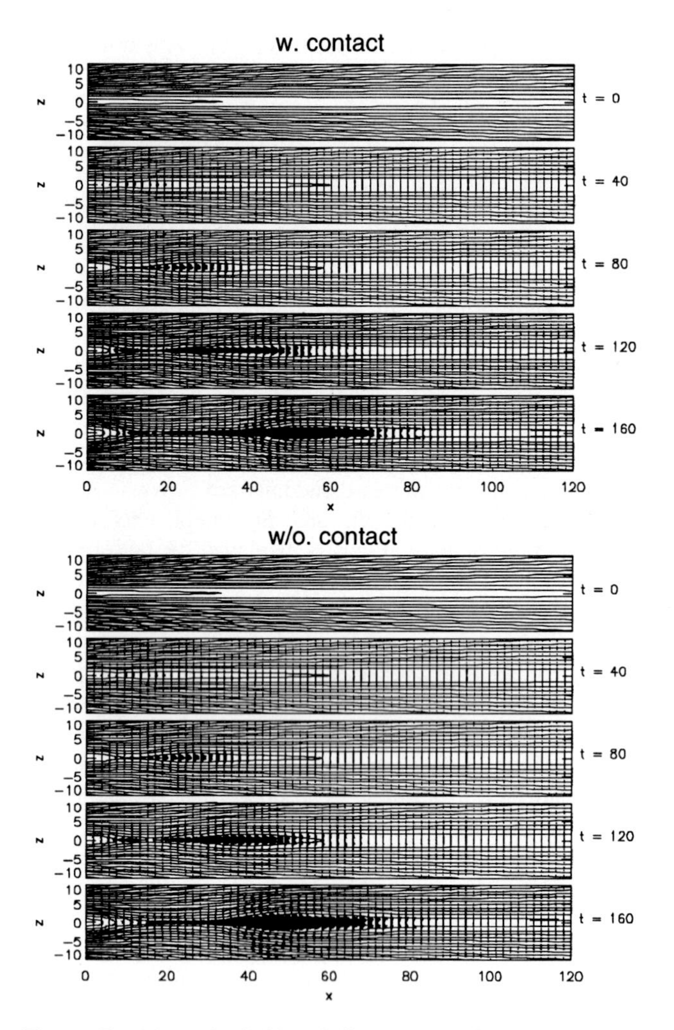

Figure 5. Magnetic field and flow vectors for the run with plasma contact (upper panel) and without (lower panel). The formation of a plasmoid can be discerned in both cases at t = 40 and is evident at t = 80.

example, Lepping et al, 1995, 1996]. They neglect effects of the plasma and field that surround the flux tube, and, in particular, do not reveal the anisotropy imposed by the pressure of the lobe field. Fits must be arbitrarily cut off at some external boundary. Moldwin and Hughes [1991] addressed these deficiencies by using a "stretching" function to describe the compression of the flux tube by the externally imposed field and plasma pressure, and also modeled the field of the surroundings by regarding the flux rope itself as an obstacle in an incompressible flowing plasma. The model is useful, but introduces strong current sheets at the flux rope boundaries that are probably unphysical. Birn [1992] modeled a plasmoid or flux rope embedded in an incompressible flow. His results are particularly relevant to the interpretation of TCRs although they have not yet been applied systematically to the interpretation of observations. Slavin et al [1995] adopted a model with a cylindrically symmetric force-free flux rope embedded in a non-force free sheath with elliptical cross-section, and also modeled the exterior region assuming incompressible flow past a rigid obstacle. The forms adopted are sufficiently flexible to provide good fits to the observations with the unknown path of the spacecraft through the structure assumed to be a straight line.

Kivelson and Khurana [1995] developed an alternate treatment of the analytical structure. Their self-consistent, two-dimensional static solution describes an array of flux ropes embedded in a Harris neutral sheet. In applying their results to observations, they considered only one elementary cell of the infinite array. Absent in this model are effects of flowing plasma in the surroundings and, like

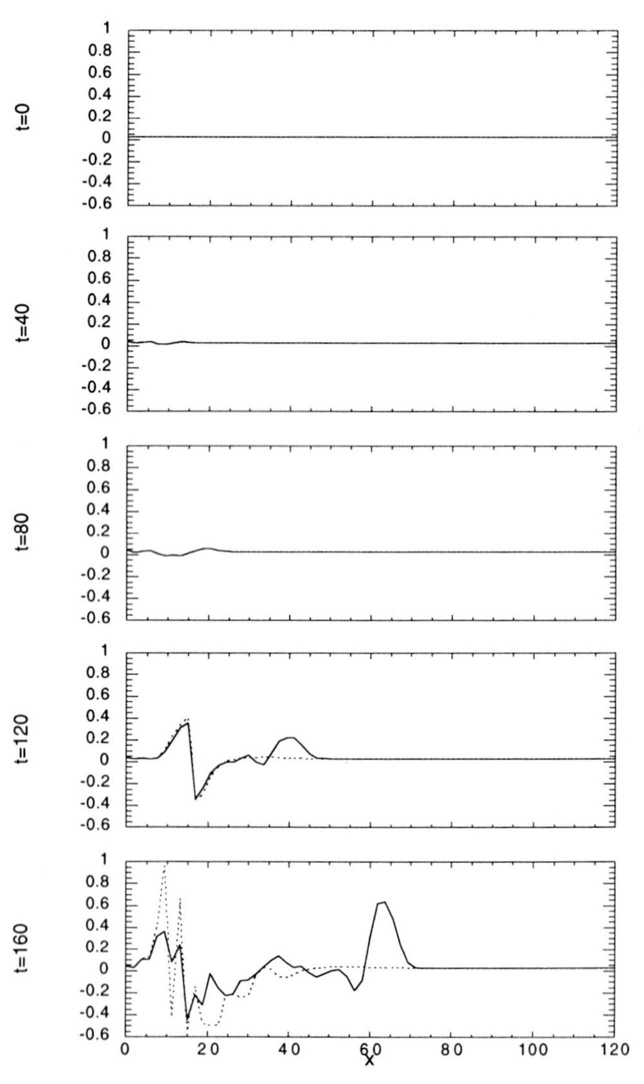

Figure 6. Variation of B_y along the x axis for the run with contact (solid line) and the run without (dash-dotted line)

other models, the form is independent of distance along the axis of the rope. However, as force balance is correctly incorporated, an elliptical cross section develops naturally. In the limit of vanishing thermal pressure, this model is electromagnetically force-free. In the model, the B_x component varies monotonically with z. In applications, the measured $B_x(t)$ serves as a proxy for $z(t)$, and the measured ion flow velocity v_x is used to determine $x(t)=v_x(t-t_0)$, thus providing a possibly meandering path through the structure. The model was successfully used to describe the features of two flux ropes observed in the magnetotail near lunar orbit during Galileo's first flyby of Earth (the field structure and the spacecraft path through the structure are shown in Figure 8). The ion pressure did not drop within either of these flux ropes and the total pressure near the centers of the structures was higher than the lobe pressure, as required by the model. The magnetic curvature force balances the excess total (thermal and magnetic) pressure in the tail.

None of the analytical models of flux ropes attempt to reflect the dynamical processes that produce flux ropes.

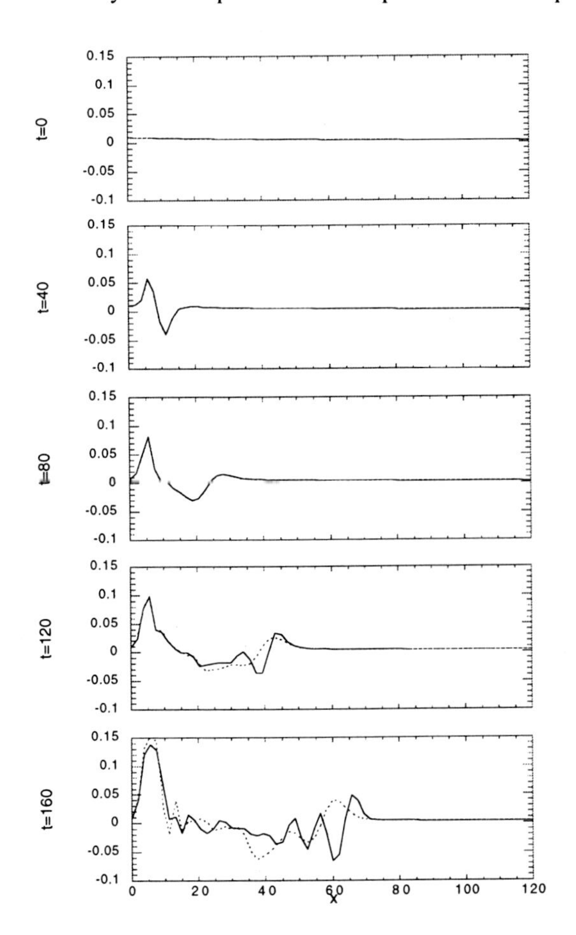

Figure 7. Variation of B_z along the *x* axis for the run with contact (solid line) and the run without (dash-dotted line)

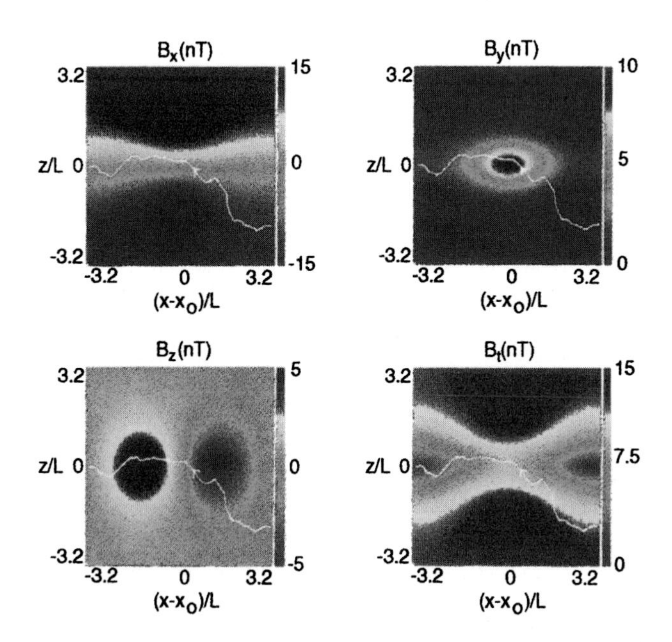

Figure 8. The Galileo trajectory through the modeled flux rope centered at 0714 UT on December 8, 1990 from Kivelson and Khurana [1995]. Color shows the three components of the magnetic field and the total field in nT. The Galileo trajectory through the flux rope is shown in white. The trajectory was inferred from measured values of B_x as discussed in the text. Distances are scaled in terms of the flux rope length scale which for this event was ~1000 km. The total current in this flux rope was 5.6×10^4.

However, by making it possible to estimate total currents and magnetic flux in the structures, the models may become useful for the study of substorms.

6. FLUX ROPE - IONOSPHERE COUPLING AND THE RELATION OF FLUX ROPES TO SUBSTORMS

Above we have remarked that in three-dimensional MHD simulations, flux ropes develop on closed plasma sheet field lines [Schindler et al., 1988; Ogino et al. 1990; Hesse and Birn, 1991; Walker and Ogino, 1996] and are therefore initially fully magnetically connected to the ionosphere. As it evolves, the flux rope may incorporate field lines linked to the solar wind. Beyond the relevance of magnetic connections to the low-latitude boundary layer (see above), it is conceivable that the magnetic connection to the ionosphere may also be important already at an early stage of the evolution. Spacecraft observations support the suggested linkage of flux ropes to the ionosphere. For example, the complex field topology explains the presence of 10 keV to 3 MeV ionospheric ions within individual flux ropes observed by Geotail [Lui et al., 1994] and the considerable variation of ion composition within a flux rope, although ionospheric ions may have been present in the plasma sheet already prior to substorm onset [e.g., Baker et al., 1982]. Tailward flowing energetic oxygen bursts observed in Geotail flux ropes by Wilken et al. [1995] are thought to occur on field lines linked at one end to the ionosphere and at the other end to the solar wind.

Kivelson et al. [1995] described how flux ropes draped across the near magnetotail would connect to the ionosphere. The currents in the core of a flux rope must flow toward dusk in order to produce the right-handed toroidal field (relative to the y-axis) required for a plasmoid-like cross section. The core field, whose sign is governed by the orientation of the IMF, as noted previously, may be parallel or antiparallel to the current. The sign of B_y in the core determines whether the duskward end of the flux rope enters the northern or southern hemisphere ionosphere. For example, when the core field has $B_y > 0$, the duskward end of the flux rope links to the ionosphere in the northern hemisphere and the dawnward end links to the ionosphere in the southern hemisphere. Thus there is unbalanced current into the northern ionosphere and out of the southern ionosphere. The closure currents have not yet been unambiguously identified, but may link to the substorm current wedge [Clauer and McPherron, 1974]. This possibility requires further study.

That there is a link between flux ropes and substorms is clear [Lepping et al., 1995; Machida et al, 1994; Moldwin and Hughes, 1991, 1992, 1993, 1994; Nagai et al, 1994; Slavin et al., 1992, 1993, 1995], but what is not yet clear is how important flux ropes may be dynamically. The difficulty with assessing the importance of flux ropes in the early stages of development is that they may be very highly localized. An example is given in Figure 9 from ISEE 1 and 2 data on March 21, 1979. ISEE 1 at (-21.32, -6.23, 1.10) was skimming the northern boundary of the plasma sheet, dipping in and out as suggested by intermittent decreases of the field magnitude between 18:30 and 18:40 UT. During this interval, ISEE 2, only 0.4 R_E away at (-21.14, -6.43, 0.75) entered the southern plasma sheet briefly between 18:34 and 18:34:30 UT and crossed back across the current sheet where it encountered a flux rope. Figure 9B shows that the total pressure within the flux rope was substantially larger than the lobe pressure, confirming the interpretation of the magnetic perturbations as a twisted field configuration. It is remarkable that no evidence of the existence of the flux rope is apparent in the ISEE 1 data despite the extremely small separation of the two spacecraft. This makes clear that the transverse scale of the flux rope structure is small. If such mesoscale structures play an important role in substorm dynamics, one can account for many of the difficulties that have plagued the field of substorm research. It is easy to understand that divergent conclusions about the underlying processes will be reached if observers spatially separated by fractions of an R_E record very different signatures.

7. OUTLOOK

While simulations of plasmoid evolution in magnetotail configurations with a net cross-tail B_y component always produce plasmoids with helical magnetic fields (as can be predicted by analytical theory, see Hesse [1995]), they do not typically develop strong core magnetic fields such as seen in ISEE 3 [e.g., Slavin et al., 1989; Moldwin and Hughes, 1992] and recently in GEOTAIL [e.g., Lepping et al., 1995] observations. An explanation of the absence or presence of such cross-tail field enhancements was recently provided by Hesse et al. [1996].

Hesse et al. suggested that core field enhancements could result from a heat loss from the plasmoid to the cold plasma of the low-latitude boundary layer or the magnetosheath, which had not been taken into account in their previous simulation models. When magnetic reconnection establishes a magnetic connection between the core regions of the plasmoid and the colder plasma at the magnetotail flanks, a plasma pressure reduction in this region could result from the combined effects of mass and heat flow. This pressure loss results in the collapse of the plasmoid core, due to magnetic tension effects in the x-direction and lobe pressure effects in the z-direction. The shrinking cross section mandates through flux conservation an increase in B_y, until, in the limiting case, the increased B_y balances magnetic pressure and tension forces. At this stage, the plasmoid would, at least in its inner regions, resemble a force free flux rope, as seen in a number of deep tail observations. Hesse et al. demonstrated this process by means of a simplified 2-1/2 dimensional MHD simulation. If this scenario proves to be correct at least some of the observed flux-ropes are aged plasmoids, i.e., evolve from initially more bubble-like structures. Future work will have to involve three-dimensional simulations to show this effect in a more complete model and will have to determine the rate at which thermal energy can be drained from the plasma within a flux rope. This work should also address the possibility that flux rope compression could result from draping of field lines behind the core [Mukai et al., 1996].

More observations will also be needed to establish fully the properties of flux ropes at different locations in the tail and to relate them to substorm phenomena. In particular, the actual location of the transition from bubble-like to flux-rope-like needs to be investigated. This is even more interesting in light of recent observations which appear to indicate that plasmoids can grow in size earthward of some $100R_E$ [Ieda et al., submitted to J. Geophys. Res., 1997]. Also, one is tempted to ask whether the mesoscale flux

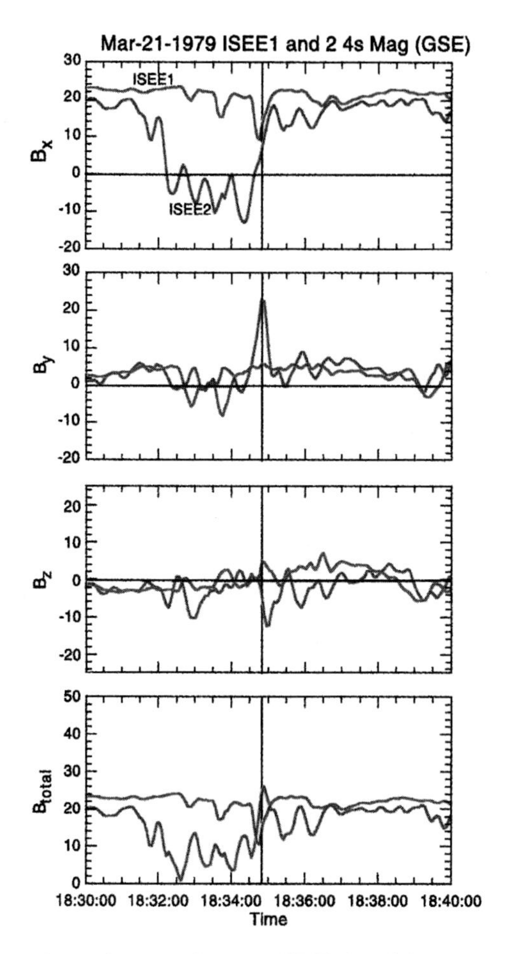

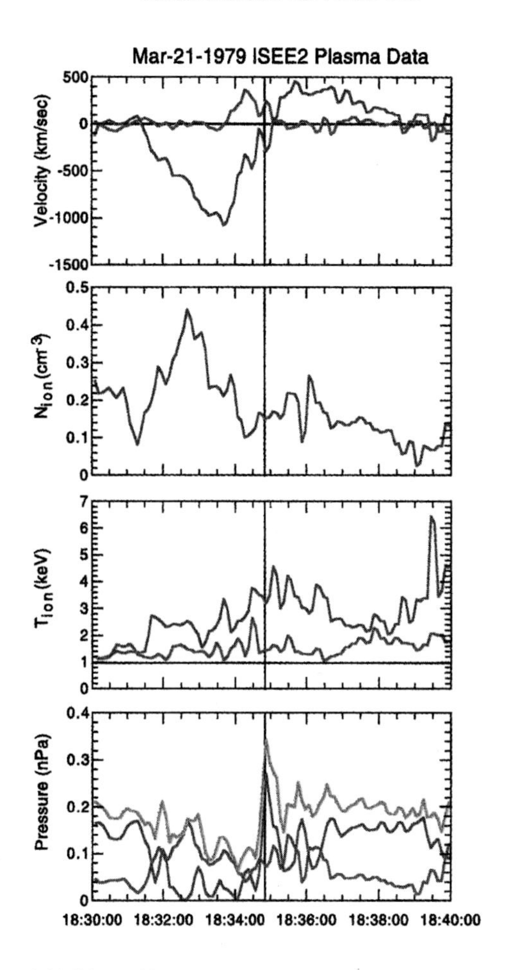

Figure 9. Data from the ISEE 1 and 2 spacecraft at GSE positions (-21.32, -6.23, 1.10) and (-21.14, -6.43, 0.75), respectively, between 18:30 and 18:40 UT on March 1, 1979. Again distances are in R_E. In Figure 9a, the three components and the total magnetic field (nT) are plotted in red for ISEE 1 and in blue for ISEE 2. A vertical line identifies the center of the flux rope. Plasma parameters measured on ISEE 2 are presented in Figure 9b. The upper panels show the x and y components of the flow velocity, the ion density, the ion temperature (blue trace) and anisotropy (red trace). The lowest panel shows the ion pressure (red), the magnetic pressure (blue) and the total pressure (green). A vertical line shows the center of the flux rope.

rope structures relate to the "bursty bulk flows" (BBFs) of Angelopoulos et al. [1992]. BBFs move rapidly earthward or tailward with a significant B_z component. They transport substantial momentum and magnetic flux along the tail. However, it has been argued that the structures must be localized across the tail or else they would transport more flux than is available in the magnetotail. However, if BBFs are similar to flux ropes in magnetic structure, they could extend far across the tail. This is because in a twisted flux rope, a single field line cuts the equator multiple times so the net flux transported need not be large even if the cross-tail extent is large. A rope-like structure moving rapidly along the tail could satisfy the conditions required for a BBF without being highly localized in the cross tail direction. The link between these two types of phenomena has not been explored and it may prove fruitful to look for a connection.

In conclusion, we find that although many features of magnetospheric flux ropes are presently understood, many interesting features remain yet to be explained and explored. Because of the ready accessibility to in-situ measurements and the progress in magnetospheric modeling, we expect that these uncertainties will eventually be removed. The present and future understanding obtained in these investigations might also prove useful to the study of other systems which exhibit flux ropes but are not as readily accessible to measurements. Thus, because of the ubiquitous presence of flux ropes in the solar wind, the solar corona, and potentially in astrophysical plasmas, the study of flux ropes in the magnetotail may be particularly rewarding.

Acknowledgments: MGK thanks R. J. Walker and L. Kepko for important contributions to the work discussed here. MGK was supported in part by the Atmospheric Sciences Division of the National Science Foundation under grant ATM-9314239 and by NASA under JPL contract 958694. MH was supported by the NASA Space Physics Theory and Supporting Research and Technology Programs.

REFERENCES

Angelopoulos, V., W. Baumjohann, C. F. Kennel, F. V. Coroniti, M. G. Kivelson, R. Pellat, R. J. Walker, H. Luhr, and G. Paschmann, Bursty bulk flows in the inner central plasma sheet, *J. Geophys. Res., 97*, 4027, 1992.

Baker, D. N., E. W. Hones, Jr., D. T. Young, and J. Birn, The possible role of ionospheric oxygen in the initiation and development of plasma sheet instabilities, *Geophys. Res. Lett., 9*, 1337-1340, 1982.

Baker, D. N., R. C. Anderson, R. D. Zwickl, and J. A. Slavin, Average plasma and magnetic field variations in the distant magnetotail associated with near-Earth substorm effects, *J. Geophys. Res., 92*, 71, 1987.

Birn, J., Magnetotail equilibrium theory: The general three-dimensional solution, *J. Geophys. Res., 92*, 11,101, 1987.

Birn, J., M. Hesse, and K. Schindler, Filamentary structure of a three-dimensional plasmoid, *J. Geophys. Res., 94*, 241, 1989.

Birn, J., The boundary value problem of magnetotail equilibrium, *J. Geophys. Res., 96*, 19,441, 1991.

Birn, J., Quasi-steady current sheet structures with field-aligned flow, *J. Geophys. Res., 97*, 16,817, 1992.

Birn, J., and M. Hesse, The magnetic topology of the plasmoid flux rope in a MHD simulation of magnetotail reconnection, in *Physics of Magnetic Flux Ropes*, Geophys. Monogr. Ser., Vol. 58, edited by C. T. Russell, E. R. Priest, and L. C. Lee, p. 655, AGU, Washington, D.C., 1990.

Birn, J., M. Hesse, and K. Schindler, Filamentary structure of a three-dimensional plasmoid, *J. Geophys. Res., 94*, 241, 1989.

Clauer, C. R., and R. L. McPherron, Mapping the local time-universal time development of magnetospheric substorms using mid-latitude magnetic observations, *J. Geophys. Res., 79*, 2811-20, 1974.

Elphic, R. C., C. A. Cattell, K. Takahashi, S. J. Bame, and C. T. Russell, ISEE-1 and 2 observations of magnetic flux ropes in the magnetotail: FTE's in the plasma sheet? *Geophys. Res. Lett., 13*, 648. 1986.

Fairfield, D. H., On the average configuration of geomagnetic tail, *J. Geophys. Res., 84,* 1950, 1979.

Fairfield, D. H., et al., Substorms, plasmoids, flux ropes, and magnetotail flux loss on March 25, 1983, CDAW 8, *J. Geophys. Res., 94,* 135, 1989.

Frank, L. A., W. R. Paterson, S. Kokubun, T. Yamamoto, and others, Direct detection of the current in a magnetotail flux rope, *Geophys. Res. Lett., 22*, 2697-700, 1995.

Hesse, M., Three-dimensional magnetic reconnection in space- and astrophysical plasmas and its consequences for particle acceleration, *Rev. Mod. Astr., 8*, 323 , 1995.

Hesse, M., and J. Birn, Plasmoid evolution in an extended magnetotail, *J. Geophys. Res., 96,* 5683, 1991.

Hesse, M., J. Birn, M. Kuznetsova, and J. Dreher, A simple model for core field generation during plasmoid evolution, *J. Geophys. Res., 101*, 10797, 1996.

Hones, E. W., Jr., Substorm processes in the magnetotail: Comments on "On hot tenuous plasma fireballs and boundary layers in the earth's magnetotail" by L. A. Frank, K. L. Ackerson and R. P. Lepping, *J. Geophys. Res., 82*, 5633, 1977.

Hones, E. W., Jr., D. N. Baker, S. J. Bame, W. C. Feldman, J. T. Gosling, D. J. McComas, R. D. Zwickl, J. A. Slavin, E. J. Smith, and B. T. Tsurutani, Structure of the magnetotail at $220R_E$ and its response to geomagnetic activity, *Geophys. Res. Lett., 11*, 5, 1984.

Hughes, W. J. and D. G. Sibeck, On the 3-dimensional structure of plasmoids, *Geophys. Res. Lett., 14*, 636, 1987.

Khurana, K. K., M. G. Kivelson, L. A. Frank, and W. R. Paterson, Observations of magnetic flux ropes and associated currents in Earth's magnetotail with the Galileo spacecraft, *Geophys. Res. Lett., 22*, 2087, 1995.

Khurana, K. K., M. G. Kivelson, and L. A. Frank, The relationship of magnetic flux ropes to substorms, *Adv. Space Res., 18*, 59, 1996.

Kivelson, M.G., C. F. Kennel, R. L. McPherron, C. T. Russell, D. J. Southwood, R. J. Walker, K. K. Khurana, P. J. Coleman, C. M. Hammond, V. Angelopoulos, A. J. Lazarus, and R. P. Lepping, The Galileo Earth encounter: The magnetometer data and allied measurements, *J. Geophys. Res., 98*, 11,299, 1993.

Kivelson, M. G., K. K. Khurana, Models of flux ropes embedded in a Harris neutral sheet: force-free solutions in low and high beta plasmas, *J. Geophys. Res., 100*, 23,637-45, 1995.

Kivelson, M.G., K.K. Khurana, R.J. Walker, L. Kepko, and D. Xu, Flux ropes, interhemispheric conjugacy, and magnetospheric current closure, *J. Geophys. Res., 101*, 27,341, 1995.

Lee, L. C., Z. W. Ma, Z. F. Fu and A. Otto, Topology of magnetic flux ropes and formation of fossil flux transfer events and boundary layer plasmas, *J. Geophys. Res., 98*, 3943, 1993.

Lepping, R. P., D. H. Fairfield, J. Jones, L. A. Frank, and others, Cross-tail magnetic flux ropes as observed by the GEOTAIL spacecraft, *Geophys. Res. Lett., 22*, 1193-6, 1995.

Lepping, R. P., J. A. Slavin, M. Hesse, J. A. Jones, and A. Szabo, Analysis of magnetotail flux ropes with strong core fields: ISEE 3 observations, *J. Geomag. Geoelectr., 4*8, 589, 1996.

Lui, A. T. Y., D. J. Williams, S. P. Christon, R. W. McEntire, V. Angelopoulos, C. Jacquey, T. Yamamoto, S. Kokubun, A preliminary assessment of energetic ion species in flux ropes/plasmoids in the distant tail, *Geophys. Res. Lett., 21*, 3019-22, 1994.

Ma, Z. W., A. Otto, and L. C. Lee, Core magnetic field enhancement in single X line, multiple X line and patchy reconnection, *J. Geophys. Res,. 99*, 6125, 1994.

Machida, S., T. Mukai, Y. Saito, T. Obara, T. Yamamoto, A. Nishida, M. Hirahara, T. Terasawa, and S. Kokubun, GEOTAIL low energy particle and magnetic field observations of a plasmoid a X_{GSM}=-142R_E, *Geophys. Res. Lett., 21*, 2995, 1994.

Moldwin, M. B., and W. J. Hughes, A 2 1/2-dimensional magnetic field model of plasmoids, in *Physics of Magnetic Flux Ropes*, Geophys. Monogr. Ser., Vol. 58, edited by C. T. Russell, E. R. Priest, and L. C. Lee, p. 663, AGU, Washington DC, 1990.

Moldwin, M. B., and W. J. Hughes, Plasmoids as magnetic flux ropes, *J. Geophys. Res., 96*, 14,051, 1991.

Moldwin, M. B., and W. J. Hughes, On the formation and evolution of plasmoids: A survey of ISEE 3 geotail data, *J. Geophys. Res., 97*, 19,259, 1992.

Moldwin, M. B., and W. J. Hughes, Geomagnetic substorm association of plasmoids, *J. Geophys. Res., 98*, 81, 1993.

Moldwin, M. B., and W. J. Hughes, Observations of earthward and tailward propagating flux rope plasmoids: expanding the plasmoid model of geomagnetic substorms, *J. Geophys. Res., 99*, 183-98, 1994.

Mukai, T., M. Fujimoto, M. Hoshino, S. Kokubun, S. Machida, K. Maezawa, A. Nishida, Y. Saito, T. Terasawa, and T. Yamamoto, Structure and kinetic properties of plasmoids and their boundary regions, *J. Geomag. Geoelec., 48*, 541, 1996.

Nagai, T., K. Takahashi, H. Kawano, T. Yamamoto, S. Kokubun, and A. Nishida, Initial GEOTAIL survey of magnetic substorm signatures in the magnetotail, *Geophys. Res. Lett., 21*, 2991, 1994.

Ogino, T., R. J. Walker, and M. Ashour-Abdalla, Magnetic flux ropes in 3-dimensional MHD simulations, in *Physics of Magnetic Flux Ropes, Geophys. Monogr. Ser.*, vol 58, edited by C. T. Russell, E. R. Priest, and . C. Lee, p. 669, AGU, Washington, D.C., 1990.

Richardson, I. G., S. W. Cowley, E. W. Hones, and S. J. Bame, Plasmoid-associated energetic ion bursts in the deep geomagnetic tail: Properties of plasmoids and the postplasmoid plasma sheet, *J. Geophys. Res., 92*, 9997, 1987.

Schindler, K., A theory of the substorm mechanism, J. Geophys. Res., 79, 2803, 1974.

Schindler, K., M. Hesse and J. Birn, General magnetic reconnection, parallel electric fields and helicity, *J. Geophys. Res., 93*, 5547, 1988.

Slavin, J. A., E. J. Smith, B. T. Tsurutani, D. G. Sibeck, H. G. Singer, D. N. Baker, J. T. Gosling, E. W. Hones, Jr., and F. L. Scarf, Substorm associated traveling compression regions in the distant tail: ISEE 3 geotail observations, *Geophys. Res. Lett., 11*, 651, 1984.

Slavin, J. A., M. F. Smith, E. L. Mazur, D. N. Baker, T. Iyemori, H. J. Singer, and E. W. Greenstadt, ISEE 3 plasmoid and TCR observations during an extended interval of substorm activity, *Geophys. Res. Lett., 19*, 825, 1992.

Slavin, J. A., M. F. Smith, E. L. Mazur, D. N. Baker, E. W. Hones Jr., T. Iyemori and E. W. Greenstadt, ISEE 3 observations of traveling compression regions in the Earth's magnetotail, *J. Geophys. Res., 98*, 15,425, 1993.

Slavin, J. A., et al., CDAW 8 observations of plasmoid signatures in the geomagnetic tail: An assessment, *J. Geophys. Res, 94*, 15153, 1989.

Slavin, J. A., C. J. Owen, M. M. Kuznetsova, M. Hesse, ISEE 3 observations of plasmoids with flux rope magnetic topologies, *Geophys. Res. Lett., 22*, 2061, 1995.

Walker, R. J., and T. Ogino, A global magnetohydrodynamic simulation of the origin and evolution of magnetic flux ropes in the magnetotail, *J. Geomag. Geoelectr., 48*, 765-780, 1996.

Wilken, B., Q. G. Zong, I. A. Daglis, T. Doke, and others, Tailward flowing energetic oxygen ion bursts associated with multiple flux ropes in the distant magnetotail: GEOTAIL observations, *Geophys. Res. Lett., 22*, 3267-70, 1995

Michael Hesse, Laboratory for Extraterrestrial Physics, NASA/Goddard Space Flight Center, Greenbelt, MD and Margaret G. Kivelson, Institute of Geophysics and Planetary Physics and Department of Earth and Space Sciences, University of California, Los Angeles, CA

Kinetic Ion Behavior in Magnetic Reconnection Region

M. Hoshino

The Institute of Space and Astronautical Science (ISAS)

A wide variety of non-Maxwellian ion distribution functions are now known to exist in the plasma sheet and the plasma sheet boundary layer (PSBL). Not only the gyrotropic ion distributions, but also the non-gyrotropic ion distribution functions are often observed during the passage of plasmoid. We discuss the formation and evolution of the ion distribution functions associated with magnetic reconnection. Using a test particle simulation, the ion dynamics in the plasma sheet is investigated. The ion distribution functions observed by the GEOTAIL satellite are compared with the test particle simulation results. We find that the basic behavior of the non-Maxwellian ions observed by GEOTAIL can be understood as the result of the meandering particle motions in a thin plasma sheet where an ion Larmor radius becomes comparable to the thickness of the plasma sheet. This kinetic ion behavior is thought to play an important role on plasma mixing and heating in magnetic reconnection process.

1. INTRODUCTION

The recent intensive spacecraft measurements of particle dynamics and magnetic fields in the magnetotail provide us with some new clues for understanding magnetotail dynamics. A wide variety of velocity distribution funcions which are not described by a simple Maxwellian distribution function are now known to exist in plasma sheet, plasmoids/flux ropes and the plasma sheet boundary layer (PSBL). A number of kinetic features observed by a satellite are thought to play a significant role on magnetotail plasma processes.

The observed velocity distribution functions often show the non-Maxwellian properties: one of them is an anisotropic, high speed ion beams, which are often observed when a satellite crosses the boundary between the lobe and the plasma sheet [e.g. Lui et al., 1977; DeCoster and Frank, 1979; Sarris and Axford, 1979; Eastman et al., 1984; Takahashi and Hones, 1988; Nakamura et al., 1991]. The boundary associated with the high speed ion beams is called as the plasma sheet boundary layer (PSBL). The PSBL ion distribution function consists of the cold lobe core ions and the tailward and/or earthward high speed beam ions along the magnetic field line. The beam speed often exceeds 1000 km/s. Another class of the observed non-Maxwellian distribution functions is the counter-streaming ions (CSIs) which is often observed inside the plasmoids/flux ropes [Mukai et al., 1996]. The CSIs are characterized by the two counter-streaming ion beams along the magnetic field line. The relative speed of two ion components often exceeds the local Alfven velocity of several hundred km/s. The ion distribution function with non-gyrotroppic bunched ions in velocity space has been observed in the current sheet [Frank et al., 1994] and in the region after the passage of the plasmoid [Hoshino et al., 1997].

There have been several attempts to provide theoretical and numerical explanations for the observed particle distributions in the magnetotail. Onsager et al. [1991] studied the anisotropic ion beam populations observed in the plasma sheet boundary layer (PSBL). They discussed that the high speed ion beam component is streaming out from the plasma sheet as one of the consequences of magnetic reconnection, and is propagating along the magnetic field line under the influence of the dawn-dusk electric field convection motion. They

New Perspectives on the Earth's Magnetotail
Geophysical Monograph 105

assumed a prescribed model electric and magnetic field, and followed the guiding center motion by assuming total kinetic energy and magnetic moment conservation.

As a self-consistent kinetic approach to the modeling of the plasma dynamics during magnetic reconnection, the numerical simulation studies have been carried out using the hybrid simulation code and the full-particle simulation code [Krauss-Varban and Omidi, 1995; Fujimoto et al., 1996; Lin and Swift, 1996; Lottermoser and Scholer, 1996; Hoshino et al., 1998]. They discussed the evolution of the velocity distribution functions as well as the plasma heating in the reconnecting current sheet with the self-consistent electric and magnetic fields. In association with the evolution of the reconnecting current sheet, the basic features such as the PSBL, slow shocks, and the magnetic diffusion region are successfully reproduced, and the plasma dynamics has been discussed. Due to the computational limitation, however, the signal to noise level of the distribution function does not seem to be enough to discuss the charged particle dynamics in details.

To know a better understanding of the structure of the distribution function, the test particle simulations have been used under prescribed electric and magnetic fields modeled after a magnetic reconnection region [e.g., Lyons and Speiser, 1982; Chen and Palmadesso, 1986; Martin and Speiser, 1988; Büchner and Zelenyi, 1989; Burkhart et al., 1990; Chen, 1992; Ashour-Abdalla et al., 1993; Büchner and Kuska, 1996; Ashour-Abdalla et al., 1996a; Ashour-Abdalla et al., 1996b]. They studied the dynamics of ions and electrons in a current sheet where the adiabatic invariants are not necessarily valid, and discussed the non-adiabatic particle energization mechanism associated with magnetic reconnection.

The recent spacecraft measurements of plasma dynamics reveal several new features on the ion dynamics, which are not discussed before. It is worth while revisiting the behavior of the test particle motion as well as the velocity distribution function. In this paper, we discuss the behavior of observed non-Maxwellian distribution functions in terms of comparison between the GEOTAIL observations and a test particle simulation. The objective of the present paper is to investigate theoretically the detailed structure of the velocity distribution function in a simple model reconnection region. We study first ion dynamics associated with magnetic reconnection using the test particle simulation under a two-dimensional steady state reconnection field geometry. We discuss the phase space structure as the functions of the dawn-dusk electric field, the thickness of the plasma sheet, and the distance from the X-type region. Next we compared the distribution function obtained by the test particle simulation with the GEOTAIL observations. We attempt to address the question of what unique signatures are produced by magnetic reconnection in the Earth's magnetotail.

2. TEST PARTICLE SIMULATION

To investigate the observed wide variety of complex and highly non-Maxwellian ion distribution function, we discuss a test particle behavior using a somewhat cleaner fields and particle source model. There have been several attempts to provide theoretical and numerical explanations for ion distributions in the magnetotail using the same framework [e.g., Lyons and Speiser, 1982]. The method used here is to integrate particles' orbits backward in time from a given position in magnetic reconnection region until the particles reach an assumed source region which corresponds to the lobe/mantle region. The Liouville theorem is then used to obtain the phase space density at the given position, and the particles at the given position are weighted according to an assumed Maxwellian distribution function at the source region. Namely, we have assumed that the phase space density is conserved along a particle's trajectory in the phase space.

2.1. *Test Particle Modeling*

We adopted the following prescribed fields,

$$\vec{B} = B_{0x}\tanh(z/\lambda_z)\vec{e_x} + B_{0z}\tanh(x/\lambda_x)\vec{e_z}, \quad (1)$$
$$\vec{E} = E_{0y}\vec{e_y}. \quad (2)$$

For easy of comparison to the magnetotail observation, we use the magnetotail coordinate system, i.e., the x, y, and z axes are respectively aligned with the sun-earth, the dawn-dusk, and south-north directions. Under the fields, a total of 30 x 30 x 30 initial velocities in the 3 dimensional velocity space at each given position are used to trace the particles backward in time until the particles reach the boundary at $z/\lambda_z = \pm 5$. We set $y = 0$ at the given position. We assume a shifted Maxwellian distribution function given below at the boundary, and the phase space density at the given position is reconstructed by assuming the conservation of the phase space density. The distribution at the source region is given by

$$f(v_x, v_y, v_z) = \quad (3)$$
$$\alpha_0 \exp\left(-\frac{(v_x - v_{0x})^2 + (v_y - v_{0y})^2 + (v_z - v_{0z})^2}{v_{th}^2}\right).$$

The drift velocities of v_{0j} are assumed to be

$$v_{0x} = v_{\text{tail}}\frac{B_x}{|B|}\text{sign}(B_x), \quad (4)$$
$$v_{0y} = 0, \quad (5)$$
$$v_{0z} = (v_{\text{conv}} + v_{\text{tail}}\frac{B_z}{|B|})\text{sign}(B_x), \quad (6)$$

Table 1. Parameters of Test Particle Simulation

RUN	B'_{0z}	λ'_x	ρ'_i	x'	z'	v'_{th}	v'_{tail}
1	0.4	2.5	0.25	-0.01	0.01	1.0	0.0
2	0.4	2.5	0.25	-6.0	0.01	1.0	0.0
3	0.4	2.5	0.25	-6.0	0.25	1.0	0.0
4	0.4	2.5	0.25	-20.0	3.0	1.0	0.0
5	0.4	2.5	0.01	5.0	1.0	25.0	-25.0
6	0.2	5.0	0.01	5.0	1.0	25.0	-25.0
7	0.2	5.0	0.01	5.0	1.0	25.0	0.0

where $v_{\rm conv} = E_{0y}/B_{0x}$ and $v_{\rm tail}$ is modeled after the tailward bulk flow in the magnetotail. The observed tailward plasma flow velocity $v_{\rm tail}$ is about 100 km/s - 300 km/s in the lobe/mantle region. For the sake of comparison to the observations in the magnetotail, the maximum phase space density α_0 in the source region is assumed to be $10^{-12.5}{\rm s}^3/{\rm m}^6$.

We use the following normalized variables defined by $v/v_{\rm conv} \rightarrow v'$, $x/\lambda_z \rightarrow x'$, $t\Omega_i \rightarrow t'$, $B/B_{0x} \rightarrow B'$, $E/(B_{0x}v_{\rm conv}) \rightarrow E' = 1$. The plasma parameters of $B'_{0z} = B_{0z}/B_{0x}$, $\lambda'_x = \lambda_x/\lambda_z$, and $\rho'_i = v_{\rm conv}/(\Omega_i\lambda_z)$ discussed in this paper are summarized in Table 1, where $\Omega_i = eB_{0x}/m$ is the ion gyro frequency and $v_{\rm conv}$ is the plasma convection speed toward the plasma sheet in the lobe.

Note that we cannot do Lorentz transform into the frame of $E_y = 0$, because the magnetic field B_z is not constant in the reconnecting current sheet. We will focus on particle acceleration near the X-type neutral region where the B_z is not uniform.

2.2. Distributions Near an X-type Reconnection Region

We first study the case that an ion Larmor radius is comparable to the thickness of the plasma sheet i.e., $\rho'_i = \rho_i/\lambda_z = 0.25$, where $\rho_i = v_{\rm conv}/\Omega_i$. Other parameters are shown in Table 1. For the sake of easy of comparison, you may take the following typical tail plasma parameters for Run 1-4: B_{0x} = 10 nT, B_{0z} = 4 nT, E_y = 2 mV/m, $v_{\rm conv}$ = 200 km/s, λ_z = 800 km, and $T_{\rm ion}$ = 200 eV. However, you may re-scale these parameters by keeping that $E_{0y}/(B_{0x}^2\lambda_z)$, $v_{\rm conv}\lambda_z$ and $T_{\rm ion}/v_{\rm conv}^2$ are constant. By re-scaling these parameters, the phase space density is also modified. For a slower convection velocity, the phase space density is enhanced.

Figure 1 shows the ion distribution function for Run 1, which is obtained near the X-type neutral region at $(x/\lambda_z, y/\lambda_z, z/\lambda_z)$ =(-0.01, 0, 0.01). The distance of the given position from the X-type neutral line is much shorter than the plasma sheet thickness. The top and left-hand panel (a) shows a slice of a 3 dimensional ion distribution function in a plane including the magnetic field and the ion convection flow vectors. The vertical and horizontal axes are aligned respectively with the magnetic field v_b and the plasma convection velocity $v_c = \vec{E} \times \vec{B}/B^2$. From this panel we find that the distribution function consists of two large bunched ion components and the additional two small bunched ion components attached to the main components.

In this region, ions are unmagnetized, namely, the local ion Larmor radius is larger than the characteristic scale length such as λ_z or λ_x. Therefore, the averaged ion velocity perpendicular to the magnetic field is not necessarily the same as the plasma convection velocity of $\vec{E} \times \vec{B}/B^2$. The unmagnetized ion distribution function becomes almost symmetric with respect to the v_x axis in the vicinity of the neutral sheet, though the distribution in Figure 1a, which is represented in the coordinate system referred to the magnetic field, slightly declines from the v_b axis. The degree of the decline is consistent with the magnetic field tilt of $\tan^{-1}(B_z/B_x) = -9.1°$ at the given observation position of $(x/\lambda_z, z/\lambda_z)$ =(-0.01, 0.01). (Note that the color scale in the distribution indicates the level of the phase space density from $10^{-50.0}$ to $10^{-12.5}$ s^3 m^{-6}, which covers the wider range than that of the plasma observations to date. It should be also noted that we used $v_{\rm tail} = 0$ in RUN 1-4. However, we find that the distributions with a tailward bulk flow do not change much their structure as long as $v_{\rm tail} \leq v_{\rm conv}$.)

Other panels (b)-(d) show the positions of particles at the source region. The particles at the given observation position are traced backward in time until the particles arrive at the source region of $z/\lambda_z = \pm 5$. The positions of the particles are shown in the color contour. In the diagram of the position z, the red (blue) color means z/λ_z = +5 (−5), namely, the particles in the red (blue) region come form the north (south) lobe/mantle. By the help of the x position diagram, one can find that the particles contributing to $v_b \sim 0$

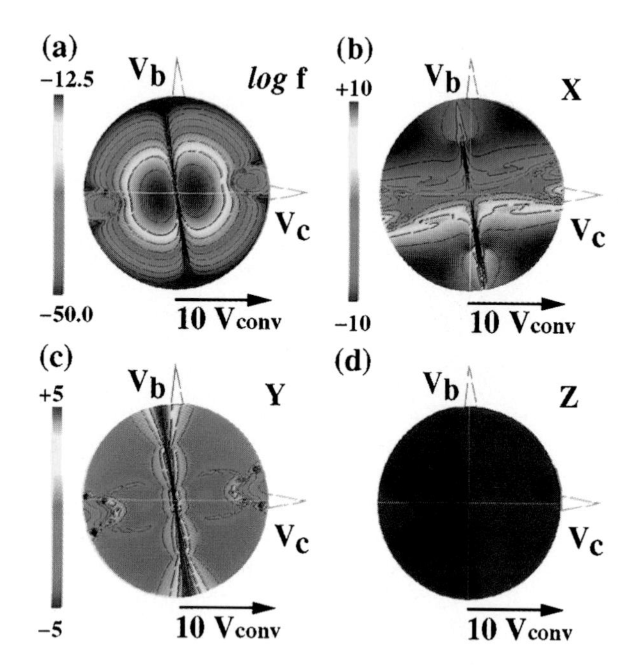

Figure 1. The ion distribution function near an X-type reconnection region and the positions of particles which are traced backward in time until the particles reach at $Z/\lambda_z = \pm 5$. The panel (a) is a slice of the 3-D distribution function in the v_b and v_c plane for $v_e = 0$, where v_b, v_c, and v_e mean respectively the magnetic field direction, the plasma convection flow direction defined by $\vec{E} \times \vec{B}$, and the electric field direction The panels (b)-(d) are the 3-D position diagrams. The color scale of the velocity distribution function shows the level of the phase space density from $10^{-50.0}$ to $10^{-12.5}$ s^3 m^{-6}, and the color scales in the position diagram x and y are respectively from -10 to +10 and from -5 to + 5, which are normalized by the thickness of the plasma sheet λ_z. The red (blue) color in the position diagram z corresponds to the north (south) lobe origin $z/\lambda_z = +5$ (-5).

in the velocity distribution function of the panel (a) are convected from above or below the given observation position, while the particles with $v_b > 0$ ($v_b < 0$) are convected from the negative x (positive x) lobe region, i.e., the tailward (earthward) lobe. From the position diagram of y, it is seen that the two small components with $(v_c, v_b) \sim (\pm 10 v_{\text{conv}}, 0)$ come from the larger negative y region, i.e., dawnward of the given observation point. Since the energy gain of particle is proportional to the distance travelling in the y direction, we find that the two small components strongly resonated with the dawn-dusk electric field E_y.

The above result can be understood by a simple idea of the meandering particle in the plasma sheet with a weak B_z magnetic field. Figure 2 shows four typical ion trajectories in our model fields. The two large components in Figure 1a are just convected from the northern and southern lobes (labelled by S0 and N0 in Figure 2), and the two small components are convected from the lobe and are once reflected back into the other hemisphere at the meandering radius where the local Larmor

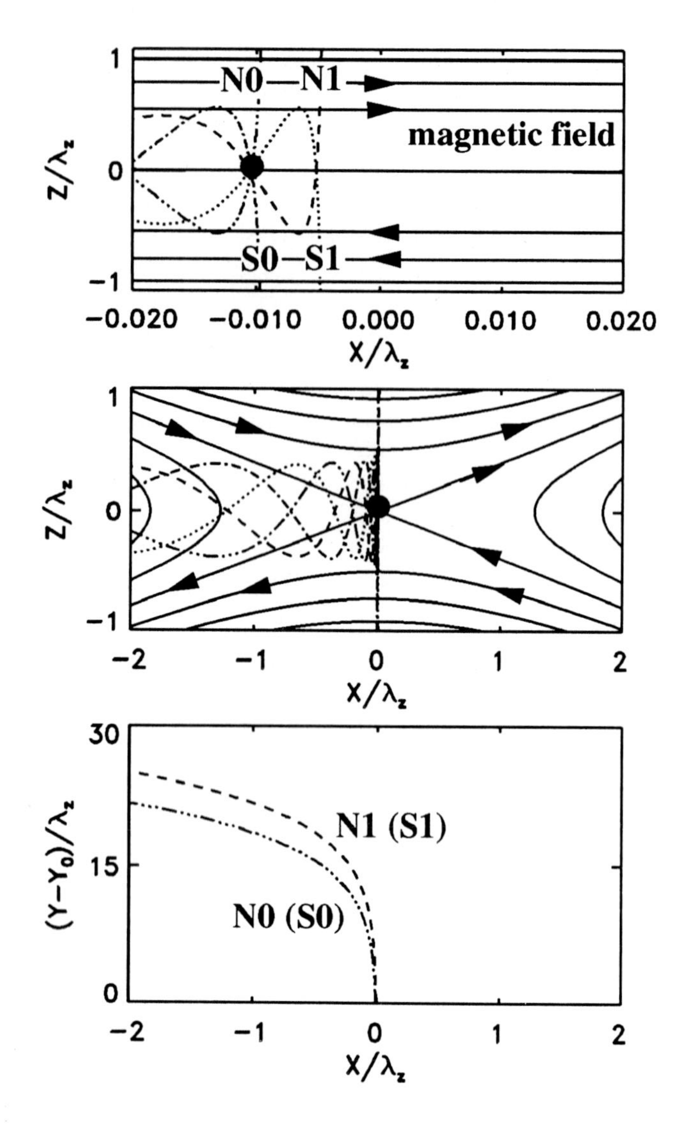

Figure 2. Selected ion trajectories in the x-z plane. The trajectores labelled by S (N) mean the particles with the south (north) lobe origin. The number after the label S or N indicates the number of reflection at the meandering boundary, where the Lamor radius becomes comparable to the distance from the neutral sheet. The closed circle shows the given position of observation. (Top) the close view of the reconnection region, (middle and bottom) the global views of the x-z plane and the x-z plane.

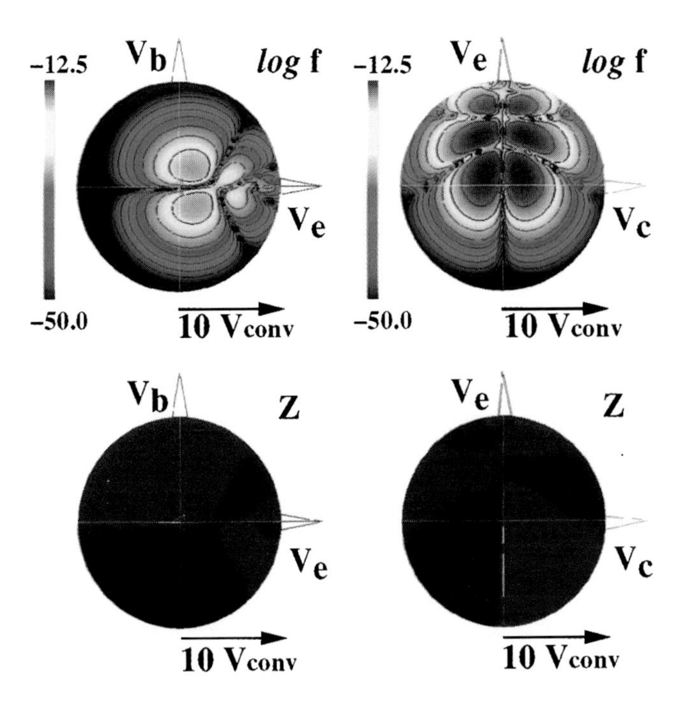

Figure 3. The ion distribution functions and the position diagrams z which are the same result as those in Figure 1, but the left-hand panels show the slices in the v_b and v_e plane, while right-hand panels are those in the v_e and v_c plane. The color scales of the velocity distribution function and of the position diagram z are the same as those in Figure 1.

radius become comparable to the distance from the neutral sheet (labelled by S1 and N1). The middle and the bottom panels in Figure 2 show the wide range of the particles' trajectories in the x-z plane and in the x-y plane, respectively. In the bottom panel, the origin of the y axis is shifted according to the lobe position, that is, $Y_0 \sim -3.1$ for N1 and S1 and $Y_0 \sim -0.22$ for N0 and S0. The trajectories of S1 and S0 in the x-z plane are almost same as those of N1 and N0. We find that the particles gain their energy near the X-type neutral region, because the distance in the y axis is proportional to the potential energy gain of the dawn-dusk electric field.

Figure 3 shows other slices of the 3-D distribution function and the position diagram. The left-hand panels show the cut including the magnetic field and the electric field, while the right-hand panels are the cut perpendicular to the magnetic field. The top panels are the velocity distribution functions, while the bottom panels are the corresponding position diagrams to know the origin of particles which come from the north or the south lobe. Based on an idea of the meandering particle orbit shown in Figure 2, we find that the two large components located around the center in the $v_e - v_b$ velocity space correspond to S0 and N0 ions, and other small components are N1 and S1. In the slice perpendicular to the magnetic field, i.e., in the $v_c - v_e$ velocity space, we can find S0 and N0 components near the center, N1 and S1 components in the middle v_e velocity range, and S2 and N2 components in the high v_e velocity region. The S2 and N2 particles are reflected twice at the south and the north meandering radii. The bunched meandering particles, i.e., S1, S2,... and N1, N2,..., are accelerated toward the electric field E_y direction, which process is recognized as the basic collisionless reconnection process in the magnetic diffusion region [e.g., Coppi et al., 1966; Laval and Pellat, 1968]. The structure of the bunched particles which are accelerated toward the electric field direction can be proposed as a signature of an X-type magnetic field diffusion region.

2.3. Distributions on Edge of Magnetic Diffusion Region

We study the distribution function in an edge of magnetic diffusion region where ions are still unmagnetized. The left-hand panels in Figure 4 shows the distribution function (top panel) and the position diagram of z (bottom panel) for the case of RUN 2 obtained at $(x/\lambda_z, y/\lambda_z, z/\lambda_z) = (-6, 0, 0.01)$. In addition to two main ion components flowing along the magnetic field, i.e., $(v_c, v_b) \sim (0, \pm 10 v_{\rm conv})$, we find additional four bunched ion components in the velocity space, i.e., $(v_c, v_b) \sim (4v_{\rm conv}, \pm 2v_{\rm conv})$ and $(v_c, v_b) \sim (6v_{\rm conv}, \pm 4v_{\rm conv})$. The phase space densities of those bunched ions decrease with increasing v_c velocity.

It is seen that the topology of the main two populations is much the same as that observed in the vicinity of the X-type region in Figure 1, if we study those distribution functions in the v_x, v_y, and v_z coordinate system. Note that the magnetic field in Figure 4 (Figure 1) is almost parallel to the z axis (the x axis). We find that the north-south B_z magnetic field does not strongly affect on the meandering motion in this region. Together with the position diagram below, we know that the main components are the S0 and N0 populations, the bunched ions with $v_c \sim 4v_{\rm conv}$ are the S1 and N1 populations, and another small components with $v_c \sim 6v_{\rm conv}$ are S2 and N2.

To understand the three dimensional structures of the bunched ions, the isosurface of the distribution function with the phase space density of 10^{-15} s^{3}m^{-6} is plotted in Figure 5. Figures 5a and 5b are the three-dimensional views of RUNs 1 and 2, respectively. We also find that the topology of those bunched ions in the $v_b - v_c$ plane is similar to the bunched ions in the $v_b - v_e$ plane and

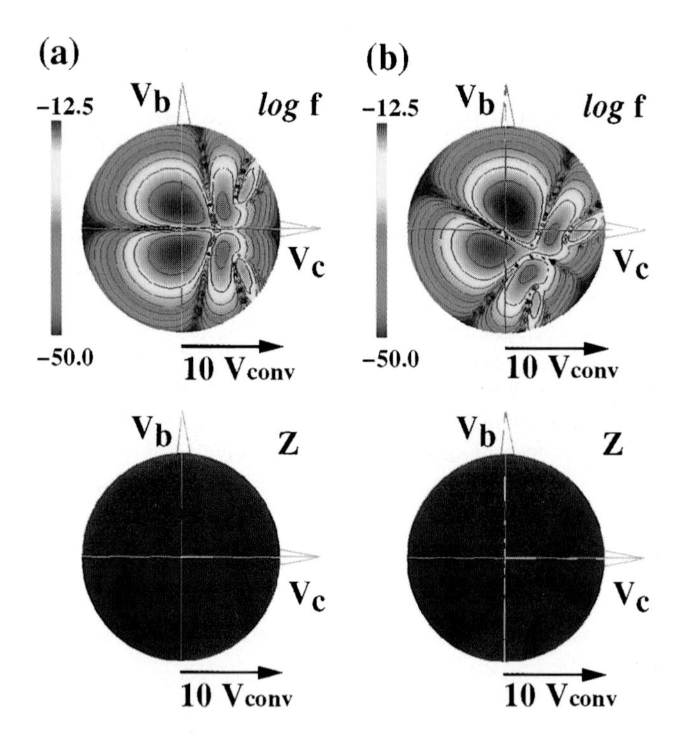

Figure 4. The ion distribution functions obtained in the edge of the magnetic diffusion region where ions are still unmagnetized into the magnetic field. The left-hand and the right-hand panels show respectively the distribution function at the neutral sheet (RUN 2) and in the region slightly above the neutral sheet (RUN 3). The bottom two panels are the corresponding position diagrams of z. The color scales of the velocity distribution function and of the position diagram z are the same as those in Figure 1.

the $v_e - v_c$ plane observed near an X-type region after a coordinate transformation. We also find the multi-reflected bunched ions (S1, N1, S2, N2, ...) form their "strings of pearl" toward the direction between the v_c and v_e axes. We find that those meandering particles start to be partially magnetized.

The distribution function and the position diagram at the same x position, but above the neutral sheet is shown in the right-hand panels in Figure 4. The given position is $(x/\lambda_z, y/\lambda_z, z/\lambda_z)$ =(-6, 0, 0.25). The bunched ion structure is similar to that in the left-hand panel in Figure 4, but the global structure is rotated toward the clockwise direction, namely, toward the x-axis. Another difference is found in the separatrix region between S0 and S1 ions. For the distribution in the left-hand panel, the bunched ions of S0 and S1 are separated, while the distribution in the right-hand panel has a "bridge" structure which connects the S0 population with the S1 population. This can be clearly confirmed from the bottom panels of the position diagrams z. We can expect that the bunched ions of S0 and S1 will merge each other as it goes toward the edge of the meandering region, which width is approximated by $\sqrt{v_{md}\lambda_z/\Omega_i}$, where v_{md} is the velocity of the meandering motion.

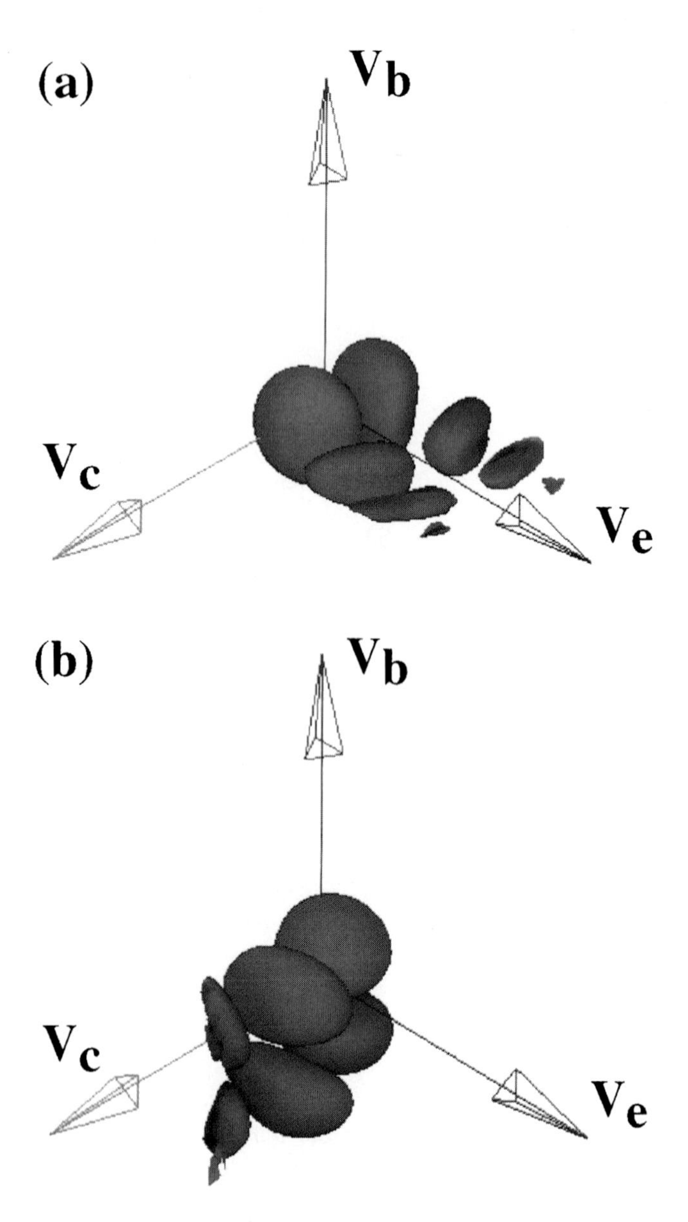

Figure 5. The isosurface of the phase space density for the 3-D ion distribution function. (a) the same distribution as RUN 1 or Figures 1 and 3, (b) the same distribution as RUN2 or Figure 4a.

2.4. Distribution Around Boundary Between Lobe and Plasma Sheet

Since the magnetic field becomes stronger with increasing distance from an X-type region, it is expected that the meandering motion starts to be magnetized in the magnetic field. The meandering ions accelerated in a weak magnetic field region are ejected from the plasma sheet due to the Speiser motion [1965], and are flowing along the magnetic field line. The velocity distribution function will become gyrotropic, i.e., symetric with respect to the magnetic field direction.

Figure 6 shows the ion distribution function and its position diagrams obtained in the boundary region between the lobe and the plasma sheet and far away from the X-type region (RUN 4), i.e., (x/λ_z, y/λ_z, z/λ_z) =(-20, 0, 3). In addition to the isotropic, core ion component, we find another beam component flowing antiparallel magnetic field direction. The beam component has a larger thermal spread perpendicular to the magnetic field. From the position diagrams, we find that the core component comes from $x/\lambda_z \approx -20$, $y/\lambda_z \approx 0$, and the north lobe. This is suggestive that these ions are convected just from the above of the given observation position without any significant particle acceleration. On the other hand, the beam ions come from $x/\lambda_z \approx 0$, $y/\lambda_z < 0$, and both north and south lobes, which means that the particles gain their energy from the dawn-dusk electric fields. By tracing the particles' trajectories (not shown here), we find that the beam ions are accelerated around the X-type region, and that are ejected from the plasma sheet after the particles gain their energy. It should be noted that the PSBL ion beam population has a fine structure as the result of the particle mixing originated from both north and south lobes.

We also studied the ion distribution function obtained in the lobe side, and we found that the beam population has a higher speed than those obtained in the plasma sheet side. This result is consistent with the interpretation of the "velocity filter" effect [e.g., Onsager et al., 1991]. Namely, the equatoward plasma motion due to $E \times B$ drift produces a spatial layering of particles according to the ion beam speed.

2.5. Dependence of E_y

Next we discuss the distributions for the case with a small convection electric field E_{0y}. The ion Larmor radius defined by v_{conv}/Ω_i is much smaller than the thickness of the plasma sheet. This parameter regime is similar to that used by Lyons and Speiser [1982], Martin et al. [1994], and Joyce et al. [1995]. For the sake of easy of comparison, one may assume the following typical tail plasma parameters for Run 5: $B_{0x} = 10$ nT, $B_{0z} = 4$ nT, $E_y = 0.08$ mV/m, $v_{conv} = 8$ km/s, $\lambda_z = 2 \times 10^4$ km, and $T_{ion} = 200$ eV. In this subsection, we also study how a finite tailward bulk plasma flow in the lobe/mantle region affects the distribution functions in the plasma sheet.

The top and left-hand panel (a) in Figure 7 shows a slice of a 3 dimensional ion distribution function in a plane including the magnetic field and the electric field E_y, while the right-hand panel (b) shows another slice in a plane including the magnetic field and the ion convection velocity vector defined by $\vec{E} \times \vec{B}$. We chose the given observation position in the earthward side of the X-type region, i.e., at (x/λ_z, y/λ_z, z/λ_z) =(5, 0, 1). We assumed that the finite tailward flow velocity $v_{tail}/v_{conv} = -25$. It is seen that the velocity distribution becomes almost an isotropic distribution compared with the large E_y and the thin current sheet case, though the bunched ion structure can be still observed in the lower velocity region around the center. In the higher velocity region, we can find the region with a large gradient of the phase space density. This behavior is similar to the "ridge" structure discussed by Martin and Speiser [1988] and Speiser and Martin [1992].

The bottom two panels in Figure 7 show the position diagrams of x and y in a plane including the magnetic

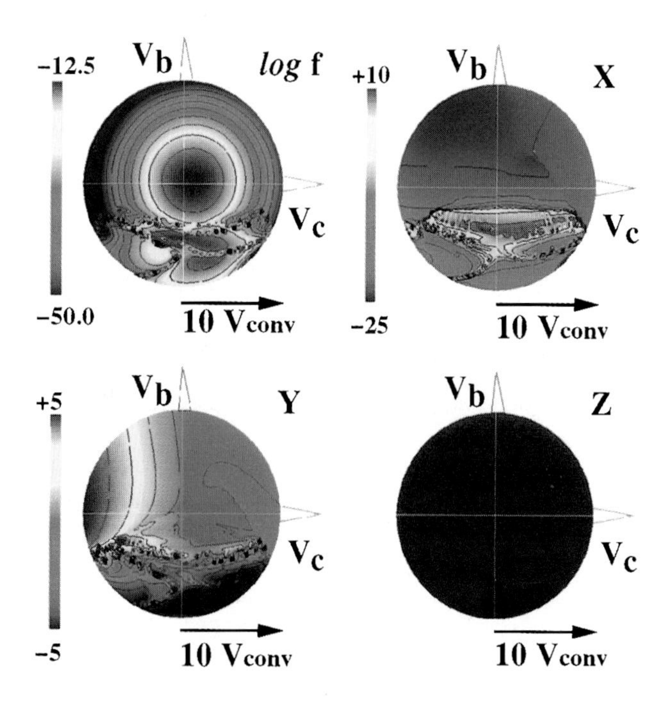

Figure 6. The ion distribution functions and the position diagrams around the boundary between the lobe and the plasma sheet (RUN 4). The same format as Figure 1.

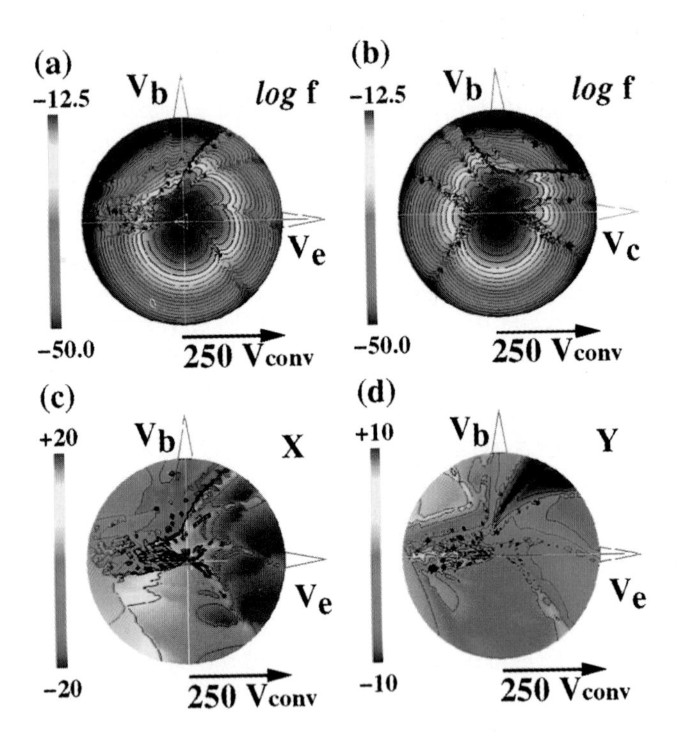

Figure 7. The ion distribution functions and the position diagrams for the small Larmor radius case compared with the thickness of the plasma sheet (RUN 5). The panel (a) shows a slice of the 3-D distribution function in the v_b and v_e plane for $v_c = 0$, and the panel (b) is another slice of the distribution in the v_b and v_c plane for $v_c = 0$. The bottom panels (c) and (d) are the corresponding position diagrams X and Y in the plane including the v_b and v_e vectors.

field and the electric field. It is seen that the particles above the ridge come from a large negative x region with $x/\lambda_z < -10$, while the particles below the ridge originate from the positive region. We also find that the particles around the ridge come from a large negative y region, which means that the particles gain their energy from the dawn-dusk electric field around the X-type neutral region.

The ridge structure becomes more clear when the amplitude of B_z becomes small. The left-hand panel in Figure 8 shows the case with $B_z' = B_{0z}/B_{0x} = 0.2$. A slice of the 3-D distribution function is shown in the top and left-hand panel, and the corresponding 2-D distribution function averaged by means of $\frac{1}{2\pi}\int f(v_\parallel, v_\perp \cos\phi, v_\perp \sin\phi) d\phi$ is shown in the bottom panel. We find the clear ridge structure associated with an isotropic component. On the other hand, the ridge structure becomes faint as the tailward plasma flow velocity relative to the reconnection structure becomes slow. We assumed no tailward flow in the right-hand panel in Figure 8. Other parameters are the same. Although a small "corn" structure instead of the ridge can be found in the almost same location appeared in the ridge distribution, the distortion of the phase space density is little.

The formation of the ridge can be understood as follows: The origin of the particles above (below) the ridge comes from the tailward (earthward) side of the X-type region with the earthward (tailward) flow velocity by passing through the X-type region. If the lobe plasma as the source region has the tailward flow, the phase space density of the source distribution with the earthward flow is smaller than that with the tailward flow. Therefore, the ridge structure is controlled by the tailward flow, which makes the difference of the phase space density in the source region [Speiser and Martin,1992]. For no tailward plasma flow case, the difference of phase space density between below and above the ridge disappears, and the corn structure with the enhancement of the phase space density can be seen as the result of the particle acceleration around the X-type region.

3. PLASMOID OBSERVATIONS

In this section, we review several common characteristics of the observed ion distribution functions in association with plasmoid/flux rope in the Earth's magnetotail, and then we discuss whether or not those observed ion distribution functions can be understood based on our test particle simulation.

We show the GEOTAIL typical plasmoid observation on September 18, 1993 in Figure 9. From the top, time series of three components of the magnetic field (nT), three components of bulk flow velocity (km/s), ion temperature (keV), and plasma density (cm^{-3}). The Geocentric Solar Magnetospheric (GSM) coordinate system is used. The spacecraft was situated in the night side magnetotail, about $70R_E$ from the Earth.

A ground-based magnetometer at Kakioka station detected a Pi-2 onset at 10:14 UT. The time series of magnetic variations show a north-to-south bipolar signature in the B_z component around 10:18 UT. Associated with the bipolar magnetic field, the tailward plasma flows are enhanced, and the plasma density and temperature increase. The evolution of the velocity distribution function during the plasmoid passage for this event has already been discussed by Mukai et al. [1996] and Hoshino et al. [1998]. We review shortly a family of the observed ion distribution functions in order to compare with the test particle simulation results obtained above.

Prior to the northward turning of the B_z magnetic field around 10:17:30 UT, the PSBL (plasma sheet boundary layer) ion beams, which consist of the cold lobe ions and the tailward high speed beam ions of

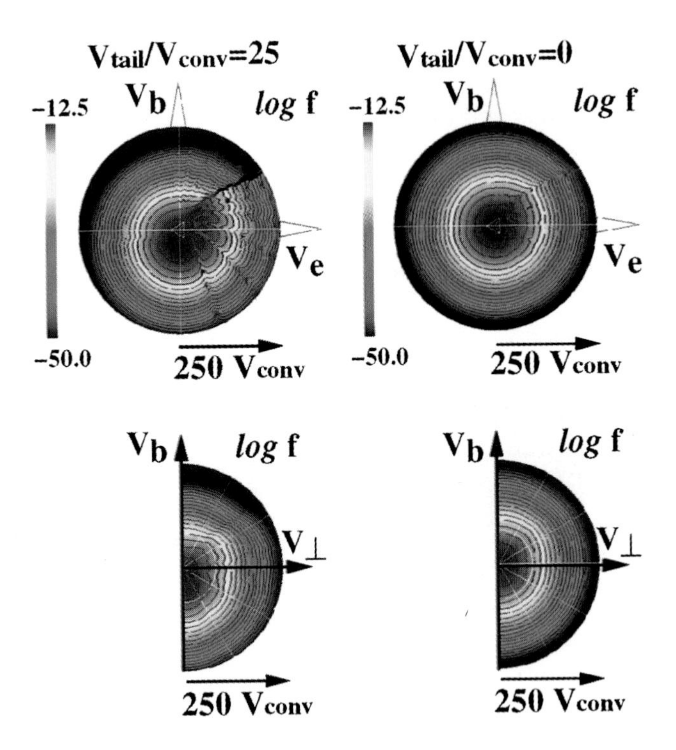

Figure 8. Effects of tailward plasma velocity on the ridge structure. The left-hand panels show the case of $v_{\mathrm{tail}}/v_{\mathrm{conv}} = 25$ (RUN 6), while the right-hand panels are the case of $v_{\mathrm{tail}} = 0$ (RUN 7). The 2-D distribution functions averaged in the phase angle ϕ perpendicular to the magnetic field are shown in the bottom panels.

$\gtrsim$2500km/s, are observed from time interval of 10:17:11-10:17:23 UT. Figure 10a shows an ion distribution function observed in the spacecraft frame at the time interval of 10:17:59 - 10:18:11 UT. The vertical and horizontal axes are respectively aligned to the directions of the magnetic field and the plasma bulk velocity perpendiculat to the magnetic field. The small red spot situated close to the center corresponds to the cold lobe plasma, and the high speed component flowing the opposite direction against the magnetic field is the PSBL ion beam. The white arrow indicates the ion bulk flow vector against the center of the spacecraft frame. The beam velocity becomes slower compared with that observed in the lobe side region due to the velocity filter effect [e.g., Onsager et al., 1991].

After the passage of the center of the plasmoid, that is, after the southward turning of the B_z polarity, the "three bunched" ions are observed at the time interval of 10:19:11-10:19:23 in Figure 10b, and the "four bunched" ions at the time interval of 10:19:35-10:19:47 in Figure 10c. The corresponding, schematic ion distribution functions are shown in Figure 11. In Figure 10b/11b we call this distribution as the three-bunched ions, because we find third ion component in addition to two counter-streaming ions along the magnetic field. We call the distribution in Figure 10c/11c as the four-bunched ions, because one may find four maxima of the phase space density embeded in a partial ring structure in the velocity space. The events (b) and (c) correspond to the stage where the southward B_z magnetic field is decreasing, which is suggestive that the satellite is approaching to the X-type neutral point. Since we find that the event (b) has the larger magnetic field B_x than the event (c), the event (b) with the three-bunched

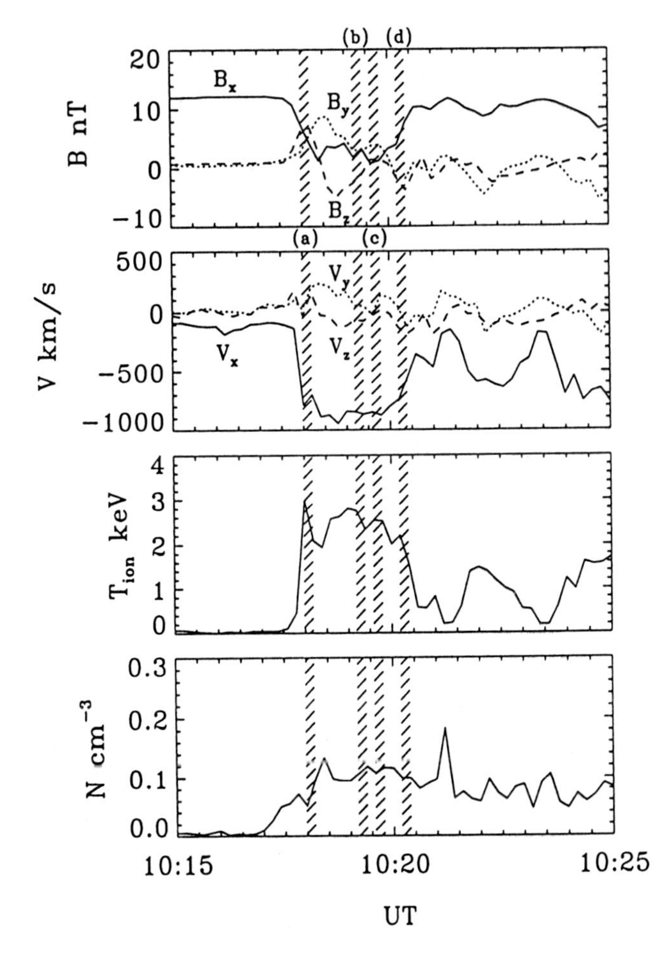

Figure 9. Typical plasmoid event observed by GEOTAIL on September 18, 1993. The spacecraft was situated in the night side magnetotail, about 70 R_E form the Earth. From the top, time series of three components of the magnetic field (nT), three component of bulk flow velocity (km s^{-1}, ion temperature (keV), and plasma density (cm^{-3}). The Geocentric Solar Magnetospheric (GSM) coordinate system is used.

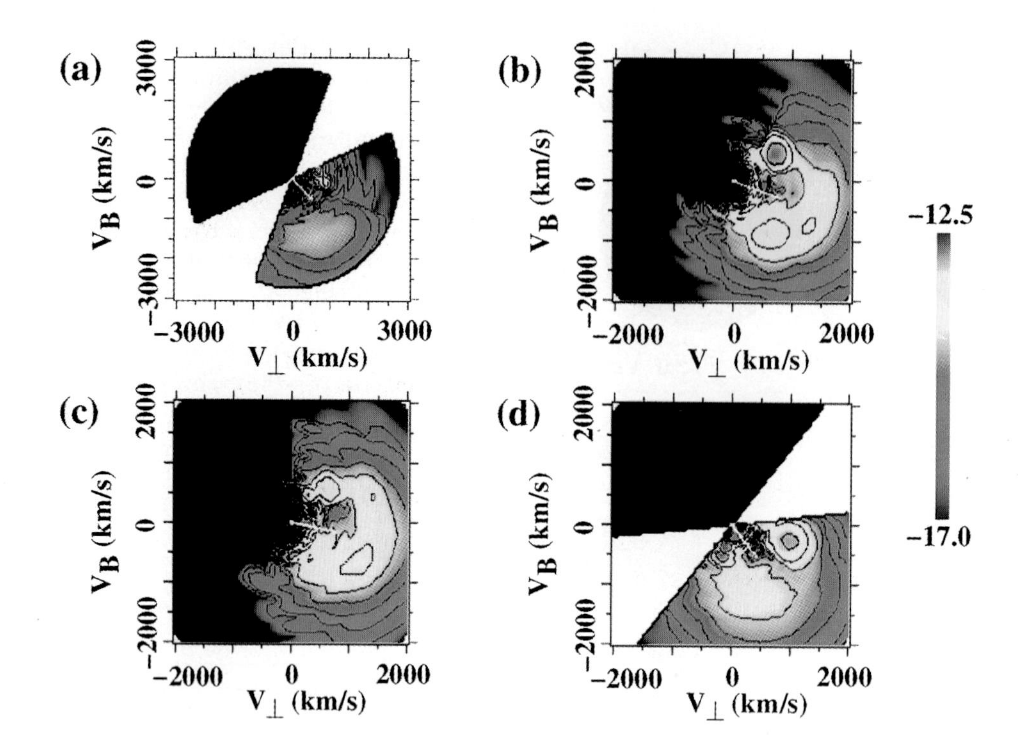

Figure 10. The ion distribution functions observed by GEOTAIL in the spacecraft frame. The time intervals for each distribution functions (a)-(f) are indicated by hatching in Figure 9. The cut of the 3 dimensional distribution function is show in the reference coordinate system to the magnetic field. v_B, $v_\perp$, and $v_{B\times v}$ correspond to the magnetic field direction, the plasma convection flow direction, and the electric field direction determined by $\vec{E}+\vec{v}\times\vec{B}=0$, respectively. The center of the cut is the origin of the spacecraft frame, and the white arrow shows the ion bulk flow velocity. The color scale show the level of the phase space density from 10^{-17} to $10^{-12.5}$ s^3 m^{-6}.

ions is thought to be observed slightly above the neutral sheet compared with the event (c).

Figure 10d shows another class of non-Maxwellian ion distribution function observed just after the passage of the plasmoid at the time interval of 10:20:11 - 10:20:23 UT. We find that the ion distribution function is characterized by two cold ion components perpendicular to the magnetic field together with a hot "halo" population. From another slice of 3-D distribution function in the plane perpendicular to the magnetic field (not shown here), we also find the bunched ions is not symmetric with respect to the magnetic field, which is suggestive that this distribution is described as the non-gyrotropic ions. The details of the distribution functions has been discussed by Hoshino et al. [1998].

In this event, we did not observe the plasma flow reversal from the tailward to the earthward flow after the passage of the plasmoid, and the satellite seemed to be located in the post-plasmoid plasma sheet (PPPS) of the tailward X-type region [Richardson and Cowley, 1985; Richardson et al., 1987]. We have also studied the distribution functions observed in the earthward side of an X-type reconnection region in order to investigate the ridge phenomena. In the near-earth magnetotail of 15 $R_E \sim 40$ R_E from the earth, we often observe the plasma flow reversal associated with the passage of the plasmoids. Roughly speaking, the observed distribution functions, however, show a similar behavior to that observed in Figure 10, and we could not find any clear evidence of the ridge structure.

4. COMPARISION BETWEEN OBSERVATIONS AND THEORY

By comparing the GEOTAIL observations with the single particle kinetic theory, we find that the basic signatures in the observed distribution functions can be

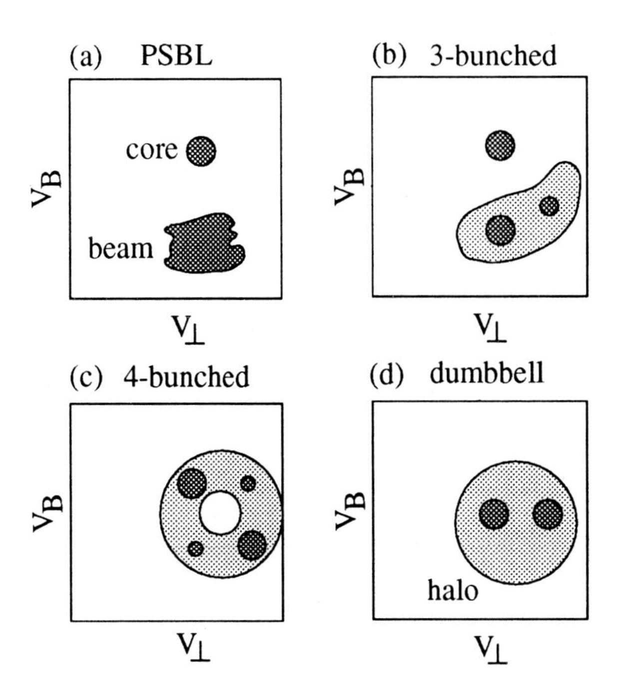

Figure 11. The schematic views of the ion distribution functions in Figure 10.

understood by the simple test particle simulation where the meandering particles play an important role on the formation and evolution of the distribution function:

The two bunched ions perpendicular to the magnetic field are found in the GEOTAIL observation just after the passage of the plasmoid (Figure 10d), while we find the similar distribution in the test particle simulation (Figure 1). (Note that the scale of the color contour of the phase space density is not same.) Since not only non-Maxwellian, gyrotropic distributions, but also non-gyrotropic distributions are often observed in the magnetotail, we think that the ion Larmor radius is not necessarily much smaller than the thickness of the plasma sheet, and that the ion thermalization time is not much shorter than the dynamic time of the reconnection evolution in the magnetotail.

During the passage of the plasmoid, we sometimes find the three-bunched ions (Figure 10b) and the four-bunched ions (Figure 10c), which are respectively compared to the test particle distributions obtained in Figures 4a and 4b. Although the bunched ions in Figure 4b show a complicated structure compared with the observation in Figure 10b, the complicated, fine structure will be smeared out if we take into account of the time and spatial resolution of a satellite observation. The agreement between the observation and the test particle simulation seems good. It should be noted that the distribution with the four-bunched ions is observed in the neutral sheet side compared with the distribution with the three-bunched ions. This behavior is also consistent with the test particle simulation.

The PSBL anisotropic ion beams obtained by GEOTAIL (Figure 10a) has almost the same feature as that obtained by the test particle simulation of Figure 6. The PSBL crescent ion beam population observed by GEOTAIL has often a fine, "wisp" structure. We interpret that the wisp structure is produced as the result of the plasma mixing of both the north and the south lobe cold ions near the X-type reconnection region.

From a point of view of the meandering particle motion, we can estimate the velocity difference between the two ion components observed near an X-type region in Figure 10d. The ions gain their energy by travelling in the y direction in the magnetic diffusion region. The energy gain in a part of the gyro-motion with respect to B_x magnetic field is equated to the change in potential energy in the y direction, i.e.,

$$\frac{1}{2} m_i v^2 \approx eEd_i, \tag{7}$$

where d_i is the meandering width which is defined by $\sqrt{v\lambda_d/\Omega_i}$. λ_d is the characteristic scale length of B_x in the magnetic diffusion region, and Ω_i is the ion cyclotron frequency. We have assumed that the distance which the particle travels in the y direction is almost equal to the meandering width [Sonnerup 1971]. Therefore, the velocity difference between non-gyrotropic ions $\Delta v = 2v$ is described as

$$\frac{\Delta v}{v_{\mathrm{conv}}} = 2^{5/3} (\frac{\lambda_d}{\rho_i})^{1/3}, \tag{8}$$

or

$$\begin{aligned} \Delta v &\approx 2^{5/3} (\frac{eE_y^2 \lambda_d}{m_i B_x})^{1/3} \\ &\approx 1700 \text{ km/s} (\frac{E_y}{4\text{mV/m}})^{2/3} (\frac{\lambda_d}{0.1\text{R}_e})^{1/3} (\frac{12\text{nT}}{B_x})^{1/3}. \end{aligned} \tag{9}$$

The maximum intensity E_y in the lobe observed after the passage of the plasmoid of Figure 9 was about 5.7 mV/m. The dawn-dusk electric field intensity is expected to become 1 $\sim$ 4mV/m during the magnetic reconnection phase [Saito et al., 1996]. The thickness of the diffusion region is an open question. If the meandering ions play an important role on the energy dissipation, λ_d may become as thin as d_i. By equating λ_d to d_i, we obtain $\lambda_d = E_y/B_x\Omega_i \approx 2.9 \times 10^2$ km. The observed velocity difference between two ion components in Figure 10d is almost consistent with the above estimation based on the meandering particle's motion.

5. DISCUSSION AND CONCLUSIONS

Reconnection is believed to be responsible for the plasma heating and acceleration associated with substorms. We discussed the evolution of ion distribution functions which often show non-gyrotropic properties. We studied how and where the non-Maxwell distributions are formed, and compared the test particle simulation result with the GEOTAIL satellite observations. In our theoretical modelling, we focused on the ion distributions observed near an X-type magnetic reconnection region where the ion thermalization time scale is smaller than the dynamic time scale of the evolution of the ion distributions.

We found that the non-gyrotropic bunched ions in the velocity space are often observed near the X-type region, and that the thickness of the plasma sheet is not much larger than the ion Larmor radius. The finite Larmor motion plays a significant role on the ion dynamics. We proposed the dumbbell-like, non-gyrotropic ion distribution function as a signature of the X-type region/the magnetic diffusion region. We have confirmed our proposal with the GEOTAIL plasma sheet observations after the passage of the plasmoid.

We also discussed that the dumbbell-like, non-gyrotropic distribution is modified as it goes toward the edge of the magnetic diffusion region where ions are still unmagnetized. The distribution consists of four bunched ion components. This kind of distribution can be often observed during the plasmoid passage if the magnetic field B_x is weak. In the region slightly above the neutral sheet where ions are though to be still unmagnetized, our test particle simulation shows the modified four bunched ion structure which is rotated according to the degree of the magnetic field direction against the x axis. The distribution seems to be different form the observed distribution given in Figure 10b. However, by taking into account the time and spatial resolution of a satellite observation, we think that a fine structure will be smeared out. The agreement between the GEOTAIL observation and the test particle theory seems good.

Our test particle simulation can also successfully reproduce most of natures of the observed PSBL ion beams in magnetotail. The high speed ion beams obtained in our test particle simulation appear in the boundary between the lobe and the plasma sheet with the "velocity filter" effect [e.g., Onsager et al., 1991].

One may find other type of the ion distribution function in the magnetotail, though we think that the distribution function could be understood in terms of a family of the bunched ions discussed in this paper. For example, the CSIs observed by Mukai et al. [1996] may be an intermediate state between the PSBL ions and the four (or three) bunched ions. In this paper, we only discussed the smiple and basic features of the family of the distribution functions.

Several other important issues still remain to be resolved:

We assumed that the ion temperature $T_{\rm ion}$ in the source region is about 200 eV in our test particle study, though the temperature seems to be higher than that of the standard temperature observed in the lobe/mantle. If we assumed the 50 eV ion temperature in the source region, the temperature of the bunched ion seems to become colder than that observed in the magnetotail. Moreover, in the GEOTAIL observations, we often observe a hot "halo" component associated with the multi-bunched ion structure in the velocity space. It seems to be difficult to explain the halo population by our simple test particle simulation alone. We will need an additional plasma heating process. For example, plasma heating due to slow shocks around the boundary between the lobe and the plasma sheet, which is not included in our simple model, may be responsible to the ion heating on the bunched ion populations as well as the formation of the hot halo population [Saito et al., 1996; Hoshino et al., 1997]. MHD turbulence observed inside the plasma sheet may also contribute to an additional plasma heating process [Hoshino et al., 1994; Zelenyi et al., 1998].

We assumed, for simplicity, the drift velocity $v_{0y} = 0$ in the source region. However, the plasmas may have a finite drift velocity in the dawn-dusk direction even in the lobe region. In the standard one-dimensional Harris plasma sheet with $\vec{E} = 0$, the drift velocity is given by $v_{0y} = -T_{\rm ion}/(eB_{0x})dN/dz$, where N is the plasma density [Harris, 1962]. However, we confirmed that the bunched ion structure discussed in this paper is not sensitive to the possible plasma parameters.

We did not discuss much on the observations of the ridge structure, because we could not find any clear evidence of the ridge in the GEOTAIL observations. By studying the effect of the normalized Larmor radius $\rho_i' = v_{\rm conv}/(\Omega_i \lambda_z)$ on the ridge structure, we find that the ridge structure can be observed for the plasma sheet with a small electric field and a thick current sheet. The plasma sheet observed during the passage of the plasmoid/flux rope, however, associates with a large convection velocity of about 100 km/s, which corresponds to about $E = 1$ mV/m for the standard tail magnetic field of $B = 10$ nT. It seems that the large convection velocity prevents from forming the ridge structure in the magnetotail if the thickness of the magnetic diffusion region does not strongly depend on the dawn-dusk electric field. However, the thickness of the plasma sheet around the magnetic diffusion region is still open question. It may be useful to study carefully the existence or the occurrence probability of the ridge structure for our better understanding of the physics around the magnetic diffusion region.

The size of the reconnection region in the dawn-dusk direction affects on the ion dynamics. Since the particles energized in the plasma sheet travelled in a large dawn-dusk distance, the velocity distribution in the high energy range function will be modified by a finite dawn-dusk dimension. However, if the dawn-dusk dimension of the reconnection region is not shorter than about 10 λ_z, we think that the modification of the distribution function is little. It is also an interesting subject to study how the distribution function is observed in the dawn or the dusk edge of the reconnection region. This topic will be deferred in another paper.

Our test particle simulation study is only a first step to understand the magnetic reconnection process in terms of a comparison to satellite observations in magnetotail. It will be required to study further the dynamics of the X-type region with the help of our findings of the family of the non-Maxwellian distribution functions. We believe that our claim provides a basis for extension of magnetic reconnection studies.

Acknowledgments. The author has enjoyed useful discussions with T. Mukai, A. Nishida, M. Scholer, D. Biskamp, F. Jamitzky, T. Terasawa, M. Fujimoto, T. Yamamoto, S. Kokubun, K. Tsuruda, D. Krauss-Varban and H. Karimabadi.

REFERENCES

Ashour-Abdalla, M., J. Berchem, J. Büchner, and L. M. Zelenyi, Shaping of the magnetotail from the mantle: Global and local structuring, *J. Geophys. Res.*. 98, 5651, 1993.

Ashour-Abdalla, M., L. A. Frank, W. R. Paterson, V. Peroomian, and L. M. Zelenyi, Proton velocity distributions in the magnetotail: Theory and observations, *J. Geophys. Res.*, 101, 2587-2598, 1996a.

Ashour-Abdalla, M., L. M. Zelenyi, V. Peroomian, L. A. Frank, and W. R. Paterson, Coarse-grained texture of ion distributions in the magnetotail: A fractal-like approach, *J. Geophys. Res.*, 101, 15287-15296, 1996b.

Büchner, J. and L. M. Zelenyi, Regular and chaotic charged particle motion in magnetotail-like field reversals, 1. Basci theory, *J. Geophys. Res.*, 94, 11821, 1989.

Büchner, J. and J.-P. Kuska, On the formation of cup-like ion beam distributions in the plasma sheet boundary layer, *J. Geomagn. Geoelectr.*, 48, 781-797, 1996.

Burkhart, G. R., J. F. Drake, and J. Chen, Magnetic reconnection in collisionless plasmas: Prescribed fields, *J. Geophys. Res.*, 95, 18833-18848, 1990.

Chen, J. and P. J. Palmadesso, Chaos and nonlinear dynamics of single-particle orbits in a magnetotail-like magnetic field, *J. Geophys. Res.*, 91, 1499, 1986.

Chen, J., Nonlinear dynamics of charged particles in the magnetotail, *J. Geophys. Res.*, 97, 15011, 1992.

Coppi, B., Laval, G., and Pellat, R., Dynamics of geomagnetic tails, *Phys. Rev. Lett.*, **16**, 1207-1210, 1966.

DeCoster, R. J., and L. A. Frank, Observations pertaining to the dynamics of the plasma sheet, *J. Geophys. Res.*, 84, 5099, 1979.

Eastman, T. E., L. A. Frank, W. K. Peterson, and W. Lennartsson, The plasma sheet boundary layer, *J. Geophys. Res.*, 89, 1553, 1984.

Frank, L. A., Paterson, W. R., and Kivelson, M. G., Observations of nonadiabatic acceleration of ions in Earth's magnetotail, *J. Geophys. Res.*, 99, 14877-14890, 1994.

Fujimoto, M., M. S. Nakamura, T. Nagai, T. Mukai, T. Yamamoto, and S. Kokubun, New kinetic evidence for the near-Earth reconnection, *Geophys. Res. Lett.*, 23, 2533-2536,1996.

Harris, E. G., On a plasma sheath separating regions of oppositely directed magnetic field, *Nuovo Cimento*, 23, 115, 1962.

Hoshino, A. Nishida, T. Yamamoto, and S. Kokubun, Turbulent magnetic field in the distant magnetotail: Bottom-up process of plasmoid formation?, *Geophys. Res. Lett.*, 21, 2935-2938, 1994.

Hoshino, M., Y. Saito, T. Mukai, A. Nishida T. Yamamoto, and S. Kokubun, Origin of hot and high speed plasmas in plasma sheet: plasma acceleration and heating due to slow shocks, *Adv. Space Res.*, 20, 973-982, 1997.

Hoshino, M., T. Mukai, T. Yamamoto, and S. Kokubun, Ion dynamics in magnetic reconnection: Comparisons between numerical simulations and Geotail observations, *J. Geophys. Res.*, in press, 1998.

Joyce, G., J. Chen, S. Slinker, D. L. Holland, and J. B. Harold, Particle energization near an X-line in the magnetotail based on global MHD fields, *J. Geophys. Res.*, 100, 19167-19176, 1995.

Krauss-Varban, D. and N. Omidi, Large-scale hybrid simulations of the magnetotail during reconnection, *Geophys. Res. Lett.*, 22, 3271, 1995.

Laval, G. and R. Pellat, Stability of the plane neutral sheet for oblique propagation and anisotropic temperature, *ESRO SP-36*, 5, 1968.

Lin, Y. and D. W. Swift, A two-dimensional hybrid simulation of the magnetotail reconnection layer, *J. Geophys. Res.*, 101, 19859-19870, 1996.

Lottermoser, R.-F. and M. Scholer, Undriven magnetic reconnection in magnetohydrodynamic and Hall magnetohydrodynamic, *J. Geophys. Res.*, 102, 4875-4892, 1996.

Lui, A. T. Y., E. W. Hones, Jr., F. Yasuhara, S.-I. Akasofu, and S. J. Bame, Magnetotail plasma flow during plasma sheet expansions: VELA 5 and 6 and IMP 6 observations, *J. Geophys. Res.*, 82, 1235, 1977.

Lyons, L. R. and T. W. Speiser, Evidence for current sheet acceleration in the geomagnetic tail, *J. Geophys. Res.*, 87, 2276-2286, 1982.

Martin, R. F., Jr., and T. W. Speiser, A predicted signature of a neutral line in the geomagnetic tail, *J. Geophys. Res.*, 93, 11521-11526, 1988.

Martin, R. F., Jr., T. W. Speiser, and K. Klamczynski, Effect of B_y on neutral line ridges and dynamical source ordering, *J. Geophys. Res.*, 99, 23623-23638, 1994.

Mukai, T. M. Fujimoto, M. Hoshino, S. Kokubun, S. Machida, K. Maezawa, A. Nishida, Y. Saito, T. Terasawa, and T.

Yamamoto, Structure and kinetic properties of the plasmoid and its boundary region,, *J. Geomag. Geoelectr.*, 48, 541, 1996.

Nakamura, M., G. Paschmann, W. Baumjohann, and N. Sckopke, Ion distributions and flows near the neutral sheet, *J. Geophys. Res.*, 96, 5631-5649, 1991.

Onsager, T. G., M. F. Thomsen, R. C. Elphic, and J. T. Gosling, Model of electron and ion distributions in the plasma sheet boundary layer, *J. Geophys. Res.*, 96, 20999-21011, 1991.

Richardson, I. G., and Cowley, S. W. H., Plasmoid-associated energetic ion bursts in the deep geomagnetic tail: properties of the boundary layer, *J. Geophys. Res.*, **90**, 12133-12158, (1985).

Richardson, I. G., Cowley, S. W. H., Hones, E. W. Jr., and Bame, S. J., Plasmoid associated energetic ion bursts in the deep geomagnetic tail: properties of plasmoids and the postplasmoid plasma sheet, *J. Geophys. Res.*, **92**, 9997-10013, (1987);

Saito, Y., T. Mukai, T. Terasawa, A. Nishida, S. Machida, S. Kokubun, and T. Yamamoto, Foreshock structure of the slow-mode shocks in the Earth's magnetotail, *J. Geophys. Res.*, 101, 13267-13274, 1996.

Sarris, E. T. and W. I. Axford, Energetic protons near the plasma sheet boundary, *Nature*, 277, 460, 1979.

Sonnerup, B. U. Ö., Adiabatic particle orbit in a magnetic null sheet, *J. Geophys. Res.*, 76, 8211-8222, 1971.

Speiser, T. W. Particle trajectories in model current sheet 1, Analytical solution, *J. Geophys. Res.*, 70, 4219, 1965.

Takahashi, K. and E. W. Hones, Jr., ISEE 1 and 2 observations of ion distributions at the plasma sheet-tail lobe boundary, *J. Geophys. Res.*, 93, 8558, 1988.

Zelenyi, L. M. et al., in this issue, 1998.

M. Hoshino, 3-1-1 Yoshinodai, Sagamihara, Kanagawa 229, Japan

Advances in the Physics of Earth's Magnetotail

L. A. Frank and W. R. Paterson
Department of Physics and Astronomy, The University of Iowa, Iowa City, Iowa

S. Kokubun
Solar–Terrestrial Environment Laboratory, Nagoya University, Aichi 442, Japan

T. Yamamoto
Institute of Space and Astronautical Science, Kanagawa 229, Japan

The combination of the tailored orbit of the Geotail spacecraft during its extensive tour of Earth's magnetotail with its state–of–the–art instrumentation has allowed many significant advances in our knowledge of this region. Three such advances are discussed. The first advance obtains from a rigorous survey for "bursty, bulk flows" of plasmas in the near–Earth plasma sheet that have been reported from observations with previous spacecraft. The near–equatorial orbit of the spacecraft at radial distances in the range of 10 to 50 R_E was ideally suitable for such a study. The plasma measurements were sufficiently accurate to separate the bulk flows into their perpendicular and parallel components. Examination of the histograms for the perpendicular flows finds that there is no evidence for the bursty bulk flow of plasmas and magnetic flux at geocentric radial distances of about 10 to 25 R_E in the central plasma sheet. On the other hand, there is substantial evidence of the presence of previously reported high–speed flows of plasmas at larger distances of about 50 R_E which were associated with "magnetotail fireballs". A second advance in the studies of magnetotail physics is the finding that the ion pitch angle distributions in the near–Earth plasma sheet are not quasi–isotropic but exhibit such features as "eyes" and "wings" that are associated with the previous history of transport and acceleration of these ions in the magnetotail. These "memories" in the complex ion velocity distributions can be used as remote sounding tools for the magnetotail when they are decoded with a realistic simulation model. Finally it is now possible to directly detect the currents flowing in the magnetotail with plasma instrumentation. Such measurements allow the in–situ determinations of current densities and the nature of the electron and ion velocity distributions which are associated with these current densities. These direct determinations of the current densities are expected to contribute significantly in revealing the current diversions and disruptions in the magnetotail that are associated with magnetic substorms. In addition the capability of determining the off–diagonal elements of the pressure tensor appears to provide new insight into the dynamics of the magnetotail current sheet.

INTRODUCTION

The Geotail spacecraft, its orbit and its sophisticated scientific instruments provide a simply splendid opportunity to greatly extend our frontiers of knowledge con-

New Perspectives on the Earth's Magnetotail
Geophysical Monograph 105

cerning the dynamics and structure of Earth's magnetotail. This mission has taken advantage of a substantial foundation of experience and findings from a series of spacecraft with access to the magnetotail for a period in excess of a quarter of a century, such as the IMP, ISEE and AMPTE spacecraft. In addition the orbit of the Geotail spacecraft was carefully tailored in order to comprehensively observe the many facets of the magnetotail as functions of downstream distance from our planet. During the first observing period during September 1992 to October 1994 the spacecraft provided a survey of the distant magnetotail with a series of orbits with apogee radial distances in the range of 73 to 211 R_E (R_E, Earth radii) and perigee distances of 2.4 to 19 R_E. Later the orbit was changed in order to study a region of different character, the midtail, during November 1994 to February 1995. The apogee distances were in the range of 39 to 64 R_E, most at 48 R_E, and the perigees were 8.7 to 10.2 R_E. The third, and current orbital phase began in March 1995 in the near-magnetotail region with perigee and apogee distances of 10 and 30 R_E, respectively. The Comprehensive Plasma Instrumentation (CPI) is capable of high-resolution determinations of the three-dimensional velocity distributions of electrons and ions within the energy/charge range of 1.3 V to 48.2 kV [*Frank et al.*, 1994]. In excess of 1000 samples for each determination of an electron or ion velocity distribution were taken with temporal resolutions of about 20 to 60 s. This intensive sampling allowed accurate determinations of such parameters as number densities, temperatures, bulk flow speeds, pressure tensors and, necessarily, the spacecraft potential. Often the velocity distributions were not quasi-Maxwellian and the fine-grained sampling of the plasmas allowed the acquisition of the actual characteristics of the distributions. The high-quality observations of the ambient magnetic fields with the Magnetic Field Experiment (MGF) [*Kokubun et al.*, 1994] were combined with the plasma measurements to provide the advances in three areas of magnetotail physics which are discussed in detail in the present paper.

Steady and Transient Plasma Flows in the Plasma Sheet

Over twenty years ago transient, high-speed flows of plasmas and magnetic flux transport were reported for plasma and magnetic field observations acquired in the plasma sheet with IMP 8 [*Frank et al.*, 1976]. The perigee and apogee radial distances for this spacecraft were 23.1 and 46.3 R_E, respectively. Within these turbulent plasma regions earthward and tailward flows at speeds up to 1000 km/s were observed. These regions of strong plasma acceleration and transport were called "fireballs". Several years later *Coroniti et al.* [1980] further documented the fields-and-particles phenomena which were associated with these transient, localized regions with observations with IMP 7 at radial distances of about 35 R_E and near the neutral sheet. It was evident that an important instability was occurring within or near the plasma sheet at these distances. Subsequent surveys of the characteristics of the central plasma sheet at somewhat lesser radial distances, < 22 R_E, found that the ion flows were typically < 50 km/s [*Huang and Frank*, 1986; *Baumjohann et al.*, 1989]. However, it is noted that *Baumjohann et al.* [1989] reported the transient appearance of flows in excess of 400 km/s.

More recently an extensive series of observations of plasmas and magnetic fields with the AMPTE Ion Release Module (IRM) spacecraft out to radial distances of 19 R_E has been employed to identify sporadic, bulk flows of ions in the central plasma sheet [*Angelopoulos et al.*, 1992]. These bulk flows can exceed 400 km/s and can possibly transport a significant amount of magnetic flux from the plasma sheet at these distances into the near-Earth plasma sheet at 10 R_E or so. This phenomenon is called "bursty bulk flows". From the paper by *Angelopoulos et al.* [1992] it is clear that "bursty bulk flows" are identified with periods which are characterized by bulk flows in excess of about 400 km/s with durations of one minute or more in an overall envelope of 10 to 20 minutes. The criteria for positioning these events in the central plasma sheet, rather than the plasma sheet boundary layer, were not secure, i.e., there were no constraints which completely eliminated the possibility that occasional samples of the plasma sheet boundary layer were included in the data set for the central plasma sheet. This is particularly important because the occurrence rate for the bursty bulk flows is low. In addition the bulk flow velocity vector was not separated into its parallel and perpendicular components, a necessity for a robust discussion of magnetic flux transport in the near-Earth plasma sheet.

The analyses of the CPI plasma measurements with the Geotail spacecraft is marvelously straightforward because this spacecraft provides a thorough survey of plasma sheet phenomena at radial distances of 10 to 212 R_E in or near the midplane of the magnetotail. Importantly, for the first time, there is a comprehensive survey of the parallel and perpendicular components of ion flows throughout this region. The time resolution for the determination of plasma parameters as presented here is 1 minute and adequate for investigation of bursty bulk flows. In order to investigate the existence of the bursty transport of magnetic flux in the central plasma sheet it is not necessary to construct or define the boundary between the central plasma sheet and its boundary layer. It is only necessary to determine whether or not there are high-speed bulk flows perpendicular to the magnetic field which are directed toward Earth in a high-β plasma. In fact the present study shows that such flows do not exist in the near-Earth plasma sheet and that the higher-speed bulk flows are parallel to the magnetic field, as expected for the plasma sheet boundary layer. A complete description of the results is given by *Paterson et al.* [1997].

Figure 1 summarizes the determinations of the projections of the X component of plasma flows parallel and

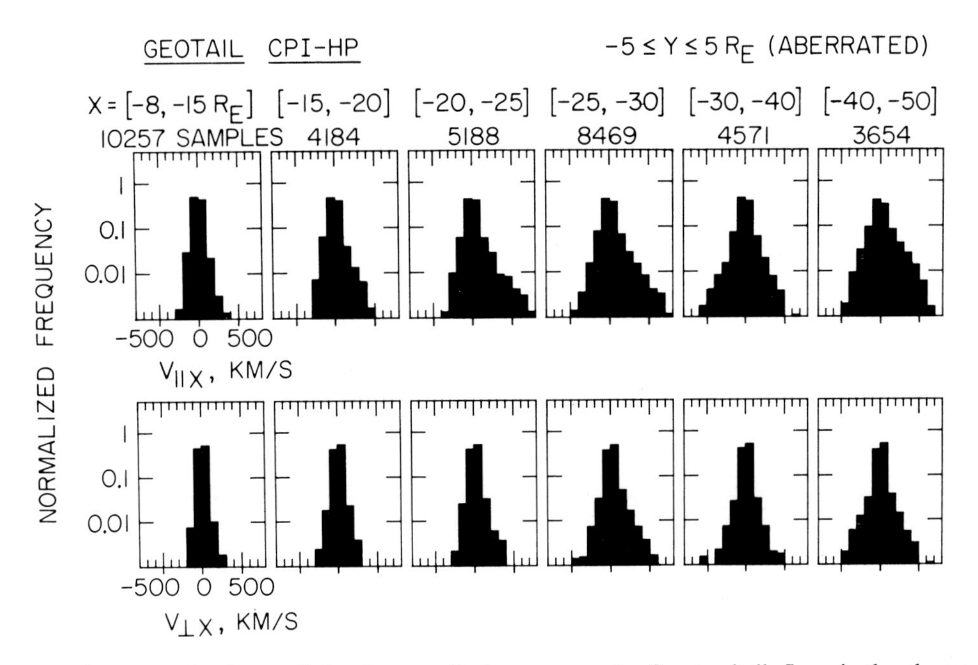

Figure 1. Histograms for the parallel and perpendicular components of proton bulk flows in the plasma sheet near the equatorial plane as a function of downstream distance from Earth. Note that the high-speed, earthward plasma motions in the downstream distance range of –8 to –25 R_E are associated with flows parallel to the magnetic field [after *Paterson et al.*, 1997].

perpendicular to the magnetic field vector, $Vx_{||}$ and $Vx_{\perp}$, respectively. The flows are computed for 1–minute accumulations of the analyzer responses. The observations were obtained during the period 1 December 1994 to 1 June 1996. These data have been placed into 6 bins along the X–axis. The number of samples within each bin are also shown in Figure 1. Positive values of the flow components are directed earthward, negative values tailward. The cross–tail dimension Y of these bins is 10 R_E, centered on the –X axis. The range of distances of the computed neutral sheet position to the spacecraft is –6 to +6 R_E. The flow determinations were limited to β values > 0.1 in order to exclude samples within the lobe, mantle, and magnetosheath boundary layer. Selection criteria for separating the central plasma sheet from its boundary layer are generally imperfect so that both plasma regimes are included in this study. Because the components of flow parallel and perpendicular to the local magnetic field are accurately determined in this study, magnetic flux transport in the magnetotail is clearly separated from flows along the magnetic field. *Huang and Frank* [1994] have previously reported perpendicular components of bulk flows in the central plasma sheet from ISEE–1 data. The time resolutions of these measurements were about 2 and 8 minutes, with most of the samples with the latter resolution. The present Geotail determinations of bulk flows with resolution of 1 minute are sufficient to detect bursty bulk flows with speeds $\gtrsim$ 400 km, if such flows exist.

The histograms for the X–component parallel to the magnetic field, $Vx_{||}$, are shown in the upper row of panels in Figure 1. For the distance range of $-8 > X > -30$ R_E there is an increasing fraction of parallel bulk flows > 200 km/s with increasing downstream distance. This distance range $X > -25$ R_E includes the orbits of AMPTE/IRM and ISEEs 1 and 2. Flow speeds as high as 700 km/s are present. For the distance range $-30 > X > -50$ R_E the occurrence of high–speed flows is divided more–or–less equally for earthward and tailward directions. This is the spatial regime of the IMP–7 and –8 orbits.

On the other hand, the histograms for the X–component of flow perpendicular to the magnetic field, i.e., the convection component $Vx_{\perp}$, are displayed in the lower row of panels in Figure 1. For the range of $X > -19$ R_E for the AMPTE/IRM orbit the occurrence frequencies for earthward and tailward flows are about equal. The magnitudes of these components of convective flow are < 300 km/s. For $-20 > X > -30$ R_E there is a small fraction of earthward flows with speeds in excess of 300 km/s. For $-30 > X > -50$ R_E the occurrences of earthward and tailward flows are about equal in frequency and magnitude.

The results shown in Figure 1 are a considerable advance in the study of plasma sheet dynamics in that the components of bulk flow parallel and perpendicular to the magnetic field are given for the first time for an extensive survey. The large convection and parallel flows of

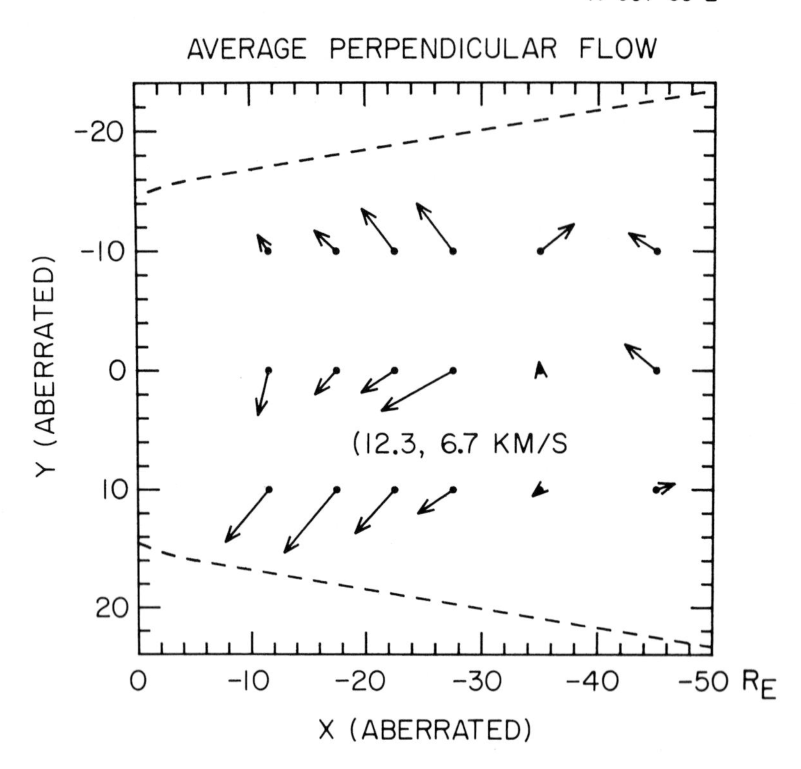

Figure 2. Average perpendicular flow velocities in the plasma sheet. The scale is given below the corresponding vector [after *Paterson et al.*, 1997].

the fireballs detected at the IMP–8 and –7 spacecraft at $X < -30\ R_E$ are clearly seen in the panels of Figure 1. However the bursty bulk flows at $X > -20\ R_E$ as observed with the AMPTE/IRM spacecraft are shown to be earthward flows parallel to the magnetic field with no evidence of high–speed convection toward Earth. The Geotail orbit was ideally positioned to achieve this latter result. Thus the bursty bulk flows are related to the field–aligned beams of ions in the plasma sheet boundary layer. These bursty bulk flows do not appear to be significant in the transport of magnetic flux toward the near–Earth plasma sheet as evidenced by the perpendicular components $V_{x\perp}$ as shown in the first two panels of the bottom row.

The ability to separate the parallel and perpendicular components of plasma bulk flows allows a determination of the average convection in the plasma sheet from the measurements shown in Figure 1. These average convection velocities are displayed in Figure 2. The arrows are directed from the centers of the spatial blocks used in binning the individual velocity samples. The convection appears to be well organized with a "stagnation point" in the local evening sector at $-30 > X > -40\ R_E$. The corresponding electric field magnitudes range from about 0 to 0.1 mV/m.

The "Memories" in the Ion Distributions in the Plasma Sheet

The Geotail spacecraft is equipped with a plasma analyzer which has a combination of improved coverage of the unit sphere for charged particle velocity vectors, angular resolution, energy range and energy resolution for hot plasmas relative to its predecessors in Earth's magnetotail. One of the surprising results of these observations with this CPI in the near–Earth plasma sheet was the complexity of the ion velocity distributions which were previously treated as Maxwellian distributions [*Huang and Frank*, 1986; *Baumjohann et al.*, 1989]. Major exceptions to the assumption of Maxwellian distributions were the structured ion velocity distributions which were reported by *Nakamura et al.* [1991] for the plasma observations with the AMPTE/IRM spacecraft and those described by *Chen et al.* [1990] as found with the ISEE–1 spacecraft. And, of course, the "lima bean" shaped isodensity contours of the ion velocity distributions in the plasma sheet boundary layers located between the central plasma sheet and the relatively barren magnetotail lobes were identified early in the studies of the magnetotail [*DeCoster and Frank*, 1979]. For the past thirty years or so there has

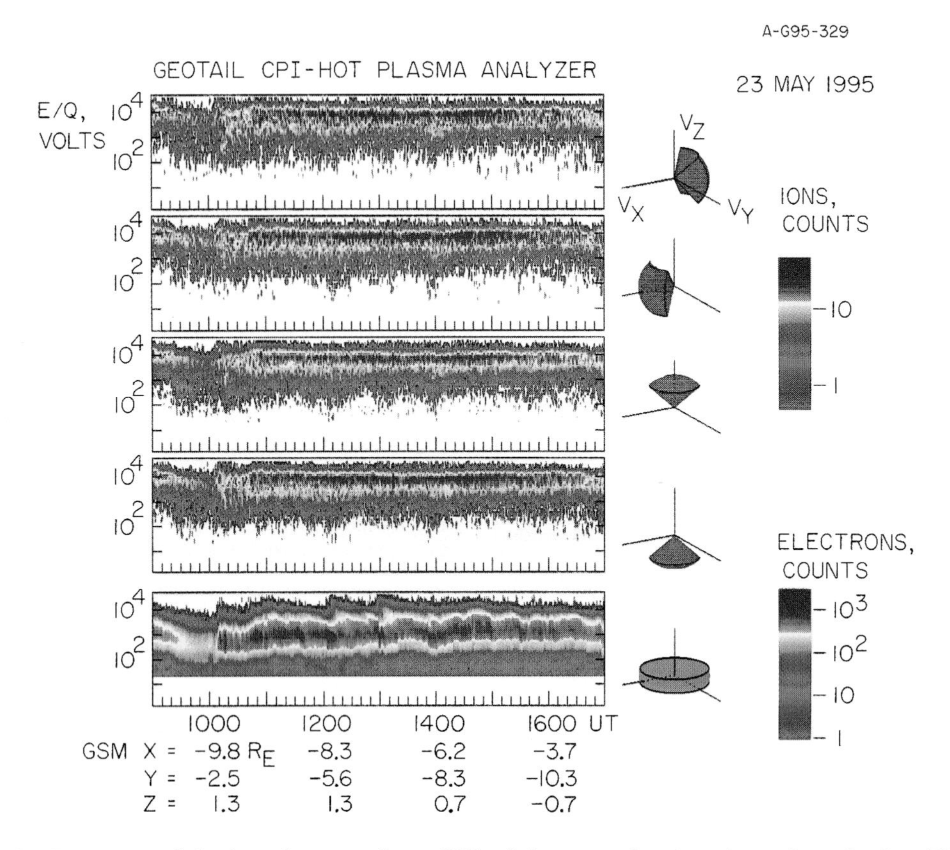

Figure 3. Responses of the hot plasma analyzer (HP) of the comprehensive plasma investigation (CPI) as functions of energy–per–unit charge (E/Q) and universal time during a crossing of the near–Earth plasma sheet on May 23, 1995. The upper four spectrograms display the responses to protons within the indicated solid angles. The bottom spectrogram records the electron responses [after *Frank et al.,* 1996].

been continuing active interest in the non–Maxwellian features of plasmas in and near the mid–plane of the magnetotail although the reports of measured velocity distributions were quite meager. A partial list of relevant publications includes *Speiser* [1965], *Lyons and Speiser* [1982], *Büchner and Zelenyi* [1989], *Ashour–Abdalla et al.* [1993; 1996] and the review by *Chen* [1992]. The interested reader is encouraged to access the reference lists given in these papers. We summarize here the remarkable features of the ion velocity distributions which have been recently detected with the Geotail spacecraft in the near–Earth plasma sheet.

A common mode of providing an overview of plasma measurements are the energy–time (E–t) spectrograms such as those shown in Figure 3. For these particular spectrograms the responses of the ion and electron sensors are color–coded and plotted as functions of energy/charge (ordinate) and Universal Time (abscissa). The color code for the sensor responses is given on the right–hand side of Figure 3. The four E–t spectrograms for the ion sensors each correspond to the four solid angles as shown to the right of the spectrograms, e.g., the upper spectrogram corresponds to ion velocity vectors directed along $-V_x$ (tailward). The bottom spectrogram in Figure 3 is a record of the electron responses summed over electron velocity directions parallel to the equatorial plane of the plasma analyzer, i.e., nearly parallel to the ecliptic plane. The positions of the spacecraft during this eight–hour series of observations are given under the abscissa in Earth–centered solar–magnetospheric coordinates. The spacecraft is located in the post–midnight sector at geocentric radial distances of about 10 R_E and at low–latitudes. The plasma energy spectra are relatively featureless as shown in Figure 3 and are typical of those previously reported with spacecraft observations in this region of the magnetotail.

The moments of the plasma distributions and the magnetic fields appear to reflect the quiescent state of the plasma sheet as exhibited in the spectrograms of Figure 3. These parameters are shown in Figures 4, 5 and 6. The fluctuations of all of these parameters indicate that the dynamical state of the near–Earth plasma sheet is calm for this period of observations. For example the average bulk flows for the ions shown in Figure 4 are only (V_x,

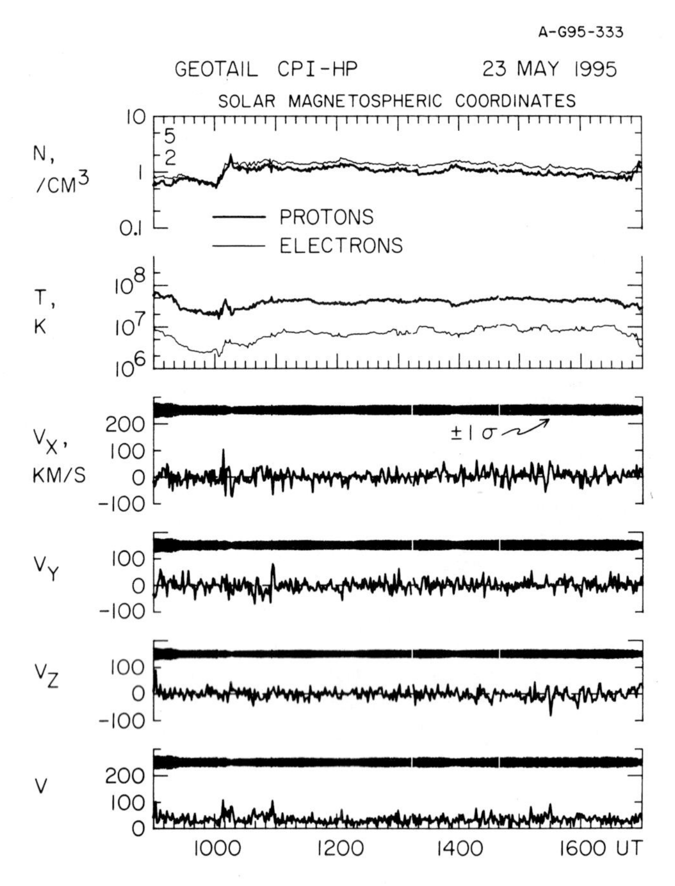

Figure 4. Plasma moments corresponding to the period of observations shown in Figure 3 [after *Frank et al.,* 1996].

V_y, V_z) = (3(±2), 1(±2), 3(±1)) km/s for the period 1300–1500 UT. Inspection of Figure 5 finds that the plasma β is about 1 or greater for most of the period and that the magnetic equator was crossed at about 1430 UT with the reversal of the B_x component of the magnetic field. A representative sample of the ion composition is given in Table 1 [*Frank et al.,* 1996]. The proton and electron temperatures perpendicular and parallel to the magnetic field, and the individual values for β corresponding to the protons and electrons, as displayed in Figure 6 further corroborate the apparent lack of activity in the near–Earth plasma sheet. However, a small hint of important plasma dynamics may be indicated by the increases of $T_{||}$ for the electrons.

It is of interest to investigate the possibility that the improved capabilities for determination of the plasma velocity distributions with the plasma analyzer might reveal features which were not observed in previous, extensive surveys of the spatial distributions and dynamical behavior in the near–Earth plasma sheet. An example of the exciting findings is shown in Figure 7. The pitch angle distributions for six consecutive samples are shown. These distributions differ greatly from those for a Maxwellian or a bi–Maxwellian distribution. The observed distributions exhibit the durable presence of "eyes" and "wings" or "ledges". These proton velocity distributions evidently offer a "memory" of their previous acceleration and transport. That is, each feature, or group of ions, in pitch angle coordinates arrived at the spacecraft from a different source and along a different path in the magnetotail. The simplest feature to interpret is the ions at $V_{||}$ = −1000 km/s. These ions are arriving at the spacecraft from the ionosphere and have been accelerated by a field–aligned potential difference at lower altitudes as observed with other spacecraft [*Shelley and Collin,* 1991]. The features in the ion velocity distributions offer a remarkably effective probe for the topology and dynamics of the global magnetotail.

A time-dependent global magnetohydrodynamic (MHD) simulation of the magnetosphere with responses to the temporal fluctuations of the solar wind parameters can be used to identify the sources and paths for the ions de-

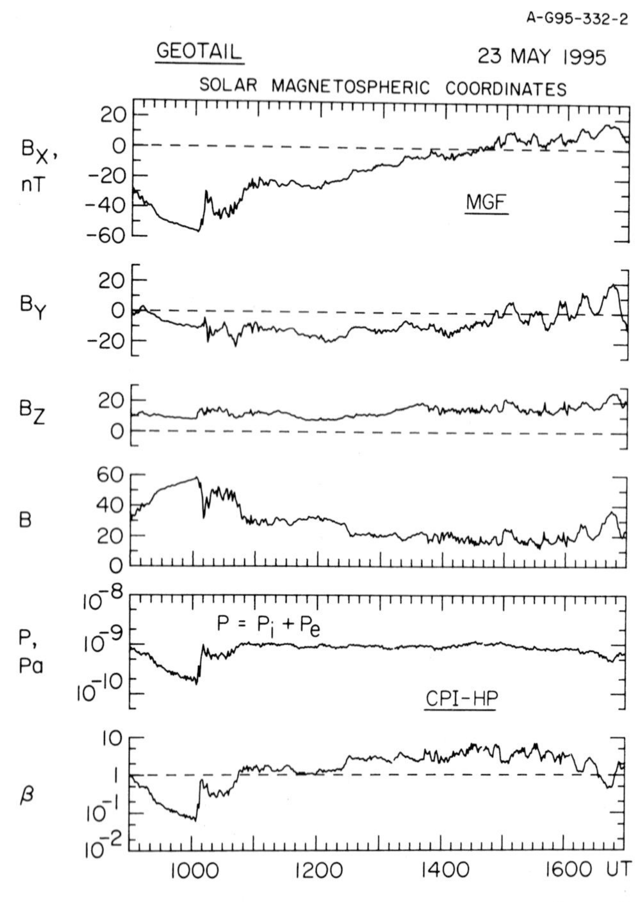

Figure 5. One–minute averages of the components and magnitude of the magnetic fields as measured with the magnetic field experiment (MGF) for the spectrograms shown in Figure 3. The total plasma pressure and the plasma β are shown in the bottom two panels [after *Frank et al.,* 1996].

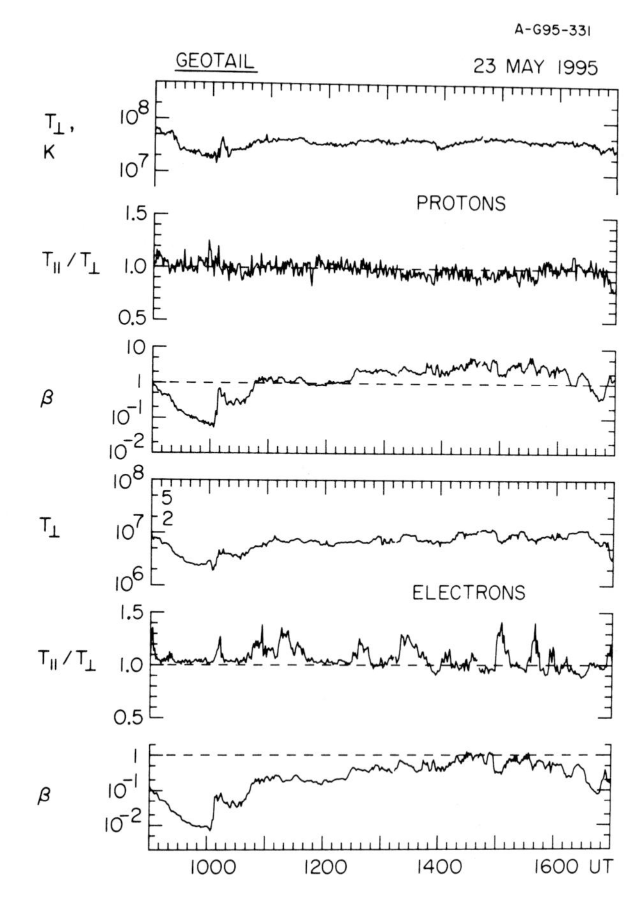

Figure 6. Perpendicular temperature $T_\perp$, temperature anisotropy $T_{||}/T_\perp$, and plasma β for protons (upper three panels) and electrons (lower three panels) during the observations in the near-Earth plasma sheet shown in Figure 3 [after *Frank et al.*, 1996].

tected at the Geotail spacecraft. Such a simulation for a measured ion pitch angle distribution has been performed by *Ashour-Abdalla et al.* [1996]. The model is not self-consistent in that the observed ion velocity distributions are not fed into the MHD model. The model utilizes the time-series interplanetary data for magnetic fields and plasma as input. The resulting magnetospheric configuration is then used to trace the paths of the various parts of the ion velocity distribution. These sources and paths were tracked by time-reversing their paths. The results of the simulation are very encouraging as ions from the ionosphere, near-Earth polar mantle, and the low-latitude boundary layer were identified in the pitch angle distribution. Ions with higher energies from nonadiabatic acceleration in the downstream current sheet were also present at the Geotail spacecraft. The measured pitch angle distributions offer a solid test for the accuracy of numerical simulations of the global magnetosphere.

Although the electron pitch angle distributions do not exhibit the complex features of those found for the ions

Table 1. Heavy Ions, 1231–1406 UT, May 23, 1995 (after *Frank et al.* [1996])

Ion	Energy Range, keV	Flux, cm^{-2} s^{-1}	Density, cm^{-3}
H^+	0.6–34.5	$8.6\ (\pm 0.03) \times 10^7$	$8.5\ (\pm 0.03) \times 10^{-1}$
He^{++}	2.4–34.5	$4.8\ (\pm 1.7) \times 10^6$	$7.5\ (\pm 2.6) \times 10^{-2}$
He^+	1.7–24.7	$1.9\ (\pm 0.9) \times 10^6$	$4.0\ (\pm 2.0) \times 10^{-2}$
O^+	1.7–24.7	$1.2\ (\pm 0.7) \times 10^6$	$5.6\ (\pm 3.0) \times 10^{-2}$

one notable feature is often seen. These electron distributions are shown in Figure 8 for a portion of the time interval for the proton measurements in Figure 7. The electron feature is the field-aligned beams at low energies which are directed both parallel and anti-parallel to the local magnetic field. The energies of these electron beams are in the range of 100 to 300 eV. The sampling of the pitch angle distributions with the multiple fields of view allows almost simultaneous sampling of the two electron beams. The differences in electron beam intensities evident in Figure 8 strongly suggests that there are significant field-aligned currents which are carried by the electrons.

A diagram which summarizes our initial results for the observations of proton and electron pitch angle distributions in the near-Earth plasma sheet is offered in Figure 9. The ion pitch angle distributions exhibit a remarkable "memory" of their initial magnetotail entry and subsequent paths through the magnetosphere to the spacecraft position. These equatorial distributions of plasmas are linked to the ionosphere at lower altitudes via two field-aligned potential drops. The ions at the spacecraft position have been accelerated by a potential in the range of 3–5 kV and the electrons by a weaker drop of 100–300 V. It is clear that the slowly convecting plasmas in the equatorial plasma sheet are strongly coupled to the ionosphere.

Direct Measurement of Currents and Pressure Tensors

The transient, large-scale currents near the midplane of the magnetotail during a substorm are believed to be associated with a diversion of the currents into the ionosphere and return of these currents to the magnetotail to form a "current wedge" [*Atkinson,* 1967; *Coroniti and Kennel,* 1972; *McPherron et al.* 1973; *Lopez and Lui,* 1990; *Ohtani et al.,* 1990; *Jacquey et al.,* 1991; *Lui,* 1996]. A broad spectrum of magnetotail phenomena have been reported for the onset including short bursts of large-amplitude electrostatic fields [*Aggson et al.,* 1983], irregular magnetic and electric pulsations [*Robert et al.,* 1984], and transient fluctuations of magnetic fields and increases in the intensities of energetic charged particles [*Lui et al.,* 1988]. Direct observations of currents with

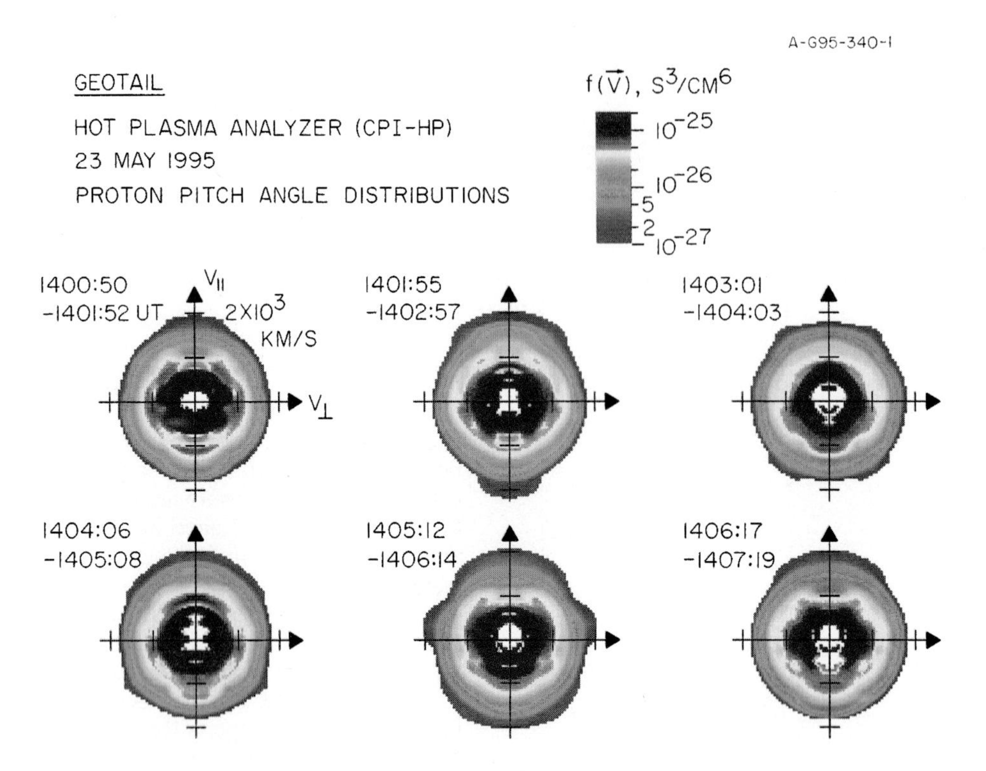

Figure 7. Pitch angle distributions for protons during six consecutive sampling intervals. The phase space densities are color-coded according to the color bar at the upper right-hand side [after *Frank et al.,* 1996].

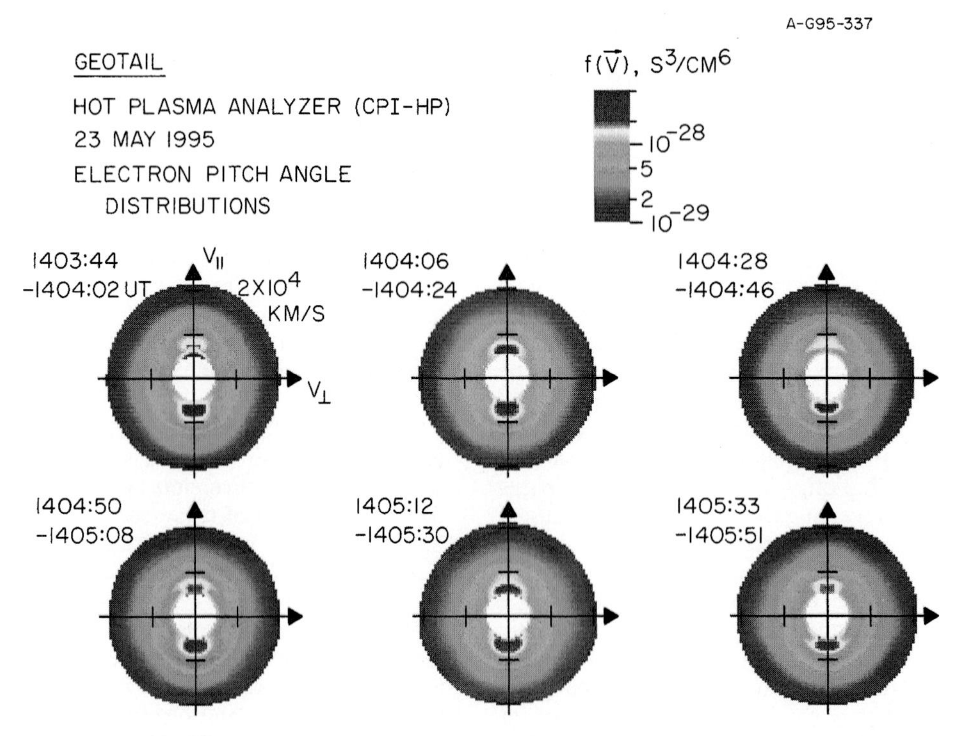

Figure 8. Pitch angle distributions for electrons during six consecutive sampling intervals. These samples were also acquired during part of the period covered by Figure 7 for the proton pitch angle distributions [after *Frank et al.,* 1996].

plasma instrumentation have not been often reported until recently. We summarize such measurements during a small isolated substorm.

Observations of plasmas and magnetic fields in the pre-midnight sector of the near-Earth plasma sheet are shown in Figure 10 for the period 0200 to 0800 UT on 9 February 1995. The plasma β in the third panel provides an effective roadmap as to the locations of the Geotail spacecraft with respect to the plasma domains. The regions with relatively high β, and high number densities, are periods during which the spacecraft was located within the central plasma sheet. Two such periods are 0320–410 UT and 0520–0545 UT. The intervals for which β < 1 exhibit features of the plasma sheet boundary layer, e.g., 0250–0320 UT and 0410–0520 UT. The Canopus magnetometer chain detected the onset of a small magnetic substorm at 0437 UT. The spacecraft position at this time was X_{GSM} = (−32.2, 5.7, −1.5 R_E) in Earth-centered, solar-magnetospheric coordinates. As shown in Figure 10 the spacecraft was located within the plasma sheet boundary layer and recorded a step-like decrease in the magnitude of the magnetic field of about 5 nT during a several-minute period during the substorm onset. The decrease in amplitude is evident in the X component of the magnetic field, with fluctuations in the considerably lesser Y and Z components which indicate the presence of substantial field-aligned currents. There is no unique signature of the substorm onset in the plasma parameters shown in the upper three panels.

On the other hand, the direct determination of the currents with the measurements of the three-dimensional velocity distributions of protons and electrons finds that a transient and large current was present at the spacecraft at substorm onset. These comprehensive plasma measure-

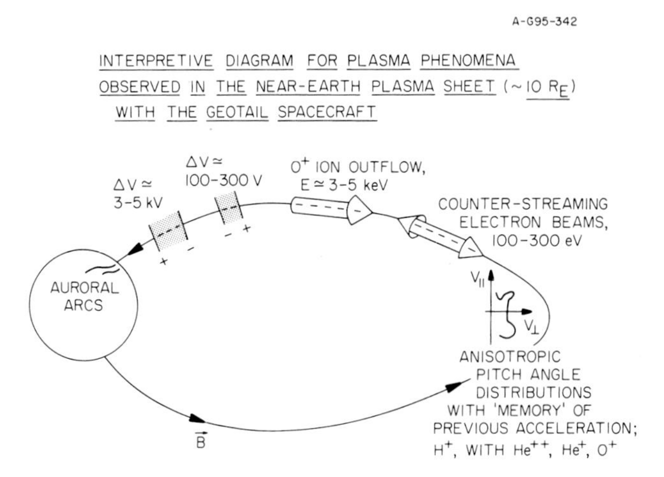

Figure 9. An interpretive and summary diagram for several of the plasma features observed in the near-Earth plasma sheet at radial distances of about 10 R_E on May 23, 1995. It is likely that the field-aligned potential drops are similarly located above the southern ionosphere [after *Frank et al.*, 1996].

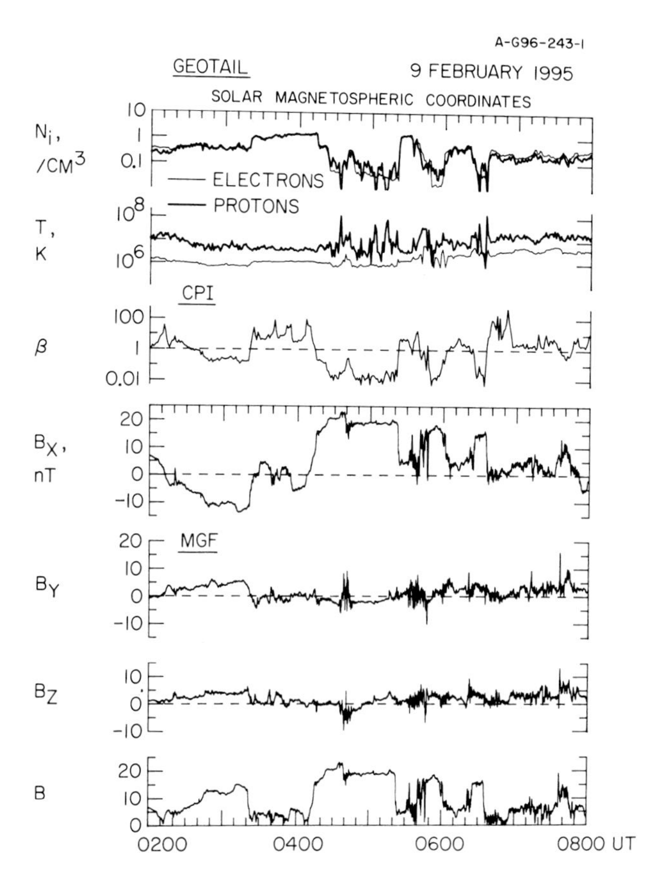

Figure 10. From top to bottom panels are displayed the proton and electron densities and temperatures, the ratio of the plasma energy density to that of the magnetic field (β), and the magnetic field components at the Geotail spacecraft position during 0200–0800 UT on 9 February 1995. The substorm onset occurred during the step-like decrease of the X component of the magnetic field of about 5 nT beginning at 0437 UT [after *Frank et al.*, 1997].

ments are sufficiently accurate to determine the contributions to the current from the electrons and protons, along with the requisite identification of the spacecraft potential. One important verification of the accuracy of the current determination is the comparison of the electron and proton number densities. A valid determination requires that these two number densities are equal. The plasma densities shown in Figure 10 show that this important condition is met for practically all of this series of observations. The field-aligned current densities for the interval 0300 to 0600 UT are shown in Figure 11. The large current impulse is readily evident for a several minute interval beginning at substorm onset at 0437 UT. The maximum current density was about 30 nA/m^2. The current is directed antiparallel to the local magnetic field. Because the X component of the magnetic field is positive the spacecraft is located northward of the magnetotail current sheet. Thus the current is directed tailward.

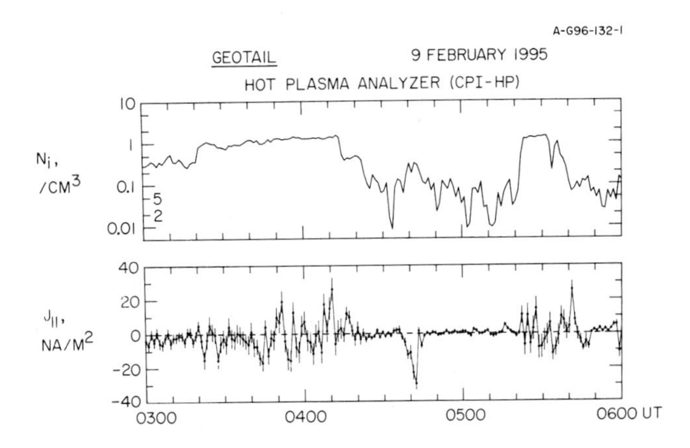

Figure 11. Magnetically field–aligned current densities during 0300–0600 UT at the Geotail spacecraft position. Note the initial appearance of the transient current pulse at about 0437 UT [after *Frank et al.*, 1997].

Because the three–dimensional ion flow velocity can be accurately determined the dimensions of the current–carrying magnetic flux tube can be estimated. The components of this velocity in the Y and Z directions are 57(±11) and −55(±9) km/s, respectively, as averaged over the period 0439:21–0441:27 UT. That is, the flux tube is moving toward local evening and downward along Z toward the current sheet. If a circular cross section is assumed for the flux tube then its diameter is about 16,000 km and the total current is in the range of 5×10^6 amperes. The magnitude of this current is in the range of the cross–tail currents in the current sheet at radial distances of 5 to 10 R_E, $2–3 \times 10^6$ A [*Kaufmann*, 1987]. The X–line can be positioned at distances either earthward or tailward of the Geotail position. There are no firm observed discriminators which resolve this issue. If the X–line is positioned inside of the Geotail position then the currents from the neutral line position must flow to the ionosphere, then transversely through the ionosphere, and subsequently out of the ionosphere in the plasma sheet boundary layer to the spacecraft position. There is no evidence in the plasma or magnetic field data that shows that the field lines threading the spacecraft position passed through the central plasma sheet or current sheet at earthward distances. If the X–line is located tailward of the spacecraft, then it is possible that the field lines at the spacecraft pass through the X–line in the current sheet. In both of the above cases the magnetic field lines at the spacecraft pass earthward into the ionosphere. A series of such direct detections of the impulsive currents associated with substorm onset will be needed to assemble a global picture of the current diversion.

The field–aligned current which was detected in the plasma sheet boundary layer during the substorm onset was contributed by both electrons and ions. This situation is different than that for the currents dominated by electrons as usually observed in the boundary layer [*Frank et al.*, 1981]. These electron velocity distributions can be generally characterized as bi–Maxwellian distributions with centroids displaced parallel or anti–parallel to the local magnetic field. The electron pitch angle distributions for the current pulse observed during substorm onset exhibit a different character as shown in the three bottom panels of Figure 12. These distributions appear to be composed of high–speed, field–aligned electron beams which are passing through a cooler electron distribution at lower speeds. The X component of the bulk flow velocity for the entire electron distribution is 298 (±37) km/s for the interval 0439:21–0441:27 UT and is directed earthward. The ion velocity distributions are more complex, as can be easily seen in the upper three panels of Figure 12. The plane used for displaying these ion velocity distributions is defined by the direction of the magnetic field and the X axis. A high–speed beam of protons at speeds in excess of 1000 km/s is passing through a cooler proton plasma which is almost at rest with respect to the Geotail spacecraft. For the above period of the electron measurements the X component of flow for the entire proton distribution is −123 km/s which contributes a substantial fraction of the current density. For the ion beam only the X component of bulk flow is −1100 (±−74) km/s with the assumption that the ions are predominantly protons. This composition is consistent with direct determinations of the mass composition with the instrument's mass spectrometers for ion beams with longer durations and outflowing from the ionosphere [*Frank et al.*, 1996]. For the present case the total ion density is 0.26 /cm^3 and that for the high–speed beam is 0.05 /cm^3. The tailward propagation of the proton high–speed beam suggests that a field–aligned potential difference of about 6 kV occurs at lesser radial distances. Such a potential difference could provide a beam of electrons with similar energies into the ionosphere. The energy influx into the ionosphere is significant, about 10 ergs/cm^2–s, and could be important in providing a conductive path for the current wedge.

There is another important region of measurable currents which are evident in Figure 11. The field–aligned currents during the interval 0330–0410 UT occur in the high–β plasmas and weak magnetic fields within and in the vicinity of the magnetotail current sheet. Inspection of Figure 10 finds that the magnitudes of the turbulent magnetic fields within this region are in the range of several nT or less. The error bars due to counting statistics have been included in the bottom panel of Figure 11. Clearly there are regions within and near the magnetotail current sheet that have significant field–aligned currents, e.g., 0350–0354 UT. The magnitudes of the currents are in the range of tens of nA/m^2. These field–aligned currents are contributed mainly by the electron velocity distributions and are directed parallel and antiparallel to the magnetic fields during various segments of the interval. Our interpretation of these field–aligned currents is that these currents provide the

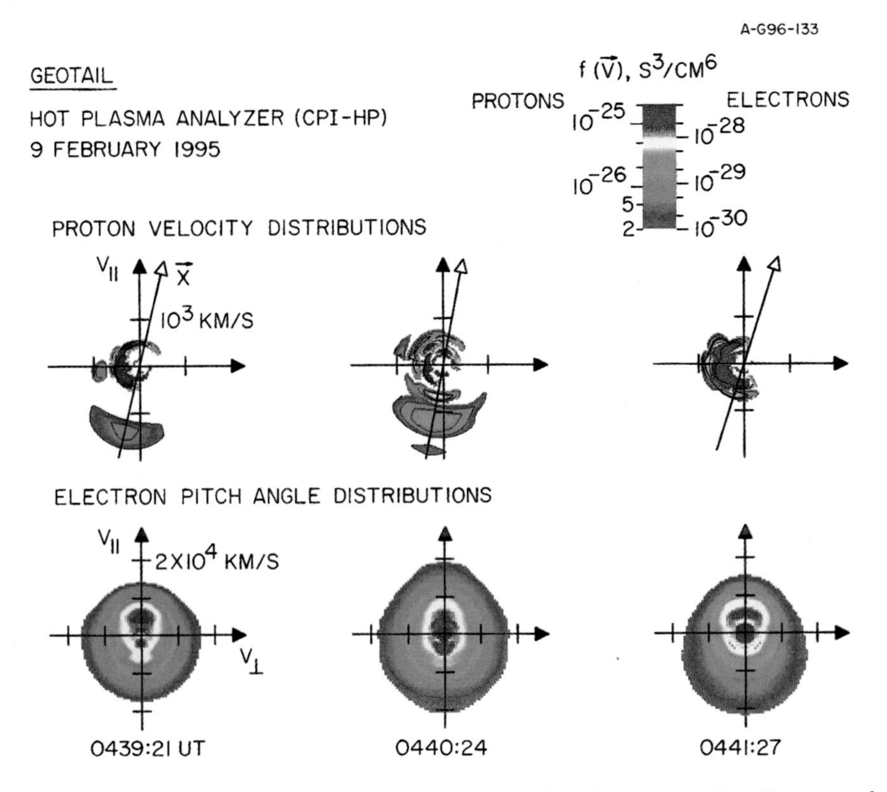

Figure 12. Proton and electron velocity distributions observed during the current pulse. Protons and electrons contribute to the current density [after *Frank et al.*, 1997].

current closure for the perpendicular currents in the magnetotail current sheet that are associated with spatial gradients of the pressure tensor. Consider the following MHD relationship for the perpendicular current [*Vasyliunas,* 1984],

$$\vec{J}_{\perp} = (c\vec{B}/B^2) \times (\rho d\vec{v}/dt + \nabla \bullet \overset{=}{P}) .$$

The gravitational term has been omitted. The pressure tensor is expected to have a significant, if not dominant, role in providing the perpendicular current.

It is customary to assume that the pressure tensor can be diagonalized into the $P_{\perp}$ and $P_{||}$ elements with the off–diagonal elements neglected. However, our recent determinations of these off–diagonal elements show that they may have an important role in determining the perpendicular current. These off–diagonal elements are shown in the bottom three panels of Figure 13 for the above interval of interest. The error bars indicate $\pm 1\sigma$ for the counting statistics. There are periods for which these elements are statistically important, e.g., 0350–0354 UT. For comparison the total pressure, magnetic pressure, the difference of parallel and perpendicular pressure and the dynamic pressures are shown in the upper three panels of Figure 13. Because the gradients of the pressure are the relevant parameters it is evident that the off–diagonal elements are important in determining the perpendicular currents and hence the dynamics of the current sheet. This possibility was previously suggested and justified on the basis of realistic numerical simulations of the global dynamics of the magnetotail [*Ashour–Abdalla et al.*, 1993]. Non–adiabatic motions of the ions in the weak magnetic fields in the magnetotail current sheet are expected to provide significant magnitudes for the off–diagonal pressure elements. These findings are provided with further support by high–resolution determinations of the proton velocity distributions in the current sheet such as those shown by the two examples in Figure 14. These ion velocity distributions were observed during the above noted interval for which significant off–diagonal elements were detected. The observational complexity of the proton velocity distributions renders difficult a simple identification of the main contributors to the off–diagonal elements, but the new ability to accurately determine the relevant plasma moments is a crucial step in understanding the origin of the magnetotail current sheet.

SUMMARY

We have discussed three important areas of research into the physics of the magnetotail which were uniquely

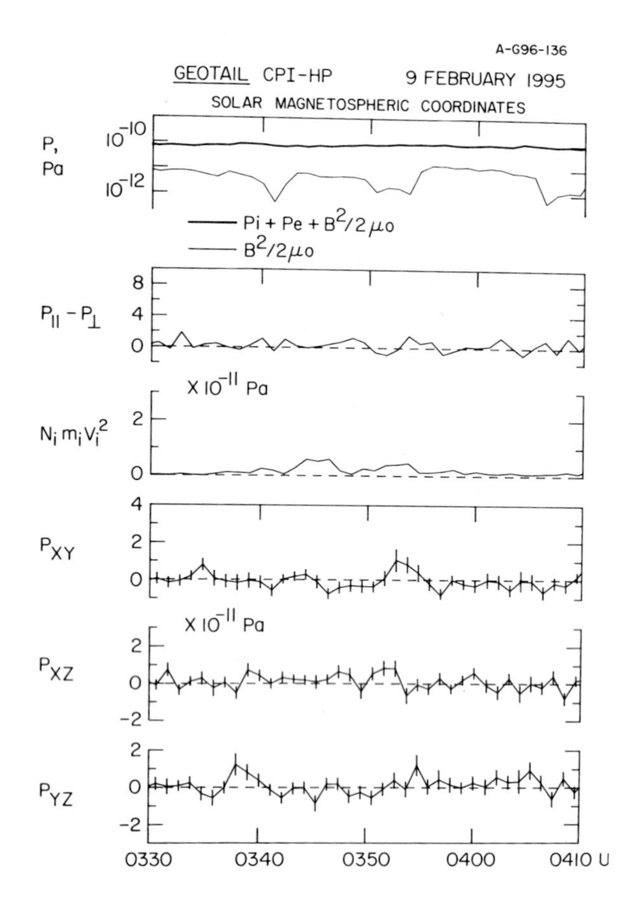

Figure 13. From top to bottom panels are shown the total pressure, pressure anisotropy due to the plasmas, the proton dynamic pressure, and the off–diagonal elements of the pressure tensor during an interval for which the Geotail spacecraft was located in the current sheet.

achieved with the Geotail mission. Namely, these three topics are (1) the nature of the plasma bulk flow components perpendicular and parallel to the magnetic field within the radial distance range of 10 to 50 R_E for which bursty bulk flows are not seen in the nearer–Earth distances and fireballs are detected at the larger distances, (2) the complex pitch angle distributions of ions in the near–Earth plasma sheet which exhibit memories of their previous magnetotail entry position, transport path and acceleration, and (3) the direct detection of intense field–aligned currents in the magnetotail during substorm onset and the direct determination of significant off–diagonal elements of the pressure tensors in the magnetotail current sheet. The contents of the present paper are intended to be exemplary because of the limitations in length. Examples of other major contributions from the plasma and magnetic field observations with the Geotail spacecraft are (1) comprehensive measurements of slow–mode shocks in the magnetotail [*Seon et al.,* 1996a,b], (2) direct comparison of a global MHD model which is responding to fluctuations of the solar wind with the plasma and magnetic field measurements with the Geotail spacecraft [*Frank et*

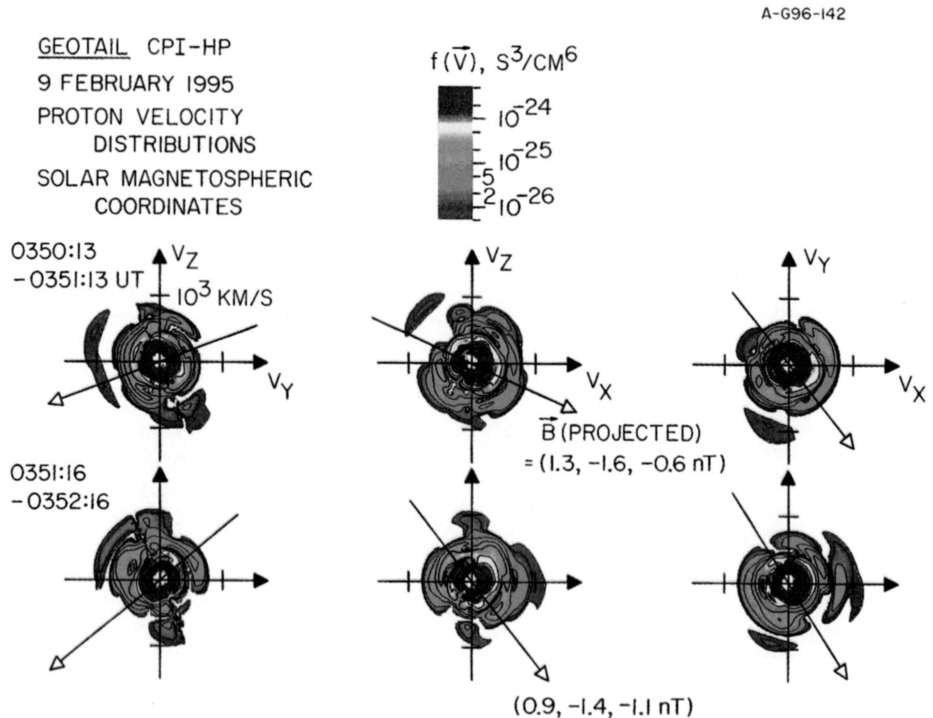

Figure 14. Two examples of the three–dimensional velocity distributions of protons in the magnetotail current sheet.

al., 1995; *Raeder et al.*, 1997; *Berchem et al.*, 1997a,b], (3) cross–tail magnetic flux ropes [*Lepping et al.*, 1995], (4) an investigation of the relevance of fractals in describing coarse–grained structure of ion velocity distributions in the magnetotail [*Ashour–Abdalla et al.*, 1996], and (5) magnetotail boundary motions such as windsock and breathing [*Siscoe et al.*, 1994a,b; *Shodhan et al.*, 1996; *Kaymaz et al.*, 1995]. Indeed the Geotail spacecraft has already provided a wealth of advances in the study of Earth's magnetosphere and can be expected to reward us with many additional findings in the future.

Acknowledgments. This research was supported in part at The University of Iowa under NASA grant NAG5–2371.

REFERENCES

Aggson, T. L., J. P. Heppner and N. C. Maynard, Observations of large magnetospheric electric fields during the onset phase of a substorm, *J. Geophys. Res.*, *88*, 3981–3990, 1983.

Angelopoulos, V., W. Baumjohann, C. F. Kennel, F. V. Coroniti, M. G. Kivelson, R. Pellat, R. J. Walker, H Lühr and G. Paschmann, Bursty bulk flows in the inner central plasma sheet, *J. Geophys. Res.*, *97*, 4027–4039, 1992.

Ashour–Abdalla, M., M. El–Alaoui, V. Peroomian, J. Raeder, R. J. Walker, R. L. Richard, L. M. Zelenyi, L. A. Frank, W. R. Paterson, J. M. Bosqued, R. P. Lepping and K. W. Ogilvie, Ion sources and acceleration mechanisms inferred from local distribution functions, *Geophys. Res. Lett.*, *24*, 955–958, 1997.

Ashour–Abdalla, M., L. A. Frank, W. R. Paterson, V. Peroomian and L. M. Zelenyi, Proton velocity distributions in the magnetotail, Theory and observations, *J. Geophys. Res.*, *101*, 2587–2598, 1996.

Ashour–Abdalla, M., L. M. Zelenyi, V. Peroomian, L. A. Frank and W. R. Paterson, Coarse–grained texture of ion distributions in the magnetotail: A fractal–like approach, *J. Geophys. Res.*, *101*, 15,287–15,296, 1996.

Ashour–Abdalla, M., L. M. Zelenyi, V. Peroomian and R. L. Richard, On the structure of the magnetotail current sheet, *Geophys. Res. Lett.*, *20*, 2019–2022, 1993.

Atkinson, G., The current system of geomagnetic bays, *J. Geophys. Res.*, *72*, 6063–6067, 1967.

Baumjohann, W., G. Paschmann and C. A. Cattell, Average plasma properties in the central plasma sheet, *J. Geophys. Res.*, *94*, 6597–6606, 1989.

Berchem, J., J. Raeder, M. Ashour–Abdalla, L. A. Frank, W. R. Paterson, K. L. Ackerson, S. Kokubun, T. Yamamoto and R. P. Lepping, Large–scale dynamics of the magnetospheric boundary: Comparisons between global MHD simulation results and ISTP observations, submitted to AGU Monograph *Encounter Between Global Observations and Models in the ISTP Era*, 1997.

Berchem, J., J. Raeder, M. Ashour–Abdalla, L. A. Frank, W. R. Paterson, K. L. Ackerson, S. Kokubun, T. Yamamoto and R. P. Lepping, The distant tail at 200 R_E: Comparison between Geotail observations and the results from a global magnetohydrodynamic simulation, *J. Geophys. Res.*, submitted for publication, 1997.

Büchner, J. and L. M. Zelenyi, Regular and chaotic charged particle motion in magnetotail–like field reversal, 1, Basic theory of trapped motion, *J. Geophys. Res.*, *94*, 11,821–11,842, 1989.

Chen, J., Nonlinear dynamics of charged particles in the magnetotail, *J. Geophys. Res.*, *97*, 15,011–15,050, 1992.

Chen, J., G. R. Burkhart and C. Y. Huang, Observational signatures of nonlinear magnetotail particle dynamics, *Geophys. Res. Lett.*, *17*, 2237–2240, 1990.

Coroniti, F. V., L. A. Frank, D. J. Williams, R. P. Lepping, F. L. Scarf, S. M. Krimigis and G. Gloeckler, Variability of plasma sheet dynamics, *J. Geophys. Res.*, *85*, 2957–2977, 1980.

Coroniti, F. V. and C. F. Kennel, Polarization of the auroral electrojet, *J. Geophys. Res.*, *77*, 2835–2850, 1972.

DeCoster, R. J. and L. A. Frank, Observations pertaining to the dynamics of the plasma sheet, *J. Geophys. Res.*, *84*, 5099–5121, 1979.

Frank, L. A., K. L. Ackerson, W. R. Paterson, J. A. Lee, M. R. English and G. L. Pickett, The comprehensive plasma instrumentation (CPI) for the Geotail spacecraft, *J. Geomag. and Geoelectr.*, *46*, 23–37, 1994.

Frank, L. A., K. L. Ackerson and R. P. Lepping, On hot tenuous plasmas, fireballs, and boundary layers in the Earth's magnetotail, *J. Geophys. Res.*, *81*, 5859–5881, 1976.

Frank, L. A., M. Ashour–Abdalla, J. Berchem, J. Raeder, W. R. Paterson, S. Kokubun, T. Yamamoto, R. P. Lepping, F. V. Coroniti, D. H. Fairfield and K. L. Ackerson, Observations of plasmas and magnetic fields in Earth's distant magnetotail: Comparison with a global MHD model, *J. Geophys. Res.*, *100*, 19,177–19,190, 1995.

Frank, L. A., R. L. McPherron, R. J. DeCoster, B. G. Burek, K. L. Ackerson and C. T. Russell, Field–aligned currents in Earth's magnetotail, *J. Geophys. Res.*, *86*, 687–700, 1981.

Frank, L. A., W. R. Paterson, K. L. Ackerson, S. Kokubun and T. Yamamoto, Plasma velocity distributions in the near–Earth plasma sheet: A first look with the Geotail spacecraft, *J. Geophys. Res.*, *101*, 10,627–10,637, 1996.

Frank, L. A., W. R. Paterson, S. Kokubun, T. Yamamoto, R. P. Lepping and K. W. Ogilvie, Observations of a current pulse in the near–Earth plasma sheet associated with a substorm onset, *Geophys. Res. Lett.*, *24*, 967–970, 1997.

Huang, C. Y. and L. A. Frank, A statistical study of the central plasma sheet: Implications for substorm models, *Geophys. Res. Lett.*, *13*, 652–655, 1986.

Huang, C. Y. and L. A. Frank, A statistical survey of the central plasma sheet, *J. Geophys. Res.*, *99*, 83–95, 1994.

Jacquey, C., J. A. Sauvaud and J. Dandouras, Location and propagation of the magnetotail current disruption during substorm expansion: Analysis and simulation of an ISEE multi–onset event, *Geophys. Res. Lett.*, *18*, 389–392, 1991.

Kaufmann, R. L., Substorm currents: Growth phase and onset, *J. Geophys. Res*, *92*, 7471–7486, 1987.

Kaymaz, Z., H. E. Petschek, G. L. Siscoe, L. A. Frank, K. L. Ackerson and W. R. Paterson, Disturbance propagation times to the far tail, *Geophys. Res. Lett., 100,* 23,743–23,748, 1995.

Kokubun, S., T. Yamamoto, M. H. Acuña, K. Hayashi, K. Shiokawa and H. Kawano, The GEOTAIL magnetic field experiment, *J. Geomag. Geoelectr., 46,* 7–21, 1994.

Lepping, R. P., D. H. Fairfield, J. Jones, L. A. Frank, W. R. Paterson, S. Kokubun and T. Yamamoto, Cross–tail magnetic flux ropes as observed by the GEOTAIL spacecraft, *Geophys. Res. Lett., 22,* 1193–1196, 1995.

Lopez, R. E. and A. T. Y. Lui, A multisatellite case study of the expansion of a substorm current wedge in the near–Earth magnetotail, *J. Geophys. Res., 95,* 8009–8017, 1990.

Lui, A. T. Y., Current disruption in the Earth's magnetosphere: Observations and models, *J. Geophys. Res. 101,* 13,067–13,088, 1996.

Lui, A. T. Y., R. E. Lopez, S. M. Krimigis, R. W. McEntire, L. J. Zanetti and T. A. Potemra, A case study of magnetotail current sheet disruption and diversion, *Geophys. Res. Lett., 15,* 721–724, 1988.

Lyons, L. R. and T. W. Speiser, Evidence for current sheet acceleration in the geomagnetic tail, *J. Geophys. Res., 87,* 2276–2286, 1982.

McPherron, R. L., C. T. Russell and M. P. Aubry, Satellite studies of magnetospheric substorms on August 15, 1968, 9. Phenomenological model for substorms, *J. Geophys. Res., 78,* 3131–3149, 1973.

Nakamura, M., G. Paschmann, W. Baumjohann and N. Sckopke, Ion distributions and flows near the neutral sheet, *J. Geophys. Res., 96,* 5631–5649, 1991.

Ohtani, S., S. Kokubun, R. Nakamura, R. E. Elphic, C. T. Russell and D. N. Baker, Field–aligned current signatures in the near–tail region, 2. Coupling between the Region 1 and the Region 2 systems, *J. Geophys. Res., 95,* 18,913–18,927, 1990.

Paterson, W. R., L. A. Frank, S. Kokubun and T. Yamamoto, Geotail survey of ion flow in the plasma sheet: Observations between 10 and 50 R_E, *J. Geophys. Res.*, in press, 1998.

Raeder, J., J. Berchem, M. Ashour–Abdalla, L. A. Frank, W. R. Paterson, K. L. Ackerson, S. Kokubun, T. Yamamoto and J. A. Slavin, Boundary layer formation in the magnetotail: Geotail observations and comparisons with a global MHD simulation, *Geophys. Res. Lett., 24,* 951–954, 1997.

Robert, P., R. Gendrin, S. Perraut and A. Roux, GEOS 2 identification of rapidly moving current structures in the equatorial outer magnetosphere during substorms, *J. Geophys. Res., 89,* 819–840, 1984.

Seon, J., L. A. Frank, W. R. Paterson, J. D. Scudder, F. V. Coroniti, S. Kokubun and T. Yamamoto, Observations of slow–mode shocks in Earth's distant magnetotail with the Geotail spacecraft, *J. Geophys. Res., 101,* 27,383–27,398, 1996.

Seon, J., L. A. Frank, W. R. Paterson, J. D. Scudder, F. V. Coroniti, S. Kokubun and T. Yamamoto, Observations of ion and electron velocity distributions associated with slow–mode shocks in Earth's distant magnetotail, *J. Geophys. Res., 101,* 27,399–27,411, 1996.

Shelley, E. G. and H. L. Collin, Auroral ion acceleration and its relationship to ion composition, in *Auroral Physics,* ed. C.–I. Meng, M. J. Rycroft and L. A. Frank, pp. 129–142, Cambridge Univ. Press, New York, 1991.

Shodhan, S., G. L. Siscoe, L. A. Frank, K. L. Ackerson and W. R. Paterson, Boundary oscillations at Geotail: Windsock, breathing and wrenching, *J. Geophys. Res., 101,* 2577–2586, 1996.

Siscoe, G. L., L. A. Frank, K. L. Ackerson and W. R. Paterson, Properties of the mantle–like magnetotail boundary layer: GEOTAIL data compared with a mantle model, *Geophys. Res. Lett., 21,* 2975–2978, 1994.

Siscoe, G. L., L. A. Frank, K. L. Ackerson and W. R. Paterson, Irregular, long–period boundary oscillations beyond ~100 R_e: GEOTAIL plasma observations, *Geophys. Res. Lett., 21,* 2979–2982, 1994.

Speiser, T. W., Particle trajectories in model current sheets, 1, Analytical solutions, *J. Geophys. Res., 70,* 4219–4226, 1965.

Vasyliunas, V., Fundamentals of current description, in *Magnetospheric Currents,* ed. T. A. Potemra, pp. 63–66, Geophysical Monograph 28, American Geophysical Union, Washington, D.C., 1984.

Louis A. Frank, Department of Physics and Astronomy, 212 Van Allen, The University of Iowa, Iowa City, IA 52242-1479

William R. Paterson, Department of Physics and Astronomy, 403 Van Allen, The University of Iowa, Iowa City, IA 52242-1479

Susumu Kokubun, Solar–Terrestrial Environment Laboratory, Nagoya University, 3–13 Honohara, Toyokawa, Aichi 442, Japan

Tatsundo Yamamoto, Institute of Space and Astronautical Science, Sagamihara, Kanagawa 229, Japan

Heavy Ion Acceleration by Reconnection in the Magnetotail: Theory and GEOTAIL observations

J. Büchner, J.-P. Kuska, B. Wilken, Q.-G. Zong

Max-Planck-Institut für Aeronomie, Katlenburg-Lindau, Germany.

The GEOTAIL project has provided rich information about the ionic composition of the magnetotail in general and about the mass-dependent spectra of energetic ions in particular. These measurements have encouraged further efforts toward a better understanding of the mechanisms of particle acceleration in the magnetosphere. In order to interpret the observations quantitatively we have modeled the mass-dependence of the spectra of energetic particles, strongly accelerated by magnetotail reconnection. In the course of strong acceleration the final energy distribution of energetic particles decouples from the source distribution. We derive mass dependent energetic particle spectra parametrized by the lower energy cutoff-velocity of the spectrum and by the ratio of reconnection electric field over the asymptotic normal magnetic field. The theoretical spectra are veryfied by test particle calculations. We compare the theoretical spectra with those, observed onboard GEOTAIL. We show that energization and spectra of ionospheric origin protons and O^+ ions as well as of solar wind origin He^{++} ions can indeed be directly explained by strong acceleration in the course of magnetotail reconnection. Our approach allows us to remotely estimate the current sheet half-thickness in the acceleration region as 1800 km and the reconnection electric field as $E = 2.7\ mV/m$.

1. INTRODUCTION

After the *in situ* discovery of the magnetosphere in the late fifties the ionosphere was considered by many researchers to be an unimportant ion source for the magnetosphere. The main argument was the higher energy of the solar wind particles, compared with the $< 1eV$ energy of ionospheric ions. Only Dessler and Hanson [1961] noticed early that, while the energy loading of the magnetosphere is perhaps due to the solar wind, the ionic content of the magnetosphere could be well provided by the ionosphere! Axford [1970] considered a possible ionospheric origin of the radiation belt and auroral precipitating ions on the basis of theoretical arguments. Using their ISEE-1 measurements Shelley et al. [1972] observed for the first time precipitating energetic O^+ ions. Later they found also outflowing O^+ ions [Shelley et al., 1976]. Ionospheric O^+ ions are extracted from the high latitude ionosphere as a result of magnetospheric substorm activity. The ISEE mission also helped to reveal an increase of the ion feeding of the inner magnetosphere and the plasma sheet *in situ* at times of enhanced magnetospheric activity [Ipavich et al., 1984; Ipavich et al., 1985]. DE (Dynamics Explorer) -1 measurements determined the outflow energy of O^+ ions as being typically less than 1 keV [Yau et al., 1985]. The mass and charge composition

New Perspectives on the Earth's Magnetotail
Geophysical Monograph 105

in the suprathermal energy range 10 keV - 230 keV near-earth magnetotail was then investigated by the AMPTE/IRM spacecraft. It detected a flux increase and spectra hardening of suprathermal ions including O^+ in close correlation with a strong storm [Möbius et al., 1987]. Based on GEOS measurements, Geiss et al. [1987] considered the ionosphere as a plasma source for the magnetosphere at least comparable with the solar wind. Chappell et al. [1987] stated that the ionosphere is a fully adequate plasma source for the Earth's magnetosphere. Since the late eighties there is no doubt left about the heavy ion feeding of the magnetosphere by the ionosphere, especially during periods of enhanced magnetospheric activity. Cladis [1986] considered the parallel acceleration of ions from the polar ionosphere to the plasma sheet. Cladis and Francis [1992] suggested that ionospheric O^+ ions in the magnetotail can trigger substorms. While AMPTE/CCE measurements revealed an increase of the energy density of ionospheric origin ions in the near Earth tail [Daglis et al., 1994] the GEOTAIL spacecraft has detected 1 - 5 keV O^+ ions in the tail lobes at essentially all explored distances (10 $R_E < |X_{GSM}| <$ 210 R_E) [Mukai et al., 1994; Hirahara et al., 1996]. The possibility of populating the tail by ionospheric origin ions was discussed based on model calculations by Baker et al. [1996]. On the other hand Lennartsson [1992] and Hirahara et al. [1996] provided ample evidence that for a good portion of time solar wind origin ions including He^{++} can dominate the plasma sheet. The charge state of helium ions in the magnetotail was shown to vary between one and two, indicating ionospheric and solar wind origin [Christon et al., 1994, 1996]. The relative abundance of the different ionic charge state species varied, depending on the relative proximity to sources and acceleration sites. Both, solar wind and ionospheric ions were observed traveling tailward at the same speed [Mukai et al., 1994; Hirahara et al., 1996]. GEOTAIL observations have also detected very energetic heavy ions in the mid- and distant tail. Ionospheric as well as solar wind origin heavy ions in the energy range 10 keV/e – 3.6 MeV/e were found by the EPIC/ICS and STICS devices onboard GEOTAIL [Williams et al., 1994; Jacquey et al., 1994; Christon et al., 1994]. Christon et al. [1994] presented some three-dimensional measurements of H^+ and He^{++} as well as of ionospheric origin heavy ions (O^+, N^+, NO^+). They demonstrated the highly non-isotropic nature of the flow in two tailward flow events, comparing them with the nearby flowing anisotropic magnetosheath ions. The level of collimation was not directly substantiated, however, e.g. by giving a number for the front to back anisotropy. The three-dimensional imaging time-of-flight spectrometer HEP-LD onboard GEOTAIL allows the measurement of mass-separated energy spectra in the range 144 keV – 4 MeV. On August 27, 1993 as well as on January 15 (five cases) and February 13, 1994 (two cases) HEP-LD measured the spectra of strongly collimated energetic helium and oxygen ion beams in the mid-tail ($40 < |X_{GSM}| < 60R_E$) [Wilken et al., 1995; Zong et al., 1997; Zong et al., 1998]. All observed ion beams correlated well with enhanced substorm activity. At distances beyond 60 R_E the tailward ion flows occured up to 40 - 50 minutes after a substorm onset. Near 40 R_E earthward directed energetic ion beams reached the spacecraft about 20 minutes after substorm onset. HEP-LD does not allow the determination of the charge state. Its observations were, therefore, compared in [Zong et al., 1997, 1998] with EPIC measurements [Williams et al., 1994; Christon et al., 1994]. This way [Zong et al., 1997, 1998] identified the beam constituents as singly charged H^+ and O^+ as well as He^{++} ions. Sample spectra of ionospheric origin energetic ions will be given in section 1.

Which acceleration process can explain the well collimated heavy ion beams in the magnetotail? The observed high energies exclude the current sheet acceleration mechanism [Cowley, 1980; Lyons and Speiser, 1982]. The observed close proximity of the beam observations to substorms suggests that the energy could have gained in reconnection electric fields. The coincidence of the beams with plasmoid-boundary traversals further supports the hypothesis of acceleration by reconnection. In order to verify this hypothesis it is useful to theoretically derive the mass-dependent spectra of heavy ions accelerated by reconnection which could be compared with the observed spectra

Here in this paper we present results of modeling the mass-dependent spectra of energetic ions accelerated by magnetotail reconnection both theoretically and by means of test particle calculations. Important conditions for any modeling effort are the initial energies of the ionospheric origin like O^+ and solar wind ions like He^{++} ions before entering the acceleration region in the tail. The temperatures of ionospheric ions which are ejected poleward of the auroral oval are of the order $1 - 2$ eV or even below [Yau, 1996]. During substorms thermal ionospheric ions gain higher energies, up to the order of several 100 eV, in parallel electric fields [Cladis, 1986]. While lower energy (< 10 eV) ions stay usually trapped near ionospheric heights due to gravitation, higher energy ions can reach the distant magnetotail, as shown for 0.075, 1.2, and 3.6 keV ions by [Baker et al., 1996]. Previously several attempts have been made to explain higher ion beam energies in the tail of dozens and hundreds of keV. Examples are given in [Fujimoto and Nakamura, 1994] who simulated the ion acceleration by reconnection numerically using a hybrid plasma code which neglects the electron mass. The maximum ion velocities obtained were, however, just of the order of the Alfvén speed in the ambient plasma. This is about 1000 km/sec instead of the 5 – 6000 km/sec beam velocities, observed by HEP-LD. Reconnection, however, can accelerate to higher beam velocities as shown in the past by means of test particle calculations. These calculations used either analytically prescribed fields (cf., e.g., [Curran et al., 1987]), MHD reconnection fields, e.g. of two-dimensional tail reconnection and plasmoids [Scholer and Jamitzky, 1989; Sachsenweger et al., 1989], two-dimensional turbulent reconnection model fields [Ambrosiano et al., 1988], fields of three-dimensional MHD reconnection [Birn and Hesse, 1994] or global MHD models [Joyce et al., 1996]. In order to parametrize the spectra, however, test particle calculations alone do not suffice, theoretical solutions are needed. The analytical treatment of particle motion in current sheets was started many years ago by Speiser [1965]. An approximate solution of the equations of particle motion near a reconnection X-line was found by Bulanov and Sasorov [1975]. It was later reconsidered by Deeg et al. [1991], Burkhart et al. [1990] and Moses et al. [1993]. The consequences of time dependent reconnection were discussed in Zelenyi et al. [1990]. Recently we have developed a theory of strong particle acceleration by reconnection [Büchner, 1995]. We have shown that the reconnection accelerated energetic ion velocity distribution decouples from the source distribution comparable to a pick-up process. As a result the observable spectra and distribution functions contain the most important information about the acceleration process – reconnection electric field and current sheet thickness. In section 2 we extend the theory of strong acceleration by reconnection to heavy ions. Then, in section 3, we verify the theoretical results by means of test particle calculations. In section 4 we apply the theoretically derived asymptotic energy distribution to analyse observed spectra.

2. OBSERVATION

Figure 1 depicts spectra observed by HEP-LD in a case, described as event B by Zong et al. [1998]. On February 13, 1994, between 18:30 UT and 19:10 UT, a tailward flowing energetic ion beam was observed at the spacecraft position $X_{GSE} = -63R_E$, $Y_{GSE} = 7R_E$ and $Z_{GSE} = -3.8R_E$. The occurence of the beam coincided with the encounter of the boundary of a plasmoid, moving tailward over the spacecraft. The B_x magnetic field evolution indicates that the spacecraft was essentially north of the neutral plane during this period, leaving the plasma sheet south bound at about 19:15 UT. The B_z magnetic field component was positive before the heavy ion beam reached the spacecraft. It turned negative at 18:30 UT when the H^+ and He^{++} components of the beam appeared and after the spacecraft left beam and boundary layer. While between 18:30 and 18:47 UT the beam ions were composed mainly of protons and helium ions, after 18:47 UT the O^+ flow became dominant. The heavy ion beam disappeared after 19:15 UT. Figure 1 depicts the spectra of three species, obtained by considering ions, reaching the instrument between 18:45 UT and 19:10 UT. Before the beam was ob-

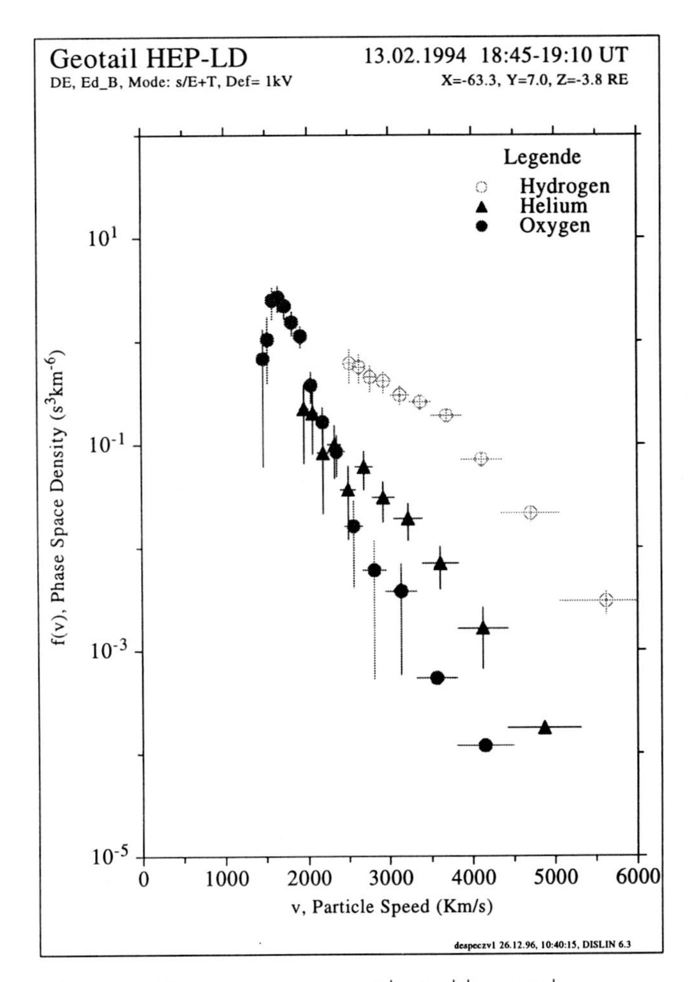

Figure 1. Sample of observed H^+, He^{++} and O^+ ion spectra $f(V)$ $vs.$ V, as measured onboard GEOTAIL by HEP-LD

served a substorm onset took place at 18:05 UT, seen in LANL geosynchronous energetic electron data as well as in ground magnetograms from Kiruna, Leirvogur, College and Valentia. The observed beams could, therefore, well be formed in a near-Earth reconnection region, created in the course of the substorm. Previous work has indeed shown that reconnection accelerates ions along its separatrices [Axford, 1984], inward of the boundaries of a newly formed plasmoid [Scholer and Jamitzky, 1987]. In the following section we describe the theoretical model of the formation of the energetic heavy ion spectra accelerated by reconnection.

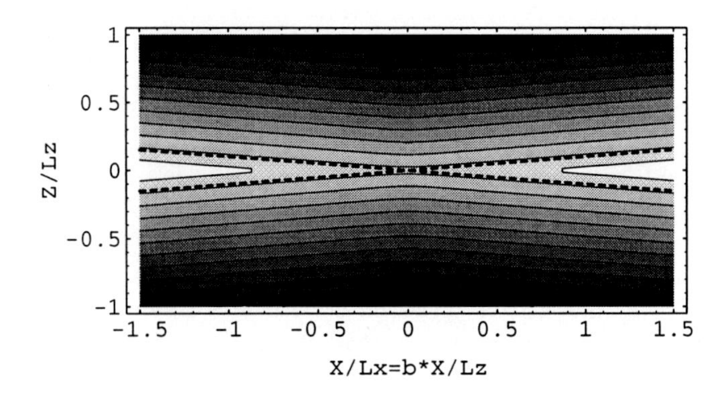

Figure 2. Magnetic field lines of the model used

3. THEORETICAL MODEL

The acceleration time by reconnection is typically less than a minute even for heavy ions. It is, therefore, shorter than the growth time of reconnection and plasmoids (several minutes). Hence, the spectrum of accelerated particles can be derived using a stationary field model. We specify the acceleration region by chosing the following field model

$$\vec{B} = B_o \cdot \tanh\left(\frac{Z}{L_z}\right) \cdot \vec{e}_x + B_n \cdot \tanh\left(\frac{X}{L_x}\right) \cdot \vec{e}_z$$

$$\vec{E} = E_y \cdot \vec{e}_y. \qquad (1)$$

The coordinate axes in expression (1) are chosen in accordance with the GSM-system, but the reconnection neutral line is shifted to the origin, being, therefore, the Y-axis of our coordinate system. As it has been shown by Cowley [1980] as well as by Lyons and Speiser [1982], characteristic values of the electric field E_y can be obtained from the lower energy cutoff velocity of observed ion beam distributions. The cutoff velocity arises from the limit determined by current sheet acceleration. It is given by $v_{cutoff} = \frac{1}{2}E_y/B_n$ [Cowley, 1980, Lyons and Speiser, 1982]. The observed cutoff velocity value was 2000 km/s. With this number and assuming a normal magnetic field at the relevant tail distances of $B_n \approx 2\ nT$ one obtains an estimate of the electric field strengths $E_y = \frac{1}{2}v_{cutoff}B_n \approx 1\ mV/m$. The typical tail-lobe magnetic field strength 20 nT provides a good lowest order estimate for B_o and 1 R_E = 6300 km is a good lowest order estimate for the spatial scale, the current sheet half-thickness L_z. Flux conservation arguments reveal $L_x = L_z B_o/B_n = 10\ R_E$.

We frther will use dimensionless variables, denoting them by small letters like $x, v_y, p_z, \ldots$ instead of coordinates, velocities, momenta etc. in physical units, denoted by capital letters $(X, V_y, P_z, \ldots)$. The time is normalized to $T_{norm} = \Omega_o^{-1} = m_p/(e \cdot B_o)$, the inverse proton gyro-frequency in a magnetic field B_o. The spatial coordinates are normalized to the half-thickness of the current sheet $L_{norm} = L_z$. The velocity unit is $V_{norm} = L_{norm}/T_{norm} = L_z \cdot \Omega_o$ and the momentum unit $P_{norm} = m_p \cdot v_{norm} = m_p \cdot L_z \cdot \Omega_o$. The energy unit is $W_{norm} = \frac{m_p}{2} \cdot V_{norm}^2$. A natural choice for the electric field unit E_{norm} is $B_o \cdot V_{norm}$, so that the normalizing electric field E_{norm} causes in the ambient magnetic field B_o an $\vec{E} \times \vec{B}$ drift velocity equal to V_{norm}. With the proton mass and current sheet parameters, introduced before, one finds $\Omega_o = 1.67\ rad/s$, i.e. $T_{norm} = 0.6\ s$. With $L_{norm} = L_z = 6300\ km$ the velocity unit is $V_{norm} = L_{norm}\Omega_o = 10,500\ km/s$ and the normalizing energy unit is $W_{norm} = \frac{1}{2}m_p V_{norm}^2 = 0.8\ MeV$. The normalizing electric field strength is $E_{norm} = B_o \cdot V_{norm} = 230\ mV/m$. The dimensionless values for current sheet electric fields $\approx 1\ mV/m$ is $\epsilon = E_y/E_{norm} \approx 0.004$ and $b = B_n/B_o = 0.1$. As a result dimensionless fields given by (1) can be written as

$$\vec{b} = \frac{\vec{B}}{B_o} = \tanh(z) \cdot \vec{e}_x + b \cdot \tanh(b \cdot x) \cdot \vec{e}_z; \qquad \vec{e} = \epsilon \cdot \vec{e}_y \quad (2)$$

The fields (equation (2)) can be represented by the dimensionless vector- and scalar potentials

$$a_y(x, z) = \ln\cosh(b \cdot x) - \ln\cosh(z)$$

and

$$\phi(y) = -\epsilon \cdot y, \qquad (3)$$

where a_y is the dimensionless Y-component of the vector potential, normalized to $A_{norm} = B_o \cdot L_z$ and the scalar potential in (3) is normalized to $\phi_{norm} = E_{norm} \cdot L_z = \Omega_o \cdot L_z^2 \cdot B_o = e \cdot (L_z B_o)^2/m_p$. A consideration of the potentials given by equation (3) in the generalized momentum equations reveals the Hamiltonian of the system. The dimensionless Hamiltonian of

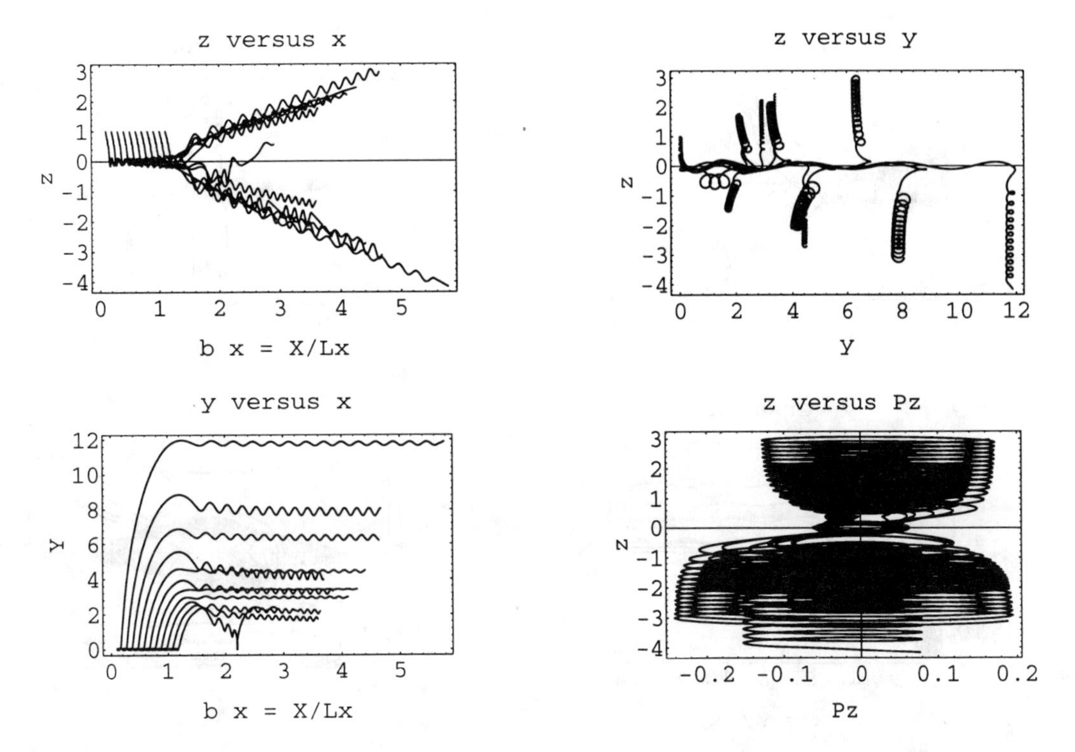

Figure 3. Projections of proton orbits in physical space and phase space (z *vs.* p_z, lower right panel)

the motion of $q = Q/e$ times positively charged ions with a mass $M = M_i/m_p$ times the proton mass in the reconnection fields of by equation (1) is given by

$$h(x, p_x, y, p_y, z, p_z) = \frac{q}{2M}p_x^2 + \frac{q}{2M}\left[p_y - \ln\cosh\left(b \cdot x\right)\right.$$

$$\left. + \ln\cosh(z)\right]^2 + \frac{q}{2M}p_z^2 - \epsilon \cdot y \tag{4}$$

Note that we normalized the Hamiltonian to $2W_{norm} = m_p \cdot V_{norm}^2$ in order to maintain the division by 2 in correspondence with the canonical form. From the Hamiltonian (4) one obtains a set of ordinary differential equations of motion:

$$\frac{dp_x}{dt} = -\frac{\partial h}{\partial x} = b \cdot \frac{q}{M} \cdot$$

$$\cdot \tanh(b \cdot x) \cdot [p_y - \ln\cosh(b \cdot x) + \ln\cosh(z)] \tag{5}$$

$$\frac{dp_y}{dt} = -\frac{\partial h}{\partial y} = \epsilon \tag{6}$$

$$\frac{dp_z}{dt} = -\frac{\partial h}{\partial z} = -\frac{q}{M} \cdot \tanh(z) \cdot$$

$$\cdot [p_y - \ln\cosh(b \cdot x) + \ln\cosh(z)] \tag{7}$$

$$\frac{dx}{dt} = \frac{\partial h}{\partial p_x} = \frac{q}{M} \cdot p_x = v_x \tag{8}$$

$$\frac{dy}{dt} = \frac{\partial h}{\partial p_y} = \frac{q}{M} \cdot \tag{9}$$

$$cdot\, [p_y - \ln\cosh(b \cdot x) + \ln\cosh(z)] = v_y$$

$$\frac{dz}{dt} = \frac{\partial h}{\partial p_z} = \frac{q}{M} \cdot p_z = v_z \tag{10}$$

Let us demonstrate the mass dependence of ion orbits by calculating test protons and test oxygen ion trajectories. In accordance with the typically low energy of the ionospheric origin ions the initial velocity is chosen to be $V_o = 100\ km/s$ which reveals the dimensionless value $v_o = 0.01$. This particle velocity corresponds to a proton kinetic energy of 70 eV and to an oxygen ion energy of 700 eV. The latter overestimates the real initial oxygen energy a little bit. It nevertheless still fulfills the necessary condition of applicability of the strong acceleration theory $v_o \ll E_y/B_n$ [Büchner, 1996]. Figures 3 and 4 depict proton and singly charged oxygen ion orbits, launched at different x_o positions between $b \cdot x_o = 0$ but constant $b \cdot x_o = 1$, $y_o = 0$ and $z_o = 1$. The parameters were chosen $\epsilon = 0.01$, $b = 0.1$ and $v_o = 0.01$. Test particle trajectories of protons (Figure 3) and oxygen ions (Figure 4) illustrate the major steps in spectrum formation.

The upper left panels of Figures 3 and 4 depict a meridian plane ($y = 0$) projection of the ion motion.

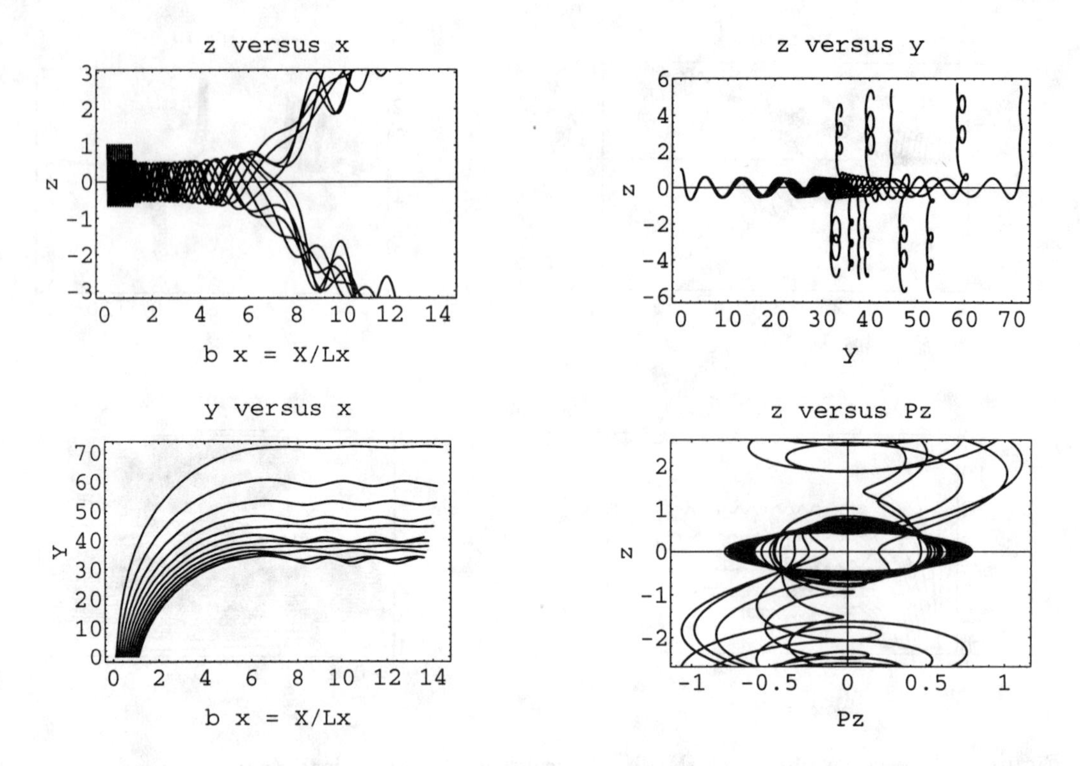

Figure 4. Projections of oxygen orbits in physical space and phase space (z *vs.* p_z, lower right panel)

After the ions are launched from their initial position in the upper left part of the Figure they first gyrate and $[\vec{E} \times \vec{B}]$ drift toward the equatorial plane ($z = 0$). However, before reaching the the equatorial plane the ions become de-magnetized. At a distance z_{de} from the neutral plane and they start to meander across the plane $z = 0$ instead of gyrating around the magnetic field direction. The particles continue to meander on so called Speiser [1965] orbits, until they get re-magnetized and ejected in the positive x direction. The upper right panels of Figures 3 and 4 depict the projection of the ion motion onto an $x = const.$ plane, the lower left panels the projection onto the equatorial plane $z = 0$. They complete the picture of the different types of motion, gyration and meandering, which is essential for modeling the spectrum theoretically.

The first transition, from gyration to meandering, takes place, when the local Larmor-radius becomes comparable to the remaining distance from the neutral plane. Since we consider the acceleration of particles with $v_o < E_y/B_n$, ($v_o = \sqrt{v_{xo}^2 + v_{yo}^2 + v_{zo}^2}$ is the normalized initial particle velocity), the $[\vec{E} \times \vec{B}]$ drift velocity near the neutral plane $v_d \approx \epsilon \cdot \sqrt{z_{de}^2 + b^2 \tanh^2(bx_{de})}^{\,-1}$ $\epsilon \cdot z_{de}^{-1}$ exceeds v_o by far. Hence, the particle velocity at de-magnetization is essentially the local $[\vec{E} \times \vec{B}]$ velocity. For the Larmor radius one finds $\rho(z_{de}) \approx v_d/z_{de} \cdot M/q \approx M \cdot \epsilon \cdot z_{de}^{-2}/q$. From the condition $\rho(z_{de}) \approx z_{de}$ one obtains

$$z_{de} \approx \sqrt{\epsilon^{2/3} - b^2 \tanh^2(bx_{de})} \approx (\frac{M}{q} \cdot \epsilon)^{1/3} \qquad (11)$$

Since $z_{de} \approx (\frac{M}{q} \cdot \epsilon)^{1/3} \ll 1$ one can replace $\ln \cosh(z)$ in the potentials and equations of motion in the non-gyotropic acceleration phase by $\frac{1}{2} \cdot z^2$.

The lower right panels of Figures 3 and 4 depict a projection of the phase space trajectory to the cutting plane z *vs.* $p_z = \frac{M}{q} \cdot v_z$. It illustrates an important feature of the reconnection acceleration process, the conservation of the action integral $I_z = \oint p_z dz$ during the meandering motion, i.e. during the main acceleration process. The upper left and the upper and lower right panels of Figures 3 and 4 illustrate that the amplitude of the z-oscillations remains practically constant $\approx z_{de}$. This fact can be used, together with the translational symmetry of the Hamiltonian in the y-direction, to reduce the dimension of the mathematical problem. From the y-invariance, i.e. the translational symmetry of the problem follows a second independent integral of motion. From equation (6) one obtains

$$p_y = \epsilon \cdot (t - t_o) + p_{yo}. \qquad (12)$$

In equation (12) p_{yo} denotes the initial p_y momentum. Replacing p_y with the help of equation (12) one finds instead of equation (5) by using $< z^2 >= \frac{1}{2} z_{de}^2$ in accordance with equation (11)

$$\frac{M}{q} \cdot \frac{d^2x}{dt^2} = \frac{dp_x}{dt} = b \cdot \frac{q}{M} \cdot \tanh(b\ x) \cdot [p_{yo} - \epsilon \cdot t_o +$$

$$+ \epsilon \cdot t - \ln\cosh(b \cdot x) + \frac{1}{4} \cdot (\frac{M}{q}\ \epsilon)^{2/3}\Big] \qquad (13)$$

In the derivation of the master equation of the acceleration model equation (13) all expressions (5)-(10) are used. Let us now try to solve it.

First, it is necessary to determine beginning and end of the acceeration, i.e. de- and re- magnetization. The upper right panels of Figures 3 and 4 indicate that after one or more periods of the meandering across the neutral plane z-oscillations the ions become re-magnetized. With the re-magnetization the non-adiabatic acceleration comes to an end. The accelerated ions can be ejected either back to the half-space, from which they were, or forward into the opposite half-space. There are different ways to determine the time of re-magnetization. One of them uses the fine structure of the meandering motion [Büchner, 1995]. For our purpose here we use the close relation of the re-magnetization condition, developed there and the properties of the projection of the average motion in the $x - y$ plane (see the lower left panels of Figures 3 and 4). The lower left panels of Figures 3 and 4 also demonstrate that the ions are re-magnetized close to a turning point of their orbits in the Y-direction, i.e. when they gained a maximum energy. This property means that re-magnetization takes place when the y direction reaches a turning point ($v_y = 0$) and v_x also stops to change, i.e. $dp_x/dt = 0$. From equation (13) follows, therefore,

$$\epsilon \cdot t_{acc} = \epsilon \cdot (t_{re} - t_{de}) = \ln\left[\frac{\cosh(bx_{re})}{\cosh(bx_{de})}\right], \qquad (14)$$

where x_{re} and x_{de} are the x-coordinates of de- and re- magnetization. Let us now determine the energy gain during the acceleration phase between de- and re-magnetization, respectively. Since the fields and the Hamiltonian are constant in time and the potential varies just in the y-direction one finds from the Hamiltonian equation (4) the following equations for the kinetic energy variation:

$$\frac{dh}{dt} = \frac{\partial h}{\partial t} = 0 \qquad \rightarrow \qquad h_o = \text{const.} =$$

$$= \frac{1}{2}\ \frac{M}{q} \cdot v_o^2 - \epsilon \cdot y_o = \frac{1}{2}\ \frac{M}{q} \cdot v^2 - \epsilon \cdot y, \qquad (15)$$

Notice that in equation (15) y_o is the y coordinate at $t = t_o$. Expression (15) was derived replacing the p_y momentum using equation (3). Although the total energy $W_{tot} = H = h \cdot 2W_{norm} = W_{kin} + W_{pot}$ is conserved due to the time-independence of the Hamiltonian, the particle kinetic energy $W_{kin} = \frac{M \cdot m_p}{2} V^2 = Mv^2 \cdot W_{norm}$ can change at the expense of the electric field by $W_{pot} = e \cdot E_y \cdot (Y - Y_o) = \epsilon \cdot (y - y_o) \cdot W_{norm}$, it is directly proportional to the shift in the Y- direction. One can determine the shift in Y by solving equation (13), which describes the particle motion in the x-direction in the non-gyrotropic phase of motion.

Next one has to solve equation (13). It is an ordinary but highly non-linear differential equation of the average motion $x(t)$, which can be solved in different limits analytically. Since HEP-LD data are restricted to higher energies we further address the most efficient reconnection-acceleration close to the neutral line, were the highest energies are gained. There, i.e. near $x = 0$, one can neglect the lowest order iteration term $\ln\cosh(bx)$ in equation (13), as long as $bx_{de} \ll 1$. Possible solutions of the resulting Airy-equation were discussed in the literature [Bulanov and Sasorov 1976; Deeg et al., 1991; Burkhart et al., 1990, Moses et al., 1993]. They basically reveal

$$t_{acc}(x_{de}, x_{re}) = \left(\frac{M^2}{q^2 \cdot \epsilon \cdot b^2}\right)^{1/3} \left[\frac{3}{2} \ln\left(\frac{x_{re}}{x_{de}}\right)\right]^{2/3} \qquad (16)$$

Solving equations (14) and (16) together one can eliminate x_{re} and determine $t_{acc}(x_{de})$ explicitely and replace x_{re}. Using $y_{re} - y_{de} = \frac{1}{2}\epsilon t_{acc}^2$ one, finally, obtains the energy gain

$$w\,(x_{de})\,|_{bx_{de} \ll 1} = \frac{1}{2}\left(\frac{M}{q}\right)^{1/3} \left(\frac{\epsilon}{b}\right)^{4/3} \left[\frac{3}{2} ln\left(\frac{x_{re}}{x_{de}}\right)\right]^{4/3}. \qquad (17)$$

The asymptotic spectrum, i.e. the velocity-distribution of the ions, can be obtained as follows. First, let us relate the ion density at the de-magnetization point to the density far from the neutral point, say at a position $\{x_o, z_o\}$. Since $b \ll 1$ one finds (cf., e.g., [Deeg et al., 1991]):

$$x_{de} \approx \alpha \cdot x_o, \qquad \text{where} \qquad \alpha \approx \left(\frac{z_o}{x_o}\right)^{b/z_o cosh^2(bx_o)} \approx 1. \qquad (18)$$

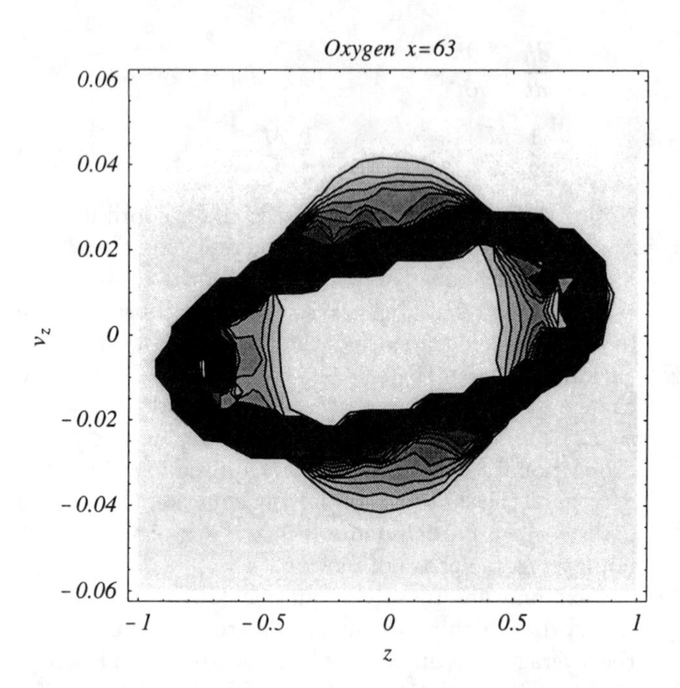

Figure 5. Phase space density of z *vs.* v_z O^+ ion orbits taken at $bx = 6.3$

Further the Liouville-theorem can be applied. It provides $n(w)\ dw = n_{lin}(x_o)\ dx_o = n_{lin}(x_o)\ dx_{de}$, where $n_{lin}(x_o) = const.$ is the linear particle number density along the x direction at $z = z_o$. Applying the Liouville-theorem and differentiating equation (17) one can derive an asymptotic expression for the velocity spectrum $f(v)$ in the higher energy limit. Note that in the derivation the identity $dN = f(w)\ dw = 4\pi\ v^2\ f(v)\ dv$ is used. One obtains the asymptotic velocity spectrum

$$f(v) \propto v^{-3/2} \cdot \exp\left\{ -\frac{2b}{3\epsilon} \cdot \left(\frac{M}{q}\right)^{1/2} \cdot v^{3/2} \right\} \quad (19)$$

The high energy velocity spectrum given by equation (19) can be verified by test particle calculations, as demonstrated in the following section.

4. RESULTS OF TEST PARTICLE CALCULATIONS

In this section we present the results of numerical calculations of 250,000 orbits of singly charged ions, M = 1, 4 and 16 times the proton mass. The charged states are chosen for protons, double charged helium ions and singly charged oxygen ions, respectively. The ions were launched into the model magnetic and electric fields given by expressions (1). The particles are initially distributed randomly between $bx_o = 0$ and $bx_o = 1$ at $z_o = 1$ above the reconnection region. Like the initial velocity of the eleven test particle in section 2 now the initial thermal velocity of the 250,000 ions was chosen to be $v_o = 0.01$. The center of their Maxwellian distribution was shifted in accordance to the local $[E \times B]$ velocity. As depicted in Figures 3 and 4 the particles first drift on gyro-orbits until they become de-magnetized and accelerated close to the neutral plane $z = 0$. After being accelerated the ions become re-magnetized. Since it is impossible to draw all orbits we use section surfaces to illustrate the re-distribution of the different ions in space and velocity space. For our purpose $z - v_z$ cuts through the space- density surfaces are appropriate, a section plane which corresponds to the projection plane of the eleven test ions to the z *vs.* p_z plane, depicted in the right lower panel of Figure 4 (notice that $p_z = v_z \cdot M/q$). Figures 5 and 6 show snapshots of the phase space cuts of oxygen ions, taken at two different distances from the neutral line, at $b \cdot x = 6.3$ (Figure 5) and $bx = 10$ (Figure 6), respectively. Figure 5 illustrates how the beam formation starts near $|z| = 0.75$ out of the plasma sheet population $|z| < 0.5$. At a distance $b \cdot x = 10$ from the reconnection site (Figure 6) the beam still flows near the plasmoid boundary, now at $|z| = 2.5$, while no particles are left near the center of the plasma sheet $z = 0$. At this position the asymptotic spectrum was diagnosed. Figure 7 depicts the resulting

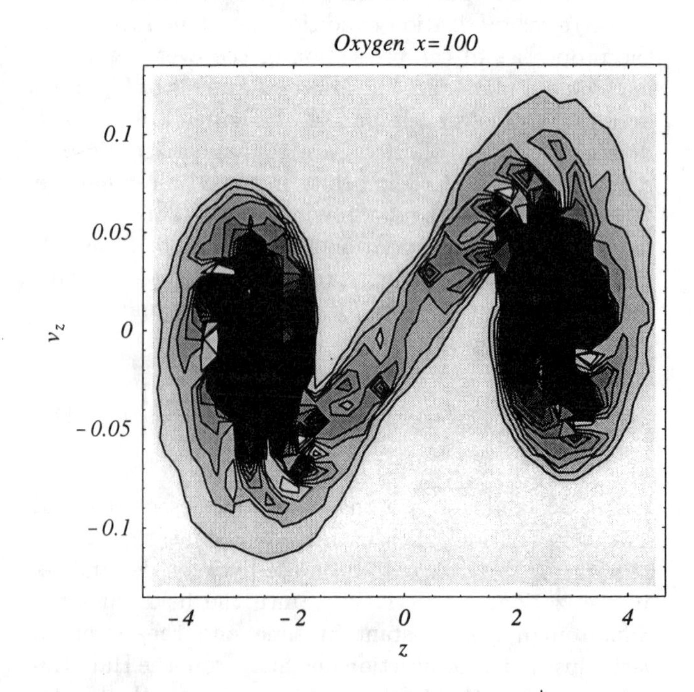

Figure 6. Phase space density of z *vs.* v_z O^+ ion orbits taken at $bx = 10$

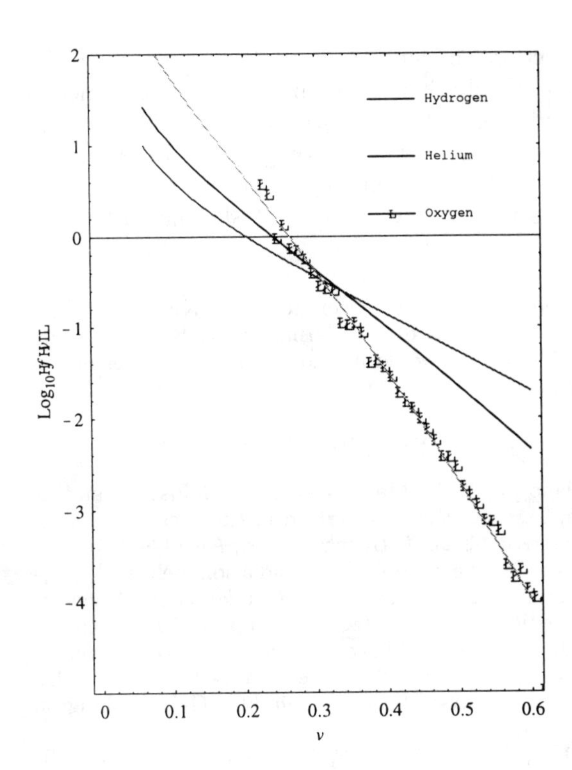

Figure 7. O^+, He^{++} and H^+ test particle spectra $f(v)$ *vs.* v for $\epsilon/b = 0.1$ and fitting curves corresponding to equation (20)

energetic O^+, He^+ and H^+ velocity spectra $f(v)$ *vs.* v, obtained at $b \cdot x = 10$.

In order to compare them with theoretically derived asymptotic spectra it is appropriate to find best fits of the test ion spectra with the velocity dependence of expression (19). The asymptotic spectrum (19) can be parametrized generally as

$$f(v) = A \cdot v^{-\frac{3}{2}} \cdot \exp\{B \cdot v^{\frac{3}{2}}\} \qquad (20)$$

The best fit of expression (20) with the test particle spectra revealed, with the confidence intervals given in parantheses, $-B = 6.2(6.0 - 6.4), 11.6(10.9 - 12.3)$ and $-24.3(23.2 - 25.3)$ for $M/q = 1, 4$ and 16, respectively. These coefficients are related to each other approximately as predicted by expression (19), i.e. $B \propto \sqrt{M/q}$. Since $\langle B \rangle = -6.1\ \sqrt{M/q}$ they revealed $\epsilon/b = 0.109$. This is within a 10 % deviation from the model parameter, the scattering being due to the exponentially decreasing number of ions reaching the highest velocities. Figure 7 depicts the test particle spectra together with the best fitting curves according to equation (20). For the experimental uncertainties , see Figure 1.

5. DATA COMPARISON

After confirming the correctness of the theoretically derived asymptotic spectra, given by equation (19), to a high degree of accuracy, one can feel confident that the asymptotic expression (20) models the high energy part of the spectrum of reconnection accelerated ions well. Its parameter B should help to determine the ϵ/b field ratio in the reconnection region. Since $b = B_n/B_o$ can be directly measured the determination of ϵ/b would allow a remote determination of the accelerating electric field. In order to obtain ϵ/b from the observed spectra one must specify the normalizing velocity $v_{norm} = L_z\Omega_o$. Since B_o is the known lobe magnetic field strength $\approx$ 20 nT, v_{norm} is mainly determined by the sheet half-thickness L_z. The normalizing velocity relates the observable cutoff velocity and the ϵ/b parameter as $V_{cutoff} = 2V_{norm}\epsilon/b$. Hence, L_z can also be determined, respectively, from ϵ/b and V_{cutoff}

The best fitting values of ϵ/b and V_{cutoff} can be obtained in an iterative way. First one guesses a most probable value ϵ/b. In the case of the the Earth's magnetotail an appropriate numer to start with is $0.01/0.1 = 0.1$. Then one calculates the normalizing velocity V_{norm} which corresponds to the chosen ϵ/b and the measured V_{cutoff} as $V_{norm} = V_{cutoff} \cdot b/2\epsilon$. Best fitting the parametrized spectrum (20) to the experimental data normalized by V_{norm} reveals the parameter B and, therefore, a best fit value for ϵ/b for every single mass per charge ratio. The mass-corrected ϵ/b must not coincide with the first assumed level. Varying the choice of the initially assumed value ϵ/b one finds the best fitting one. In this case the cutoff velocity was about $V_{cutoff} = 1600\ km/s$. In the case of these spectra the method, described above, revealed $V_{norm} = 3000\ km/s$ and $\epsilon/b = 0.27$. For these parameters the best fitting values $-B$ were $-B = 3.2\ (2.8 - 3.3), 4.4\ (3.8 - 5.1)$ and $12.9(10.5 - 15.4)$ for the H^+, He^{++} and O^+ spectra, respectively. Hence, the sheet half-thickness in the reconnection region was closer to 1800 km rather than 6300 km and the electric field was about 2.7 mV/m instead of the originally assumed 1 mV/m. Figure 8 depicts the test particle spectra together with the best fitting curves according to equation (20). Again, for the experimental uncertainties , see Figure 1.

6. SUMMARY AND DISCUSSION

Based on theoretical considerations we have derived a mass-per-charge-dependent asymptotic expression for the velocity spectrum of energetic ions, accelerated by reconnection. The model spectrum exhibits a charac-

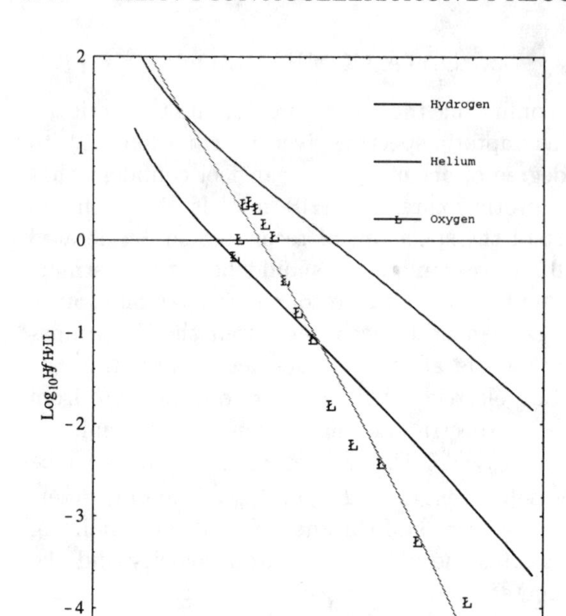

Figure 8. Observed H^+, He^{++} and O^+ ion spectra $f(v)$ *vs.* v, (velocity normalized to $V_{norm} = 3000\ km/s$) and the fitting curves corresponding to equation (20)

teristic $\propto \exp\left\{\sqrt{M/q}\right\}$ dependence on the mass-per-charge ratio. The validity of the derived spectrum has been verified by test particle calculations. We fit the asymptotic form of the $f(v)$ distributions to the spectra, measured by HEP-LD onboard GEOTAIL. We found that the observed spectra agree well with the ones, derived for reconnection acceleration. In particular they exhibit the characteristic square-root mass-per-charge dependence predicted theoretically. Our results show that the acceleration of ionospheric origin ions, like O^+, and solar wind origin ions like He^{++}, to high energy beams, observed in the tail can be well explained by a one-step strong reconnection acceleration process. Since the acceleration takes place in a cross-tail electric field the maximum energy, which can be gained this way,is given by $\Delta W = e \cdot E_y \cdot \Delta Y$, where ΔY is limited by the size of the tail. The limits for proton energies are of the order of 100 keV and for oxygen ion energies of the order of 1 MeV. The observations, we have used, are below those levels. By the way, the observation of highest energy ions at the dusk-side of the tail agrees well with our explanation. We have derived an asymptotic energy-per-charge dependent spectrum of reconection accelerated ions. It allowed us to determine remotely, just from the measured parameters V_{norm} and ϵ/b, a sheet half-thickness of 1800 km and and an electric field strength of $E = 2.7\ mV/m$ in the reconnection region by fitting the theoretically derived to the spectra, observed by HEP-LD onboard GEOTAIL.

Acknowledgments. The work of J.-P. Kuska has been supported by a grant of the German DFG Research Community. The Editor would like to thank the referees of this manuscript.

REFERENCES

Ambrosiano, J., W.H. Matthaeus, M.L. Goldstein, and D. Plarte, Test particle acceleration in turbulent reconnecting magnetic fields, *J. Geophys. Res.*, *93*, 14,383, 1988.

Axford, I., On the origin of the radiation belt and auroral primary ions, in *Particles and Fields in the Magnetosphere*, edited by B.M. McCormac, pp. 46-59, 1970.

Axford, I., Magnetic field reconnection, in *Reconnection in space and Laboratory Plasma,* edited by E.W. Hones Jr., pp. 4-14, Geophysical Monograph 30, AGU, Washington D.C., 1984.

Baker, D.N., T.I. Pulkkinen, P. Toivanen, M. Hesse, and R. McPherron, A possible interpretation of cold ion beams in the earth's tail lobe, *J. Geomagn. Geoelctr,*, 48, 699, 1996.

Büchner, J., The Nonlinear Dynamics of Strong Particle Acceleration by Reconnection, in *Physics of Space Plasmas, 1993,* ed. T. Chang, G.B. Crew and J.R. Jasperse (eds.), Scientific Publishers Inc., Cambridge, MA, 1995.

Büchner, J., and S. Teselkin, Signatures for remote sensing of reconnection in the distant magnetotail, *Adv. Space Res.*, *18,* 45, 1996.

Büchner, J. and L.M. Zelenyi, Regular and chaotic charged particle motion in magnetotail-like field reversals, 1. Basic theory, *J. Geophys. Res.*, *94,* 11,821, 1989.

Birn J., and M. Hesse, Particle acceleration in the dynamic magnetotail: Orbits in self-consistent three-dimensional MHD fields, *J. Geophys. Res.*, *99*, 109, 1994.

Bulanov, S.V., and P.V. Sasorov, Energetic spectrum of particles accelerated in the vicinity of a magnetic field zero, *Russian J. Astronomy*, *52*, 763-769, 1975.

Burkhart, G.R., J.F. Drake, and J. Chen, Magnetic reconnection in collisionless plasmas: Prescribed fields, *J. Geophys. Res.*, *95*, 18,833, 1990.

Chappell, C.R., T.E. Moore, and J.H. Waite, The ionosphere as a fully adequate source of plasma for the Earth's magnetosphere, *J. Geophys. Res.*, *92*, 5696-5910, 1987.

Christon, S.P., G. Gloeckler, D.J. Williams, T. Mukai, R.W. McEntire, C. Jacquey, V. Angelopoulus, A.T.Y. Lui, S. Kokubun, D.H. Fairfield, M. Hirahara, and T. Yamamoto, Energetic atomic and molecular ions of ionospheric origin observed in distant magnetotail flow reversals, *Geophys. Res. Lett.*, *21*, 3023-3026, 1994.

Christon, S.P., G. Gloeckler, D.J. Williams, R.W. McEntire, and A.T.Y. Lui, The downtail distance variation of energetic ions in the Earth's magnetotail region: GEOTAIL measurements at $X > -208\ R_E$, *J. Geomag. Geoelectr.*, *48*, 615-627, 1996.

Cladis, J.B., Parallel acceleration and transport of ions from polar ionosphere to plasma sheet, *Geophys. Res. Lett.*, *13*, 893-896, 1986.

Cladis, J.B., and W.E. Francis, Distribution in the magnetotail of O^+ ions from the cusp/cleft ionosphere: A possible substorm trigger, *J. Geophys. Res.*, *97*, 123-130, 1992.

Cowley, S.W.H., Plasma populations in a simple open model magnetosphere, *Space Sci. Rev.*, *26*, 217, 1980.

Curran, D.B., C.K. Goertz, and T.A. Whelan, Ion distributions in a two-dimensional reconnection field geometry, *Geophys. Res. Lett.*, *14*, 99, 1987.

Daglis I.A., S. Livi, E.T. Sarris, and B. Wilken, Energy density of ionospheric and solar wind origin ions in the near-Earth magnetotail during substorms, *J. Geophys. Res.*, *99*, 5691-5703,1994.

De Hoffman, F., and E. Teller, Magnetohydrodynamic Shock, *Phys. Rev.*, *80*, 692, 1950.

Deeg, H.J., J.E. Borovsky, and N. Duric, Particle acceleration near X-type magnetic neutral lines, *Phys. Fluids*, *3*, 2660, 1991.

Dessler, A.J., and W.B. Hanson, Possible energy source for the aurora, *Astroph. J.*, *134*, 1024-1025, 1961.

Fujimoto, M. and M. Nakamura, Acceleration of heavy ions in the magnetotail reconnection layer, *Geophys. Res. Lett.*, *21*, 2955-2958, 1994.

Geiss, J., H. Balsinger, P. Eberhardt, H.P. Walker, L. Weber, D.T. Young, and H. Rosenbauer, Dynamics of magnetosphericion composition as observed by the GEOS mass spectrometer, *Space Sci. Rev.*, *22*, 537-566, 1987.

Hirahara, M., T. Mukai, T. Terasawa, S. Machida, Y. Saito, T. Yamamoto, and S. Kokobun, Cold dense ion flows with multiple components observed in the distant tail lobe by GEOTAIL, *J. Geophys. Res.*, *101*, 4, 7769, 1996.

Ipavich F.M., A.B. Galvin, M. Scholer, G. Gloeckler, D. Hovestadt, and B. Klecker, Energetic (¿ 100 keV) O^+ ions in the plasma sheet, *Geophys. Res. Lett.*, *11*, 504-507, 1984.

Ipavich F.M., A.B. Galvin, M. Scholer, G. Gloeckler, D. Hovestadt, and B. Klecker, Suprathermal O^+ and H^+ behavior during the March 22, 1979 (CDAW 6) substorms, *J. Geophys. Res.*, *90*, 171,105-17,119, 1985.

Jacquey, C., D.J. Williams, R.W. McEntire, A.T.Y. Lui, V. Angelopoulus, S.P. Christon, S. Kokubun, T. Yamamoto, G.D. Reeves, and R.D. Belian, Tailward energetic ion streams observed at about 100 R_E by GEOTAIL-EPIC associated with geomagnetic activity intensifications, *Geophys. Res. Lett.*, *21*, 3015-3018, 1994.

Joyce G., J. Chen, S. Slinker, and D.L. Holland, Particle energization near an X-line in the magnetotail based on global MHD fields, *J. Geophys. Res.*, *100*, 1995.

Lennartsson W., A scenario for solar wind penetration of Earth's magnetic tail based on ion composition data from the ISEE 1 spacecraft, *J. Geophys. Res.*, *97*, 19,221, 1992.

Lyons, L. and T.W. Speiser, Evidence for current sheet acceleration in the geomagnetic tail, *J. Geophys. Res.*, *87*, 2276, 1982.

Lundin, R., and L. Eliasson, Auroral energization process, *Ann. Geophys.*, *9*, 202-223, 1991.

Möbius, E., M. Scholer, B. Klecker, D. Hovestadt, G. Gloeckler, and F.M. Ipavich, Acceleration of ions of ionospheric origin in the plasma sheet during substorm activity, in *Magnetotail Physics*, ed. by A.T.Y. Lui, pp 231-234, Johns Hopkins Univ. Press, Baltimore, MD, 1987.

Moses, R.W., J.M. Finn, and K.M. Ling, Plasma heating by collisionless magnetic reconnection: Analysis and computation, *J. Geophys. Res.*, *98*, 4013, 1993.

Mukai, T., M. Hirahara, S. Machida, Y. Saito, T. Terasawa, and A. Nishida, Geotail observation of cold ion streams in the medium distance magnetotail lobe in the course of substorms, *Geophys. Res. Lett.*, *21*, 1023, 1994.

Sachsenweger D., M. Scholer, E. Möbius, Test particle acceleration in a magnetotail reconnection configuration, *Geophys. Res. Lett.*, *16*, 1027, 1989.

Sato, T., H. Matsumoto, and K. Nagai, Particle acceleration in time-developing magnetic reconnection process, *J. Geophys. Res.*, *87*, 6089, 1982.

Scholer, M., and F. Jamitzky, Particle orbits during the development of plasmoids, *J. Geophys. Res.*, *92*, 12.181, 1987.

Shelley, E.G., R.G. Johnson, and R.D. Sharp, Satellite observations of energetic heavy ions during a geomagnetic storm, *J. Geophys. Res.*, *77*, 6104, 1972.

Shelley E.G., R.D. Sharp, and R.G. Johnson, Satellite observations of an ionospheric acceleration mechanism, *Geophys. Res. Lett.*, *3*, 654-656, 1976.

Sonnerup, B.U.Ö., Adiabatic particle orbits in a magnetic null sheet., *J. Geophys. Res.*, *76*, 8211, 1971.

Speiser, T., Particle trajectories in model current sheets, 1. *J. Geophys. Res.*, *70*, 4219, 1965.

Vasyliunas, V.M., Theoretical models of magnetic field line merging, 1, *Rev. Geophys. Space Phys.*, *13*, 303, 1975.

Williams, D.J., A.T.Y. Lui, R.W. McEntire, V. Angelopoulus, C. Jacquey, S.P. Christon, L.A. Frank, K.L. Ackerson, W.R. Paterson, S. Kokubun, T. Yamamoto, D.H. Fairfield, Magnetopause encounters in the magnetotail at distances of about 80 R_E, *Geophys. Res. Lett.*, *21*, 3007-3010, 1994.

Wilken, B., Q.-G. Zong, I.A. Daglis, T. Doke, S. Livi, K. Maezawa, Z.Y. Pu, S. Ullaland, and T. Yamamoto, Tailward flowing energetic oxygen ion bursts associated with multiple flux ropes in the distant magnetotail: geotail observations, *Geophys. Res. Lett.*, *22*, 3267, 1995.

Yau, A.W., E.G. Shelley, W.K. Peterson, and L. Jenchyshyn, Energetic auroral and polar ion outflow at DE-1 altitudes, *J. Geophys. Res. Lett.*, *90*, 8417-8432, 1985.

Yau, A.W., T. Abe, and B.A. Whalen, Cold plasma source of upflowing ionospheric origin ions in the nightside auroral ionosphere, *J. Geomag. Geoelectr.*, *48*, 947-957, 1996.

Zelenyi, L.M., J.G. Lominadze, and A.L. Taktakishvili, Generation of the energetic proton and electron bursts in the planetary magnetotail, *J. Geophys. Res.*, *95*, 3883, 1990.

Zong, Q.-G., B. Wilken, G.D. Reeves, I.A. Daglis, T. Doke, T. Iyemori, S. Livi, K. Maezawa, T. Mukai, S. Kokubun, Z.-Y. Pu, S. Ullaland, J. Woch, R. Lepping and T. Yamamoto, Geotail observations of energetic ion species and magnetic field in plasmoid-like structures in the course of an isolated substorm event, *J. Geophys. Res.*, *102*, 11,409-11,428, 1997.

Zong, Q.-G., B. Wilken, T. Mukai, T. Doke, G.D. Reeves, S. Livi, K. Maezawa, D.J. Williams, Q. S. Kokubun, S. Ullaland, and T. Yamamoto, Energetic oxygen ion bursts in the distant magnetotail as a product of intense substorms: Three case studies, *J. Geophys. Res.*, 103, 1998, in press.

J. Büchner, J.-P. Kuska, B. Wilken, Q.-G. Zong, Max-Planck-Institut für Aeronomie, Max-Planck-Str. 2, D-37191 Katlenburg-Lindau, Germany. (e-mail: buechner@linmpi.mpg.de)

Particle Dynamics in the Near-Earth Magnetotail and Macroscopic Consequences

D. C. Delcourt

Centre d'étude des Environnements Terrestre et Planétaires, Saint-Maur des Fossés, France

G. Belmont

Centre d'étude des Environnements Terrestre et Planétaires, Vélizy, France

We examine the transport of plasma sheet ions in the near-Earth (~10-15 R_E) magnetotail using single-particle trajectory simulations. From a dynamical viewpoint, this region of space corresponds to a transition between the distant tail where particles meander about the midplane and the dipolar region at low L- shells where the particle motion is adiabatic (magnetic moment conserving). In the near-Earth tail, because of Larmor radii comparable to the field line curvature radius, plasma sheet ions experience nonadiabatic motion in a manner which differs from that in the distant tail. This nonadiabatic motion can be viewed as the result of an impulsive centrifugal force perturbing the gyromotion near the field minimum. Three regimes of magnetic moment variations can be distinguished: (i) at large pitch angles, particles exhibit negligible magnetic moment change. (ii) At intermediate pitch angles, particles experience magnetic moment damping at specific phase angles; hence, filling of the atmospheric loss cone and subsequent precipitation. (iii) At relatively small pitch angles (up to ~30°), particles are subjected to systematic magnetic moment enhancement and prominent bunching in gyration phase. Within the limits of the calculations which do not treat the particle dynamics in a self-consistent manner, the phase bunching effect is found to lead to a distinctive current distribution at the earthward termination of the magnetotail current sheet, namely: a striated pattern of currents both in the dawn-dusk and in the Earth tail directions, the latter current leading to field line inclination in the dawn-dusk direction at low latitudes. The non-gyrotropic distributions resulting from this bunching in gyration phase yield off-diagonal terms in the pressure tensor, the Z-gradient of which appears essential to achieve stress balance.

1. INTRODUCTION

In the Earth's magnetotail, particles may not conserve their magnetic moment (first adiabatic invariant) due to significant field variations on the length scale of their cyclotron turn (see, e.g., *Chen* [1992] for a review). This nonadiabatic particle behavior can be characterized by the parameter κ defined as the square root of the minimum curvature radius- to- maximum Larmor radius ratio [e.g., *Büchner and Zelenyi*, 1989]. In the distant tail, the Larmor radius of plasma sheet ions is significantly larger than the field line curvature radius and the κ parameter is much smaller than unity. In this regime, particles may experience meandering motion about the field minimum as initially

New Perspectives on the Earth's Magnetotail
Geophysical Monograph 105

shown by *Speiser* [1965]. Even though these meandering trajectory sequences are not adiabatic stricto senso, they do have some regularity with the action integral $I_Z = \int \dot{z}\, dz$ as an approximate invariant [e.g., *Speiser*, 1968; *Büchner and Zelenyi*, 1989]; hence their denomination as quasi-adiabatic. In the $\kappa << 1$ dynamical regime, *Chen and Palmadesso* [1986] demonstrated that the phase space is partitioned into distinct regions (viz., transient or Speiser-type orbits, stochastic or quasi-trapped orbits, and trapped orbits), each of them being characterized by different time scales. *Burkhart and Chen* [1991] showed that the phase space partitioning considerably favors Speiser-type orbits at specific values of the κ parameter or, equivalently, for a series of resonant energies. Also, *Ashour-Abdalla et al.* [1994] put forward that meandering motion at $\kappa << 1$ leads to the formation of a thin current sheet throughout most of the magnetotail, with thickness comparable to the Z-oscillation length scale. On the other hand, in the innermost dipolelike magnetosphere where the parameter κ is much larger than unity, the particle motion is adiabatic (magnetic moment conserving) and can be described with the help of the guiding center approximation.

Between these current sheet ($\kappa << 1$) and adiabatic ($\kappa >> 1$) limits, plasma sheet ions exhibit κ of the order of unity in a narrow (a few R_E) region of space at the transition between taillike and dipolelike field configurations. In this $\kappa \sim 1$ regime, the particle motion has been shown to be chaotic [e.g., *Büchner and Zelenyi*, 1986] and cannot be characterized by any invariant. This regime which is thus distinct from the adiabatic one at low L- shells and the quasi-adiabatic one in the distant tail, has important dynamical implications. For instance, it has been suggested that the magnetic moment scattering occuring in this regime and the consequent filling of the loss cone may be responsible for intense precipitation over the auroral zone [e.g., *Sergeev et al.*, 1983; *Zelenyi et al.*, 1990a]. This conversely allows remote sensing of the magnetotail magnetic field [e.g., *Sergeev and Gvozdevsky*, 1995]. In a recent study [*Delcourt et al.*, 1994], we demonstrated that the nonadiabatic particle behavior in the $\kappa \sim 1$ regime can be viewed as the result of an impulsive centrifugal force perturbing the gyromotion near the field minimum, an interpretation framework referred to as the centrifugal impulse model. The analytical description of magnetic moment change obtained within this model reproduces well the results of numerical trajectory calculations, and *Delcourt and Martin* [1997] ("Extension of the centrifugal impulse model to particle dynamics in sharp field reversal", manuscript submitted for publication in Geophysical Research Letters; hereinafter referred to as DM-97) showed that, to some extent, this centrifugal impulse model can be applied to $\kappa < 1$ regimes. In subsequent studies, we also demonstrated that de-trapping of plasma sheet ions at $\kappa \sim 1$ can be easily understood within the centrifugal impulse model [*Delcourt et al.*, 1996a] and that particles which are scattered toward large pitch angles are subjected to

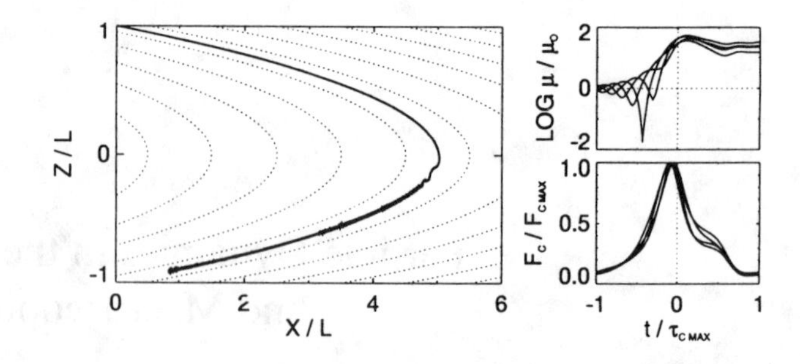

Figure 1. Model proton trajectories at $\kappa = 1.3$ in the parabolic field (1): (left) trajectory projection in the X-Z plane, (top right) magnetic moment (normalized to the initial value) versus time, (bottom right) centrifugal force (normalized to the maximum value) versus time. The time is normalized to the cyclotron period at the field minimum and measured from the time of the first $Z = 0$ crossing. Dotted lines in the left panel depict magnetic field lines.

prominent bunching in gyrophase; hence, the formation of a thin current sheet made of phase-bunched current carriers [*Delcourt et al.*, 1996b; *Delcourt and Belmont*, 1998]. The present paper provides a summary of these various studies.

2. PARTICLE DYNAMICS IN THE $\kappa \sim 1$ REGIME

2.1. The Centrifugal Impulse Model

To illustrate nonadiabatic behaviors at $\kappa \sim 1$, Figure 1 shows an example of particle orbits in the parabolic field defined as

$$\mathbf{B}(X,Z) = B_o \frac{Z}{L}\mathbf{x} + B_n \mathbf{z} \qquad (1)$$

Here L is a reference scale length representing the half-thickness of the current sheet and $\mathbf{x}$ and $\mathbf{z}$ are unit vectors in the X and Z directions, respectively. The magnetic field thus defined is invariant by translation in X and Y, and the minimum curvature radius at $Z = 0$ is: $R_C = LB_n/B_o$. In Figure 1, we adopt the following parameter values: $B_o = 10$ nT, $B_n = 1$ nT, and $L = 3\ R_E$. In this figure, test protons were launched from the edge of the current sheet ($Z = L$) with distinct gyration phases (from 0° to 360° by steps of 90°) and an energy E such that: $\kappa = [R_C/\rho_{Lmax}]^{1/2} = [qLB_n^2/B_o]^{1/2}[2mE]^{-1/4} = 1.3$ (ρ_{Lmax} being the maximum Larmor radius, q the particle charge, and m its mass). The pitch angle at the field minimum in the adiabatic limit (i.e., as derived from magnetic moment conservation) is set to 5°. These test protons were traced during one interaction with the current sheet, that is, we consider trajectory sequences where the particles approach the field reversal in an adiabatic (magnetic moment conserving) manner, experience nonadiabatic motion near $Z = 0$, and then

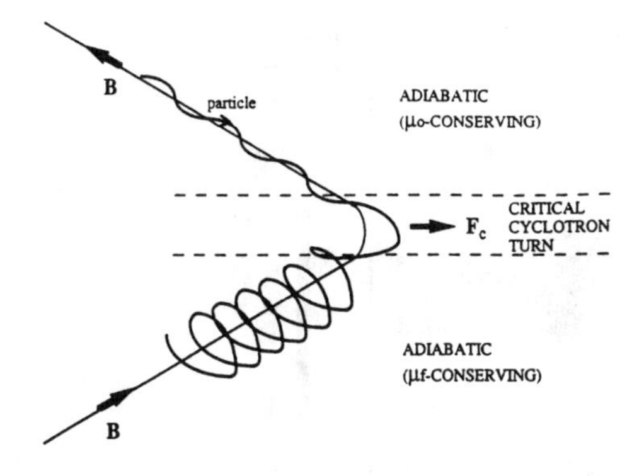

Figure 2. Schematic representation of the centrifugal impulse model. **B** denotes the magnetic field, $\mathbf{F}_C$ the impulsive centrifugal force, and μ_o and μ_f the magnetic moments prior to and after crossing of the field reversal, respectively.

recover an adiabatic behavior until they reach their mirror point (i.e., **V.B** = 0).

The left panel of Figure 1 presents the particle orbits in the *X-Z* plane, whereas the right panels show the instantaneous magnetic moment (normalized to its initial value) and the centrifugal force acting upon the particles (normalized to its maximum value) as a function of time (as measured from the time of the first $Z = 0$ crossing and normalized to the maximum gyroperiod at $Z = 0$). It can be seen in Figure 1 that, regardless of their gyration phase, the particles intercept the $Z = 0$ plane once and experience significant magnetic moment enhancement upon crossing of the current sheet (top right panel). This enhancement occurs as the particle is subjected to a large centrifugal impulse near the field minimum (bottom right panel), and it leads to mirroring at low Z-height at exit (left panel). In particular, it is of interest to note in the bottom right panel that the timescale of the centrifugal impulse is of the order of the cyclotron period at the field minimum.

In view of these results, *Delcourt et al.* [1994] (hereinafter referred to as paper 1) made the following assumption: the motion is adiabatic prior to and after crossing of the neutral sheet, and these two adiabatic sequences are separated by a "critical" cyclotron turn near the field minimum where an impulsive centrifugal force (due to the locally enhanced field line curvature) perturbs the gyromotion. The centrifugal force is thus viewed as being responsible for the scattering of the particle magnetic moment, as is schematically illustrated in Figure 2. In paper 1, a simple functional form was chosen for the impulsive centrifugal force, namely: $F_C(t) = (mV_\parallel^2/2R_C)\,[1 - \cos(2\pi t/\tau)]$ where $V_\parallel$ is the particle parallel speed, and τ, the impulse time scale. After solving the equation of motion in the gyration plane (assuming the magnetic field **B** to be approximately constant), and after deriving the various coefficients from boundary conditions at $t = 0$, a first order description of the two gyrovelocity components during the impulse was obtained (a brief account of this calculation is given in Appendix). If we define two orthogonal axis *T* and *Y* in the gyration plane and consider that the centrifugal force $\mathbf{F}_C$ (perpendicular to **B**) acts along axis *T* which coincides with the *X* axis of Figure 1 at $Z = 0$, the two gyrovelocity components V_T and V_Y at a given time *t* of the impulse are (equations (A8) and (A9) of the Appendix):

$$V_T(t) = V_\perp \sin\left(\frac{2\pi t}{\tau_C} + \psi_o\right) - \frac{mV_{//}^2}{2qBR_C}\frac{1}{\chi^2 - 1}\left[\sin\left(\frac{2\pi t}{\tau_C}\right) - \chi \sin\left(\frac{2\pi t}{\tau}\right)\right] \quad (2)$$

$$V_Y(t) = V_\perp \cos\left(\frac{2\pi t}{\tau_C} + \psi_o\right) + \frac{mV_{//}^2}{2qBR_C}\frac{1}{\chi^2 - 1}\left[1 - \cos\left(\frac{2\pi t}{\tau_C}\right) - \chi^2 + \chi^2 \cos\left(\frac{2\pi t}{\tau}\right)\right] \quad (3)$$

where τ_C is the particle gyroperiod, χ, the τ/τ_C ratio, and where $V_\perp$ and ψ_o are the initial gyration velocity and phase, respectively. Note that the origin of phase adopted lies along the *T* axis in the direction opposite that of the centrifugal impulse as illustrated in Figure A1 of the Appendix. From equations (2) and (3), one may derive the instantaneous magnetic moment (normalized to its initial value) according to

$$\mu(t) = \frac{V_T^2(t) + V_Y^2(t)}{V_\perp^2} \quad (4)$$

and the instantaneous phase of gyration according to:

$$\psi(t) = \mathrm{Arctan}[V_T(t) / V_Y(t)] \quad (5)$$

As will be discussed further hereinafter, it should be noted that the χ ratio introduced in (2) and (3) is related to the κ parameter since κ is also the ratio of the minimum Larmor frequency- to- maximum bounce frequency ($\kappa^2 = \omega_C/\omega_B$) [*Büchner and Zelenyi*, 1989].

2.2 Magnetic Moment Variations

If $\tau \approx \tau_C$ (i.e., $\chi \approx 1$) as displayed in Figure 1, equation (4) yields the following magnetic moment after the "critical" cyclotron turn:

$$\mu(\tau) = 1 + \frac{1}{\kappa_a^4} - \frac{2}{\kappa_a^2}\sin\psi_o \quad (6)$$

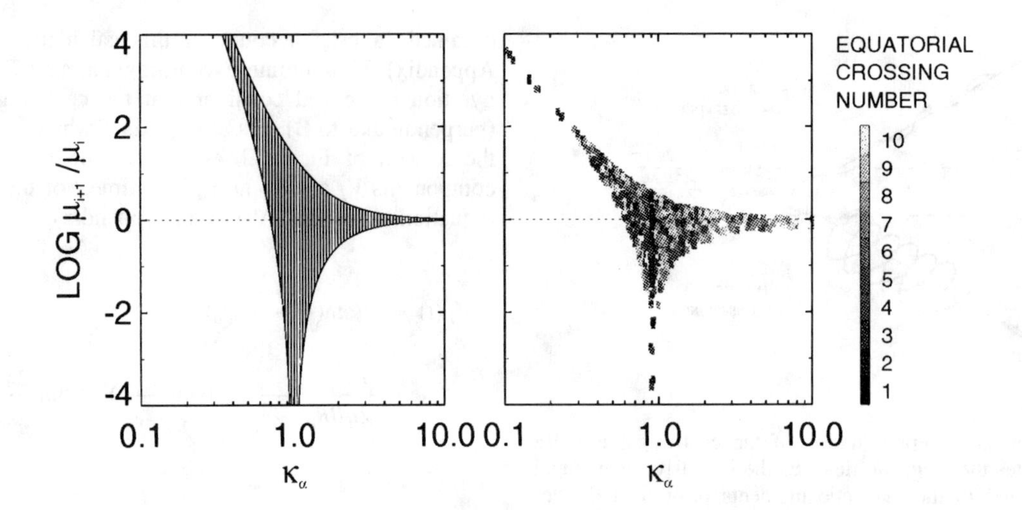

Figure 3. (left) Analytical magnetic moment variations (equation (7)) as a function of κ_α. (right) Computed magnetic moment variations in the parabolic field (1) as a function of κ_α. In these computations, $\kappa = 1.3$ particles are launched from the edge of the current sheet with 5° pitch angle and distinct gyrophases (from 0° to 360° by steps of 2°). They are traced during 10 successive interactions (coded according to the gray scale at right) with the current sheet.

Equivalently, one has for the envelope of magnetic moment:

$$\mu_\pm(\tau) = [1 \pm \frac{1}{\kappa_\alpha^{\ 2}}]^2 \qquad (7)$$

In these equations, the quantity κ_α is related to the κ parameter according to $\kappa_\alpha = \kappa \sin^{1/2}\alpha / \cos\alpha$. On a given magnetic field line, one may thus derive from magnetic moment conservation the pitch angle α at the field minimum, calculate the corresponding value of κ_α, and then estimate the amount of μ scattering using (7).

The left panel of Figure 3 shows the analytical variations of $\mu_\pm(\tau)$ as obtained from (7) as a function of κ_α. This panel puts forward a characteristic three-branch pattern of magnetic moment variations, namely: (i) at small pitch angles, systematic μ enhancements regardless of gyration phase, (ii) at large pitch angles, negligible μ change, and (iii) in between, either magnetic moment enhancement or damping depending upon gyration phase. These variations are in good agreement with those obtained numerically, as can be seen by comparison with the right panel of Figure 3 which shows the results of trajectory calculations. In this latter panel, the test protons considered in Figure 1 were followed during 10 successive interactions with the neutral sheet and their magnetic moment after each crossing is presented as a function of κ_α (note that the magnetic moment is gray-coded according to the crossing number and normalized to the previous value, i.e., μ_{i+1}/μ_i denoting by i the crossing number). It can be seen in the right panel of Figure 3 that the computed μ variations exhibit a three-branch pattern similar to the analytical one (left panel). In the long term, it is also apparent that the vertical branch near $\kappa_\alpha = 1$ is responsible for chaotization of the particle motion [e.g., *Büchner and Zelenyi*, 1986]. Indeed, since the μ change in this latter branch directly depends upon gyration phase, particles which experience μ damping subsequently find themselves in the oblique branch at small pitch angles, whereas those experiencing μ enhancement are injected toward the horizontal branch. In the long term, these particles initially sampled about a single guiding center cover the full area of the three-branch pattern. Also, it should be pointed out that the above κ_α parameter bears a quite specific meaning since one has $\kappa_\alpha^{\ 2} = qV_\perp B/(mV_\parallel^{\ 2}/R_C)$. In other words, κ_α simply is the ratio of the Lorentz force to centrifugal force near the field minimum. The large μ enhancements obtained at small pitch angles are due to predominant centrifugal effects, whereas the μ damping obtained near $\kappa_\alpha = 1$ corresponds to a situation where one force balances the other.

Equation (7) which is at the root of the three-branch μ variation pattern in Figure 3 was obtained in the limit $\chi \approx 1$. In a manner similar to *Büchner and Zelenyi* [1989], we may approximate the length scale of the field reversal by a half-circle of radius R_C. The χ ratio then becomes:

$$\chi = \frac{\pi R_c}{V}\frac{qB_n}{2\pi m} = \frac{\kappa^2}{2} \qquad (8)$$

Hence, taking $\chi \approx 1$ amounts stricto senso to considering $\kappa \approx 2^{1/2}$. Still, *Delcourt et al.* [1996a] demonstrated that the three-branch dynamics portrayed in Figure 3 is a quite general feature of nonadiabatic behavior between $\kappa \sim 1$ and the adiabatic limit at $\kappa \sim 3$. In fact, the functional form (7) can be used to globally characterize the μ changes by modifying somewhat the definition of κ_α. That is, we rather consider a nonlinear dependence of κ_α upon κ and write $\kappa_\alpha = f(\kappa)\sin^{1/2}\alpha/\cos\alpha$. Equation (7) becomes:

$$\mu_\pm(\tau) = [1 \pm \frac{\cos^2\alpha}{f(\kappa)^2\sin\alpha}]^2 \quad (9)$$

and a fit to the numerical results gives:

$$f(\kappa) = [0.65 + \exp(\kappa^{1.7}-1)]^{1/2} \quad (10)$$

In a like manner to (6), one has for the phase-dependent form of (9):

$$\mu(\tau) = 1 + [\frac{\cos^2\alpha}{f(\kappa)^2\sin\alpha}]^2 - 2\frac{\cos^2\alpha}{f(\kappa)^2\sin\alpha}\sin\psi_o \quad (11)$$

Figure 4 presents the results of systematic trajectory calculations at different κ values within the 1-3 interval. Here, in a like manner to Figure 1, test protons were launched within the parabolic field (1) with distinct pitch and phase angles and they were traced during one interaction with the neutral sheet. In Figure 4, their final magnetic moment (normalized to the initial value) is shown as a function of equatorial pitch angle. As mentioned above, Figure 4 displays that, at any value of κ between 1 and 3, the μ variations are organized according to a three-branch pattern. Also it is apparent that the analytical envelope given by (9) (solid line) is in good agreement with the computations. As κ becomes of the order of 3, the three-branch pattern vanishes and one has $\mu_\pm(\tau) \approx 1$ over most of the pitch angle range, as expected from transition toward adiabatic behavior. On the other hand, it can be seen in Figure 4 that as κ decreases toward unity, the three-branch pattern spreads over a wider pitch angle interval. This is consistent with the above analysis based upon the κ_α parameter since, given the particle speed, decreasing κ amounts to decreasing R_C and thus to increasing the centrifugal force acting upon the particle. Hence, as $\kappa \to 1^+$, the situation where Lorentz force and centrifugal force are comparable (i.e., $\kappa_\alpha \sim 1$) occurs at larger and larger pitch angles.

Finally, it should be pointed out that equations (6) and (11) bear some resemblance to the calculation of *Birmingham* [1984] (see equation (23) of that study). As examined by *Martin and Delcourt* [1996], both the Birmingham's result and the present model exhibit a sinusoidal dependence upon gyration phase. However, Birmingham's estimate does not feature a phase-independent term as in (6) and (11) which is of importance at small pitch angles and yields systematic magnetic moment enhancements. For this reason, Birmingham's result does not give the above three-branch pattern correctly when κ approaches unity (see, e.g., Figure 5 of *Martin and Delcourt* [1996]).

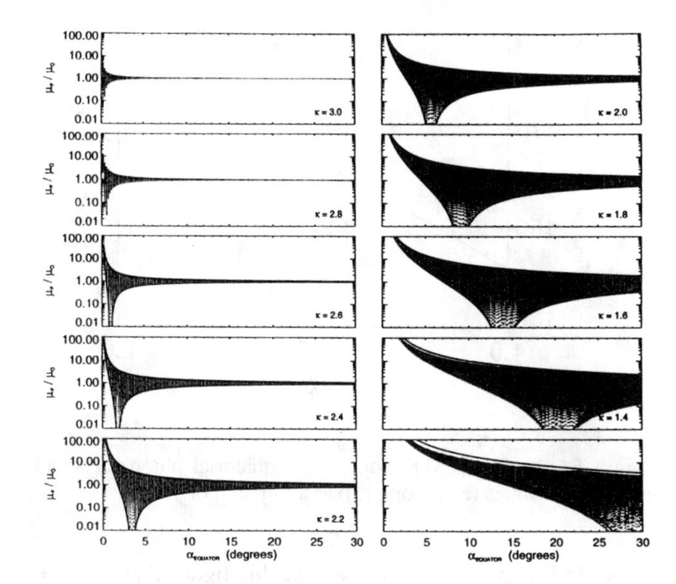

Figure 4. Computed μ variations in the parabolic field (1) as a function of initial pitch angle (at equator and in the adiabatic limit) for distinct values of the κ parameter. Magnetic moments are normalized to the initial value. The solid lines in each panel show the analytical μ variations obtained from (9). From *Delcourt et al.* [1996a].

2.3. Phase Variations

Figure 1 shows that the impulsive centrifugal force is maximum near $Z = 0$, as expected from the locally enhanced field line curvature. In our analytical description (equation (A5) of the Appendix), this maximum centrifugal impulse occurs at $t = \tau/2$. In other words, we may estimate the gyration phase at $Z = 0$ by taking $t = \tau/2$ in equation (5). This yields the following expression for the equatorial gyration phase:

$$\psi(\tau/2) = \mathrm{Arctan}\{ [\sin(\pi\chi+\psi_o) - \frac{1}{2\kappa_\alpha^2(\chi^2-1)}\sin\pi\chi] / [\cos(\pi\chi+\psi_o) + \frac{1}{2\kappa_\alpha^2(\chi^2-1)}(1-\cos\pi\chi-2\chi^2)] \} \quad (12)$$

For $\kappa_\alpha << 1$ (i.e., at small pitch angles), this equation simply amounts to:

$$\tan\psi(\tau/2) \approx -\frac{\sin\pi\chi}{1-\cos\pi\chi-2\chi^2} \quad (13)$$

In other words, regardless of their initial gyrophase, particles with small pitch angles have a nearly identical phase at equator, which is solely controlled by the χ ratio. The variation of this equatorial phase as a function of χ is

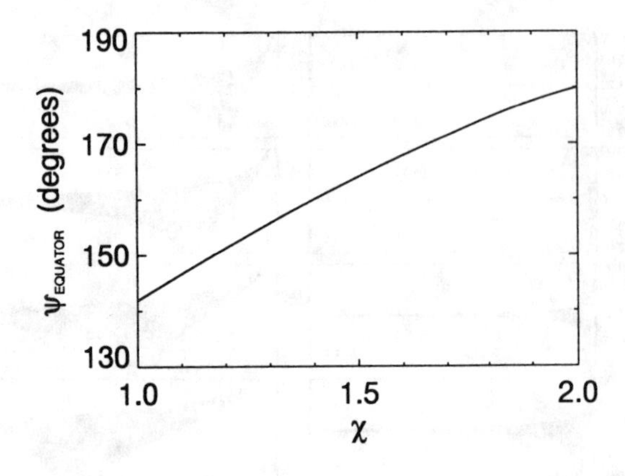

Figure 5. Analytical variation of the equatorial phase for small initial pitch angles (equation (13)) as a function of χ.

shown in Figure 5. It can be seen in this figure that, for χ = 1, the equatorial phase is of the order of 140°, and that it increases at larger χ values. This χ-dependent (or equivalently κ-dependent; equation (8)) phase bunching effect is schematically illustrated in Figure 6. This figure shows the orientation of the velocity vector in the equatorial plane for two distinct values of the κ parameter. Keeping in mind that ψ represents the angle between the particle velocity and the Y axis, it can be seen in Figure 6 that, at equator, particles with larger energies (smaller κ values) have a more X-oriented velocity, consistently with enhanced centrifugal effects.

The variability of this phase bunching effect with pitch angle can be appreciated in Figure 7. The left panel of this figure presents the analytical variation of $\psi(\tau/2)$ derived from (12) (with χ = 1) as a function of κ_α. As displayed in Figure 5, it can be seen in this panel that particles with small pitch angles systematically exhibit an equatorial phase of ~140°, regardless of their initial phase. As κ_α becomes larger, the phase spread gradually increases and ultimately covers an angular sector of 2π for pitch angles such that $\kappa_\alpha \geq 1$. In other words, whereas the phase rate of change $\dot{\psi}$ differs from the Larmor frequency at small pitch angles, these frequencies become comparable at large pitch angles as described by equation (A10) of the Appendix. Returning to the three-branch pattern of Figure 3, it is apparent that the prominent bunching in gyration phase goes together with the systematic μ enhancements at small pitch angles, whereas the negligible μ changes at large pitch angles are associated with negligible phase change. As for the vertical branch with possible μ damping, it lies at the transition between these distinct $\dot{\psi}$ regimes.

The right panel of Figure 7 shows the equatorial phases obtained numerically using a format similar to that of Figure 3. Test protons are traced during 10 successive interactions with the neutral sheet, and their phase (gray-coded according to the crossing number) at $Z = 0$ is presented as a function of κ_α. Keeping in mind that the particles are initially sampled about a single guiding center and launched with a small pitch angle, it can be seen that they have a nearly identical phase at the first Z = 0 crossing. Subsequently, as these particles cover the full area of the three-branch pattern (Figure 3), they exhibit phase variations consistent with the analytical ones in the right panel. In particular, it is clearly apparent that $\kappa_\alpha \sim 1$ sharply delineates the region where prominent phase bunching occurs ($\kappa_\alpha << 1$) and that where phases are distributed over 2π ($\kappa_\alpha >> 1$). The quantity κ_α^2 being the ratio of the Lorentz force-to-centrifugal force, this result simply demonstrates that phase bunching occurs when centrifugal effects prevail ($\kappa_\alpha << 1$) and that it becomes negligible if the Lorentz force dominates ($\kappa_\alpha >> 1$).

Finally, it should be pointed out that, in the limit $\chi \approx 1$, equation (5) yields the following phase at the end of the "critical" cyclotron turn:

$$\psi(\tau) = \text{Arctan}\left[\frac{\sin\psi_0 - \dfrac{1}{\kappa_\alpha^2}}{\cos\psi_0}\right] \tag{14}$$

For small pitch angles ($\kappa_\alpha << 1$), one has $\tan\psi(\tau) \rightarrow -\infty$ or, equivalently, $\psi(\tau)$ of the order of $3\pi/2$ regardless of initial gyrophase. At the end of the centrifugal impulse, the particle velocity thus systematically points earthward, that is, in the direction opposite that of the centrifugal impulse.

2.4. Extension of the Centrifugal Impulse Model to $\kappa < 1$ Regimes

In the analytical model described above, the nonadiabatic behavior of the particles is viewed as the result of an

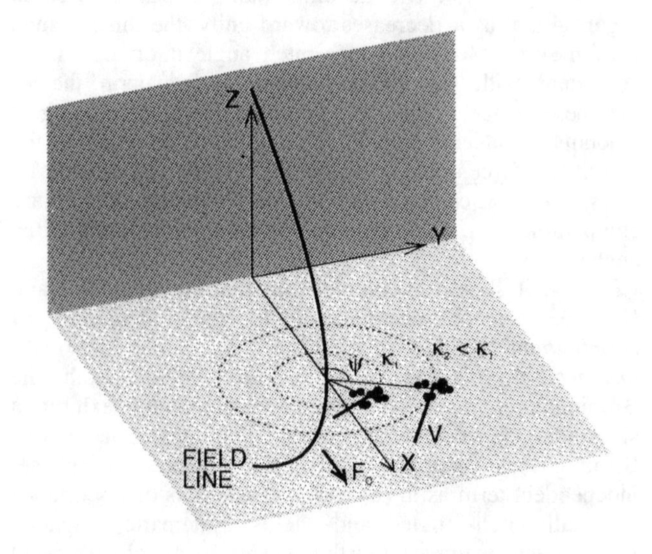

Figure 6. Schematic representation of the κ-dependent phase bunching effect near the $Z = 0$ plane.

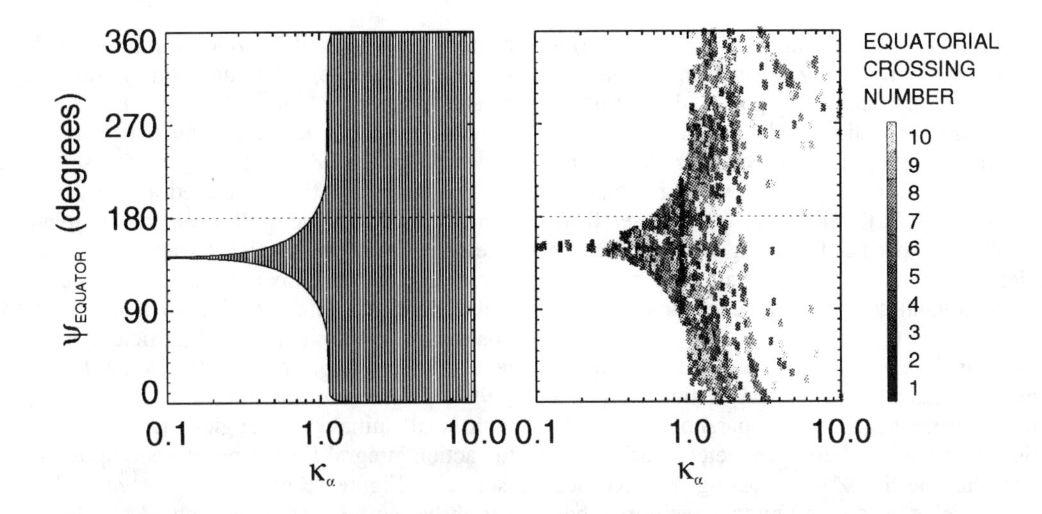

Figure 7. (left) Analytical equatorial phases (equation (12)) as a function of κ_α. (right) Computed equatorial phases as a function of κ_α. The trajectory calculations are identical to those described in Figure 3. Equatorial phases are coded according to the gray scale at right.

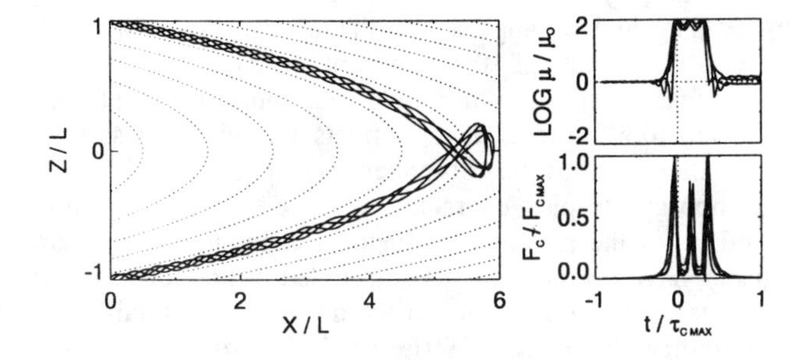

Figure 8. Identical to Figure 1 but for $\kappa = 0.315$.

enhanced centrifugal force perturbing the gyromotion near the field minimum. As such, we expect this interpretation to apply only to particles which execute full cyclotron turns near the field minimum and we exclude particles which experience fast oscillations about the $Z = 0$ plane like those examined by *Speiser* [1965]. Equivalently, for this interpretation to be valid, the maximum Larmor radius should be comparable to or smaller than the minimum field line curvature radius so that particles do not meander about the midplane, i.e., $\kappa \geq 1$. It appears however that the centrifugal impulse model reproduces some essential features of the particle dynamics even when κ is smaller than unity.

To provide insights into this result, Figure 8 presents model particle orbits in the parabolic field (1) with a κ parameter set to 0.315. In this figure, we again consider one interaction with the neutral sheet, that is, trajectory sequences where the particles approach the field reversal in an adiabatic manner, experience nonadiabatic motion near $Z = 0$, and then recover an adiabatic behavior until they reach their mirror point (**V.B** = 0). In Figure 8, it can be seen that the particles cross the $Z = 0$ plane three times before escaping with a magnetic moment nearly identical to the initial one (top right panel). This behavior is characteristic of Speiser-type orbits, the κ value considered here corresponding to the second energy resonance envisionned by *Chen and Palmadesso* [1986]. It is also apparent from the bottom right panel of Figure 8 that the particles are subjected to repeated centrifugal impulses. The time scales of these impulses are of the same order of magnitude and much smaller than the cyclotron period. The comparison of the top and bottom panels at right also shows that the first impulse gives rise to a large μ enhancement (as in the oblique branch of the three-branch pattern) and the last one, to μ damping back to the initial value (as in the vertical branch of the three-branch pattern), whereas the impulse in between yields negligible μ change.

On this basis, DM-97 extended the centrifugal impulse description to small κ regimes by noting that, during one "critical" cyclotron turn, the particles may cross the $Z = 0$ plane many times (i.e., $\tau << \tau_C$) and that the successive phases at $Z = 0$ can be readily derived from (5). These phases approximately occur at $t = (n - 1/2)\,\tau$ where n is an integer comprised between 1 and some upper limit n_{max} such that $(n_{max} - 1/2)\,\tau \leq \tau_C$. Equivalently, one has

$$n_{max} = \left\lfloor \frac{1}{\chi} + \frac{1}{2} \right\rfloor \tag{15}$$

For incoming particles with small pitch angles, the first term on the right handside of (2) and (3) can be neglected, and one has from (5):

$$\psi[(n - 1/2)\tau] = \mathrm{Arctan}\left[\frac{-\sin 2\pi\chi(n-1/2)}{1-2\chi^2 - \cos 2\pi\chi(n-1/2)}\right] \tag{16}$$

Thus, for a given value of χ, one may derive from (16) a series of equatorial phases by letting n vary from 1 to n_{max} as given by (15). In a similar manner to Figure 5, the left panel of Figure 9 shows these ψ values (labeled according to n number) as a function of χ. Let us consider decreasing values of χ in this panel. First, a single ψ value is obtained for χ between 1 and 2/3. In this interval, ψ decreases from 140° toward 90° as expected from Figure 6. When χ becomes smaller than 2/3, two ψ values are obtained and, more generally, additional ψ is obtained each time one has $\chi = 2/(2n_{max} -1)$ (equation (15)).

The right panel of Figure 9 shows the equatorial phases obtained from systematic trajectory calculations. In this panel, test protons were traced in the parabolic field (1) considering distinct values of the κ parameter. During one cyclotron period after the first $Z = 0$ crossing, we recorded the phase of the particles (as defined by (5)) each time they intercept the $Z = 0$ plane. In the right panel of Figure 9, these computed phase series are presented as a function of κ. The comparison of the two panels in this figure displays a remarkable similarity between analytical and numerical results. As mentioned above, this suggests a close relationship between the χ ratio and κ. In particular, the vertical dotted lines in the right panel of Figure 9 show the location of energy resonances as derived by *Burkhart and Chen* [1991], namely:

$$\kappa_i \approx \frac{0.8}{i+0.6} \tag{17}$$

where i is an integer ≥ 1. At these resonances, particles preferentially execute Speiser-type orbits, that is: they are turned back from the $+X$ to $-X$ direction while oscillating about the midplane and subsequently escape from the current sheet with a magnetic moment nearly identical to that at entry. It is apparent from the right panel of Figure 9 that, at resonance, the ψ values are equally distributed about 180°. A similar analysis of the left panel of Figure 9 allows to establish a one-to-one correspondence between the resonant χ values and κ_i. On this basis, DM-97 derived the following relationship between χ and κ:

$$\chi = \frac{2.5\kappa}{2+\kappa} \tag{18}$$

The analysis of DM-97 also shows that the analytical magnetic moment obtained from (4) at $t = \tau_c$ does become comparable to unity whenever $\kappa \approx \kappa_i$, quite consistent with Speiser-type orbits. It follows from this result that energy resonances correspond to specific situations where the centrifugal impulse time scale nears a submultiple of the cyclotron period.

It should be pointed out here that the study of DM-97 is restricted to (nearly) resonant particle orbits, that is, particles which experience periodic Z- oscillations upon crossing of the field reversal and which subsequently escape. At this stage, it is not clear to which extent the centrifugal impulse description can be applied to the various $\kappa << 1$ regimes, including for instance quasi-trapped behavior [e.g., *Ashour-Abdalla et al.*, 1991]. Still, regardless of the κ value (smaller than 1), DM-97 put forward that each $Z = 0$ crossing gives rise to a three-branch pattern similar to that described above, consistent with an enhanced centrifugal pull near the midplane. Also it is interesting that, for particles with $\kappa << 1$ and small pitch angles at equator, numerical trajectory calculations reveal a phase bunching effect similar to that discussed above. In this regard, in their analysis of particle motion in sharp field reversals, *Savenkov et al.* [1991] and *Zelenyi and Savenkov* [1993] put forward an intriguing feature of particle orbits with small initial pitch angles, namely, that the change in the action integral I_Z does not depend upon initial gyrophase (see, e.g., Figure 4a of *Savenkov et al.* [1991]). Prominent variability of I_Z with gyrophase is achieved only at subsequent interactions. This behavior clearly resembles that obtained in the centrifugal impulse model at $\kappa \sim 1$. The physical reason for this may be that nearly field-aligned particles execute full cyclotron turns during approach (that is, denoting by ρ_L the Larmor radius, these particles verify $\rho_L << Z$ throughout approach of the neutral sheet), in which case perturbation of the gyromotion by an impulsive centrifugal force near the field minimum is a relevant approximation. In fact, several effects that will be described hereinafter (such as the consequences of phase bunching for the non-gyrotropic character of the ion distribution functions or the role of non-diagonal terms in the pressure tensor), have been extensively discussed in previous studies of the $\kappa << 1$ ion dynamics in the distant magnetotail [e.g., *Ashour-Abdalla et al.*, 1991, 1994; *Holland and Chen*, 1993; *Zelenyi and Savenkov*, 1993]. Our intent here is to examine these effects in the $\kappa \sim 1$ limiting case of the inner plasma sheet.

3. IMPLICATIONS OF $\kappa \sim 1$ DYNAMICS FOR THE NEAR-EARTH MAGNETOTAIL

In the Earth's magnetotail, plasma sheet ions exhibit $\kappa \sim 1$ at the transition between dipolar and taillike field configurations. This $\kappa \sim 1$ situation is achieved in a region of rapid increase of the equatorial magnetic field and it is thus limited to a narrow X interval, typically of the order of a few R_E near $L = 10$. Note that, because of the large gradient of the equatorial magnetic field, the $\kappa \sim 1$ dynamics of near-Earth plasma sheet ions is complicated by substantial gradient drift. Still, in the trajectory calculations discussed hereinafter, we focus on ions with relatively small pitch angles at equator (up to 30°-40°) and, for these ions, the curvature term largely exceeds (by at least a factor 10) that due to $\nabla_\perp B$ in the region of space considered. Also it should be pointed out that, in contrast to the far tail which is essentially one-dimensional (X-invariant), the parameter κ is

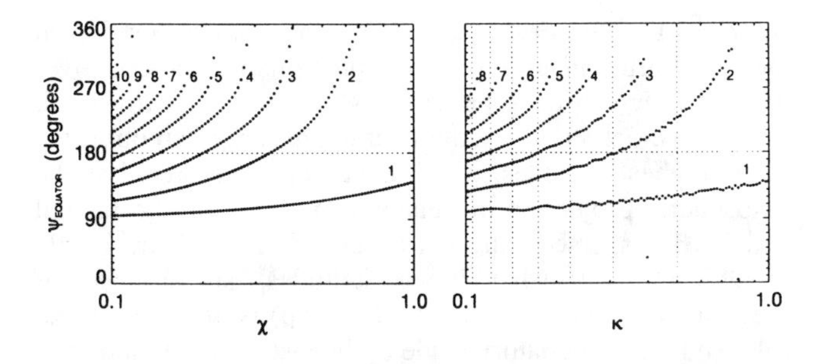

Figure 9. (left) Analytical phases at $Z = 0$ crossing (equation (16)) as a function of χ. (right) Computed phases at $Z = 0$ crossing as a function of κ. The crossing number n is indicated in both panels. Vertical dotted lines in the right panel show the resonant κ values (equation (17)). From DM-97.

not truly appropriate in the two-dimensional near-Earth tail [e.g., *Burkhart et al.*, 1991]. In this region, the κ sampling adopted is obtained from a local calculation along the various guiding center field lines that we consider. That is, for given particle energy and given guiding center field line, we compute the corresponding κ value from the curvature radius of this magnetic field line at equator. In deriving these κ values, we implicitly assume negligible variation of the minimum curvature radius on the length scale of the Larmor radius. As a matter of fact, in the energy range examined, this variation was found to lead to changes in the computed κ of at most 10%. We will first examine the consequences of magnetic moment damping (vertical branch of the three-branch pattern), and then those related to phase bunching (oblique branch).

3.1. Ion Precipitation from the Near-Earth Plasma Sheet

It was stated above that, because the μ change in the vertical branch of the three-branch pattern critically depends upon gyration phase, this branch is responsible for chaotisation of the particle motion in the long term (see, e.g., right panel of Figure 3). The fact that particles experience here magnetic moment damping at specific pitch and phase angles is of particular importance since this leads to significant filling of the loss cone and subsequent precipitation into the ionosphere. Previous studies [e.g., *Sergeev et al.*, 1983; *Zelenyi et al.*, 1990a] already suggested that the strong pitch angle diffusion occurring in the $\kappa \sim 1$ regime in the near-Earth tail provides a potential de-trapping mechanism for plasma sheet ions. To further investigate this issue, Figures 10 and 11 show the analytical flux variations obtained after one crossing of the neutral sheet and for distinct values of κ between 1 and 3. To compute these variations, we developed the $\mu(\tau)$ ratio in equation (11) and derived the following relationship between equatorial pitch angles prior to (α_o) and after (α_f) crossing of the neutral sheet (see equation (7) of *Delcourt et al.* [1996a]):

$$\sin^2\alpha_f = \sin^2\alpha_o + \frac{\cos^4\alpha_o}{f(\kappa)^4} - 2\frac{\cos^2\alpha_o \sin\alpha_o}{f(\kappa)^2}\sin\psi_o \quad (19)$$

Given the angular distribution at entry, equation (19) allows a direct evaluation of the distribution at exit. In Figure 10, we have first considered a beam incident upon the field reversal with an isotropic distribution between the loss cone boundary (arbitrarily set to 2.5°) and some upper limit (arbitrarily set to 8.5°). This initial distribution is represented by the shaded area in Figure 10. On the other hand, the solid line in this figure depicts the flux profile achieved at exit using (19) to relate initial and final pitch angles.

As a first comment, it can be seen in Figure 10 that the analytical results are in good agreement with the results of systematic trajectory calculations in the parabolic field (1) (dashed lines). The profiles achieved also are in agreement with those of *Sergeev et al.* [1983] (dotted lines), though minor deviations are noticeable due to the distinct field models adopted (hence, slightly different timescales of the centrifugal impulse). Most notably, it is apparent from Figure 10 that, whereas a negligible filling of the loss cone

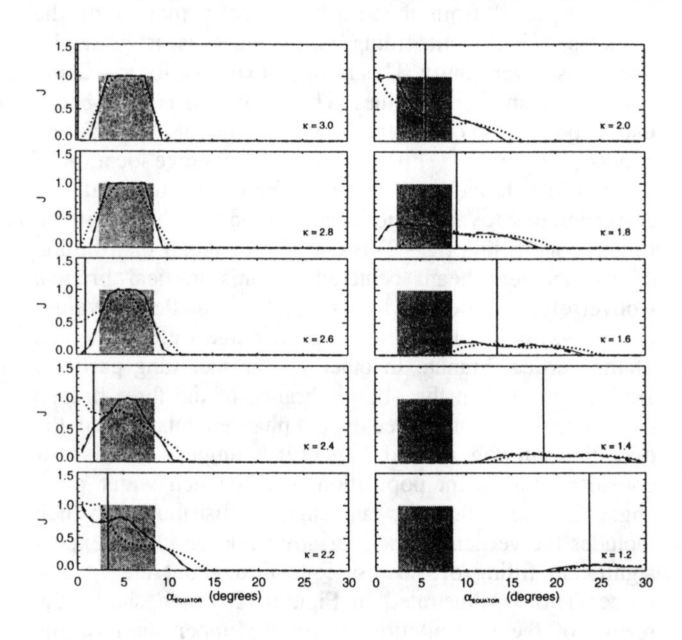

Figure 10. Directional differential flux (normalized to unity) as a function of pitch angle for distinct values of the κ parameter. Analytical variations are shown in solid lines, computations in the parabolic field (1) in dashed lines, and computations in the *Sergeev et al.* [1983] model in dotted lines. The initial isotropic distribution (shaded area) extends from 2.5° up to 8.5°. Vertical lines show the average location of the vertical branch of the three-branch pattern. From *Delcourt et al.* [1996a].

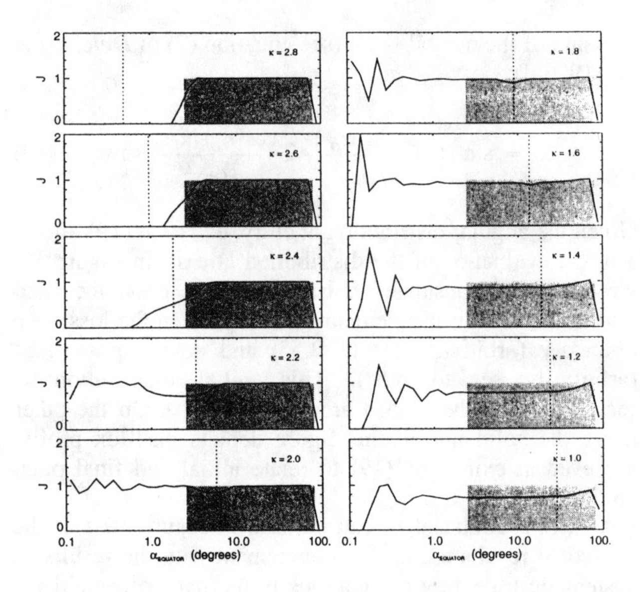

Figure 11. Similar to Figure 10 but the initial isotropic distribution extends from 2.5° up to 80°. From *Delcourt et al.* [1996a].

(below 2.5° pitch angle) is achieved when κ is of the order 3 (as expected from the nearly adiabatic motion of the particles), significant filling is obtained as κ gradually decreases toward unity. This filling maximizes for κ ~ 2 and vanishes at smaller κ values. This evolution can be better understood if one considers the vertical line in the different panels of Figure 10. This line shows the average location of the vertical branch of the three-branch pattern at the corresponding κ value, and it can be seen that the maximum in loss cone filling (κ ~ 2) is due to an angular distribution of the incident beam centered on this vertical branch. Conversely, the depleted loss cone at smaller κ values (below ~2) is due to a beam angular range well below that of the vertical branch. In other words, incoming particles are here located in the oblique branch of the three-branch pattern and thus subjected to μ enhancements; hence the overall scattering toward large pitch angles. If we now consider an incident population over a much wider pitch angle interval (that is, an angular distribution which includes the vertical branch for any value of κ), we expect significant filling of the loss cone throughout the ~1-3 κ range. This is illustrated in Figure 11 which shows the results of the computations when the upper limit of the angular distribution is set to 80°. It is clearly apparent from this figure that a prominent loss cone filling is obtained regardless of the κ value.

These results put forward nonadiabatic motion in the near-Earth plasma sheet as a significant de-trapping mechanism. In particular, they confirm the interpretation of *Sergeev et al.* [1993] which identifies the equatorward boundary of intense isotropic precipitation (referred to as the Isotropic Boundary) as the onset (κ ~ 3) of nonadiabatic motion in the near-Earth tail. In typical ion precipitation patterns over the nightside auroral zone (see, e.g., Plate 1 of *Senior et al.* [1994]), velocity dispersed ion structures are often observed at the highest latitudes, which have been interpreted as the ionospheric projection of beams launched from the far tail and **E**×**B** dispersed during transport [e.g., *Zelenyi et al.*, 1990b]. At somewhat lower latitudes, an absence of precipitation (referred to as the Ion Gap) is often observed, followed on the equatorial side by intense precipitation. The results presented above demonstrate that the κ ~ 1 dynamical regime cannot be related to an absence of precipitation as featured in the Ion Gap, unless some unrealistic assumption is made on the Central Plasma Sheet populations (such as to be fairly collimated along the magnetic field as in Figure 10). As schematically illustrated in Figure 12, it appears that the Ion Gap must rather be related to the distant magnetotail where the loss cone is made very small. In fact, it may well be that the κ ~ 1 scattering process is responsible for the main body of auroral ion precipitation, as envisaged for instance by *Newell et al.* [1996] (see, e.g., the *b2i* boundary in their Plate 1).

3.2. Phase Bunching and Related Current Sheet Structure

It was shown above that particles incident upon the field reversal may be subjected to bunching in gyration phase. This bunching is effective at relatively small pitch angles (i.e., in the oblique branch of the three-branch pattern where particles experience systematic μ enhancements; see Figures 3 and 7) and is due to predominant centrifugal effects (i.e., $\kappa_\alpha < 1$). Still, it is apparent from Figure 4 that the domain of velocity space affected by phase bunching significantly depends upon the value of κ, the oblique branch extending only over of a few degrees in pitch angle for κ ~ 2-3 and up to 30° for κ ~ 1. This result is not surprising if one considers that, given the minimum field line curvature radius, decreasing κ amounts to increasing the particle speed; hence, a predominant centrifugal force over a larger pitch angle interval.

To better illustrate this bunching effect, Figure 13 shows an example of particle orbits at L = 10 in the *Tsyganenko* [1989] model (hereinafter referred to as T-89). In this figure, test protons were launched from a given off-equator position with distinct phases of gyration (from 0° to 360° by steps of 45°) and a pitch angle of 170° at Z = 2.5 R_E. An initial energy of 5 keV was used which corresponds to κ ~ 1.45 at this distance in the magnetotail (the dawn- to- dusk convection electric field was not considered in these calculations). The particles were traced during one crossing of the neutral sheet, until they reach their mirror point (**V.B** = 0) in the opposite hemisphere. It is apparent from the two right panels of Figure 13 that, upon approach of the field

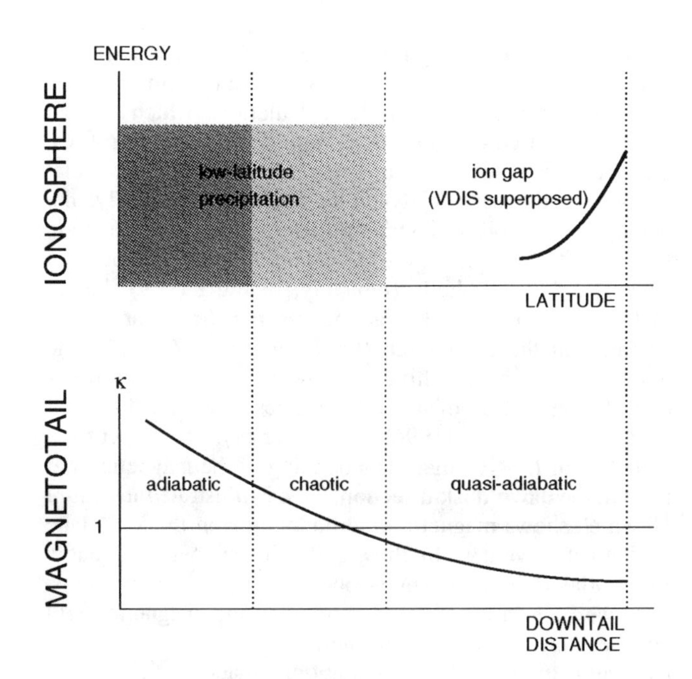

Figure 12. Schematic representation of the relationship between auroral precipitation structures and particle dynamics in the magnetotail. From *Delcourt et al.* [1996a].

reversal, velocity distributions are centered about zero, the particles being uniformly distributed in gyration phase. In contrast, in the vicinity of the $Z = 0$ plane, it can be seen that the particles travel together in the $+T$ or $-T$ (with T perpendicular to both **B** and Y; see Appendix), $+Y$ or $-Y$ directions. This effect rapidly vanishes as the particles move away from the equator due to small differences in Larmor frequency; hence a progressive mixing in gyration phase. The Z interval where the particles are bunched together is thus restricted to the immediate vicinity of the magnetotail midplane. In particular note that, at $Z = 0$, the V_T and V_Y distributions point toward tail and dusk, respectively, as schematically shown in Figure 6. These velocities subsequently point in the opposite directions when the centrifugal force relaxes.

The phase bunched velocity distributions portrayed in Figure 13 have important implications for the net current carried by the particles. This can be appreciated in Figure 14 which shows the results of systematic trajectory calculations in the T-89 model. Here, test protons were launched from a Z height of 2 R_E along distinct L- shells (from $L = 8$ up to $L = 14$ by steps of 1) within the T-89 model, using distinct energies, pitch angles and gyration phases and assuming that the initial distribution outside the loss cone is an isotropic Maxwellian (f_M). The particles were then traced during one interaction with the field reversal as described above. Once all the calculations were performed, the current in a given Z bin along a given L-shell was computed by assuming symmetrical inflows from

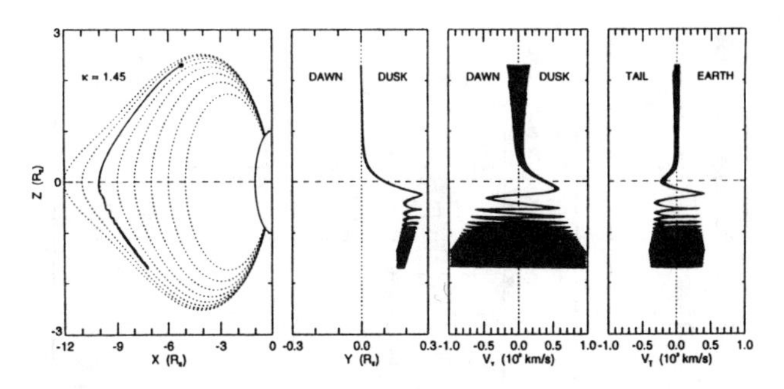

Figure 13. Model trajectory of protons in the T-89 model: (from left to right) projection in the X-Z plane, projection in the Y-Z plane, V_Y velocity component versus Z, and V_T velocity component versus Z (T denoting the direction perpendicular to both **B** and Y and pointing toward Earth ; see Figure A1). The ions are launched from $Z = 2.5$ R_E along the $L = 10$ field line with 5-keV energy, 170° pitch angle, and distinct phases of gyration (from 0° to 360° by steps of 45°). Dotted lines in the leftmost panel show selected T-89 field lines.

both hemispheres and by summing over the particles occuring within this bin. Inside this Z bin, the net current in the Y direction was thus obtained as:

$$J_Y = \frac{q}{\Omega} \{ \textstyle\sum_{n_N} [F < V_Y > (T_{out} - T_{in})]_{E,\alpha} + \sum_{n_S} [F < V_Y > (T_{out} - T_{in})]_{E,\alpha} \} \quad (20)$$

where Ω is the bin volume, E the particle energy and α its pitch angle, and where n_N and n_S denote particles originating from northern and southern hemispheres, respectively, $<V_Y>$, the average Y velocity of the particle within the bin, and T_{in} and T_{out}, its respective time of entry and time of exit in the bin. Also, $F_{E,\alpha}$ is the flow rate (in ions s^{-1}) at the corresponding pitch angle α and energy E, viz., $F_{E,\alpha} = f_M\ E\ \delta E\ \delta S\ \cos\alpha\ \delta(\cos\alpha)$ (where δS is a surface element perpendicular to **B**). Substituting $<V_T>$ for $<V_Y>$, an expression similar to (20) can be obtained for the net current in the T direction. In addition, whereas the ion density (set to 1 ion cm^{-3} [e.g., *Lennartsson and Shelley*, 1986]) was taken constant in first approximation over the entire line source at $Z = 2$ R_E, the temperature was varied with L- value in such a way that the computed Z-integrated current intensity (as obtained by summation of (20) over Z) matches that in T-89. This leads to a temperature of ~2.5 keV at $L = 14$ up to ~8 keV at $L = 8$ (see Figure 2 of *Delcourt and Belmont* [1998]), in rough agreement with in-situ measurements [e.g., *Lui and Hamilton*, 1992].

The right panels of Figure 14 show the J_Y and J_T distributions thus obtained versus Z for the distinct L values considered (note that Y and T point here in directions opposite to those in Figure 13). Also, since the parameter κ depends upon energy, there exists, at each L, a κ

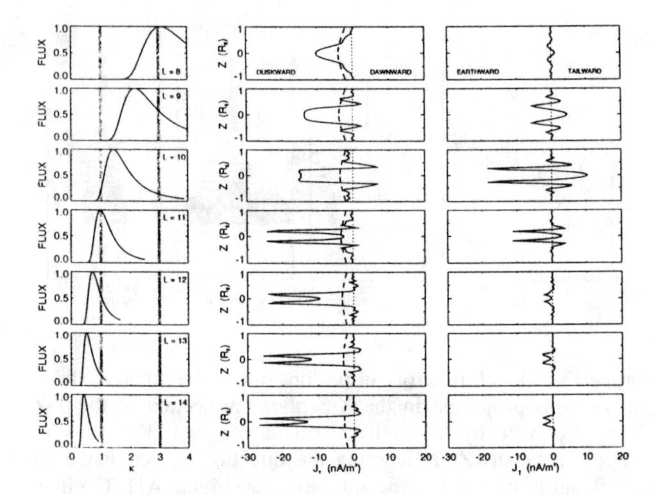

Figure 14. (left) Differential flux (normalized to the maximum value) versus κ, (center) computed J_Y versus Z, (right) computed J_T versus Z (note that Y and T are pointing in directions opposite to those in Figure 13). These variations are shown for distinct L values, from L = 8 (top) up to L = 14 (bottom). The dashed line in the center panels shows the J_Y current density in T-89. From *Delcourt and Belmont* [1998].

distribution in the plasma which directly depends upon the local ion temperature. This κ distribution is illustrated in the leftmost panels of Figure 14 which present the differential flux, $E f_M$, as function of κ. It can be seen in these panels that, with a maximum flux near κ ~ 3, the L = 8 shell corresponds to an overall transition from nonadiabatic to adiabatic plasma transport. As L increases (from top to bottom in Figure 14), the ion population is gradually shifted toward smaller κ values, yielding increasing nonadiabatic effects near Z = 0 and the build-up of a thin current sheet in the Y direction (center panels of Figure 14). Interestingly, it should be noted that the J_Y distributions achieved here closely resemble those obtained in the study of *Ashour-Abdalla et al.* [1994] where a large number of particles was traced from a mantle source (see, e.g., Plate 5 of *Ashour-Abdalla et al.* [1994] and Plate 1 of *Delcourt and Belmont* [1998]). Even though the total current intensity is by construction identical to that in T-89, the computations thus put forward a Z-distribution of the current significantly different from that in T-89. In particular, it was shown in *Delcourt et al.* [1996b] that such a thin current sheet appears as a persistent feature if iterative computations are performed.

Moreover, it is apparent from Figure 14 that the details of the computed current sheet significantly depend upon the dynamical regime experienced by the particles. For instance, at L = 10 where the bulk of the ion population has κ ~ 1.5, the thin sheet of duskward current near Z = 0 is delineated by substantial dawnward currents on either side. Simultaneously, a complicated current pattern can be seen in the T direction (rightmost panel of Figure 14) with tailward current near Z = 0 and earthward current on either side. Within the limits of the calculations which do not provide a self-consistent description of the current profiles, the structuring portrayed in Figure 14 emphasizes special features of the particle dynamics near κ = 1, namely: the deployment of phase bunching effects along the magnetic field line (see, e.g., Figure 13). This structuring does not appear at low L- shells (typically, below L ~ 9) due to global evolution toward adiabatic transport (κ ~ 3 or above) whereas, in the distant tail (typically, above L ~ 11), the effect of phase bunching is smeared by the complex meandering motion of κ << 1 particles (see, e.g., Figure 2 of *Delcourt et al.* [1996b]). Conversely, the J_T current pattern near L = 10 suggests a bending of the magnetic field lines in the dawn-dusk direction. This is illustrated in Figure 15 which shows magnetic field hodograms in the X-Y plane at distinct L values. In this figure, it can be seen that a substantial B_Y component is obtained at L = 10 whereas, earthward or tailward of this region, the magnetic field reverses in a nearly linear manner.

Because of the various assumptions made in the present computations (most notably, the fact that they do not treat the particle dynamics in a self-consistent manner), it is difficult to thoroughly assess the role of phase bunching for the magnetotail current sheet structure. Still, it is interesting that, in the self-consistent calculations of *Burkhart et al.* [1992], a B_Y component similarly emerges when the bulk ion population evolves from κ << 1 up to κ ~ 1. The above calculations put forward that this peculiar feature which should occur near the earthward termination of the magnetotail current sheet, directly follows from centrifugally-driven phase bunching. On the other hand, neither the present calculations nor those of *Burkhart et al.* [1992] account for the dynamics of electrons and whether these latter particles will significantly alter the current profiles portrayed in Figure 14 is an issue which remains to be investigated.

4. MACROSCOPIC CONSEQUENCES OF NON-GYROTROPIC ION DISTRIBUTIONS

In the above trajectory calculations, a fairly large amount (up to ~70%) of the ion population launched from Z = ±2 R_E at L = 10 was found to be subjected to phase bunching (e.g., Figure 5 of *Delcourt et al.* [1996b]). In the distant magnetotail, it has been shown that the κ << 1 dynamics leads to non-gyrotropic ion distribution functions [e.g., *Ashour-Abdalla et al.*, 1991]. The above phase bunching results actually suggest a similar non-gyrotropic character of the ion distribution functions in the near-Earth tail where κ ~ 1. The build-up of such distributions leads us to briefly revisit some common ideas about the magnetotail, since these ideas are usually derived from MHD concepts which

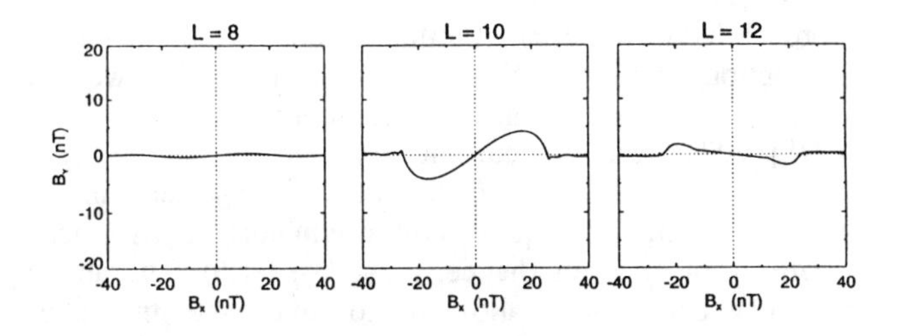

Figure 15. Hodograms showing B_Y versus B_X along distinct magnetic field lines: (from left to right) $L = 8$, $L = 10$, and $L = 12$. From *Delcourt and Belmont* [1998].

exclude non-gyrotropic effects. In particular, in a like manner to the distant tail where off-diagonal elements of the pressure tensor have been shown to be essential to maintain stress balance [e.g., *Holland and Chen*, 1993; *Ashour-Abdalla et al.*, 1994], we will show that these off-diagonal elements are of importance in the $\kappa \sim 1$ dynamical regime of the inner plasma sheet. As a preliminary comment, it should be kept in mind that the two primary conditions for MHD theory to be valid have to do with the slowness of variations, yielding $\partial_t << \omega_C$ and $\partial_r << \omega_C / V_{th}$. For a static equilibrium, the first condition is always verified. As for the second condition, it requires that most particles (those carrying the pressure) have a Larmor radius everywhere much smaller than characteristic length scales of the system, and in particular much smaller than the minimum curvature radius of the magnetic fied lines. These necessary (but not sufficient) conditions therefore meet the "adiabatic" criterion $\kappa >> 3$ introduced for the description of individual trajectories. The pressure being proportional to the integral of $V^2 f(V)$, it can be seen in the leftmost panels of Figure 14 that this pressure is essentially carried by particles at $\kappa >> 1$ for $L < 8$ and by particles at $\kappa << 1$ for $L > 12$. As for $L \sim 10$, it corresponds to a transition region with $\kappa \sim 1$. Thus, whereas MHD may apply to the inner plasmasheet, it clearly is questionable near $L \sim 10$ and beyond.

4.1. Magnetic Field / Pressure Equilibrium

For a static equilibrium to exist, the Lorentz force has to be balanced by a plasma force according to the usual equality:

$$\nabla.(\bar{\bar{\mathbf{P}}}) = \mathbf{J} \times \mathbf{B} \qquad (21)$$

We use here the notation $\bar{\bar{P}}$ for the "global pressure", i.e., the non-centered second moment of the distribution function: $\bar{\bar{P}} = \int m\mathbf{ww}f(\mathbf{w})d^3\mathbf{w}$. When necessary, this quantity can be separated as usual in two parts, namely, the dynamical pressure (related to the mean velocity) and the ordinary pressure (centered second moment): $\bar{\bar{P}} = m\mathbf{VV} + \bar{\bar{\pi}}$. Note that, generally speaking, $\mathbf{J}$ and $\bar{\bar{P}}$ are due to all species present in the plasma, but we neglect here the contribution of electrons and only consider that of protons. Using a local reference frame with unit vectors $\mathbf{t}$ and $\mathbf{n}$ tangential and normal to the field line, respectively, the magnetic force can also be written as a function of the curvature radius R_C of the field line and the scale length L_g of the magnetic field modulus gradient:

$$\mathbf{J} \times \mathbf{B} = \frac{B^2}{\mu_0}[\frac{\mathbf{n}}{R_C} - \frac{\mathbf{g}_\perp}{L_g}] \qquad (22)$$

Here $\mathbf{g}$ is a unit vector in the $\nabla(B)$ direction and $\mathbf{g}_\perp$ its projection perpendicular to $\mathbf{B}$, whereas R_C and L_g are defined by: $\mathbf{n}/R_C = \mathbf{t}.\nabla(\mathbf{t})$ and $\mathbf{g}/L_g = \nabla(B)/B$, respectively. The second term in (22) is usually small at equator and even null in the simplest models where the magnetic field is X-invariant such as in the parabolic model (1).

At the magnetic equator, it is clear from symmetry arguments that the magnetic force is in the X-direction (and, more precisely, earthward since J_Y points toward dusk). This force thus has to be balanced by the X-component of $\nabla.(\bar{\bar{P}})$, which is the sum of three terms: $\partial_X(P_{XX}) + \partial_Y(P_{XY}) + \partial_Z(P_{XZ})$. Here, we neglect the second of these terms by assuming a two-dimensional field geometry with invariance in the Y-direction. In MHD theory, the ordinary pressure is supposed to be scalar (i.e., $\bar{\bar{\pi}} = p\,\bar{\bar{I}}$), so that $p_{XX} = p$ and $p_{XZ} = 0$. Under this assumption (and as long as the dynamical pressure can be neglected in the reference frame adopted, viz., the convection frame), the only term able to balance the magnetic stress is thus a $\partial_X(p)$ term. In all MHD models of the magnetotail, knowledge of J_Y and B_Z at equator is then sufficient to uniquely determine the radial increase of pressure toward the Earth. If we approximate the magnetotail by a one-dimensional field (with minimum curvature radius of the order of a few tenths of R_E), this increase of pressure is linear. MHD models cannot then account for a very long (several tens of R_E) tail, since this would lead to unrealistically large pressure differences between distant and near-Earth regions. Conversely, MHD models with realistic values of the equatorial pressure exhibit unrealistically small cross-tail currents and, accordingly, weakly elongated magnetotails [e.g., *Schindler and Birn*, 1982].

One could a priori expect anisotropic MHD to solve these difficulties. Assuming the ion distribution to be everywhere gyrotropic (i.e., exhibiting cylindrical symmetry around $\mathbf{B}$), the distribution function does not depend upon gyration phase (i.e., $f(\mathbf{V}) = f(V_\parallel, \mathbf{V}_\perp)$), but the pressure tensor relies on two scalars $p_\parallel$ and $p_\perp$ as well as on the direction $\mathbf{t}$ of the local magnetic field, viz., $\bar{\bar{\pi}} = p_\parallel \mathbf{tt} + p_\perp(\bar{\bar{I}} - \mathbf{tt})$. Its divergence can therefore be written:

$$\nabla.(\bar{\bar{P}}) = \nabla_\parallel(p_\parallel) + \nabla_\perp(p_\perp) + (p_\parallel - p_\perp)\nabla.(\mathbf{tt}) \qquad (23)$$

with

$$\nabla.(\mathbf{tt}) = \frac{\mathbf{n}}{R_C} - \frac{\mathbf{g}_{//}}{L_g} \tag{24}$$

Let us use (23) to derive the X-component of $\nabla.(\bar{\bar{\pi}})$ at equator in anisotropic MHD. This component is now the sum of two terms: $\partial_X(p_\perp) + (p_\parallel - p_\perp)/R_C$, the second of these terms resulting from a non-zero p_{XZ} (viz., $p_{XZ} = (p_\parallel - p_\perp)t_X t_Z$). Accordingly, the variation of the perpendicular pressure along the tail axis, $\partial_X(p_\perp)$, now differs from the above isotropic result. However, a thin current sheet with a small $\partial_X(p_\perp)$ would still require $p_\parallel >> p_\perp$ [e.g., *Cowley*, 1978], which is not the case in the central plasma sheet.

Finally, let us consider the above non-gyrotropic distributions due to centrifugally-driven phase bunching. For simplicity, we assume the particles to be perfectly bunched at equator so that they exhibit identical perpendicular velocities V_{XO} and V_{YO} at $Z = 0$, whatever their V_Z component. In the close vicinity of $Z = 0$, the magnetic force imposes that the V_X component varies as $V_X = V_{XO} + \omega_{CO}V_{YO}Z/V_Z$. The two pressure components P_{XX} and P_{XZ} able to balance the magnetic force at the equator are then such that:

$$\partial_X(P_{XX}) = \partial_X(nmV_{XO}^2) \tag{25}$$

and

$$\partial_Z(P_{XZ}) = nm\omega_{CO}\, V_{YO} \tag{26}$$

With $L_X >> \rho_L$, it is easy to see that the gradient described by (26) prevails over that in (25), even if $V_X \sim V_Y$. In other words, because of the sharp gradient of P_{XZ} in the Z direction, equilibrium can now be achieved over an X-interval much larger than in the above isotropic and anisotropic MHD cases. These crude estimates suggest that, whereas MHD can account for the large earthward increase of pressure in the near-Earth plasma sheet (typically, within $L \sim 8$ where the bulk of the ion population has $\kappa >> 1$), equilibrium further out in the magnetotail has a different origin, in particular near $L = 10$ where the ion phase bunching seems to play a major role.

4.2. Origin of Cross-Tail Current

If we neglect the role of electrons and consider ions as the only current carriers, the very existence of a perpendicular current implies that the ions have a mean perpendicular velocity. Thus, regardless of the physical mechanism at the origin of the current, the local ion distribution function, $f(\mathbf{V})$, is by essence nongyrotropic. In the above trajectory calculations, particles were launched with the same guiding center along a given L- shell. Such a "guiding-center" based approach led us to define at initialization a "guiding-center" distribution function $f_{gc}(\mathbf{V})$ (the above isotropic Maxwellian f_M) to locally describe the ion population (see, e.g., *Catto* [1978]). The resulting current can then be due to (i) a perpendicular gradient of f_{gc} (hence, a $\partial_X(P_{XX})$ term in the above description of equilibrium at equator) or (ii) a non-gyrotropy of f_{gc} itself (hence, a $\partial_Z(P_{XZ})$ term). The above trajectory calculations allow us to investigate the latter effect. Examination of the former effect would have required us to trace particles from various X values with different densities f_{gc}, to reconstruct the local distribution function f, and finally to calculate the current as a moment of f.

In the near-Earth plasma sheet, most particles follow adiabatic trajectories, gyration phases (ψ) remain equally distributed, and f_{gc} is gyrotropic (i.e., independent of ψ). The current in this region is then due to the first mechanism mentioned above, namely: a perpendicular gradient of f_{gc}. It is the so-called "magnetization" current, $\mathbf{J}_\perp = \mathbf{t}\times\nabla(p_\perp)/B$ (with $\mathbf{B} = B\mathbf{t}$). In contrast, at large geocentric distances, we have shown that particles at $\kappa \sim 1$ or below can be subjected to prominent bunching in gyration phase. This effect is responsible for the non-gyrotropy of f_{gc} and thus for a current. In particular, it was demonstrated in the preceeding section that, near $L = 10$, the deployment of phase bunching effects along the magnetic field line yields a complicated J_T pattern whereas, further out in the magnetotail ($\kappa < 1$), this J_T pattern is smeared by the complex meandering motion of the particles near the field minimum (the net result being here a large Y excursion of the particles about the small B_Z component).

To interpret the results of the computations, we must consider two types of particles passing at $Z = \pm 2\ R_E$, namely: those participating to the current through a gradient effect and those participating to the current through a non-gyrotropy effect. In Figure 16, current intensities integrated between $Z = +2\ R_E$ and $Z = -2\ R_E$ are presented as a function of L. These intensities are obtained considering either the only non-gyrotropy effect (dotted line) or both non-gyrotropy and gradient effects (solid line). In the former case, the current intensity is calculated by summation of (20) over Z, whereas in the latter case, it is estimated from the divergence of $\bar{\bar{\mathbf{P}}}$:

$$J_Y = -\left[\frac{\nabla.(\bar{\bar{\mathbf{P}}})\times\mathbf{t}}{B}\right].\mathbf{y} \tag{27}$$

hence:

$$J_Y = \frac{1}{B}\{[\partial_X(P_{XX}) + \partial_Z(P_{XZ})]\cos\phi - [\partial_X(P_{XZ}) + \partial_Z(P_{ZZ})]\sin\phi\} \tag{28}$$

(ϕ being the angle between $\mathbf{B}$ and the Z axis). At each L-value in Figure 16, the current intensity in dotted line is thus identical to that in T-89 since, by construction, the ion temperature used at $Z = \pm 2\ R_E$ has been chosen to meet this equality. It should be stressed here that this intensity is obtained with a better accuracy than that in solid line, since derivatives with respect to X in (28) are computed within the L- sampling of the computations (i.e., from 8 to 14 by steps of 1). Figure 16 clearly demonstrates that, at $L \sim 10$ and beyond, the profiles in dotted and solid lines nearly coincide. In other words, in this region of space, the effect of X-gradient is negligible and the cross-tail current can essentially be attributed to the non-gyrotropy effect as put forward in previous studies [e.g., *Ashour-Abdalla et al.*, 1994]. In contrast, below $L \sim 10$, the current profiles rapidly diverge, indicating an increasing role of the gradient effect. As mentioned above, it should be kept in mind that particles which mirror between $Z = +2\ R_E$ and $Z = -2\ R_E$ (i.e., particles with fairly large pitch angles at the equator) were not considered in the trajectory calculations, and the current (of the gradient type) contributed by these particles is thus not accounted for in Figure 16. This contribution is likely to be small beyond $L \sim 10$ because of weak X-gradients but should become significant at low L- shells; hence an increased divergence of the two current profiles in Figure 16. As for electrons, their role in this interpretation framework is unclear and will be examined in a future study.

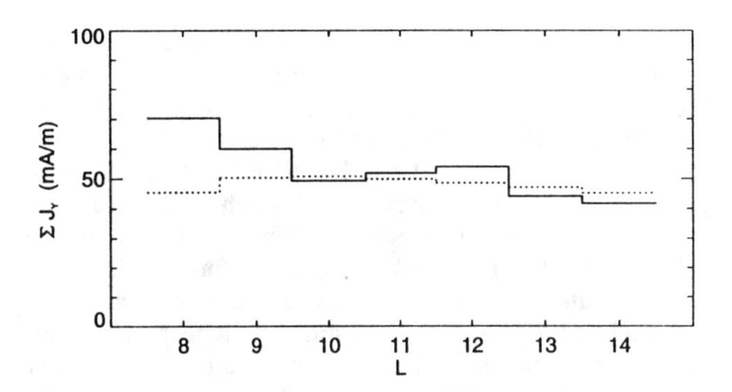

Figure 16. Integrated current intensity versus L- value. The solid line shows the current due to both non-gyrotropy and gradient effects, the dotted line, that due to non-gyrotropy alone.

4.3. Non-Gyrotropic Distribution Functions and Magnetic Reconnection

Magnetic reconnection in the Earth's magnetotail is a key issue in magnetospheric physics, both from the viewpoint of global equilibrium and regarding the stability of the magnetosphere (substorms) [e.g., *Vasyliunas*, 1975]. The local possibility of reconnection can conveniently be quantified by the vector:

$$\Omega = \frac{1}{\delta X\ B}\ d_t(\delta \mathbf{X} \times \mathbf{B}) \tag{29}$$

where $\delta\mathbf{X}$ is the vector difference between two points arbitrarily close, aligned along a magnetic field line and convected with the mean velocity $\mathbf{V}$. The vector Ω is zero if $\delta\mathbf{X}$ and $\mathbf{B}$ remain aligned (i.e., if $d_t(\delta\mathbf{X}\times\mathbf{B}) = 0$). Otherwise, $\delta\mathbf{X}$ and $\mathbf{B}$ depart from alignment with angular velocity Ω. With the notation $\varepsilon = \mathbf{E} + \mathbf{V}\times\mathbf{B}$, it is easy to put Ω under the form:

$$\Omega = \frac{[\nabla\times(\boldsymbol{\varepsilon})]\times\mathbf{t}}{B} \tag{30}$$

The usual result that the magnetic diffusion rate is zero in ideal MHD is obviously obtained by setting ε to zero. When taking into account the ion kinetic effects for a stationary equilibrium (due to finite Larmor radius), ε is given by :

$$\varepsilon = \frac{\nabla.(\bar{\bar{\mathbf{P}}})}{nq} \tag{31}$$

The rate of magnetic diffusion therefore only involves:

$$\nabla\times(\varepsilon) = \frac{1}{nq}\{\nabla\times[\nabla.(\bar{\bar{\mathbf{P}}})] - \frac{\nabla(n)}{n}\times\nabla.(\bar{\bar{\mathbf{P}}})\} \tag{32}$$

At equator, we assume, for symmetry reasons, that the density gradient is directed in the X direction so that the second term on the right handside of (32) vanishes. Under such conditions, the only term responsible for reconnection in the equatorial plane is $\nabla \times [\nabla.(\bar{\bar{\mathbf{P}}})]$. This term is obviously zero when the pressure tensor is scalar (i.e., for isotropic ion distributions), and we can conclude that finite Larmor radius effects are unable to allow for magnetic reconnection at equator as long as distribution functions remain isotropic. As for anisotropic but gyrotropic distributions, Ω is proportional to $\partial_{XZ}{}^2(p_\parallel - p_\perp)$ and the magnetic diffusion again remains zero since the equatorial plane must be a symmetry plane for the anisotropy. On the other hand, as put forward in previous studies [e.g., *Büchner and Zelenyi*, 1987; *Cai et al.*, 1994], it is apparent from the above arguments that non-gyrotropic distributions (both for ions and for electrons) and related non-diagonal pressure terms may provide more degrees of freedom for building such effects.

5. SUMMARY

The present study puts forward that, between the adiabatic limit ($\kappa >> 1$) which prevails in the innermost magnetosphere and the quasi-adiabatic limit ($\kappa << 1$) of the distant magnetotail, there exists an intermediate dynamical regime ($\kappa \sim 1$) in the near-Earth tail where the motion of plasma sheet ions is perturbed by an impulsive centrifugal force near the equator. In this regime, significant filling of the loss cone may be achieved because of magnetic moment damping in a limited domain of velocity space. The $\kappa \sim 1$ regime is also characterized by prominent bunching in gyration phase at relatively small pitch angles, due to enhanced centrifugal effects. Within the limits of the calculations, the phase bunching effect is found to lead to a striated pattern of currents at low latitudes, with a thin (a few tenths of R_E) sheet of duskward current near the equator and a significant current in the Earth-tail direction on either side. This latter current leads to field line inclination in the dawn-dusk direction at the earthward edge of the magnetotail current sheet. From a macroscopic viewpoint, the non-gyrotropic distributions produced via centrifugally-driven phase bunching are responsible for off-diagonal elements in the pressure tensor, the Z-gradient of which appears essential to achieve stress balance.

APPENDIX

We define two orthogonal axis (hereinafter labeled as T and Y) in the gyration plane and consider that the centrifugal force F_C (perpendicular to **B**) acts along axis T which, at $Z = 0$, coincides with the X axis of Figure 1. We also choose an origin of phase along this same axis in the direction opposite that of the centrifugal impulse as illustrated in Figure A1. With such definitions, the equation of motion in the gyration plane becomes

$$\dot{V_T} = \omega_C V_Y + \frac{F_c}{m} \quad \text{(A1)}$$

$$\dot{V_Y} = -\omega_C V_T \quad \text{(A2)}$$

ω_C being the Larmor frequency which, in first approximation, we assume constant. Derivation of (A1), (A2), and substitution of $\dot{V_T}$, $\dot{V_Y}$ yield the following equations:

$$\ddot{V_T} + \omega_C^2 V_T = \frac{\dot{F_c}}{m} \quad \text{(A3)}$$

$$\ddot{V_Y} + \omega_C^2 V_Y = -\omega_C \frac{F_c}{m} \quad \text{(A4)}$$

We then consider an impulsive centrifugal force of the form:

$$F_C(t) = \frac{mV_{//}^2}{2R_c}\left[1 - \cos\left(\frac{2\pi t}{\tau}\right)\right] \quad \text{(A5)}$$

where m is the particle mass, τ the impulse timescale, and where $V_{\|}$ and R_C are the particle parallel speed and magnetic field line curvature radius, respectively, which we assume constant. Though obviously an oversimplification, the good agreement between the numerical results and the analytical predictions suggests that such an assumption is not unreasonable as a first approximation. Note that the form (A5) is such that: $F_C(0) = F_C(\tau) = \dot{F_c}(0) = \dot{F_c}(\tau) = 0$. Making use of (A5), general solutions of (A3) and (A4) are readily found as

$$V_T(t) = C_1 \sin(\omega_C t) + C_2 \cos(\omega_C t) + \frac{V_{//}^2}{2R_c}\frac{\omega}{\omega_c^2 - \omega^2}\sin(\omega t) \quad \text{(A6)}$$

$$V_Y(t) = C_3 \sin(\omega_C t) + C_4 \cos(\omega_C t) - \frac{V_{//}^2}{2R_c}\left[\frac{1}{\omega_c} - \frac{\omega_c}{\omega_c^2 - \omega^2}\cos(\omega t)\right] \quad \text{(A7)}$$

where ω is the impulse frequency and the C_1, C_2, C_3, C_4 coefficients are evaluated with respect to the boundary conditions at $t = 0$. Denoting by χ the ω_C/ω ratio, and by ψ_o and $V_\perp$, the gyration phase and perpendicular velocity at $t = 0$, respectively, we obtain at a given time t of the impulse

$$V_T(t) = V_\perp \sin\left(\frac{2\pi t}{\tau_c} + \psi_o\right) - \frac{mV_{//}^2}{2qBR_c}\frac{1}{\chi^2 - 1}\left[\sin\left(\frac{2\pi t}{\tau_c}\right) - \chi \sin\left(\frac{2\pi t}{\tau}\right)\right] \quad \text{(A8)}$$

$$V_Y(t) = V_\perp \cos\left(\frac{2\pi t}{\tau_c} + \psi_o\right) + \frac{mV_{//}^2}{2qBR_c}\frac{1}{\chi^2 - 1}\left[1 - \cos\left(\frac{2\pi t}{\tau_c}\right) - \chi^2 + \chi^2\cos\left(\frac{2\pi t}{\tau}\right)\right] \quad \text{(A9)}$$

With the help of (5), equations (A8) and (A9) yield the following rate of change for the gyration phase:

$$\frac{\dot{\psi}(t)}{\omega_c} = 1 + \frac{mV_{//}^2}{qBR_c V_\perp}\frac{V_\perp V_Y(t)}{V_T^2(t) + V_Y^2(t)}\sin^2\left(\frac{\pi t}{\tau}\right) \quad \text{(A10)}$$

On the other hand, one has by combination of (4), (A8) and (A9):

$$\frac{\dot{\mu}(t)}{\omega_c} = \frac{2mV_{//}^2}{qBR_c V_\perp}\frac{V_T(t)}{V_\perp}\sin^2\left(\frac{\pi t}{\tau}\right) \quad \text{(A11)}$$

Equations (A10) and (A11) give $\dot{\psi}(0) = \dot{\psi}(\tau) = \omega_C$ and $\dot{\mu}(0) = \dot{\mu}(\tau) = 0$ as they should.

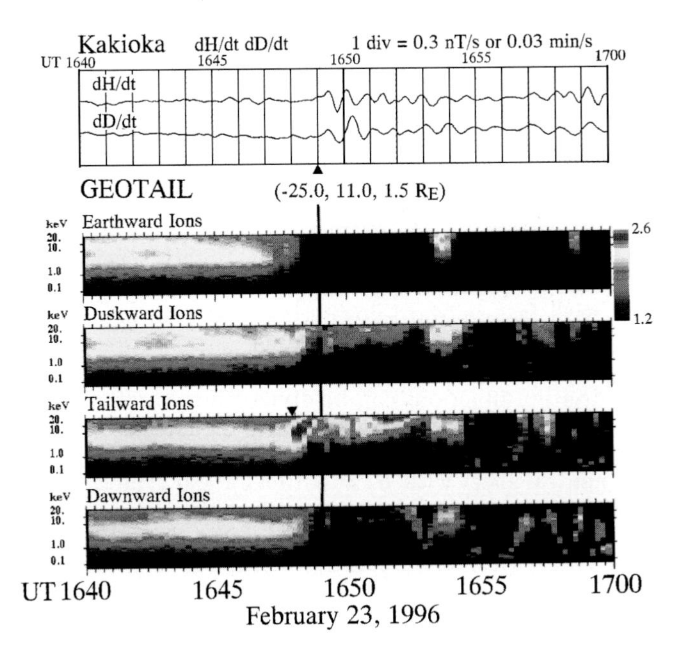

Figure 8. Pi 2 activity in the magnetic field data dH/dt (the upper trace) and dD/dt (the lower trace) at Kakioka for the period 1640-1700 UT on February 23, 1996. One box corresponds to one-min interval. Color panels are directional ion energy-time diagrams observed by Geotail for the same period. One tick mark corresponds to 1/3-min interval. Counts/sample are color-coded according to the logarithmic color bar from 1.2 to 2.6 at the right-hand side.

near (-25.0, 11.0, 1.5 R_E). The onset of Pi 2 pulsation was found at 1649 UT. Tailward-flowing ions with energies of keV appeared at 1648 UT and B_z became strongly southward (B_z was southward after 1645 UT) [see *Nagai et al.* 1998, Figure 2]. This event indicates that magnetic reconnection started al least one minute before the ground onset (V_x started to decrease 2 minutes before the ground onset in this event).

Figure 9 shows Pi 2 activity at Kakioka and ion energy-time diagrams from Geotail for the period 1155-1215 UT on October 15, 1995. Geotail was located near (-17.2, -0.2 -1.4 R_E). The onset of Pi 2 pulsation was found at 1204 UT. Earthward flowing ions with energies of higher than 10 keV appeared at 1202.5 UT, indicating that magnetic reconnection started at least 1.5 minutes before the ground onset.

Figure 10 shows time delays for the 43 tailward flows and the 17 earthward flows. Here, the time delay is start time of flow (V_x) minus substorm onset. It is important to note that nine flows start before onsets in the premidnight region at 20-30 R_E (panel(a)). In this region, five tailward flows start almost simultaneously with ground onsets (within 0-1 min) and only four tailward flows start with positive time delay. Beyond 30 R_E, most tailward flows start a few minutes after onsets. Four earthward flows start before onsets (panel(c)) and most of other earthward flows start almost simultaneously with ground onset (panel(d)).

4.2. Plasmoid Analysis

We used the data from September 1993 through October 1994. In this analysis, the method of substorm selection was the same as that used in the substorm analysis. We obtained 43 well-defined substorms in the periods when Geotail was in the magnetotail. We excluded events in which Geotail entered from the magnetosheath into the magnetotail.

In Figure 11, the delay times between the onset of substorm activity and the arrival of the plasmoid/flux rope at Geotail are plotted versus X_{GSM} (X in the aberrated GSM coordinate system). The arrival of the plasmoid/flux rope is defined as that of tailward flows. The arrival time of the plasmoid/flux rope relative to the onset is progressively delayed as the satellite distance from the Earth increases. The delay is typically 11 min at 100 R_E and 25 min at 200 R_E. A linear fit to data points (the correlation coefficient is 0.83) indicates that plasmoids start at X_{GSM} = -20 R_E and have a tailward speed of 760 km/s.

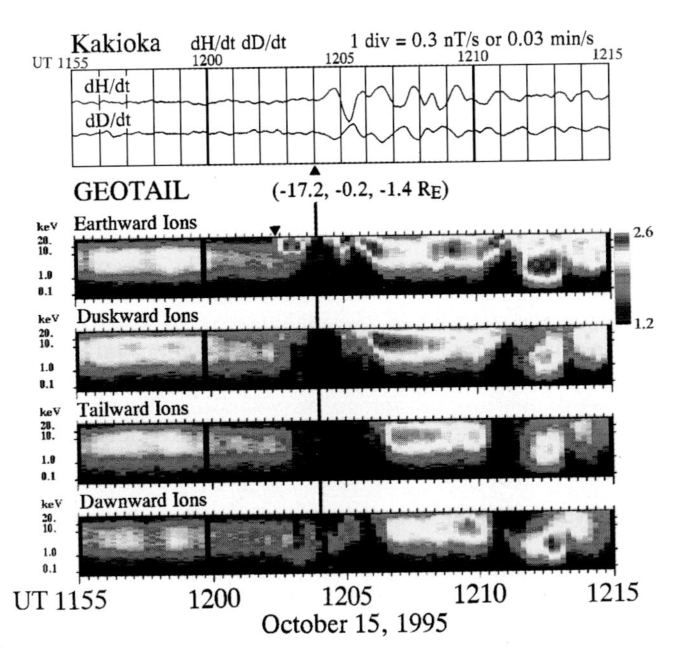

Figure 9. Pi 2 activity in the magnetic field data dH/dt (the upper trace) and dD/dt (the lower trace) at Kakioka for the period 1155-1215 UT on October 15, 1995. One box corresponds to one-min interval. Color panels are directional ion energy-time diagrams observed by Geotail for the same period. One tick mark corresponds to 1/3-min interval. Counts/sample are color-coded according to the logarithmic color bar from 1.2 to 2.6 at the right-hand side.

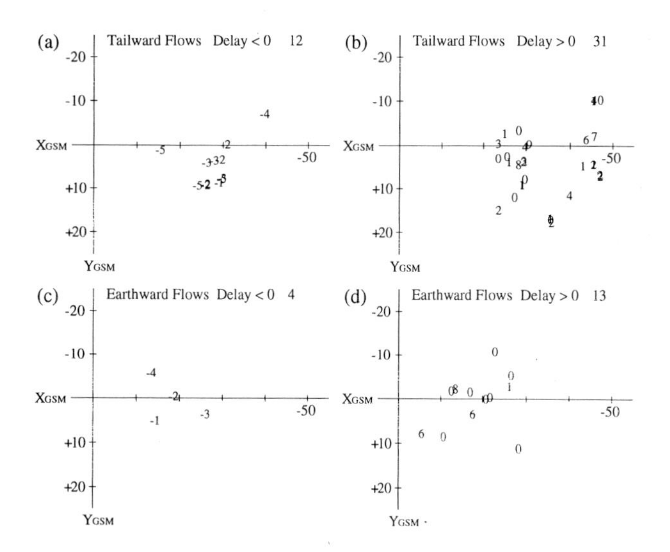

Figure 10. Start times of tailward flows and earthward flows relative to substorm onset presented at Geotail positions in the aberrated GSM XY plane. Negative time delays mean that flows start prior to ground onsets [(a) and (c)], whereas positive time delays mean that flows start at and after ground onsets [(b) and (d)].

5. MAGNETIC FIELD AND PLASMA FLOW STRUCTURE

In several tailward ion flows identified in the above analyses, highly accelerated electrons occasionally appear. The highest-energy (10-keV) component of these electrons shows tailward anisotropy. Simultaneity in the appearance of high-speed tailward-escaping electrons and tailward-flowing ions means that the observations were carried out in the immediate vicinity of the acceleration region, that is, the magnetic reconnection site; otherwise, we would first observe high-energy tailward-escaping electrons and then tailward-flowing ions because of the large difference in velocities. This section examines ion and electron distribution functions as well as magnetic field variations for a tailward flow event on January 27, 1996 [*Nagai et al.*, 1998]. We delineate the structure of the magnetic field and plasma flow near the magnetic reconnection site. Magnetic reconnection produces tailward flows and earthward flows. The global evolution of these flows can be visualized by taking a snapshot of the XZ-plane cross section of the plasma sheet. Although there are only a small number of Geotail data sets for this purpose, we present a preliminary attempt to explore the temporal and spatial development of tailward/earthward flows.

5.1. *The January 27, 1996 Event*

Geotail was located at (-28.9, +5.8, -2.6 R_E) in the GSM coordinates at 1400 UT on January 27, 1996, and a moderate-size substorm took place near the Geotail meridian. Figure 12 shows magnetic field and plasma flow velocity variations for the period 1355-1415 UT on January 27, 1996. Ion and electron energy-time diagrams are also presented. Pi 2 pulsation started at 1355 UT, and it intensified near 1404 UT (the Pi 2 activity is presented for the period 1350-1420 UT in the bottom panel of Figure 12). Tailward flows started near 1359 UT and persisted for 12 min, although tailward convection flows ended near 1407 UT. Highly accelerated (5-10 keV) electrons appeared in the interval 1404-1406 UT with fast tailward-flowing ions, as seen in the ion and electron energy-time diagrams.

During the period 1404-1406 UT, ion and electron distribution functions can be divided into three classes, each with a distinct magnetic field structure: type N (near the neutral sheet with $B_x < 5$ nT), type A (off the neutral sheet with $B_x = 5\text{-}10$ nT), and type B (near the plasma sheet/tail lobe boundary with $B_x = 10$ nT). Here, distribution functions are presented in the BCE coordinate system. In this system, the B-C plane contains the magnetic field. The B direction is along the local magnetic field vector, and the C direction is the direction of the plasma flow component perpendicular to the local magnetic field. E completes the right-hand coordinate system. Therefore, B is the field-aligned direction, C is the convection direction ($\mathbf{E} \times \mathbf{B}$ drift direction), and E is the assumed electric field direction.

Panels (a) and (d) of Figure 13 present a type N distribution taken in the 12-s sample period beginning at 1404:59 UT. In this period, the B direction was approximately southward and the C direction was almost tailward. Ions show a strong convection motion at a speed of > 2000 km/s (a significant number of ions have energies higher than the 40-keV instrument energy limit). Electron distribution is basically isotropic. As seen in the energy-time diagrams, the low-energy (< 1 keV) part of the distribution decreases, whereas the high-energy (> 1 keV) part of the distribution increases, in comparison with those seen before 1403 UT

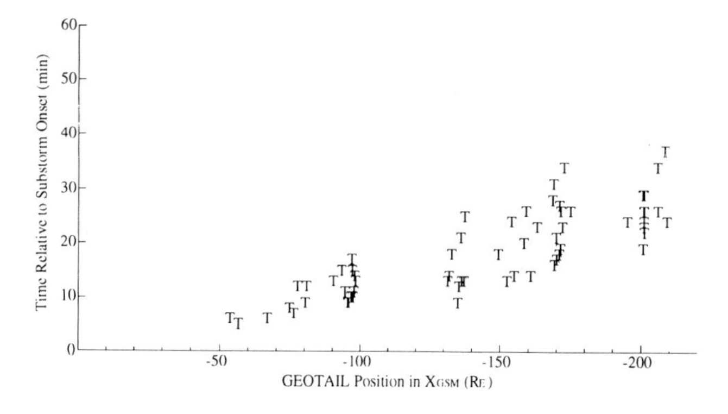

Figure 11. Delay times between substorm onset and arrival of tailward flow as a function of Geotail distance in the aberrated X_{GSM} (R_E).

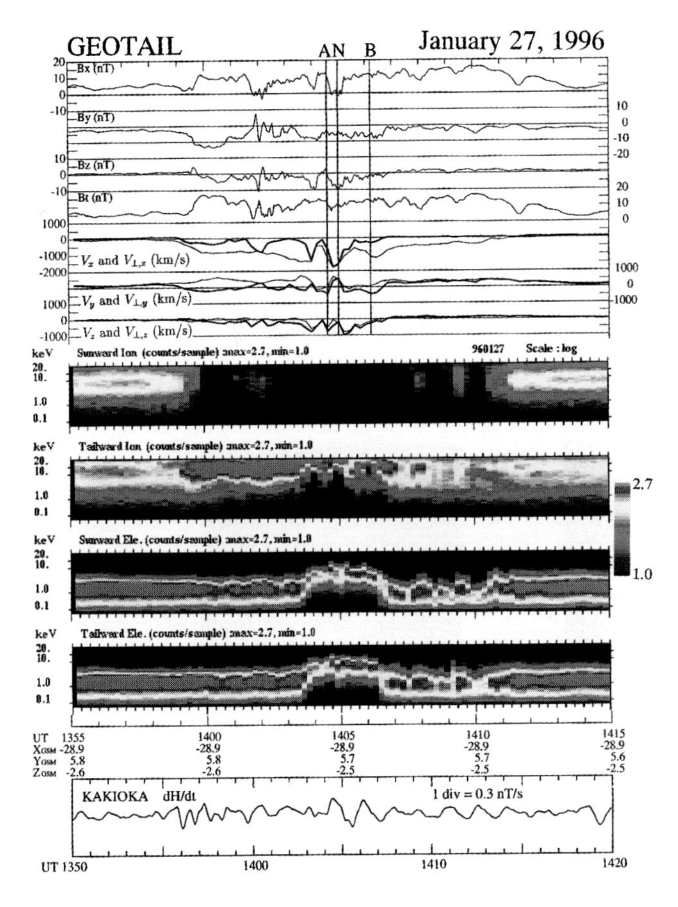

Figure 12. Magnetic field variations (B_x, B_y, B_z, and B_t), plasma velocity variations (V_x, V_y, and V_z are shown by thin curves, while $V_{\perp,x}$, $V_{\perp,y}$, and $V_{\perp,z}$ are shown by thick curves), and energy-time diagrams of ions and electrons for the period 1355-1415 UT on January 27, 1996. $V_{\perp,x}$ is the x component of the plasma velocity perpendicular to the magnetic field. In the energy-time diagrams, ion counts/sample and electron counts/sample are color-coded according to the logarithmic color bar from 1.0 to 2.7 at the right-hand side. Type A distribution is seen at 1404:34 UT (the vertical line labeled A), type N distribution is seen at 1404:59 UT (the vertical line labeled N), and type B distribution is seen at 1406:12 UT (the vertical line labeled B). In the bottom panel, Pi 2 pulsation activity at Kakioka is presented for the period 1350-1420 UT.

or after 1407 UT. Therefore, these electrons undergo heating and/or acceleration.

Panels (b) and (e) of Figure 13 present a type A distribution seen in the 12-s sample period beginning at 1404:34 UT. During this period, B_x decreased almost monotonically from 10 nT to 0 nT, and B_z decreased from 0 nT to -5 nT (in the 1/16-s time resolution data). The B direction is roughly tailward and the C direction is toward the neutral sheet. Ions have a field-aligned component and a convection component. The field-aligned component shows a fast tailward flow almost along the field line, and its energy seems to be beyond the instrument's energy limit (40 keV). The convection component is distributed in a wide range in the +C direction. The convection speed is high in type N distribution near the neutral sheet (small B_x), but low in type B distribution (which is presented below) near the boundary region (large B_x). Presumably, the high convection speed part is sampled in the small-B_x region, whereas the slow convection speed part is sampled in the large-B_x region. The temperature of the convection component increases with the convection velocity. The electron distribution is an almost "flat-top" distribution, indicating that electrons are highly accelerated. An important feature is an excess of high-energy (> 10-keV) electrons in the -B direction, which means tailward escaping of high-energy electrons.

Panels (c) and (f) of Figure 13 present a type B distribution taken in a 12-s sample period beginning at 1406:12 UT. In this period, although the magnetic field is inclined slightly southward, it is almost earthward. Ions have two components: a field-aligned component and a convection component. The field-aligned component forms a fast tailward flow, whereas the convection component is mostly concentrated in one position of the C axis, and its speed is not high. The electrons show a new feature. Tailward escaping of high-energy (10-keV) electrons is evident as the population in the -B direction. There is a peak in the electron distribution in the +B direction, near V_B = +32,000 km/s. This peak indicates that medium-energy (3-keV) electrons form a field-aligned beam in the direction opposite to that of ion flows. Therefore, high-energy electrons stream tailward, whereas medium-energy electrons stream earthward.

Type N distribution was seen at 1403:57, 1404:46, and 1404:59 UT, when B_x was small. Type A distribution was seen at 1403:45, 1404:10, 1404:34, and 1405:11 UT, just prior to and just after type N distribution. Type B distribution was seen in the period 1405:23 to 1406:24 UT, when the magnetic field is almost earthward.

5.2. Evolution of Tailward/Earthward Flows

To study the dynamics of tailward/earthward flows, we made a superposed epoch analysis using the plasma moment and magnetic field data in the magnetotail at X_{GSM} = -10 to -40 R_E. We examined the data for the time interval from -8 min to +10 min relative to onset that was determined with Pi 2 pulsations. The average flow pattern of V_x was obtained in each 2-min interval and presented in the X_{GSM} - ion plasma β plane. The distance from the neutral sheet is considered to be measured well by the ion plasma β value rather than the spacecraft Z_{GSM} value. The results are given in Figure 14. Earthward flow patterns (V_x) and tailward flow patterns (-V_x) are presented separately to easily distinguish the earthward flows from the tailward flows. We did not discriminate the data in the northern hemisphere from those in the southern hemisphere, so that the results are

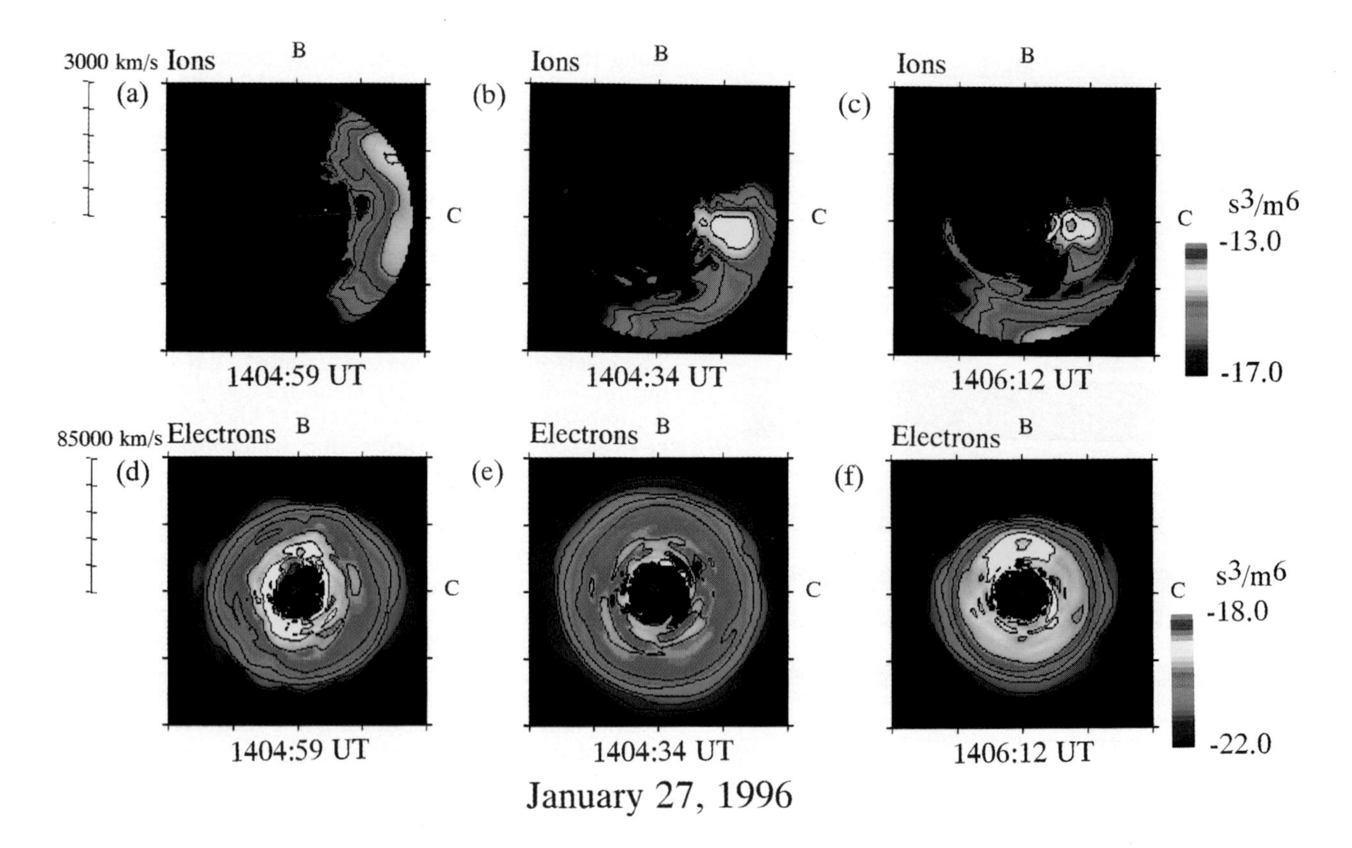

Figure 13. Ion and electron velocity distributions observed in the period 1404-1406 UT on January 27, 1996. The distribution functions are presented in the BCE coordinate system (B is the magnetic field direction, C is the convection velocity direction, and E is the assumed electric field direction). Ion phase space densities from -17.0 to -13.0 s^3/m^6 are color-coded according to the color bar, and electron phase space densities from -22.0 to -18.0 s^3/m^6 are color-coded according to the color bar.

symmetric relative to the equatorial plane (ion plasma β value = 100).

Earthward flows are seen near X_{GSM} = -40 R_E in the time interval beginning at -8 min. These flows exist in the region $0.1 < \beta < 3$, which corresponds to the plasma sheet/tail lobe boundary, and they are mostly field-aligned. An extended analysis covering the tail up to X_{GSM} = -200 R_E shows that these flows are seen even before -8 min, and they appear to originate in the distant neutral line beyond 100 R_E [see *Nishida et al.*, 1996]. It is not clear, at this stage, whether these flows are a signature of the growth phase or remnants of preceding substorm activity.

Enhancement of tailward flows start near X_{GSM} = -25 R_E in the time interval beginning at -2 min. The tailward flows originate in the high-β (10 to 100) region. As time proceeds, the tailward flow region extends to the lower-β region of the down tail and tailward flow speed increases. Earthward flows are enhanced in the time interval beginning at -2 min, as seen mostly in the low-β (around 1) region. It is evident that the earthward flows are less pronounced than the tailward flows.

6. SUMMARY AND DISCUSSION

6.1. *Magnetic Reconnection Site*

One of the important issues in substorm physics is where magnetic reconnection begins for substorm onsets. An answer based on the Geotail results is that magnetic reconnection most likely starts somewhere in the premidnight region of the plasma sheet between X_{GSM} = -22 and -30 R_E. In the substorm analysis, tailward flows are observed at X_{GSM} < -16 R_E, but only three tailward flows are seen at X_{GSM} > -22 R_E. Earthward flows are observed only at X_{GSM} > -30 R_E. No earthward flows are observed at X_{GSM} < -30 R_E in the expansion phase, and tailward flows are observed for all substorms in the central tail at X_{GSM} < -30 R_E. Near X_{GSM} = -22 R_E, there are two earthward flows and three tailward flows. Furthermore, three reversal events occur at X_{GSM} = -22 to -30 R_E. In the flow analysis, the frequency of occurrence of tailward flows is small at X_{GSM} > -22 R_E, and it increases with distance from the Earth between X_{GSM} = -22 and -

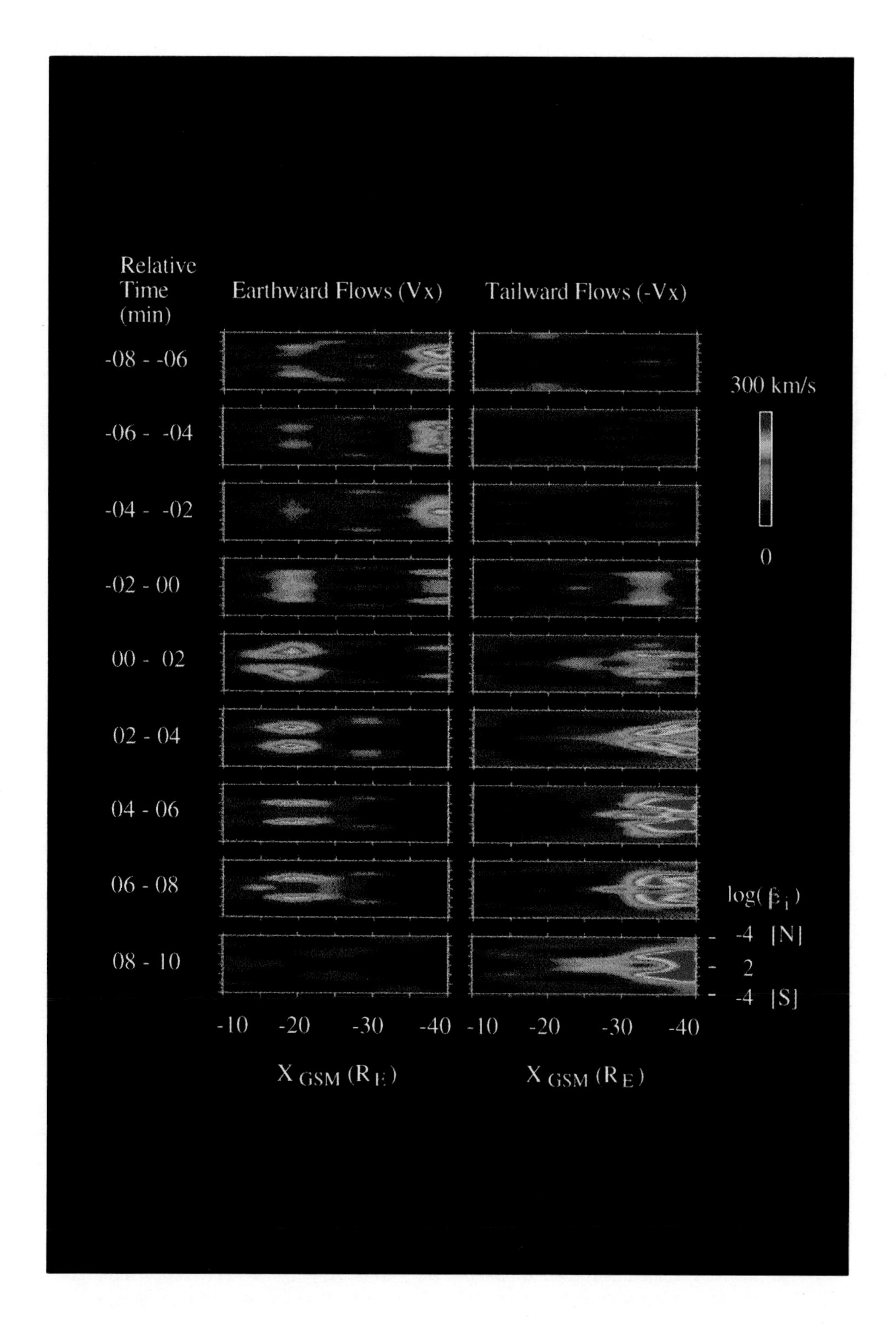

Figure 14. Evolution of earthward flows and tailward flows in the plasma sheet at X_{GSM} = -10 to -40 R_E. Ion plasma β is used for the vertical axis value, indicating distance from the neutral sheet. Zero epoch is substorm onset time.

30 R_E. It appears to be saturated near X_{GSM} = -30 R_E. The concentration of fast convection flows in the premidnight sector is evident in both substorm analysis and flow analysis. In the onset analysis, flow events in which tailward flows start prior to the ground onset time are concentrated in the premidnight region at X_{GSM} = -22 to -30 R_E. The start of tailward flows is delayed for a few minutes at X_{GSM} < -30 R_E. Hence, the three analyses assign the same premidnight region at X_{GSM} = -22 to -30 R_E for magnetic reconnection in association with substorm onsets.

In early observations with VELA satellites [*Hones et al.*, 1971], magnetic reconnection was thought to form at 15-20 R_E. *Hones and Schindler* [1979] examined flows in association with substorm onsets with Imp 6 and Imp 8 at X_{GSM} = -18 to -35 R_E and found that 80% of substorm-associated flows were tailward and 20% earthward. *Cattell and Mozer* [1984] examined **E** × **B** flows with the ISEE 1 satellite (with an apogee at 22.3 R_E) and concluded that magnetic reconnection usually formed tailward of X_{GSM} = -20 R_E. *Baumjohann et al.* [1989; 1990] examined flow characteristics inside 18.7 R_E with the IRM satellite and found that fast flows were earthward in this region. *Angelopoulos et al.* [1994] made an extensive survey of flows with the IRM satellite and the ISEE satellite, and confirmed the results of *Cattell and Mozer* [1984] and *Baumjohann et al.* [1989; 1990]. These past observations, therefore, indicate that the inner boundary for the magnetic reconnection site is near X_{GSM} = -20 R_E. There were no conclusive analyses of the outer boundary for the magnetic reconnection site, because there were few observations beyond 30 R_E. The work of *Hones and Schindler* [1979] suggested that the outer boundary is near 30 R_E, since earthward flows were rare beyond that point. Hence, the present results are fairly consistent with those of the past except for the results from the VELA satellites. The VELA results are doubted by *Lyons and Huang* [1994] because of the low time resolution of those measurements.

There are, however, a non-negligible number of fast tailward flows inside 22 R_E in the ISEE observations [e.g., *Cattell and Mozer,* 1984; *Angelopoulos et al.*, 1994]. The average Kp value was 2_0 in the present Geotail observations. The average Kp value was 3_- in the ISEE tail survey period (January-June 1978 and January-June 1979). The low frequency of occurrence in the near-Earth region seen with Geotail might be an effect of geomagnetic activity levels or solar wind conditions.

The concentration of the convection flows is evident in our analyses. The dawn-dusk asymmetry of the magnetic reconnection site has not been addressed in past studies. *Hones* [1979] indicated that magnetic reconnection took place at Y_{SM} = -12 to +12 R_E; however, no rate of occurrence was reported inside this dawn-dusk extent. At synchronous orbit, signatures of substorm onset start in a longitudinally limited region that is centered near the 23 MLT meridian [*Nagai,* 1982; 1991]. The present results are consistent with near-Earth onset signatures. In the substorm analysis, fast nonconvection flows tend to be observed in the postmidnight sector, and in the flow analysis, they are frequently observed near the midnight meridian and east of it. Convection flows are easily observed near the neutral sheet, whereas field-aligned flows are easily observed near the plasma sheet/tail lobe region. It is possible that the difference in flow characteristics is caused by that in plasma sheet dynamics during substorms. This issue should be studied further.

6.2. Start Time of Magnetic Reconnection

Nishida et al. [1981] examined the relationship between fast tailward flows with negative B_z observed by Imp 6 (with an apogee at 33 R_E) and Pi 2 pulsations. They found that Pi 2 pulsations started almost simultaneously with, or within 1 or 2 min of, the onset of fast tailward flow. *Hones et al.* [1986] reported that the fast tailward flows started a few minutes after the clear ground onset at X_{GSM} = -20 R_E for the substorm of April 24, 1979. *Sergeev et al.* [1995] observed fast tailward flows a few minutes before the distinct onset signatures on the ground in the substorm on April 15, 1979. The present study provides a number of plasma flows that start prior to ground onsets. It is reasonably concluded that magnetic reconnection starts in the limited extent of the plasma sheet prior to an onset signature as identified by Pi 2 pulsations on the ground. The observed time delays in the tail are considered to be a propagation effect and/or longitudinal expansion of the magnetic reconnection site. Since onset of a Pi 2 pulsation is a good indicator for a substorm onset, the present results provide strong evidence that various substorm phenomena originate from magnetic reconnection.

The ISEE 3 survey in the distant tail provided a variety of information on substorm dynamics. Magnetic reconnection at substorm onsets produces plasmoids/flux ropes in the plasma sheet, and they can be observed as Traveling Compression Regions (TCRs) in the tail lobes. In the plasmoid analysis, we examined tailward flows in the magnetotail at X_{GSM} = -50 to -210 R_E. In the ISEE 3 survey, the observations were done mainly near X_{GSM} = -80 R_E and beyond X_{GSM} = -200 R_E, because of the orbital configuration [e.g., *Hones et al.*, 1984; *Moldwin and Hughes,* 1993; *Slavin et al.*, 1993]. In the plasmoid analysis, we can obtain plasmoid observations in various locations between X_{GSM} = -50 R_E and X_{GSM} = -210 R_E. The good fitting of the data points with one line suggests that the substorms start in the limited X_{GSM} range and that tailward flows start almost simultaneously with ground onset. Therefore, the results of the plasmoid analysis support those obtained in the other analyses.

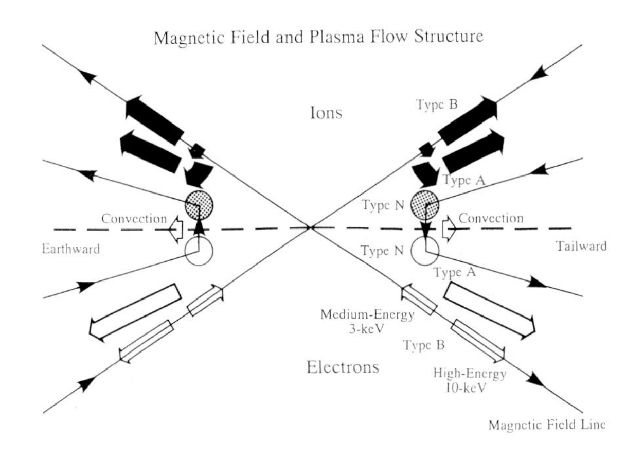

Figure 15. Structure of the magnetic field and plasma flow of magnetic reconnection based on the Geotail observations. Ion behavior is presented in the upper part, and electron behavior in the lower part.

6.3. *Structure of Magnetic Reconnection*

Recently, simulations of magnetic reconnection have been carried out with hybrid code [e.g., *Krauss-Varban and Omidi,* 1995; *Lin and Swift,* 1996; *Nakabayashi and Machida,* 1997; *Nakamura et al.,* 1998]. These simulations allow the characteristics of ion distribution functions to be studied. An important finding is that ion distributions largely depend on observed locations relative to the neutral sheet.

Figure 15 summarizes schematically the structure of the magnetic field and plasma flow in the vicinity of magnetic reconnection, on the basis of the January 27, 1996, event [*Nagai et al.,* 1998]. Near the neutral sheet, ions convect away from the X-type neutral line. High-energy isotropic electrons exist in this region. Off the neutral sheet, one ion component convects toward the neutral sheet and is heated and accelerated tailward. Another component has passed near the X-type neutral line and has been strongly accelerated and then ejected along the magnetic field line. Electrons are highly accelerated and have almost isotropic distributions; however, the highest energy electron component shows escape. Near the boundary region, ions also have a field-aligned component and a convection component, although heating and acceleration during convection become weak. An unexpected feature is two-beam structure in the electrons. High-energy electrons stream away from the neutral line, while medium-energy electrons stream into the neutral line region. Since the outflow velocity of electrons is significantly larger than that of ions, a potential drop can form along the magnetic field line. Low-energy electrons can be accelerated by this potential drop and stream into the neutral line region as medium-energy electrons in order to keep charge neutrality there. Although the basic characteristics of the present observations are those expected from the simulation results, the Geotail observations certainly provide more detailed information on the magnetic reconnection process.

The earthward side of the X-type neutral line is not exactly symmetric to the tailward side of the line. If the convection electric field is constant in the plasma sheet, the flow speed, which is E/B, should become small in the near-Earth plasma sheet ($< 20\ R_E$). Furthermore, ion energies are generally high inside 15 R_E so that any plasma instrument cannot observe the whole ion population (the upper energy limit of LEP is 40 keV). The calculated moment values might not be correct. Therefore, the flow characteristics presented in Figure 14 cannot represent substorm onset signatures correctly in the near-Earth plasma sheet. Information on energetic particles should be included in the figure.

In the substorm study, the onset signatures can be detected for all substorms in the region $Y_{GSM} = 0$ to $+10\ R_E$ inside 30 R_E. The localization of the onset signatures in the premidnight sector is well established in the vicinity of synchronous orbit [e.g., *Nagai*, 1982; 1991]. In the near-Earth plasma sheet inside 20 R_E, plasma pressure is fairly high. It is likely that the effect of magnetic reconnection is transmitted earthward in the narrow "channel" located in the premidnight region of the plasma sheet. Beyond 30 R_E, the plasma pressure is lower and the magnetic field structure is not firm in the plasma sheet. It is possible that fast plasma flows generated by magnetic reconnection expand in a large area of the plasma sheet tailward side of the X-type neutral line.

7. CONCLUSIONS

The Geotail mission has added significantly to our understanding of magnetic reconnection for substorm onsets. When and where, exactly, does magnetic reconnection initiate? These questions have been answered by Geotail. We are witnessing magnetic reconnection in the near-Earth plasma sheet. The findings presented in this paper and other studies should be combined with more sophisticated and realistic simulation studies to gain a further understanding of the microphysics of magnetic reconnection. Furthermore, we must answer the important question "why does magnetic reconnection initiate?"

Acknowledgments. The digital magnetic field data from Kakioka were provided by Kakioka Magnetic Observatory. We thank K. Yumoto and S. I. Solovyev for supplying magnetic field data from the 210° magnetic meridian network stations. Other magnetic field data were supplied by WDC-C2, Kyoto University. The authors thank D. H. Fairfield and T. Mukai for informative discussion.

REFERENCES

Angelopoulos, V., C. F. Kennel, F. V. Coroniti, R. Pellat, M. G. Kivelson, R. J. Walker, C. T. Russell, W. Baumjohann, W. C. Feldman, and J. T. Gosling, Statistical characteristics of bursty bulk flow events, J. Geophys. Res., *99,* 21257-21280, 1994.

Baker, D. N., T. I. Pulkkinen, V. Angelopoulos, W. Baumjohann, and R. L. McPherron, Neutral line model of substorms: Past results and present view, J. Geophys. Res., *101,* 12975-13010, 1996.

Baumjohann, W., G. Paschmann, and C. A. Cattell, Average plasma properties in the central plasma sheet, J. Geophys. Res., *94,* 6597-6606, 1989.

Baumjohann, W., G. Paschmann, and H. Lühr, Characteristics of high-speed ion flows in the plasma sheet, J. Geophys. Res., *95,* 3801-3809, 1990.

Cattell, C. A., and F. S. Mozer, Substorm electric fields in the Earth's magnetotail, in *Magnetic Reconnection in Space and Laboratory Plasmas,* edited by E. W. Hones Jr., pp. 208-215, AGU, Washington, D. C., 1984.

Fairfield, D. H., T. Mukai, A. T. Y. Lui, C. A. Cattell, G. D. Reeves, T. Nagai, G. Rostoker, H. J. Singer, M. L. Kaiser, S. Kokubun, A. J. Lazarus, R. P. Lepping, M. Nakamura, J. T. Steinberg, K. Tsuruda, D. J. Williams, and T. Yamamoto, Geotail observations of substorm onset in the inner magnetotail, J. Geophys. Res., *193,* 103-117, 1998.

Hones, E. W., Jr., Transient phenomena in the magnetotail and their relation to substorms, *Space Sci. Rev., 23,* 393-410, 1979.

Hones, E. W., Jr., and K. Schindler, Magnetotail plasma flow during substorms: A survey with Imp 6 and Imp 8 satellites, J. Geophys. Res., *84,* 7155-7169, 1979.

Hones, E. W., Jr., J. R. Asbridge, and S. J. Bame, Time variations in the magnetotail plasma sheet at 18 R_E determined from concurrent observations by a pair of Vela satellites, J. Geophys. Res., *76,* 4402, 1971.

Hones, E. W., Jr., D. N. Baker, S. J. Bame, W. C. Feldman, J. T. Gosling, D. J. McComas, R. D. Zwickl, J. A. Slavin, E. J. Smith, and B. T. Tsurutani, Structure of the magnetotail at 220 R_E and its response to geomagnetic activity, Geophys. Res. Lett., *11,* 5-7, 1984.

Hones, E. W., Jr., T. A. Fritz, J. Birn, J. Cooney, and S. J. Bame, Detailed observations of the plasma sheet during a substorm on April 24, 1979, J. Geophys. Res., *91,* 6845-6859, 1986.

Krauss-Varban, D., and N. Omidi, Large-scale hybrid simulations of the magnetotail during reconnection, Geophys. Res. Lett., *22,* 3271-3274, 1995.

Kokubun, S., T. Yamamoto, M. H. Acuña, K. Hayashi, K. Shiokawa, and H. Kawano, The GEOTAIL magnetic field experiment, *J. Geomagn. Geoelectr., 46,* 7-21, 1994.

Lin, Y., and D. W. Swift, A two-dimensional hybrid simulation of the magnetotail reconnection layer, J. Geophys. Res., *101,* 19859-19870, 1996.

Lyons, L. R., and C. Y. Huang, Plasma sheet expansion at r = 15-20 R_E: A recovery phase or expansion phase phenomenon? J. Geophys. Res., *99,* 10995-11004, 1994.

McPherron, R. L., Physical processes producing magnetospheric substorms and magnetic storms, in *Geomagnetism Volume 4,* edited by J. A. Jacobs, pp. 593-739, Academic Press, London, 1991.

Moldwin, M. B., and W. J. Hughes, Geomagnetic substorm association of plasmoids, J. Geophys. Res., *98,* 81-88, 1993.

Mukai, T., S. Machida, Y. Saito, M. Hirahara, T. Terasawa, N. Kaya, T. Obara, M. Ejiri, and A. Nishida, The low energy particle (LEP) experiment onboard the GEOTAIL satellite, *J. Geomagn. Geoelectr., 46,* 669-692, 1994.

Nagai, T., Observed magnetic substorm signatures at synchronous altitude, J. Geophys. Res., 87, 4405-4417, 1982.

Nagai, T., An empirical model of substorm-related magnetic field variations at synchronous orbit, in *Magnetospheric Substorms,* edited by J. R. Kan, T. Potemra, S. Kokubun, and T. Iijima, pp. 91-95, AGU, Washington, D. C., 1991.

Nagai, T., M. Fujimoto, Y. Saito, S. Machida, T. Terasawa, R. Nakamura, T. Yamamoto, T. Mukai, A. Nishida, and S. Kokubun, Structure and dynamics of magnetic reconnection for substorm onsets with Geotail observations, J. Geophys. Res., *103,* 4419-4440, 1998.

Nakabayashi, J., and S. Machida, Electromagnetic hybrid-code simulation of magnetic reconnection: Velocity distribution functions of accelerated ions, Geophys. Res. Lett., *24,* 1339-1342, 1997.

Nakamura, M. S., M. Fujimoto, and K. Maezawa, Ion dynamics and resultant velocity space distributions in the course of magnetic reconnection, J. Geophys. Res., *103,* 4531-4546, 1998.

Nishida, A., Reconnection in Earth's magnetotail: An overview, in *Magnetic Reconnection in Space and Laboratory Plasmas,* edited by E. W. Hones Jr., pp. 159-167, AGU, Washington, D. C., 1984.

Nishida, A., The GEOTAIL mission, Geophys. Res. Lett., *21,* 2871-2873, 1994.

Nishida, A., H. Hayakawa, and E. W. Hones Jr., Observed signatures of reconnection in the magnetotail, J. Geophys. Res., *86,* 1422-1436, 1981.

Nishida, A., T. Mukai, T. Yamamoto, Y. Saito, and S. Kokubun, Magnetotail convection in geomagnetically active times 1. Distance to the neutral lines, *J. Geomagn. Geoelectr., 48,* 489-501, 1996.

Sergeev, V. A., V. Angelopoulos, D. G. Mitchell, and C. T. Russell, In situ observations of magnetotail reconnection prior to the onset of a small substorm, J. Geophys. Res., *100,* 19121-19133, 1995.

Slavin, J. A., M. F. Smith, E. L. Mazur, D. N. Baker, E. W. Hones Jr., T. Iyemori, and E. W. Greenstadt, ISEE 3 observations of traveling compression regions in the Earth's magnetotail, J. Geophys. Res., *98,* 15425-15446, 1993.

Yumoto, K., and the 210° MM magnetic observation group, The 210° magnetic meridian network project, *J. Geomagn. Geoelectr., 48,* 1297-1309, 1996.

Tsugunobu Nagai, Department of Earth and Planetary Sciences, Tokyo Institute of Technology, Tokyo 152-8551, Japan. (e-mail: nagai@geo.titech.ac.jp)

Shinobu Machida, Department of Geophysics, Kyoto University, Kyoto 606-8502, Japan. (e-mail: machida@kugi.kyoto-u.ac.jp)

Traveling Compressions Regions

James A. Slavin

NASA/GSFC, Laboratory for Extraterrestrial Physics, Greenbelt, Maryland

Traveling compression regions (TCRs) are short duration, ~1-3 min, small amplitude, $\Delta B/B \sim 10\%$, perturbations of the lobe magnetic field caused by the downstream passage of plasmoids. Their characteristic signature, a north-then-south tilting of the lobe field centered upon a smooth compression, was first noted in the Explorer 35 observations about a decade prior to the first definitive observations of plasmoids by ISEE 3. Since that time, TCRs have been studied extensively using observations from the ISEE 3, IMP 8, Galileo and GEOTAIL missions. In addition to providing a unique example of magnetic field draping in a low β plasma (i.e., tail lobes), TCRs have been used very successfully to determine the number and size of plasmoids being ejected down the tail during substorms and to infer the time of near-Earth neutral line formation relative to substorm onset. In this article we review the previous experimental and theoretical investigations of traveling compression regions and discuss new research directions and applications involving this space physics phenomenon.

INTRODUCTION

A very interesting set of magnetic field observations analyzed by Maezawa (1975) is displayed in Figure 1. They consist of magnetic field measurements taken by Explorers 33 and 35 and ground-based stations about 30 years ago on September 20, 1967. These 4 hrs of simultaneous observations are plotted relative to substorm onset which was determined from the positive bay in the H component of the magnetic field observed at Guam shown in the bottom panel. This onset identification, marked with a solid vertical line, is supported by a rapid decrease in the auroral AL index which commenced at about the same time. The top panel plots the B_z component of the interplanetary magnetic field (IMF) as observed by Explorer 33. As shown, the IMF had a significant southward component for over an hour prior to the substorm onset. Finally, the magnetic field in the north lobe of the tail at $X \sim -60$ R_e was monitored by Explorer 35 with the magnitude and latitude angle displayed in the middle panels.

New Perspectives on the Earth's Magnetotail
Geophysical Monograph 105

Published in 1998 by the American Geophysical Union

Maezawa (1975) was primarily interested in the loading of magnetic flux into the tail lobes prior to expansion phase onset and its unloading afterwards as a source of energy to power substorms. This is a corner stone of the near-earth neutral line (NENL) model (see review by Baker et al., 1996). The build-up in lobe magnetic flux prior to onset is evident in Figure 1 from the increase in lobe field intensity in the second panel and the increased flare angle of the field lines as indicated by the latitude angle of the lobe field in the third panel. After substorm onset the process reverses with the lobe field strength decreasing and the elevation angle tending toward zero. Finally, about 40 min after onset, the unloading of the tail is further signaled by the decrease in the diameter of the tail which leaves Explorer 35 in the magnetosheath; the magnetopause being indicated by a vertical dashed line.

What is of interest to us here was Maezawa's observation of a several min long increase in lobe field strength about 10 min following substorm onset in the second panel. Examination of the third panel indicates that it was accompanied by a north-then-south variation in the latitude angle of the lobe field. Maezawa stated in his paper (italics added for emphasis) "...just after onset the field

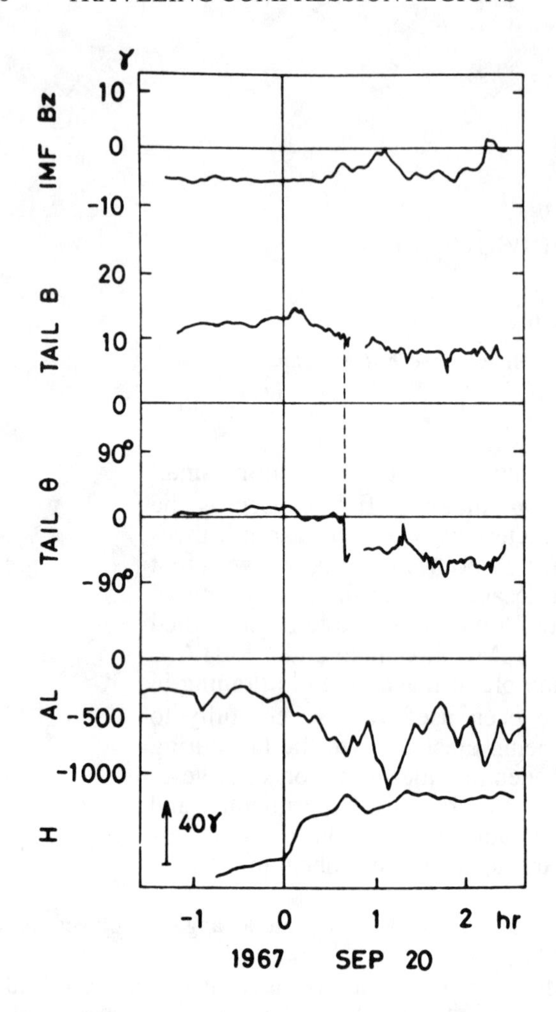

Figure 1. IMF B_z observations from Explorer 33 (top panel), tail lobe magnetic field magnitude and polar angle at Explorer 35 (middle panels) and the auroral AL index and the H component of the magnetic field measured at a mid-latitude station (bottom panel) are displayed for 4 hr interval on September 20, 1967 (adapted from Maezawa, 1975).

magnitude was further increased…by a few gammas…*such a small hump in the field is often seen immediately after expansion phase onsets*…therefore it is not of solar wind origin. A possible interpretation is *that the tail field lines are temporarily compressed from inside* as…hot plasma streams through the plasma sheet and down the tail…". At the time of the Maezawa (1975) work, the foundations of the modern NENL model of substorms were just being laid by McPherron et al. (1973), Schindler (1974), Hones (1977) and others. Central to this class of model is the formation of an x-type neutral line in the near-Earth (i.e., X > -30 R_e) plasma sheet where reconnection effectively disconnects the downstream plasma sheet and leads to its being ejected downtail. The flux tubes comprising the north and south lobes of the tail, no longer being separated by the plasma sheet, then come together and reconnect. It is this reconnection of the open flux tubes in the lobes which liberates stored magnetic energy to power the substorm expansion phase (e.g., Baker et al., 1996). The disconnected downstream plasma sheet, which is ejected near the time of substorm onset, is termed a "plasmoid" and it rather closely resembles Maezawa's "streaming of hot plasma" down the tail.

Slavin et al. (1984) re-discovered Maezawa's "humps" while analyzing the ISEE 3 distant tail magnetic field measurements and came to very similar conclusions concerning their origin. Keying on their observation that the time delay between substorm onset and the arrival of these perturbations at ISEE 3 was proportional to the downtail distance of the spacecraft, Slavin et al. termed these features "traveling compression regions", or TCRs, because they appeared to travel down the tail. Furthermore, they noted the close correspondence between the durations of TCRs and plasmoids, their very similar downtail propagation speeds and their close associations with substorm activity and concluded that TCRs and plasmoids are causally linked. Figure 2, taken from Slavin et al. (1984), depicts the draping and compression of field lines about a plasmoid to form the TCR. Although indepently derived, this is effectively the same interpretation as that of Maezawa (1975), but up-dated with the source of the compression being identified as a plasmoid.

Since the initial investigations using the Explorer 33/35 and ISEE 3 observations, TCRs have been studied in the

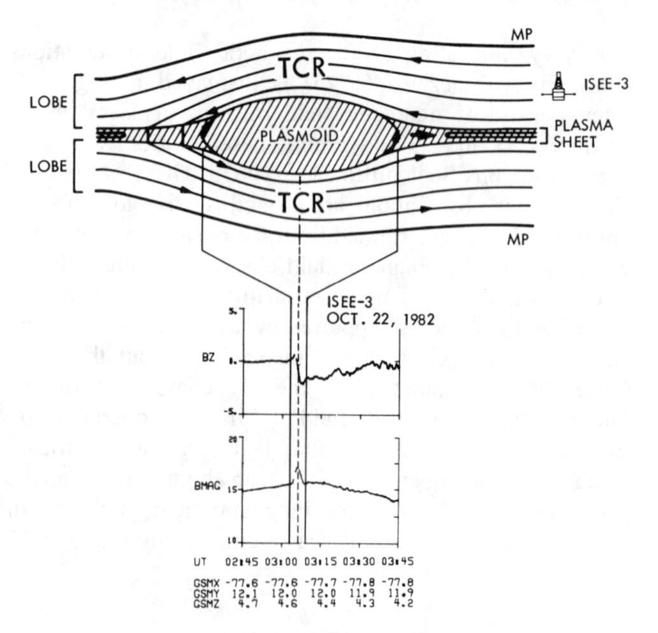

Figure 2. A schematic view of a traveling compression region is presented along with actual magnetic field magnitude and B_z observed by ISEE 3 during a TCR on October 22, 1982 (adapted from Slavin et al., 1984).

magnetic field measurements returned by the IMP 8 (Taguchi et al., 1996; 1997a,b,c; Moldwin and Hughes, 1994; Slavin et al., 1990), Galileo (Kivelson et al., 1993) and GEOTAIL (Kawano et al., 1994; Nagai et al., 1994) missions. As will be discussed in the sections to follow, the interpretation that these field signatures are caused by the draping of lobe flux tubes about plasmoids moving rapidly tailward has been supported by numerous subsequent studies. Other measurements such as energetic particle anisotropies and electric fields (Kawano et al., 1994; Owen and Slavin, 1992; Murphy et al., 1987) as well as theoretical modeling studies (Walker and Ogino, 1996; Raeder, 1994) have confirmed the draping model. In addition, two spacecraft studies by Moldwin and Hughes (1992a) and Slavin et al. (1998) have directly observed both the plasmoid and its accompanying compression region, albeit at different downtail distances. Finally, the TCR signature has been used as a readily available proxy for the more localized plasmoid signature in studies of substorm dynamics and energetics and, more recently, to study the time of NENL formation relative to substorm onset (Taguchi et al., 1996; 1997c; Moldwin and Hughes, 1994; Slavin et al., 1993).

MAGNETIC FIELD SIGNATURE

An example of a TCR in the ISEE 3 magnetic field measurements at X = -79 R_e taken on April 11, 1983 is displayed in Figure 3a (Courtesy of the ISEE 3 Magnetometer P.I. – E.J. Smith). Nearly centered on 03:00 UT is a ~ 2 min long compression of the lobe magnetic field with an amplitude of ~1 nT. The lobe field is largely oriented in the X direction so that the B_x and B_{total} perturbations are essentially the same. The next largest variation is found in the B_z signature. Here we see the defining northward-then-southward tilting of the field with the inflection point near the peak in the field magnitude. However, as the observations are being made Earthward of the distance where flaring usually ceases, X ~ -100 to –120 Re (Kokubun et al., 1996; Yamamoto et al., 1994; Slavin et al., 1985) there is still some flaring present and B_z is biased slightly negative. Hence, when one refers to "north-then-southward tilting" which is "symmetric", these terms are used in a relative sense with respect to the "baseline" or mean B_z level. Following the north/south B_z tilting, this component gradually recovers toward zero over the next 6-8 min.

For this event there is little B_y variation although B_y perturbations with amplitudes of ~30–60% that of the B_z component are common (Slavin et al., 1993). The B_y perturbation is thought to arise from the draping of the lobe field lines about a 3-D plasmoid bulge which is localized in its east–west extent (Slavin et al., 1989). When the spacecraft passes directly over the centerline of the

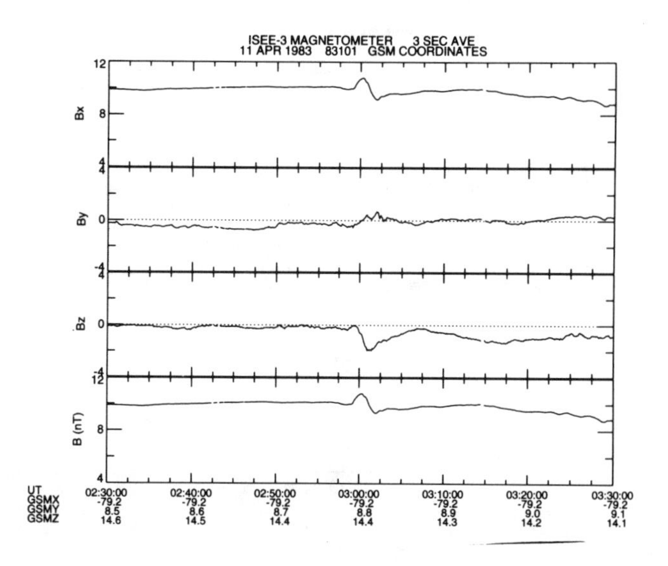

Figure 3a. A close-up of a typical TCR in the ISEE 3 magnetic field observations (GSM coordinates).

plasmoid, or the plasmoid has a very large extent in the east–west direction, then there should be little or no B_y signature. However, when the plasmoid bulge is localized in its Y extent and passes to the east or west of the spacecraft, a significant east–west field line tilting would be expected whose sense should tell us on which side of the spacecraft the plasmoid passed (e.g., Slavin et al., 1989). Simultaneous observations of both the plasmoid and TCR by an array of spacecraft in the tail would, of course, be necessary to demonstrate this in a definitive manner, but overall this draping model of TCRs is well supported by a large body of indirect and statistical evidence.

Figure 3b shows the TCR from the previous figure in principal axis coordinates (Sonnerup and Cahill, 1967). As is apparent from the large ratios of the maximum to intermediate and intermediate to minimum variance eigenvalues, the principal axes are very well defined for this event. The constant field component in the minimum variance direction indicates that the TCR field perturbation is essentially two-dimensional. It consists of a simple "rocking" in the B2-B1 plane which, from inspection of the eigenvectors, is very close to the GSE X – Z plane. The lack of any B3 variation also strongly suggests that this magnetic field perturbation is not associated with a more complex magnetic structure. In the case of a flux ropes, for example, a planar magnetic field signature is only possible if the spacecraft trajectory intersects the central axis of the structure on a largely radial chord (e.g., see diagrams in Moldwin and Hughes, 1991; Slavin et al., 1989). From previous studies we know that "wind-sock" variations in tail orientation with changing solar wind flow direction produces great variations in the spacecraft trajectories

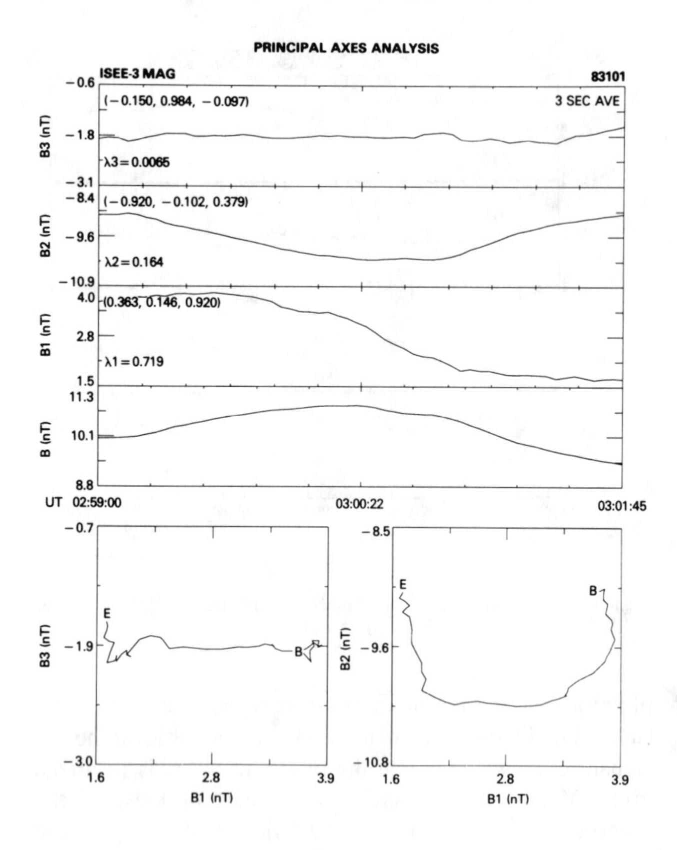

Figure 3b. The same TCR as in the previous figure, but displayed in principal axes coordinates. The directions of the maximum, intermediate and minimum variance axes (B1, B2, B3) in GSE coordinates are displayed in the upper left-hand corner of each panel. The associated eigen vectors are shown in the lower left-hand portion of each panel. Hodograms of the field variations are plotted in the bottom panels.

relative to the central axes of these structures from event to event (e.g., Slavin et al., 1995; Moldwin and Hughes, 1992). For TCRs with larger B_y values the perturbation remains two dimensional, but the normal to the B1 – B2 plane (i.e., the B3 direction) in which the field line rotation takes place will tilt in the + or – Z direction. Additional case studies and statistical studies of principal axis eigenvectors for TCRs can be found in Slavin et al. (1993; 1989; 1984).

An example of a TCR observed by IMP 8 much closer to the Earth on March 10, 1994 is displayed in Figure 4 (Courtesy of the IMP 8 Magnetometer P.I. - R.P. Lepping). In this instance the spacecraft was at $X = -31\ R_e$, $Y = -5\ R_e$, and $Z = 7\ R_e$. The difference in location between IMP 8 and a spacecraft in the trans-lunar tail is important for several reasons. First, a spacecraft in the distant tail is most likely observing a TCR associated with a mature plasmoid moving tailward as opposed to detecting a compression which is caused by a nascent plasmoid growing in the nearby plasma sheet just prior to or just following its ejection. Hence, at IMP 8 lobe compressions associated with plasmoids may reflect not only the downtail motion of the plasmoid, but also its temporal development as the plasmoid bulge grows in the plasma sheet beneath the observing spacecraft (Taguchi et al., 1997c). Second, the tail is flaring far more strongly at IMP 8 distances as compared with most ISEE 3 and GEOTAIL events. The superposition of strong tail flaring and the plasmoid induced compression can produce some subtle, but significant modifications of the simpler TCR signatures seen in the more distant tail as shown by Taguchi et al. (1997a,b).

No magnetospheric plasma measurements were available as late in the IMP 8 mission as March 10, 1994, but the strong B_x component and the low level of the fluctuations clearly indicate that IMP 8 was in the north lobe of the magnetotail during this event. The TCR is identified by the several minute long compression in the field magnitude centered on 01:49 UT, the weak northward then southward variation in B_z coincident with the field maximum, and, finally, the extended interval of depressed B_z afterwards. Again, the tilting of the magnetic field is not truly bipolar in B_z because the strong flaring of the field in the lobes causes a several nT negative shift in the background magnetic field.

The B_z component of the TCR perturbation in Figure 4 is not symmetric and quasi-sinusoidal as observed in Figure 3a. Instead, only a vestigial northward tilting is observed just prior to its well-defined southward excursion. It is this asymmetry that Slavin et al. (1990) and Taguchi et al. (1996) have attributed to lobe flux tube draping over a plasmoid forming in the adjacent plasma sheet so that the draping of the lobe field over the anti-sunward end of the plasmoid cannot be observed. The extended interval of enhanced southward B_z (i.e., below the baseline southward tilting due to flaring) following the compression region signature is, again, thought to be caused by the reconnection of lobe field lines following the release of the plasmoid.

Another interesting study of TCRs (and plasmoids) utilizing the IMP 8 magnetic field measurements was carried out by Moldwin and Hughes (1994). In addition to identifying 37 TCRs similar to the event displayed in Figure 4, they found 19 lobe compression regions with south-then-north (SN) tilting of B_z. These SN events appeared to be somewhat smaller in duration and amplitude than their north-then-south (NS) counterparts and they were associated with weak, very high latitude geomagnetic activity as opposed to the strong correlations with substorm expansion phase onset found for the NS TCRs. Two hypotheses for the origin of these SN TCRs have been considered. First, unlike the much more common north-then-south TCRs, the SN perturbations cannot be interpreted in a unique manner because they can also be produced externally by regions of enhanced solar wind pressure as they convect past the magnetosphere (see Fig. 2 in Slavin et al., 1993). Indeed, Moldwin and Hughes did find that about half of their SN events were associated with transient increases in the H component of the geomagnetic field at low latitude ground stations consistent with the occurrence of brief solar wind pressure enhancements. Their favored interpretation, however, was that the SN TCRs might be caused by "proto-plasmoids" forming in the cis-lunar plasma sheet and being convected Earthward until such a time as a substorm occurs and ejects them tailward as part of a fully formed plasmoid. This interesting and, if correct, important result represents another good example of how lobe compressions can be used to "sound" for bulges in the plasma sheet relating to its internal dynamics.

SPATIAL DISTRIBUTION OF TCRS

Another interesting aspect of TCRs is their rate of occurrence as a function of distance down the tail. As shown in Figure 5 (from Slavin et al., 1993), the number of TCRs per unit time is not independent of downtail distance. Rather, it increases until reaching a broad maximum at X ~ -60 to -130 R_e. The gradual increase is consistent with plasmoids having lengths of several tens of Earth Radii and generally forming sunward of X ~ -100 R_e (Ieda et al., 1997; Moldwin and Hughes, 1992c). For this reason, most examples of TCRs in the literature are not from the very distant tail, e.g., X ~ -100 to –200 R_e, but rather at ~ -60 to –130 R_e where they are more frequent. As an aside, it should be noted that while tail flaring is becoming weak at these distances, it is not absent and most examples of TCR B_z variations are biased slightly negative, as in Figure 3a, due to residual tail flaring.

Following the maximum near X ~ -100 R_e, TCR occurrence frequency then falls by about a factor of ~3 between X ~ -140 and –240 R_e. No corresponding decrease in the rate of plasmoid occurrence has been observed (Ieda et al., 1997; Moldwin and Hughes, 1992c). It was suggested by Slavin et al. (1993) that some of the observed decrease may be due to the obscuring of TCR signatures by frequent, large amplitude tail motion and a growth in plasma sheet boundary layer (PSBL) wave amplitudes at large downstream distances. However, this explanation is difficult to test quantitatively and may, at best, be only a partial explanation. As will be discussed in a later section, a second explanation has been put forward involving an evolution in the nature of the plasmoid – lobe interaction as conditions in the tail lobes change with increasing distance.

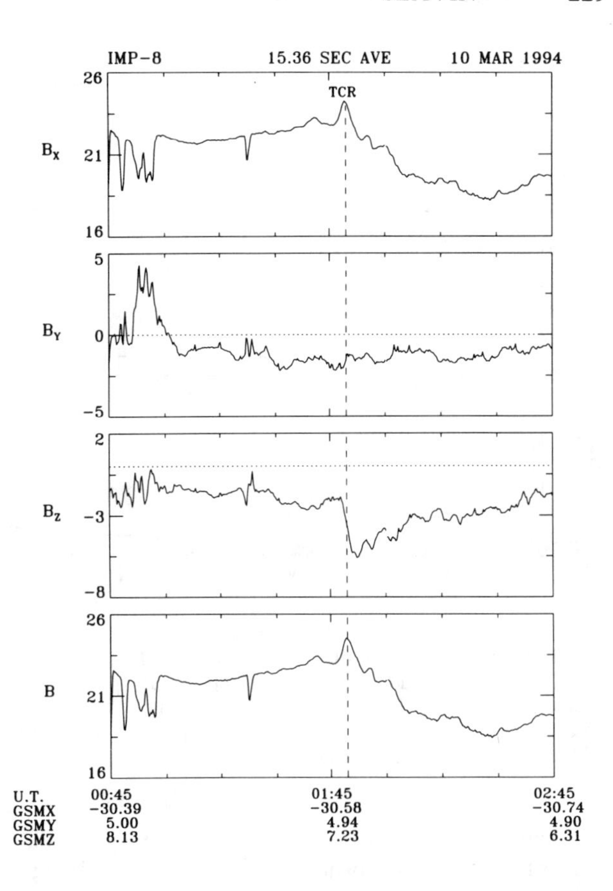

Figure 4. Two hours of IMP 8 magnetic field measurements (15.36 sec averages) taken on March 10, 1994 are displayed in GSM coordinates. Note the correlated field compression and north-south field perturbation centered on 01:50 UT which constitute the traveling compression region.

PLASMOIDS AND TCRS

The magnetic field and plasma measurements made in vicinity of plasmoids are known to be heavily influenced by the path taken by the spacecraft relative to the center of the plasmoid (Moldwin and Hughes, 1992b; Slavin et al., 1989). At one extreme, a TCR is observed if the spacecraft never encounters the plasmoid directly. At the opposite extreme, very strong "core" magnetic fields are sometimes observed when a spacecraft passes near the center of plasmoids (e.g., Machida et al., 1994). This latter result is consistent with at least some plasmoid possessing flux rope-type magnetic field topologies (Ieda et al., 1997; Slavin et al., 1995; Lepping et al., 1995; Moldwin and Hughes, 1991). Recently, progress has been made with respect to inferring the relative trajectory of spacecraft through plasmoids by assuming a specific force-free magnetic field topology and then determining the spacecraft trajectory which yields the best fit between the measured magnetic field and the model (Kivelson and

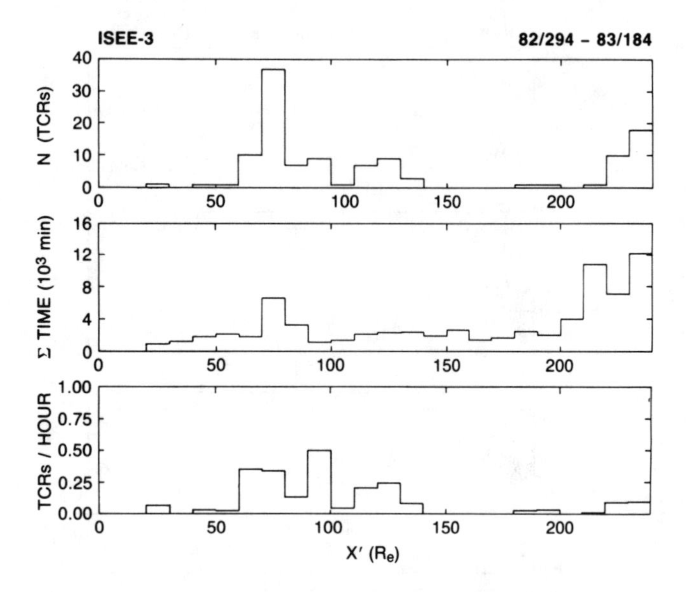

Figure 5. This histogram displays the number of TCRs observed by ISEE 3 (top), the amount of time the spacecraft spent in the tail lobes (middle), and the ratio of these two quantities (bottom) all as a function X', of the solar wind aberrated GSM X coordinate of the spacecraft (adapted from Slavin et al., 1993).

Khurana, 1995). While such an approach is very useful, the lack of a model independent means of determining the trajectory of the spacecraft through plasmoids with arbitrary magnetic structure is still one of the factors which greatly complicate the analysis and modeling of plasmoids and the traveling compression in the surrounding lobe region.

The observation of plasmoids and TCRs in close association with one another has also been offered as additional evidence for the former being the cause of the latter (e.g., Slavin et al., 1989). An example of 2 plasmoids and 3 TCRs observed by GEOTAIL during the space of only 2 hours, i.e., 18:00 to 20:00 UT on September 26, 1994 (Courtesy of the LEP and MGF P.I.s, T. Mukai and S. Kokubun, respectively) is presented in Figure 6. At this time GEOTAIL was located near the nominal, aberrated east-west center of the tail at X = -169 R_e. The first plasmoid around 18:15 UT apparently passed well below Geotail which was located in the north lobe of the tail. Accordingly, a TCR was observed with the characteristic several minute long compression and a north-then-south tilting of the field with no significant signature in the plasma measurements. Note, that for this example of a TCR in the rather distant tail the B_z perturbation is not biased negatively as is the case in the presence of tail flaring, but slightly positive probably due to some windsock southward pointing of the tail due to the natural variations in solar wind flow direction.

Following a brief neutral sheet encounter, a several minute long rotation in the B_z component of the magnetic field centered on 18:36 UT occurred coincident with the appearance of high β plasma sheet-type plasma flowing tailward at a peak speed of 1,000 km/s. The high plasma ion β, >10, indicates that GEOTAIL penetrated deeply into this plasmoid, while the weak magnetic field and irregular B_y signature suggest that it did not possess a flux rope-like structure. While tailward plasma flow had been observed since GEOTAIL entered the PSBL/plasma sheet around 18:30 UT, the tailward flow speed peaks at ~ 1000 km/s in the center of the plasmoid near the center of the B_z rotation. A second, weaker TCR is evident at 18:59 UT. This event is qualitatively the same as the first TCR, but much smaller in amplitude and shorter in duration presumably due to a smaller plasmoid with less height in the north-south direction and less length in the X direction than the first TCR.

Around 19:30 UT another clear, large amplitude bipolar B_z signature is observed. The small decrease in total field strength and modest increases in ion temperature and

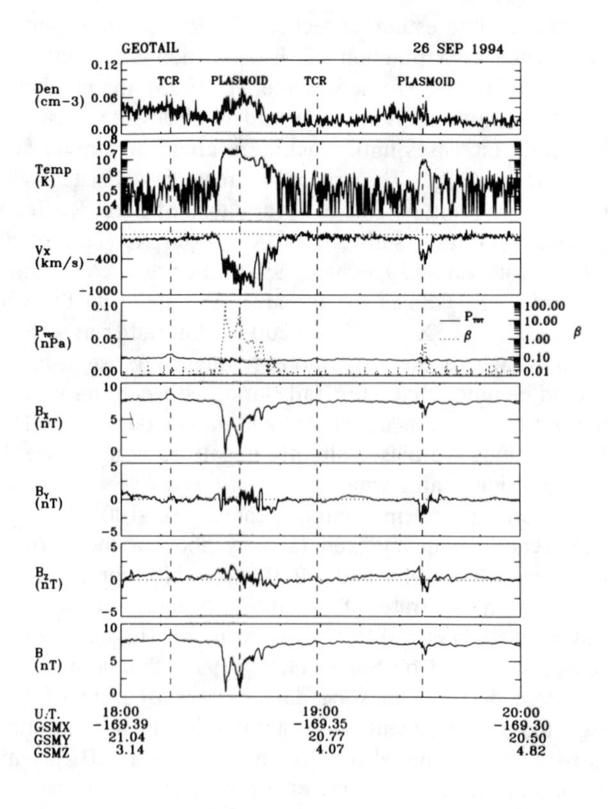

Figure 6. Two hours of Geotail magnetic field and plasma parameters (24 sec averages) taken on September 26, 1994 are displayed in GSM coordinates. As indicated a series of TCRs and plasmoids were observed during this interval with the last plasmoid being observed while GEOTAIL was located in the plasma sheet boundary layer.

density accompanying the bipolar B_z signature identify this as a plasmoid encounter for which the spacecraft did not penetrate deeper than the plasma sheet boundary layer which overlays the northern and southern faces of the plasma sheet and, hence, plasmoids. Interestingly, the B_y magnetic field signature is most pronounced for this event and, hence, suggests more of a helical-type magnetic structure than the 18:36 UT plasmoid despite the grazing trajectory of GEOTAIL. Moldwin and Hughes (1992b) have called attention to such events and discussed the variation in plasmoid signatures as a function of the height of the spacecraft trajectory relative to the center of the plasmoid.

Finally, there is a very weak TCR signature at 19:39 UT which has an amplitude of only ~0.2 nT. Statistical studies such as Slavin et al. (1993) have required that the compression signature be no smaller than 1% of the background lobe field in order for it to be identified as a TCR in a reliable and reproducible manner. This compression signature just exceeds this threshold. Given their modest amplitudes, it is clear that many TCRs, especially the weaker ones such as the last event in Figure 6 are easily overlooked in the data unless care is taken in the analysis process.

Examination of the B_z perturbations in Figure 6 nicely demonstrates the comparability of the temporal durations of the TCRs and plasmoids. Measured from the maximum $+B_z$ to the minimum $-B_z$, they have temporal durations of ~2 to 4 min in agreement with previous surveys. If the underlying plasmoids associated with all of the events were moving down the tail with the ~800 km/s suggested by the LEP plasma measurements taken during the 18:36 UT event, then their lengths in X are ~15 to 30 R_e in reasonable agreement with previous surveys (e.g., Ieda et al., 1997).

The amplitudes of the TCRs, ~2 to 10%, are very typical and imply, via magnetic flux conservation (see Slavin et al., 1993), a total plasmoid height in the $\pm$Z direction of approximately ~ 8 to 15 R_e. Although dependent upon modeling assumptions regarding tail diameter, flux content, and the shape of the plasmoid bulge in the Y-Z plane, the amplitudes of the TCR perturbations are presently our most direct source of information on the height of plasmoids in the Z direction.

A better method for deducing the height of plasmoids may be to directly measure and integrate the V_z perturbation associated with the TCR which corresponds to the speed with which the plasmoid bulge displaces lobe flux tubes first outward and then inward as it moves down the tail. In the TCR studies utilizing the ISEE 3 data, it has been noted that no significant variations were observed in the plasma electron measurements (e.g., Slavin et al., 1984), but detailed analyses were never presented. The GEOTAIL plasma measurements displayed in Figure 6 show very little variation during the TCR events unless the plasmoid bulge is large enough to push the PSBL over the spacecraft. Kawano et al. (1994), however, took a different approach in that they used electric and magnetic field observations to infer the V_z component of local plasma convection speed during a series of TCR events at X = -78 R_e. In each case Kawano et al. observed the expected north-then-south perturbations in V_z coincident with the TCRs as the underlying plasmoids displaced and compressed the lobe flux tubes in its path. The amplitudes of these north-then-south velocity perturbations were several hundred km/s consistent with the typical plasmoid Z dimensions and durations inferred from analysis of the magnetic field perturbations alone.

ENERGETIC PARTICLE SIGNATURES

Whereas plasma measurements have been used primarily to separate TCRs from boundary layer plasmoids, observations of more energetic ions, > 30 keV, have been used to remotely sense the movement of the plasma sheet as it is displaced outward toward spacecraft in the lobe by the passage of plasmoids and to determine the location of reconnection X-lines relative to the underlying plasmoid.

Murphy et al. (1987) first analyzed and modeled the arrival time and direction of energetic ions of differing energies, and, hence, gyro-radius, during some selected TCR events. These analyses confirmed that TCR signatures in the lobe magnetic field are indeed coincident with outward bulging of the plasma sheet and PSBL in the manner originally suggested by Maezawa (1975) and Slavin et al. (1984). A typical example of the energetic particle measurements taken by ISEE 3 during TCR events is presented in Figure 7a from a study by Owen and Slavin (1992). The magnetic field measurements in the bottom panels indicate the presence of 3 TCRs marked with vertical dashed lines. As indicated at the bottom of the figure, ISEE 3 was located at a downstream distance of X = -78 R_e. The top panel displays the spin averaged fluxes in the 3 lowest ion energy channels, 35-56, 56-91, and 91-147 keV. As shown, only the first TCR has associated with it a clear enhancement in the energetic ions. The second panel separates the ion flux in the lowest energy channel into the ions moving tailward (solid line), Earthward (dashed line) and perpendicular to the X axis (dotted line).

Overall, the Owen and Slavin analysis of the EPAS energetic ion observations revealed that about half of the TCRs in ISEE 3 data set had associated with them energetic ion enhancements. The TCR population with energetic ion enhancements was similar in duration to those lacking the enhancements, but the former possessed a mean amplitude approximately twice that of the latter. This result is consistent with the energetic ions being largely

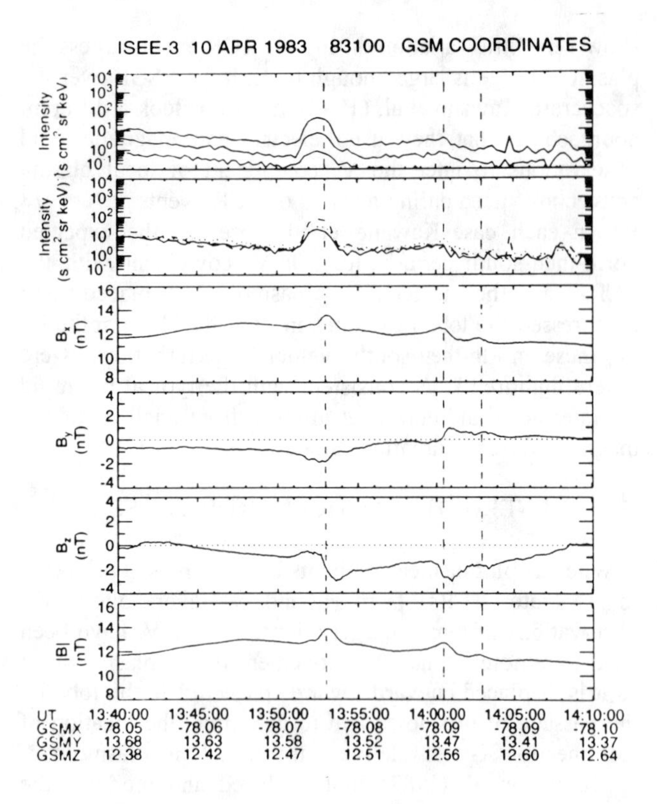

Figure 7a. ISEE 3 energetic particle and magnetic field measurements for a multiple TCR event on April 10, 1983. As described in the text, the top panels provide the energetic ion flux in various energy channels and look directions. The bottom panels display the magnetic field in GSM coordinates (adapted from Owen and Slavin, 1992).

confined to a PSBL with finite width in the ± Z direction which is more likely to be displaced past the spacecraft as the height of the plasmoid in Z and, hence, the amplitude of the TCR increases.

The anisotropy of the energetic ions was also examined. Owen and Slavin found that nearly all of the TCR associated energetic ion enhancements observed at X < -100 R_e possessed tailward anisotropies. This was interpreted as evidence that beyond ~-100 R_e plasmoids are essentially fully formed, disconnected from the Earth and in the process of being ejected. At X > -100 R_e, approximately half of the TCR associated energetic ion enhancements possessed no clear anisotropy while the majority of the remaining events exhibited Earthward anisotropy. They interpreted this result as a further indication that Earthward of X ~ –100 R_e plasmoids are still under the influence of reconnection both at the NENL and a "distant" neutral line. Further examples of Earthward PSBL flows just prior to plasmoid encounters and the role of the distant neutral line in plasmoid formation can be found in Mukai et al. (1996) and Slavin et al. (1998).

Figure 7b, again taken from Owen and Slavin (1992), summarizes their results in a schematic manner. They found that for about half of the TCRs the spacecraft was too deep in the lobes, i.e. trajectory A-A', to penetrate into the PSBL even when the plasmoid was directly beneath the spacecraft. The other half of the events corresponded to measurements taken along trajectories similar to B-B' which does intersect the PSBL. The nature of the anisotropies in the PSBL is determined by the location of the spacecraft relative NENL and the distant neutral line and their respective reconnection rates. Hence, just as we discussed with respect to the TCR and plasmoid events shown in Figure 6, the spacecraft trajectory in the frame of the plasmoid is a major factor in determining the nature of the returned measurements.

MULTI-SPACECRAFT OBSERVATIONS

Two spacecraft studies of plasmoid downtail motion between IMP 8 and ISEE 3 or GEOTAIL in the distant tail have been conducted by Moldwin and Hughes (1992b) and Slavin et al. (1998). In both studies the observation of a TCR at IMP 8 was used as the "start time" and the arrival of a plasmoid at ISEE 3 or GEOTAIL as the "stop time" for time-of-flight (TOF) determinations of average plasmoid speed. The TOF plasmoid speeds of 200 to 700 km/s determined by these two studies are in good agreement with the larger statistical studies.

One of the IMP 8/Geotail events analyzed by Slavin et al. (1998) occurred on April 16, 1994 and is shown in Figure 8. It displays 90 min of simultaneous, 3 sec averaged magnetic field observations from IMP 8 at X = -29 R_e and Geotail at X = -197 R_e. The low variance, largely X-directed magnetic fields in the lower panels indicate that IMP 8 was located in the south lobe of the tail

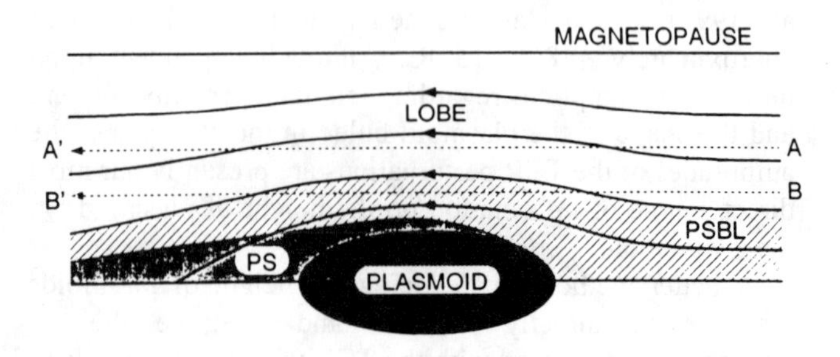

Figure 7b. A schematic view of two different spacecraft trajectories in the rest frame of a tailward moving plasmoid. Along path A the spacecraft never comes close enough to the plasmoid to enter the PSBL and observe the energetic particles populating this region. In the case of path B, however, the energetic particles in the PSBL are briefly observed in the immediate vicinity of the plasmoid (adapted from Owen and Slavin, 1992).

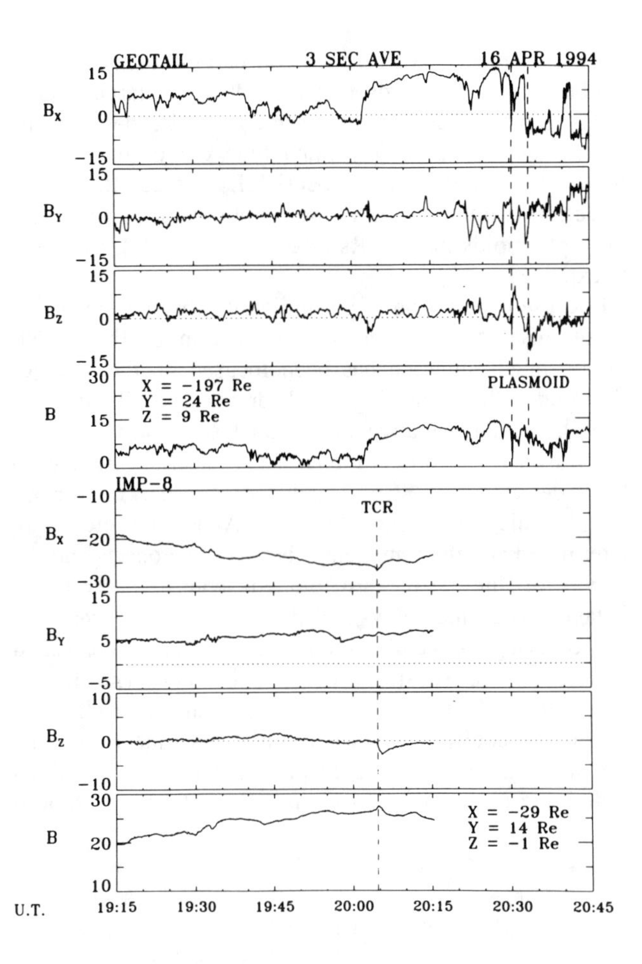

Figure 8. Simultaneous 3 sec averaged magnetic field measurements taken by GEOTAIL and IMP 8 from 19:15 to 20:45 UT on April 16, 1994 are displayed in GSM coordinates. For this event the plasmoid at GEOTAIL follows by 27 min the observation of the TCR at IMP 8 (adapted from Slavin et al., 1998).

throughout this interval. At 20:05 UT a TCR is clearly visible in the IMP 8 measurements as a brief field compression accompanied by a sharp southward tilting in B_z followed by a slower recovery. The lack of a significant B_y variation suggests that the spacecraft passed directly over the plasmoid as opposed to being off to the east or west (e.g., Slavin et al., 1989) or that the plasmoid extended across much of the tail (e.g., Walker and Ogino, 1996). During the 50 min preceding the TCR the lobe field intensity at IMP 8 steadily increased by about 25% in a very typical substorm growth phase. The TCR coincides with the broad maximum in lobe field strength which is seen to decrease afterwards.

The GEOTAIL magnetic field measurements in the top panels of Figure 8 show the spacecraft first residing in the plasma sheet, where the field is weak and variable, and then the PSBL or north lobe where the field is much stronger. The most salient feature in the GEOTAIL measurements is the plasmoid centered on ~20:32 UT. It is indicated by the large amplitude north-then-south B_z variation accompanied, in contrast with the TCR in the lower panels, by a weakening of the field strength as the high β plasmoid envelopes the spacecraft. Note also the slow recovery of the southward B_z field following the plasmoid due to the reconnection of lobe flux tubes following plasmoid ejection. The time-of-flight speed for this plasmoid from IMP 8 to GEOTAIL was 168 R_e/27 min ~ 664 km/s. Given this speed and the ~3.5 min duration of the plasmoid, the estimated length of the plasmoid was ~22 R_e. The multi-spacecraft events analyzed by Moldwin and Hughes and Slavin et al. have shown 1) that the TCRs at IMP 8 are indeed associated with the ejection of plasmoids down the tail, 2) that, at least in some cases, the measured bulk speeds in the distant tail are significantly larger than the time-of-flight speeds suggesting some additional acceleration after their initial release, and 3) that plasmoid release appeared to have taken place at or within a few minutes of substorm expansion phase onset.

SUBSTORM ASSOCIATION

Beginning with the initial studies by Maezawa (1975) and Slavin et al. (1984) clear substorm associations have been found for most TCR events. Figure 9 displays a "best typical case" from the ISEE 3 database for December 25, 1982. In this example the tail lobe magnetic field at X = -56 R_e shows a strong build-up of flux during a ~30 min growth phase. Pi2 events were first detected in the AFGL mid-latitude magnetometer chain beginning at 07:46 UT (W.J. Hughes, private communication, 1995) about 5 min before the first TCR and 6 min before the AL index responds to the substorm onset. The peak in lobe field intensity at 07:51 UT coincided with the observation of the first of 5 TCRs (indicated by the solid arrow heads) which were observed as the tail lobe field intensity decreased to levels similar to those observed prior to the growth phase.

This "Christmas Day" event displays many of the TCR substorm associations which have been reported in other studies. 1) For events observed Earthward of about X = -100 R_e, where magnetopause flaring ceases, there is a clear tendency for the lobe field strength to increase prior to TCRs and decrease afterwards (Slavin et al., 1984). 2) There is a delay between the time of substorm expansion phase onset and TCR arrival which increases from ~5-10 min at the orbit of the moon (i.e., X = -60 R_e) to about 30 min at X = -200 R_e (Nagai et al., 1994; Slavin et al., 1993). 3) Substorms, especially more intense ones, frequently produce multiple TCRs during their expansion phase (Slavin et al., 1993).

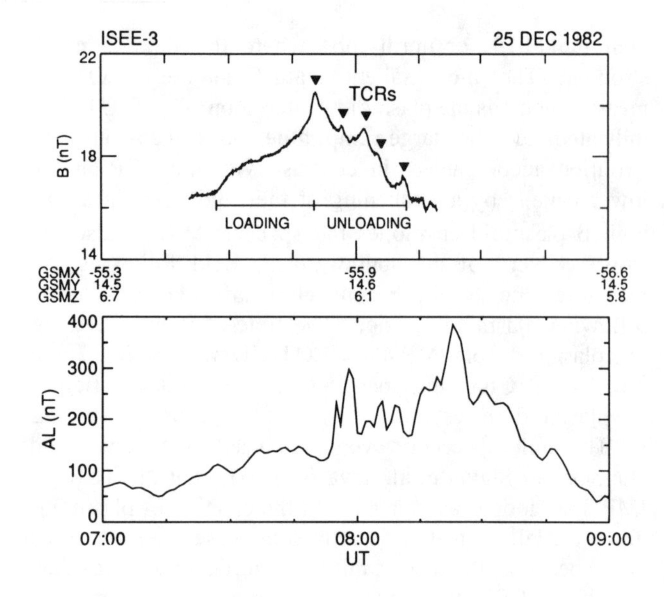

Figure 9. A particularly clear example of a tail lobe "loading" with magnetic flux prior to substorm onset followed by the release of a series of plasmoids, inferred from the presence of TCRs in the lobe regions at ISEE 3, and then a decrease in the lobe field strength as the tail "unloads".

The very close association of TCRs and plasmoids with substorm activity is demonstrated further in Figure 10 taken from Slavin et al. (1992). During this unusual 36 hr interval on April 9-11, 1983 a total of 12 substorms were observed in the AL index (bottom panel), auroral kilometric radiation measured at ISEE 3 (panel 8), energetic electrons measured at geosynchronous orbit (panels 5-7), and Pi2 pulsation events measured at Memambetsu (Japan) and Wingst (Germany) (arrows beneath panel 6). The individual substorms are numbered in the bottom panel. The ISEE 3 magnetic field observations displayed in the top panels of Figure 10 were taken in the tail at downstream distances of X = -76 to 80 R_e. Except for a few strong dips in field magnitude, signifying entries into the plasma sheet or plasmoids, the spacecraft was located in the north lobe of the tail most of the time. Associated with nearly all of the substorms is the characteristic loading/unloading signature in the field magnitude. Solid and dashed vertical lines mark the locations of TCRs and plasmoids, respectively, determined from inspection of the higher time resolution plasma and field data. In cases where more than 2 plasmoids or TCRs were released during a given substorm, only the first and the last are marked with vertical lines. These observations show that all 12 substorms produced plasmoids either observed directly or through the TCRs they produced in the lobes. A more comprehensive analysis of the individual substorms and plasmoid events on this day with an emphasis on the ISEE 3 energetic particle measurements can be found in Richardson et al. (1996). The large ratio of TCRs to plasmoids is common in the ISEE 3 data set due to its trajectory spending much time out of the ecliptic plane. In contrast, the GEOTAIL spacecraft, whose trajectory is much closer to the ecliptic, appears to see far more plasmoids than TCRs (Nagai et al., 1994), as would be expected.

Because the ejection of plasmoids at substorm expansion phase onset is a fundamental prediction of the NENL model, much attention has been focused on the issue of plasmoid/TCR timing. Figure 11, drawn from Slavin et al. (1993), is representative of the analyses using TCRs as proxies for direct observations of the plasmoids. For each TCR detected at ISEE 3 substorm onsets in various substorm diagnostics (i.e., AL index, AKR, Pi2s, etc.) were determined and the time delays between the onsets and the arrival of the TCRs are plotted against downstream distance. The time delays and the downstream distance of the observing spacecraft were found to be highly correlated for all of the substorm onset indicators. The results displayed in Figure 11 using the AL index produced a correlation coefficient of 0.8 and a best linear fit slope indicating a mean tailward speed for the TCR and its underlying plasmoid of 589 km/s. These findings are also

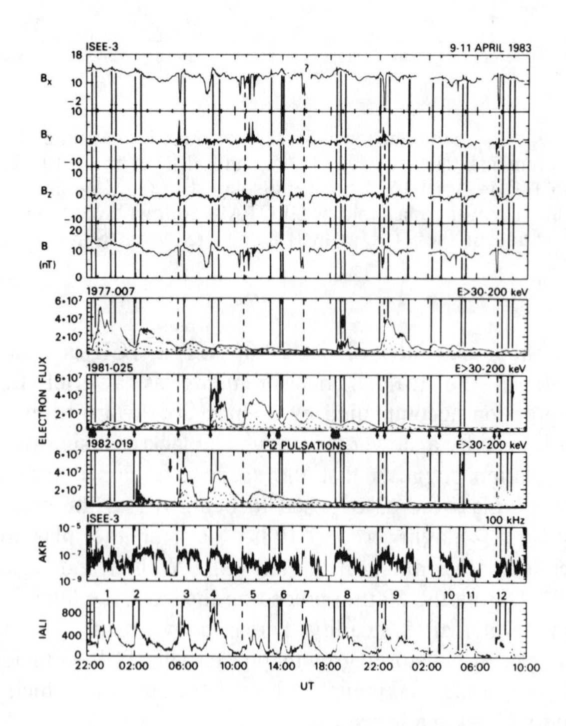

Figure 10. A 36 hr interval on April 9-11, 1983 when ISEE 3 was continuously in the tail and observing plasmoids and TCRs associated with 12 substorms (adapted from Slavin et al., 1992).

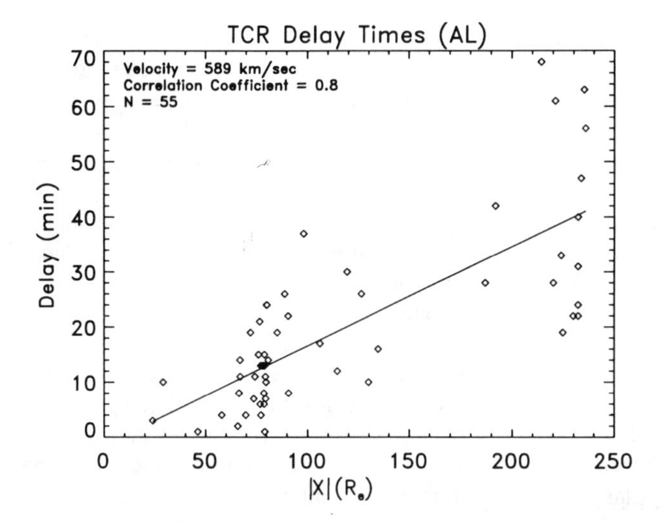

Figure 11. An example of the relationship between the downtail location of ISEE 3 and the time delay between the onset of substorm expansion phase in the AL index and the arrival of the TCR at the spacecraft. The linear regression fit displayed as a solid line indicate a mean downtail speed of 589 km/s for the plasmoids producing these TCRs (adapted from Slavin et al., 1993).

consistent with the results of similar analyses by Nagai et al. (1994) using both plasmoids and TCRs from the Geotail data base. Superposed epoch studies (Baker et al., 1987) have produced similar results demonstrating a strong tendency for plasmoids to be released near expansion phase onset.

Recently, Taguchi et al. (1996) have carried out a new study of TCR substorm associations in the middle tail using IMP 8 observations. Observed at X ~ -30 to -40 R_e, the TCRs at IMP 8 are detected in the same general region where plasmoids are formed and released. Hence, as noted earlier, the time delays associated with the downtail motion of the plasmoids should be small, ~ 1-2 min. In addition, the TCRs at IMP 8 often lack or possess only a vestigial northward B_z as was discussed earlier (see Figure 4). For this reason it has been suggested that the TCRs at IMP 8 are often produced by a plasmoids forming near the location of the spacecraft. In this case the time delays would be near zero or perhaps even slightly negative as plasmoid formation is caused by closed field reconnection which precedes open field line reconnection and the onset of substorm expansion phase. A histogram from the study by Taguchi et al. (1996) of the time differences between the substorm onset determined by Pi2s and these "B_z lacking TCRs" is displayed in Figure 12. As shown, it has a peak very near zero time delay with 74% of the events having delays of +/-2 min consistent with the interpretation that this special class of TCRs is produced by plasmoids still in the very early stages of formation/ejection. Additional studies by Taguchi and co-workers utilizing other measures of substorm onset have produced very similar results (Taguchi et al., 1997c).

PLASMOID – LOBE INTERACTION

The interaction between the tailward moving plasmoid and the flux tubes of the lobe region was treated in only the most cursory of manners by Maezawa (1975) and Slavin et al. (1984). However, it is important to remember that the draping of lobe flux tubes about the plasmoid bulge is carried out by the launching of fast mode wave fronts. This process is displayed schematically by Slavin et al. (1994) in Figure 13. These compressive waves, indicated by the arrows in the figure, radiate outward from the plasmoid and cause the plasma and embedded lobe field lines to become compressed and move aside as the plasmoid moves down the tail. When these wave fronts reach the magnetopause they cause it to move outward slightly (see Slavin et al., 1994; 1993). Complementary rarefaction waves are launched from the trailing end of the plasmoid to fill-in behind this rapidly moving body and return the tail to its previous state.

If the fast mode speed in the lobes were everywhere large compared to the downtail speed of the plasmoid, then the case depicted in the top portion of Figure 13 would always hold. In this instance the wave speed in the lobes is sufficiently high that there is time for the fast mode wave fronts to travel from the plasmoid to the magnetopause before the plasmoid escapes down that tail. In this situation

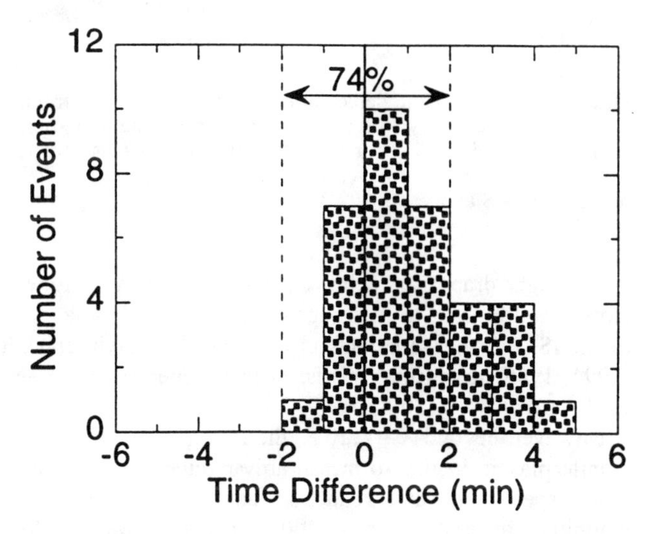

Figure 12. A histogram of the delay times between substorm onset, determined using ground-based Pi2 events, and the observation of TCRs at IMP 8 in the mid-tail (adapted from Taguchi et al., 1996).

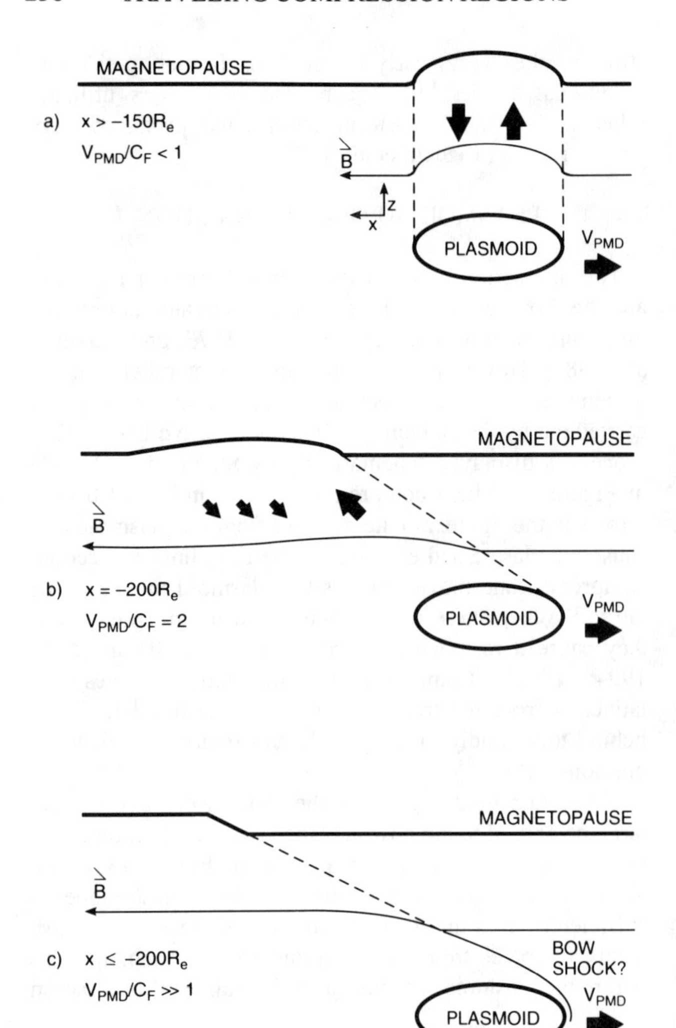

Figure 13. A schematic depiction of the fast mode wave fronts which are launched by the plasmoid to produce the lobe field compression and draping for the different ratios of plasmoid to fast mode wave speed observed at various distances down the tail (adapted from Slavin et al., 1994).

a very tight draping is achieved with the lobe flux tubes closely conforming to the shape of the plasmoid. Analysis of the ISEE 3 measurements in the lobes by Slavin et al., (1994; 1985) found that this is, in fact, generally the case earthward of X ~ -150 R_e.

However, beyond X ~ -150 R_e the filling of the lobe with mantle plasma begins to measurably reduce the local fast mode speed and that reduction continues to grow in magnitude at least out to ~ –200 R_e (Slavin et al., 1985). Initially, as the lobe fast mode speed is reduced only modestly, the primary result is a sweeping back of the region of lobe compression so that it trails the plasmoid as a result of the latter body's high relative speed. This situation is depicted in the middle panel of Figure 13. Invariably such fluid interactions can be expected to lead to some steepening at the leading edge of the compression and a spreading out of the trailing rarefaction. Eventually, beyond ~ -200 R_e, the plasmoid speed may at times exceed the fast mode speed in the lobes as they continue to fill with mantle plasma. In this case Slavin et al. speculated that an incipient bow or limb shock might form very near the leading end of the plasmoid, but only in the lobes as the plasmoid speed appears to be always sub-magnetosonic with respect to the plasma sheet plasma ahead of the plasmoid (see Slavin et al., 1989; 1985). However, even if the evolutionary conditions for the formation of such a shock were satisfied, it would be very weak and quickly decay into a "bow wave" with increasing distance from the plasmoid as shown in the bottom panel of Figure 13.

Aside from providing a unique example of flux tube draping in a very low beta plasma, this examination of the plasmoid – lobe interaction is important because of the puzzling result noted earlier that the numbers of TCRs observed per unit time spent in the lobes decreases with increasing downstream distance beyond X ~ –130 R_e. As there is no corresponding downturn in the numbers of plasmoids detected in the distant tail (Ieda et al., 1997; Moldwin and Hughes, 1992c) we can conclude that the reason is not related to the coalescence or dissipation of the plasmoids themselves. Slavin et al. (1994) suggested instead that in the very distant tail the TCR waveform is steepening and becoming distorted by the decreased fast mode speed in the lobes to the point where they are no longer being recognized as such in surveys. For this reason the region where the TCR signature is of most utility in such investigations may be limited to X > -150 R_e.

MHD SIMULATIONS

The compression of the lobes by the plasmoid bulge has been noted in a number of simulation studies (e.g., Walker and Ogino, 1996; Hesse and Birn, 1991; Ogino et al., 1990). Raeder (1994), in particular, explicitly examined the relationship between plasmoids and TCRs in the distant magnetotail during simulations of a series of substorms. In this global MHD simulation three substorms were observed to develop. The plasmoids and their accompanying TCRs were then observed by a pair of "virtual spacecraft" located in the plasma sheet and north lobe at X = -100 R_e. Figure 14, taken from Raeder (1994), displays observations from the lobe "spacecraft" with vertical lines indicating the times of substorm onset. The arrival times of these plasmoids at the plasma sheet spacecraft (not shown) were 310, 320, 335, 405, 430, 460, and 515 min into the simulation. Inspection of the magnetic field data collected by the lobe spacecraft displayed shows a clear TCR

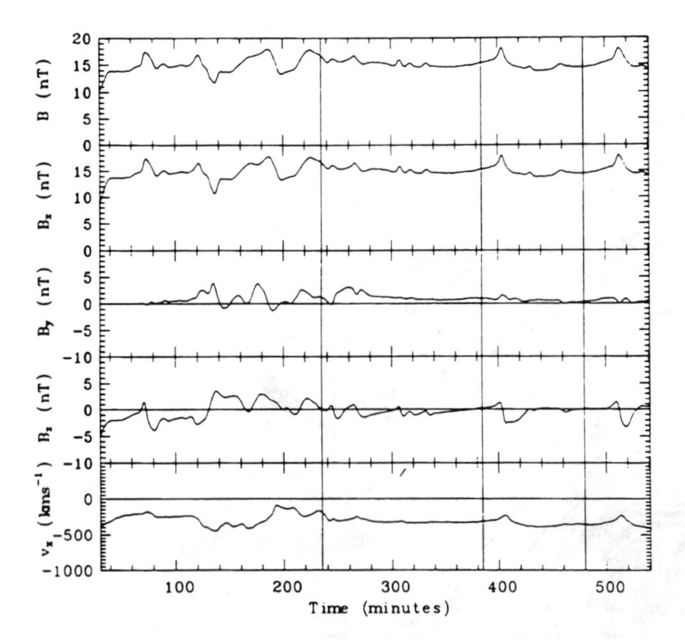

Figure 14. The lobe magnetic field at X = -100 R_e produced by a MHD simulation in which a total of 7 plasmoids were released. Solid vertical lines mark the onset of the substorms in the simulation (adapted from Raeder, 1994).

perturbation for each of these plasmoids. Furthermore, the largest plasmoids, the 405 and 515 min events, produced the largest TCR amplitudes and durations. Although there is still considerable theoretical work to be done in deconvolving TCR perturbations to yield better measures of the dimensions of the underlying plasmoids, simulations such as this one provide strong support for the relatively simple analytical analyses which has been performed on these structures in previous studies.

SUMMARY

A schematic view of plasmoid formation with the surrounding compression regions explicitly marked is presented in Figure 15. As with most such diagrams it somewhat exaggerates the overall dimensions of the plasmoid for the purposes of visibility (e.g., Baker et al., 1987; Hones et al., 1977). The diameter of the lobes have also been under-represented in the interest of producing a compact depiction of the flow fields and magnetic fields associated with plasmoid formation and ejection. Paying particular attention to aspects of the formation/ejection process observable using TCRs, the top diagram depicts the presence of a plasmoid developing underneath IMP 8 just tailward of the NENL within several min of substorm onset. In fact, recent studies by Taguchi et al. (1997c) indicate that the initial compression of the lobes by the rapidly forming plasmoid can sometimes be detected even a few min prior to expansion phase onset while reconnection at the NENL is still confined to closed flux tubes. Within 10-15 min a mature plasmoid has been ejected and is about to envelop GEOTAIL near X ~ -100 Re. Whether a plasmoid or TCR is observed will depend upon the size of the plasmoid and the relative location of the spacecraft. GEOTAIL's location in Figure 15, between the two X-lines, will result in it observing Earthward, tailward, or bi-directional energetic particle anisotropies in the PSBL depending upon the respective reconnection rates at the two X-lines. If the plasmoid is encountered, complex magnetic signatures are often observed as they attempt relax toward a force-free flux rope geometry. This is especially true in the core region where plasma β may drop below 1 as plasma escapes by means of the plasmoid's connectivity to the rest of the tail and, possibly, the magnetosheath (Hesse et al., 1996). Finally, GEOTAIL may encounter the post-plasmoid plasma sheet (PPPS). This region is composed of north and south lobe flux tubes which have reconnected at the NENL after the plasmoid was ejected. The southward magnetic field in this region is so pronounced that a southward tilting of the lobe magnetic field is produced which is clearly observed for ~15-30 min following TCRs. Note that this region is not shown to scale in Figure 15. Given that its duration can be an order of magnitude longer than that of the plasmoid, the length of the PPPS could reach 100 Re or more before it terminates with the tailward retreat of the NENL.

As described in this article, traveling compression regions are now well understood as a magnetic draping phenomenon driven by the plasmoid – lobe interaction. Studies of TCRs and their properties have been very significant with respect to: 1) confirming the high downtail speed of plasmoids, 2) establishing the approximate dimensions of plasmoids (especially their height in Z), 3) determining that the plasmoid bulge is often localized in its east – west extent, 4) demonstrating that plasmoids are typically ejected downtail at or within a few min of substorm expansion phase onset, 5) confirming that plasmoid formation and ejection is followed by a ~15 – 30 min interval of lobe field line reconnection, and 6) showing that it is common for a single substorm to produce multiple plasmoids.

As the modeling of magnetotail dynamics becomes more comprehensive in terms of the physical processes they can represent, and more sensitive with respect to spatial resolution and amplitude so that waves and other dynamic features can be detected, the TCR signature should merge naturally with the outermost layers of the plasmoid. Questions to be addressed with such simulations and modeling in the future will include 1) the time-dependent compression of the lobes in the mid-tail region

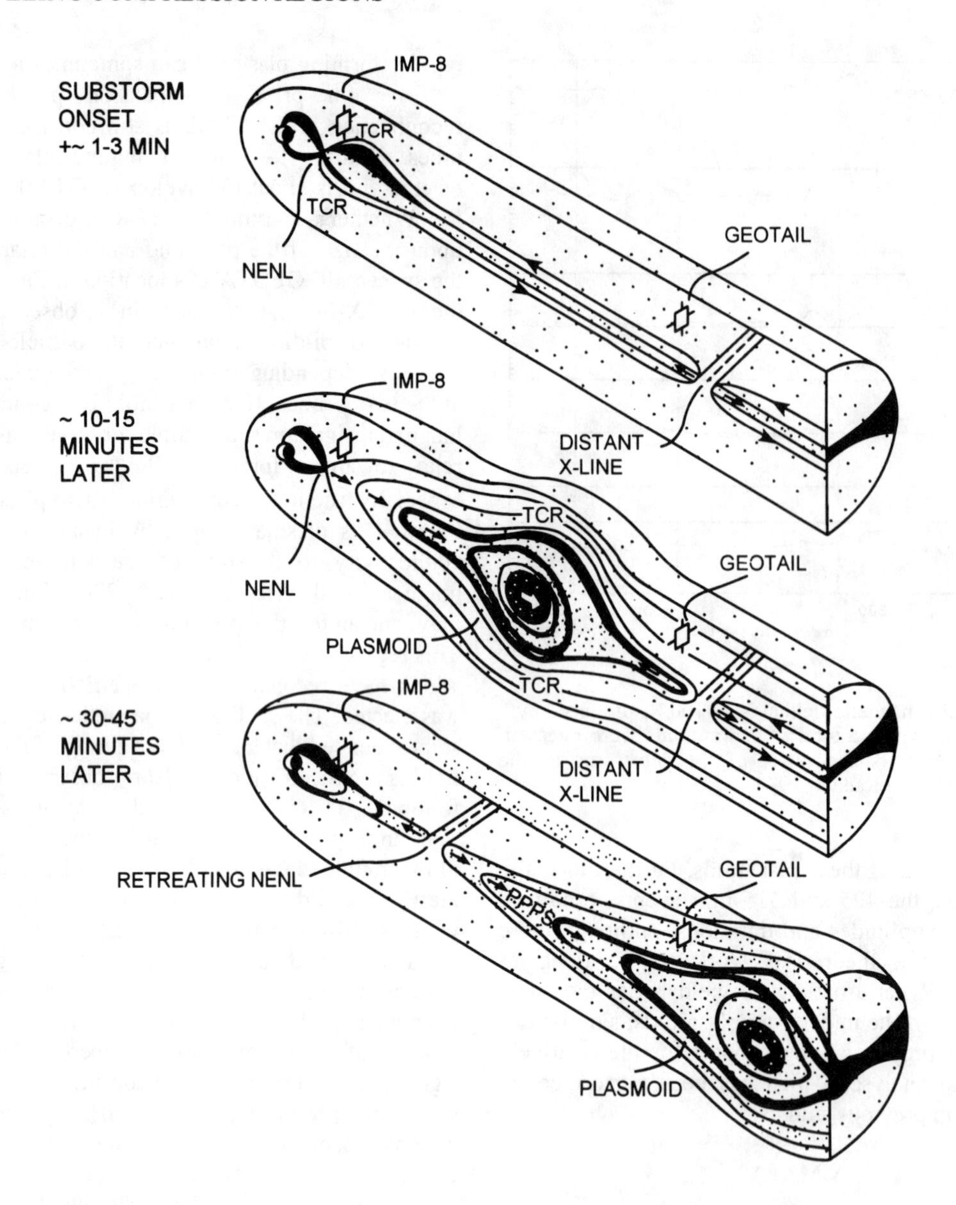

Figure 15. A schematic view of plasmoid formation and ejection as a result of the occurrence of a near-Earth neutral line inside the orbit of IMP-8.

as the plasmoid is forming and just beginning to move tailward and 2) the effect of the increasing lobe plasma content with increasing downtail distance on the plasmoid – lobe interaction in general and the possible steepening of the TCR compression signature as the lobe fast mode speed decreases.

Future studies of TCRs and their associated plasma and energetic particle signatures should offer many further opportunities for significant contributions. As knowledge of reconnection in the tail becomes more sophisticated, an increased emphasis may be placed upon the separatrix layers they produce in the PSBL. Analysis of their associated TCR signatures, when present, may provide important contextual information on the size of the underlying plasmoids and, indirectly, the rates of reconnection at the near and distant X-lines. If the V_z perturbation associated with the displacement of lobe flux tubes by plasmoids can be reliably detected with either electric field or plasma instruments, then simple integration may yield a more direct means of measuring plasmoid height than the analysis of TCR compression amplitude. In addition, the "south-then-north B_z" TCRs which have been reported at IMP 8 distances need to be re-examined using observations for which simultaneous upstream solar wind pressure measurements exist so that their origins can be attributed to external or internal causes in a definitive manner. If they are associated with Earthward convecting "proto-plasmoids" as suggested by Moldwin and Hughes (1994), then their study many lead to some significant refinements in the NENL with respect to tail dynamics during the growth phase and, possibly, psuedo-onsets. Finally, TCRs will continue to be used as proxies for plasmoids in substorm studies and as start or stop "pulses" for gauging their downtail motion and estimating plasmoid dimensions.

Acknowledgments. The author wishes to acknowledge the many colleagues who have participated and consulted in his studies of TCRs and their relationship to substorms with special thanks to D.N. Baker, J. Birn, M. Hesse, E.W. Hones, Jr., W.J. Hughes, M. Kuznetsova, M.B. Moldwin, C.J. Owen, I.G. Richardson, D.G. Sibeck, H.J. Singer, G.L. Siscoe and S. Taguchi. The author also gratefully acknowledges the use of measurements from the ISEE 3 (P.I. – E.J. Smith), IMP 8 (P.I. – R. Lepping) and GEOTAIL (P.I. - S. Kokubun) magnetometer investigations and the GEOTAIL LEP investigation (P.I. - T. Mukai). The extremely thoughtful comments contributed by both referees are also gratefully acknowledged.

REFERENCES

Baker, D.N., R. C. Anderson, R. D. Zwickl, and J. A. Slavin, Average plasma and magnetic field variations in the distant magnetotail associated with near-Earth substorm effects, *J. Geophys. Res.*, *92*, 71, 1987.

Baker, D.N., T.I. Pulkkinen, V. Agnelopoulos, W. Baumjohann, and R.L. McPherron, Neutral line model of substorms: Past results and present view, *J. Geophys. Res.*, *101,* 12, 975, 1996.

Hesse, M., and J. Birn, Plasmoid evolution in an extended magnetotail, *J. Geophys. Res.*, *91*, 5,683, 1991.

Hesse, M., J. Birn, M.M. Kuznetsova, and J. Dreher, A simple model of core field generation during plasmoid evolution, *J. Geophys. Res.*, *101*, 10,797, 1996.

Hones, E.W., Jr., The magnetotail: Its generation and dissipation, *Physics of solar planetary environments*, ed. by D.J. Williams, pp. 559-571, Washington D.C., 1976.

Hones, E.W., Jr., Substorm processes in the magnetotail: Comments on "On hot tenuous plasmas, fireballs, and boundary layers in the earth's magnetotail" by L.A. Frank et al., *J. Geophys. Res.*, *82,* 5,633, 1977.

Hones, E.W., Jr., D. N. Baker, S. J. Bame, W. C. Feldman, J. T. Gosling, D. J. McComas, R.D. Zwickl, J. A. Slavin, E. J. Smith, and B. T. Tsurutani, Structure of the magnetotail at 220 R_e and its response to geomagnetic activity, *Geophys. Res. Lett.*, *11*, 5, 1984.

Ieda, A., S. Machida, T. Mukai, Y. Saito, T. Yamamoto, A. Nishida, T. Terasawa, and S. Kokubun, Statistical analysis on the plasmoid evolution with Geotail observations, *J. Geophys. Res.*, in press, 1997.

Kawano, H., T. Yamamoto, S. Kokubun, K. Tsuruda, A.T.Y. Lui, D.J. Williams, K. Yumoto, H. Hayakawa, M. Nakamura, T. Okada, A. Matsuoka, K. Shiokawa, and A. Nishida, A flux rope followed by recurring encounters with traveling compression regions, *Geophys. Res. Lett., 21, 2,891, 1994.*

Kivelson, M.G., C.F. Kennel, R.L. McPherron, C.T. Russell, D.J. Southwood, R.J. Walker, K.K. Khurana, P.J. Coleman, C.M. Hammond, V. Angelopoulos, A.J. Lazarus, R.P. Lepping and T.J. Hughes, The Galileo Earth encounter: Magnetometer and allied measurements, *J. Geophys. Res.*, *100*, 11,299, 1993.

Kivelson, M.G., and K. Khurana, Models of flux ropes embedded in a Harris neutral sheet: Force-free solutions in low and high beta plasmas, *J. Geophys. Res.*, *100*, 23,637, 1995.

Kokubun, S., L.A. Frank, K. Hayashi, Y. Kamide, R.P. Lepping, T. Mukai, R. Nakamura, W.R. Paterson, T. Yamamoto, and K. Yumoto, Large field events in the distant magnetotail during magnetic storms, *J. Geomagn. Geoelectr.*, *48*, 561, 1996.

Lepping, R.P., D.H. Fairfield, J. Jones, L.A. Frank, W.R. Paterson, S. Kokubun, and T. Yamamoto, Cross-tail magnetic flux ropes as observed by the Geotail spacecraft, *Geophys. Res. Lett.*, *22,* 1,193, 1995.

Machida, S., et al., Geotail low energy particle and magnetic field observations of a plasmoid at X = 142 R_e, *Geophys. Res. Lett.*, *21,* 2,295, 1994.

Maezawa, K., Magnetotail boundary motion associated with geomagnetic substorms, *J. Geophys. Res.*, *80,* 3,543, 1975.

McPherron, R.L., C.T. Russell, and M.P. Aubry, Satellite studies of magnetospheric substorms on August 15, 1968, 9, Phenomenological model for substorms, *J. Geophys. Res.*, *78,* 3,131, 1973.

Moldwin, M.B., and W.J. Hughes, Plasmoids as flux ropes, *J. Geophys. Res.*, *96*, 14,051, 1991.

Moldwin, M.B., and W.J. Hughes, Multi-satellite observations of plasmoids: IMP 8 and ISEE 3, *Geophys. Res. Lett.*, *19*, 1,081, 1992a.

Moldwin, M.B., and W.J. Hughes, Plasmoid observations in the distant plasma sheet boundary layer, *Geophys. Res. Lett.*, *19*, 1,911, 1992b.

Moldwin, M.B., and W.J. Hughes, On the formation and evolution of plasmoids: A survey of ISEE 3 data, *J. Geophys. Res.*, *97*, 19,259, 1992c.

Moldwin, M.B., and W.J. Hughes, Observations of Earthward and tailward propagating Flux rope plasmoids: Expanding the plasmoid model of geomagnetic substorms, *J. Geophys. Res.*, *99*, 183, 1994.

Mukai, T., M. Fujimoto, M. Hoshino, S. Kokubun, S. Machida, K. Maezawa, A. Nishida, Y. Saito, T. Terasawa, and T. Yamamoto, Structure and kinetic properties of plasmoids and their boundary regions, *J. Geomag. Geoelectr.*, *48*, 541, 1996.

Murphy, N., J.A. Slavin, D.N. Baker, and W.J. Hughes, Enhancement of energetic ions associated with traveling compression regions in the deep geomagnetic tail, *J. Geophys. Res.*, *92*, 64, 1987.

Nagai, T., K. Takahashi, H. Kawano, T. Yamamoto, S. Kokubun, and A. Nishida, Initial Geotail survey of magnetic substorm signatures in the magnetotail, *Geophys. Res. Lett.*, *21*, 2,991, 1994

Nagai, T., M. Fujimoto, Y. Saito, S. Machida, T. Terasawa, R. Nakamura, T. Yamamoto, T. Mukai, A. Nishida, and S. Kokubun, Structure and dynamics of magnetic reconnection for substorm onsets with Geotail Observations, submitted to *J. Geophys. Res.*, 1996.

Ogino, T., R.J. Walker, and M. Ashour-Abdalla, Magnetic flux ropes in 3-D MHD simulations, *Physics of Magnetic Flux Ropes*, eds. C.T. Russell, E.R. Priest, pp. 669, AGU, Washington, D.C., 1990.

Owen, C.J., and J.A. Slavin, Energetic ion events associated with traveling compression regions, *Proc. International Conf. on Substorms*, ESA SP-335, pp. 365-370, May, 1992.

Raeder, J., Global MHD simulation of the dynamics of the magnetosphere: Weak and strong solar wind forcing, Substorms 2, Proc. 2nd Int'l Conf. on Substorms, eds. J.R. An, J.D. Craven, and S.-I. Akasofu, pp. 561-568, Fairbanks, 1994.

Richardson, I.G., S.W.H. Cowley, E.W. Hones, Jr., and S.J. Bame, Plasmoid-associated energetic ion bursts in the deep geomagnetic tail: Properties of plasmoids and the post-plasmoid plasma sheet, *J. Geophys. Res.*, *92*, 9,997, 1987.

Richardson, I.G., C.J. Owen, and J.A. Slavin, Energetic (>0.2 MeV) electron bursts in the deep geomagnetic tail observed by ISEE 3: Association with substorms and magnetotail structures, *J. Geomag. Geoelectr.*, *48*, 657, 1996.

Schindler, K., A theory of the substorm mechanism, *J. Geophys. Res.*, *79*, 2,803, 1974.

Slavin, J.A., E. J. Smith, B. T. Tsurutani, D. G. Sibeck, H. J. Singer, D. N. Baker, J. T. Gosling, E. W. Hones, and F. L. Scarf, Substorm-associated traveling compression regions in the distant tail: ISEE 3 observations, *Geophys. Res. Lett.*, *11*, 657, 1984.

Slavin, J.A., E. J. Smith, D. G. Sibeck, D. N. Baker, R. D. Zwickl, and S.-I. Akasofu, An ISEE 3 study of average and substorm conditions in the distant magnetotail, *J. Geophys. Res.*, *90*, 10,875, 1985.

Slavin, J.A., D.N. Baker, J.D. Craven, R.C. Elphic, D.H. Fairfield, L.A. Frank, A.B. Galvin, W.J. Hughes, R.H. Manka, D.G. Mitchell, I.G. Richardson, T.R. Sanderson, D.J. Sibeck, H.J. Singer, E.J. Smith, and R.D. Zwickl, CDAW 8 observations of plasmoid signatures in the geomagnetic tail: An assessment, *J. Geophys. Res.*, *94*, 15,153, 1989.

Slavin, J.A., R.P. Lepping, and D.N. Baker, IMP 8 observations of traveling compression regions: New evidence for near-Earth plasmoids and neutral lines, *Geophys. Res. Lett.*, *17*, 913, 1990.

Slavin, J.A., M. F. Smith, E. L. Mazur, D. N. Baker, T. Iyemori, H. J. Singer, and E. W. Greenstadt, ISEE 3 plasmoid and TCR observations during an extended interval of substorm activity, *Geophys. Res. Lett.*, *19*, 825, 1992.

Slavin, J.A., M. F. Smith, E. L. Mazur, D. N. Baker, T. Iyemori, and E. W. Greenstadt, ISEE 3 observations of traveling compression regions in the Earth's magnetotail, *J. Geophys. Res.*, *98*, 15,425, 1993.

Slavin, J. A., C. J. Owen, and M. Hesse, The Evolution of the Plasmoid-Lobe Interaction with Downtail Distance, *Geophys. Res. Lett.*, *21*, 2,765, 1994.

Slavin, J. A., C. J. Owen, M. M. Kuznetsova, and M. Hesse, ISEE 3 Observations of Plasmoids with Flux Rope Magnetic Topologies, *Geophys. Res. Lett.*, *22*, 2,061, 1995.

Slavin, J.A., D.H. Fairfield, M. Kuznetsova, C.J. Owen, R.P. Lepping, S. Taguchi, T. Mukai, Y. Saito, T. Yamamoto, S. Kokubun, A.T.Y. Lui, and G.D. Reeves, ISTP Observations of Plasmoid Ejection: IMP 8 and Geotail, *J. Geophys. Res.*, *103*, 119 , 1998.

Sonnerup, B.U.O., and L.J. Cahill, Magnetopause structure and attitude from Explorer 12 observations, *J. Geophys. Res.*, *72*, 171, 1967.

Taguchi, S., J.A. Slavin, R.P. Lepping and M. Nose, Traveling compression regions observed in the mid-tail lobes near substorm expansion phase onset, *Proc. Int'l Conference on Substorms (ICS-3)*, ESA SP-389, pp. 603-607, Versailles, May 12-17, 1996.

Taguchi, S., J.A. Slavin, and R.P. Lepping, Characteristics of mid-tail traveling compression regions, *Geophys. Res. Lett*, *24*, 353, 1997a.

Taguchi, S., J.A. Slavin and R.P. Lepping, Traveling compression regions in the mid-tail: 15 years of IMP 8 observations, *J. Geophys. Res.*, in press, 1997b.

Taguchi, S., J.A. Slavin, M. Kiyohara, M.Nose, R.P. Lepping and G. Reeves, Temporal relationship between mid-tail TCRs and substorm onset: Evidence for near-Earth neutral line formation in the late growth phase, submitted to *J. Geophys. Res*, 1997c.

Walker, R.J., and T. Ogino, A global MHD simulation of the origin and evolution of magnetic flux ropes in the magnetotail, *J. Geomag. Geolectr.*, *48*, 765, 1996.

Yamamoto, T., K. Shiokawa, and S. Kokubun, Magnetic field structure of the magnetotail as observed by GEOTAIL, *Geophys. Res. Lett.*, *21*, 2,875, 1994.

James A. Slavin, NASA/GSFC, Laboratory for Extraterrestrial Physics, Greenbelt, Maryland, USA 20771

Near Earth Plasma Sheet Penetration and Geomagnetic Disturbances

L. R. Lyons[1,2], G. T. Blanchard[1], J. C. Samson[3], J. M. Ruohoniemi[4],

R. A. Greenwald[4], G. D. Reeves[5], and J. D. Scudder[6]

The plasma sheet has a distinct inner boundary that moves inward during periods of, and as a result of, enhanced convection. This leads to dramatic increases in plasma pressure and cross-tail current in the near-Earth (equatorial radial distances r ~ 6-10 R_E) region of the nightside magnetosphere that are associated with geomagnetic activity. During quiet times, the plasma sheet remains beyond r ~ 10 R_E and precipitation from the plasma sheet is weak and peaks at invariant latitudes $\Lambda \geq 68°$. Weak auroral disturbances, here called "separatrix disturbances", are occasionally seen to move equatorward from near the magnetic separatrix. During the growth phase of isolated substorms, the plasma sheet moves into the near-Earth night side region leading to large increases in plasma pressure. Precipitation expands equatorward from its quiet-time location to Λ = 64°-67°. At the onset of the substorm expansion phase, active aurora initiates within the inner plasma sheet at an invariant latitude ~5-6° equatorward of the separatrix. The region of active aurora then moves several degrees in latitude poleward during the expansion phase. This corresponds to a large tailward motion of the equatorial region that maps to the active aurora from its initial location in the near-Earth region. After the expansion phase, the plasma sheet precipitation re-forms well poleward of its location at onset and near where it is located during quiet times, implying a significant decrease in near-Earth plasma sheet pressure. Particle injections are juxtaposed with a decrease in plasma pressure within the inner plasma sheet during the expansion phase of substorms. Separatrix disturbances are seen during all phase of substorm activity. During convection bays, the plasma sheet remains within the near-Earth region with enhanced pressures. Auroral activity during convection bays is dominated by separatrix disturbances that can be quite strong (ground magnetic perturbations up to ~300 nT) and move towards the inner plasma sheet from near the separatrix. Poleward moving active aurora that characterize

[1] Department of Atmospheric Sciences, University of California, Los Angeles, Los Angeles, California
[2] Space and Environment Technology Center, The Aerospace Corporation, Los Angeles, California
[3] Department of Physics, University of Alberta, Edmonton, Canada
[4] Johns Hopkins University, Applied Physics Laboratory, Laurel, Maryland
[5] Los Alamos National Laboratories, Los Alamos, New Mexico
[6] Department of Physics and Astronomy, University of Iowa, Iowa City, Iowa

New Perspectives on the Earth's Magnetotail
Geophysical Monograph 105

INTRODUCTION

Early magnetospheric observations revealed a broad region of low energy (thermal energies of a few hundred eV to several hundred keV) plasma within the magnetotail [e.g., *Gringauz et al.*, 1960; *Freeman*, 1964; *Bame et al.*, 1967; *Frank*, 1967]. This region, known as the plasma sheet, extends throughout the portion of the magnetotail that lies within a few R_E of the magnetic neutral sheet. Plasma sheet particles are the primary carriers of plasma energy within the magnetotail and are responsible for the current that crosses the tail.

Plasma sheet dynamics lead to dramatic changes in plasma pressure and cross-tail current in the near-Earth (equatorial radial distances r ~ 6-10 R_E,) region of the

traditional substorm expansions, and associated reductions in plasma pressure in the near-Earth region, are absent. The separatrix disturbances are associated with equatorward flows in the ionosphere that should correspond to bursts of earthward flow (BBF's) within the high-altitude plasma sheet that propagate earthward toward the inner plasma sheet. During a substorm expansion, the tailward expanding reduction of cross-tail current gives a dipolarization of the magnetic field and associated induced electric field that propagates tailward through the plasma sheet. This should give a tailward propagating BBF that does not map to the ionosphere.

nightside magnetosphere. In addition, plasma sheet dynamics in this region are associated with current-wedge formation and its mapping along field lines to the region of auroral breakup at substorm onset [e.g., *Samson et al.*, 1992]. Despite its clear importance to magnetospheric dynamics and geomagnetic disturbances, the near-Earth plasma sheet has received only limited study because of a lack of measurements. While there have been many satellites in geosynchronous orbit, very few satellites have traversed the r ~ 6-10 R_E region. However, ground based and satellite measurements of the plasma sheet are now being made that can be used to address the important physics of this region.

This paper reviews what is known about the dynamics of the near-Earth plasma sheet and its relation to geomagnetic activity. Because so few studies have been performed, an initial analysis is also performed on data that relates the dynamics to geomagnetic activity. The analysis emphasizes data from the CANOPUS (Canadian Auroral Network for the OPEN Program Unified Study) ground-based, meridian-scanning photometers. Conditions of steady weak, steady enhanced, and variable convection are investigated. It is shown that the plasma sheet responses to each are unique and identifiable, and that the relation between geomagnetic disturbances and each of these plasma sheet conditions can be identified. The relationship between the disturbances observed within the ionospheric mapping of the plasma sheet map and their counterpart within the high altitude plasma sheet is also addressed.

BACKGROUND

The importance of the dynamics of the near-Earth plasma sheet was initially addressed by *Vasyliunas* [1968]. He found that plasma sheet electrons have a sharp inner boundary at a location that varies dramatically with geomagnetic activity. During geomagnetically quiet times, the inner boundary was found at r ~ 11 R_E. However, the boundary was found at r ~ 6-8 R_E during periods of geomagnetic activity, and it has been seen as close in as r = 5 R_E [*Smith and Hoffman*, 1974]. Based on the continuity of electron spectra as a function r during geomagnetically active periods, *Vasyliunas* [1968] concluded that "the plasma sheet actually moves inward until its inner boundary has been displaced to the observed close distances," and he suggested that the inward motion was due to the cross-tail electric field.

Vasyliunas's idea that the inward motion of the inner boundary of the plasma sheet moves inward in response to enhancements in convection is supported by calculations of the trajectories of plasma sheet particles [e.g., *Kavanagh et al.*, 1968; *Jaggi and Wolf*, 1973; *Wolf*, 1995]. Figure 1 illustrates trajectories (based on the calculations of *Chen et al.* [1993]) of equatorially mirroring protons in the equatorial plane for conditions of weak and enhanced convection. The trajectories are sketched for a magnetic moment of 20 MeV/Gauss, which is representative of plasma sheet protons (~20 keV at r = 7 R_E). In general, plasma sheet particles convected earthward until they reach the radial distance where magnetic drift deflects them around azimuthally around the Earth, electrons toward the dawn side and positive ions toward the dusk side. When convection is weak, the deflection prevents plasma sheet particles from reaching the < 10 R_E portion of the nightside magnetosphere. However, enhanced convection brings particles well into this region.

The strong effect of convection strength on the radial distance to which plasma sheet particles have access ought to have very significant effects on particle distributions in the ~6-10 R_E region of the nightside magnetosphere, and this is the region where the largest plasma-sheet pressure changes are observed. To measure the response of the near-Earth plasma sheet to variations in convection and the relation of this response to geomagnetic activity requires measurements of plasma sheet particles as a function of equatorial radial distance. Such measurements have been very limited in the past. However the data of *Schield and Frank* [1970] show that the inner edge of the plasma sheet is displaced inward during the main phase of moderate storms, and *Kivelson et al.* [1973] inferred that the boundary moves inward during the growth phase of substorms. Both the main phase of storms and the growth phase of substorms are conditions of enhanced electric fields, for which inward motion of the inner edge of the plasma sheet is expected.

The expected response of the plasma sheet to changes in magnetospheric convection can be seen in the particle data

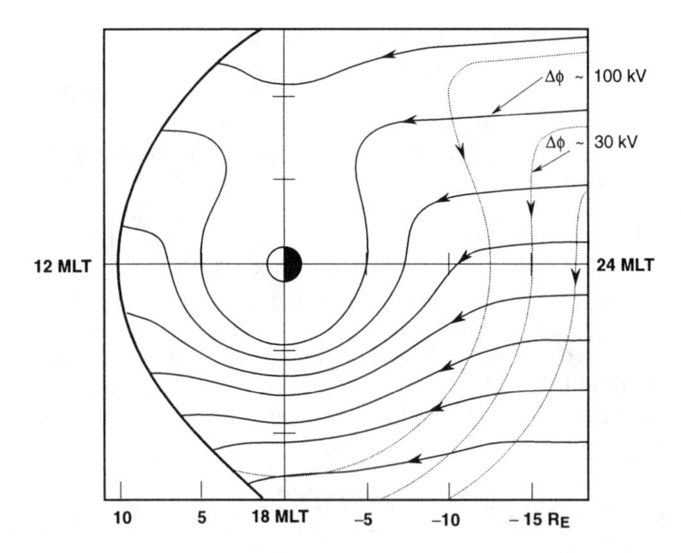

Figure 1. Illustration of 20 MeV/G proton trajectories in the equatorial plane for weak convection (lighter lines, $\Delta\phi \sim 30$ kV) and for enhanced convection (heavier lines, $\Delta\phi \sim 100$ kV). The heavy solid curve on the left hand side of the illustration is the dayside magnetopause.

from the POLAR satellite in Figure 2. Figure 2 shows spectrograms of ion and electron counts from ~0.01 to ~15 keV as obtained from the HYDRA instrument [*Scudder et al.*, 1995] on POLAR on May 9 (top) and May 12, 1996 (bottom). For both passes, the satellite was near 2245 MLT and moved equatorward and earthward from the northern hemisphere lobe, entering the plasma sheet at an invariant latitude $\Lambda \sim 69°$ on May 9 and ~70° on May 12. The poleward boundary of the plasma sheet is clearly identified by an abrupt increase in ion and electron count rates at energies from a few hundred eV to the highest energy of detectable counts. As the satellite continued equatorward and earthward, the inner boundary plasma sheet was encountered. This boundary is particularly abrupt for electrons, and is seen at significantly different locations for the two orbits.

The May 12 orbit was during an extended period of positive interplanetary magnetic field (IMF) B_z (~5 nT) so that convection is expected to have been weak. The inner boundary of plasma sheet electrons on this orbit was located at a model invariant latitude of $\Lambda \sim 68.5°$. During the period of the May 9 orbit, the IMF B_z was negative (−2 to −5 nT). As expected from enhanced convection during a period of negative IMF B_z, the inner edge of the plasma sheet was encountered at significantly lower Λ (~63°) on this orbit than on the May 12 orbit. While the invariant latitudes in Figure 2 were obtained from a magnetic field model which does not consider magnetic field changes due to plasma sheet dynamics, Figure 2 shows a clear difference in the location of the inner edge of the plasma sheet between the two orbits that relates to the expected strength of convection. For both orbits, the inner edge of plasma sheet ions is located very close to the inner edge of plasma sheet electrons. The ions, however, extend further earthward over a narrow range of energies near 8 keV, This earthward extension was referred to as a "nose structure" and explained as a particle drift effect by *Smith and Hoffman* [1974]. At these energies, the westward magnetic drift speeds and eastward corotation speeds for ions are nearly equal, so that the total azimuthal drift speed is very low. Thus azimuthal drift, which limits the earthward penetration of most particles, is less effective over this energy range and allows enhanced earthward penetration via convection. This existence of this nose structure supports the idea that it is the balance between inward convection and magnetic drift that determines the inward penetration of the nightside plasma sheet.

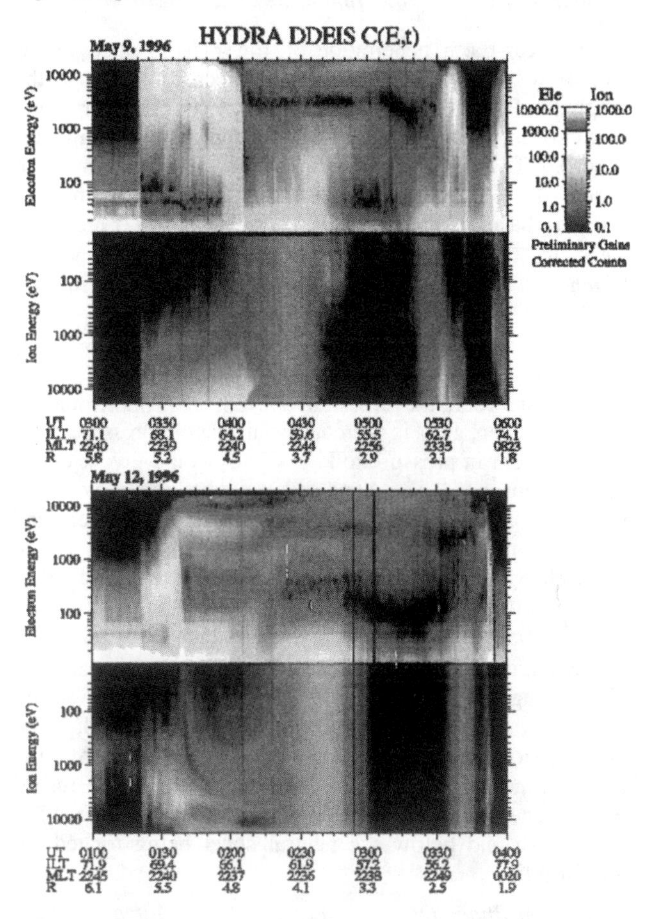

Figure 2. Spectrograms of electron and ion count rates from the HYDRA detector for inbound passes of the POLAR satellite on May 9 and May 12, 1996. UT, MLT, and the invariant latitude (ILT) and radial location (R) of the satellite are annotated along the bottom of the spectrograms for each pass.

The SuperDARN radars [*Greenwald et al.*, 1995] measures plasma flows in the ionosphere which can be related to the POLAR observations. Figure 3 shows evening sector plasma flows from two of the SuperDARN radars (MLT ~ UT − 7 hr for data in Figure 3) as a function of magnetic latitude and UT during the 00-06 UT period of southward IMF on May 9 (which includes the times of the May 9 POLAR orbit in Figure 2) and during the 00-06 UT period of northward IMF on May 12 (which includes the times of the May 12 POLAR orbit). It can be seen that when flows were observable between 65° and 69° magnetic latitude, they were generally significantly higher on May 9 than on May 12. This is consistent with the inward penetration of the plasma sheet as seen on May 9 being the result of enhanced convection.

PLASMA SHEET DISTURBANCES

Growth Phase Pressure Increases

The near-Earth penetration of the plasma sheet gives enhancements in plasma pressure and perpendicular (to the magnetic field **B**) current that, except for the stormtime ring current, are the largest that occur within the magnetosphere. These enhancements in pressure are shown in Figure 4. Here equatorial pressures versus r during the growth-phase of substorms [*Kistler et al.*, 1992] are compared to the equatorial pressures during quiet periods [*Spence et al.*, 1989]. Convection is enhanced during the southward IMF conditions that lead to the growth phase substorms, and this can be seen to lead to plasma pressure increases of ~2-4 nPa in the inner magnetosphere.

To first order, the cross-tail current is proportional to $[Pl_{z=0}]^{1/2}$, where $Pl_{z=0}$ is the equatorial plasma pressure. An enhancement in pressure to $Pl_{z=0} = 5$ nP would give a cross-tail current of 0.18 A/m and a lobe magnetic field perturbation of 110 nT. This perturbation is larger than the unperturbed magnetic field magnitude at $r \geq 6\ R_E$. The large changes in magnetic field geometry resulting from such a perturbation in the $r \geq 6\ R_E$ region were observed by *Thomas and Hedgecock* [1975]. Figure 5 shows sketches of the magnetic field based on their observations near midnight along the trajectory of the HEOS 1 spacecraft during a disturbed (top panel) and a quiet (bottom) period. The large distortion of the magnetic field caused by the enhanced near-Earth cross-tail current during the disturbed period can be seen in the figure, the magnetic field direction just above and below the neutral sheet being rotated by almost 90°.

Expansion Phase Dipolarizations and Associated Earthward Flows

The increases in pressure and current in the near-Earth plasma sheet are associated with geomagnetic activity as illustrated by the magnetic field data in Figure 6. Figure 6 [from *McPherron and Manka*, 1985] shows the typical relation between the magnetic field observed at geosynchronous orbit (by the GOES 3 satellite) and the IMF (observed by IMP 8) and ground magnetometers. Times of the onsets of two large substorm expansion phases, apparently triggered by northward turnings of the IMF (i.e., reductions in the magnitude of a negative IMF B_z) are indicated in the figure. The GOES data is shown for times surrounding the first onset, when the satellite was in the vicinity of magnetic midnight. Notice that the southward turning of the IMF at 1008 UT leads to an ~70 nT increase in the magnitude of the V (radial) component of the magnetic field observed on GOES. This increase occurred over the ~35 min growth-phase period from ~1015 to 1050 UT, and it must have been associated with a corresponding increase of plasma pressure in the vicinity of

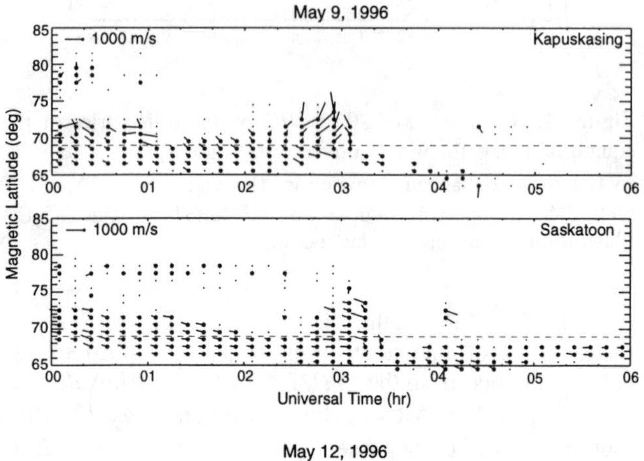

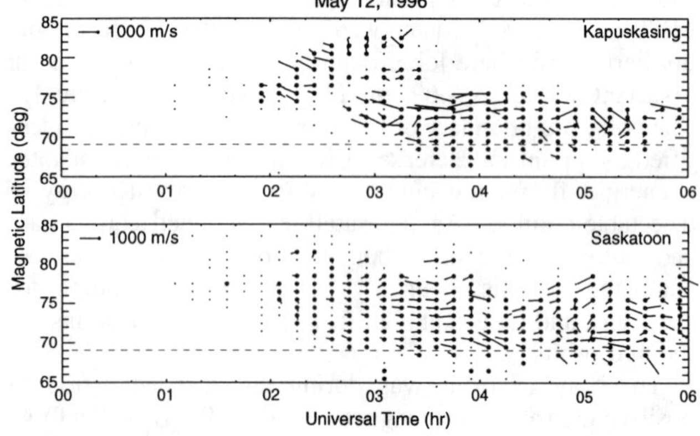

Figure 3. Plasma flows from the Kapuskasing and Saskatoon SuperDARN radar. Magnetic latitude is given in the Adjusted Corrected Geomagnetic Coordinate system, which is the updated version of the PACE (Polar American Conjugate Experiment) geomagnetic coordinate system [*Baker and Wing*, 1989]. The L-shell fitting procedure described by *Ruohoniemi et al.* [1989] has been used to obtain two-dimensional velocities from the line-of-sight radar flow measurements made as a function of latitude and longitude. A light dashed line demarcates 69° magnetic latitude. Points indicate locations from which radar echoes were received, and flow vectors are shown only where the data is of high quality .

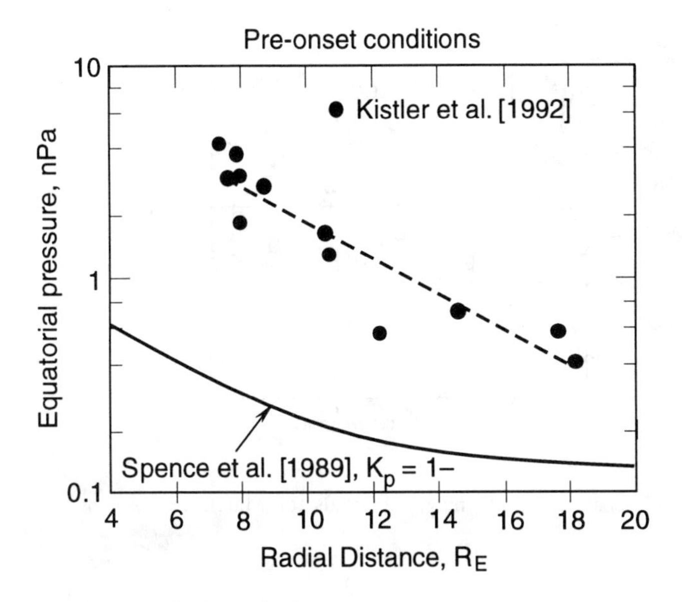

Figure 4. Equatorial plasma pressure versus radial distance for conditions preceding substorm expansion phase onset and for quiet conditions.

synchronous orbit. The pressure must have increased everywhere between the inner edge of the plasma sheet that formed following the enhancement in convection associated with the southward IMF and the inner edge of the plasma sheet that existed beforehand.

The decrease in magnitude of the V-component following the onset was even more rapid than was the increase during the substorm growth phase. This decrease in the V-component is associated with an increase in the H-component (parallel to the geomagnetic dipole and positive northward). These changes reflect the return of the magnetic field to a more dipolar configuration following its more stretched configuration. This "dipolarization" is associated with a decrease in the cross-tail current, and implies a decrease in equatorial plasma pressure. This decrease in plasma pressure must occur despite the well-know particle injections in the vicinity of synchronous orbit, and decreases in equatorial plasma sheet pressure following substorm onset have been observed by [*Lui et al.*, 1992] and discussed by *Lyons* [1996a] using observations from *Roux et al.*, [1991]. This illustrates an interesting question concerning plasma sheet dynamics during substorms that has not been adequately answered and warrants further study: How can particle injections occur in regions where the expansion-phase reduction of cross-tail current requires a reduction in plasma pressure?

The dipolarization of the magnetic field during the expansion phase, gives a strong induced electric field that displaces particles inwards throughout the region of reduction in cross-tail current [e.g., *Shepherd et al.*, 1980; *Aggson et al.*, 1983; *Mauk and Meng*, 1987]. These induced electric fields are short-lived (one to a few minutes) and have magnitudes (~10 mV/m) which are significantly greater than growth-phase electric fields. An example of the measured y-component of the electric field E_y during a substorm dipolarization observed on April 8, 1977 at L= 7.5 and a magnetic latitude of −20° is shown is in Figure 7 [from *Aggson et al.*, 1983]. The figure shows the decrease in $|B_x|$ associated with dipolarization at 1857-1858 UT, and $|B_z|$ was observed to increase at the same time. The electric field can be seen to be enhanced while the magnetic field is dipolarizing, and the dotted lines show that E_y is proportional to the time rate of change of B_x when the rapid changes in B_x and E_y are neglected. Such electric fields cause particles to move earthward and towards the magnetic equator during the short interval that the magnetic field is dipolarizing. The burst of earthward flow speed associated with such an induced electric fields will be ~100 km/s in the vicinity of synchronous orbit, which is capable of displacing particles earthward by ~ 1 R_E earthward [*Pedersen*, 1992]. Note that "earthward convection by the

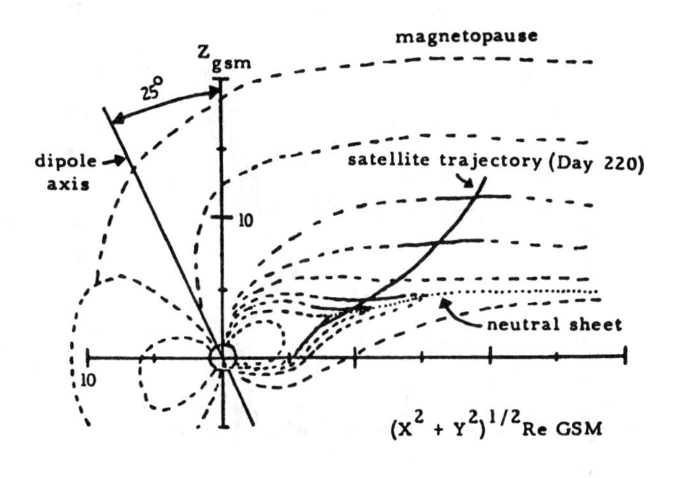

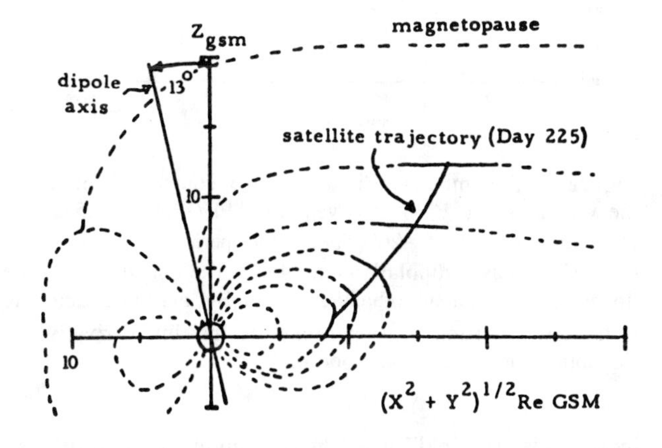

Figure 5. Local meridian sketches of magnetic field lines consistent with field directions observed in 1969 [*Thomas and Hedgecock*, 1975] by HEOS 1 along the trajectories drawn in the figure for days 220 (disturbed) and 225 (quiet).

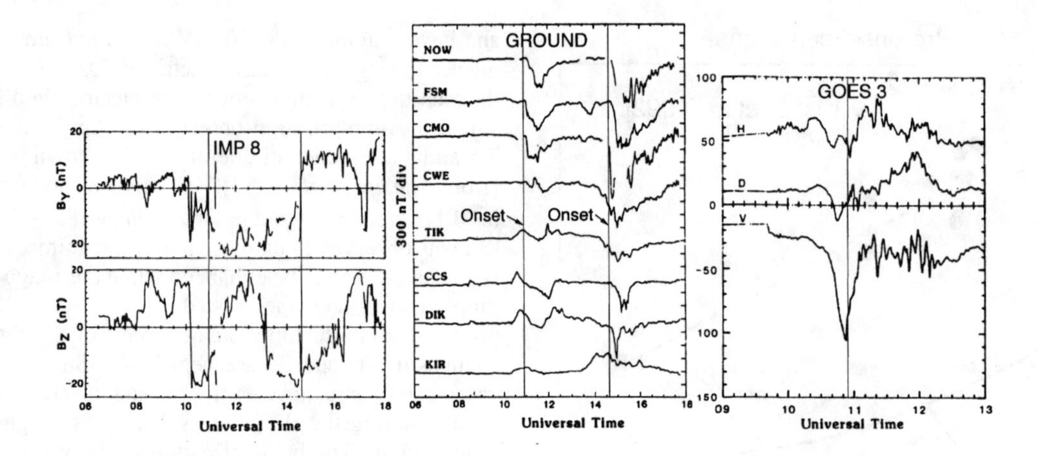

Figure 6. Interplanetary, ground H-component, and geosynchronous magnetic fields for March 22, 1979 [from *McPherron and Manka*, 1985].

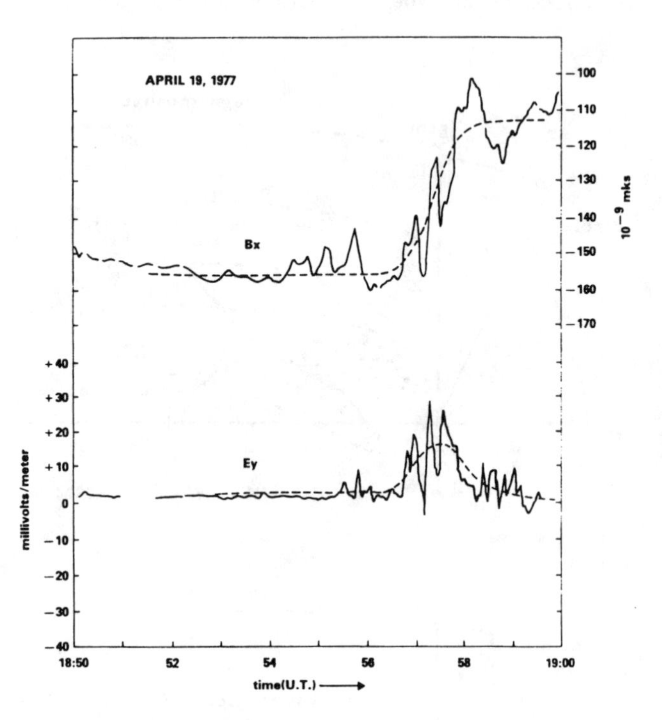

Figure 7. A comparison between the x component of **B** and the y component **E** as observed by ISEE 1 at L= 7.5 and a magnetic latitude of –20° during the period of a substorm expansion phase dipolarization of **B**. The dotted line in the upper portion is a smoothed version of B_X and the dotted line in the lower portion is proportional to the time derivative of the dotted line in the upper portion.

induced electric field" and "dipolarization of the magnetic field" are different ways of viewing the same process.

Because of the large electric-field magnitude, there should be a displacement of energetic particles to regions earthward of where they have access during the growth phase. Thus the largest expansion-phase flux increases at a particular location should occur at energies for which there was no access during the growth phase. Consistent with this, the particle fluxes of *Roux et al.* [1991] show increases at higher energies where the growth-phase energy spectra are very steep and consistent with a lack of access during the growth phase. This may offer a partial explanation for particle injections in the region where the cross-tail current decreases. However, this possibility requires further observational verification as well as quantitative understanding of how the particle fluxes at the lower energies decrease as the induced electric field moves particles earthward.

An additional part of the explanation may be that the inward displacement of particles as the magnetic field dipolarizes can move plasma earthward of where the bulk of the plasma sheet had access during the preceding growth phase. In these regions, particle injections without a concurrent decrease in total plasma pressure are possible. Such injections have been found at synchronous orbit by *Birn et al.* [1997], as would be expected if synchronous orbit is often somewhat earthward of the inner edge of the growth-phase plasma sheet.

After originating in the near Earth plasma sheet (r~ 6-10 R_E), the reduction in cross-tail current expands tailward during the expansion phase at velocities of a few hundred km/s [*Jacquey et al.*, 1991, 1993; *Ohtani et al.*, 1992]. The dipolarization of the magnetic field and associated induced electric field should propagate tailward with the expansion of the current reduction region, giving a tailward propagating burst of earthward plasma flow Two satellite observations of the magnetic field dipolarization and associated induced electric field show that it takes ~ 5 min for the burst of earthward flow to reach r~ 20 R_E [*Pedersen*, 1992]. At r~ 20 R_E the electric fields reported by Pedersen were several mV/m, which give earthward flow speeds of several hundred km/s, and a flow speed of ~200 km/s was observed by *Nishida et al.* [1997b] as a substorm-associated magnetic field dipolarization expanded past x = – 19 R_E.

The tailward expanding reduction in cross-tail current is also associated with the poleward moving region of active aurora that is observed during the substorm expansion phase [see *Lyons*, 1996b and references therein]. The active aurora results from the upward field-aligned currents in the region of divergence of the cross-tail current on the duskward side of the current reduction region. We thus expect a clear association between the regions of active auroral (and westward electrojet) during the expansion phase of substorms, that are observed from the ground to move poleward at ~1 km/s, and bursts of earthward flow observed in the tail. Bursts of enhanced flow (known as "BBF's") are now known to be a common feature of the plasma sheet [*Baumjohann et al.*, 1990; *Angelopoulos et al.*, 1992], and the tailward propagating bursts of flow during the substorm expansion phase should constitute a particular type of BBF. Since the induced electric field associated with substorm magnetic field dipolarizations do not map to the ionosphere, these BBF's should not be seen in the ionosphere.

Convection Bays

Relatively steady enhanced convection can lead to periods that are refereed to as convection bays. During convection bays, convection remains enhanced without the large tail-current reductions that are associated with substorms [*Kokubun et al.*, 1977; *Pytte et al.*, 1978], so that the magnetic field in the inner plasma sheet remains highly stretched as it is during the growth phase of substorms [*Sergeev et al.*, 1994]. This implies that the plasma sheet penetrates inward during convection bays as it does during the substorm growth phase. The important difference is that the enhanced plasma pressures are maintained for extended periods in the inner plasma sheet without the large reductions of cross-tail current, and the associated reductions in plasma pressure, that occur during substorms.

Separatrix Disturbances and Earthward Flows

Another important type of geomagnetic disturbance on the night side has been identified from the ground measurements of ionospheric flow and auroral intensities during periods of enhanced convection [*Sergeev et al.*, 1990] and during quiet periods [*de la Beaujardière*, 1994]. This type of disturbance consists of few minute bursts of enhanced equatorward flow at speeds of ~500 m/s that are observed to propagate equatorward and westward from a location initially detected near the polar cap boundary, an example of which is shown in Figure 8. These flow bursts have been observed to repeat after a time of ~10 - 30 min, and they have been found to be associated with significant enhancements in auroral emissions and with ground magnetic disturbances of ~20-100 nT [*de la Beaujardière et al.*, 1994; *Kauristie et al.*, 1996; *Yeoman and Lühr*, 1997]. Figure 8 shows that the arcs are seen on the western edge

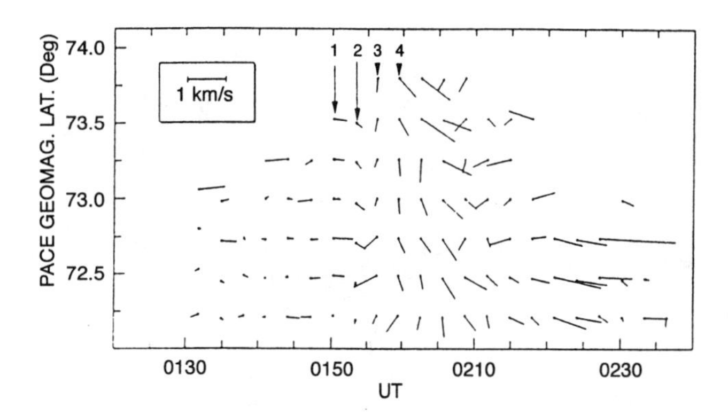

Figure 8. Sondrestrom radar velocity vectors from January 13, 1989 displayed as a function of time and magnetic latitude. The numbers at the top refer to four scans of the radar that measured an equatorward drifting auroral arc that was observed during scan 1, reached maximum intensity during scan 3, and drifted equatorward of the radar field of view after scan 4.

(since the flow enhancement propagates westward) of the flow enhancement, which corresponds to the region of converging ionospheric electric field and Pedersen current.

These equatorward flow enhancements are seen in the ionosphere where they are associated with potential electric fields that are expected to map along field line to the high-altitude plasma sheet. They thus should be seen within the high-altitude plasma sheet. *Sergeev et al.* [1990] found an associated between the ionospheric flow enhancements and BBF's observed at r~ 20 R_E, and *Kauristie et al.* [1996] found associations with BBF's as close in as x ~ –11 R_E. Also, *Nishida et al.* [1997a] found repeated earthward particle bursts at r = 10-15 R_E during an extended geomagnetically quiet interval which would be expected to have been associated with equatorward flow enhancements in the ionosphere. Thus this second class of disturbance is most likely associated with BBF's that map to the ionosphere and propagate earthward within the plasma sheet. These type of disturbances can be referred to as "separatrix disturbances", since they are associated with auroral disturbances that initiate near the boundary between open and closed field lines, and they can occur during all levels of geomagnetic activity.

USE OF CANOPUS PHOTOMETERS

The substorm expansion phase is clearly related to an inward penetration of the plasma sheet that occurs during the preceding growth phase. However, the changes in the plasma sheet following the expansion phase have not been well evaluated. Also, while the separatrix disturbances have been observed during geomagnetically quiet and disturbed periods, their relation to plasma sheet penetration has not previously been determined. Additionally, the temporal evolution of the plasma sheet during convection bays and as a function of substorm phase has not been evaluated. To evaluate the evolution of the inner plasma

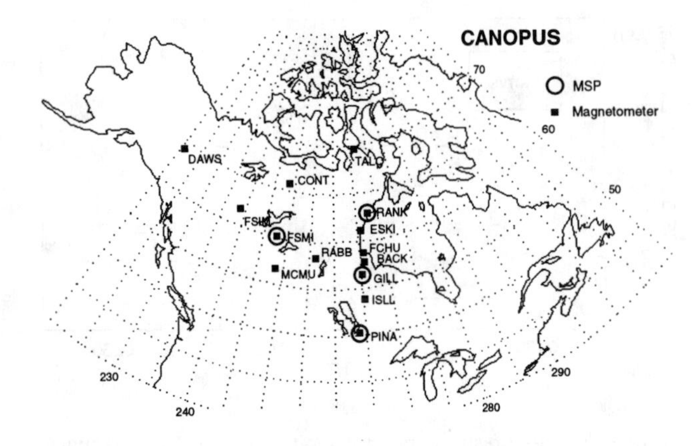

Figure 9. Locations of CANOPUS ground magnetometers and meridian scanning photometers (MSP). Geographic latitude and longitude are indicated. See *Rostoker et al.*, [1995] for station names, exact station coordinates, and other CANOPUS ground instrumentation.

sheet and its relation to the IMF, convection, and geomagnetic disturbances, it is desirable to be able to monitor the plasma sheet as function of time. This cannot be done using in-situ satellite observations. However, the isotropy of plasma sheet particles leads to precipitation, which causes atmospheric emissions that can be measured with ground-based photometers. Emissions from plasma sheet particle precipitation are now being routinely monitored with the CANOPUS meridian-scanning photometers, and data from these photometers can be used to monitor the intensity, motion, and location of the plasma sheet as a function of time. The locations of the CANOPUS photometers and ground magnetometers are shown in Figure 9 [*Rostoker et al.*, 1995]. Here we use CANOPUS photometer data to perform an initial analysis of the temporal evolution of the plasma sheet and its relation to convection and geomagnetic activity. We use data from Gillam (GILL) and Rankin Inlet (RANK), which are at approximately the same longitude, and ground magnetic field data from the meridian chain of magnetometers extending from Pinawa to Rankin Inlet. The photometers at Pinawa (PINA) and Fort Smith (FSMI) give measurements at lower latitudes and approximately 17° in longitude to the west of Gillam, respectively, but are not used in this preliminary study.

Figure 10 shows a merging of meridian-scanning-photometer data from Gillam and Rankin Inlet, the magnetic x-component from the meridian chain of magnetometers, and IMF measurements from IMP 8. Data is shown for a 12 hr period of low geomagnetic activity from 01-13 UT on December 18, 1990. Magnetic midnight is at ~07 UT. Emission intensities from the photometers are shown for three different wavelengths as a function of magnetic latitude and UT. The photometers measure emission intensities as a function of elevation angle along the magnetic meridian. Latitude profiles have been obtained from these measurements by assigning a fixed altitude to each of the emissions. The 6300 Å emission reflects precipitation of plasma-sheet electrons at energies $\lesssim$ 1 keV. This emission shows a clear poleward boundary which tracks the poleward boundary of the plasma sheet and provides a good identification of the separatrix that separates the region of open polar-cap field lines from the lower latitude region of closed field lines [*Samson et al.*, 1992; *Blanchard et al.* 1995]. The location of this boundary as determined by an automated procedure developed by *Blanchard et al.* [1996] is shown in the figure as a black and white line. The 4861 Å emission is due to proton precipitation. Plasma sheet protons precipitate via scattering into the loss cone that results from the violation of the guiding center approximation in the region of the cross-tail current sheet. The most intense 4861 Å emissions correspond to the innermost portion of the plasma sheet proton distribution [*Samson et al.*, 1992], so that the band of most intense emissions tracks the temporal evolution of the inner plasma sheet as mapped to the ionosphere. The latitude of the peak in the emission profile is shown versus UT by a black and white line. The 5577 Å emission responds primarily to $\gtrsim$ 1 keV electron precipitation, but also responds weakly to protons. This emission is particularly useful for identifying and tracking auroral disturbances.

ANALYSIS OF PLASMA SHEET DYNAMICS

To perform an initial analysis, we obtained photometer data from the period of December 1990 to February 1991. From this data, we identified 15 clear nights with good quality photometer data. IMF data is available for 7 of these nights.

Four of the nights with good data had mostly very weak auroral activity (December 18; January 5, 10, 23). IMF data was available for three of these quiet nights, and showed that the IMF B_z remained mostly positive during the quiet periods. Convection is expected to be weak for such northward IMF conditions. Figure 10 shows an example of data for one of these quiet periods (on December 18) that had predominantly northward IMF from ~02 UT until ~ 09 UT. Note that the peak of the 4861 Å emission remained poleward of invariant latitude Λ = 68° and relatively weak (as compared to what is seen during more disturbed periods), and that the poleward boundary of the plasma sheet remained relatively stable at Λ ~ 73° until over an hour past magnetic midnight. Consistent with the lack of significant auroral activity, the magnetometer data for this period are extremely quiet. The above features concerning the location of the peak and the intensity of the 4861 Å emission, the stability of the of the polar cap boundary near 73°, and the quiet magnetometer data are characteristic of all four of the extended quiet periods in our data set. Note, however, that some weak auroral enhancements can be seen near the polar-cap boundary in Figure 10, particularly in the 5577 Å emissions, and stronger

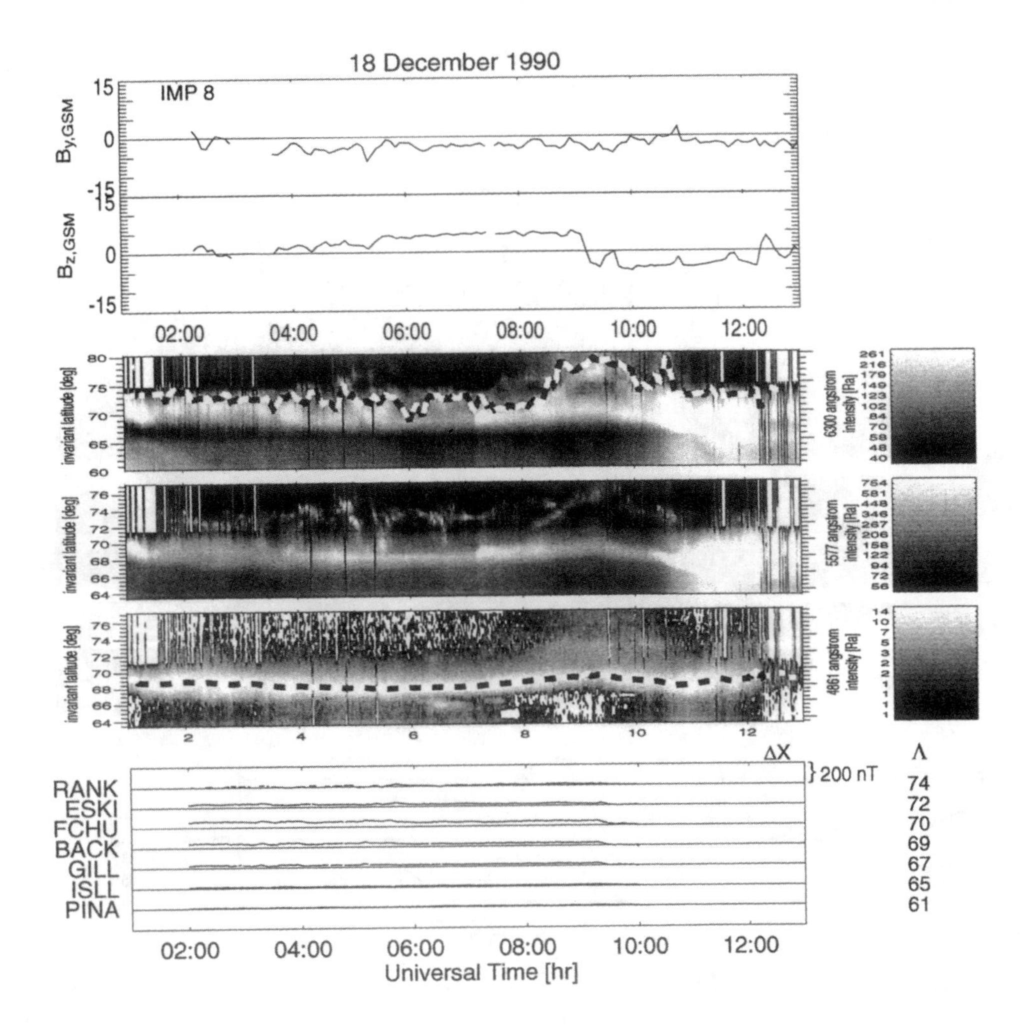

Figure 10. Interplanetary magnetic field from IMP 8 (top panels), merged meridian scanning photometer data from the CANOPUS stations Rankin Inlet and Gillam (middle panels), and ground magnetic field data from the meridian chain of magnetometers extending from Pinawa to Rankin Inlet (lower panels) for 01-13 UT on December 18. 1990. Black and white lines give the latitude of peak 4861 Å emissions and the poleward boundary of the plasma sheet as determined from the 6300 Å photometer *data [Blanchard et al.,* 1996].

separatrix disturbances are seen in the photometer from one of the other quiet nights (January 5).

Figure 11 shows IMF, photometer, and ground magnetometer data for 1-13 UT on January 12, 1991. This includes a substorm period with onsets (indicated by dotted gray lines) at ~06 and ~0630 UT. The photometer data shows important differences from that during quiet times. Following a southward turning of the IMF at ~0450 UT, the 4861 Å emissions intensified and moved equatorward. The peak of the emissions moved from its quiet-time location at $\Lambda > 68°$ to $\Lambda \approx 65°$, implying that the plasma sheet moved earthward during this growth-phase period of southward IMF. During the same period, the poleward boundary of the plasma sheet moved equatorward from $\Lambda = 74°$ to $\Lambda = 71°$. After the expansion phase activity, the band of proton precipitation reappeared at higher latitudes (centered at $\Lambda \approx 67°$), consistent with an expansion-phase reduction in plasma pressure in the inner portion of the growth-phase plasma sheet. Expansion phase auroral activity begins with an abrupt intensification of 5577 Å emissions at $\Lambda = 65°$, which is 6° equatorward of the poleward boundary of the plasma sheet and within the band of intense proton precipitation. The major substorm onset is at 0630 UT and is followed by a large poleward motion of the region of active aurora, though the structure of the poleward-moving active aurora during the expansion phase is somewhat masked by higher than usual atmospheric scattering on this night. This onset is accompanied by injections of energetic particles seen by the Los Alamos (LANL) geosynchronous satellites (data not shown).

The data in Figure 11 are typical of what is seen during isolated substorm periods and show the plasma sheet and

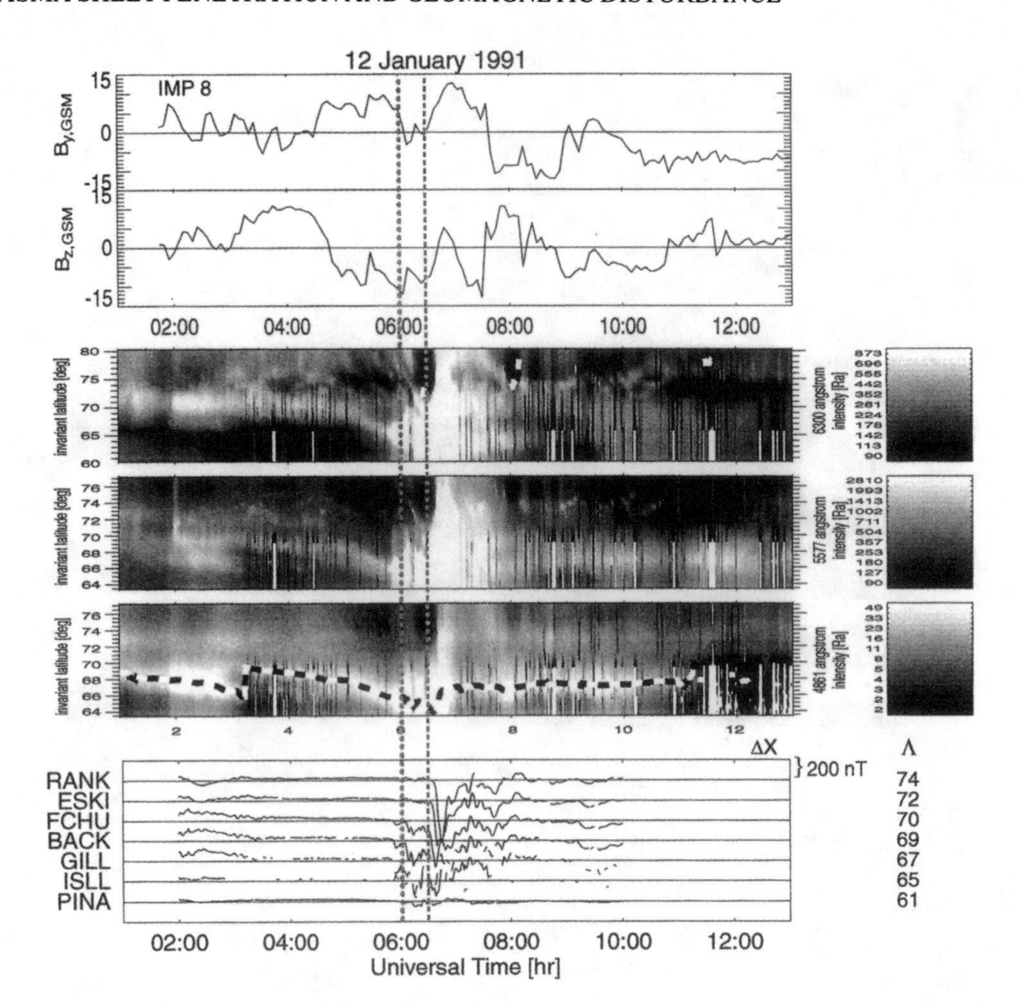

Figure 11. Interplanetary magnetic field from IMP 8 (top panels), merged meridian scanning photometer data from the CANOPUS stations Rankin Inlet and Gillam (middle panels), and ground magnetic field data from the meridian chain of magnetometers extending from Pinawa to Rankin Inlet (lower panels) for 01-13 UT on January 12, 1991. Substorm onsets are indicated by dotted gray lines.

geomagnetic activity response to large-scale variations in the IMF. The substorm growth phase is associated with an extended (> 30 mn) period of southward IMF and equatorward motion of the peak in the 4861 Å emissions, and the major onset is associated with a large northward turning of the IMF. The smaller onset seen a half before the larger onset is associated with a smaller northward turning of the IMF. Another example of photometer observations during a typical substorm has been discussed by *Samson* [1995] and *Lyons* [1996b], and two further examples from February 12, 1991 are shown in Figure 12 (no IMF measurements available). The primary onsets on February 12 were at ~0215 and ~0800 UT, and the poleward-moving expansion-phase aurora are clearly seen following the onsets. Also, geosynchronous injections can be seen in the LANL geosynchronous energetic electron data (Figure 13) for both substorms. The large enhancement seen in the 1987-097 satellite data at ~0215 UT is characteristic of substorm injections seen near midnight and in the early morning, whereas the more gradual and weaker injection seen at the other satellites and in association with the 0800 UT substorm are characteristic of injections seen at other MLT's.

Figure 12 shows a number of separatrix disturbances from ~0330 to ~0930 UT, which indicates that these disturbances can occur during the growth, expansion, and recovery phases of substorms as well as during quiet times. The disturbances during the period of substorms in Figure 12 appear to be more intense than those during the quiet nights. Separatrix disturbances are not as prevalent in Figure 11 as in Figure 12; however the lack of many observable disturbances during that night could be do to hazy observing conditions that caused considerable atmospheric scattering.

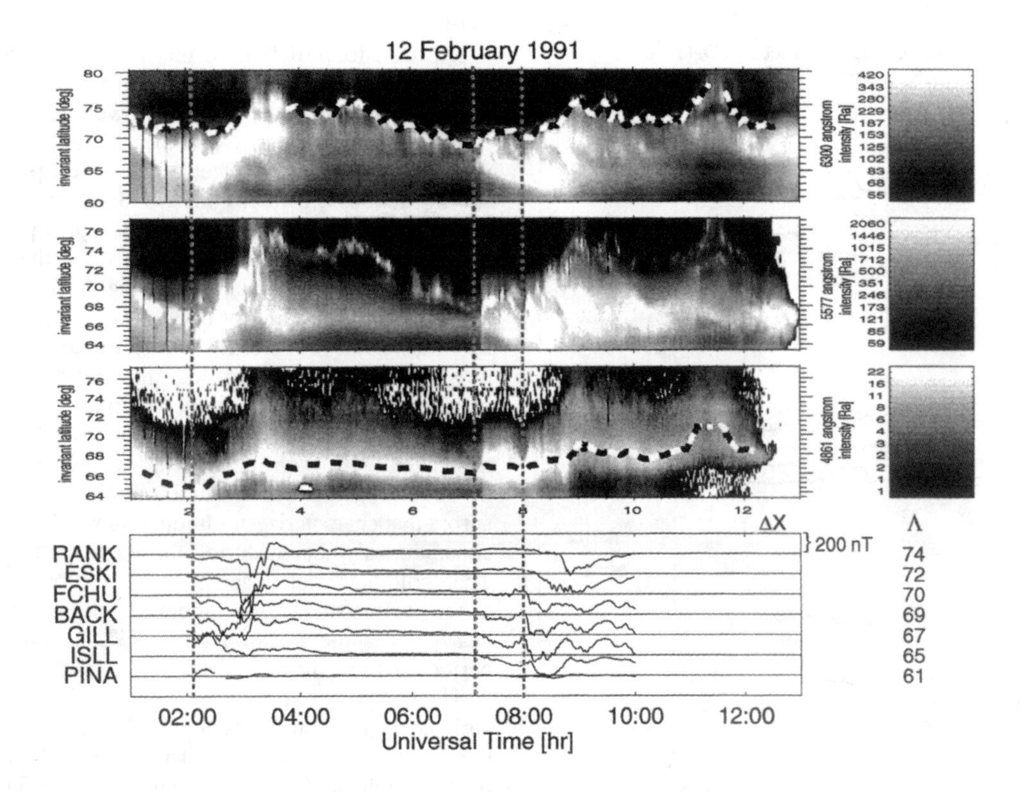

Figure 12. Merged meridian scanning photometer data from the CANOPUS stations Rankin Inlet and Gillam for 01-13 UT on February 12, 1991. Substorm onsets are indicated by dotted gray lines.

Relatively isolated traditional substorms were seen during a total of five of the nights (December 15, 24; January 9, 12; February 12) for which we have good photometer data. In all these cases, the inward motion of the plasma sheet during the growth phase carried the peak of the plasma sheet precipitation from near its quiet to location to $\Lambda = 64°\text{-}67°$. At substorm expansion phase onset, active aurora initiates with the inner plasma sheet at a latitude ~5-6° equatorward of the magnetic separatrix. The region of active aurora then moves several degrees in latitude poleward during the expansion phase. After the expansion phase, the band of plasma sheet precipitation reforms well poleward of its location at onset and near where it is located during quiet times. Strong separatrix disturbances, as seen in Figure 12, were seen during three of these nights.

Figure 14 shows IMF, photometer, and ground magnetometer data for 1-13 UT on January 24, 1991, when the IMF was predominantly southward. Such conditions should lead to relatively steady enhanced convection as has been associated with convection bays. This period can be seen to have had a persistent, enhanced, inner plasma sheet with a peak that stayed near 66° latitude, which is definitively equatorward of the quiet time location of the peak. Such an equatorward location for the ionospheric mapping of the inner plasma sheet implies that the inner plasma sheet indeed remains close to the earth, with enhanced pressures, during convection bays, and that the pressure reductions that follow substorm expansions are absent. The intensity of the ion precipitation and its low latitude extension decreased noticeably during the ~0430 to 0600 UT period when the IMF became temporarily northward, but gradually intensified and moved equatorward after the IMF again became steadily southward. The plasma sheet can be seen to have moved poleward at ~09 UT, which may be partially a local time effect. The ground magnetometer data for this period were far more stable than during the substorm periods and do not show evidence of substorms, except during the weak activity after 08 UT. The LANL geosynchronous energetic electron data for this period (Figure 15) also are relatively stable, distinct injections of the type associated with substorms being absent except in association with the activity after 08 UT.

Strong separatrix disturbances can be seen throughout much of the period in Figure 14. These enhancements appear to move equatorward from the polar cap boundary, are associated with up to 200 nT perturbations in the ground magnetic field, and are the most distinct type of auroral activity during this period.

A second example of a convection bay period (December 13, 1990) is shown in Figure 16. While there was no IMF data for this period, the inner plasma sheet remained

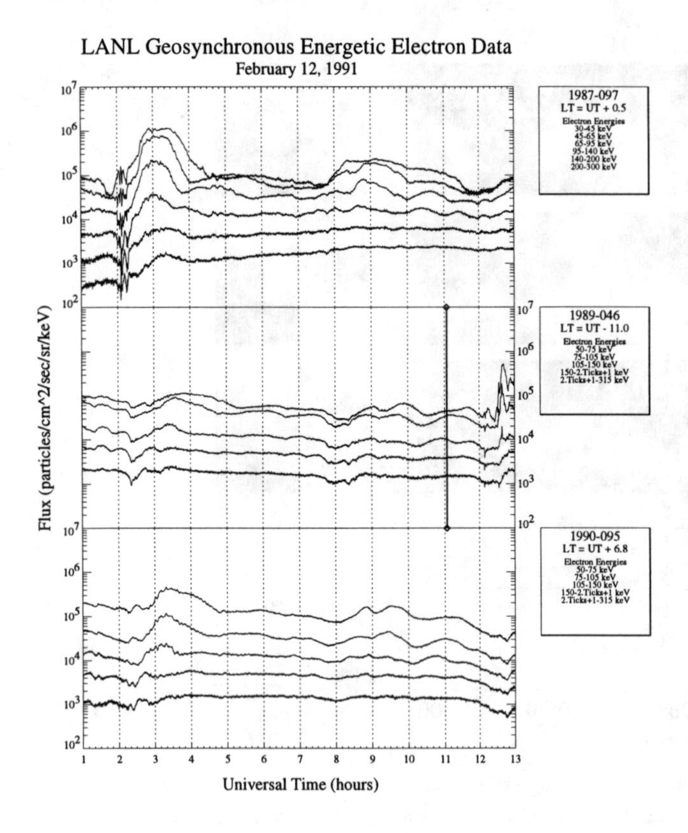

Figure 13. LANL geosynchronous energetic electron data for 01-13 UT on February 12, 1991.

centered at $\Lambda \lesssim 66°$ from 01 to 09 UT as expected for a prolonged period of relatively steady, enhanced convection. As with the January 24 convection bay, the photometer data in Figure 16 show strong separatrix disturbances throughout the period of apparently enhanced convection. These enhancements are associated with significant perturbations in the ground magnetic field. The perturbations reached ~300 nT at ~0415 UT and after 08 UT. The perturbation that began at ~08 UT looks very much like that associated with a substorm expansion phase onset. However the auroral disturbance moved equatorward, and there is no evidence of the region of poleward-moving active aurora that is associated with traditional substorms and is seen in Figures 11 and 12. The LANL geosynchronous data for the December 13 period are shown in Figure 17, and consistent with the ground and optical data, distinct signatures of substorm injections are absent, except after 08 UT. A significant particle enhancement occurred after 08 UT. The flux increase associated with this enhancement is somewhat more gradual than typically associated with substorms, but nevertheless resembles a substorm injection. Clearly the differences and similarities between substorms and the earthward moving enhancements requires further study, and the possibility that the equatorward moving enhancements can be associated with particle injections in the inner plasma sheet may need to be considered.

During a total of four of the nights with good data, the inner plasma sheet remained much more intense than during quiet times and centered at lower Λ's. Two of these convection bay periods (December 24, and January 8) were weaker and had plasma sheet precipitation peaked between $\Lambda = 66°$ and $67°$, whereas the two stronger convection bays discussed above had plasma sheet precipitation peaked between $\Lambda = 64°$ and $66°$. We expect that the IMF is relatively stable with $B_z < 0$ during these periods, though IMF data are available for only one of them. Auroral activity during these periods is dominated by equatorward-moving separatrix disturbances that move towards the inner plasma sheet from near the separatrix. These disturbances are particularly intense during convection bays and can be associated with significant (up to ~300 nT) magnetic perturbations on the ground. Poleward-moving active aurora that characterize traditional substorm expansions are conspicuously absent during these convection bay periods. The photometer data also show that the poleward boundary of the plasma sheet during the convection bay periods is at nearly the same location ($\Lambda \sim 73°$) as during the quiet periods with positive IMF B_z.

The two remaining nights (February 7 and 11) had quite variable IMF B_z and moderately active aurora and ground magnetic fields. We have not interpreted the data for these more complicated situations, which may be further complicated by IMP 8 being well away from the Earth-sun line ($|y| > 23\ R_E$).

CONCLUSIONS

The inward penetration of the plasma sheet is clearly a crucial aspect of magnetospheric dynamics and disturbances. It is related to the important changes in nightside magnetic fields and currents that occur at r ~ 6-10 R_E, and is uniquely related to geomagnetic disturbances. Based on what has been published in the literature and the preliminary study described in this paper, we can reach a number of tentative conclusion. Since these conclusions are based on a limited number of previous studies and on the small data set considered here, their validity should be evaluated with much larger studies. The tentative conclusions are as follows:

During periods when convection is weak, as it is when the IMF B_z remains positive, the inner edge of the plasma sheet remains considerably farther from the Earth than synchronous orbit. The 4861 Å emissions observed from the ground show that the proton precipitation from the inner plasma sheet precipitation is weak during these periods, and that, in contrast to more active periods, precipitation from the inner plasma sheet remains peaked at $\Lambda \geq 68°$. Auroral and magnetic activity remain low during such periods. However, weak equatorward moving disturbances that initiate near the magnetic separatrix can occur during quiet periods. These separatrix disturbances are associated with few minute enhancements of equatorward flow in the ionosphere that propagate westward and equatorward within the poleward portion of the plasma

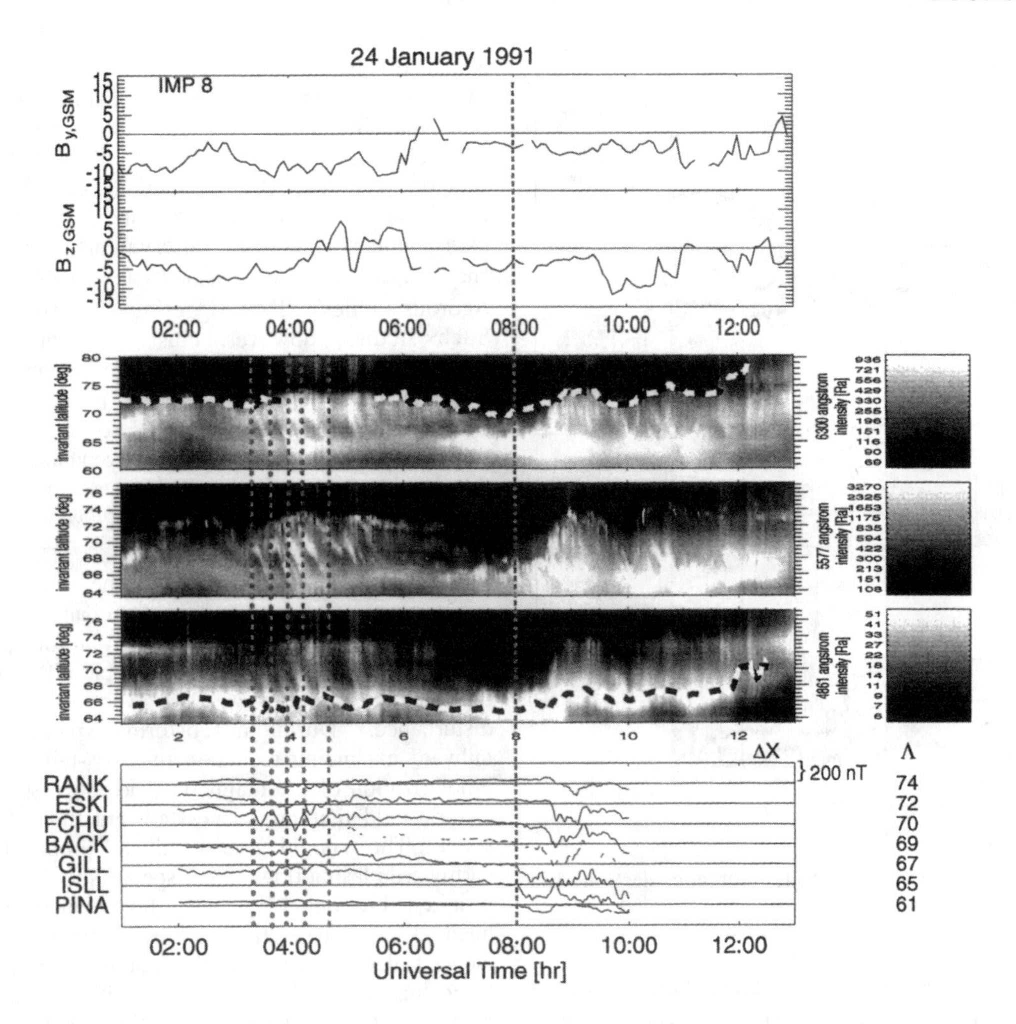

Figure 14. Interplanetary magnetic field from IMP 8 (top panels), merged meridian scanning photometer data from the CANOPUS stations Rankin Inlet and Gillam (middle panels), and ground magnetic field data from the meridian chain of magnetometers extending from Pinawa to Rankin Inlet (lower panels) for 01-13 UT on January 24 1991. A series of equatorward moving auroral enhancements is indicated by gray dashed lines, and a possible onset of a weak substorm is indicated by a dotted gray line.

sheet. Auroral enhancements are on the western edge of the flow enhancements.

Enhanced convection moves the inner edge of the plasma sheet earthward to near synchronous orbit as postulated by *Vasyliunas* [1968], and can occasionally move the plasma sheet earthward of synchronous obit. Enhanced convection that persists for an extended period of time ($\gtrsim$ 2 hr) leads to what is referred to as a convection bay. Convection bay periods are associated with an inner plasma sheet that remains much more intense than during quiet times and the peak in precipitating proton fluxes is centered at lower Λ's (64° - 67°) than during quiet periods. We expect that the IMF is relatively stable with $B_z < 0$ during these periods, though IMF data are only available for one of the four convection bay periods that we considered. During convection bays, enhanced plasma pressures are maintained for extended periods in the inner plasma sheet without the large reductions of cross-tail current, and the associated reductions in plasma pressure, that occur during substorms. Auroral activity during these periods is dominated by separatrix disturbances that move towards the inner plasma sheet from near the separatrix. These disturbances are particularly strong during convection bays and can be associated with significant (up to ~ 300 nT) magnetic perturbations on the ground. Poleward-moving active aurora that characterize traditional substorm expansions are not seen during these convection bay periods.

It is also interesting that the magnetic separatrix during the convection bay periods is at nearly the same location (L ~ 73°) as during the quiet periods with positive IMF B_z. A similar lack of sensitivity of the nightside separatrix location to the sign of the IMF B_z has been noted from

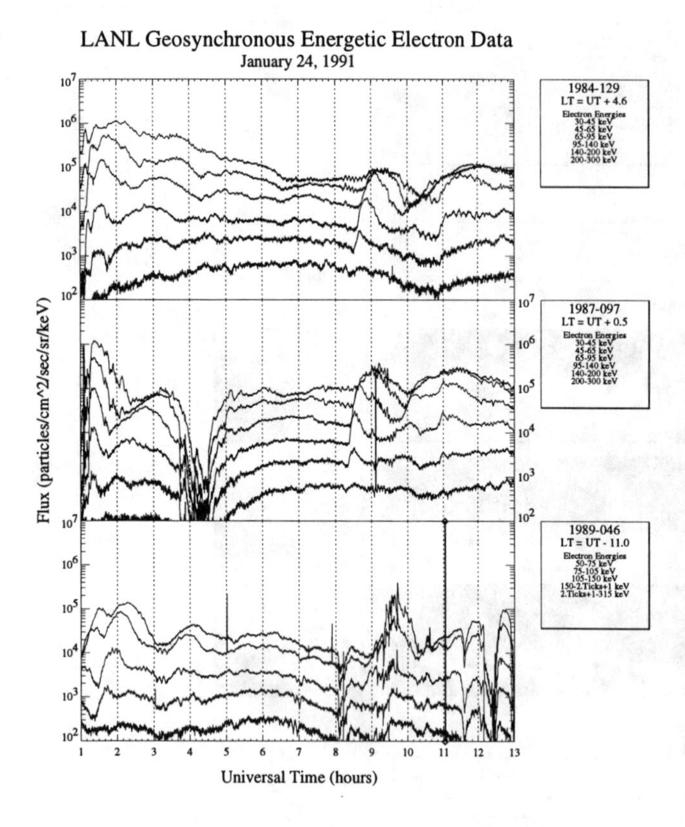

Figure 15. LANL geosynchronous energetic electron data for 01-13 UT on January 24, 1991.

low-altitude measurements of precipitating particles [*Lyons et al.*, 1996]. In contrast, the separatrix is seen to move equatorward during the growth phases of the substorms considered here (Figure 11 and 12). This topic of the separatrix location relative to the IMF and to substorms clearly warrants further study.

Shorter periods of enhanced convection ($\gtrsim 0.5$ hr) are associated with a much more dynamic inner edge of the plasma sheet. The inner plasma sheet intensifies and the peak in proton precipitation moves equatorward to Λ = 64°-67° during the period of enhanced convection, which constitute the growth phase of substorms. The equatorward motion of the plasma sheet precipitation reflects the earthward penetration of the plasma sheet associated with enhanced convection. At substorm expansion phase onset, active aurora initiates within the inner plasma sheet at an invariant latitude ~5-6° equatorward of the separatrix, and the region of active aurora then moves poleward several degrees in latitude. After the expansion phase, the band of inner plasma sheet precipitation reforms well poleward of its location at onset and near where it is located during quiet times, implying a reduction in plasma pressure in the near-Earth plasma sheet. This behavior of the plasma sheet is what is expected from an enhancement of convection during the growth phase and a reduction in convection during the expansion phase. It is consistent with the expansion phase being triggered by IMF changes that lead to a reduction in convection (i.e., northward turnings or reduction in the IMF $|B_y|$) following a period of enhancement in convection. The IMF data for the one substorm for which it is available (Figure 11) shows such IMF behavior, and it has been found to be characteristic of many published substorm observations [See *Lyons*, 1996b, and references therein, and *Lyons*, 1997]. Additional rigorous studies of IMF triggering of substorms are needed. Such studies, however, must take into account the considerable IMF structure that often occurs in the plane perpendicular to the Earth-sun line [e.g., *Crooker et al.*, 1982]. This structure can lead to uncertainties in the timing of IMF triggers, and it can even cause IMF changes that impinge upon the magnetosphere to occasionally be missed by an IMF monitor $\gtrsim 10$ R_E away from the Earth-sun line [*Sotirelis et al.*, 1997].

It is suggested here that both the expansion phase of substorms and the separatrix disturbances are associated with BBF's observed within the high-altitude plasma sheet. However, there should be considerable differences between the BBF's during these two very different types of disturbance. During a substorm expansion phase, the tailward expanding reduction of cross-tail current gives a dipolarization of the magnetic field and associated induced electric field that propagates tailward through the plasma sheet. This should give a tailward propagating burst of earthward plasma flow. Flow speeds should be ~100 km/s near synchronous orbit and should increase to several hundred km/s at r~ 20 R_E. Since these BBF's are due to induced electric fields in the magnetotail, corresponding flow bursts should not occur within the ionosphere. However, the tailward propagation of the flow bursts should be associated with the poleward moving region of active aurora and westward electrojet that is observed in the ionosphere during the expansion phase.

The separatrix disturbances on the other hand, are associated with potential electric fields that should cause flow within the equatorial plasma sheet and within the ionosphere. The flows propagate equatorward in the ionosphere and the ionospheric flow speed are ~500 m/s. The earthward flow speeds in the magnetotail should be several hundred km/s, and the flow bursts should propagate earthward towards the inner plasma sheet. These BBF's do not generally reach the inner edge of the plasma sheet, but they have been seen as close to the Earth as x ~ -11 R_E. These flow bursts can occur during all geomagnetic conditions, and should be particularly common and intense during the enhanced convection conditions associated with convection bays. They should thus be most common when inner plasma sheet pressures are enhanced. However, the frequency of occurrence of these flow bursts as function of geomagnetic activity has not yet been definitively determined.

In addition to the above plasma sheet responses to large-scale convection, the plasma sheet has important dynamical responses to substorms that have not been adequately

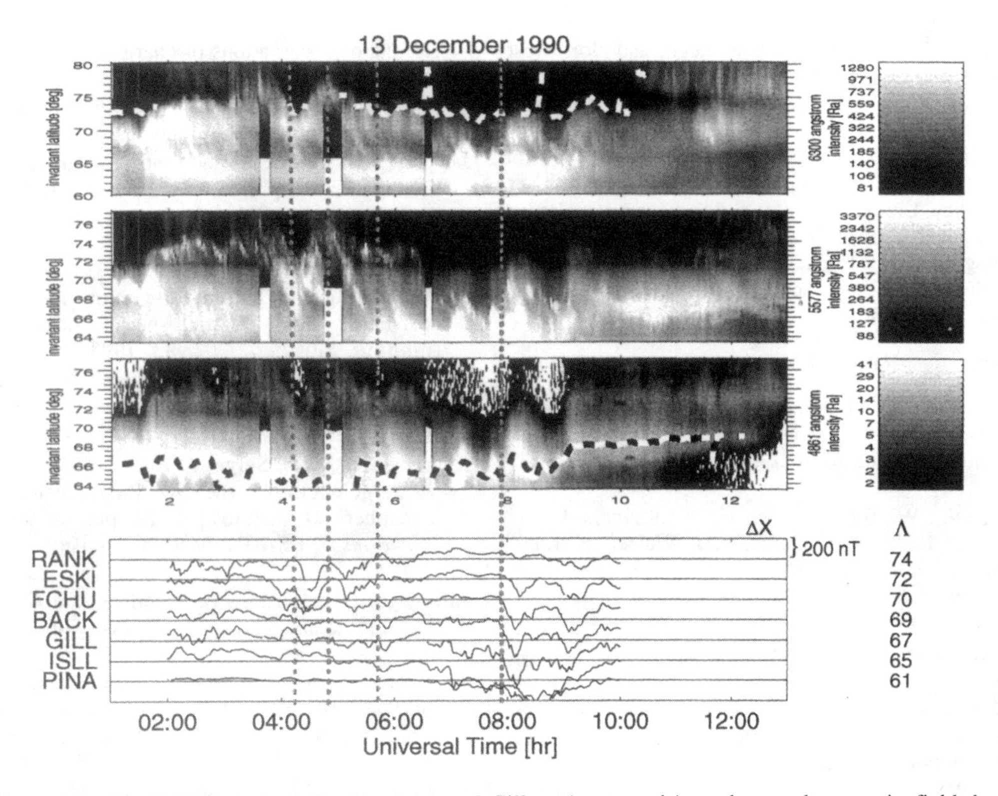

Figure 16. CANOPUS stations Rankin Inlet and Gillam (top panels), and ground magnetic field data from the meridian chain of magnetometers extending from Pinawa to Rankin Inlet (lower panels) for 01-13 UT on December 13, 1990. Three equatorward moving auroral enhancements are indicated by gray dashed lines, and such an enhancement with a magnetic perturbation resembling that associated with substorms is indicated by a dotted gray line.

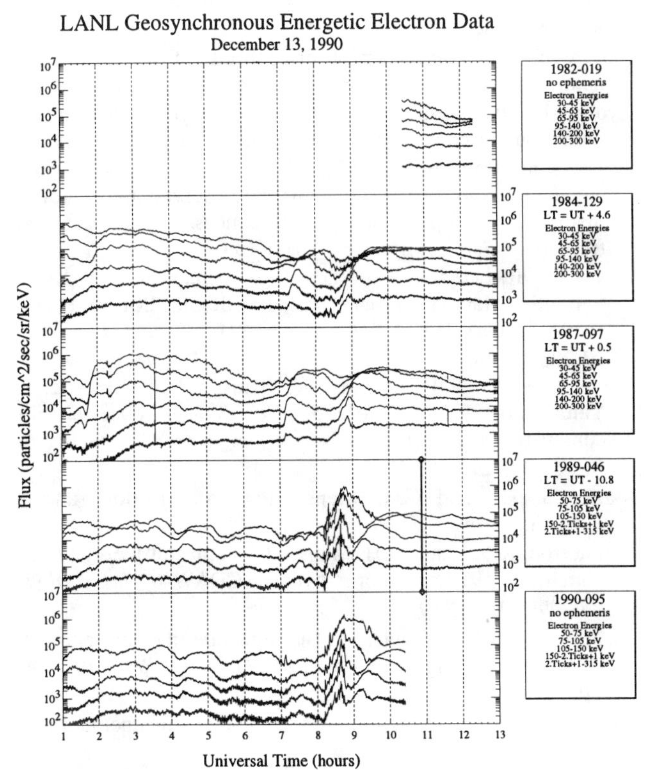

Figure 17. LANL geosynchronous energetic electron data for 01-13 UT on December 13, 1990.

studied. For example, plasma sheet particle pressures must decrease as the cross-tail current decreases during the expansion phase of substorms. At the same time, however, the inward motion of particles as the magnetic field dipolarizes gives rise to particle injections. A possible partial resolution to this dilemma is that the inward motion due to magnetic field dipolarization gives particle flux increases at energies higher than the thermal energy, while the fluxes at lower energies decrease giving a decrease in total plasma pressure. However, there are currently only very limited published observations showing that this occurs, and a quantitative explanation of how particle fluxes at the lower energies decrease as the magnetic field dipolarizes has yet to be obtained.

Acknowledgments. This work has been supported by NSF grants ATM-9120072 and OPP-9423409 and NASA grant NAGW-3968, and the Aerospace Sponsored Research Program. CANOPUS data has been obtained with support of the Canadian Space Agency. GGS-Polar-Hydra data have been made possible by contributions from the University of Iowa, NASA-GSFC, the University of New Hampshire, Max-Planck Institute

Lindau, the University of California San Diego, and Northwest Space Enterprises and were funded by NASA grant NAG5-2231 to the University of Iowa. The Kapuskasing SuperDARN radar is operated by the Johns Hopkins University Applied Physics Laboratory with the support of NASA grant NAG5-1099 and NSF grant ATM-9502993. The Saskatoon SuperDARN radar is operated by the University of Saskatchewan under a Collaborative Special Project Grant from the Natural Sciences and Engineering Research Council of Canada. G. D. Reeves would like to thank the Department of Energy Office of Basic Energy Science for support.

REFERENCES

Aggson, T. L., J. P. Heppner, and N. C. Maynard, Observations of large magnetospheric electric fields during the onset phase of a substorm, *J. Geophys. Res.*, *88*, 3981, 1983.

Angelopoulos, V., W. Baumjohann, C. F. Kennel, F. V. Coroniti, M. G. Kivelson, R. Pellat, R. J. Walker, H. Lühr, and G. Paschmann, Bursty bulk flows in the central plasma sheet, *J. Geophys. Res.*, *97*, 4027, 1992.

Baker , K. B., and S Wing, A new magnetic coordinate system for conjugate studies of high latitudes, *J. Geophys. Res.*, *94*, 9139, 1989.

Bame, S. J., J. R. Asbridge, H. E. Felthauser, E. W. Hones, and I. B. Strong, Characteristics of the plasma sheet of the earth's magnetotail, *J. Geophys. Res.*, *72*, 113, 1967.

Baumjohann, W. J., G. Paschmann, and H. Lühr, Characteristics of high-speed flows in the plasma sheet, *J. Geophys. Res.*, *95*, 3801, 1990.

Birn, J., et al., Characteristic plasma properties during dispersionless substorm injections at geosynchronous orbit, *J. Geophys. Res.*, *102*, 2309, 1997.

Blanchard, G. T., L. R. Lyons, J. C. Samson, and F. J. Rich, Locating the polar cap boundary from observations of 6300 Å auroral emission, *J. Geophys. Res.*, *100*, 7855, 1995.

Blanchard, G. T., L. R. Lyons, and J. C. Samson, Accuracy of 6300 Å auroral emission to identify the separatrix on the night side of the Earth, *J. Geophys. Res.*, *102*, 9697, 1997.

Chen, M. W., M. Schulz, L. R. Lyons, and D. J. Gorney, Stormtime transport of ring current and radiation belt ions, *J. Geophys. Res.*, *98*, 3835, 1993.

Crooker, N. U., G. L. Siscoe, C. T. Russell, and E. J. Smith, Factors controlling degree of correlation between ISEE 1 and ISEE 3 interplanetary magnetic field measurements, *J. Geophys. Res.*, *87*, 2224, 1982.

de la Beaujardière, O., L. R. Lyons, J. M. Ruohoniemi, E. Friis-Christensen, C. Danielsen, F. J. Rich, and P. T. Newell, Quiet-time intensifications along the poleward auroral boundary near midnight, *J. Geophys. Res.*, *99*, 287, 1994.

Frank, L. A., Several observations of low-energy protons and electrons in the Earth's magnetosphere, *J. Geophys. Res.*, *72*, 1905, 1967.

Freeman, J. W., Electron distribution in the outer radiation zone, *J. Geophys. Res.*, *69*, 1691, 1964.

Greenwald, R. A., et al., DARN/SuperDARN: A global view of the dynamics of high latitude convection, *Space Sci. Rev.*, *71*, 761, 1995.

Gringauz, K. I., V. G. Kurt, V. I. Moroz, and I. S. Shklovsky, Results of observations of charged particles out to 100,000 km with the aid of charged particle traps on Soviet space probes, *Astron. Zh.*, *37*, 716, 1960 (English translation: *Soviet Astronomy-AJ*, *4*, 680, 1961.)

Jacquey, C., J. A. Sauvaud, and J. Dandouras, Location and propagation of the magnetotail substorm expansion: Analysis and simulation of an ISEE multi-onset event, *Geophys. Res. Lett.*, *18*, 389, 1991.

Jacquey, C., J. A. Sauvaud, and J. Dandouras, Tailward propagating tail current disruption and dynamics of the near-Earth tail: A multi-point measurement analysis, *Geophys. Res. Lett.*, *20*, 983, 1993.

Jaggi, R. K., and R. A. Wolf, Self-consistent calculation of the motion of a sheet of ions in the magnetosphere, *J. Geophys. Res.*, *78*, 2852, 1973.

Kauristie, K., V. A. Sergeev, T. I. Pulkkinen, R . J. Pellinen, V. Angelopoulos, and W. Baumjohann, Study of the ionospheric signatures of the plasma sheet bubbles, in *Substorms 3*, ESA Publications Division, Noordwijk, p.93, 1996.

Kavanagh, L. D., Jr., J. W. Freeman, and A. J. Chen, Plasma flow in the magnetosphere, *J. Geophys. Res.*, *73*, 5511, 1968.

Kistler, L. M., E. Mobius, W. Baumjohann, G. Paschmann, and D. C. Hamilton, Pressure changes in the plasma sheet during substorm injections, *J. Geophys. Res.*, *97*, 2973, 1992.

Kivelson, M. G., T. A. Farley, and M. P. Aubry, Satellite studies of magnetospheric substorms on August 15, 1968, 5. Energetic electrons, spatial boundaries, and wave-particle interactions at OGO 5, *J. Geophys. Res.*, *78*, 3079, 1973.

Kokubun, S., R. L. McPherron, and C. T. Russell, Triggering of substorms by solar wind discontinuities, *J. Geophys. Res.*, *82*, 74, 1977.

Lui, A. T. Y., et al., Current disruptions in the near-Earth neutral sheet region, *J. Geophys Res.*, *97*, 1461, 1992.

Lyons, L. R., Theory for substorms triggered by sudden reductions in convection, in *Substorms 3*, ESA Publications Division, Noordwijk, p.126, 1996a.

Lyons, L. R., Substorms: Fundamental observational features, distinction from other disturbances, and external triggering, *J. Geophys. Res.*, *101*, 13,011, 1996b.

Lyons, L. R., G. Lu, O. de la Beaujardière, and F. J. Rich, Synoptic Maps of Polar Caps for Stable IMF Intervals during January 1992 GEM Campaign, *J. Geophys. Res.*, *101*, 27,283, 1996.

Lyons, L. R., G. T. Blanchard, J. C. Samson, R. P. Lepping, T. Yamamoto, and T. Moretto, Coordinated observations demonstrating external substorm triggering, *J. Geophys. Res.*, *102*, 27,039, 1997.

Mauk, B. H., and C.-I. Meng, Plasma injection during substorms, *Physica Scripta*, *T18*, 128, 1987.

McPherron, R. L., and R. H. Manka, Dynamics of the 1054 UT March 22, 1979, substorm event: CDAW 6, *J. Geophys. Res.*, *90*, 1175, 1985.

Nishida, A., et al., Traversal of the nightside magnetosphere at 10 to 15 R_E during northward IMF, *Geophys. Res. Lett.*, *24*, 939, 1997a.

Nishida, A., et al., Response of the near-Earth magnetotail to a northward turning of the IMF, *Geophys. Res. Lett.*, *24*, 943, 1997b.

Ohtani, H., S. Kokubun, and C. T. Russell, Radial expansion of the tail current disruption during substorms: A new approach to the substorm onset region, *J. Geophys. Res.*, *97*, 3129, 1992.

Pedersen, A., Substorm electric and magnetic field signatures on GEOS-1, GEOS-2, and ISEE-1, in *Substorms 1*, ESA SP-335, p. 237, European Space Agency, Paris, France, 1992.

Pytte, T., R. L. McPherron, E. W. Hones, Jr., and H. I. West, Jr., Multiple-satellite studies of magnetospheric substorms: Distinction between polar magnetic substorms and convection-driven magnetic bays, *J. Geophys. Res.*, *83*, 663, 1978.

Rostoker, G., et al., CANOPUS-A ground-based instrument array for remote sensing the high latitude ionosphere during the ISTP/GGS program, *Space Sci. Rev.*, *71*, 743, 1995.

Roux, A., et al., Plasma sheet instability related to the westward traveling surge, *J. Geophys. Res.*, *95*, 17,697, 1991.

Ruohoniemi, J. M., R. A. Greenwald, K. B. Baker, J.-B. Villain, C. Hanuise, and J. Kelly, Mapping high-latitude plasma convection with coherent HF radars, *J. Geophys. Res.*, *94*, 13,463, 1989.

Samson, J. C., Mapping substorm intensifications from the ionosphere to the magnetosphere, in *Substorms 2*, edited by J. R. Kan and S.-I. Akasofu, p. 237, University of Alaska, Fairbanks, Alaska, 1995.

Samson, J. C., L. R. Lyons, B. Xu, F Creutzberg, and P. Newell, Proton Aurora and Substorm Intensifications, *Geophys. Res. Lett.*, *19*, 2167, 1992.

Schield, M. A., and L. A. Frank, Electron observations between the inner edge of the plasma sheet and the plasmasphere, *J. Geophys. Res.*, *75*, 5401, 1970.

Scudder, J., et al., HYDRA-A 3-dimensional electron and ion hot plasma instrument for the POLAR spacecraft of the GGS mission, *Space Sci. Rev.*, *71*, 459, 1995.

Sergeev, V. A., O. A. Aulamo, R. J. Pellinen, M. K. Vallinkoski, T Bösinger, C. A. Cattell, R. C. Elphic, and D. J. Williams, Non-substorm transient injection events in the ionosphere and magnetosphere, *Planet. Space Sci.*, *38*, 231, 1990.

Sergeev, V. A., T. I. Pulkkinen, R. J. Pellinen, and N. A. Tsyganenko, Hybrid state of the tail magnetic configuration during steady convection events, *J. Geophys. Res.*, *99*, 23,571, 1994.

Shepherd, G. G., et al., Plasma and field signatures of poleward propagating auroral precipitation observed at the foot of the GEOS 2 field line, *J. Geophys. Res.*, *85*, 4587, 1980.

Smith, P. H., and R. A. Hoffman, Direct observations in the dusk hours of the characteristics of the stormtime ring current during the beginning of magnetic storms, *J. Geophys. Res.*, *79*, 966, 1974.

Sotirelis, T., P. T. Newell, and C.-I. Meng, Polar rain as a diagnostic of recent rapid dayside merging, *J. Geophys. Res.*, 1997 (in press).

Spence, H. E., M. G. Kivelson, R. J. Walker, and D. J. McComas, Magnetospheric plasma pressures in the midnight meridian: Observations from 2.5 to 35 R_e, *J. Geophys. Res.*, *94*, 5264, 1989.

Thomas, B. T. and P. Hedgecock, Substorm effects on the neutral sheet inside 10 Earth radii, in *The Magnetospheres of Earth and Jupiter*, ed. by V. Formisano, pp. 55-70, D Reidel Publishing Co., Dordrecht-Holland, 1975.

Vasyliunas, V. M., A survey of low energy electrons in the evening sector of the magnetosphere with OGO -1 and OGO-3 , *J. Geophys. Res.*, *73*, 2839, 1968.

Wolf, R. A., Magnetospheric configuration, in *Introduction to Space Physics*, ed. by M. G. Kivelson and C. T. Russell, p. 288, Cambridge Univ. Press, New York, 1995.

Yeoman, T. K., and H. Lühr, CUTLASS/IMAGE observations of high-latitude convection features during substorms, *Ann. Geophysicae*, *15*, 692, 1997.

G. T. Blanchard and L. R. Lyons, Department of Atmospheric Sciences, University of California, Los Angeles, CA 90095-1565

R. A. Greenwald and J. M. Ruohoniemi , Johns Hopkins University, Applied Physics Laboratory, Johns Hopkins Road, Laurel, MD 20723-6099

G. D. Reeves, Los Alamos National Laboratories, NIS-2 Mail Stop D436, Los Alamos, NM 87545

J. C. Samson, Department of Physics, University of Alberta, Edmonton, Alberta, Canada T6G 2E9

J. D. Scudder, Department of Physics and Astronomy, University Iowa, Iowa City, IA 52252-1479

Plasma Waves in Geospace: GEOTAIL Observations

H. Matsumoto, H. Kojima, and Y. Omura

Radio Atmospheric Science Center, Kyoto University, Kyoto, Japan

I. Nagano

Department of Electrical and Computer Engineering, Kanazawa University, Kanazawa, Japan

This article presents an overview of plasma wave observations in Geospace by the Plasma Wave Instrument (PWI) onboard the GEOTAIL spacecraft. The GEOTAIL spacecraft covers a wide area in Geospace from the solar wind down to the distant geomagnetic tail. Accordingly the plasma wave data obtained in the vast area show a very rich collection of plasma waves and related phenomena. To overview and understand the GEOTAIL plasma wave data, this article covers not only the plasma wave data themselves, but also the details of the PWI instrument and related computer experiments. The plasma wave data are discussed from two aspects: from the wave modes and related wave-particle and wave-wave interactions, and from their geophysical roles in the different regions of Geospace.

1. INTRODUCTION

Plasma waves in geospace have been studied extensively over the past 30 years by both ground-based observations and in situ satellite observations. The literature of these observations and related theoretical works is quite large and not listed here (instead, references are cited in the following chapters when needed). Though interesting and stimulating plasma wave data from other planetary space are now available [e.g., see *Gurnett et al.*, 1980; *Kurth*, 1988 and references therein], the geospace provides the most detailed information partly because of the large number of available satellites and partly of higher telemetry capacity with richer available instruments on plasma particles and fields.

New Perspectives on the Earth's Magnetotail
Geophysical Monograph 105

GEOTAIL was launched on July 24, 1992 as one element of the ISTP fleet [*Acuña et al.*, 1995]. The expected role of GEOTAIL is to observe the deep tail for the first two years after its launching and then cover near-Earth geospace such as the outer magnetosphere, the magnetosheath, the magnetopause, the bow shock and the upstream solar wind as well as the near-tail region for the rest of its mission life after the orbital change. The GEOTAIL Plasma Wave Instrument (PWI) is dedicated to observation of the frequency spectra and waveforms (for a limited frequency range below 4kHz though) of plasma waves. The instrument was carefully designed and installed as a satisfactory quiet receiver set free from artificial noise from the spacecraft. Since the successful extension of its sensors (two electric and three magnetic sensors), the GEOTAIL PWI became functioning since September 1992. Since then it has been observing plasma waves almost continuously.

This article provides an overview of plasma wave research by GEOTAIL over the past five years after the launch attempting to describe the observed plasma waves from both interests of geophysics and plasma physics. We have therefore designed to construct our chapters in such a way that we discuss plasma waves from two aspects; First in terms of the wave modes and relevant instabilities and second in terms of regional characteristics.

The outline of this article is as follows. Chapter 2 is prepared to give the background information to have better understanding of the presented wave data. It provides a brief theoretical background of linear dispersion relation of plasma waves in cold and hot magnetized plasmas. Also described is the instrumental setup of both sensors and receivers. The GEOTAIL orbits are also discussed in this chapter. Chapter 3 and Chapter 4 are devoted to description of electromagnetic and electrostatic waves observed by GEOTAIL, respectively. Their description is based on the wave modes or wave names in principle. Chapter 5, on the contrary, focuses on the description of plasma waves in different regions of geospace. Plasma waves associated with transient phenomena such as CME(Coronal Mass Ejection), CIR(Corotating Interaction Region) in the solar wind and plasmoid passage in the tail are then presented in Chapter 6. In Chapter 7, we present a summary of our efforts of computer experiments using particle codes on nonlinear evolution of electron beams in plasmas. The computer experiments have been carried out to explain newly discovered natures of the broadband electrostatic noise(BEN) by GEOTAIL PWI. The GEOTAIL observation disclosed a fact that BEN is the result of a series of solitary bi-polar electric field (or equivalently a mono-humped potential) running along the geomagnetic field in the plasma sheet boundary layer(PSBL). Chapter 8 summarizes this article and provides some discussion on the role and future direction of the plasma wave research.

2. GEOTAIL PLASMA WAVE OBSERVATIONS

2.1. Fundamental Characteristics of Plasma Waves

Waves in magnetized plasmas generally exhibit a rich variety in their characteristics. Any motion of charged particles cannot be free from the control of magnetic and electric fields as well as inter-particle forces. Plasma particles in general are highly coupled with magnetic and electric wave fields through their self-consistent motion. Therefore the dynamic nature of the collective motion of plasmas is embedded in the wave data. Plasma waves in geospace are therefore valuable for the research on dynamic processes in geospace and thereby provide useful and sometimes unique information, which otherwise is not obtainable.

As the plasma wave phenomena observed in geospace exhibit a highly variable and diverse nature both spatially and temporally, it is often useful to give thoughts on fundamental characteristics of plasma waves. In this section, therefore, we start with a review of the plasma dispersion characteristics which are often used to determine the wave mode. The wave mode identification is normally the first step of understanding the observed waves. However, the information of wave number magnitude is not available with a single space observation. Therefore the mode identification is basically attempted with the use of the relative frequency of the observed waves in reference to the plasma and cyclotron frequencies. Occasionally other measurable information such as the polarization and k-vector direction are used as well. The possibility of Doppler shift effects is also examined when the identified wave mode has a slow phase velocity relative to the plasma flow velocity. The most useful and popular format to show the plasma dispersion relation is the $\omega - k$ diagrams which we use in the following discussion.

Waves in magnetized and homogeneous plasmas can be classified into two categories; electromagnetic waves and electrostatic waves. The electromagnetic waves are those having both electric and magnetic components, and usually propagate with a high velocity and thereby can be observed at far distance from the source region where they are generated. The electrostatic waves, on the other hand, are those without the magnetic components. Their wave numbers are in most cases large and therefore their phase velocity ω/k is small. Their group velocity is much slower than that of electromagnetic waves and hence electrostatic waves strongly reflect the local nature of the plasma which maintains the waves.

2.1.1. Four Electromagnetic Modes in Magnetized Cold Plasmas. We start with a review of the electromagnetic wave nature in the homogeneous and magnetized cold plasma. The wave nature of this category was studied intensively in 1960's by radio scientists for the study of wave propagation in the ionosphere [e.g., see a textbook by *Budden*, 1985]. Figures 1 and 2 show the $\omega - k$ diagrams for the plasma waves in a homogeneous magnetized cold plasma for two different plasma parameters; $\Pi_e/\Omega_e = 5$ and $\Pi_e/\Omega_e = 0.2$, respectively. Note that Π_e and Ω_e represent plasma and cyclotron angular frequencies of electrons, respectively. As the ion mass is much heavier than the electron mass, the dispersion curves relating to the ion motion, i.e., the ion modes, are not seen in the linear scale as used in these figures. Therefore the modes appearing in these figures are essentially electron modes. As far as we stay in the frequency range much above the characteristic frequency of the ion modes such as the ion plasma frequency or ion cyclotron frequency, we have four electron modes

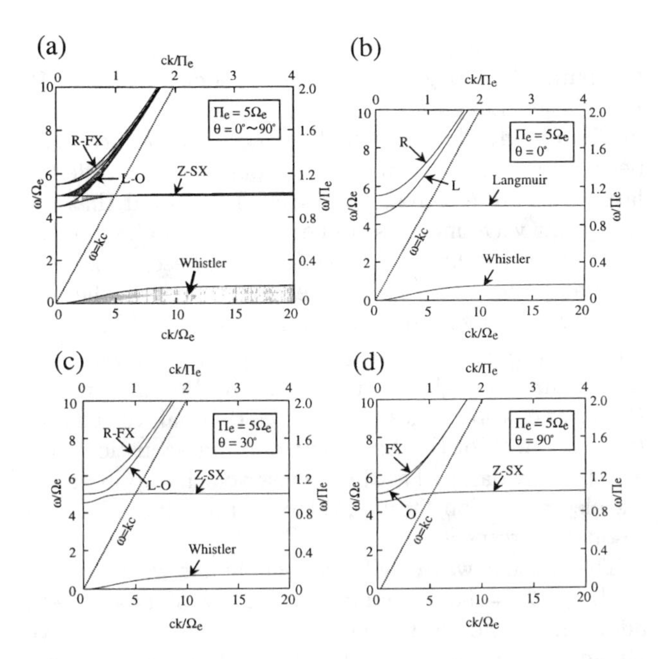

Figure 1. Dispersion ($\omega - k$) diagram for the electromagnetic waves in a homogeneous magnetized cold plasma with $\Pi_e = 5\Omega_e$. The four frames are the four propagation angles (a) $\theta = 0 \sim 90°$, (b) $\theta = 0°$, (c) $\theta = 30°$, and (d) $\theta = 90°$.

for the k vector range displayed in these $\omega - k$ diagrams. The dispersion characteristics of the four modes change as the wave normal angle θ (defined as an angle between the magnetic field and the wave k vector) changes. Three panels (b), (c) and (d) in Figures 1 and 2 show the dispersion curves for $\theta = 0°$, $30°$ and $90°$, respectively. The dispersion curves for an arbitrary wave normal angle between 0° and 90° drop in the shaded areas shown in the $\omega - k$ diagram (a). It is noted that the cutoff and resonance frequencies are given at $k = 0$ and $k \to \infty$, respectively.

The four modes represented by the four shaded areas are called R–FX mode, L–O mode, Z–SX mode, and whistler mode, respectively. The nomenclatures of R–FX, L–O and Z–SX stands for the Right-hand polarized and Fast X mode, the Left-hand polarized and Ordinary mode, and the Z and Slow X mode, respectively. It is noted that the L–O mode has a special limiting branch called Langmuir oscillation at $\theta = 0°$. This oscillation mode has no magnetic component and hence is not an electromagnetic wave. However it becomes a propagating electromagnetic mode as seen in the case for $\theta \neq 0°$.

The whistler mode appears always below the electron cyclotron frequency and is the most frequently observed electromagnetic wave mode in geospace. In the case of $\Pi_e > \Omega_e$, the uppermost resonance frequency $k \to \infty$ of whistler waves decreases smoothly as θ increases and finally reaches the lower hybrid frequency. It is noted, however, the whistler wave mode in Figure 1(d) seems to be vanishing as the lower hybrid frequency is so close to $\omega = 0$ axis. In the case where $\Pi_e < \Omega_e$, on the other hand, the whistler mode branch is split into two different shaded areas as seen in Figure 2. The whistler mode branch for $\theta = 0°$, which starts from $\omega = 0$ and reaches the electron cyclotron frequency, is split into two different branches; the higher and lower frequency branches. The higher frequency branch keeps similar dispersion shape to that for the whistler mode wave at $\theta = 0°$, but is not the whistler mode but the Z–SX mode. Different from the normal whistler characteristics, it now has the lower cutoff frequency at f_{Z-SX} defined by

$$f_{Z-SX} = -\frac{f_{ce}}{2} + \sqrt{f_{pe}^2 + \frac{f_{ce}^2}{4}} \tag{1}$$

Therefore it changes to the Z–SX mode with the upper resonance frequency which exceeds the electron cyclotron frequency, and no more whistler mode. The lower frequency branch suffers from a drastic change. It now has the upper resonance frequency below Π_e. The resonance frequency decreases from Π_e down to the lower hybrid frequency as θ increases. This whistler mode is sometimes referred to as magnetosonic wave or compressional Alfven wave mode in the limit of $\theta \to 90°$. It is noted that the whistler mode can be

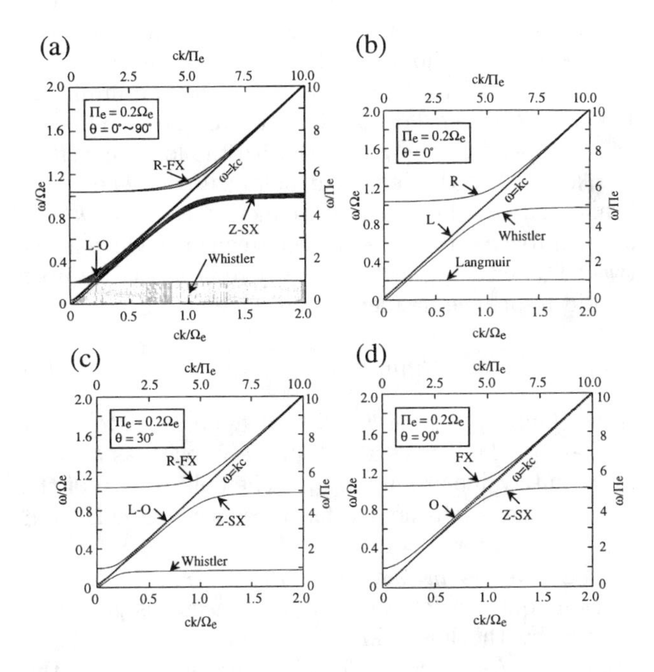

Figure 2. The dispersion ($\omega - k$) diagram for a homogeneous magnetized cold plasma with $\Pi_e = 0.2\Omega_e$. Others are the same as in Figure 1.

found only below the electron plasma frequency except for the purely parallel propagation, (i.e., $\theta = 0°$) in the case where $\Pi_e < \Omega_e$.

The other three modes are essentially free space propagation modes modified by the plasma presence. In free space (i.e., in the limit of the zero plasma density), these three modes shrink into the free space electromagnetic mode propagating with a speed of light. Due to the modification by the plasma presence, they have a cutoff frequency. Below the cutoff frequency, each mode becomes an evanescent mode and does not propagate any more. The lower cutoff frequencies of these modes are given by Eq. (1) and by

$$f_{\mathrm{L-O}} = f_{\mathrm{pe}} \tag{2}$$

$$f_{\mathrm{R-FX}} = \frac{f_{\mathrm{ce}}}{2} + \sqrt{f_{\mathrm{pe}}^2 + \frac{f_{\mathrm{ce}}^2}{4}} \tag{3}$$

The L–O mode is not influenced by the magnetic field and is influenced only by the plasma presence. In contrast, the two X modes appear only under the influence of the external magnetic field, and thereby shrink into the L–O mode when the magnetic field vanishes. The dispersion relation of the R–FX mode does not much change with the increasing θ, but the Z–SX and L–O modes do show a drastic change as θ changes. It should be kept in mind that the L mode propagation in the limit of $\theta = 0°$ is a very special case. It is a special branch starting from the $f_{\mathrm{Z-SX}}$ and smoothly increases its frequency without any frequency gap as $k \to \infty$. The L branch for $\theta = 0°$ is split into the two branches, L–O and Z–SX as θ deviates from zero. The L–O mode is the upper frequency branch with the lower cutoff at $f_{\mathrm{L-O}}$, and has the left-hand polarization. The Z–SX branch has the lower cutoff frequency $f_{\mathrm{Z-SX}}$ at $k = 0$ and approaches to the resonance frequency which is between Ω_e and ω_{UHR} as $k \to \infty$, where ω_{UHR} is the upper hybrid resonance frequency that is defined by

$$\omega_{\mathrm{UHR}} = \sqrt{\Pi_e^2 + \omega_e^2}. \tag{4}$$

In the limiting case of $\theta = 90°$, it approaches to ω_{UHR} as $k \to \infty$. This gives a slow phase velocity as $k \to \infty$ for a finite value around ω_{UHR}. The polarization of the Z–SX mode is left-handed for a frequency $\omega < \Pi_e$ and right-handed for $\omega > \Pi_e$.

2.1.2. Waves in Magnetized Hot Plasmas. The dispersion equation for waves in hot plasmas is generally treated by the linear kinetic theory [e.g., Stix, 1962; Stix, 1992; Akhiezer, 1975]. Numerical solutions of the linear dispersion equation in hot plasmas are shown in Figure 3. Note that both of the frequency and wave number axes are given in logarithmic scales. Figure 3 provides four cases of $\theta = 0°$, $30°$, $80°$, and $90°$. In the parallel propagation case, five modes are seen; R, L, whistler, Langmuir and ion acoustic modes as shown in Figure 3(a). The former three modes are electromagnetic waves discussed in the previous section, while the latter two are electrostatic waves. It is noted that the Langmuir wave mode is connected to the Langmuir oscillation in the limit to $k \to 0$. The phase velocity of the Langmuir waves approaches to the electron thermal velocity as k increases. The phase velocity V_{ph} of the ion acoustic wave mode is given by $V_{\mathrm{ph}} = \sqrt{k_{\mathrm{B}} T_e / m_i}$ in the small k region, where k_{B} is the Boltzmann constant, and approaches to the ion thermal velocity as $k \to \infty$. Both of the Langmuir and ion acoustic waves are strongly damped as their phase velocity approach to the electron or ion thermal velocity through the Landau resonance, respectively.

The plasma waves with perpendicular wave normal angles ($\theta = 90°$) exhibit a variety of modes. In addition to the electromagnetic waves which we have already discussed, a new family of electrostatic waves appears in the large k region. The O mode and X mode are the same as discussed before. Their phase velocity approaches to the light velocity as k increases. The SX mode, however, approaches once to the upper hybrid frequency as already discussed as k increases but its frequency finally drops smoothly as $k \to \infty$ to the integer multiple frequency of the electron cyclotron frequencies which are closest to the upper hybrid frequency. The new family is called Electron Cyclotron Harmonic (ECH) waves. They have a harmonic structure in frequency with a spacing approximately equal to the electron cyclotron frequency. They are sometimes called Bernstein modes [*Bernstein*, 1958; *Stix*, 1962, 1992]. The magnetosonic wave mode (the limiting case of whistler mode wave as $\theta \to 90°$) has a phase velocity equal to the Alfven velocity V_{A} given by

$$V_{\mathrm{A}} = \frac{B_o}{\sqrt{\mu_o m_i N_i}} \tag{5}$$

where μ_o, m_i, N_i, and B_o are the magnetic permeability in vacuum, the ion mass and density, and the external magnetic field, respectively. However, the phase velocity of the magnetosonic wave decreases as its frequency approaches to the lower hybrid frequency.

2.1.3. Discussion on Plasma Wave Features. We have quickly reviewed the dispersion relation of the electron modes. A similar feature is found for the ion modes but is not discussed here simply because the frequency range of the ion modes is much lower than the frequency coverage of the GEOTAIL wave receivers. Another point we must mention is that the dispersion features discussed above are limited to the Maxwellian plasmas. The wave modes in the non-Maxwellian plasmas show different dispersion. However, a superposition of the Maxwellian velocity distribution function can provide a fairly good approximation of non-Maxwellian plasmas. One example of such case will be discussed in Chapter 7 for the cases of plasmas composed of multiple electron species with different drift and/or different temperature. Another approach has been attempted to approximate non-Maxwellian plasmas with a modified Olbertian velocity distribution function [*Vasyliunas*, 1968; *Matsumoto*, 1972] which is given by

$$\begin{aligned} g_s(v_\perp, v_\parallel) &= N_s \sqrt{\frac{m_s}{(2\lambda-3)\Pi\kappa T_{s\parallel}}} \\ &\times \frac{m_s}{(2\lambda-3)\Pi\kappa T_{s\perp}} \frac{\Gamma(\lambda+1)}{\Gamma(\lambda-\frac{1}{2})} \\ &\times \left\{1+\frac{m_s V_{s\parallel}^2}{(2\lambda-3)\kappa T_{s\parallel}}+\frac{m_s V_{s\perp}^2}{(2\lambda-3)\kappa T_{s\perp}}\right\} \end{aligned} \quad (6)$$

where Γ is the gamma function and λ is an arbitrary integer.

A similar attempt has been made recently by Summers and Thorne [1992] and Summers et al. [1996]. They call their velocity distribution function a Kappa function. The plasma dispersion relation can also be found by numerical computations even in these non-Maxwellian distribution cases.

Another important discussion related to plasma waves is, of course, concerned with instabilities and dampings. Though both are interesting and important, they are beyond the scope of this chapter. In the following chapter, we touch upon the instability and damping issues when necessary.

2.2. PWI onboard GEOTAIL

The PWI onboard the GEOTAIL spacecraft consists of the following three different sets of receivers [*Matsumoto et al.*, 1994a]: (1) Sweep Frequency Analyzer (SFA), (2) Multi-Channel Analyzer (MCA) and (3) Wave-Form Capture (WFC). The block diagram of the GEOTAIL PWI is shown in Figure 4. The first two sets of receivers are devoted to measuring frequency spectra of plasma waves, while the last one was designed to capture wave forms of two electric and three magnetic field components, simultaneously. In order to measure weak electric and magnetic fields of plasma waves in the geomagnetic tail region, we designed the two electric sensors and tri-axial search coils with a special attention to reduction of spacecraft noise. Figure 5 shows the configuration of the GEOTAIL spacecraft along with the definition of the antenna axes. Two sets of long dipole antennas with a length of 100 m tip-to-tip are radially deployed and are dedicated to the electric field measurement. They are wire- and probe-antennas called 'WANT' and 'PANT', respectively. The PANT is a top-loaded antenna with a tip sphere load with a diameter of 10 cm. The WANT measures the electric field component E_U, while the PANT measures E_V component.

The assembly of PWI tri-axial search coils (which is hereafter called PWI-SC) and their pre-amplifiers are mounted on the top of a mast called MST–S which has 6 m in length. The tri-axial search coils associated with the MGF (MGF-SC) are mounted on the same mast but 2 m closer to the spacecraft from the PWI-SC. The PWI-SC is normally connected to the PWI measurement system. However, we can select the MGF-SC by a telemetry command if necessary. The three axes of the PWI search coils are defined in a cylindrical coordinate system as follows: α is in the tangential direction, β is in the radial direction, and γ is parallel to the spin Z axis of the spacecraft, respectively.

The highest sensitivities of the electric and magnetic sensors are $5 \times 10\text{nV}/\sqrt{\text{Hz}}$ and $1.5 \times 10^{-5}\text{nT}/\sqrt{\text{Hz}}$, respectively. Further, in order to reduce the effects of the noise interference from other instruments, the PWI investigators had required the other instruments to reduce the noise current on the power lines below a specified level, and not to radiate electromagnetic noise from the harness cables [*Tsutsui et al.*, 1992]. We performed the ElectroMagnetic Compatibility (EMC) tests before the launch and checked if each instrument satisfies the requirement from the PWI. After the launching of GEOTAIL, we could confirm on orbits the successful achievements of the reduction of the electromagnetic noise from other instruments.

2.2.1. PWI Receivers. The SFA provides spectral information of plasma waves over the frequency range from 24 Hz to 800 kHz for the electric field and 24 Hz to 12.5 kHz for the magnetic field. The SFA consists of eight independent receivers covering 5 frequency bands for the electric fields and 3 frequency bands for the magnetic fields. The receiver specifications are listed in Table 1.

Each receiver has a very good frequency resolution of 1/128 of the receiver frequency band although their time resolution is somewhat coarse (64 sec for Bands 1 and 2, and 8 sec for Bands 3 $\sim$ 5). The effective dynamic range of the SFA receivers is $\sim$ 90 dB. The electric field receivers are normally set in the High gain mode which is 30 dB higher than the Low gain mode.

The MCA is provided by R. R. Anderson at the University of Iowa. This subsystem contains two spectrum analyzers with fixed frequency channel filters. It provides high time resolution data to complement the coarser time resolution data by the SFA. However, their frequency resolution is coarse because they have only four frequency channels per decade in frequency. One multi-channel spectrum analyzer is used to measure the

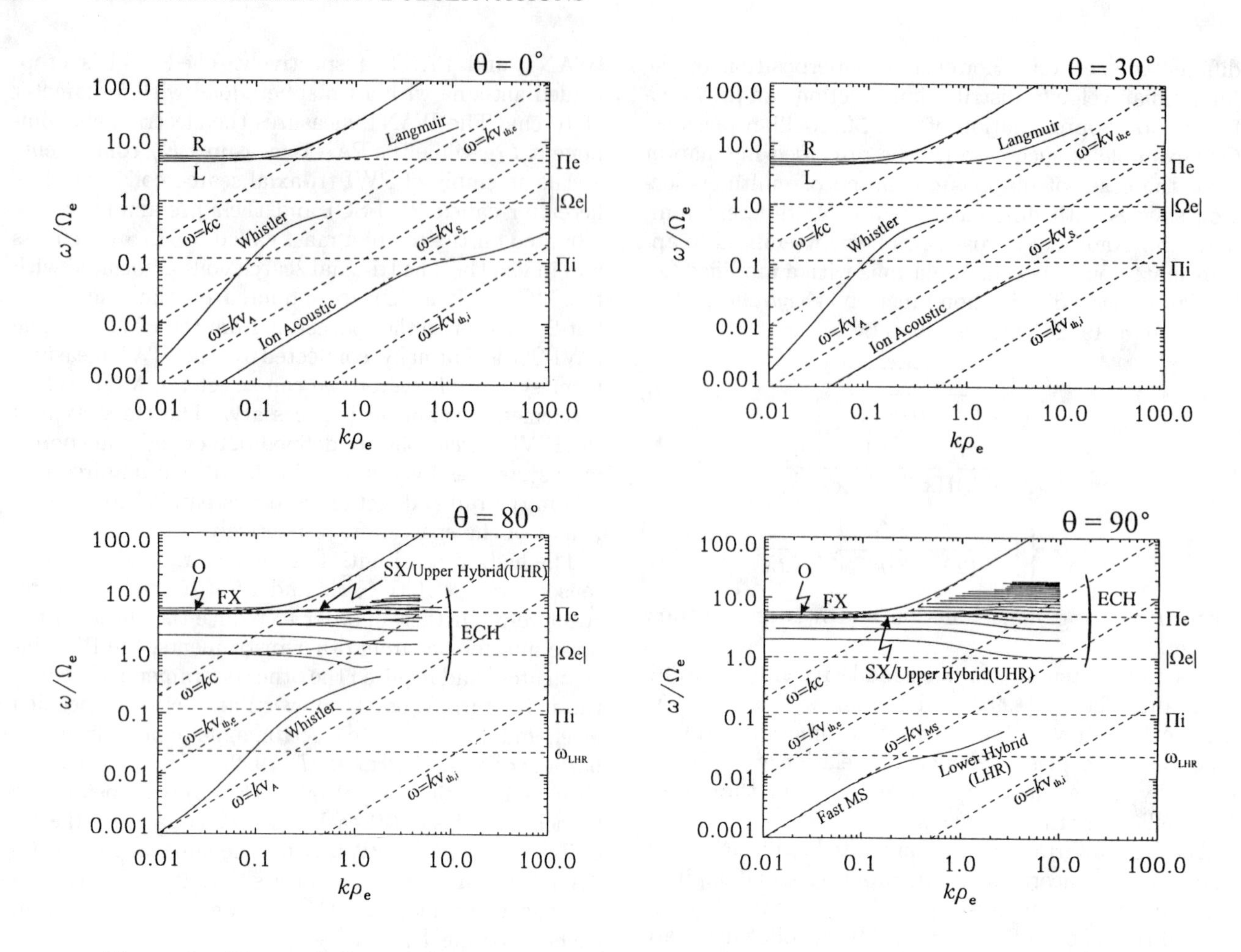

Figure 3. The $\omega - k$ diagram obtained from the numerical solutions to the plasma dispersion equation in a magnetized hot plasmas for four different wave normal angles; (a) $\theta = 0°$, (b) $\theta = 30°$, (c) $\theta = 80°$, and (d) $\theta = 90°$. The quantity ρ_e is electron larmour radius. $v_{th,e}$, and $v_{th,i}$ denote electron and ion thermal velocity, respectively. v_s, v_{MS}, and v_A denote ion sound velocity, magnetosonic velocity, and Alfven velocity, respectively.

wave electric field and is composed of 20 channels covering the frequency range from 5.62 Hz to 311 kHz. The other spectrum analyzer is used for wave magnetic field measurements and has 14 channels covering the frequency range from 5.62 Hz to 10 kHz. The bandwidths of the filters are ±15% of the channel center frequency in the frequency range below 10 kHz and ±7.5% of the center frequency for frequency above 10 kHz. The MCA instrument measures the wave electric field of either the E_U or E_V component (depending on which antenna is used) with a dynamic range of $\sim$ 110 dB and the wave magnetic field of either the B_α or B_γ component with a dynamic range of $\sim$ 100 dB.

Signals from all channels are sampled simultaneously so that the ratio of the electric to magnetic field strength can be calculated accurately. The signals are sampled once (data acquisition time is 1.037 msec) every 250 msec in the highest time resolution mode.

The WFC data are used for the detailed analyses of wave characteristics, such as determination of the wave vector, polarization, Poynting flux, and antenna sheath impedance. The WFC system has five (two electric and three magnetic) receivers each connected to one sensor (E_U, E_V, B_α, B_β, or B_γ), and has two different operation modes: Memory mode and Direct mode.

In the Memory mode, the observed wave signals are measured simultaneously as E_U E_V, B_α, B_β, and B_γ. These are fed to a gain-controller followed by an anti-aliasing filter (LPF) with an upper frequency cutoff of 4 kHz, then through high pass filters (HPF) which have a low frequency cutoff of 10 Hz. Waveforms of the analog signal are sampled and then converted into 12 bit

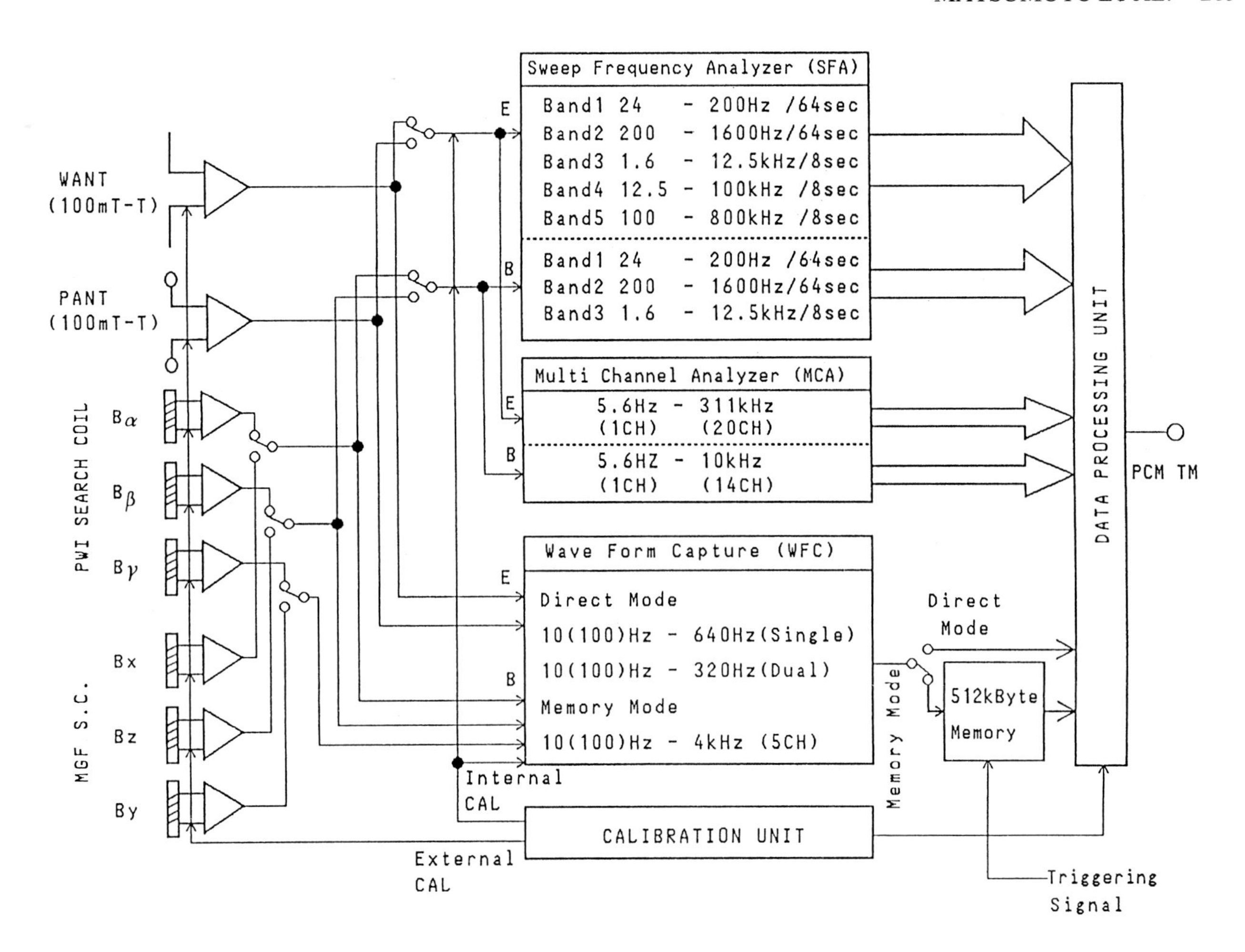

Figure 4. Block diagram of the Plasma Wave instrument (PWI) onboard GEOTAIL spacecraft. PWI consists of three different sets of receivers : (1) Sweep Frequency Analyzer (SFA), (2) Multi-Channel Analyzer (MCA), and (3) Wave-Form Capture (WFC) [*Matsumoto et al.*, 1994a].

data by an A/D converter with a sample frequency of 12 kHz. The 12 bit digital data are compressed into 8 bit by a quasi-logarithmic compression method. The compressed 8 bit data are then stored into the onboard memory with a storage of 512 kBytes for a period of 8.7sec. The stored waveform data are read out of the memory and telemetered to the ground using the PCM telemetry during an uninterrupted period of 275 seconds.

Besides the Memory Mode, the WFC can be operated in the Direct mode, with two selectable options: one is the Single channel mode in which only one field component is measured continuously and telemetered to ground on a real-time basis. In this operation mode the instrument can measure continuous waveforms of one electric or one magnetic component with an upper limit frequency of 640 Hz. The other is the Dual channel mode where two field components are measured and telemetered simultaneously. In this mode the upper limit frequency is 320 Hz. For both of these modes, the data sampling frequency is set to three times the upper limit in frequency.

The WFC system can measure the wave amplitude with a dynamic range of 66 dB. The receiver gain, keeping its dynamic range, can be stepped up by 40 dB or 20 dB by the gain controller.

Since the WFC receiver allows us to analyze the wave phase as well as the wave amplitude, it is very important to calibrate the phase rotation through the electric circuit. The phase rotation from the pre-amplifiers to the output of the receiver had been measured by the ground test. However, we need to measure the antenna impedance of the electric field antennas during the flight in space environment as the antenna

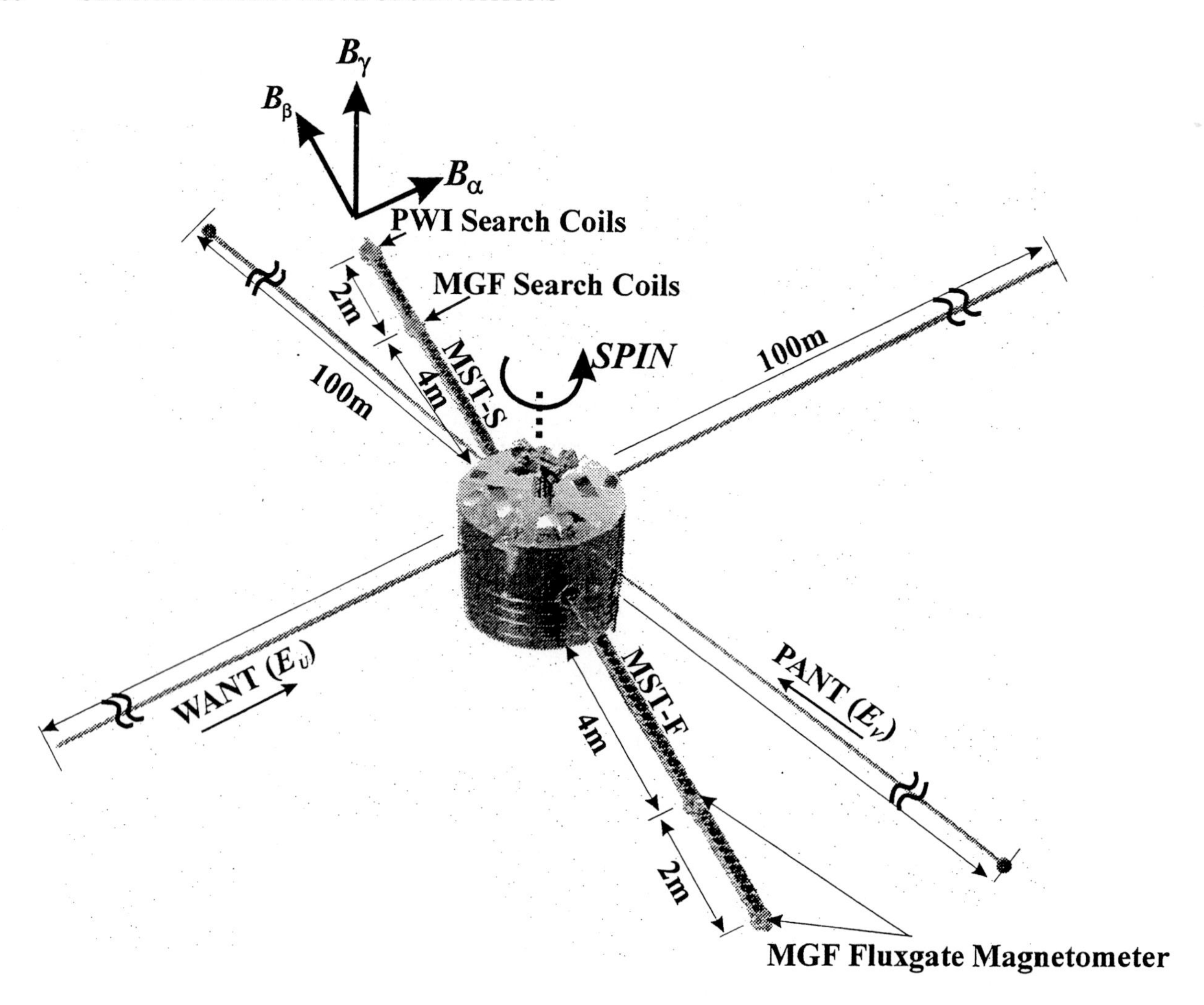

Figure 5. External view of the GEOTAIL spacecraft and antenna configuration.

impedance strongly depends on the surrounding plasma parameters. The PWI has a capability to measure the antenna impedance during the flight [*Matsumoto et al.*, 1994a; *Tsutsui et al.*, 1997]. The detailed measurement technique is described in *Tsutsui et al.*, [1997]. On the basis of the measurement during the flight, *Tsutsui et al.* [1997] obtained the complex antenna impedance and showed that it strongly depends on the plasma density which, as we will introduce in Chapter 5, depends on the location of the spacecraft in the geospace. Therefore we make use of the typical value of the antenna impedance measured in each region of geospace for the detailed analysis of plasma wave data.

2.3. GEOTAIL Orbits

GEOTAIL was launched on July 24, 1992. After the initial spacecraft check, the dipole antenna and the masts for the magnetic field sensor were deployed on August and September, 1992. The plasma wave observation started immediately after the deployment of the antennas. After the initial check of the PWI instrument, the full-fledged observation of plasma waves was initiated and has been carried on without any serious problems up to now. The operation mode of the PWI instrument is controlled at ISAS (Institute of Space and Astronautical Sciences) on the real time basis by the members of the PWI team.

The GEOTAIL orbit was so designed as to cover the wide area of the geospace. It was first injected into the orbits to cover the geomagnetic tail region for the first two year period. Then it was maneuvered into much lower apogee of about $30R_E$ on November 12, 1994. Since then the spacecraft entered into the near Earth orbit phase. The orbits of the GEOTAIL for the distant

Table 1. Specification of SFA

Band	Frequency Range	Freq. Steps	Band Width	Field component	Sweep time
1	24 Hz ~ 200 Hz	128	2.6 Hz	B and E	64 sec
2	200 Hz ~ 1600 Hz	128	10 Hz	B and E	64 sec
3	1.6 kHz ~ 12.5 kHz	128	85 Hz	B and E	8 sec
4	12.5 kHz ~ 100 kHz	128	680Hz	E only	8 sec
5	100 kHz ~ 800 kHz	128	5.4 kHz	E only	8 sec

The SFA consists of eight independent receivers covering 5 frequency bands for the electric fields and 3 frequency bands for the magneticn fields. Each receiver has a fine frequency resolution of 1/128 of the receiver frequency band.

tail phase and the near Earth phase are plotted in Figures 6 (a) and (b), respectively. In Figure 6, we make use of the modified Geocentric Solar-Magnetospheric Coordinates (GSM'), which include the effect of the solar wind aberration of 4 °, the hinging distance of 10 R_E and the tilt angle of the geomagnetic dipole axis.

The orbits in the distant tail phase cover a wide area reaching 210 R_E in the x (anti-solar) direction and $\pm 80 R_E$ in the y-direction. Note that the orbits in the distant tail phase also cover the middle range of the tail from 50 to 120 R_E. These orbits have provided a good opportunity of surveying various regions such as the magnetosheath, the tail lobe, the plasma sheet and the boundary regions between them. On the contrary, the near Earth orbits are limited between $\pm 30 R_E$ in both x and y directions with a perigee of about 10 R_E. However, the spacecraft regularly repeats the rotation around the Earth except for its initial phase of the orbit maneuvering, giving suitable orbits for the study of plasma waves in the upstream solar wind, the bow shock, the magnetosheath and the magnetopause boundary.

3. ELECTROMAGNETIC WAVES OBSERVED BY GEOTAIL

3.1. *Whistler Mode Waves*

3.1.1. Whistlers. Whistlers are well-known phenomena since their discovery in 1950's by the ground observation [*Eckersley*, 1925; *Storey*, 1953] and have also been detected by in-situ satellite observations [*Smith and Angerami*, 1968; *Dunckel and Helliwell*, 1969]. The source of whistlers is lightening discharge which radiates electromagnetic energy into the magnetosphere through the ionosphere. The plasma dispersion characteristics explain the reason for the gliding tone as the group velocity is slower for lower frequency component below the nose frequency [*Ellis*, 1956; *Smith and Carpenter*, 1961; *Helliwell et al.*, 1956]. The GEOTAIL spacecraft occasionally enters into the dayside magnetosphere near its perigee. It traverses the dayside magnetosphere from the magnetopause to the inner magnetosphere. On such occasions, the PWI WFC picked up whistler phenomena. Figure 7 shows one example of a whistler [*Nagano et al.*, 1996; *Nagano et al.*, 1998] observed by GEOTAIL on April 29, 1993. The upper left panel is a dynamic spectrum for B_Z component in the spacecraft coordinates. The k vector direction relative to the ambient magnetic field direction is shown in the upper right panel. The magnitude and direction of the Poynting vector are shown in the lower left and right panels, respectively. The spacecraft was located on a field line of

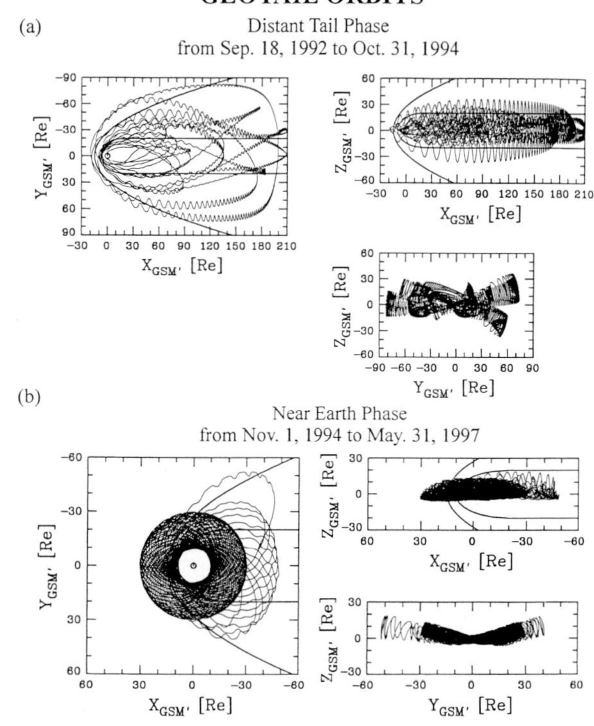

Figure 6. GEOTAIL orbits plotted in the modified GSM coordinate system. (a) Distant tail phase and (b) Near tail phase.

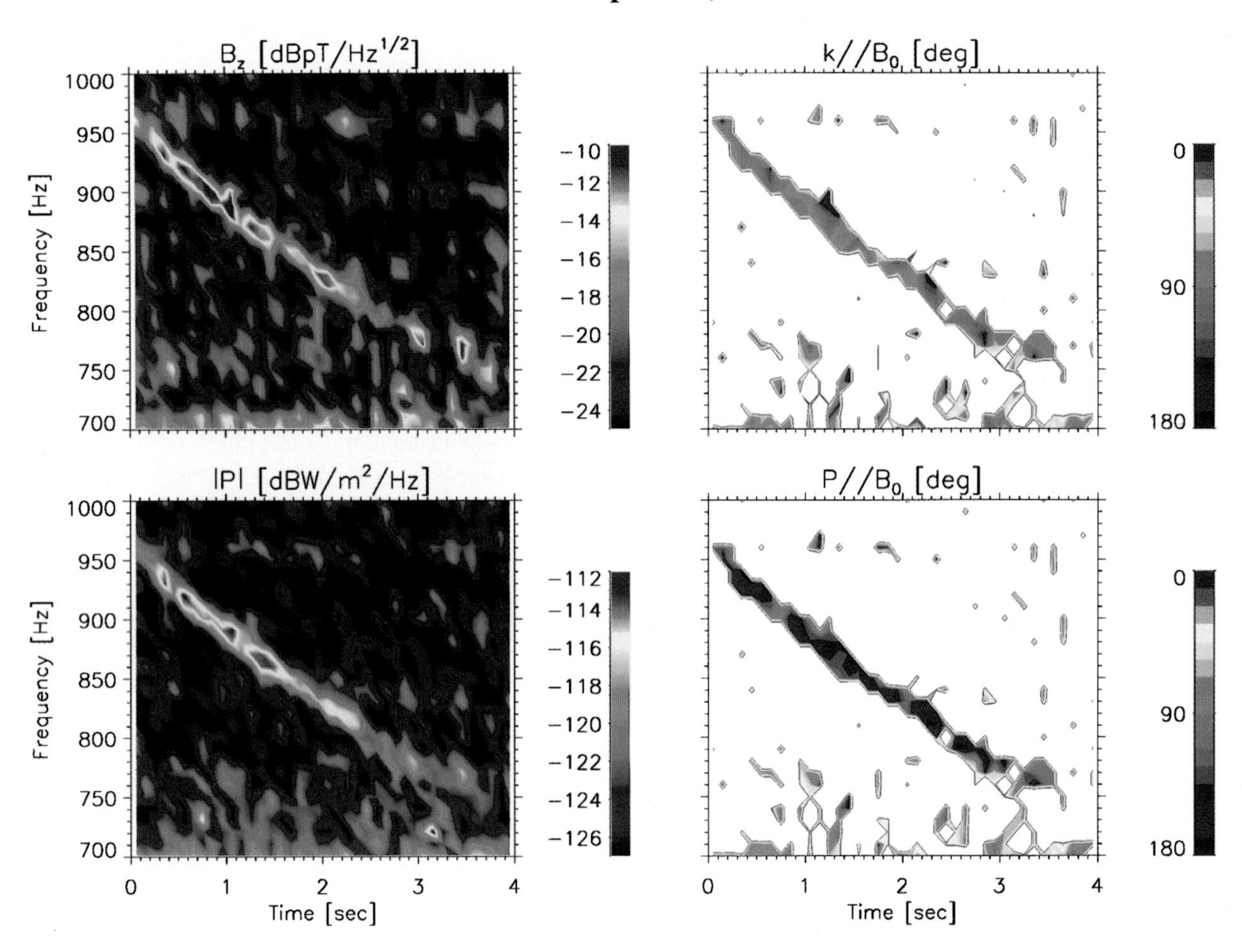

Figure 7. Example of a whistler wave observed by GEOTAIL on April 29, 1993. Upper two panels display the frequency-time spectrogram (left) and frequency-time dependence of k vector direction (right). Lower two panels show the frequency-time dependence of the magnitude and direction of the Poynting vector [*Nagano et al.*, 1998].

about $L = 10$ at 7.73 R_E with the latitude and longitude of 7.9 deg. and 153.3 deg, respectively. The observed whistler element shows a very large dispersion of $D = 860s^{1/2}$. The local electron density is about 3/cc, which was determined from the lower cutoff frequency of the continuum radiation (see Section 5). The K_p index was 5. Whistlers with such a large dispersion have rarely been observed in the past satellite observations. The WFC was used to measure three magnetic components of these whistlers by which we could deduce the k vector and the Poynting vector of the whistler element. The result indicates that the k vector is almost perpendicular to the ambient geomagnetic field B_o of the magnetosphere, while the Poynting vector is parallel to B_o. *Nagano et al.* [1998] have conducted a ray tracing in an attempt of explaining such unusual large dispersion. They concluded that the whistler with such a large D-value can only be explained in terms of a special non-ducted ray path. They have revealed the special path is created if and only if the k vector of the whistler at the injection point at the ionospheric level is inclined from the vertical direction with an angle of a few degrees.

3.1.2. Chorus and Hiss. Chorus is also a well-known phenomenon which is often observed in the magnetosphere. Though it has been known for a long time since its discovery by the ground observation [e.g., see *Helliwell*, 1965], detailed information on the chorus element has not fully been provided in the past studies. The statistical studies of the chorus distribution in the

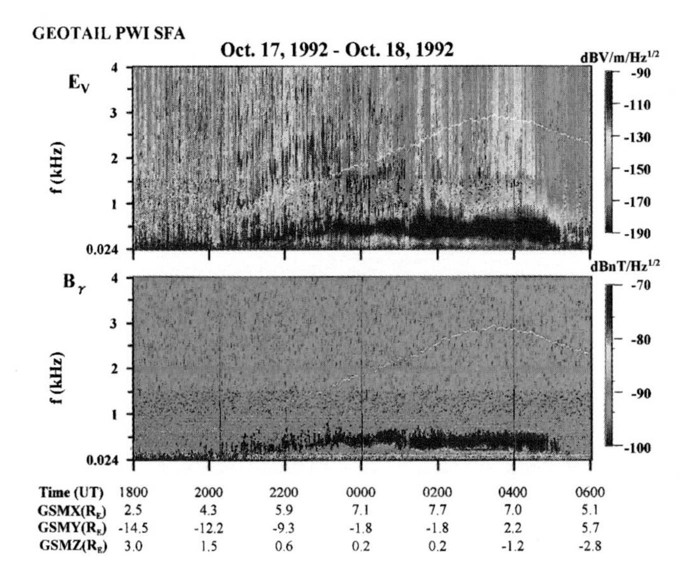

Figure 8. Typical dynamic spectrum of chorus observed on the GEOTAIL orbits passing through the magnetopause from the magnetosheath to the magnetosphere. Upper and lower panels display the frequency-time spectrogram of the electric and magnetic components, respectively. The intense electromagnetic emission observed around a few hundreds of Hz are the chorus emissions.

magnetosphere have revealed that these whistler mode waves are commonly observed in the outer magnetosphere beyond the plasmapause [*Tsurutani and Smith*, 1977]. However, the generation mechanism of chorus element is still not fully understood. The clue to this open problem can undoubtedly be provided by in-situ observations of chorus elements. Especially information on its k vector and Poynting vectors in addition to the simultaneous frequency spectra is valuable in understanding the chorus generation mechanism. The GEOTAIL PWI has actually provided such clues to this issue.

Figure 8 shows a typical dynamic spectrum of chorus observed on the GEOTAIL orbits passing through the magnetopause from the magnetosheath to the magnetosphere. These waves were examined by WFC in details and turned out to be composed of multiple risers and hooks as shown in Figure 9. In this example GEOTAIL traversed the magnetopause around 21:00 UT on October 17, 1992 and did enter from the magnetosheath into the magnetosphere and traversed in the outer magnetosphere towards the nightside. As seen in the figure, the chorus starts to appear around 2200 UT at the frequency of approximately 250 Hz around 1/4 of the local electron cyclotron frequency f_{ce} which is indicated by the white solid line in the figure, but by about one hour later, its relative frequency to the local f_{ce} decreases as the spacecraft enters into the magnetosphere. It is interesting to observe the crossing of the magnetopause is not a single event but multiple ones, though not clearly seen in the time scale in Figure 8. The chorus spectra disappear around 5:00 UT suddenly. The reason for this is not clear. One possibility is due to the spacecraft leaving from the outside magnetosphere into the plasma sheet.

As the chorus is normally quite stably observed, even the rare sampling of the waveforms by the WFC can capture its detailed waveforms with the WFC high time resolution. Figure 9 shows three such examples [*Nagano et al.*, 1996]. The three panels from the top to the bottom show typical examples of a long-lasting falling tone, a hook and risers, respectively. To obtain further information, we have carried out an analysis of k vectors and the Poynting vectors of these chorus elements. The results are shown in Figure 10, which shows the k vector direction in reference to the geomagnetic field. It is evident that the k vectors of all of these chorus elements are aligned along the geomagnetic field line except for

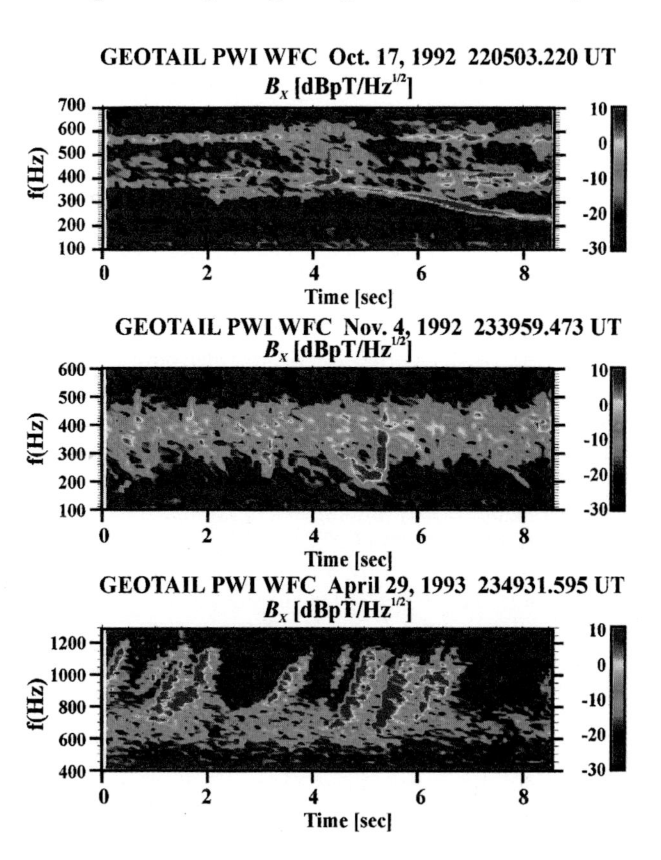

Figure 9. Examples of the chorus spectra observed by GEOTAIL on the dayside outer magnetosphere. The three panels from the top to the bottom show typical examples of a long-lasting falling tone, a hook and risers, respectively.

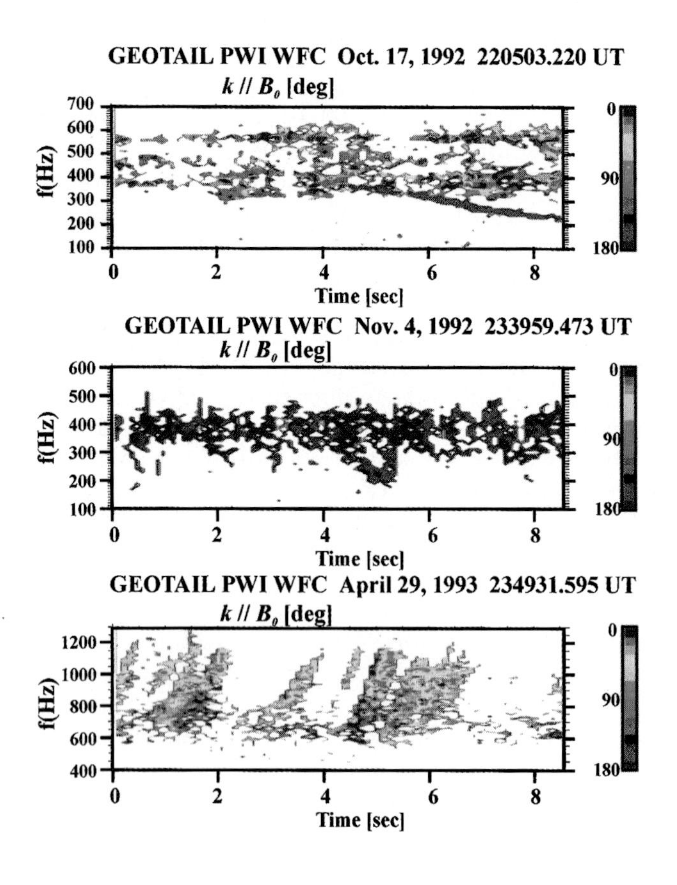

Figure 10. Wave normal angles relative to the ambient magnetic field in three cases of the chorus shown in Figure 9

the high frequency component of the risers. The Poynting vectors of the hook and risers turn out to be also parallel to the geomagnetic field. However we could not analyze the Poynting vector of the long-lasting falling tone because of the instrumental saturation due to its high intensity.

Nunn et al. [1997] attempted to simulate the generation process of the long lasting chorus element shown in the top panel of Figure 9 by a Vlasov code computer experiment. Their model for the computer experiment is based on the nonlinear phase trapping theory [*Nunn*, 1974; *Matsumoto et al.*, 1974] and could reproduce the dynamic spectra quite reasonably as shown in Figure 11 [*Nunn et al.*, 1997]. It is noted that the in-situ information on the frequency as well as its k vector is indispensable for such computer experiment. Such in-situ information at large L-shell location in the magnetosphere is first brought in by the GEOTAIL PWI. Another important information has been given from the particle measurement (CPI by L. A. Frank team in this case). It did show that the energetic electrons at the resonance velocity of the falling tone do exist and their flux is not negligibly small. The computer experiment thus adopted a model for the electron flux based on the GEOTAIL particle observation. They have demonstrated that the contribution of nonlinear phase trapping of resonant electrons is essential for the generation of the chorus element. This point has already been studied for the case of VLF triggered emission [e.g., *Nunn*, 1971, 1974; *Matsumoto*, 1972; *Omura and Matsumoto*, 1982].

As per a direct comparison of energetic electrons with the chorus event, *Yagitani et al.* [1996] surveyed a data set on Oct. 17, 1992. They found that positive correlation between the chorus event and the electron flux near the resonant energy. Figure 12 shows one example of the correlation between the chorus and the energetic resonant electron flux. They computed the growth rate of the whistler mode waves based on the CPI electron data and compared it with the SFA and MCA chorus data. For the calculation of the growth rate, they assumed that the flux measured by the CPI data is a result of a self-consistent pitch angle diffusion caused by the excitation of the whistler chorus elements. On the basis of this assumption they estimated that the temperature anisotropy of the resonant electrons is three times larger than the observed value. It is not realistic to introduce the "three times" factor. Therefore they reexamine the particle data in much detail and found that the correlation is still good without introduction of this artificial factor [Yagitani et al., 1997]. The uppermost panel of Figure 12 shows the linear growth rate as a function of frequency through the cyclotron resonance condition. Though the correlation between the calculated growth rate and the frequency band of the observed chorus event does not exactly agree, they show a good correlation for most of the observation time. This

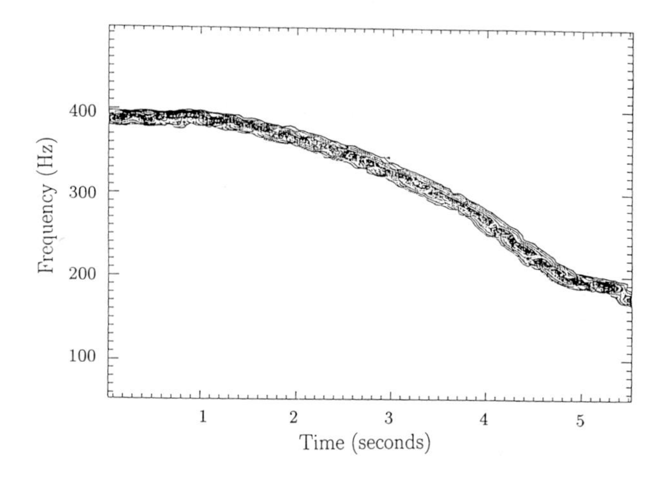

Figure 11. Chorus spectra reproduced by a computer experiment [*Nunn et al.*, 1997].

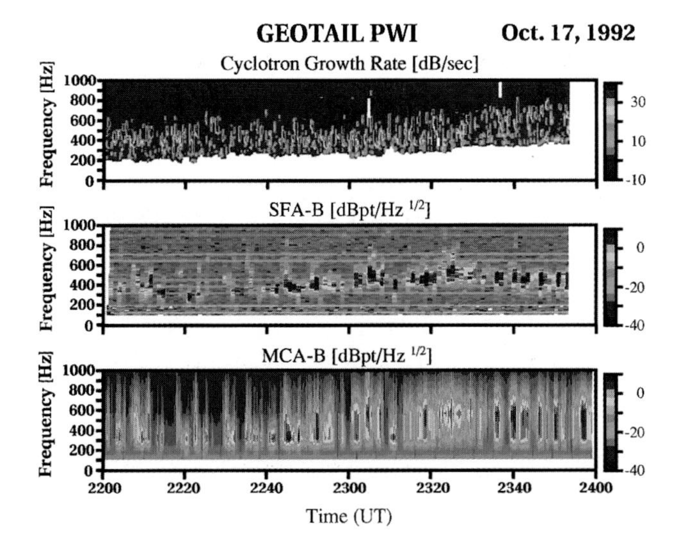

Figure 12. Correlation between the intensities of chorus emissions and growth rates expected from the linear theory. Lower two panels show the frequency-time spectrograms of the magnetic component observed by the SFA and MCA, respectively. The top panel shows the linear growth rate of the electron cyclotron resonance calculated based on the measurements of the ambient magnetic field and plasmas.

provides good evidence to support the theories and hypothesis of the electron cyclotron resonance for the chorus generation. However, it is noted that the generation mechanism may be far more complex and may involve nonlinear effects. In such case, the observed distribution function of resonant electrons have been averaged out by the nonlinear interaction and may not be appropriate for the growth rate calculations.

As the perigee of the GEOTAIL is about 10 R_E, the spacecraft rarely enters into the plasmasphere. Therefore, we have so far very few examples of hiss in the GEOTAIL observations. However, it is noted that hiss-like emission is usually observed even beyond the plasmapause as a background of chorus emission. Such an example is seen in the middle panel of Figure 9. This type of wideband and banded emission with weak intensity is sometimes called "structureless chorus." However, it is generally believed that the generation of wideband and continuous emission of this sort should be different from that of discrete emission such as chorus and VLF triggered emission.

3.1.3. Lion Roars. It has been known that strong monochromatic whistler waves exist in the magnetosheath. Due to their audio sound replayed on the audio amplifier, they are named as "Lion roars" [*Smith et al.*, 1969, 1971]. GEOTAIL also picks up the lion roars quite often when it enters into the magnetosheath. The lion roars are characterized by its monochromatic nature in the frequency range from a few tens of Hz to a couple of hundreds of Hz. Figure 13 shows an example of the lion roars observed on Aug. 24, 1996 in the format of the SFA dynamic spectrum. The spectrum is for the magnetic component. As seen in the low frequency range from about 30 Hz to 200 Hz, patchy dots are seen with a short duration and with a monochromatic nature. These dots in the dynamic spectrum are nothing but the lion roars. In Figure 14, we show the wave forms at two different times, at 0715:59 UT (case (a)) and 0739:38 UT (case (b)) on the same day. The upper panel corresponds to case (a) and shows a typical wave packet of the lion roars with a duration of several tens of cycles. In addition to the waveforms, a hodograph of the magnetic component, two projection frames ($X-Y$ and $X-Z$ planes) of the k vector (solid arrow) and magnetic field line (dotted line)) are presented for each case. Three magnetic components of the lion roars are used to find the k vector direction with the use of the minimum variance method. As seen in the figure, the lion roars have right-handed polarization. On the other hand, case (b) shows a typical example of amplitude-modulated quasi-monochromatic wave with a long duration. The polarization is also right-handed. The result shows that both types of the lion roars propagate along the external magnetic field [*Zhang et al.*, 1997]. Such short lived whistler wave packets of case (a) are often observed near the bow shock in the magnetosheath. The other type of whistler wave as represented by case (b) is often seen near the magnetopause.

Though the generation mechanism of the lion roars is not well understood, it is generally believed that these waves are generated by resonant interaction with energetic electrons which are produced via acceleration at the bow shock and penetrate into the magnetosheath region. The short duration of the lion roars may be explained in two possible ways. In either case, resonant electrons with temperature anisotropy are required in the generation region. One possible interpretation is simply due to the short life time or short production time of the resonant energetic electrons at the bow shock or the magnetopause. The other possible interpretation is that a magnetic flux tube which contains the energetic resonant electrons and whistler waves passes through the spacecraft in a short time due to the flux motion carried by the tailward plasma flow in the magnetosheath.

Another interesting point which GEOTAIL found concerning the lion roars is the high variability of its center frequency. This may be a direct reflection of high variability of central energy of the accelerated resonant electrons. If so, we can extract valuable information on physical mechanism for electron acceleration at the

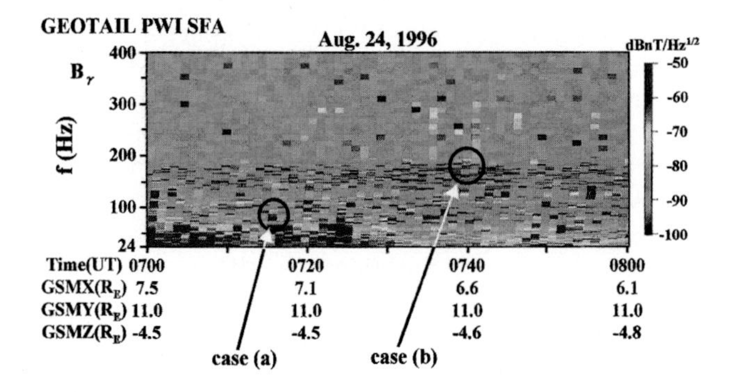

Figure 13. Example of the Lion roars observed on Aug. 24, 1996 in the format of the SFA dynamic spectrum.

bow shock and/or the magnetopause. This point is now under investigation and will be published shortly in the near future.

3.1.4. Magnetic Noise Bursts. The term "magnetic noise bursts (MNB)" is not clearly defined. *Holzer et al.* [1966] used the terminology of "broad band magnetic noise" to describe a wide band magnetic noise in the magnetosheath observed by OGO-1. *Russel* [1972], on the other hand, discussed a variety of magnetic wave phenomena in the tail. His interest was for waves with frequencies below 1 Hz, and he did not use the term magnetic noise explicitly. Unfortunately the GEOTAIL PWI can observe the magnetic waves only above 5.6 Hz, and therefore cannot confirm these waves.

The first usage of the explicit name "magnetic noise burst" is seen in *Gurnett et al.* [1976]. They found brief bursts of whistler mode magnetic noise near the neutral sheet. The MNB they discussed was a broad band noise from 40 Hz to 400 Hz.

GEOTAIL has traversed the central plasma sheet quite often when it was on the distant tail orbits. The MCA data show that the MNB has the broad band spectrum extending from the lowest detectable frequency (5.6 Hz) up to about 31.1 Hz (Such example will be discussed in section 5.6 in Figure 54.).

At the same time we also see a monochromatic wave packet in the high frequency about 100 Hz. Figure 15 shows one example of the plasma sheet MNB. The high frequency short-lived wave packet in the plasma sheet has not been discussed in the literature as far as the authors are aware of. It is noted that the high frequency whistler wave packet in the plasma sheet is very similar to the lion roars in the magnetosheath.

3.1.5. Whistler Wave Packet in the Solar Wind. In the upstream region of the bow shock in the solar wind, we occasionally observe a quasi-monochromatic whistler wave packet. Their waveforms are similar to those of the lion roars in the magnetosheath. An example of such whistler wave packet in the solar wind is shown in Figure 16. The uppermost panel shows the SFA data in the form of the dynamic spectra for both E

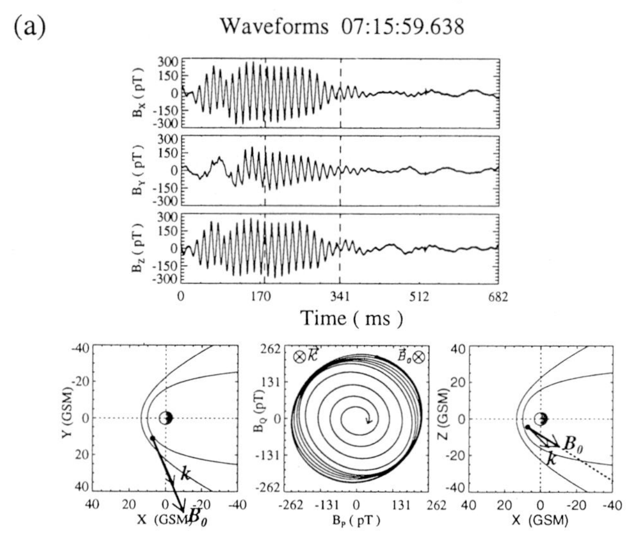

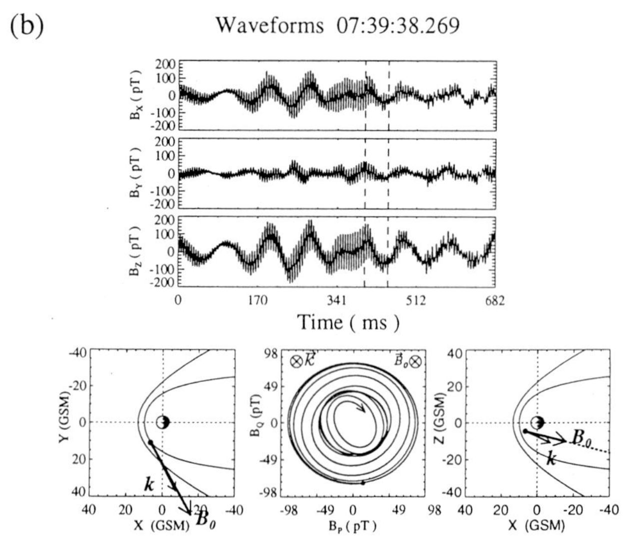

Figure 14. Waveforms at two different times, at 7:15:59 UT (case (a)) and 7:39:38 UT (case (b)) on August 24, 1996. The upper panel corresponds to case (a) and shows a typical wave packet of the lion roars with a duration of several tens of cycles. On the other hand, case (b) shows a typical example of amplitude-modulated quasi-monochromatic wave with a long duration. This type of wave is often seen near the magnetopause. In addition to the waveforms, a hodograph of the magnetic component, two projection frames of the k vector (solid arrow) and magnetic field line (dotted line) are presented.

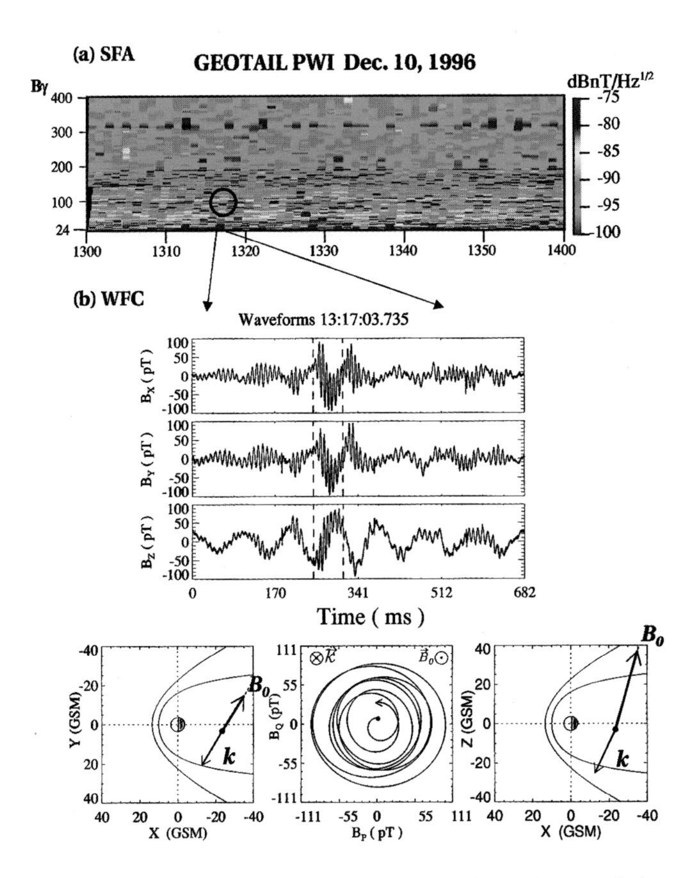

Figure 15. Dynamic spectrum and wave forms of the plasma sheet MNB. The wave forms of the MNB show packet-like structure with a frequency of about 100Hz. Note that their amplitude is modulated. The left and right panels show k vector and the bottom center panel shows its polarization (R-wave).

and B components. Note that the E component shows purely whistler waves and is not contaminated by otherwise frequently observed electrostatic waves in the upstream region. The middle panel shows a blow-up of the wave forms which shows the characteristics of quasi-monochromatic feature lasting a few hundreds of cycles of oscillation. The lowest panel in Figure 16 shows the result of the hodograph and k vector analyses. It clearly shows the right-hand polarization indicating the wave packet is in the whistler mode. The generation mechanism of this whistler wave packet is probably due to the cyclotron interaction of electrons in the solar wind but has not been proven experimentally with the particle data set from the GEOTAIL.

3.2. Nonthermal Continuum Radiation

3.2.1. Trapped Continuum Radiation. Continuum radiation (CR) is the most commonly observed electromagnetic wave in the magnetosphere [e.g., see a review by *Kurth,* 1991]. *Brown* [1973] and *Gurnett and Shaw* [1973] first discovered the radiation and *Gurnett* [1975] first introduced the term of "nonthermal continuum radiation", while *Jones* [1980; 1981a, b; 1982] used a different term of "terrestrial myriametric radiation". The nonthermal CR is a wide band and weak electromagnetic radiation often with a sharp lower cutoff frequency at the local plasma frequency, f_{pe}. It sometimes has an upper cutoff frequency as well. In such cases, the upper cutoff frequency is determined by the remote plasma frequency of the magnetosheath where the waves are reflected back into the magnetosphere or the magnetotail region. The trapped nonthermal CR is found within the magnetopause boundary due to the wave trapping between the plasmapause (or the plasma sheet boundary layer) and the magnetopause. Figure 17 is one of such typical nonthermal trapped CR observed in the tail lobe. The example is taken from the GEOTAIL PWI/SFA data observed on December 15, 1995. The

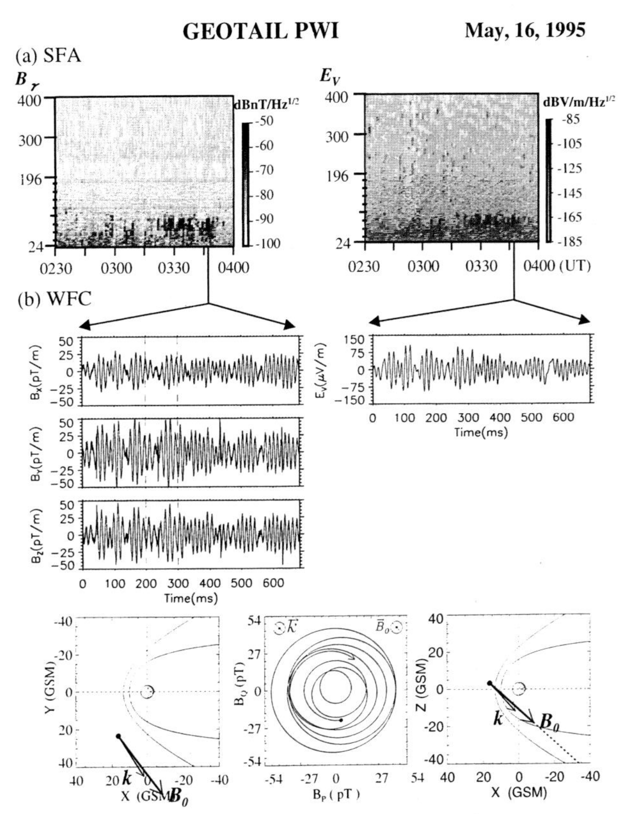

Figure 16. Dynamic spectrum and wave forms of quasi-monochromatic whistler wave pocket with an amplitude modulation. Note that the E component also does show a clear wave spectrum similar to the B spectrum, and is free from the electrostatic noise. The bottom three frames are the same as in Figure 14 and Figure 15.

spacecraft was in the tail lobe near a point of (GSM X, GSM Y, GSM Z) = (-28, -11.5, 0.0 R_E). The trapped nonthermal CR is seen in the frequency range between about 4 kHz and about 40 kHz and has both the lower and the upper cutoff frequencies. The upper cutoff frequency near 40 kHz is the electron plasma frequency of the magnetosheath above which these waves are not reflected back to the tail lobe. The lower cutoff frequency of CR is often used to read out the local electron density. This is basically based on the fact that the lower cutoff frequency of the trapped nonthermal CR is equal to the local electron plasma frequency.

The reason for the lower cutoff frequency being the local plasma frequency is explained by the cutoff nature of dispersion characteristics. Figure 1(a) is a typical $\omega - k$ diagram showing the dispersion of electromagnetic waves in a magnetized plasma with the plasma frequency larger than the cyclotron frequency. The shaded area between the two dispersion curves for parallel ($\theta = 0°$) and perpendicular($\theta = 90°$) propagation with respect to the external magnetic field is filled by the dispersion curves for waves with oblique propagation angles. The trapped continuum radiation is believed to be generated at the plasmapause by the linear wave conversion mechanism (LWCM) through the radio window [*Jones*,1980; 1981a, b; 1982]. According to the LWCM theory, the seed of the trapped nonthermal continuum radiation is generated as a result of mode conversion from the electrostatic upper hybrid waves to the L-O electromagnetic waves through the conversion window near the intersection of two dispersion curves for the Z–SX mode branch (which is connected to the electrostatic electron cyclotron harmonic waves in the vicinity of the upper hybrid frequency) and the L-O electromagnetic branch. Therefore the resultant electromagnetic waves should have the left-handed polarization at least near the cutoff frequency.

Figure 18 shows a typical dynamic spectrum of the trapped nonthermal CR observed by the GEOTAIL PWI WFC in the lobe of the geomagnetic tail on November 15, 1992 [*Nagano et al.*, 1994]. As seen in the figure, the lower cutoff frequency changes as the receiving dipole antenna rotates. The arrows below the horizontal scale in Figure 18 show the timing when the dipole antenna is aligned along the geomagnetic field. The two alternating lower cutoff frequencies turn out to be f_{R-FX} and f_{L-O} rather than the combination of f_{L-O} and f_{Z-SX}. This was confirmed by a comparison between the calculated electron cyclotron frequency from the difference of the two alternating cutoff frequencies and the observed local electron cyclotron frequency by the MGF magnetic sensor onboard the GEOTAIL. The cyclotron frequency computed from the difference of $f_{R-FX} - f_{L-O}$ agrees much better with the measured electron cyclotron frequency than that deduced from the difference of $f_{L-O} - f_{Z-SX}$. Furthermore the polarizations of the observed waves with frequency between the f_{R-FX} and f_{L-O} clearly show the left-hand polarization. Therefore the lower frequency of the modulated cutoff frequency is f_{L-O} which is equal to the electron plasma frequency. This is also confirmed by a fact that the lowest cutoff frequency is observed synchronized with timing when the dipole antenna becomes parallel to the external magnetic field. *Gurnett and Shaw* [1973] had already found similar spin modulation of the lower cutoff frequency of the trapped continuum radiation by IMP6 in the vicinity of the plasmapause. They identified that the upper and lower cutoff frequencies are f_{R-FX} and f_{L-O}, respectively. GEOTAIL found the similar nature in the vicinity of the plasma sheet boundary layer as shown in Figure 18 [*Nagano et al.*, 1994].

Figure 19 shows a good example how the trapped nonthermal CR changes its frequency band width as GEOTAIL traverses the magnetopause from the magnetosheath to the tail lobe. We could see a rapid drop and rise of the lower cutoff frequency of the continuum, i.e., the drop and rise of the electron density as the spacecraft moves into the tail lobe and back to the magnetosheath around 2000 and 2130 UT, respectively. The upper cutoff frequency, on the other hand, remains the same at the traversal of the magnetopause.

It is noted that some lower frequency components may be lost due to the enhanced density on the way of the pass. The possibility of such case would be very small. This is because the continuum radiation in the tail could reach the spacecraft from various directions via multiple reflections, most of them through the lobe where the density is very low. Therefore it is very likely that the lower cutoff frequency of the CR indicates the local electron density.

Kurth et al. [1981] and *Kurth* [1991] discussed an apparent paradox that the wide band nature of the trapped nonthermal CR is inconsistent with the discrete nature of the L-O waves directly converted from the upper hybrid waves through the Z–SX mode at the conversion window. However, they concluded that the wide band nature of the trapped nonthermal CR is a result of the smoothing process in frequency via Fermi-Compton scattering [*Kurth*, 1991] in the process of the multiple reflection at the density wall at the magnetopause boundary. Therefore the frequency band is nominally between the local plasma frequency and the plasma frequency at the magnetopause or at some place in the magnetosheath. Thus the trapped continuum is believed to be composed of superimposed

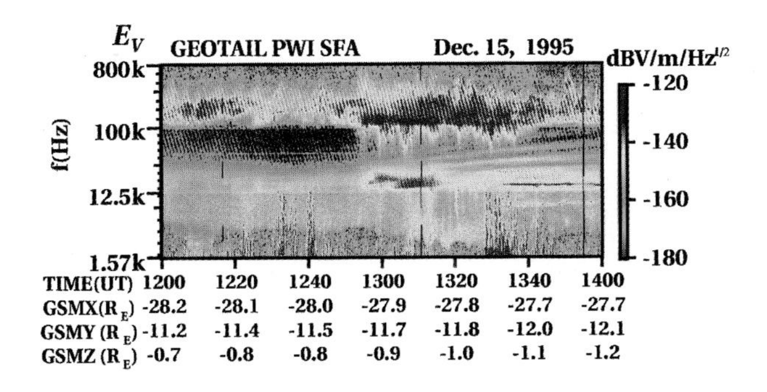

Figure 17. Typical example of Continuum Radiation (CR) seen in the frequency range between 4KHz and 40 – 100KHz. Both trapped and escaping CR are observed. The latter is clearly separated from the former after 1300UT.

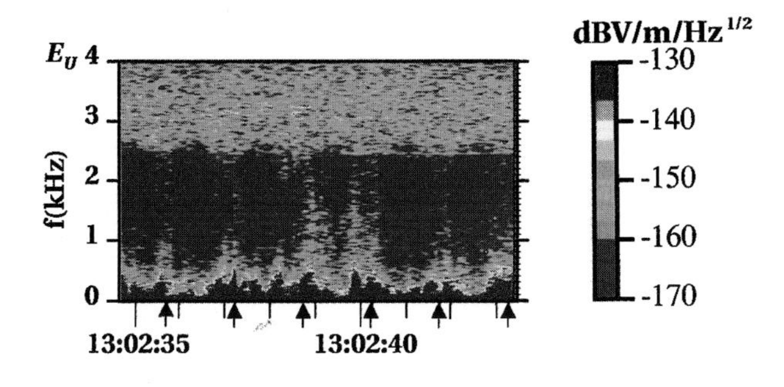

Figure 18. Expanded dynamic spectrum of trapped CR observed in the tail lobe at the spacecraft location of (GSM X, GSM Y, GSM Z) = (-120,-3.5,4.7 R_E). The lower cut off frequency around 2.6KHz is spin-modulated. The arrows along the time axis show the timings when the dipole antenna is aligned.

waves captured in a sort of magnetospheric and magnetotail cavity surrounded by the high density magnetosheath. However, the generation location of the original seed of the CR is not limited only to the plasmapause. It can be generated wherever the density gradient is large. Therefore it is highly possible that the trapped nonthermal CR is generated at the plasma sheet boundary layer and trapped between the PSBL and the magnetopause as schematically illustrated in Figure 20 [*Nagano et al.*, 1994]. This type of trapped continuum can be called "lobe trapped CR" and the classical trapped CR with the plasmapause origin can be called MS(magnetosheath) trapped CR. We will discuss the possible source region of the lobe trapped CR in the next section.

3.2.2. Source of the Trapped Nonthermal Continuum. It is noted that the source of the trapped nonthermal continuum radiation observed within the magnetopause is believed to be located at the plasmapause. It is generally agreed upon that intense electrostatic electron cyclotron waves, especially those at the local upper hybrid frequency generated by electron temperature anisotropy and/or electron beam are converted to the electromagnetic mode at the sharp density gradient at the plasmapause [e.g., *Jones*, 1981; *Gough*, 1982], though there are several nonlinear theories for the generation of CR, which involve wave-wave coupling [Melrose, 1981]. However, an interesting case was found in the GEOTAIL plasma wave data set which shows a continuum radiation generated not at the plasmapause but

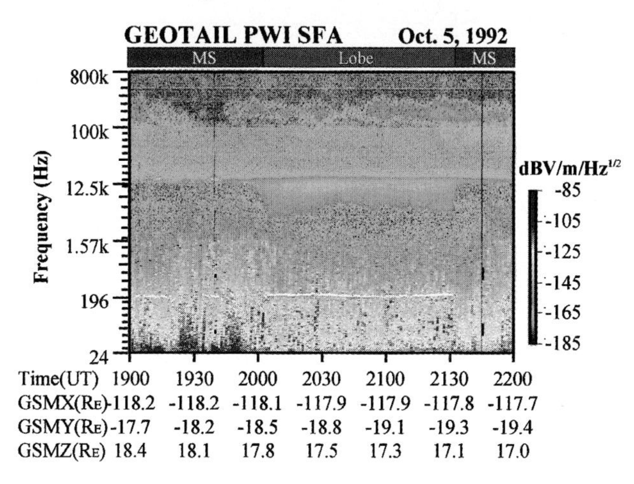

Figure 19. Example showing the change of the CR cutoff frequency as the spacecraft moved from the magnetosheath to the tail lobe passing through the magnetopause boundary around $X = -118R_E$ and moved back to the magnetosheath.

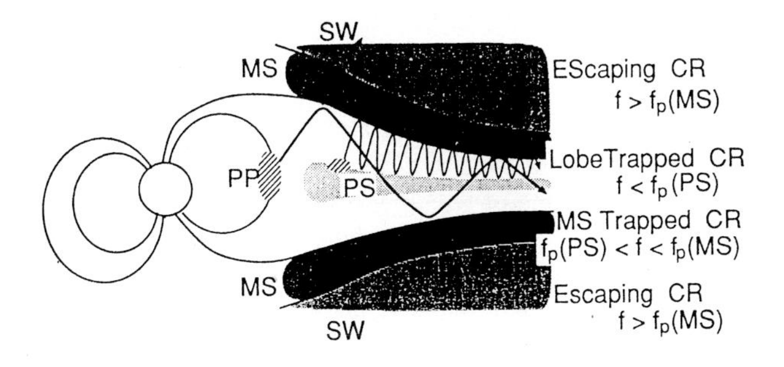

Figure 20. A schematic illustration of the propagation of the MS(magnetosheath) trapped CR and of the Lobe trapped CR. The source region of the former is near the plasma pause, while the source of the latter is near the PSBL [*Nagano et al.*, 1994].

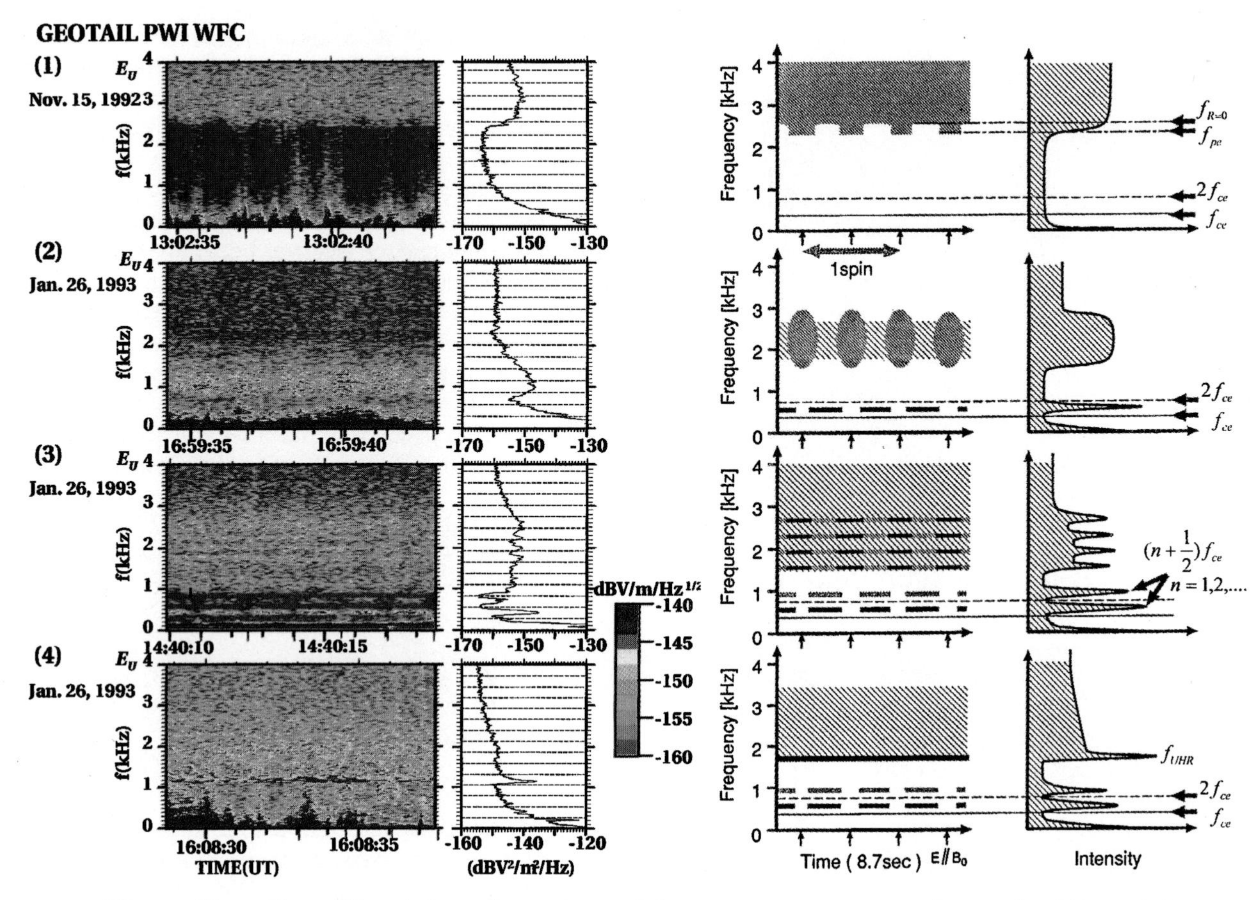

Figure 21. Four different dynamic spectra and frequency spectra averaged over the 8.7sec of the lobe trapped CR, (Left Column) observed by the GEOTAIL PWI WFC, and their schematic illustration (Right Column). The top CR is the typical lobe trapped CR with diffused spectrum above the cutoff frequency, while the rest three show the CR with more or less discrete frequency components(see text). The locations (GSM X, GSM Y, GSM Z) spacecraft were (a) (-120, -3.5, 4.7 R_E), (b) (-90.4, -1.7, -6.5 R_E), (c) (-90.5, -2.0, -6.4 R_E) and (d) (-90.7, -2.3, -6.2 R_E), respectively.

at the plasma sheet boundary layer (PSBL) where the density gradient also exists [*Nagano et al.*, 1994]. Figure 21 shows four different types of the trapped non-thermal CR observed in the tail lobe. The left two show the observed dynamic spectra and the frequency spectrum averaged over the 8.7 sec, respectively. The right two panels show the corresponding schematic illustrations. The four different types are observed at different locations of the spacecraft relative to the generation region at the plasma sheet boundary layer. The uppermost type, called type (1) here, is most often observed and can be called "continuum" because its dynamic spectrum shows continuous spectrum above the cutoff frequency with a smoothly varying intensity. It is noted that the lower cutoff frequency is not constant over the spin period but changes as the spacecraft rotates as discussed before. The small arrows on the horizontal scale in the schematic illustration, and the longer vertical lines on the horizontal scale in the observed dynamic spectra indicate the timing when the dipole antenna used for the observation of these waves is parallel to the geomagnetic field. As discussed before the lower cutoff frequency of the continuum can be described either f_{pe} or f_{R-FX} depending on the k vector of the observed waves. The second type in Figure 21, called type (2), is a banded continuum with some discrete structures in the dynamic spectrum. This type also suffers from the spin modulation, and has the spin-modulated lower cutoff frequency. The third type in Figure 21, called type (3), shows much clearer structure of discrete emissions with a harmonic structure superimposed on weak continuum radiation. Notice that both strong line emissions near 1.5 f_{ce} and weak 2.5 f_{ce} emission are seen below the local electron plasma frequency (900 Hz in this case), where f_{ce} is the electron cyclotron frequency. The horizontal dash lines in the second left column (the averaged frequency spectrum) shows the harmonics of the local electron cyclotron frequency. These two line emissions below the electron plasma frequency are the electron cyclotron harmonic (ECH) waves. The fourth type, type (4), is mainly composed of discrete emissions with a peak intensity at the local upper hybrid frequency accompanied by weaker ECH waves at lower frequencies. These ECH waves are of electrostatic nature generated near the PSBL and may well be the source of the lobe trapped electromagnetic continuum through the linear mode conversion at the density gradient near the PSBL.

As the lobe trapped continuum and related discrete emissions have frequencies of 1-2 kHz, even the name of "myriametric radiation" may not be appropriate. However the physical mechanism of these waves is very similar to that of the myriametric discrete emissions and nonthermal continuum generated at the plasmapause and trapped by the cavity surrounded by the magnetosheath.

3.2.3. Myriametric Discrete Emission (Escaping Continuum Radiation). In contrast to the diffuse or continuum nature of the spectrum of the trapped nonthermal CR, there have been observed another type of electromagnetic radiation in the similar myriametric frequency range. It is a class of radiation composed of a number of discrete emission lines with a highly structured frequency spectrum. These are called either "escaping continuum" or "myriametric discrete emissions." The former name of escaping continuum is not appropriate as the emission spectrum is no more continuum or diffuse in frequency. These emissions are the escaping component of the electromagnetic waves converted from the electrostatic waves at the plasmapause. The terminology of "escaping" continuum for these discrete emissions came from the fact that the emissions are generated somewhere within the magnetosphere and penetrate through the high density region of the magnetosheath into the solar wind. One example of the escaping myriametric discrete emission is shown in Figure 17 together with the trapped nonthermal CR. The electromagnetic components with a frequency above the plasma frequency of the magnetosheath are not reflected at the magnetopause and hence can escape into the solar wind. Such escaping components with rising discrete emissions are seen above 40 kHz extending up to about 100 kHz for a period from 1300 UT to 1400 UT.

3.2.4. Continuum Enhancement (Continuum Storm). In the previous two sections, we have discussed the nonthermal CR of two different types with both "continuum" and "discrete" spectra. These are, as described already, believed to be generated through the linear mode-conversion process from electrostatic waves near the upper hybrid frequency at the equator of the plasmapause or at the PSBL with a large density gradient. *Gurnett and Shaw* [1973], *Brown* [1973] and *Gurnett* [1975] extensively studied the CR generated at the plasmapause and found that the CR is observed in the local time zone of 0400LT – 1400LT, and is positively correlated with large-scale geomagnetic storms which last for several days [*Gurnett and Frank*, 1976]. We could call this type of continuum radiation as "classical continuum" even though the CR is mainly composed of multiple discrete emission. The classical continuum lasts for 2$\sim$12 hours with a smooth variation of both intensity and frequency. Its frequency range is generally from the local electron plasma frequency up to 100-500 kHz. However, when it is observed in the solar wind, it loses the frequency components below the frequency of 2-4 times of the local solar wind electron plasma frequency. This is caused by the reflection and refraction of these frequency components on their pass in the magnetosheath which normally has the electron plasma frequency of 2 $\sim$ 4 times of the solar wind plasma frequency.

Gough [1982] and *Filbert and Kellog* [1989] pointed out that there exists another type of nonthermal continuum radiation with much shorter duration time less than a few hours in association of individual electron injection from the tail into the magnetosphere at the time of substorms. Here we call this shorter life CR as "continuum enhancement". The continuum enhancement has a banded frequency spectrum from the local electron plasma frequency up to 100-500 kHz but lasts only for 1-3 hours. It is noted that the frequency range of the continuum enhancement is overlapping with those of the LFR (Low Frequency Radiation) which is the low frequency part of Auroral Kilometric Radiation (AKR) [*Filbert and Kellogg*, 1989] and of the LF band radio bursts [*Kaiser et al.*, 1996]. Though the LFR and LF bursts have a good correlation with substorms as well as the continuum enhancement, their duration time is much shorter than the continuum enhancement. *Kasaba* [1997a] and *Kasaba et al.* [1997c] discuss the nature of these continuum enhancement. The waves dealt by these authors are essentially the same. Figure 22 shows examples of the classical continuum and the continuum enhancement (storm) observed by GEOTAIL. The top panel of Figure 22 shows the classical continuum observed at two intervals of 0-2 UT and 5-10 UT on April 6, 1995. The GEOTAIL was located in the dayside zone as shown by the orbit (a) in the bottom panel of Figure 22. The second panel from the top in Figure 22 shows the continuum enhancement observed at three time intervals of 13-14 UT, 15-16 UT and 19-21

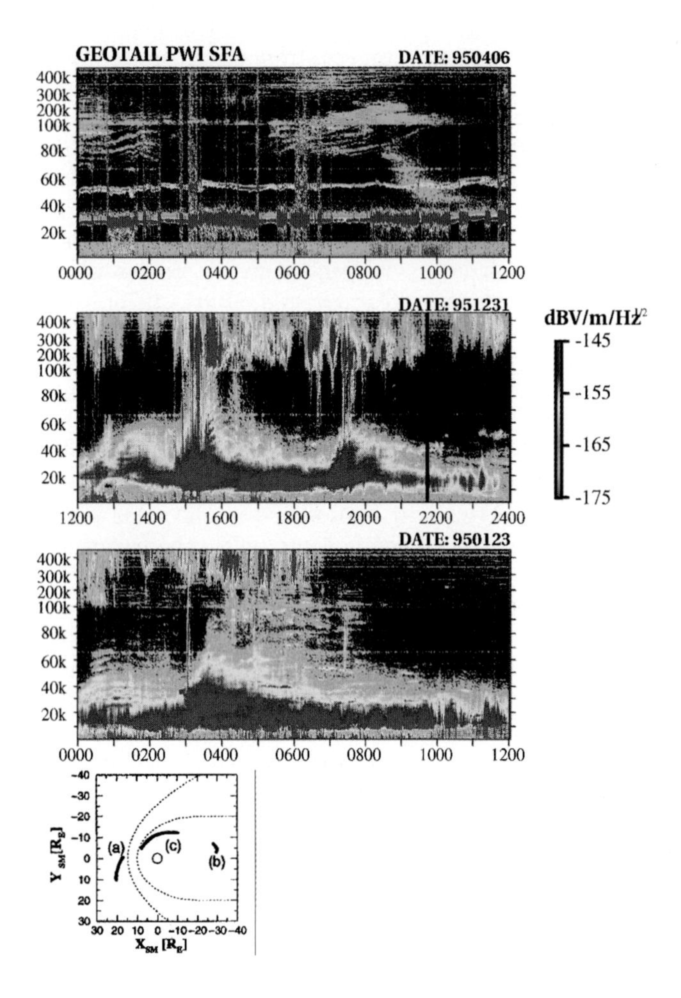

Figure 22. Dynamic Spectra observed by GEOTAIL PWI of (a) Classical continuum (00 – 02 UT and 05 – 10 UT) (b) Continuum enhancement (13 – 14 UT, 15 – 16 UT, 19 – 20 UT and 20 –21 UT), and (c) Classical continuum (00 – 03 UT) and sequential event of continuum enhancement (03 – 04 UT) followed by classical continuum (04 – 08 UT). The respective locations and orbits of GEOTAIL corresponding to each case are shown by (a), (b) and (c) in the bottom panel.

UT on December 31, 1995 when GEOTAIL is located on the nightside zone as shown by the orbit (b) in the bottom panel of Figure 22. Interesting but quite commonly observed example of the continuum enhancement is shown in the third panel of Figure 22. The continuum enhancement is occasionally followed by the classical continuum. In this example, the continuum enhancement appearing from 0300 UT to 0350 UT is followed by the classical continuum which lasts until 0800 UT. The orbit of the GEOTAIL for this case is depicted by the orbit (c) in the bottom panel of Figure 22. After being injected into the near Earth orbit, the GEOTAIL orbits cover the near Earth region within $30R_E$ quite uniformly as shown in Figure 6(b). This uniform orbital coverage has given a good opportunity of studying the statistical distribution of the classical continuum and the continuum enhancement[*Kasaba*, 1997a; *Kasaba et al.*, 1997c]. The bottom panel of Figure 23 shows the result of the occurrence histogram of the classical enhanced continuum for a period of about one year from November, 1994 to December, 1995. The black diamonds and white square symbols on the top panel of Figure 23 indicate the locations where the classical continuum and the enhanced continuum are observed by GEOTAIL, respectively. It is clearly seen in the histogram that the classical continuum is observed mainly on the late dawn side and on the dayside, while the continuum enhancement is observed mainly on the early dawnside or on the night side. This distribution is consistent with a fact that the continuum enhancement observed in the dawn time zone is sometimes followed by the classical continuum as shown in the third panel in Figure 22.

In other words these two types of the continuum are the time sequential phenomena caused by the electron injection events associated with substorms. From these facts, we could naturally draw a physical picture of the substorm electron injection into the magnetosphere. At the time of substorm event, electrons are injected from the tail into the midnight section of the magnetosphere. Then the injected electrons suffer from the $\boldsymbol{E} \times \boldsymbol{B}$ drift and the curvature and gradient drift. The former results in the inward drift motion, and the latter the azimuthal drift motion, respectively. As the curvature and gradient drift is proportional to the kinetic energy of the injected electrons, higher energy ($\geq$ 10keV) electrons suffer little azimuthal drift and cannot reach the plasmapause without a large azimuthal drift motion toward the dawn side. Low energy electrons, on the other hand, can penetrate into the higher density plasmasphere in a short time through the steep density gradient at the plasmapause, producing discrete emissions with a fast rate of frequency rise. Such fast frequency risers are seen in the example shown in the second panel of Figure 22 around 1530 UT. They exhibit a fast frequency rise within 5 minutes. The fast frequency risers are normally followed by the main part of the continuum enhancement with much slower rate of the frequency increase as seen from 1530 UT to 1550 UT in the same figure. It is noted that we need a sharp density gradient for mode conversion to overcome Landau and cyclotron damping which would otherwise absorb the source ECH waves and that the most efficient conversion occurs close to the magnetic equator since ECH waves are strongest there [e.g., *Horne* , 1989; *Horne*, 1990].

The slower frequency change is a result of slow azimuthal drift motion of higher energy electrons (≥ 10 keV) toward the dawn side due to the curvature and gradient drift through the dawn side plasmapause regions. As the average drifting time of the injected electrons from the midnight to dawnside is about 1 hour [*Filbert and Kellogg*, 1989], the time interval between the onset of the continuum enhancement and that of the classical continuum should be of the same order. The GEOTAIL observation of the time interval of the onsets of the continuum enhancement and the classical continuum shows a good agreement with this predicted value. The time history of the dynamic spectrum of the continuum enhancement thus provides information of the drift motion of the injected electrons at the time of the substorm. Usefulness of the continuum radiation for remote sensing of the plasmapause shape and distance at the time of substorms has been suggested by *Gough* [1982] and the details of the application of the GEOTAIL data to this problem are discussed in *Kasaba at al.* [1997c].

3.3. Auroral Kilometric Radiation (AKR) and Auroral Myriametric Radiation (AMR)

3.3.1. Auroral Kilometric Radiation (AKR). The Auroral Kilometric radiation (AKR) is one of the most commonly observed electromagnetic waves in geospace. At the early phase of the AKR research, it was called Terrestrial Kilometric Radiation (TKR) from a view point that the Earth is also a radio planet which radiates electromagnetic waves toward the outside universe [*Gurnett*, 1974]. The AKR has a frequency spectrum extending from several tens of kHz up to several hundreds of kHz. The average radiation power of the AKR is $10^7 - 10^8$ watts [*Gallagher and Gurnett*, 1979]. The AKR is believed to be generated through the cyclotron maser instability [*Wu and Lee*, 1979] in the plasma cavity [*Calvert*, 1981] on the field lines of discrete auroras in the nightside (22-00 MLT) auroral region [*Green et al.*, 1977]. Five dynamic color spectra of the AKR observed by GEOTAIL are shown with different time scales in Figure 24. The AKR is the emission seen in the frequency range between several tens of kHz and several hundreds of kHz on the orbits of the GEOTAIL. The reddish and yellowish parts in the dynamic spectra are the AKR segments. The uppermost dynamic spectrum indicates a typical six-hour variation of the AKR. In this frame, AKR intensification and frequency expansion are seen at three times; 1850, 1945 and 2140 UT. These three instants are close to those of the onsets of three substorms. Each of the three AKR blobs lasts for about 60 to 90 minutes which are the time scale of the substorms. During the lasting time of one blob, the

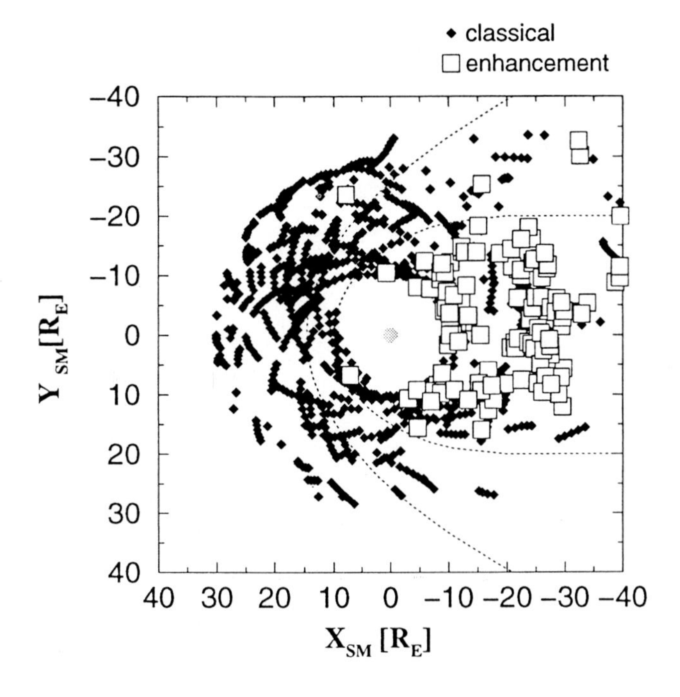

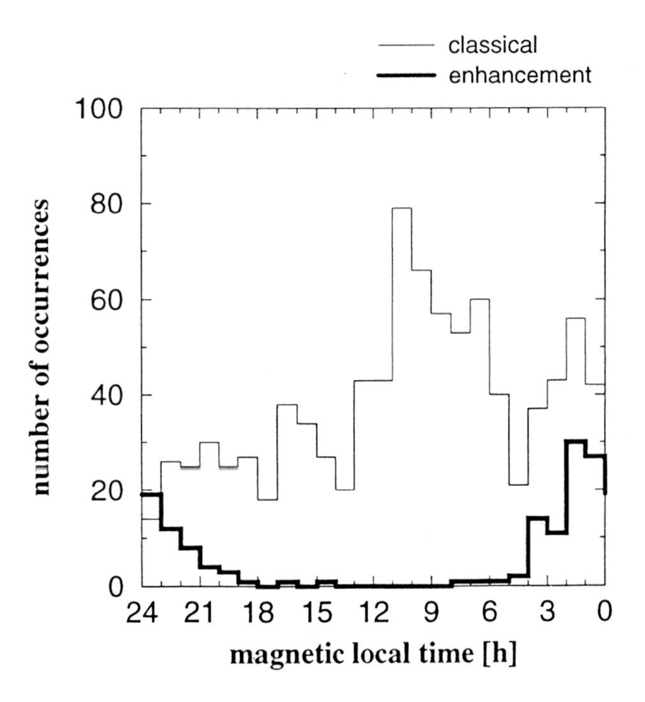

Figure 23. Mapping of locations of observed classical continuum (represented by solid diamond symbols) and of the continuum enhancement (represented by open square symbols) are shown in the upper panel. Occurrence histograms of the two types of continuum versus the magnetic local time are shown in the bottom panel. The thick and thin lines represent the histogram for the continuum enhancement and classical continuum, respectively [*Kasaba*, 1997a].

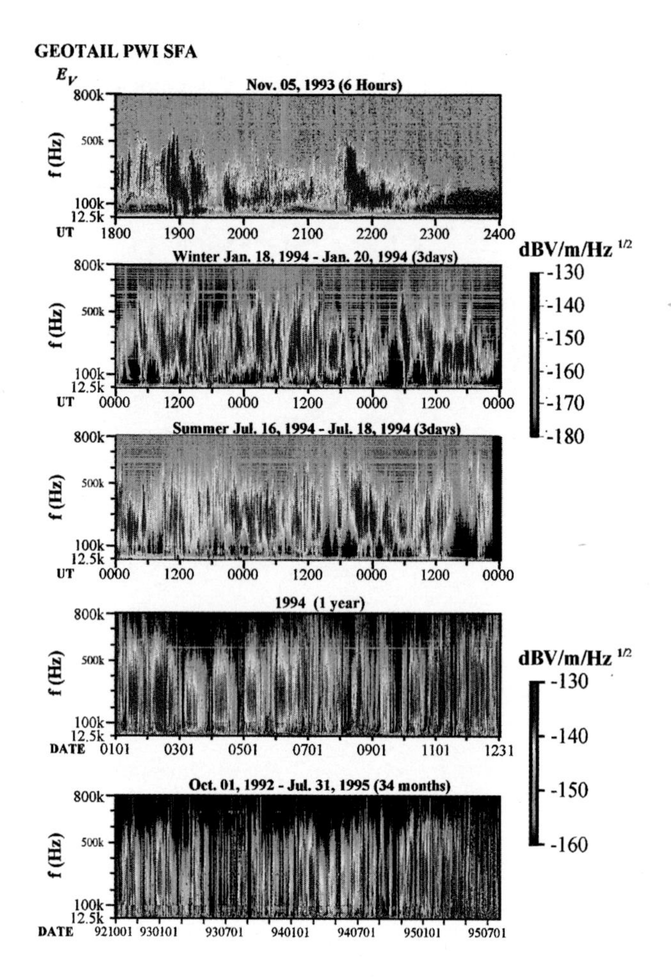

Figure 24. Dynamic spectra of AKR observed by GEOTAIL PWI over different time scales from 6 hours to about 3 years. The second and the third panels show the AKR daily variation, while the fourth and fifth panels show the seasonal variations (see text).

AKR upper cutoff frequency generally decreases with time and its intensity becomes weaker as time elapses. In addition to these characteristics, a shorter time scale variation of 5 to 10 minutes is generally observed in each blob. The reason of such short time scale modification of the AKR emission is not well understood at the present stage. One possible reason may be attributed to the modification of the precipitation of energetic electrons into the plasma cavity in the auroral region. Interrelation of the AKR intensification and frequency band expansion with substorms will be discussed in more detail in the next section.

The second and third panels from the top of Figure 24 show the daily variation of AKR in two different seasons, winter (January) and summer (July) in the northern hemisphere. These two panels are the dynamic spectra of the AKR over three days. The visible AKR frequency band shows a clear daily variation. In the case for northern winter (the second panel), the AKR frequency band is shifted to the higher frequency around 0500 UT, while in the northern summer (the third panel) the shift toward the higher frequency takes place around 2000 UT. This daily variation can be explained by the daily variation of the visibility of AKR at the location of the spacecraft. As the AKR is generally shielded by the high density plasmasphere, the visibility of AKR depends on the spacecraft magnetic latitude. The higher the latitude is, the better AKR visibility is attained. The spacecraft reaches its highest magnetic latitude when the magnetic dipole axis tilts to its maximum angle. As the intersection local time of the magnetic dipole axis with the surface of the northern hemisphere is 5 hours behind the universal time, the observed universal time of 0500 UT for the highest frequency in the second panel corresponds to the time of the maximum magnetic latitude of the GEOTAIL. The local time characteristics of the AKR frequency in the northern summer (the third panel) can be understood similarly. The fourth and fifth panels from the top in Figure 24 show the annual and triennial variation of the AKR dynamic spectra respectively. The annual variation shows a clear seasonal frequency variation. The uppermost observable frequency of AKR is maximum in both summer and winter. At equinox the uppermost frequency of AKR becomes minimum. The reason for this seasonal variation is again interpreted by the relative latitude of GEOTAIL. The triennial variation also shows the repeated seasonal variation except for the year after November, 1994. This is because the GEOTAIL orbit was changed from the deep tail orbits to the near Earth orbits. On the near Earth orbits, the spacecraft regularly crosses the bow shock where strong electrostatic waves with the similar frequency range of AKR are observed and contaminate the dynamic spectra.

3.3.2. Correlation of AKR with other geophysical phenomena. The AKR intensity and frequency bandwidth show a good correlation with the K_p index variation [e.g., *Kaiser and Alexander*, 1977]. Figures 25 and 26 are the examples to show such correlation. Figure 25 shows an example of the variation of the AKR dynamic spectra and corresponding K_p index over a period of two months covering the dayside to the deep tail orbits. The top, middle and bottom panels show the dynamic spectra of the electric component measured by the PWI SFA, the K_p index and the location of GEOTAIL, respectively. The AKR intensity is stronger on the location nearer to the Earth, and weaker at the deeper tail location. A clear one-to-one correspondence of the K_p index variation with that of the AKR activity is clearly recognized comparing the top and middle panels. Such

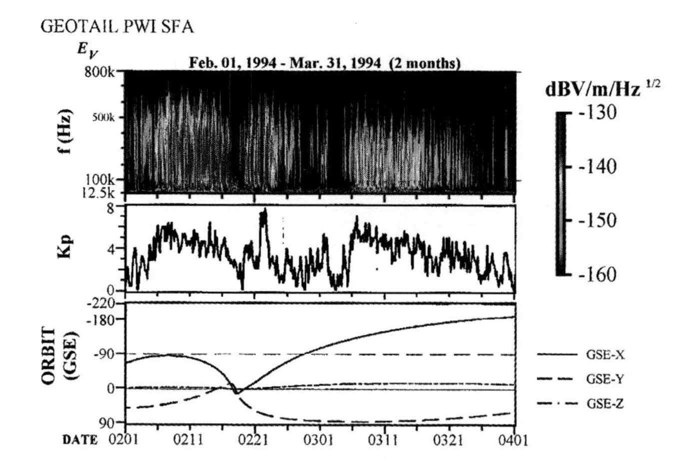

Figure 25. Comparison of AKR activity with K_p index over 2 month period. Also shows the intensity dependence of AKR on the radial distance along the Sun-Earth line (see the orbit along GSM X in the bottom).

good correlation between the K_p index and the AKR activities is also confirmed by a longer term data over one year as shown in Figure 26. The K_p index shows a quasi-periodic variation equal to 27 day rotational period of the sun. Accordingly the AKR activity changes responding to the variation of the K_p index.

It is well recognized that the AKR is closely related with the substorm onsets. *Murata* [1995] and *Murata et al.* [1997] have studied the interrelation between the AKR activities and the well-established measures of substorms. *Murata* [1995] examined the functional dependence of the AKR intensity on the distance from the Earth. Figure 27 shows that the intensity of AKR decreases as the geocentric distance towards the tail increases and that it is proportional to $\frac{1}{r}$ beyond 25 R_E far from the Earth independing on the geomagnetic activities, which are represented by K_p index.

A measure of the AKR activity can be expressed as an index defined by

$$\epsilon_{\mathrm{AKR}} = 10 \log \frac{\int_{f_l}^{f_h} E_o^2 df}{f_h - f_l} \quad [\mathrm{dBV^2/m^2/Hz}], \quad (7)$$

$$E_o = E(r)\frac{r}{r_o} \quad [\mathrm{dBV/m/\sqrt{Hz}}], \quad (8)$$

where $r_o = 25R_E$, $f_l = 50$ kHz and $f_h = 800$ kHz [*Murata et al.*, 1997]. The quantity of ϵ_{AKR} is called AKR index. This index is proportional to AKR power flux at a given time of the observation. The AKR index is computed and shown in the top panel of Figure 28 for a case observed on Jan., 1993. The second panel shows the corresponding magnetogram records from the stations in the auroral latitudes except for those located between 130° and 180° [Data Catalogue, 1995]. The superimposed magnetogram records in the second panel do not give the exact AE index but still are a good measure of substorms. It is clear that the three AKR activities labeled "a", "b" and "c" correspond to the three substorms represented by the second panel. The onsets of these substorms are well represented by the sudden increase of the AKR index.

3.3.3. Auroral Myriametric Radiation (AMR). In the SFA dynamic spectra, we occasionally find an enhancement of wave activities in the frequency range between 2 kHz and 50 kHz. The wavelength of most of these waves drops in the myriametric range. The nonthermal trapped continuum radiation observed in the magnetosphere and tail region is also observed in the same frequency range as already shown in Figure 17. Actually a terminology of "Terrestrial Myriametric Radiation (TMR)" was used for the nonthermal continuum radiation in some of the literatures [e.g., *Jones*, 1981a, b; 1982; *Hashimoto et al.*, 1994]. In this section, however, we discuss another class of radiation in the same frequency range. Figure 29 shows an example of the new class of radiation, which we call Auroral Myriametric Radiation (AMR) [*Hashimoto et al.*, 1994]. The AMR is always observed coincidentally with the AKR and has stronger intensity and shorter duration than the nonthermal continuum radiation. It is, however, noted that not all AKR is associated with AMR. In other words, AMR is associated with the injection of high energy electrons into the auroral region at the time of some substorms but not all substorms produce AMR.

In order to identify the source region of the AMR, we examined the MLT (magnetic local time) dependence of the observation of the AMR. The result of the statistical analyses is shown in the right panel of Figure 29 together with the similar analyses for AKR. The AMR is preferentially observed between 20 LT and 24 LT and show much narrower MLT distribution than AKR, though the locations of their peak occurrence rate around 22 LT is the same.

As the AMR is a new type of radiation discovered by GEOTAIL, its generation mechanism has not been well studied. Judging from the fact that the AMR frequency is much lower than the AKR frequency, it is inferred that the cyclotron maser instability in the plasma cavity is not a candidate because the cyclotron frequency in the plasma cavity is higher than the AMR frequency. Another candidate as a characteristic frequency close to the AMR frequency is the electron plasma frequency in the plasma cavity. Therefore, the electron beam instability would be a more plausible candidate because it

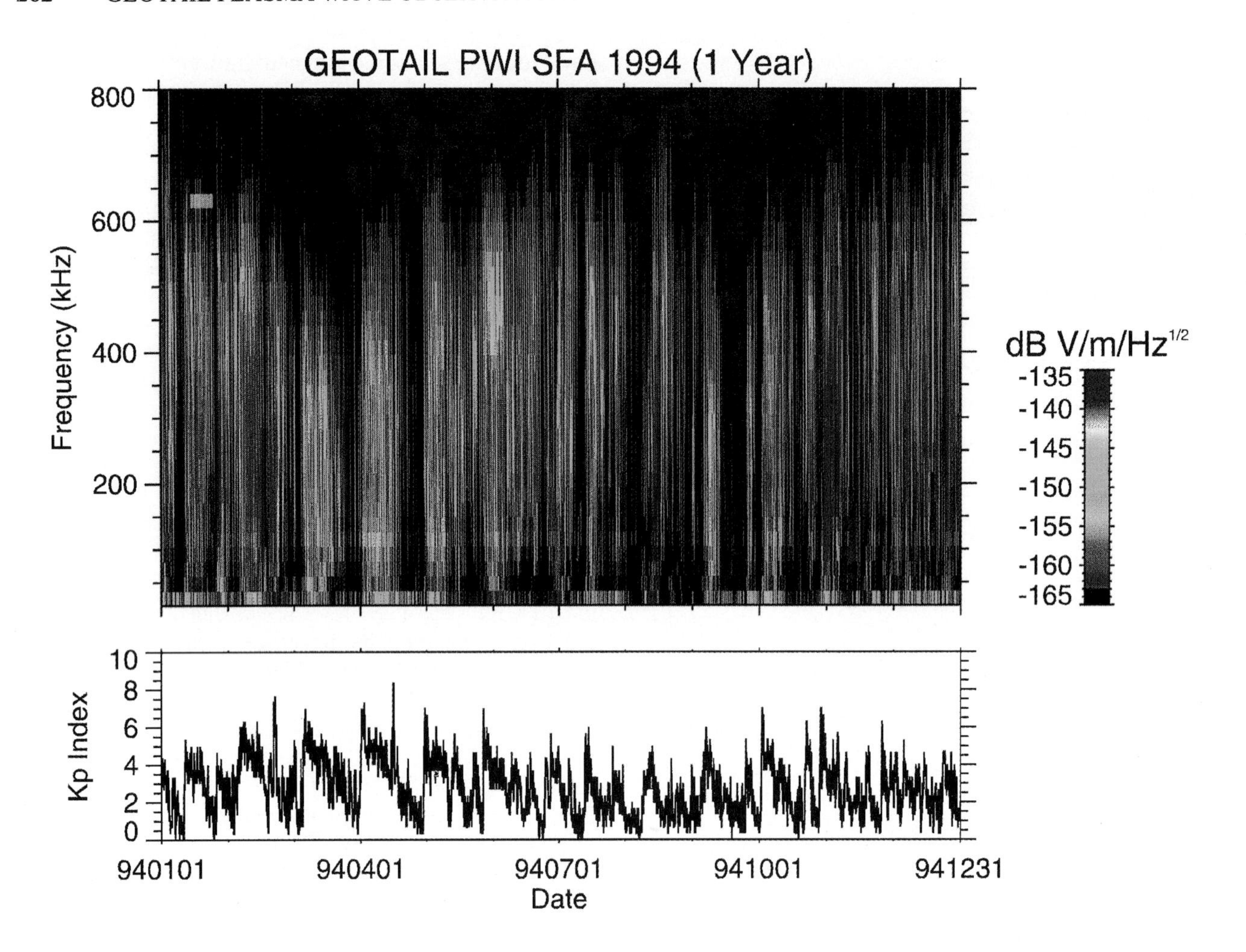

Figure 26. Correlation of AKR activity with K_p index over 2 years.

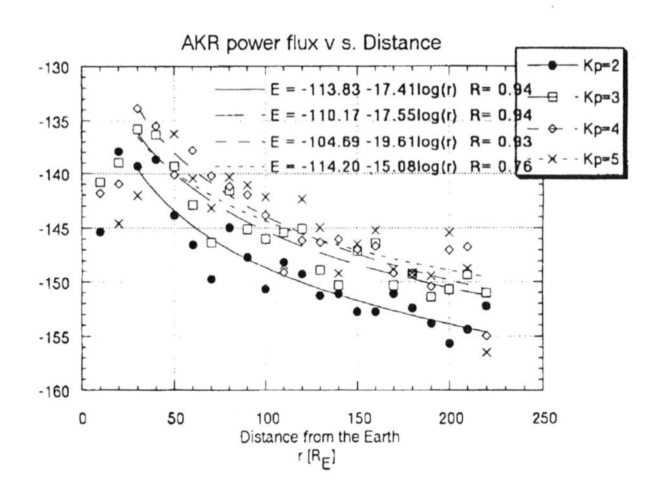

Figure 27. Dependence of AKR intensity on the geocentric distance from 0 to –220 R_E.

easily excites Langmuir waves which may well be converted to L-O electromagnetic mode near the wall of the steep density gradient in the plasma cavity along the auroral field line.

3.4. Type III Solar Burst

Type III solar radio burst is a common electromagnetic radiation observed not only in the interplanetary space but also in geospace. It is an electromagnetic wave with a smooth gliding frequency that falls with time. Figure 30 is a typical example of the type III solar radio burst observed by GEOTAIL. As seen in the figure, the lower frequency components of the type III radio burst arrive later than the high frequency elements resulting in the smooth frequency dispersion. This specific example was used to calibrate the gains of the wave receivers onboard the GEOTAIL and WIND spacecraft because the two spacecraft picked up this

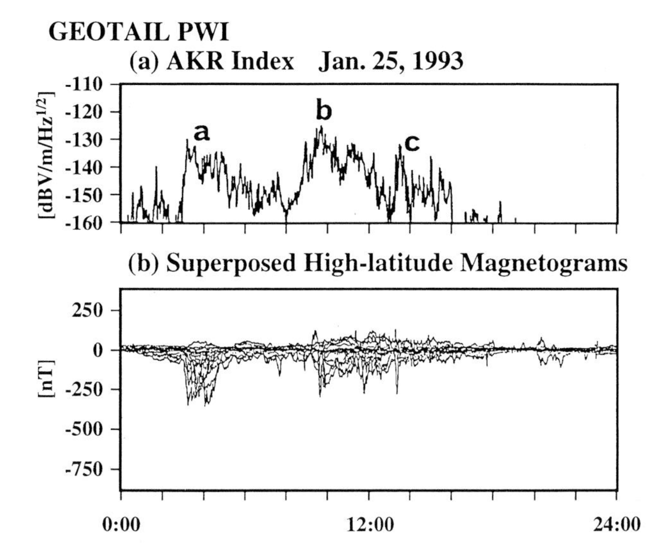

Figure 28. Comparison of the AKR index (ϵ_{AKR} defined by Eq. (8)) with a superimposed high-latitude magnetograms which gives a measure of substorm. A good correlation of the AKR index with the substorm activity is clearly found [*Murata*, 1995].

type III burst simultaneously without being disturbed by intense AKR.

Type III solar radio burst is believed to be generated by nonlinear wave-wave interactions of electron plasma waves excited by energetic streaming electrons ejected by solar flares [*Ginzberg and Zheleznyakov*, 1958; *Sturrock*, 1961; *Papadopoulos et al.*, 1974]. An extensive survey of Langmuir waves associated with type III radio bursts has been conducted using 9 years of observation with the IMP 6, IMP 8, Helios 1, Helios 2, Voyager 1 and Voyager 2 [*Gurnett et al.*, 1978]. They found that the intensity of Langmuir waves drops drastically as the heliocentric distance from the Sun increases. This fact explains why intense Langmuir waves associated with type III radio bursts are seldom observed near the orbit of the Earth. The chance of detecting the Langmuir waves simultaneously observed with type III radio bursts turns out to be only 12 during the 9-year observation. The GEOTAIL PWI has observed quite a number of type III radio bursts since the launching of the spacecraft. Among them we have picked out only those type III bursts which show a well-defined low frequency tails during the period from September, 1992 to October, 1995. Out of 141 events of such type III bursts, the lower cutoff frequency is twice the local electron plasma frequency, $2f_p$ in 28 cases, while the cutoff is seen at the local electron plasma frequency, f_p in 72 cases. The rest are the cases where neither the cutoff frequency is clear nor the local electron plasma frequency is well defined from the data. Figure 31 shows the examples of the two cases. Note that two different frequency scales are used in each panel to expand the low frequency part below 100kHz. The upper panel shows the case where the lower cutoff frequency is at the local f_{pe} and the first type III burst in the lower panel shows the case where the cutoff frequency is at $2f_{\mathrm{pe}}$. The case with the lower cutoff frequency at f_{pe} was observed in the deep tail around 200 $\mathrm{R_E}$. The type III radio burst in this case is associated with intense Langmuir waves. The PWI SFA receiver for a period from 0020 to 0130 UT was saturated with the intense Langmuir waves. This fact indicates that the intensity of the observed Langmuir waves exceeds the saturation level of about 2 mV/m.

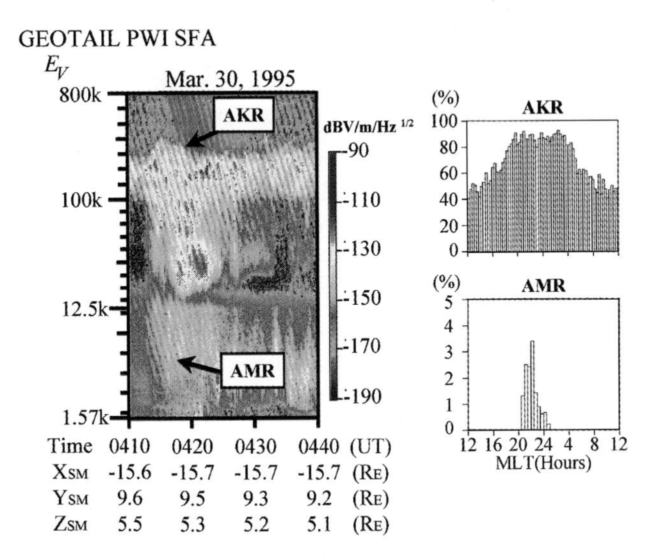

Figure 29. Typical example of dynamic spectrum of AMR which appears below 20 kHz (left panel). The right panel shows the distribution of the source location in terms of the magnetic local time for both AKR and AMR. The source location was decided from the k vector direction deduced from the spin modulation of AKR and AMR intensities.

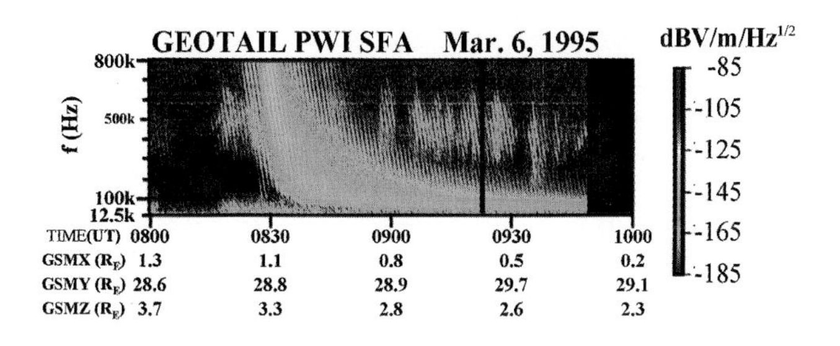

Figure 30. Typical example of type III solar radio burst observed by GEOTAIL PWI.

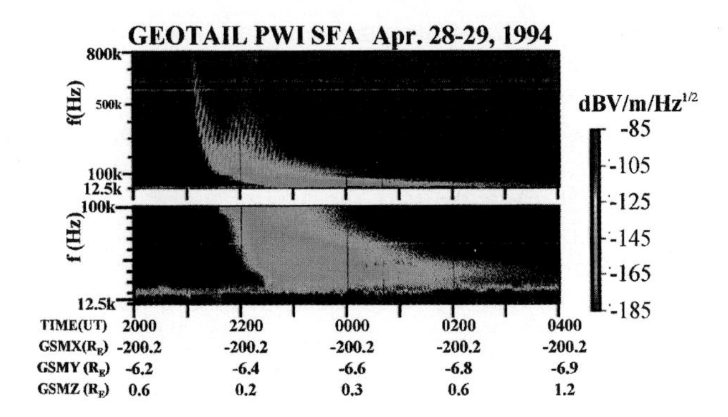

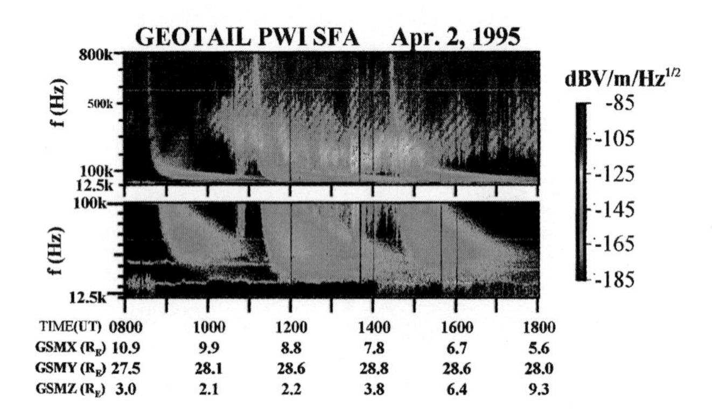

Figure 31. Type III solar burst with the lower cutoff frequency (a) at the f_{pe} (Upper panel). Type III solar burst sometimes shows the lower cutoff frequency at f_{pe} as shown in the first burst in the lower panel.

This fact seems to be contradictory with the results by *Gurnett et al.* [1978]. However, it is noted that the location of observation for weaker intensities of Langmuir waves by previous spacecraft observations was in the tail lobe and was not in the solar wind.

In some cases, consecutive type III solar radio bursts show the different cutoff frequency either at f_{pe} or $2f_{pe}$. The lower panel of Figure 31 shows such an example. The first type III burst in the figure shows the cutoff at $2f_{pe}$ though faint frequency components are seen below $2f_{pe}$ down to f_{pe}. The second and third bursts have the lower cutoff frequency at f_{pe}, though the frequency components above $2f_{pe}$ are stronger than those below $2f_{pe}$. The reason of the coexistence of the two different cutoff frequencies at f_{pe} and $2f_{pe}$ needs to be investigated. The enhanced line emission of Langmuir waves around f_{pe} is simultaneously observed with the second and the third type III bursts around 1200, 1430 and 1600 UT. However, it is noted that we need further discussions on the generation of these Langmuir waves, because we need to separate the Langmuir waves generated in the upstream region of the bow shock.

3.5. Electromagnetic discrete emission at twice the electron plasma frequency ($2f_p$ emission)

The $2f_p$ electromagnetic radiation is frequently observed in the upstream region of the Earth' bow shock. It is a discrete (narrow band) emission with twice the solar wind electron plasma frequency. Direction findings of the $2f_p$ emission by *Gurnett* [1975] and *Gurnett and Frank* [1976] showed that the source of the $2f_p$ emission is located outside the magnetosphere. It is generally believed that the $2f_p$ emission is generated by a nonlinear coupling involving intense Langmuir waves in the same manner as the generation of the type III solar burst. *Hoang et al.* [1981], using the ISEE date set, pointed out that the source region of the $2f_p$ emission is on the interplanetary magnetic field line tangent to the bow shock. Therefore it is widely recognized that there is an electron beam streaming backward from the bow shock into the upstream region along the tangent field line, and that the electron beam excites intense Langmuir waves which eventually produce the $2f_p$ emission through the nonlinear wave coupling process. Typical dynamic spectrum of the $2f_p$ emission observed by GEOTAIL is Figure 32. The horizontal line emissions starting at about 40 kHz and 80 kHz are Langmuir waves (or enhancement of thermal noise at the local electron frequency) and $2f_p$ emission, respectively. We could see the frequency variation of the $2f_p$ emission is very similar to that of the Langmuir waves. However, the frequency of $2f_p$ emission is not always equal to the double frequency of the local f_{pe}. This non-double $f_{pe_{local}}$ emission is often associated with the bifurcation of the spectrum of the $2f_p$ emission [*Lacombe et al.*, 1988; *Reiner et al.*, 1996; *Kasaba*, 1997a; *Kasaba et al.*,

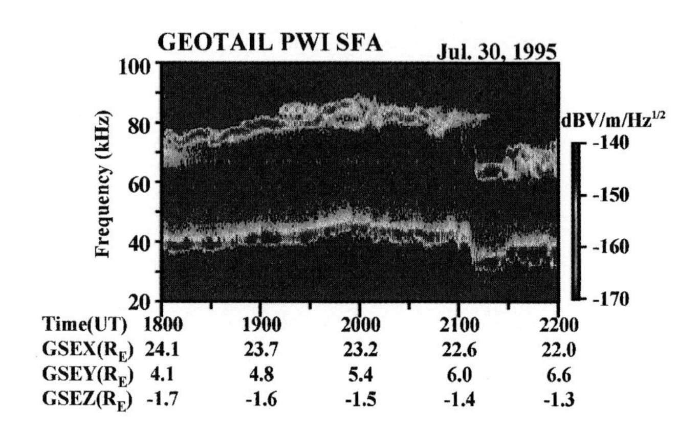

Figure 32. Dynamic spectra of $2f_p$ emission observed by GEOTAIL PWI. The lower frequency discrete emission is the Langmuir waves or enhanced noise level at the electron plasma frequency. The higher discrete emission is the electromagnetic $2f_p$ radiation. Note that the frequency of the $2f_p$ emission is not always the double of the local f_{pe}.

1997b]. The bifurcation is a phenomenon of coexistence of two $2f_p$ emissions with different frequencies for a certain time of period. An example of the bifurcation is shown in the dynamic spectrum in Figure 33 [*Kasaba*, 1997a]. It is associated with a solar wind density jump which propagates through the spacecraft and the upstream region [*Lacombe et al.*, 1988]. The lower line emission is the Langmuir line which jumps at around 2224 UT (labeled $T_{s/c}$ in the figure frame) from $(f_{pe})_1$ to $(f_{pe})_2$. The sudden jump of the local electron frequency indicates that a boundary between the lower and higher density regions in the solar wind passed through the GEOTAIL at a time labeled $T_{s/c}$. In spite of the jump in the local electron plasma frequency, the frequency of the $2f_p$ emission does not change synchronously but jumps up later to $2(f_{pe})_2$ from $2(f_{pe})_1$ around 2229 UT, though a faint emission precedes even before the $T_{s/c}$. The faint $2(f_{pe})_2$ starts to appear at a time T_{up} when the old $2(f_{pe})_1$ starts to change its frequency and intensity slightly. The old $2(f_{pe})_1$ gradually diminishes and vanishes around 2242 UT, which is labeled as T_{down} in the figure. Therefore the spacecraft observed two different electromagnetic emissions at $2(f_{pe})_1$ and $2(f_{pe})_2$ for a period from T_{up} to T_{down}.

This time-serial change of the observed emission can be explained by a schematic illustration shown below the dynamic spectra in Figure 33. The three vertical lines in the illustration indicate three locations ($X_{up}, X_{s/c}$ and X_{down}) of the density jump in the solar wind at three different times, $T_{up}, T_{s/c}$ and T_{down}, respectively. A thick bar on the tangent field line of the IMF shows the limited source region (between X_{up} and X_{down}) of the $2f_p$ emission. In the source region, streaming electrons produced at the bow shock are fertile of Langmuir waves through a beam instability. Outside of the source region, the streaming electrons are diffused and stabilized due to the velocity space diffusion by the excited Langmuir waves [*Matsumoto et al.*, 1997]. Before the time of T_{up}, the streaming electrons are all embedded in the low density solar wind, and hence produced only $2(f_{pe})_1$. After T_{down}, the streaming electrons are all covered by the higher density solar wind, and hence produce only $2(f_{pe})_2$. In the intermediate period from T_{up} to T_{down}, the streaming electron region indicated by the thick bar along the tangent field line is partially in the lower density and partially in the higher density solar wind. Therefore both of the emission at $2(f_p)_1$ and $2(f_p)_2$ are produced and observed. The information of the three times, T_{up}, $T_{s/c}$ and T_{down}, can be used to estimate the size of the source region of the $2f_p$ emission. *Kasaba* [1997a] estimated the source length based on the GEOTAIL observation and found that the typical extension of the upstream side of the source region projected on the Sun-Earth line is about 5-10 R_E while the downstreaming size is about 10-20 R_E projected on the Sun-Earth line.

Another interesting point which the example shows is the existence of a faint emission at $(f_p)_2 \sim 50$ kHz between T_{up} and $T_{s/c}$. The emission provides a proof of generation of electromagnetic waves at the electron plasma frequency in addition to the $2f_{pe}$ radio emission as a result of mode conversion from the electrostatic to electromagnetic waves. The fact that the lower cutoff of the type III solar radio bursts observed by GEOTAIL often comes down to local f_{pe} can be well interpreted based on the proof shown in the bifurcation of the $2f_p$ emission. Figure 34 shows mappings of the Langmuir waves and $2f_p$ emission observed by GEOTAIL in the upstream region of the bow shock during a period from November, 1994 to April, 1996. The result is consistent with the estimation of the source region of the $2f_p$ emission based on the bifurcation analysis stated above.

4. ELECTROSTATIC WAVES OBSERVED BY GEOTAIL

Electrostatic waves hardly propagate over a long distance from their generation region. Instead, they reflect well the nature of the plasmas in the vicinity of spacecraft and relevant wave-particle interactions nearby.

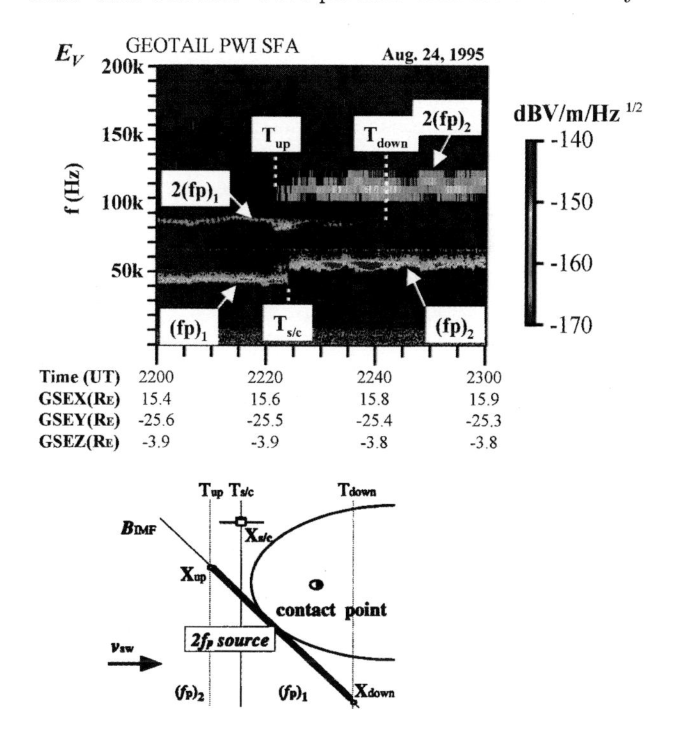

Figure 33. Example of bifurcation of the $2f_p$ emission observed by GEOTAIL PWI. The local f_p jumps at 2224UT but the jump of the $2f_p$ is delayed (see text). The lower illustration shows possible situation to explain the bifurcation phenomenon [*Kasaba*, 1997a].

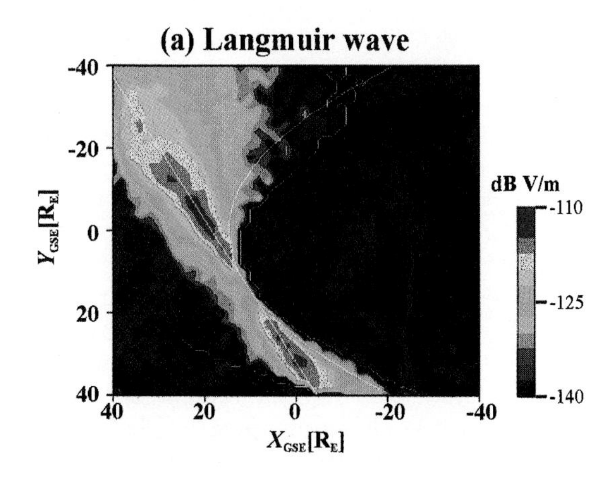

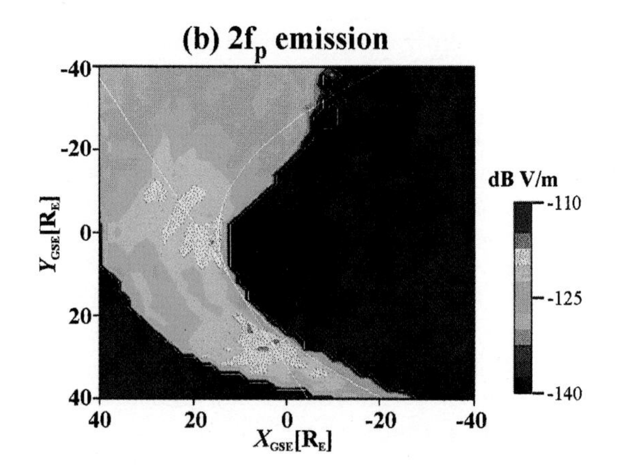

Figure 34. Statistical intensity mappings of the Langmuir waves (left) and of the $2f_p$ EM radiation (right). The mapping was made by projecting observed location relative to the tangent field line making an angle of 120 ° with the solar wind stream line [*Kasaba*, 1997a].

The GEOTAIL PWI frequently observed various electrostatic waves in geospace in addition to electromagnetic waves as discussed in Chapter 3. In this chapter, we will show an overview of the electrostatic waves observed by GEOTAIL.

4.1. Electron plasma waves

Though the terminology of "electron plasma waves" is used as an equivalent name to Langmuir waves in some literatures, we use it to describe the electrostatic waves with their frequencies close to the electron plasma frequencies in a broader sense. The electron plasma waves with amplitude modulations have been observed in the various regions. *Gurnett et al.* [1981] reported the modulated electron plasma waves observed in the upstream region of the Jovian bow shock. Some of their observed waveforms are amplitude-modulated quasi-monochromatic waves, while the others are isolated wave packets. Similar waveforms were observed associated with the electron beam in the solar wind at 1 AU, the upstream region of the terrestrial bow shock, and in the polar region of the magnetosphere as well [*Gurnett et al.*, 1993; *Kellog et al.*, 1996; *Stasiewicz et al.*, 1996]. Assuming that the observed modulated waves are Langmuir waves, they attempted to explain their generation mechanism in terms of the nonlinear parametric wave-wave coupling. However, there still remain a problem of a poor correlation between the observed Langmuir and ion acoustic waves which are needed for the parametric nonlinear wave coupling.

The GEOTAIL PWI has also observed similar modulated electron plasma waves. Figure 35 shows a typical example of such modulated electron plasma waves observed by GEOTAIL in the tail lobe region [*Kojima et al.*, 1997b]. Such electron plasma waves are frequently observed at the edge of the lobe close to the PSBL. The similar electrostatic waves were reported by *Scarf et al.* [1984] in connection with the slow shock observation in the tail. However, they did not examine their waveforms.

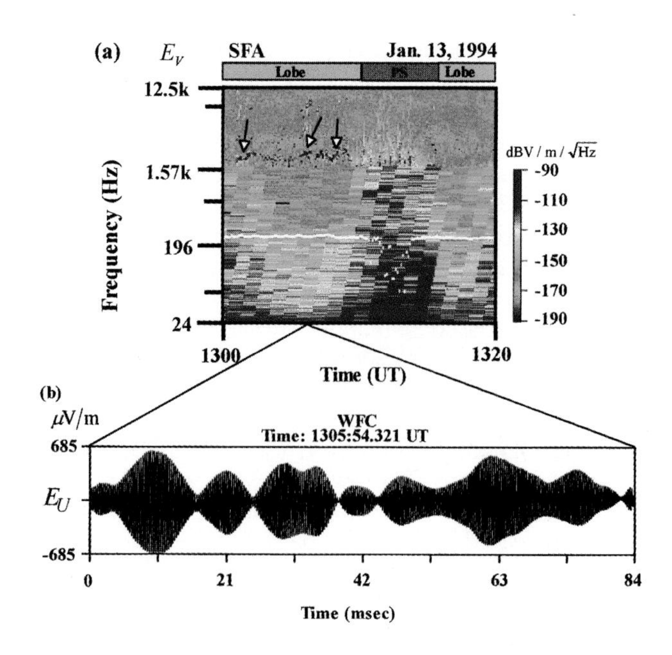

Figure 35. Typical dynamic spectrum (a) and corresponding waveforms (b) of the modulated electron plasma waves observed by GEOTAIL in the tail lobe region [*Kojima et al.*, 1997b].

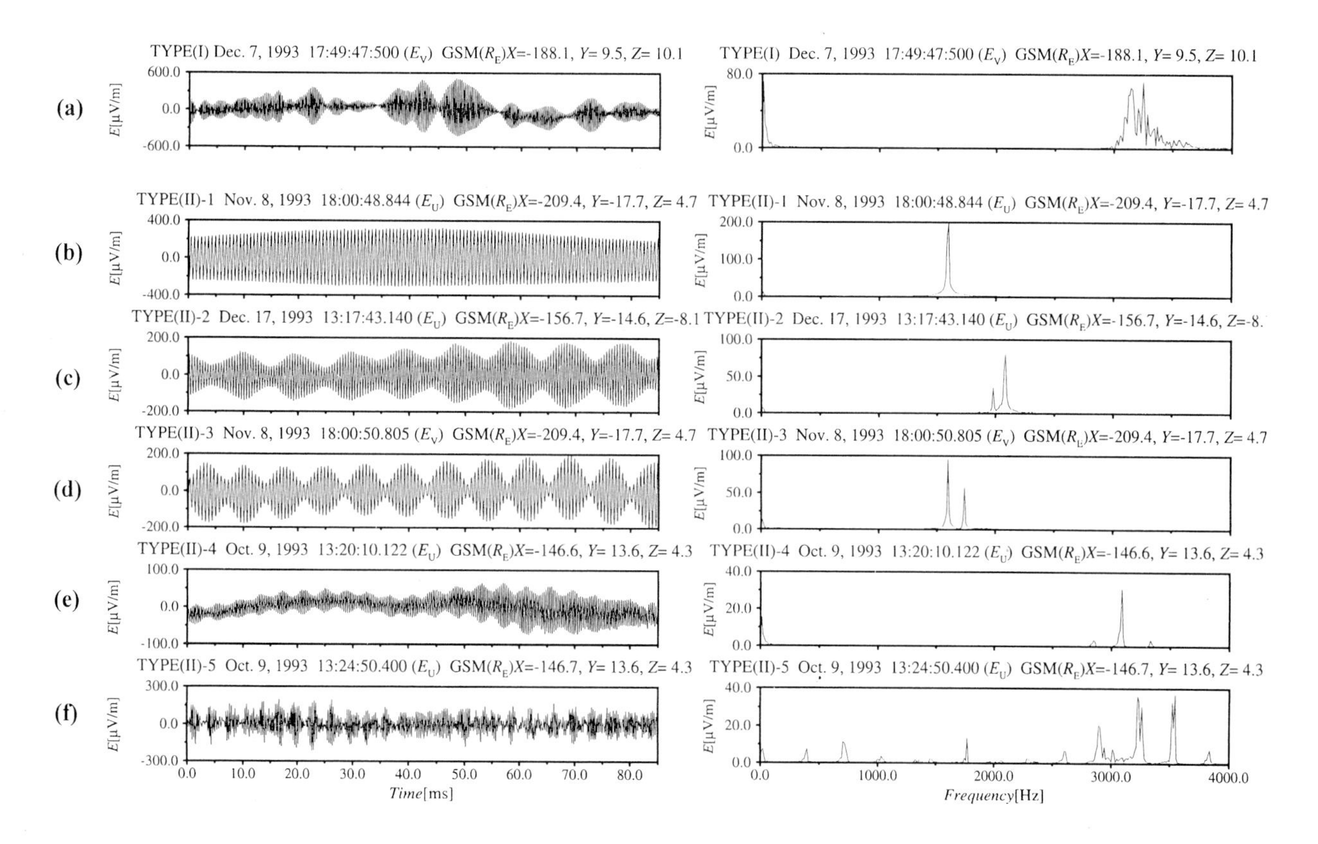

Figure 36. Classification of the electron plasma waves observed in the tail lobe region. The observed modulated waves are classified into 2 types: Langmuir waves (Type (I)) and Electron Cyclotron Harmonic waves (Type (II)). The Type (II) waves are further classified into 5 subgroups according to the spectral structures [*Kojima et al.*, 1997b].

The identification of the spacecraft location is displayed in the top bar of Figure 35(a). The spacecraft moved into the plasma sheet (PS) from the lobe and returned back to the lobe. The intense monochromatic waves displayed by white arrows are the electron plasma waves at frequencies close to the plasma frequency. Note that the electron plasma waves are seen just before the plasma sheet entry.

Figure 35(b) shows the waveforms of the modulated electron plasma waves observed at 1305:54.321 UT. The observed waveforms are strongly modulated, which are very similar to those in *Gurnett et al.* [1981], *Gurnett et al.* [1993], *Kellogg et al.* [1996], and *Stasiewicz et al.* [1996]. Such modulated waveforms can be the result of superposition of several waves with different frequencies. This possibility led these authors to the idea of the nonlinear wave-wave coupling of the Langmuir waves with waves in the lower frequency such as ion acoustic waves.

Kojima et al. [1997b] examined the polarizations of the modulated electron plasma waves observed in the lobe region. As a result, they found that there exist two different types of waves in the observed modulated electron plasma waves ; Langmuir waves (LW) and Electron Cyclotron Harmonic (ECH) waves. The LW and ECH waves are totally different modes having the parallel and perpendicular wave normal directions, respectively. Figure 36 shows the classified waveforms (left panels) and corresponding Fourier spectra (right panels). The observed modulated electron plasma waves are classified further into 6 types; The Type (I) is the modulated Langmuir waves. As we can see in the left panel (a), the envelope of the modulated waves is asymmetric and changes its frequency in a short time. The Type (II) waves are perpendicularly propagating waves. They are further classified into 5 groups. The subclassification is based on the spectrum structure shown in the right panels. Different from Type (I), the frequency spectra of Type (II) consist of discrete spectrum components. The Type (II)-1 is monochromatic and has only one spectral component. The spectra of the Type (II)-2 and -3 consist of one main and single upper or lower sideband components, respectively, while that of the Type (II)-4 has one main and 2 sideband components. The waveforms of the Type (II)-5 consists of multiple discrete components.

Kojima et al. [1997b] examined the frequency gap between the discrete spectrum components and found that their frequency gaps are almost equal to the local electron cyclotron frequency. Thus they concluded that the Type II waves are the perpendicularly propagating ECH waves. In the tail lobe region, the typical electron cyclotron frequency ($f_{ce} \sim 300$ Hz) is much smaller than the typical electron plasma frequency ($f_{pe} \sim 2$ kHz). Therefore, the typical upper hybrid resonance (UHR) frequency ($f_{UHR} \sim 2.02$ kHz), where the ECH waves are most unstable, is very close to the electron plasma frequency. This fact makes it difficult to distinguish the Langmuir waves and the ECH waves only by frequency spectral analysis. They, therefore, stressed the importance of the polarization analyses for the mode identification in such cases. Likewise in the solar wind at 1 AU the typical electron plasma frequency (about 20 kHz) is much higher than the typical electron cyclotron frequency (about 200 Hz). Therefore, the modulated waveforms observed in the solar wind at 1AU by *Gurnett et al.* [1993] may have included the ECH waves because they have not shown the polarization analysis. If so, we can understand the reported poor correlation between the modulated waves and the ion acoustic waves, because we do not need the nonlinear coupling mechanism, but can explain the modulation by a simple superposition of multiple spectral lines of ECH waves.

Nonetheless, we need to explain the modulation of Langmuir waves by some nonlinear mechanism. Previous attempts to explain the modulated Langmuir waves were to consider nonlinear wave-wave couplings. However, there still remains the problems of the poor correlation of the Langmuir waves with the ion acoustic waves and of a fact that the observed wave amplitude is much smaller than the amplitude required by the theory in the nonlinear interaction [Gurnett et al., 1993]. Another possible candidate for the generation of the modulated Langmuir waves is proposed by *Akimoto et al.* [1996]. They performed a full-particle simulation without including ion dynamics on the nonlinear Langmuir waves excited by an electron beam and demonstrated that the modulation of the Langmuir waves is possible without the nonlinear coupling with the ion waves. In their computer simulation, the Langmuir waves excited by an electron beam on the linear stage start to trap beam electrons. Their nonlinear evolution causes a phase space modulation of the trapped electrons resulting in the spatial amplitude modulation of the excited Langmuir waves. The spatially modulated Langmuir waves in the simulation are converted to the temporal amplitude modulation observed at a fixed point in the simulation. Therefore this modulation mechanism is not related to the ion dynamics, but solely to the nonlinear dynamics of electrons. In this case, we do not need to involve ion acoustic waves at all, which is consistent with observation.

In addition to the amplitude-modulated Langmuir waves observed in the tail lobe, GEOTAIL PWI observes Langmuir waves in the upstream region as well. The statistical analysis of occurrence of Langmuir events in the upstream region has already been presented in Figure 34(a). The Langmuir waves are most frequently observed on or near the tangent field line. It is noted that the Langmuir wave activity is very weak near the contact point (CP) of the tangent IMF field line with the bow shock. This can be explained [*Matsumoto et al.*, 1997] by a beam forming mechanism along the tangent field line. However the Langmuir wave activity is enhanced beyond a certain distance from the CP along the tangent field line. This tendency is also shown by Greenstadt et al. [1995]. Matsumoto et al. [1997] explained this spatial distribution of Langmuir waves in terms of a velocity filter effect. The electrons reflected back from the bow shock cannot form a bump on the tail in their velocity distribution function being masked by the abundance of the background solar wind electrons. However, slower electrons in the reflected electron population are swept away toward the Earth with the solar wind and thereby only the high speed component remains on the tangent field line. This process can produce a bump on the tail in the electron velocity distribution function. A schematic illustration of the formation of the electron beam along the tangent field line is given in Figure 37. Further downstream along the tangent field line, the Langmuir wave becomes weak and vanishes beyond about 50 Re from the CP along the tangent field line. This is due to a nonlinear diffusion caused by the Langmuir waves. Such beam destruction process will be seen in the simulation results in Chapter 7 on the nonlinear evolution of the beam instability.

4.2. Electron Cyclotron Harmonic (ECH) Waves

The first ECH waves in geospace were observed by the OGO-5 satellite in the near earth region of $4 < L < 10$ [*Kennel et al.*, 1970; *Fredricks and Scarf*, 1973]. The ECH emissions are observed at $f \sim (n+1/2)f_{ce}$, where f_{ce} and n denote the electron cyclotron frequency and a positive integer value, respectively. The observed ECH waves were almost electrostatic with the frequency spectrum with a structure of multiple harmonics. Many theoretical attempts on the generation of the ECH waves

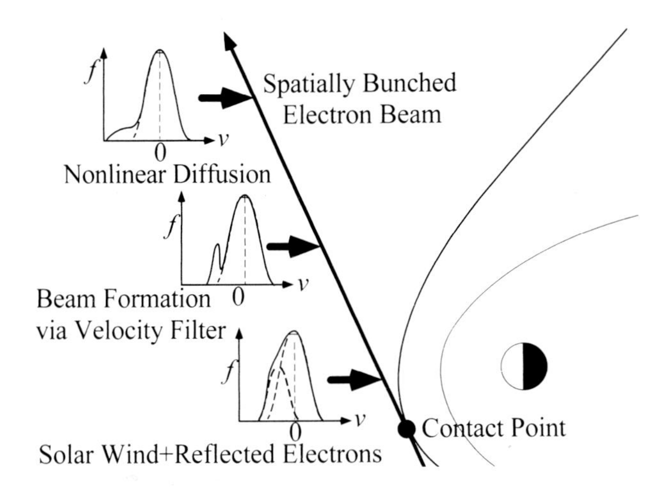

Figure 37. Schematic illustration of the formation of an electron beam along the tangent field line [*Matsumoto et al.*, 1997].

followed and concluded that the most possible candidate for the generation mechanism is an instability caused by the temperature anisotropy of electrons [see a review by *Ashour-Abdalla et al.*, 1979].

ECH waves are observed by GEOTAIL not only in the tail lobe but also in the dayside outer magnetosphere and even in the deep tail regions. Figure 38 shows five examples of the ECH waves observed by GEOTAIL in the dayside magnetosphere and in the tail region. The superposed white lines display the electron cyclotron frequency and its harmonics. Figure 38(a) shows the ECH waves often observed on the dayside of the outer magnetosphere. In the figure, two types of the ECH waves can be identified. They are: continuous and weak ECH waves and intermittent and intense ECH waves. The continuous ECH waves belong to the diffuse ECH waves in the classification by *Hubberd and Birmingham* [1978]. Usui et al. [1997] showed that the continuous ECH waves broadly appear along the dayside magnetopause, based on the statistical analyses. They have the multiple harmonic spectral structure with a weak amplitude of the order of $\sim 0.3\mu\mathrm{V/m}/\sqrt{\mathrm{Hz}}$. The latter type of ECH emissions is seen at 2216, 2227, 2234, 2257, and 2259 UT. This type of ECH emissions is observed intermittently and sporadically in time. Usui et al. [1997] showed that such intense and bursty ECH waves are mainly observed at the dawnside region near the magnetopause. Further, *Matsumoto and Usui* [1997] studied these bursty ECH waves using the WFC data as well as the spectrum data by the SFA and MCA. They called these ECH waves 'Totem Pole emissions (TP emissions)' after their spectral forms in the dynamic spectra. Though the TP emissions seem to be simply the intensified version of the diffuse type ECH waves, *Matsumoto and Usui* [1997] using more detailed spectral structure based on the high time WFC data, revealed the existence of two spectral peaks in the first electron cyclotron harmonic band from f_{ce} to $2f_{ce}$. Their maximum amplitude is about $-100\mathrm{dBV/m}/\sqrt{\mathrm{Hz}}$. The similar fine structure between f_{ce} harmonics was observed by the SCATHA satellite [*Koons and Fennell*, 1984]. *Matsumoto and Usui* [1997] stressed that such two spectral peaks are not special but are often found associated with the intense ECH waves near the dayside magnetopause. Since we cannot expect such two spectral peaks in the first harmonic band from the linear dispersion for the Maxwellian plasma, we need to consider other effects including nonlinear phenomena.

Figure 38 (b) to (e) illustrate the frequency-time spectrograms of ECH emissions observed by GEOTAIL in the tail region. ECH waves in panels (b) and (c) are observed in the near earth region ($|X| < 30\mathrm{R_E}$), those in panels (d) in the middle tail region ($|X| \sim 60\mathrm{R_E}$), and those in panel (e) in the distant tail region ($|X| > 100\mathrm{R_E}$). The ECH waves in the near and distant tail regions have already been reported by *Roeder and Koons* [1989] and *Gurnett et al.* [1976], respectively.

The ECH waves in the near earth region had been discussed in relation to the diffuse auroral electron precipitation [e.g., *Kennel and Ashour-Abdalla*, 1982]. However, *Belmont et al.* [1983] reported their intensities observed by the geostationary spacecraft GEOS 2 are too small for the precipitation. Further, using the data from AMPTE IRM and SCATHA satellites, *Roeder and Koons* [1989] also showed that the occurrence and intensity of ECH waves observed below $L < 20$ is too small to account for the continuous precipitation of magnetospheric electrons in the diffuse aurora. The GEOTAIL observations shown in Figure 38 (b) and (c) are consistent with the above results. As far as we examined our data, the observed ECH wave intensities are in most cases less than $1\ \mathrm{mV/m}/\sqrt{\mathrm{Hz}}$. Their amplitudes are not large enough to cause the diffuse electron precipitation due to pitch angle diffusion [Lyons, 1974].

The ECH emissions observed in the middle tail region at radial distances ranging from about 23.1 to 46.3 $\mathrm{R_E}$ are briefly reported by *Gurnett et al.* [1976]. They showed that the ECH waves can be observed very close to the neutral sheet. They reported that the ECH waves have a correlation with high temperature in the central plasma sheet but their occurrence rate is very low. However, GEOTAIL data displayed in Figure 38 (d) and (e) show that the ECH emissions are observed in the lobe region close to the plasma sheet but not observed near the neutral sheet. In section 4.1, we have shown that the ECH waves are observed in the form of the modulated electron plasma waves. Actually most of the ECH

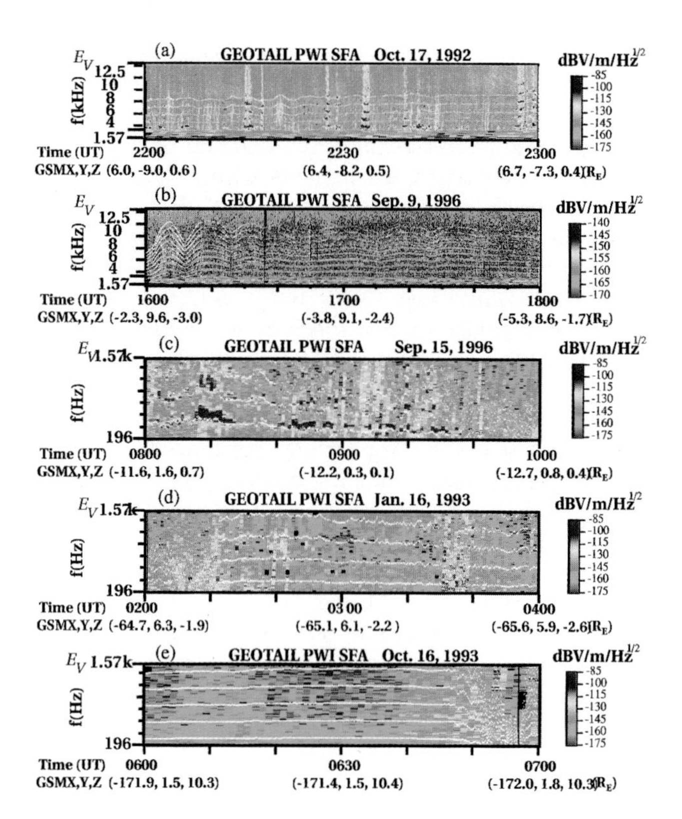

Figure 38. Typical example dynamic spectrum of the Electron Cyclotron Harmonic waves (ECH) observed by GEOTAIL in the dayside outer magnetosphere ((a)) and in the tail region ((b) $\sim$ (e)). The intense bursty spectra shown in panel (a) were addressed as "Totem Pole Emissions" by Matsumoto and Usui [1997]. The tail ECH waves shown in panel (b) to (e) are observed in the lobe region close to the plasma sheet.

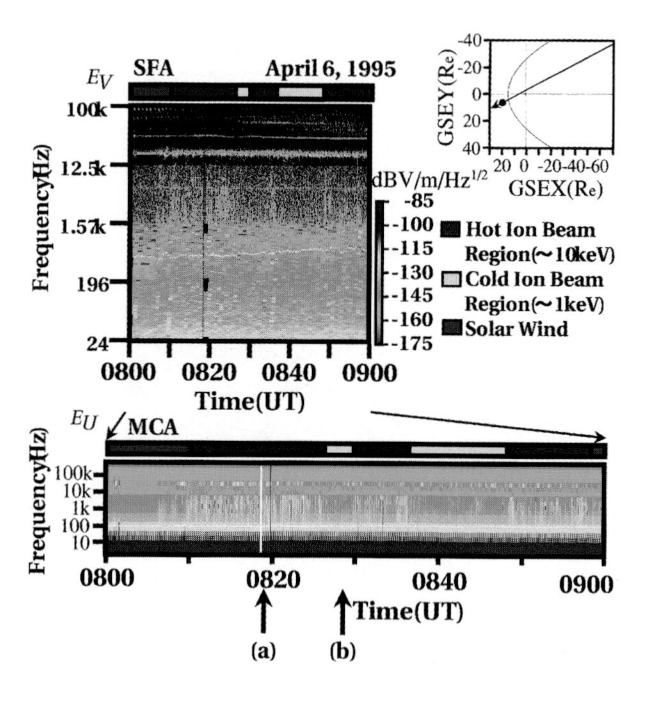

Figure 39. Dynamic Spectra of the ion acoustic-like waves observed in the ion foreshock region. The upper and lower dynamic spectra are generated from the SFA and MCA, respectively. The spectral features shown in both panels are very similar to those of the EQMW (Electrostatic Quasi-Monochromatic Waves) observed in the lobe and magnetosheath regions [*Matsumoto et al.*, 1997].

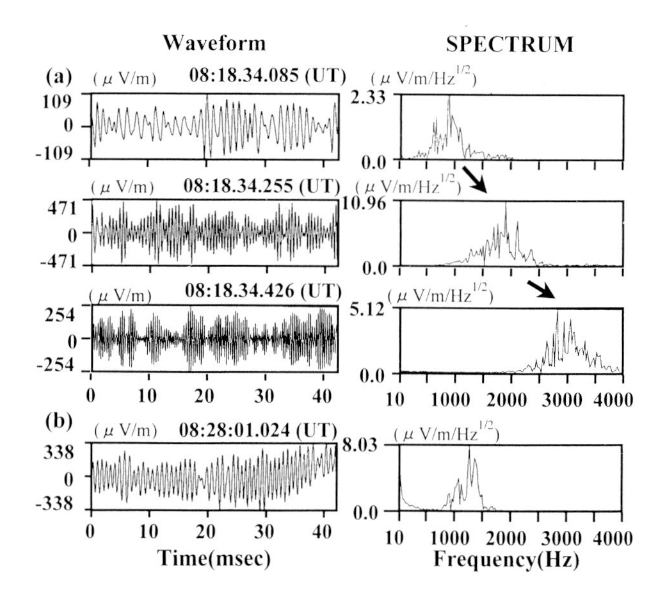

Figure 40. Waveforms of the ion acoustic-like waves observed (a) in the hot ion beam region and (b) in the cold ion beam region of the ion foreshock. [*Matsumoto et al.*, 1997]

waves observed in the middle and distant tail lobe are amplitude-modulated.

4.3. Ion Acoustic-like Waves

It is well known that ion acoustic-like waves are often observed in the ion foreshock region [*Scarf et al.*, 1970; *Anderson et al.*, 1981; *Rodriguez*, 1981; *Gurnett*, 1985]. *Matsumoto et al.* [1997] showed the detailed wave spectra and waveforms of the ion acoustic-like waves observed in the ion foreshock region. Figures 39 and 40 show that the frequency-time spectrogram of the ion acoustic-like waves and their corresponding waveforms [*Matsumoto et al.*, 1997].

The electrostatic waves which we address in this paper as ion acoustic-like waves shown in Figure 39 appear in the frequency range around a few kHz. Since the typical ion plasma frequency in the ion foreshock region is a few hundreds of Hz, we need to explain the frequency range of the ion acoustic-like waves in terms of the Doppler shift effect. *Rodriguez* [1981] suggested a possibility that these waves are not only the Doppler shifted ion acoustic waves but also could be waves generated by Buneman instability due to slow electron beam. However, no definite distinction has not been made up to the present time.

In Figure 39, we can find very patchy red or yellow dots in the SFA plot (upper panel). As the SFA sweeps the frequency slowly, these patchy natures show that the frequency of these waves varies very quickly in time. This nature is very similar to that of the NEN(Narrowband Electrostatic Noise) observed in the lobe and in the magnetosheath regions, which we discuss later. Such bursty nature of the ion acoustic-like waves can be seen well in the MCA data (lower panel) shown in Figure 39.

The waveforms at different times (a) and (b) indicated in Figure 39 are shown in Figure 40. The waveform sampled in the hot ion beam region (see Section 5.1) shows that the wave is a quasi-monochromatic wave with the amplitude modulation and its central frequency increases with time. The frequency rises rapidly, i.e., changes from about 700 Hz to 4 kHz in 0.35 sec. The possibility of such rapid change in frequency has been pointed out by spectral change in the previous observation of ISEE-2 [*Anderson et al.*, 1981]. The present observation has given its confirmation with additional information of waveforms, and with much finer time scale. It is surprising that the waves in the hot ion beam region consist of discrete riser elements. It may not be easy to understand their very fast frequency variation simply by the Doppler shift of the ion acoustic waves as considered in the past. In contrast, the waves in the cold ion beam region (see Section 5.1), the frequency and amplitude are not much variable as shown by one example in panel (b) of Figure 40. The wave is quasi-monochromatic with an almost constant frequency

4.4. Broadband Electrostatic Noise (BEN) and Electrostatic Solitary Waves (ESW)

Intense electrostatic emissions are commonly observed in the plasma sheet boundary layer (PSBL). They were referred to as broadband electrostatic noise (BEN) by *Gurnett et al.* [1979]. The wave features of the BEN reported by *Gurnett et al.* [1979] are as follows: (1) The noise usually occurs over a broad range of frequency up to the local electron plasma frequency. (2) They consist of many discrete bursts lasting from a few seconds to several minutes. (3) Their spectrum shows a marked decrease in intensity at the electron cyclotron frequency. (4) They have a low frequency cutoff corresponding to the local lower hybrid resonance frequency. (5) The electric field is oriented within ±20° from perpendicular to the magnetic field.

Gurnett et al. [1976] also reported good correlation of the BEN with ion streamings observed in the PSBL. This good correlation leads many theoretical studies on the BEN generation mechanism to the ion beam instabilities such as ion-ion two stream instability [e.g., *Akimoto and Omidi*, 1986; *Grabbe*, 1987], the ion beam acoustic instability [e.g., *Dusenbery and Lyons* , 1985; *Akimoto and Omidi* , 1986; *Ashour-Abdalla and Okuda*, 1986; *Dusenbery*, 1986; *Grabbe*, 1987; *Burinskaya and Meister*, 1989, 1990], and the Buneman instability [*Grabbe*, 1989]. *Tsutsui et al.* [1991] considered the ion flow as the plasma bulk flow to provide the Doppler shift and they attempted to explain the broad frequency characteristics of BEN by the Doppler shift of ion acoustic potential bubbles convecting with the plasma bulk flow.

On the other hand, *Onsager et al.* [1993] showed that the BEN can be observed in the electron layer of the outer PSBL without energetic ions. They showed that the BEN can be observed even in the electron layer without ion streams at just the outside of the plasma sheet. They pointed out that the ion streaming is not essential in the excitation of the BEN, and stressed that the BEN have the close relation to the electrons dynamics.

In spite of the above observational and theoretical efforts, the clear answers to the generation mechanism of the BEN have not been obtained. The main difficulty for explaining the BEN excitation mechanism is originated from the broadness of its spectrum. In order to explain the broadness of the BEN spectra, we need to combine different instabilities.

The GEOTAIL waveform observations provided a breakthrough in the study of the BEN. On the basis of the waveform observations by GEOTAIL, *Matsumoto et al.* [1994b] revealed that the BEN includes a series of solitary bipolar waveforms. The observed waveforms are addressed as PSBL Electrostatic Solitary Waves (PSBL ESW) [*Matsumoto et al.*, 1994b; *Kojima et al.*, 1997a]. Figure 41 shows a typical frequency-time spectrogram and a snapshot of the ESW waveforms observed by GEOTAIL at (GSM X, GSM Y, GSM Z) = (-95, 11, -4 R_E) on Jan. 13, 1994.

We can find the similar features of the BEN spectra shown in Figure 41 to those reported by *Gurnett et al.* [1979]. They have no corresponding magnetic components and are purely electrostatic waves. Their uppermost frequencies reach the local electron plasma frequencies ($f_{pe} \sim 2$ kHz), though, they intermittently exceed the electron plasma frequencies. Figure 41(b) shows the expanded frequency-time spectrogram observed by the MCA. We can see the bursty nature which was reported by *Gurnett et al.* [1979]. The bursty spectra are conspicuous especially in the high frequency range above the electron cyclotron frequency which is shown by a black line.

Figure 41(c) shows the representative waveforms of the PSBL ESW corresponding to the BEN shown in Figure 41(a) and (b). In this case, the pulse widths of the observed PSBL ESW are of the order of a few milliseconds. *Matsumoto et al.* [1994b] pointed out that such solitary wave structures are the main reason for the BEN broadband spectra. Since the pulse widths decide the uppermost frequency of the corresponding spectra, such solitary waves contribute especially to the high frequency parts of the broadband spectra above a few hundreds of Hertz.

The detailed features of the PSBL ESW have been described by *Matsumoto et al.* [1994b] and *Kojima et al.* [1997a] as follows; (1) Pulse widths are from a few milliseconds to a few tens of milliseconds; (2) The orientation of the electric field is parallel to the ambient magnetic field; (3) The ESW waveform corresponds to a huge potential hump flowing along the ambient magnetic field as shown in Figure 42; (4) The PSBL ESW appear very bursty and intermittently. The bursty nature of the BEN especially in a high frequency range is originated from the bursty nature of the PSBL ESW.

The above mentioned natures of the ESW can well explain the BEN natures, which were reported by *Gurnett et al.* [1979] except for one point. The difference from Gurnett et al. [1979] is the electric field orientation. *Gurnett et al.* [1979] reported that the orientation of the electric field of the BEN is almost perpendicular to the ambient field, while *Kojima et al.* [1997a] showed that the polarization of the ESW is parallel to the ambient magnetic field. *Onsager et al.* [1993] showed that the high frequency portion of the BEN spectra are polarized along the ambient magnetic field, however, even in their results, when the spectral intensity of the high frequency portions is weak, their wave polarizations are perpendicular to the ambient magnetic field. Their results on the electric field orientation are not consistent with ours. All of the ESW which we have examined are polarized along the ambient magnetic field. We do not have clear explanation for this inconsistency.

Matsumoto et al. [1994b] and *Omura et al.* [1994] showed that the ESW waveforms can be reproduced by the BGK mode which is produced as a result of the nonlinear evolution of beam excited Langmuir waves by full particle computer experiments. They set up a simple simulation model of electron two stream instabilities and showed that excited electrostatic potentials are merged into an isolated BGK potential as a result of the nonlinear evolution. The detailed results of the computer experiments will be discussed in Chapter 7.

As we will introduce in Chapter 5, BEN can also be observed in the magnetosheath region. Surprisingly, the BEN in the magnetosheath also includes the ESW waveforms. *Kojima et al.* [1997a] referred them to as MS ESW(MagnetoSheath ESW). The pulse width of the MS ESW is of the order of a few milliseconds and has a very similar nature to the PSBL ESW. Their electric field orientation in the spin plane is parallel to the ambient magnetic field. However, the occurrence rate of the MS ESW is much lower than that of the PSBL ESW [*Kojima et al.*, 1997a]. Because of the similar natures of the PSBL ESW and the MS ESW, we may be able to apply the BGK potential flow model shown in Figure 42 to the MS ESW. Further, *Matsumoto et al.* [1997] also showed that the similar solitary waves are observed in the terrestrial upstream region and the bow shock region. Figure 43 shows one example of the waveforms of the upstream waves which contain the ESW (pulses).

In the generation model of the ESW proposed by *Matsumoto et al.* [1994b] and *Omura et al.* [1994], the existence of energetic electron beams is essential. The existence of the electron beam in the PSBL, the magnetosheath and the terrestrial upstream regions can be proved through the existence of the Langmuir waves. As we have shown in Section 4.1, the Langmuir waves are observed in the lobe region close to the PSBL, the magnetosheath, and the upstream regions. This fact is consistent with our proposed ESW generation model. However, it is still questionable that the BGK potential can stably exist even in the turbulent region such as the magnetosheath, the upstream region and the bow shock.

4.5. Narrowband Electrostatic Noise (NEN)

Narrowband electrostatic noise (NEN) was initially reported in relation to the upstream waves of the slow shock around the plasma sheet [*Scarf et al.*, 1984]. Further, similar emissions were also observed in the region close to the magnetopause [*Coroniti et al.*, 1990]. The NEN is a purely electrostatic emission propagating parallel to the ambient magnetic field. Its spectrum band width is narrower than the BEN bandwidth and the center frequency of NEN is between the local electron and ion plasma frequencies. This characteristic frequency range leads us to an idea that the NEN is a Doppler-shifted ion acoustic wave, because there exists no normal electrostatic mode propagating parallel to the external magnetic field between the local electron and ion plasma frequencies. However, we have a difficulty that the condition of $T_i < T_e$ is not always satisfied in the tail region, if we apply the model of the Doppler shifted ion acoustic wave to the NEN.

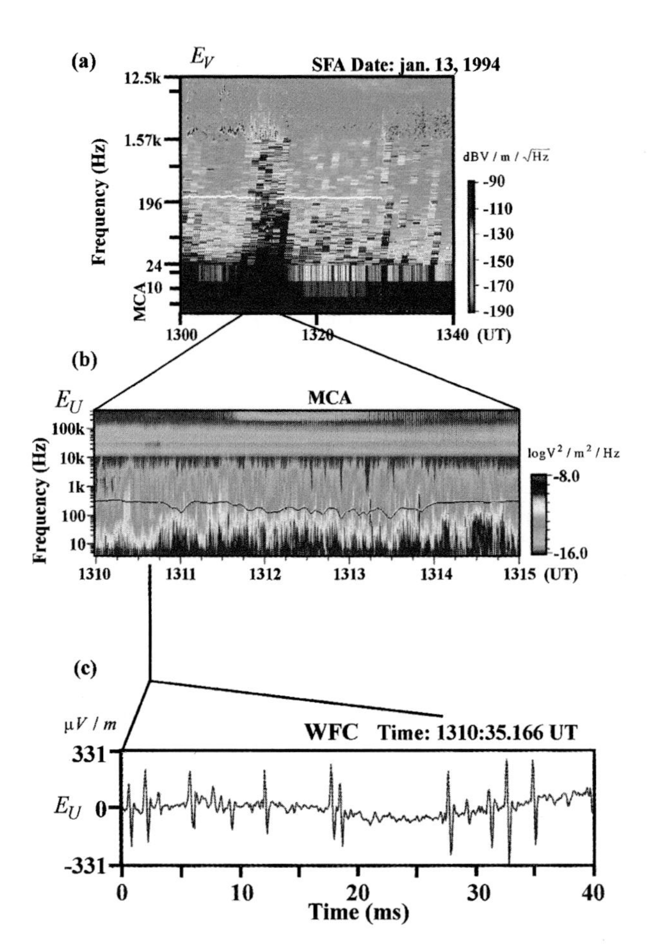

Figure 41. Typical dynamic spectra ((a) and (b)) of BEN (Broadband Electrostatic Noise) and corresponding waveforms ((c)) of the PSBL Electrostatic Solitary Waves (PSBL ESW) [*Kojima et al.*, 1997a].

Another generation mechanism of the NEN was proposed by *Coroniti and Ashour-Abdalla* [1989] *Ashour-Abdalla et al.*, [1989]. They showed that superposition of two Maxwellian distributions shifted slightly in the parallel direction unstabilize electrostatic waves with frequencies between electron and ion plasma frequencies. They referred this mode to as "Electron hole mode". The electron hole mode can be one of the possible candidates of the NEN, because the boundary region is characterized by a mixture of two different plasmas.

Kojima et al. [1997a] performed detailed analyses of the NEN using GEOTAIL plasma wave data. They showed that the NEN are not 'Noise' but are a quasi-monochromatic 'wave" based on the GEOTAIL waveform observations. Figure 44 shows the typical dynamic spectrum and waveforms of the NEN observed by GEOTAIL in the tail lobe region. The patchy dots shown in the SFA plot in Figure 44(a) are the NEN. The coarse time resolution and fine frequency resolution of the SFA result in the patchy appearance of the NEN spectra, because of the quick change of the NEN frequency. During the time duration shown in Figure 44(a), the electron and ion plasma frequencies (f_{pe}, and f_{pi}) are approximately 2 kHz and 46 Hz, respectively. Therefore the center frequency of the NEN, which is about a few hundreds of Hertz, is between the electron and ion plasma frequencies. The bursty nature of the NEN can be seen in Figure 44(b). Though we cannot see well the frequency change of the NEN spectra in the MCA data, their burstiness with the order of a few seconds are clearly seen. Figure 44(c) shows the waveforms of the NEN. The observed waveforms are quasi-sinusoidal and not noises at all. *Kojima et al.* [1997a] called NEN observed in the lobe as "Lobe Electrostatic Quasi-Monochromatic Waves (Lobe EQMW)". They showed that the Lobe EQMW are purely electrostatic and that their electric fields are oriented to the parallel direction relative to the ambient magnetic field. The frequencies of the Lobe EQMW fluctuate quickly. The frequency fluctuation of the sinusoidal waveforms leads to the patchy feature of the NEN wave spectra in the dynamic spectra. It was also pointed out that the Lobe EQMW are observed in the lobe region close to the plasma sheet and/or the magnetopause boundary regions. This result is consistent with the results on

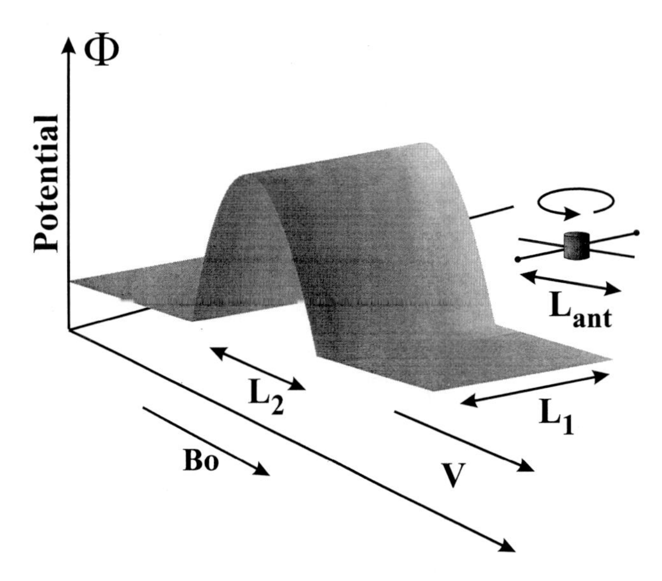

Figure 42. Schematic illustration of the ESW potential. The ESW potential flows along the ambient magnetic field B_o. Its scale (L_1) perpendicular to B_o is almost uniform relative to the scale of the GEOTAIL antenna length (L_{ant}). The potential width (L_2) is much larger than the antenna length (L_{ant}) [*Kojima et al.*, 1997a].

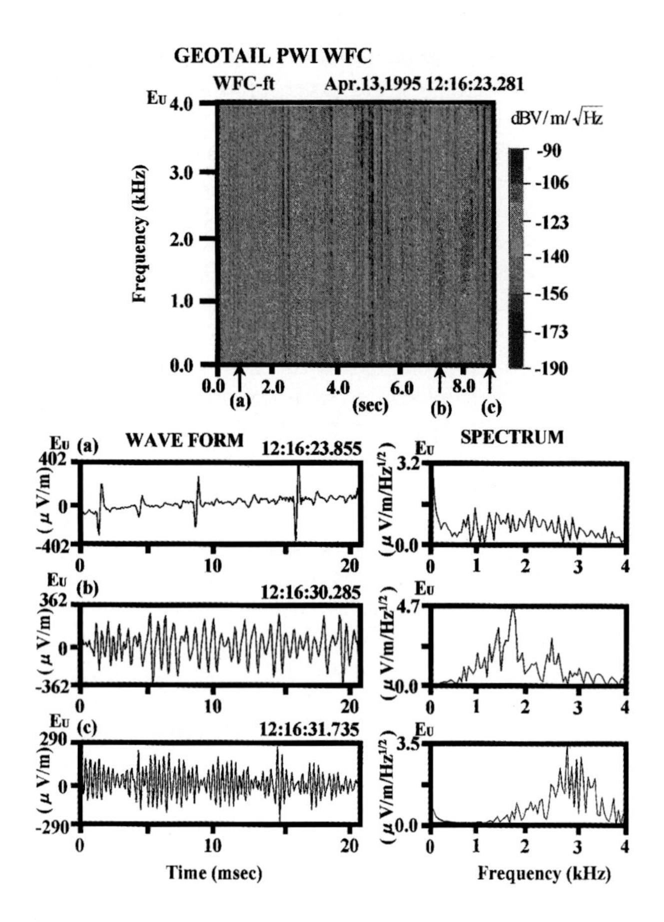

Figure 43. ESW observed in the upstream region (a). Panels (b) and (c) are the waveforms of the ion acoustic-like waves [*Matsumoto et al.*, 1997].

the NEN reported by *Scarf et al.* [1984], and *Coroniti et al.* [1990]. Further, the result seems to fit with the theory of the electron hole mode.

The similar waveforms are also observed in the magnetosheath region. *Kojima et al.* [1997a] showed that they have the similar features to those of the Lobe EQMW. Therefore, they referred to these NEN in the magnetosheath as "MagnetoSheath Electrostatic Quasi-Monochromatic Waves (MS EQMW)". Typical center frequency of the MS EQMW is about 2 kHz, which is ten times higher than those of the Lobe EQMW. The interesting point is that the frequency of the Lobe EQMW is proportional to the plasma frequency. For example, the typical NEN frequency and the electron plasma frequency in the lobe are 200 Hz and 2 kHz, respectively, while those in the magnetosheath are 2 kHz and 20 kHz, respectively. This fact is very important for a clue to the NEN generation mechanism. However, since both Doppler-shifted ion acoustic waves and electron hole mode waves are very sensitive to the local plasma parameters, we cannot judge which mechanism is more plausible.

5. PLASMA WAVE FEATURES IN GEOSPACE

Plasma wave signatures provide an important clue for identifying the spacecraft location. Figure 45 schematically illustrates the typical wave signatures observed by GEOTAIL in geospace. Each characteristic region in geospace is associated with respective wave signatures. By combining the density information obtained from the lower cutoff frequency of continuum radiation (CR) with the wave signatures shown in Figure 45, we can identify the plasma regions in geospace.

In this chapter, after summarizing the typical wave signatures in individual regions, we will show some statistical studies on the global structure of the geomagnetic tail.

5.1. *Solar wind and Electron/Ion foreshock regions*

The solar wind in the interplanetary space is one of the regions where the plasma wave activities are very

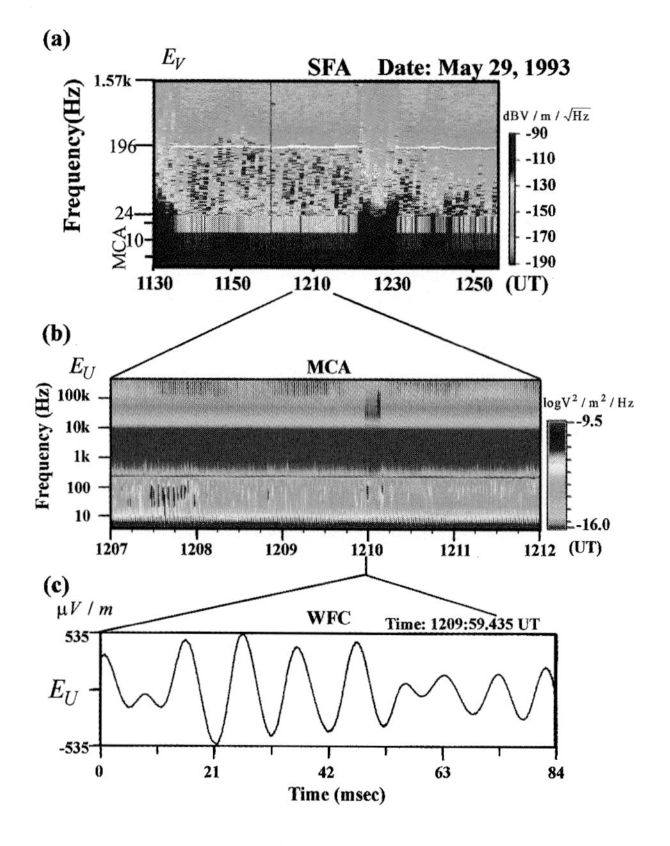

Figure 44. Typical dynamic spectra ((a) and (b)) and corresponding waveforms (c) of the Lobe Electrostatic Quasi-Monochromatic Waves (Lobe EQMW) [*Kojima et al.*, 1997a].

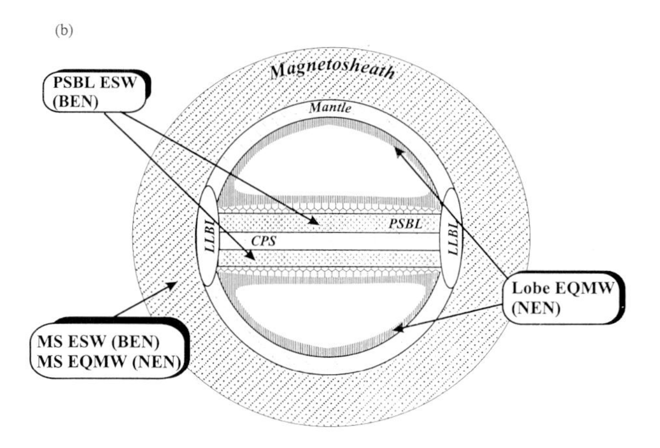

Figure 45. Schematical drawing of the plasma wave signatures observed in each characteristic region of the geospace: (a) Meridian plane, and (b) Tail cross section [Matsumoto et al., 1997; Kojima et al., 1997a]

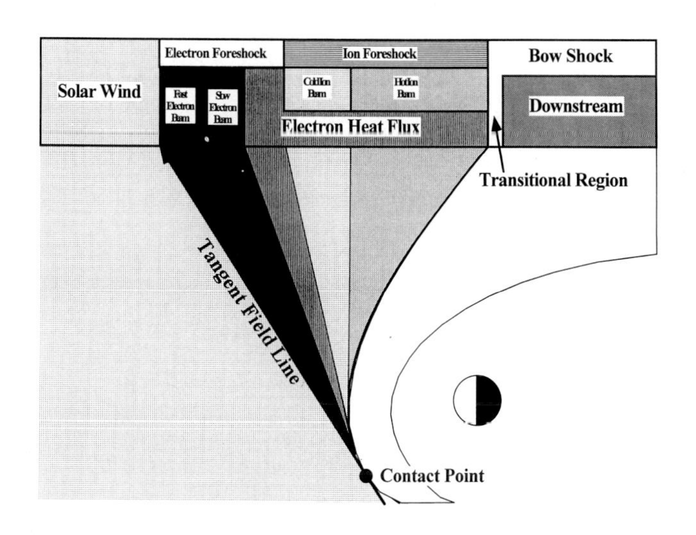

Figure 46. Schematical illustration of proposed subregion in the terrestrial upstream region [Matsumoto et al., 1997].

low. In the nominal state of the solar wind, we observe only weak UHR noise around the local plasma frequency (~20 kHz). However, in the special case such as the encounter with the interplanetary shocks, a drastic change of the plasma wave activities is found. Detailed wave features in the encounter with the interplanetary shock will be discussed in Chapter 6.

The terrestrial foreshock has been extensively surveyed by the previous spacecraft. It is now well known that the foreshock region is mainly classified into two different characteristic regions called the electron foreshock and the ion foreshock [*Gurnett*, 1985].

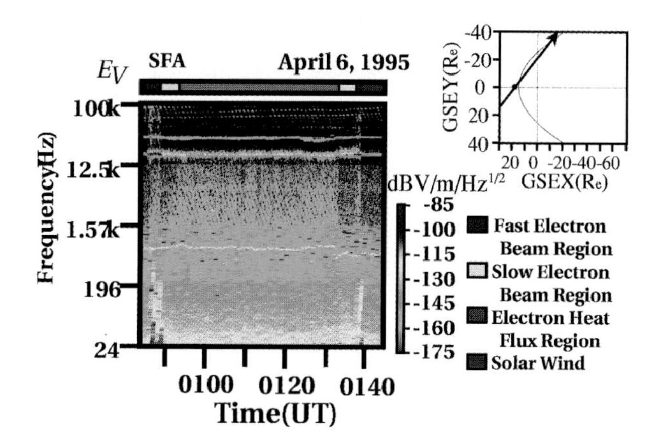

Figure 47. Frequency-time spectrogram observed in the electron foreshock. Equivalent sub-regions are indicated in the top panel. The representative wave signature of the electron foreshock is the intense electron plasma waves as described in Chapter 4 [Matsumoto et al., 1997].

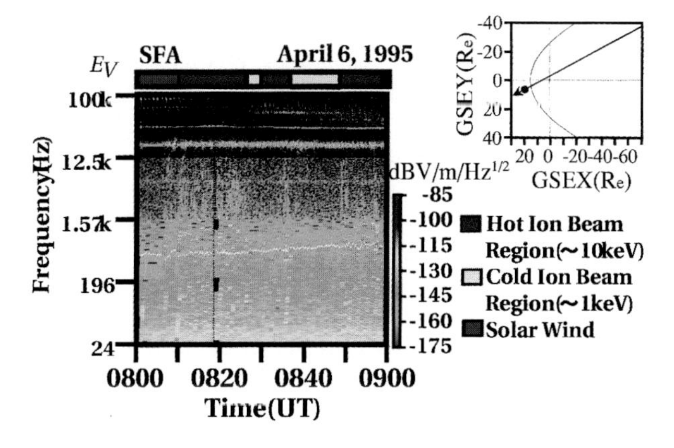

Figure 48. Frequency-time spectrogram observed in the ion foreshock. Equivalent sub-regions are indicated in the top panel. The representative wave signature of the ion foreshock is the ion acoustic-like waves, which we have introduced in Chapter 4 [Matsumoto et al., 1997].

Matsumoto et al. [1997] proposed a further division of the electron/ion foreshock regions based on the GEOTAIL low energy particle and plasma wave observations as shown in Figure 46. Figures 47 and 48 show examples of the frequency-time spectrogram observed in the electron and ion foreshock regions, respectively. The proposed subregion are also displayed on the top bar and its color captions in each figure. The spacecraft location and its relative relation to the IMF direction are shown in the upper-right panels.

Matsumoto et al. [1997] classified the electron foreshock into 3 characteristic subregion. They are: Fast electron beam region; Slow electron beam region; and Electron heat flux region (see Figure 46). The fast electron beam region is characterized by strong Langmuir waves ($\sim$ 25 kHz) associated with enhanced low frequency waves below 200 Hz. They are excited by electron beams running along the tangent field line which contacts the bow shock as discussed in the previous chapter. In the slow electron beam region, we see weaker Langmuir waves but see no wave activities below the electron plasma frequency. In the electron heat flux region, we do not find well-defined electron beam, however, there exists a high energy tail with higher temperature in the electron distribution. As shown in Figure 47, the electron heat flux region is characterized by weak Langmuir waves and simultaneous weak waves from below 10 Hz up to approximately 20 kHz.

Figure 48 is a similar plot with Figure 47 for the ion foreshock. The sub-regions in the ion foreshock proposed by [*Matsumoto et al.*1997] (see Figure 46) are hot and cold ion beam regions, which are displayed by the horizontal color bar in the top of Figure 48.

In Figure 48, we could find patchy red or yellow dots. They are very similar to the NEN emissions observed in the magnetosheath and in the tail lobe region. The patchy spectra suggest the quick change of the wave frequency and its intensity relative to the sweep periods of the SFA. The difference between waves in the hot and cold ion regions is the uppermost frequency of these waves. In the hot ion beam region it reaches approximately 10 kHz or more, while in the cold ion beam region, it is below 1 kHz.

5.2. Bow shock

Gurnett [1985] gave a good review of the plasma wave signatures around the bow shock. He pointed out that the bow shock transition layer is characterized by an abrupt burst of electric and magnetic wave components associated with the jump of the ambient magnetic field. *Matsumoto et al.* [1997] showed further detailed plasma wave structures in the bow shock using the GEOTAIL PWI data.

Figure 49 shows a typical wave spectrum observed in the vicinity of the bow shock observed by GEOTAIL. The spacecraft experienced the bow shock crossings 6 times (displayed by blue arrows) during the 2 hour period. The intense broadband emission up to 12 kHz seen in the electric field component (upper panel) indicates the bow shock crossing followed by the entry into the downstream region. The edge of such broadband emissions corresponds to the timing of the bow shock crossing. The wave intensity in the downstream of the transition region of the bow shock is almost constant in the frequency range from about 200 Hz to 2 kHz. It is, however, weakened as the distance from the bow shock increases in the frequency range from 5 Hz to about 200 Hz.

The magnetic field component waves are also enhanced in the bow shock and downstream regions. Especially in the bow shock transition, their uppermost frequencies reach the local electron cyclotron frequency which is shown by the white solid line, while in the downstream region, they remain below 100 Hz and seem to be similar to the MNB spectra.

Matsumoto et al. [1997] performed the waveform analyses on the plasma waves observed in the vicinity of the bow shock. They showed that in the ramp of the magnetic field of the bow shock both ESW and EQMW types of waveforms are observed and that their wave features rapidly change in a short time scale of a few msec. Figure 50 shows one example of waveforms of electrostatic waves at the ramp region of the bow shock [Matsumoto et al., 1997]. They show a very rapid variation of their frequency and intensity.

5.3. Dayside outer magnetosphere

During the distant tail phase of the GEOTAIL orbit, its perigee was located around the dayside region of $10R_E$ from the Earth. Its orbits are very suitable for the observation of the outer magnetosphere on the dayside and occasionally provided chances of skimming the dayside magnetopause.

GEOTAIL experienced its first dayside magnetopause skimming on October 17-18, 1992. The spacecraft moved from the dawn side of the northern hemisphere to the dusk side of the southern hemisphere. Figure 8 shows the frequency-time spectrogram observed in this first magnetopause skimming.

The gradual increase of the electron cyclotron frequency indicated by a white line shows that the spacecraft moved into the magnetosphere during the interval from 2000 UT on October 17 to 0500 UT on October 18. We find bursty electromagnetic waves below 200 Hz during the interval from 1800 UT to 2040 UT

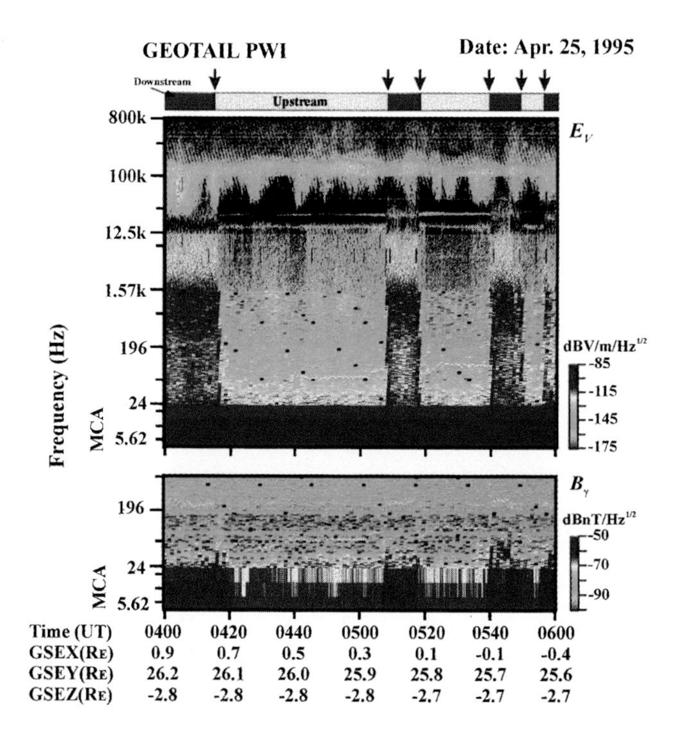

Figure 49. Plasma wave signatures in the bow shock crossing. The transition timings of the bow shock are indicated by arrows in the top panel.

(see B component). These low frequency electromagnetic waves indicate that the spacecraft was located in the magnetosheath region. They are switched on and off repeatedly, which suggests that the spacecraft repeatedly crossed the magnetopause and moved in and out of the magnetosphere several times.

The wave activities in the outer dayside magnetosphere are characterized by the chorus emissions and electron cyclotron harmonic waves. The electromagnetic waves with the frequency range of 200 Hz to 500 Hz observed in the interval from 2040 UT to 0500 UT shown in Figure 8 are the chorus emissions. The detailed structure and characteristics have already been discussed in Chapter 3. The chorus emissions disappear around 0500 UT and in turn the broadband electrostatic emissions start to appear in the E-component. This fact suggests that the spacecraft moved into the plasma sheet region around 0500 UT.

The electron cyclotron harmonic waves (ECHW) observed in the dayside outer magnetosphere are classified into the bursty ECHW and the diffuse ECHW, as discussed in the previous chapter. The bursty Totem Pole emissions (TP emissions) [Matsumoto and Usui, 1997] were repeatedly observed during the interval from 2130 UT to 2300 UT. The diffuse ECHW were continuously observed during the period from 2050 UT to 0500 UT. This diffuse ECHW also disappeared at 0500 UT simultaneously with the chorus emissions.

5.4. Magnetosheath

The magnetosheath is characterized by very high activities of plasma waves. Typical plasma wave activities in the magnetosheath are represented by BEN, NEN, electron plasma waves, and MNB. The typical plasma densities are 5/cc corresponding to the CR lower cutoff frequency (electron plasma frequency) of 20 kHz.

Figure 51 shows a representative frequency-time spectrogram generated from the SFA data observed in the magnetosheath at (GSM X, GSM Y, GSM Z) = (-86, 39, 1.0 R_E). The above mentioned typical waves are denoted by white arrows in the figure.

The BEN spectra observed in the magnetosheath are very similar to those in the PSBL except for its uppermost frequency. Though the uppermost frequency of the PSBL BEN always reaches the local electron plasma frequency ($\sim$ 2 kHz), the frequency of the magnetosheath BEN almost always stays at much lower

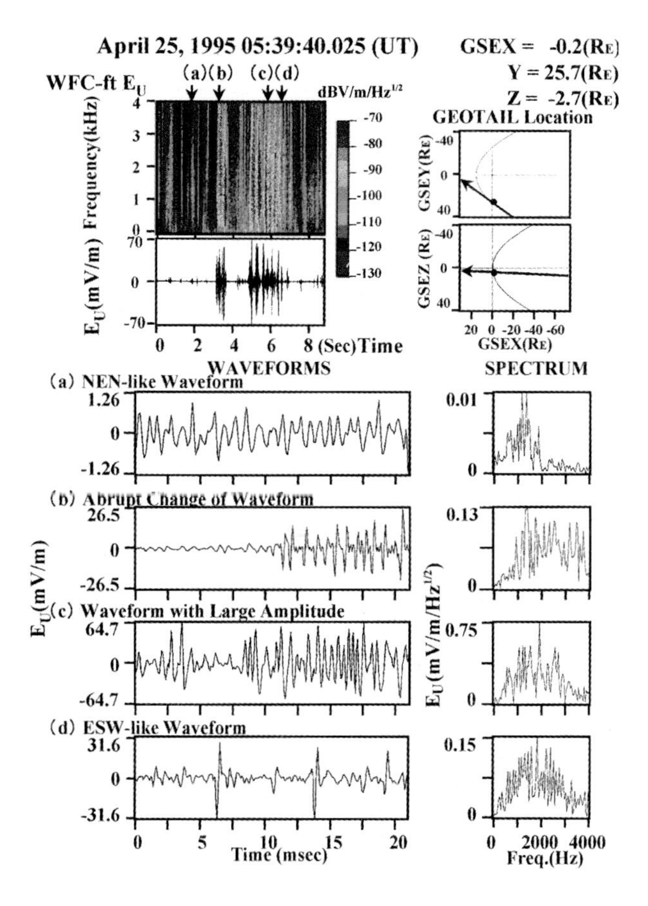

Figure 50. Typical waveforms of electrostatic waves observed in the ramp region of the bow shock [Matsumoto et al., 1997].

than the local electron plasma frequency ($\sim$ 20 kHz). This is also proved by the waveform observations of the BEN. Both BENs, called as PSBL ESW and MS ESW, have almost the same pulse widths indicating the uppermost frequency of the MS ESW in relative to the local electron plasma frequency is much smaller than in the case of the PSBL ESW.

On the other hand, the magnetosheath NEN has similar features to the lobe NEN. The interesting point, however, is that the ratio of the local electron plasma frequency and the center frequency of the NEN is almost conserved between the NENs in the lobe and in the magnetosheath. The typical center frequencies ($f_{\rm NEN}$) of the NEN in the lobe and in the magnetosheath are 200 Hz and 2 kHz, respectively. Since the typical electron plasma frequencies ($f_{\rm pe}$) in both regions are 2 kHz and 20 KHz respectively, we find that the ratio ($f_{\rm pe}/f_{\rm NEN}$) is almost equal to 10. As we have shown in Chapter 4, the waveform characteristics of the NENs in the lobe and in the magnetosheath are very similar to each other. The above observations suggest generation mechanisms of the NENs in the lobe and the magnetosheath are the same or similar.

We have discussed the electron plasma waves observed in the lobe region in Chapter 4. Our waveform observations have revealed that they are continuous waves but are strongly modulated in amplitude. Further, we have shown that they include both Langmuir waves and Electron Cyclotron Harmonic waves. The separation of these two waves can be done only by the polarization analysis based on the high time waveform observations, because the UHR frequency is too close to the electron plasma frequency in the lobe region. This situation is similar in the magnetosheath. In order to identify the plasma wave mode of the electron plasma waves observed in the magnetosheath, we need to consult the waveform observations. However, unfortunately, the frequencies of the electron plasma waves in the magnetosheath are normally higher than 4 kHz which is the upper limit frequency of the GEOTAIL WFC. This prevents us to identify the wave mode in the magnetosheath.

The MNB is the most useful wave signature for the identification of the magnetosheath. It can be clearly seen in the magnetic component (lower panel) of Figure 51. As shown in Section 3.1.5, the MNB are observed in the upstream and downstream of the bow shock as well as in the magnetosheath. The upper frequency of the MNB occasionally reaches the local electron cyclotron frequency, and its lower frequency limit is well below the lowest observable frequency (5.62 Hz) of the PWI. The detailed waveform analyses of the MNB are described in Section 3.1.

5.5. Magnetopause Boundary

In the view point of the plasma wave signatures, we can classify the magnetopause and its boundary layer into 3 types. They are: Type 1: Magnetopause with a wide and smooth boundary layer, Type 2: Magnetopause with a thin and sharp boundary layer, and Type 3: the low latitude boundary layer (LLBL) in the near tail. Typical wave signatures in these 3 types of the magnetopause boundary are shown in Figures 52 and 53.

The frequency-time spectrogram shown in Figure 52 illustrates the above mentioned first two kinds of the typical magnetopause and its boundary. In this figure, we can easily identify the clear magnetopause crossings in 2 intervals from 0048 to 0055 UT and from 0135 to 0140 UT judging from the rapid drop or rise of the plasma frequencies and the change in the magnitude of the ambient magnetic field, which is indicated by the change of the electron cyclotron frequency shown by the white solid line. Further, the low frequency MNB indicates that the spacecraft stayed in the magnetosheath during the periods from 0000 to 0048 UT and from 0135 to 0200 UT.

Two magnetopause crossings displayed in Figure 52 have the different features in the time scale. In the first crossing, the spacecraft passed through the magnetopause boundary layer in 15 minutes. The gradual decrease of the CR lower cutoff frequency and the gradual increase of the cyclotron frequency from 0048 to 0055 UT show that the spacecraft moved into the lobe region through the magnetopause boundary layer of Type 1.

On the other hand, in the second crossing, the spacecraft moved back to the magnetosheath through the Type 2 magnetopause boundary. Although the time difference between the first and second cases is only 30 min, we find that the spacecraft moved across the magnetopause boundary in the second crossing more quickly than in the first crossing. The passage time in the second crossing is only one or two minutes. The time difference can be clearly seen in the difference of the time change of the CR lower cutoff frequency around 0055 UT and around 0138 UT. Since the velocity of the spacecraft is negligible relative to the dynamic motion of the tail structure, the following two hypotheses can be inferred to explain the time scale difference between these 2 events. (1) The speed of tail flapping or oscillation motions changed during the interval between the 2 crossings. (2) The thickness of the magnetopause boundary is different at the 2 different locations the spacecraft passed through.

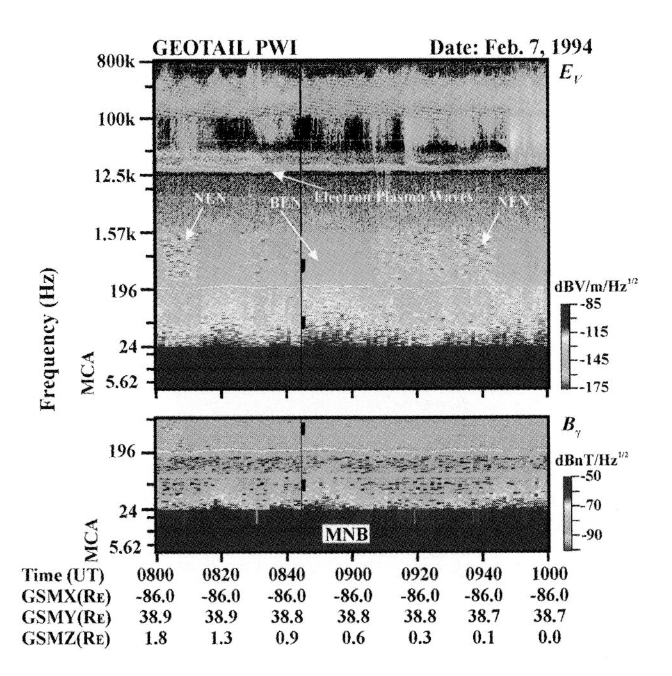

Figure 51. Typical plasma wave signatures observed in the magnetosheath region.

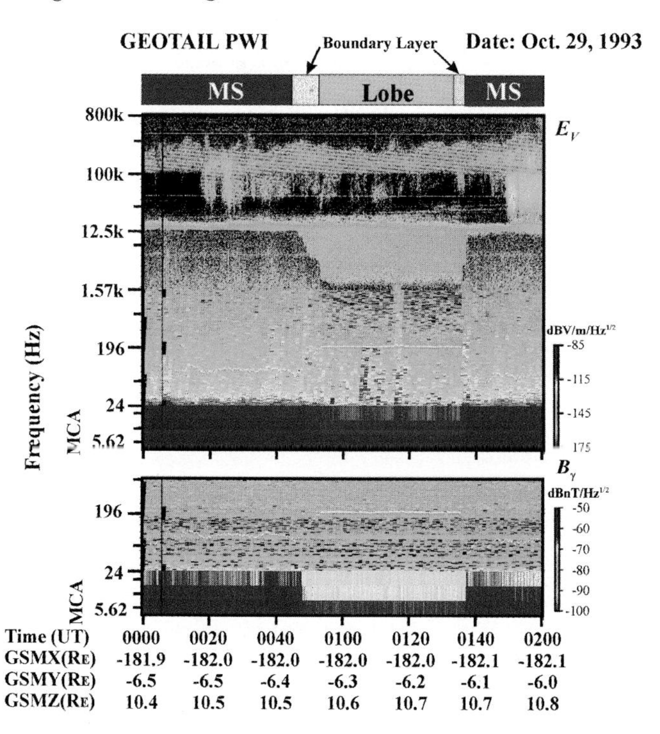

Figure 52. Magnetopause crossings of the type 1 and 2. Judging from the time variation of the CR lower cutoff frequency, and of the electron cyclotron frequency displayed by white lines, the spacecraft experienced the very smooth magnetopause crossing (type 1) around 0055 UT, while the abrupt magnetopause crossing (type 2) can be observed around 0138 UT.

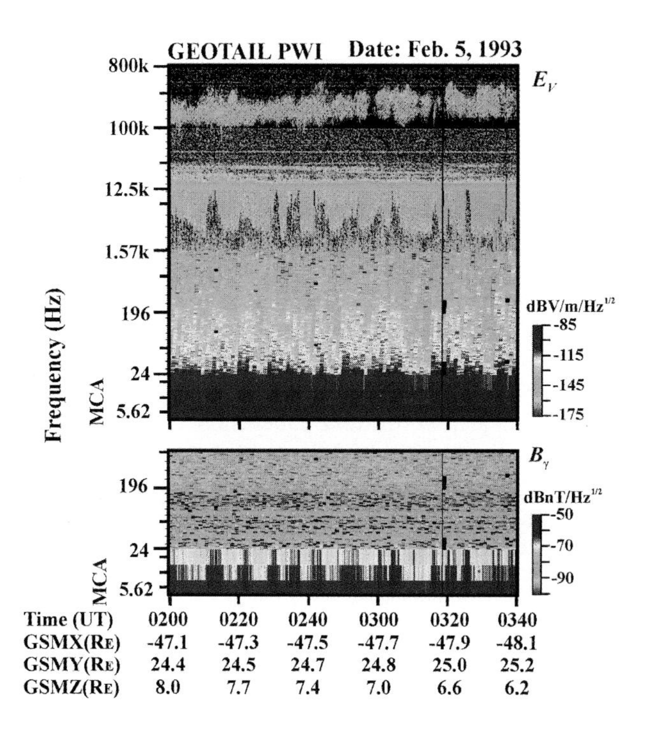

Figure 53. Magnetopause crossing of the type 3. We can find that the local plasma density changes repeatedly from the time variation of the CR lower cutoff frequency. Consulting the time variation of the ambient magnetic field, and of the plasma temperature, we conclude that the spacecraft moved around between the plasma sheet and Low Latitude Boundary Layer (LLBL).

The GEOTAIL PWI occasionally observes such events with different features during a short period as the spacecraft passes through the magnetopause. Unfortunately, since the single spacecraft observation cannot distinguish between the temporal and spatial phenomena, we need to wait for the multiple spacecraft observations in the future to verify the above two hypotheses.

Figure 53 shows the third type of the magnetopause crossing observed at (GSM X, GSM Y, GSM Z) = (-48, 25, 7 R_E) on February 5, 1993. The electron density, the intensity of MNB, and the magnetic field change quasi-periodically during the 100 min period. The time period of their variation is about five to ten minutes. The BEN type emissions are also observed simultaneously. The interesting point is that the periods of higher density coincide with the periods of the appearance of the MNB and BEN. On the other hand, in the low density periods, plasma wave activities are low. Consulting with the magnetic field, we found that the azimuthal magnetic field direction often reverses (not shown) and that their amplitudes are lower in the periods of lower density. The above wave signatures in the periods of higher density are quite similar to those in the magnetosheath. However, since the plasma densities in the periods of higher density are about 0.4/cc to 2/cc, these densities are smaller than the average density in the magnetosheath. On the other hand, the signatures of the wave and the ambient magnetic field in the preiods of lower density are typical ones in the central plasma sheet. From these results, we infer that the spacecraft stayed in the magnetopause boundary in the low latitude region into which the magnetosheath plasma invades and forms a boundary layer between the magnetosheath and the plasma sheet.

From the observations of the plasma waves, and the magnetic fields, we draw a conclusion that the spacecraft periodically repeated a motion between the LLBL and the plasma sheet for the period of present interest. During this period, the spacecraft was located in the dusk side of the near tail region, which is consistent with the conclusion we have drawn.

5.6. *Tail Lobe, Plasma Sheet and its Boundary Layer*

Figure 54 shows the plasma wave signatures of the plasma sheet entry from the lobe observed at (GSM X, GSM Y, GSM Z) = (-205.3, 16.7, -1.9 R_E) on June 8, 1993.

The activities of the plasma waves in the core lobe are low. However, as we have discussed in section 4.5, the NEN (lobe EQMW) are observed near the magnetopause/plasma sheet boundaries in the tail lobe. In Figure 54, the patchy spectra below 200 Hz observed during the periods from 0100 to 0115 UT and from 0142 to 0200 UT are the NEN in the lobe. The stable magnetic field corresponding to the electron cyclotron frequency of 200 Hz displayed by a white line is also an indication of the lobe during this period.

The intense broadband spectra with uppermost frequencies around 2 kHz seen in the two periods from 0118 to 0122 UT and from 0135 to 0141 UT are the BEN. These BEN suggest that the spacecraft was located in the PSBL during the above two periods.

During the interval between the two PSBL crossings, the wave activities below 2 kHz in the electric component channels (upper panel) become very quiet during the interval from 0125 to 0135 UT. The magnetic field measurements (not shown) indicate that the spacecraft crossed the neutral sheet crossing at 0125 UT. Therefore, the spacecraft entered into the central plasma sheet (CPS) around this time. However, one interesting point is that the lower cutoff frequency of the CR remain constant or even lower than the surrounding PSBL and the lobe. This suggests that the plasma temperature must be high in this region to maintain the pressure balance.

Though the plasma wave activities are very low in the central plasma sheet especially around the neutral sheet, we occasionally observe very low frequency component of the electromagnetic waves there. Such electromagnetic waves can be seen in Figure 54. The bursty waves below 24 Hz are seen at 0125, 0132, 0135, 0137, and 0139 UT as shown in the lower panel. The spectral intensity of these EM bursts was was very low, but we occasionally observe intense electromagnetic waves even in the central plasma sheet as those in the magnetosheath region.

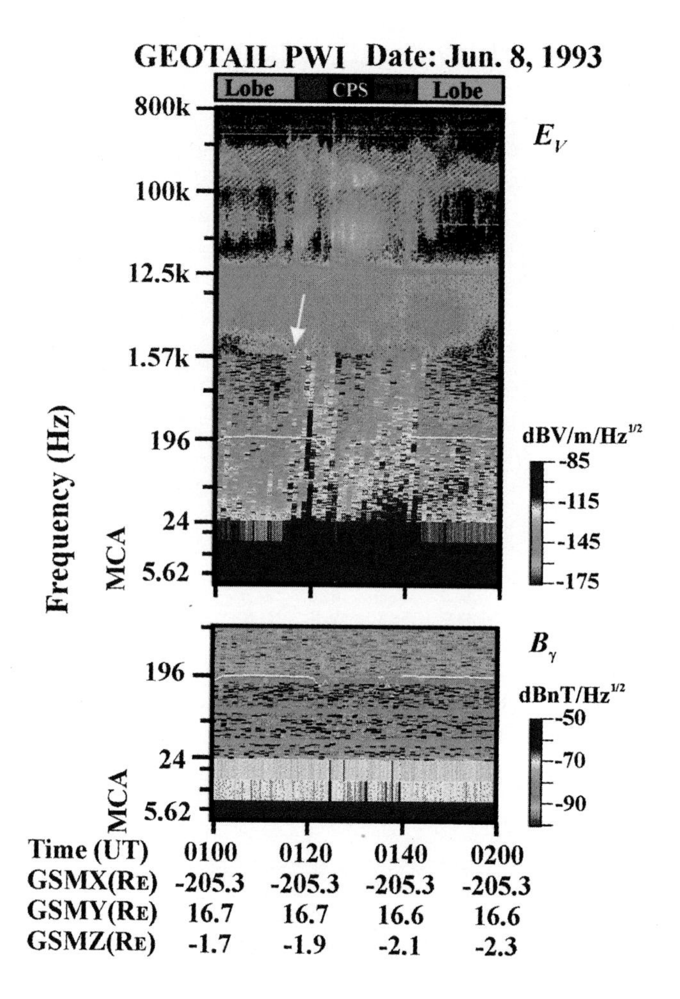

Figure 54. Plasma sheet entry from the lobe region. As introduced in Chapter 4, we frequently observe the electron plasma waves at the lobe edge close to the plasma sheet (indicated by a white arrow). The typical wave signature of the plasma sheet boundary layer is the BEN, while in the central plasma sheet especially around the neutral sheet, the plasma wave activity is very low.

Another significant wave signature is the electron plasma waves shown by a white arrow in Figure 54. As we have discussed in section 4.1, the electron plasma waves with their frequencies equal to the local electron plasma frequency are observed in the lobe edge contacting the PSBL where BEN are actively observed. Actually the electron plasma waves displayed by the white arrow are observed just before the appearance of the BEN at 0119 UT. Therefore, the electron plasma waves provide a good indication to identify the edge of the plasma sheet.

5.7. *Tail structure in the view point of the plasma*

In the previous sections, we have reviewed the plasma wave features in geospace revealed by GEOTAIL plasma wave measurements. Since each characteristic region in geospace exhibits respective plasma wave signatures, we can identify the region where the spacecraft is located by consulting these plasma wave signatures. In addition, the plasma density information obtainable from the CR lower cutoff frequency is also very useful for the region identification.

The orbits of the GEOTAIL spacecraft are designed to survey the geomagnetic tail region over the wide area. Therefore, the plasma density information along the orbits is useful for the study of the global structure of the geomagnetic tail.

Figure 55 shows one example of the distribution of the local electron plasma density deduced from the lower cutoff frequency of the continuum radiation in four different distance ranges from the Earth ($20 < |X| < 50 R_E$, $50 < |X| < 100 R_E$, $100 < |X| < 150 R_E$, and $150 R_E < |X|$). The coordinate system called GSM' (the modified Geocentric Solar-Magnetospheric Coordinates) is used to include the effect of the solar wind aberration of 4 °, the hinging distance of 10 R_E and the tilt angle of the geomagnetic dipole axis.

We have sampled the lower cutoff frequencies of continuum radiation with the time resolution of one hour. The obtained densities are mapped as follows. First we divide the $Y - Z$ plane into 120×80 cells with a grid spacing of $1 R_E$. The density measured by GEOTAIL (via the lower cutoff frequency of CR) on a specific location of GEOTAIL is distributed to the adjacent nearest 4×4 grid points in the $Y - Z$ plane. Thus the electron density data over the entire period of measurement from September 18, 1992 to Feb. 17, 1994 are distributed to the whole grid points in the $Y - Z$ plane at a fixed X point. Then an average is taken over the times of data assignments. Finally, the averaged density is further averaged over the X coordinates in each range of X.

Figure 55 thus provides the electron density map in the $Y_{GSM'} - Z_{GSM'}$ plane. The superposed white dotted circle in each panel displays a reference size of the radius of 30 R_E.

In the near tail region shown in Figure 55(a), the low density plasmas less than 1/cc are concentrated in the area within the radius of 20 R_E. The high density area which is seen beyond 20 R_E corresponds to the magnetosheath. Thus we conclude that the magnetopause is located around 20 R_E from the center axis of the tail. In other words, the averaged tail radius in the near tail region is about 20 R_E. Similar tendency is seen in Figure 55(b). However, in Figure 55(c), the radius of the low density plasma area in the distant tail region is about 25 R_E and is larger than that in the near tail region.

The interesting point seen in Figure 55(c) and (d) is that the high density areas shown by red or yellow contours can be found even in the central region of the tail with $|Y|$, $|Z| \sim 10$ R_E. Observation of these high density areas near the center of the distant tail must have been either of the following two cases: 1. Observation of the cold dense plasmas; 2. Observation of the magnetosheath plasmas. The cold dense plasma in the central distant tail was initially reported by *Mukai et al.* [1994]. The origin of the cold dense plasma is believed to be the magnetosheath cold plasmas. The cold dense plasma with the density of about 1/cc $\sim$ 0.1/cc is frequently observed in the lobe region and always flows in the tailward direction.

The case 2 is more interesting. We observe the magnetosheath plasma signature even near the center of the geomagnetic tail. Figure 56 shows a good example to indicate an abrupt observation of the magnetosheath plasma signature almost in the central geomagnetic tail region. During this period, the spacecraft was located at (GSM' X, GSM' Y, GSM' Z) = (-207, 3, 0.4 R_E). Figure 56 shows the frequency-time spectrograms of the electric field component (upper) and the magnetic field component (lower). We see the abrupt observation of the magnetosheath signature in both components. From the view point of plasma waves, the magnetosheath region is characterized by the higher CR lower cutoff frequency around 20 kHz and by the intense MNB emissions. Using these characteristics of the plasma waves, we can easily identify the spacecraft location. From 1000 to 1020 UT, the lower cutoff frequency was about 13 kHz, which is equivalent to the plasma density of 2/cc, and intense MNB below the electron cyclotron frequency was also found in the B component. Therefore the spacecraft seems to have been in the magnetosheath during this interval. However, just after 1020 UT, the lower cutoff frequency abruptly decreased to about 4 kHz, which is equivalent to the density of 0.2/cc. Synchronizing with the above abrupt change of the density, the intense MNB disappears at 1020 UT (see lower panel). Additionally, we start to see the NEN in the frequency range from 100 Hz to 200 Hz. The above change of the wave signatures shows that the spacecraft moved across

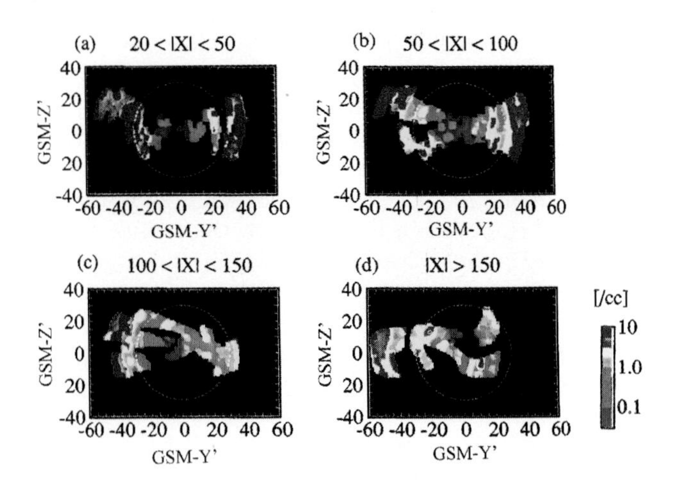

Figure 55. Result of the statistical analyses of the plasma density profile in the geomagnetic tail. The density data are read out from the lower cutoff frequency of the continuum radiation observed by GEOTAIL.

the magnetopause at 1020 UT and entered into the tail lobe region. The change of the lower cutoff frequency and the return of the MNB at 1055 UT show that the spacecraft moved back to the magnetosheath. Judging from the spacecraft location in the modified GSM coordinates the spacecraft should have stayed in the central tail region, but actually the spacecraft moved in and out of the magnetosheath. This strongly suggests that the stationary model of the geomagnetic tail is invalid. The tail in the distant region flaps or oscillates drastically. The high density areas shown in Fig. 55(d), therefore, indicate such events of the tail flapping/oscillation.

We show another example of the statistical analyses of the plasma wave data. In Figure 57 we show the distribution of the intensities averaged over 10 minutes of MNB in the GSM' X-Y plane. We have surveyed the averaged intensity of the MNB using the GEOTAIL PWI data from September 18, 1992 to October 10, 1994. The dataset of the 10 minute average of the MNB intensity was generated from the magnetic field components of the lowest three channels of the MCA (5.62 Hz, 10.0 Hz, and 17.8 Hz). The outer and inner white solid lines indicate the location of the bow shock and the magnetopause calculated by the model proposed by *Howe and Binsack* [1972].

As the intense MNB is observed usually in the magnetosheath, the inner edge of the intensive MNB region obtained from the statistical analyses must be a good indication of the magnetopause location. Actually the inner edge of the area with the intense MNB shown in Figure 57 agrees well with the inner solid line which shows the modeled magnetopause location. The agreement is especially good in the near tail of $|X| < 60$ R_E.

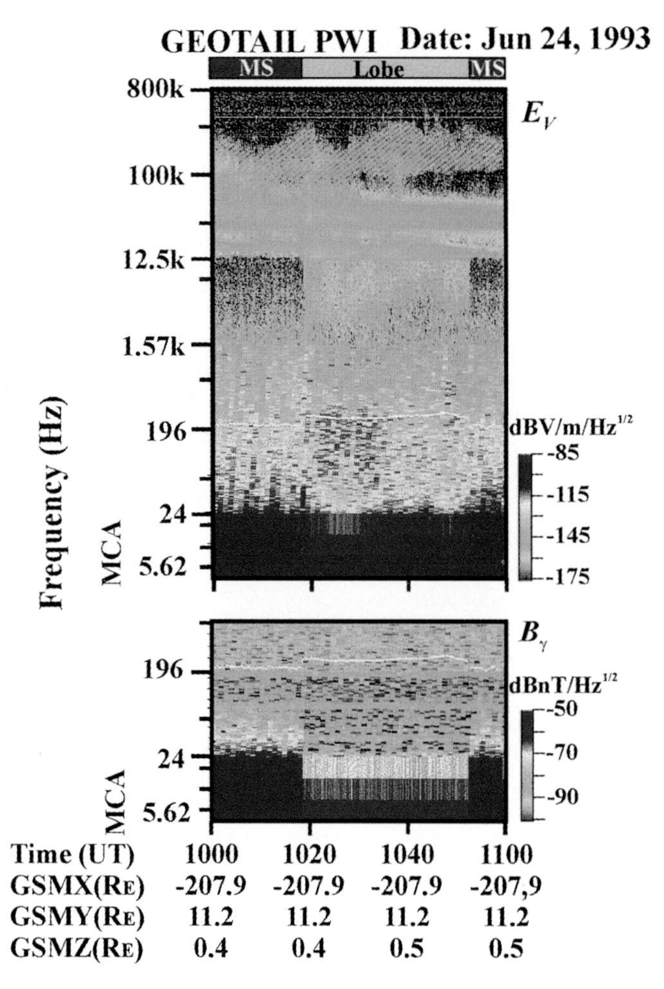

Figure 56. Sudden observation of the magnetosheath plasma in the center of the tail region.

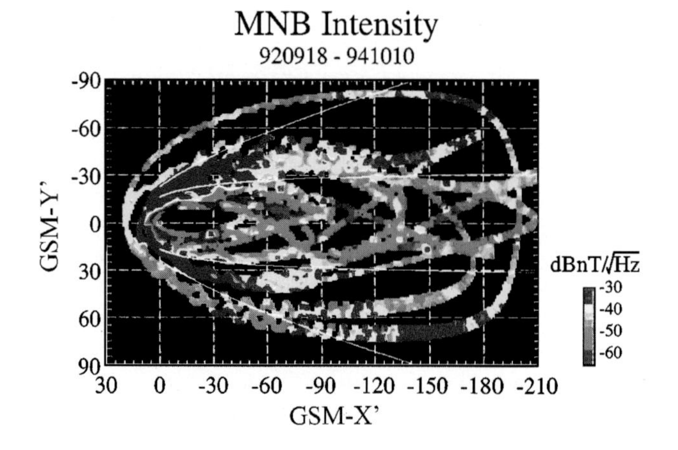

Figure 57. Contour plot of the MNB intensity projected in GSM'-X-Y plane. The solid lines show the location model of the bow shock and magnetopause proposed by Howe and Binsack [1972].

The result shown in Figure 57 is consistent with the result from the statistical analysis on the plasma density shown in Figure 55.

Another interesting point which can be drawn from Figure 57 is the fact that MNB is also occasionally found in the central plasma sheet. In Figure 57, we see intense MNB's with yellow or red color codes localized in the center area of the tail. These MNB's in the center of the tail are found in the following two special cases: 1. when the magnetosheath plasma is met in the center of the tail, 2. when the spacecraft encounters plasmoids. The above case 1 has already been discussed in this chapter. The case 2 will be discussed in Chapter 6 where we demonstrate that intensified MNB waves are observed inside or around the plasmoids.

Plasma wave data observed by spacecraft have been mainly dedicated in the past to the study of microscopic phenomena such as wave-particle interactions. However, we have shown that the plasma wave data are also very useful for studying the macroscopic structure of the geospace. The macroscopic global structure of the geospace is controlled by the macroscopic plasma phenomena such as the Kelvin-Helmholtz (K-H) instability and the magnetic reconnection process. Once the magnetic field and plasma structures are determined, conditions for microscopic instabilities decide the leading mode of plasma waves. Therefore, each characteristic region in geospace has its own plasma wave characteristics. On the other hand, the excited plasma waves scatter or thermalize the background plasmas and/or play a key role in the macroscopic instabilities such as the K-H instability, the bow shock formation and the magnetic reconnection. This point will be discussed in Chapter 8.

6. PLASMA WAVE FEATURES IN ASSOCIATION WITH TRANSIENT PHENOMENA

In the previous Chapter, we have discussed the plasma wave features and characterized signatures in different regions defined by the quasi-stationary geospace structure. The discussions there are valid for most of the time. However, we unexpectedly observe several kinds of transient phenomena in geospace, because the solar wind and its interaction with the magnetosphere are not always stationary. The typical transient phenomena in geospace are the interplanetary shock in the solar wind, and the plasmoid formation in the geomagnetic tail region. The GEOTAIL frequently observes such transient phenomena. In this section, we will discuss the plasma waves associated with such transient phenomena.

6.1. CME and CIR

Though the main objective of the GEOTAIL spacecraft is to survey the geomagnetic tail region, the spacecraft spends most of the time in the magnetosheath region, specially at far distant toward the deep tail. In the deep tail the difference between the magnetosheath plasma and the solar wind is very small. As we have shown in Chapter 5, there exist only very weak plasma wave activities in the solar wind except for special events such as the interplanetary shocks, CME (Coronal Mass Ejection) and CIR (Corotating Interaction Region). In this section, we introduce the plasma waves associated with such transient phenomena in the solar wind.

Coronal Mass Ejection (CME) is a phenomenon causing large interplanetary disturbances by the ejection of a sizable amount of the solar material [e.g., Gosling et al., 1974]. These events have origins near the solar surface in the complex magnetic structure of the coronal streamer belt known to be the origin of the slow solar wind. There are two fundamental categories of CME's: fast CME's and slow CME's. Fast CME's have velocities ranging from 450 to 1200 km/s and are easily identified by a forward propagating shock preceding the CME by approximately 4 to 8 hours. On the other hand, the slow CME is identified by the presence of a bi-directional electron heat flux [e.g., Gosling, 1990].

Figure 58 shows the wave dynamic spectrum associated with the slow CME event observed in the deep magnetosheath (GSE $X = -152$ R_E) by GEOTAIL on December 27 and 28, 1992. *Hammond et al.* [1995] examined the correlation of this slow CME and the CME event observed by Ulysses spacecraft at ~20°S of GEOTAIL and 5AU from the Sun on January 9 and 10, 1993. From the similarities of the time variation of the magnetic field, they concluded that the same CME was observed by both GEOTAIL and Ulysses. Further, they discussed the latitude dependence of the CME global structure using observation results of GEOTAIL and Ulysses spacecraft.

The plasma wave activities associated with the encounter with this CME are similar to those in the magnetosheath. The period for the CME event in Figure 58 is displayed by the red arrow along the horizontal axis. This period is decided by the rotation of the magnetic field direction, which suggests the passage of the magnetic cloud. Before the encounter with the CME, we see the strong density enhancement (from 11/cc up to 60/cc) associated with the enhancement of the ambient magnetic field amplitude at 2100 UT on December 27, 1992. This seems to correspond to the compressed portion as the precursor of the CME. We find bursty NEN-like and BEN-like emissions at 1950, 2130, 2355 and

0030 UT. The density enhancement appears again at 0200 UT when the spacecraft encounters the CME. The electron plasma waves observed from 0500 to 0630 UT on December 28 suggest the existence of the electron beam in the CME magnetic cloud. Further, the intense NEN-like emissions are observed in the latter portion of the CME from 0730 to 1130 UT on December 28. As we have discussed in Chapter 4, the center frequency of the NEN is closely related to the local plasma density. This is one of important clues for studying the excitation mechanism of the NEN. However, these NEN-like emissions associated with the CME shown in Figure 58 do not seem to have a correlation with the local density. This is very clear when we compare the NEN-like emissions in the precursor with those in the inner CME. The center frequencies of the two NEN-like emissions are almost the same in spite of the big difference of the density. This is the reason why we do not address these waves in the CME event NEN emissions. We need more detailed analyses for the NEN-like emissions comparing with the real NEN observed in the tail lobe and the magnetosheath.

Another type of the solar wind disturbance arises from the interaction between high and low speed solar wind streams as they propagate radially outward from the solar surface. These are termed Corotating Interaction Regions (CIR's). The mechanism for forming CIR is provided by the interaction between the non-interpenetrable high and low speed solar wind streams. The high speed solar wind overtakes and compresses upstream slow solar wind plasma. Therefore, the stream interface (SI) between these two regions will steepen and form a pair of shocks called the forward shocks and the reverse shock [*Gosling et al.*, 1976]. The forward and reverse shocks propagate radially outward and inward from the SI, respectively. The CIR is characterized by an enhancement in the bulk velocity, density, temperature, and magnetic field amplitude at the forward shock followed by a decrease of these at the reverse shock or reverse compression.

Figure 59 shows the frequency-time spectrogram of plasma waves associated with the CIR observed by GEOTAIL for a period from December 6 to December 8, 1992 at GSM $X = -180R_E$ behind the earth. The periods of the forward shock, stream interface and reverse shock which are judged from the magnetic field and plasma measurements are displayed in the bottom of Figure 59. This CIR event was also confirmed by Ulysses at 5 AU after 20 days from the observation by GEOTAIL (*private communication with C. M. Hammond*).

The UHR noise, from which we could read the approximate local plasma densities, shows the gradual density increase by the encounter with the foreshock. At the foreshock encounter, the density is abruptly enhanced at 0800 UT on Dec. 7, 1992. The enhanced density gradually increases until the encounter of the streaming interface. It reaches about 120/cc at maximum. The increase of the density is followed by an abrupt decrease when the spacecraft passes the streaming interface. The arrival of the reverse shock is indicated by the small density decrease at about 0400 UT on Dec. 8, 1992. The weak monochromatic waves in the frequency range 70 kHz to 200 kHz are the $2f_p$ emissions, which have been discussed in Chapter 3. They trace well the variation of the UHR frequency.

Bursty BEN-type electrostatic waves are observed at the encounter with the foreshock, for about 4 hours before the encounter with the streaming interface, and for about 6 hours before the encounter with reverse shock. Unfortunately, no waveform data corresponding to the above BEN-type waves are available. However, judging from their spectral features, they seem to consist of a series of the ESW. The existence of ESW suggests the nonlinear interaction, which will be shown in Chapter 7, of possible electron beam instabilities.

6.2. Plasmoid

The neutral line model for the substorm needs the near earth neutral point at the substorm onset time [e.g., *Russell and McPherron*, 1973]. The reconnection process at the near earth neutral point results in the ejection of the plasma cloud called '*plasmoid*'. The ejected plasma cloud travels tailward. *Machida et al.* [1994] performed detailed analyses on the plasmoid observed by GEOTAIL in the deep tail (GSM X=-142 R_E). Based on the magnetic field and plasma measurements, they showed the detailed plasma signatures in

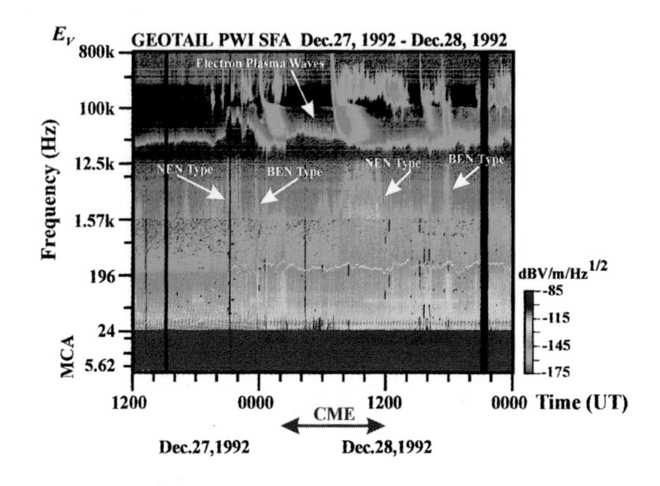

Figure 58. Plasma waves associated with the encounter of the CME observed by GEOTAIL.

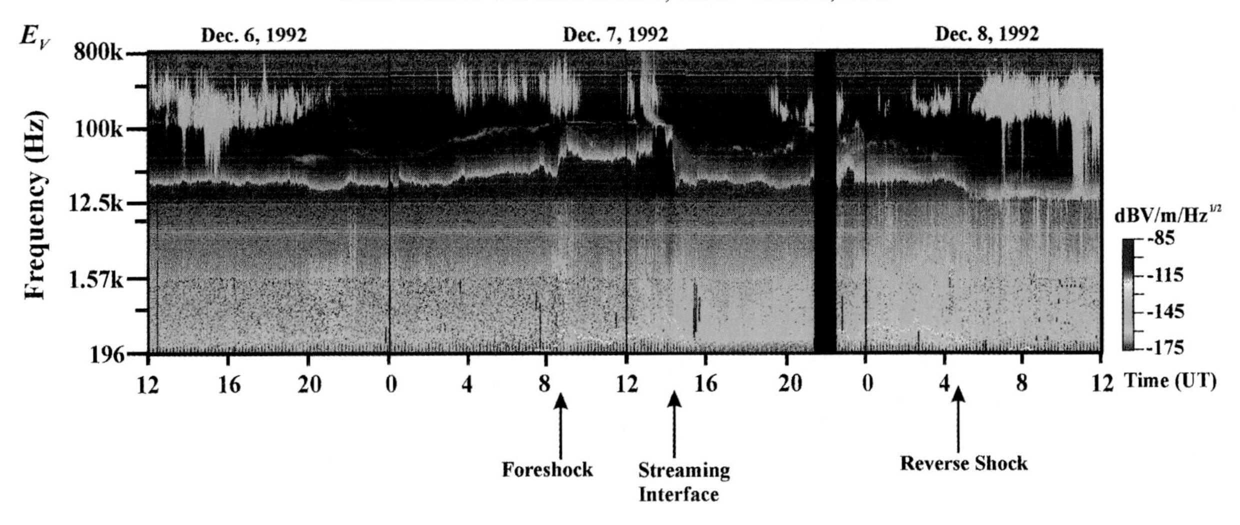

Figure 59. Plasma waves associated with the encounter of the CIR observed by GEOTAIL.

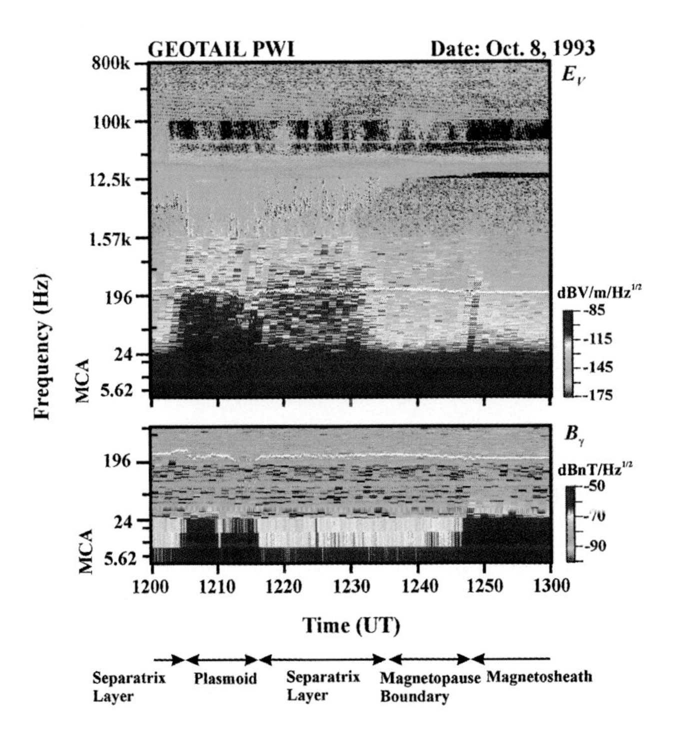

Figure 60. Plasma waves associated with the encounter of the plasmoid observed by GEOTAIL.

and around the plasmoid. We examine the frequency-time spectrogram of plasma waves associated with the plasmoid event. The result is shown in Figure 60 The spacecraft location is displayed in the bottom.

We see the intense BEN emissions during the period corresponding to the plasmoid. In the magnetic field component, we also find the low frequency MNB during the same period. These BEN and MNB are the typical wave signature in the body of the plasmoid. These signatures are very close to those of the PSBL (plasma sheet boundary layer). However, in general, the wave intensity of these waves in the plasmoid is higher than those in the plasma sheet boundary.

Machida et al. [1994] reported that the tailward electron and ion beams were observed in the separatrix layers which are located at both leading and trailing edges of the plasmoid. As shown in Figure 60, the wave activities in the leading and trailing edges are significantly different. The leading separatrix layer during the period from 1200 to 1205 UT is represented by the electron plasma waves with frequencies equal to 7 kHz. The equivalent plasma density is about 0.6 /cc. This density is higher than the average tail lobe density. This high density could be a result of the compressional effects due to the plasmoid. The observation of the electron plasma waves in this layer is very consistent with the existence of the electron beam. On the other hand, the latter separatrix layer in the trailing edge of the plasmoid shows significantly different wave features. During the period from 1215 to 1235 UT, we see rather intense but scattered waves in addition to the electron plasma waves. The scattered and sparse spectra below 1 kHz seem to represent NEN. It is noted that such space spectra are not observed in the leading edge of the plasmoid. *Machida et al.* [1994] showed the velocity distributions of electrons and ions at 6 time intervals around an in the plasmoid event. We compared them with our data, but could not find the decisive difference in the electron velocity distributions between the leading and trailing separatrix layers.

Another plasmoid event observed by GEOTAIL PWI was shown by *Matsumoto et al.* [1994a]. By the comparison of the event shown in Figure 60 with that in *Matsumoto et al.* [1994a], we find that the common wave activities during the plasmoid event are the BEN in the body of the plasmoid, and the electron plasma waves in the edge of the plasmoid.

The intense and sparse spectral components below 1 kHz shown in Figure 60 would be observed only on the special orbit. Similarly monochromatic whistler mode waves are sometimes observed in association with the plasmoid event. The occurrence of these special wave signatures associated with the plasmoid could depend on the spacecraft path through the plasmoid.

7. COMPUTER EXPERIMENTS OF FORMATION OF ESW AND NEN

The waveforms of ESW and NEN emissions presented in Chapter 7 shows close similarity with those found in electrostatic particle simulations of an electron beam instability [Matsumoto et al., 1994b]. In this chapter, we study the basic properties of electrostatic electron beam instabilities based on the linear theory, and we further study nonlinear evolution of the instabilities by computer experiments. The linear analyses and computer experiments with various parameters clarify the necessary physical conditions for generation of ESW and NEN emissions.

7.1. Model for Computer Experiments

One dimensional electromagnetic particle code named "KEMPO1" [*Omura and Matsumoto*, 1993] has been used to study the nonlinear evolution of plasma wave instability driven by an electron beam. The electron beam can excite both electrostatic waves and electromagnetic whistler mode waves through Landau resonance and cyclotron resonance, respectively. The competing process of these electromagnetic and electrostatic instabilities has been studied by the linear dispersion analysis and the particle simulation [*Omura and Matsumoto*, 1987; *Zhang, et al.*, 1993; *Borda de Agua et al.*,1995]. For the purely electrostatic waves propagating parallel to the ambient magnetic field, there is no coupling with electromagnetic modes. Therefore, we can use an electrostatic particle code, simplified version of the KEMPO1, where we can reduce the computations, and can perform longer runs up to much larger time steps. The simulation model is a one-dimensional system taken along the static magnetic field with the periodic boundary condition. We assume two groups of electrons and one group of ions. As the initial condition, each of the species is given a Maxwellian velocity distribution function with a drift velocity V_{ds} and a thermal velocity V_{ts}. In the following, we describe various kinds of electrostatic instabilities for the electrostatic waves propagating parallel to the ambient magnetic field.

7.2. Electrostatic Dispersion Relation

The kinetic description of the electron beam instability is given by the dispersion relation derived from the Poisson and Vlasov equations.

$$1 = \sum_s \frac{\Pi_s^2}{k^2} \int_L \frac{dg_v/dv}{v - \omega/k} dv \tag{9}$$

where Π_s is the plasma frequency of particle species "s", and $g_v(v)$ is the normalized velocity distribution function whose integral over the velocity v is unity. The wave number is assumed to be positive and the frequency is complex $\omega = \omega_r + i\omega_i$. The integration over the velocity is taken along the Landau contour by which the dispersion relation is analytically continued from the upper half of the complex ω plain to the lower half and defined as an analytic function over the entire complex ω plain.

For the Maxwellian plasmas consisting of the distribution function of the form:

$$g_v = \frac{1}{2\sqrt{\pi} V_{ts}} \exp(-\frac{(v - V_{ds})^2}{2V_{ts}}) \tag{10}$$

we can rewrite the dispersion relation as

$$1 + 2\sum_s (1 + \zeta_s Z(\zeta_s)) = 0 \tag{11}$$

where $Z(\zeta_s)$ is the plasma dispersion function [*Fried and Conte*, 1961].

$$\zeta_s = \frac{\omega - kV_{ds}}{\sqrt{2} k V_{ts}} \tag{12}$$

For the electrostatic dispersion relations for the parallel propagation, there is no effect of the external magnetic field. As is obvious from the dispersion relation given above, the property of the plasma is specified by the combination of the density specified by the plasma frequency $\Pi_s = \sqrt{q_s^2 n_s / \epsilon_o m_s}$, thermal velocity V_{ts} and drift velocity V_{ds} of each species "s", where q_s, n_s, m_s and ϵ_o are the charge, density and mass of the s-th particle and the permittivity in vacuum. The various combinations of these parameters could give a positive imaginary frequency ω_i, i.e., the growth of the unstable waves.

When the thermal velocities are small enough, i.e., $\zeta_s >> 1$, the dispersion function $Z(\zeta_s)$ is approximated by

$$Z(\zeta_s)) \simeq -\frac{1}{\zeta_s} - \frac{1}{2\zeta_s^2} \qquad (13)$$

The dispersion relation is then simplified to the form

$$1 = \sum_s \frac{\Pi_s^2}{(\omega - kV_{ds})^2} \qquad (14)$$

It is noted that the dispersion relation has no imaginary part. The instability as a solution to the dispersion equation without the imaginary term is called a reactive instability.

For the plasma consisting of warmer particle species with finite thermal velocities ($\zeta_s << 1$), the dispersion function is approximated by

$$Z(\zeta_s) \simeq -2\zeta_s + i\sqrt{\pi}\exp(-\zeta_s^2) \qquad (15)$$

The dispersion equation has a finite imaginary part. Rewriting the dispersion relation as

$$D_r(k,\omega) + iD_i(k,\omega) = 0 \qquad (16)$$

we can obtain an approximate expression for the growth rate under the assumption of $|\omega_i| << |\omega_r|$.

$$\omega_i = -\frac{D_i(k,\omega_r)}{\partial D_r(k,\omega_r)/\partial\omega_{\omega=\omega_r}} \qquad (17)$$

The instability given by the contribution of the imaginary part D_i is called a resistive instability.

7.3. *Reactive Instability*

Generally speaking, the reactive instabilities give rise to larger growth rates than the resistive instabilities. Because of the large growth rates, the saturation level of the electric field of the unstable modes is very large to trap a substantial part of the electron population in the system. Typical examples of the linear dispersion relations of the reactive instabilities are shown in Figure 61. We assumed two electron beams and one ion beam with densities n_1, n_2 and n_i, respectively ($n_1 + n_2 = n_i$). We assume a reduced mass ratio $m_i/m_e = 100$, where m_i and m_e are the masses of an ion and an electron, respectively. With the reduced mass ratio, the charge neutrality condition gives the ratio of ion and electron plasma frequencies as $\Pi_i/\Pi_e = 0.1$, respectively. The thermal velocity of these electrons and of ions are $V_{te} = 0.05V_d$ and $V_{ti} = 0.005V_d$, where V_d is the drift velocity between the two electron beams. The ions are assumed to be drifting with one of the electron beams, the one with density n_1 unless $n_1 = 0$. When the electron beams

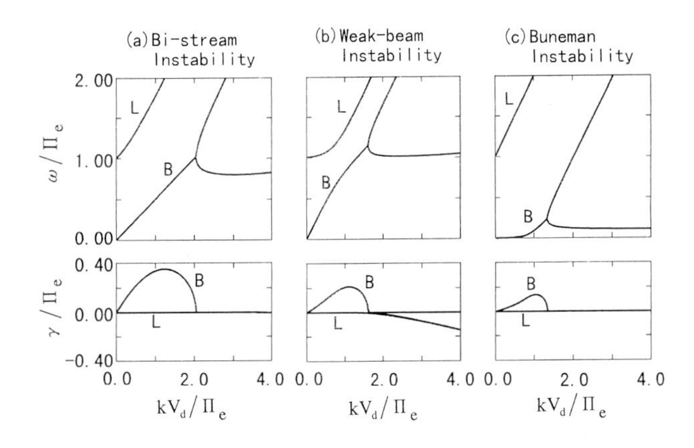

Figure 61. Typical examples of the linear dispersion relations of the reactive instabilities. We assumed two electron beams and one ion beam with densities n_1, n_2, and n_i, respectively. The ions are assumed to be drifting with one of the electron beams, the one with density n_1, unless $n_1 = 0$. Three panels correspond to three cases for the different density ratios: (a) $n_2/n_i = 0.5$, (b) $n_2/n_i = 0.05$, and (c) $n_2/n_i = 1.0$. The imaginary part of ω is represented by $\gamma(= \omega_i)$

have similar number densities, i.e., $n_1 \sim n_2$, the instability is called bi-stream instability Figure 61 (a) shows the case with $n_2/n_i = 0.5$. If the electron beam (n_2) drifting against the other electron beam and the drifting beam have much smaller density, i.e., $n_2 << n_1$, the instability is called weak-beam instability. Figure 61(b) shows the case with $n_2/n_i = 0.05$. As an opposite case, if the drifting electron beam is a majority species, $n_1 << n_2$, then the instability is called Buneman instability Figure 61 (c) is an extreme case with $n_2/n_i = 1.0$, i.e., $n_1 = 0.0$.

7.4. *Resistive Instability*

When the thermal velocity of the majority electrons is large comparable to the phase velocities of the electrostatic waves, the dispersion relation has a finite imaginary part, and the nature of the instability becomes resistive as discussed before.

Figure 62 shows examples of the resistive instabilities. The bump-on-tail instability shown in Figure 62(a) is a modification of the weak-beam instability (Figure 61(b)). We just increased the thermal velocity of the majority electrons from $V_{te} = 0.05V_d$ to $V_{te1} = 0.5V_d$, while keeping other parameters the same. The structure of the dispersion curves changes, and we do not have bifurcation of the curve any more. The most unstable mode "B" appears near the electron plasma frequency. In the low frequency range, there appears a branch "I" of ion acoustic waves.

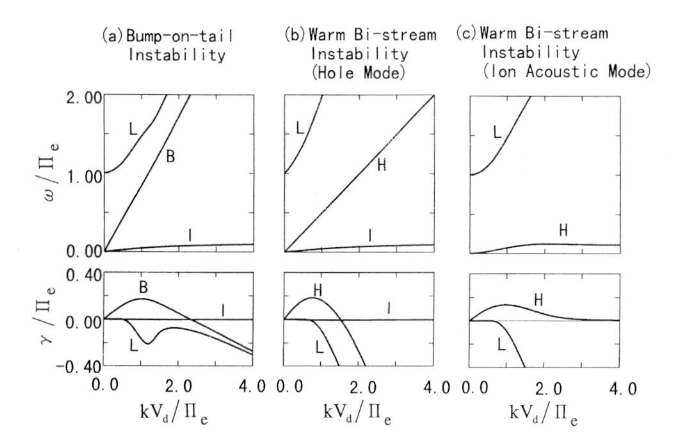

Figure 62. Same as Figure 61, but for the Resistive instability.

The warm bi-stream instability shown in Figure 62(b) is also a modification of the cold bi-stream instability in Figure 61(a). The frequency of the most unstable mode "H" is substantially lower than the electron plasma frequency. The mode is called "hole mode" because of the velocity distribution function with the depletion at the center velocity. The frequency range of the unstable mode is a function of the center velocity of the two warm electron beams. If we take the center velocity at the velocity of the ion beam, the dispersion curve "H" of the hole mode merges with that of the ion acoustic waves as shown in Figure 62 (c). Therefore, depending on the drift velocities of the warm bi-streaming electrons, waves in the frequency range between the ion plasma frequency Π_i and the electron plasma frequency Π_e can be excited. The frequency range of NEN emissions can be explained by this property of the linear dispersion relation.

7.5. Nonlinear Evolution of Electron Beam Instability

A series of computer experiments has been performed with various combinations of the parameters for the study of ESW associated with BEN [*Omura et al.*, 1994]. In all of these simulations, waves grow according to the prediction of the linear dispersion relations, and saturate at a certain amplitude of the electric field corresponding to the formation of electrostatic potentials. The evolution at much later time after the nonlinear effects comes in effect is quite different depending of the density ratio of the electrons and the thermal velocities of the electrons and ions.

We first performed two runs of bi-stream instability (Figure 61(a)) with different initial ion thermal velocities $V_{ti} = 0.1V_d$ and $0.005V_d$. As the initial condition, we put two cold electron beams with the same thermal velocity $V_{te} = 0.05_d$ at $v_x/V_d = 0.0$ and 1.0. We also placed an ion beam at $v_x/V_d = 0.0$. The evolution of the instability is shown by the phase diagrams for particles in the $x - v_x$ space in Figure 63. At $t = 26\Pi_e^{-1}$, we find the growth of the most dominant mode around Mode 10 (10 spatial wave cycles in the system) has reached saturation. The frequency of the mode is less than the plasma frequency in accordance with the linear theory. We find no significant difference between the two cases. However at $t = 26\Pi_e^{-1}$ we find that the number of vortices of trapped electrons decreases through coalescence in the left panel ($V_{ti} = 0.1V_d$), while the right panel ($V_{ti} = 0.005V_d$) shows a slightly disordered structure of vortices. Later at $t = 102\Pi_e^{-1}$, $205\Pi_e^{-1}$, and $307\Pi_e^{-1}$, we observe the formation of larger vortices through the coalescence of smaller vortices on the left panel, while the vortices seem to disappear in the right-hand panel.

In both runs, the electrons are thermalized over the velocity range between $v_x = 0$ and the drift velocities $v_x = 20$ through the strong nonlinear trapping of the whole electron beams and subsequent coalescence or diffusions of the coherent potential structures as shown in Figure 63. Thus we can estimate the electron thermal velocity V_e in the nonlinear stage as $\sim 0.5V_d$. If V_{ti} is small, the ion acoustic waves can propagate without the ion Landau damping at the phase velocity V_s given by

$$V_s \sim \sqrt{\frac{m_e}{m_i}} \frac{V_d}{2} \tag{18}$$

Since we assumed the mass ratio $m_i/m_e = 100$, we have $V_s = 0.05V_d$ on the nonlinear stage of the electron trapping. In the case presented in the right panel ($V_{ti} = 0.005V_d$), the ion acoustic wave can exist without strong ion Landau damping. The initial nonlinear electrostatic waves due to the bi-stream instability are converted into the ion acoustic waves through nonlinear wave-wave interaction (see Figure 2(b) in Omura et al.[1994]).

However, in the case presented in the left panel ($V_{ti} = 0.1V_d$), the ion acoustic wave is subject to heavy ion Landau damping because $V_{ti} > V_s$. Since there is no other normal mode for the electrostatic waves to decay into, the coalescence of the trapping potentials leads to formation of ESW. It is noted that the ESW are flowing with the average drift velocity of $v_x = 0.5V_d$ with respect to the ions at rest. The ESW are formed by the electrons trapped in the positive potentials. These potentials are very stable in the frame of reference moving with the ESW drift velocity (see Figure 4 in Omura et al.[1994]).

The stable structure forms a BGK mode which is a time independent solution of the Vlasov – Poisson equations. The density gradient forming the potential is balanced by the gradient of the velocity distribution function in the velocity phase space.

$$v_x \frac{\partial f_e(x, v_x)}{\partial x} + \frac{e}{m_e} \frac{\partial \phi}{\partial x} \frac{\partial f_e(x, v_x)}{\partial v_x} = 0 \qquad (19)$$

Since the trapping occurs only for the electrons, the equilibrium is taken only by the dynamics of the electron. Between the adjacent ESW's, there is a valley of potentials that could trap the ions and disturb the equilibrium produced by the electrons. Therefore, one of the necessary condition of the stable ESW is that the relative drift velocity V_{i-ESW} between ESW and ions must satisfy the following relation.

$$\frac{m_i}{2}(V_{i-ESW} - V_{ti})^2 > |e\Delta\phi| \qquad (20)$$

where $\Delta\phi$ is the amplitude of the ESW potentials.

We next change the density ratio between the two streaming electron beams $R = n_2/n_i$. In the above two cases, the densities of electron beam were the same, i.e., $R = 0.5$. Assuming the hot ions with $V_{ti} = 0.05V_d$, we performed 5 other runs with different density ratios $R = 0.1, 0.3, 0.7, 0.9$, and 1.0. The phase diagrams at $t = 410\Pi_e^{-1}$, which is a much later time of the nonlinear evolution, is shown in Figure 64 for the runs with different density ratios. For $R = 0.1$, the instability is the weak beam instability shown in Figure 61 (b). Although Langmuir waves are excited, the trapped electrons at the saturation of the instability become diffused via a quasi-linear diffusion process thereby exciting higher wave number modes which diffuse any structures in the phase space as shown in Panel (a). For $R = 0.3$, the characteristics of the bi-stream instability are retained, forming small ESW as seen in Panel (b). In the case of $R = 0.5$ the evolution of the instability was previously shown in Figure 63. The later time of the nonlinear evolution is shown in Panel (c). Comparing the phase diagrams at $t = 307\Pi_e^{-1}$ (Figure 63) and $t = 410\Pi_e^{-1}$(Figure 64), we can say that the potential structure does not change except that it moves to the right.

For the higher beam density ratios of $R = 0.7$ and 0.9, we can find an enhanced ESW due to the free energy of the ion beam. For $R = 1.0$, the instability is purely a Buneman instability shown in Figure 61 (c). Since we assume $V_{ti} > V_s$, we find an evident ESW in Panel (f) of Figure 64 instead of ion acoustic waves which occurs for the case with colder ions $V_{ti} << V_s$.

In the above run with $R = 0.1$, we could not find formation of ESW. It was because of the quasi-linear diffusion that excite many Langmuir waves with different phase velocity. We further investigated the case with the weak electron beam [*Omura et al.*, 1996]. The time evolution of the quasi-linear diffusion process is shown in Figure 65 for a weaker electron beam of $R = 0.05$. We show the phase diagrams on the left column and the electron velocity distribution function on the right column. We performed two different runs with hot and cold ions, but the ion temperature does not give any significant difference as we found in the case of the bi-stream shown in Figure 63. We only show the case

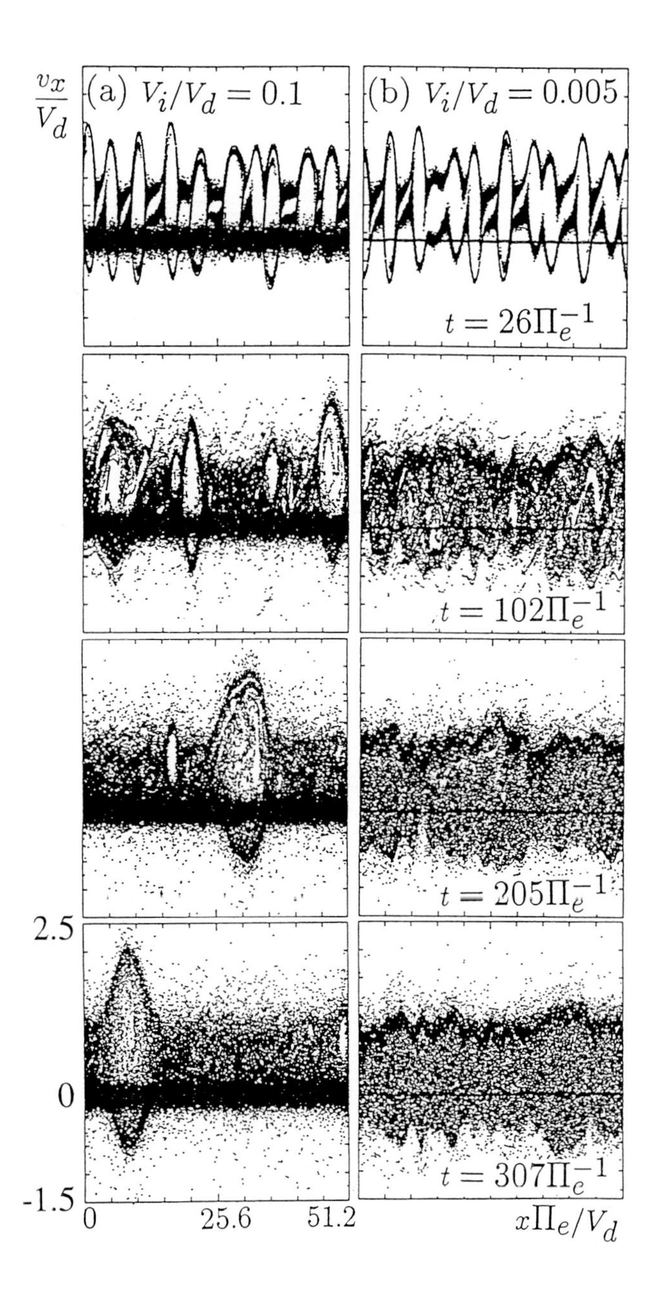

Figure 63. Time evolution in the $x - v_x$ phase space of the bi-stream instabilities with different initial ion thermal velocities ($V_{ti} = 2.0$ and 0.1) [*Omura et al.*, 1996].

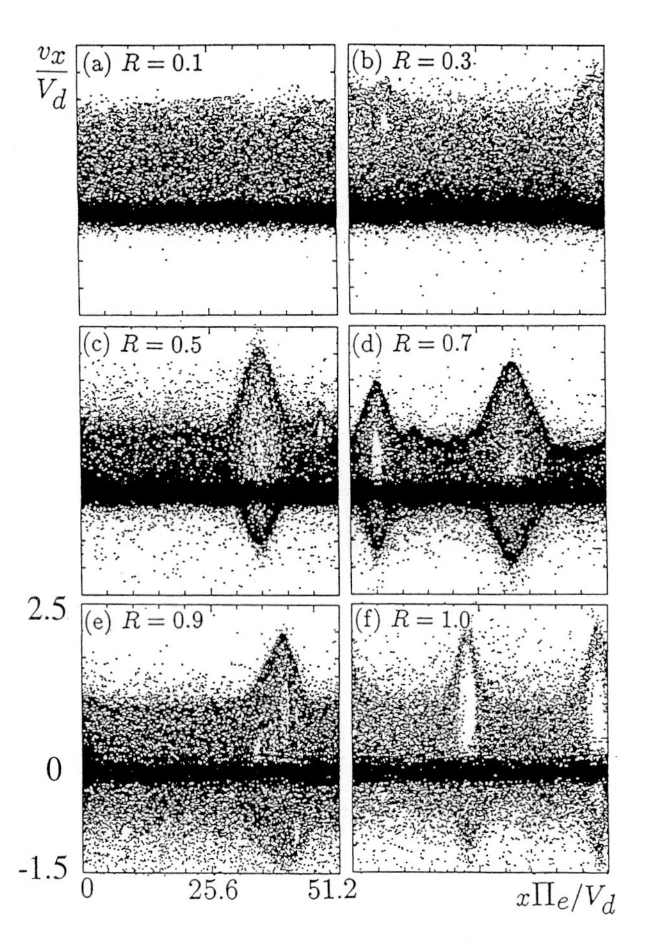

Figure 64. Density ratio dependence of the nonlinear evolution. Six panels display the phase diagrams at $t = 410\Pi_e^{-1}$, which is a much later time of the nonlinear evolution, for the density rations, $R = 0.1$, 0.3, 0.7, 0.9, 1.0, respectively [*Omura et al.*, 1996].

with cold ions ($V_{ti} = 0.005V_d$) in Figure 65. The electron beams spread over the velocity space, forming a plateau in the distribution function over the velocity range between the drift velocities of the weak electron beam and the majority electrons. The majority electrons are not much affected by the diffusion process as seen in the distribution functions in the right column in Figure 65.

In the weak-beam instability, Langmuir waves are excited as found in the ω-k diagram Figure 66 (a). Although the dispersion is modified by the electron beam, the spectra are close to the plasma frequency Π_e. The nonlinear trapping occurs only for the weak electron beam, and the majority electrons supporting the Langmuir waves are not involved in the nonlinear trapping dynamics. The dispersion relation of the Langmuir waves can be approximated by the linear dispersion relation. The electron beams are diffused gradually filling the gap in the velocity space between the initial two electron beams. As is well known, this process is described by the quasi-linear theory. Langmuir waves

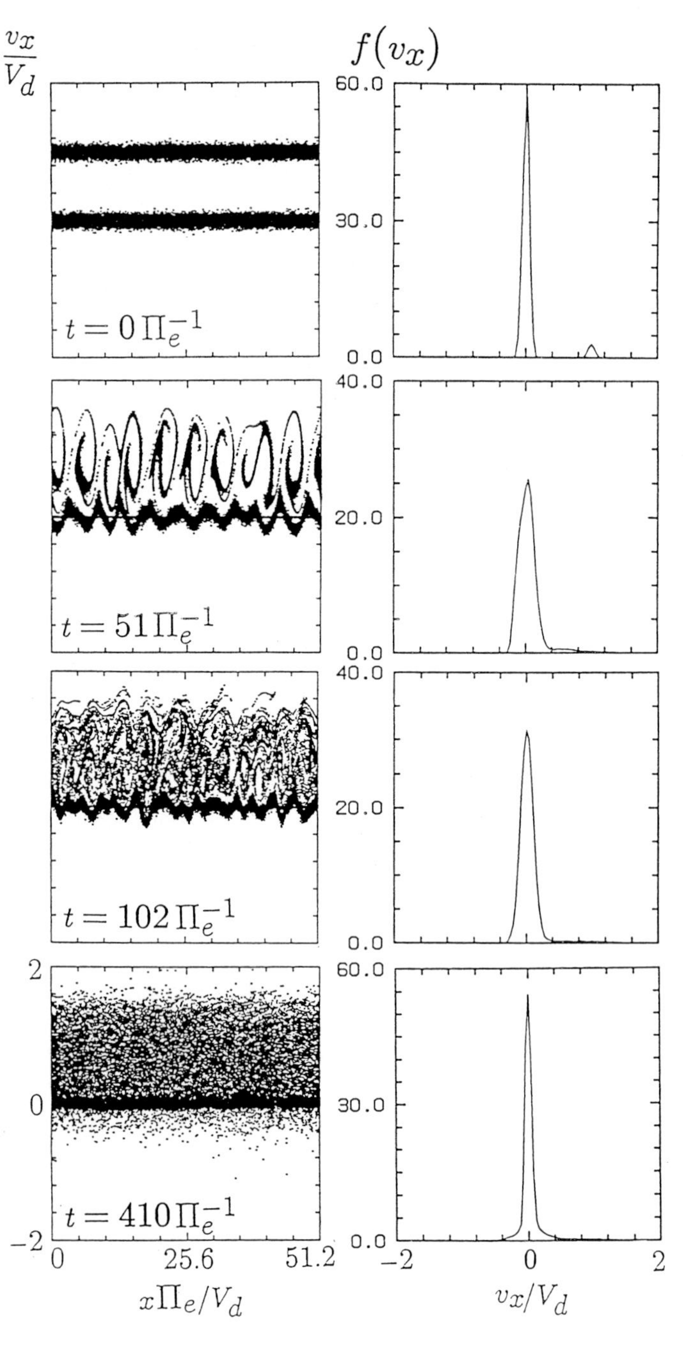

Figure 65. Time series of the phase diagrams in the quasi-linear diffusion process for a weaker electron beam of $R = 0.05$ [*Omura et al.*, 1996].

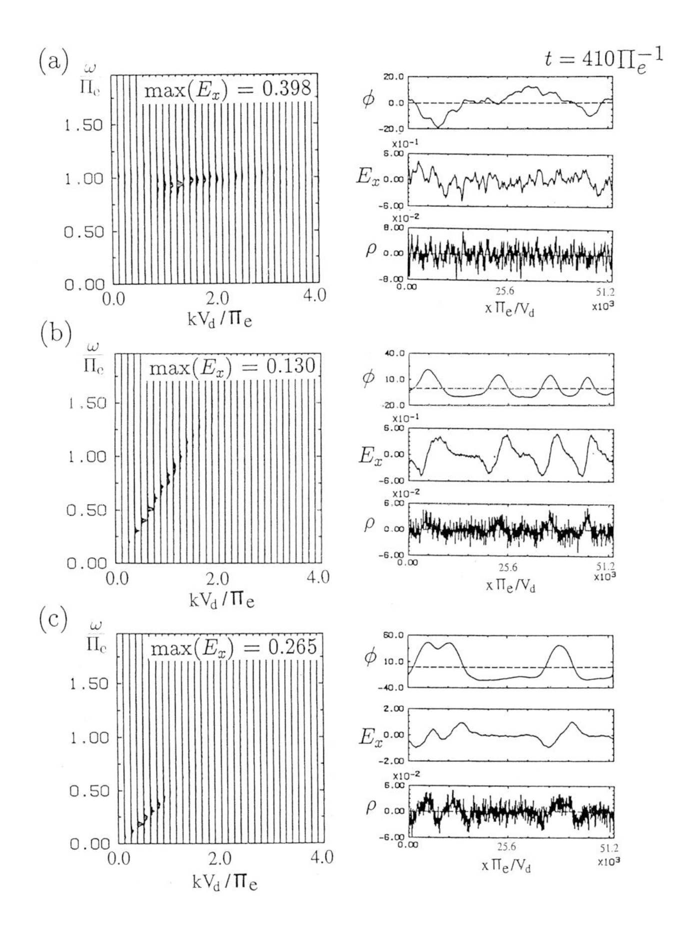

Figure 66. $\omega - k$ diagrams (left) and corresponding spatial profiles of potential ϕ, electric field E_x, and charge density ρ (right) for (a) weak-beam instability, (b) bump-on-tail beam instability, and (c) electron hole mode instability [*Omura et al.*, 1996].

with smaller phase velocities are subsequently excited. These Langmuir waves are essentially linear waves with small amplitudes, and we do not find nonlinear decay processes even in the presence of the cold ions.

Keeping the same density ratio $R = 0.05$, we just increase the thermal velocity of the majority electron beam so that the gap of the initial velocity distribution functions is filled in the velocity space. The nature of the instability changes from the reactive to the resistive one as shown in Figure 62 (a). As shown in Figure 67, the positive gradient of the velocity distribution function is localized at the small bump on the high-energy tail of the majority electron beam. A coherent Langmuir wave, whose phase velocity is slightly smaller than the drift velocity of the bump electron beam, grows and traps the bump electrons, diffusing the bump to a flat beam distribution. Contrary to the previous weak-beam instability, there remains no positive gradient that yields a positive growth rate at a different phase velocity. Therefore, there occurs no further excitation of Langmuir waves.

At the saturation of the instability, the potential structure of the Langmuir waves coalesces with each other to form ESW. After this coalescence process, we

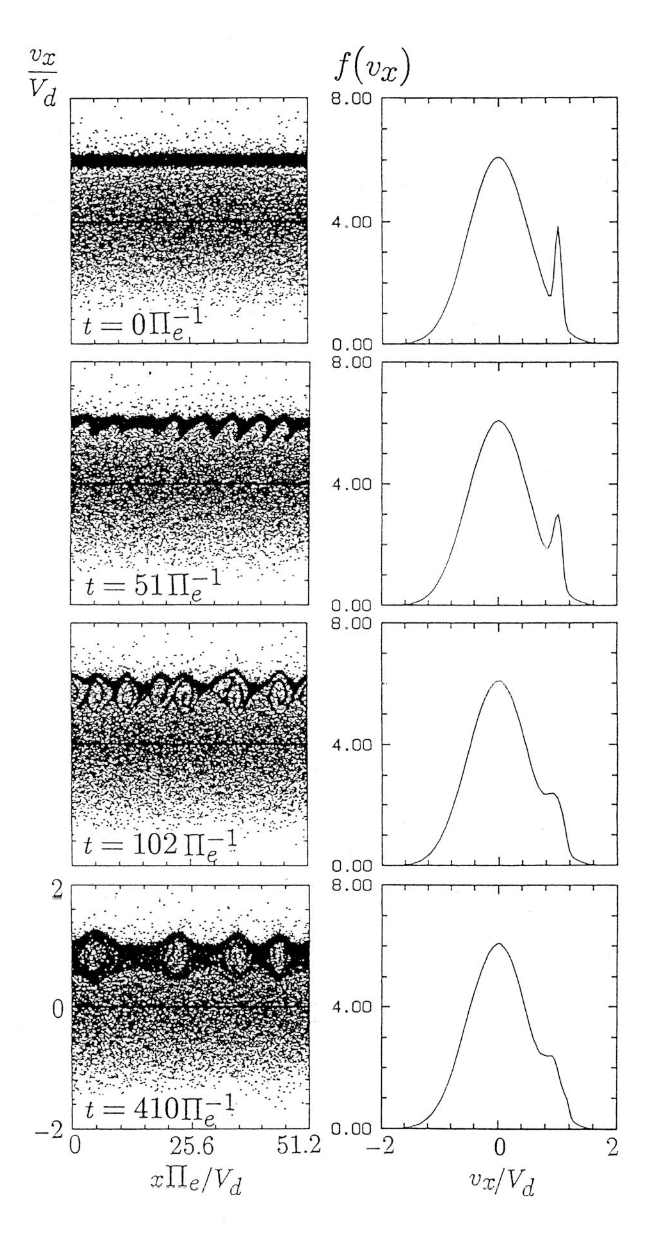

Figure 67. Time series of the $x - v_x$ phase diagrams and corresponding electron velocity distribution functions for the bump-on-tail instability [*Omura et al.*, 1996].

find a stable BGK mode, as we found in the bi-stream instability with hot ions. The temperature of the ions is not important, because the amplitude of the Langmuir wave is relatively small and the wave does not decay into the ion acoustic waves.

The solitary wave structure gives a broadband spectrum as shown in Figure 66 (b). Since the ESW are formed by the minority electron beam, it moves with the electron beam velocity slightly decreased from the initial value. Correspondingly, the spectra of the ESW extend clearly up to the plasma frequency.

We performed another run for the warm bi-stream instability where we increased the thermal velocities of the bi-streaming electron beams, making the instability weaker. The linear dispersion is given in Figure 62 (b). The time evolution of the phase diagrams and the velocity distribution function are shown in Figure 68. The wave grows up to a level that causes a nonlinear trapping of the particles. The trapping causes a slight diffusion of the two-hump velocity distribution function. We can find the phase velocity of the wave from the $\omega - k$ spectra of Figure 66. The gradient of the $\omega - k$ spectra gives the phase velocity as $\sim$ 10, which corresponds to the velocity at the valley of the two-hump velocity distribution function. After the saturation, the electron distribution becomes marginally stable and the wave exists stably as a normal mode of the plasma for a long time.

The longtime evolution of these potentials is different depending on whether the center velocity V_c at the valley of the two-hump distribution function is close to the ion acoustic velocity. As shown in Figure 62 (b) and (c), the electron hole mode is found above the ion plasma frequency in the case of $V_c >> V_s$. As we find on the right panel in Figure 66 (c), we also observe a kind of ESW. If the velocity V_c of the potentials is large enough to be free from the ion dynamics, the coalescence of the electrostatic trapping potentials also occurs even in the warm bi-stream instability, gradually forming ESW. The emission in this case may correspond to the narrowband electrostatic emissions (NEN). In the case of $V_c < V_s$, however, the dispersion relation becomes similar to that of ion acoustic waves. The strong coupling with ion dynamics results in ion acoustic waves with different phase velocities.

The nonlinear evolutions of these various instabilities are summarized in Figure 69. The left column represents the initial condition and the right column represents the final state for each instability. In the bi-stream instability, the nonlinear trapping occurs for a wide velocity range that engulfs the whole population of the electrons. The wave amplitudes are large enough to decay into the ion acoustic waves in the presence of cold ions. Langmuir waves (LW) and ESW are excited by the weak beam and bump-ontail instabilities, respectively. Electron hole modes (EHM) corresponding to NEN emissions are excited by the warm bi-stream instability. In these resistive instabilities with smaller growth rates, the wave amplitudes are not large enough to cause the nonlinear decay to the ion acoustic waves.

7.6. Comparison with GEOTAIL Observation

We have shown that the ESW move with different drift velocities from the ion flow. We can also estimate the drift velocity of the ESW from the waveform measurement by the WFC of GEOTAIL/PWI. The ESW observed by the WFC have a very clear potential structure free from the thermal fluctuations with small wavelengths. This suggests that the potential structure is larger than the thermal fluctuation due to the thermal motion of the electrons. To generate the electrostatic potential structure without thermal fluctuation, the trapping of thermal bulk electrons must take place. Since the minimum thermal energy of the electrons is about 10 eV, the depth of the potential well $\Delta\phi$ must be 10 eV at least. The maximum electric field E_{max} in the observation is about $E_{\text{max}} \sim 300\ \mu\text{V/m}$. Approximating the ESW with a sinusoidal waveform with a wavelength λ, we have the relation

$$\Delta\phi = \frac{\lambda}{\pi} E_{\text{max}} \tag{21}$$

From the WFC data, we can obtain the duration time t_{ESW} of the ESW, which give the drift velocity of the ESW as

$$V_d = \frac{\lambda}{t_{\text{ESW}}} = \frac{\pi\Delta\phi}{E_{\text{max}} t_{\text{ESW}}} \tag{22}$$

Therefore the wavelength is about 100 km for the potential depth of 10 eV. Since the typical duration time t_{ESW} is about 5 ms, we obtain a drift velocity of 20,000 km/s. The charged particles that form the BGK mode cannot be ions, because it is unlikely to observe such a high-energy ion flow in the magnetotail.

The electron drift velocity of 20,000 km/s corresponds to the energy of 1 keV electrons. If we assume a smooth velocity distribution function close to the Maxwellian distribution, the trapping potential of 10 eV electrons suggests only small fractions of the electron population are trapped by the BGK potentials. This implies that the ESW is likely to be generated by the bump-on-tail instability rather than the bi-stream instability.

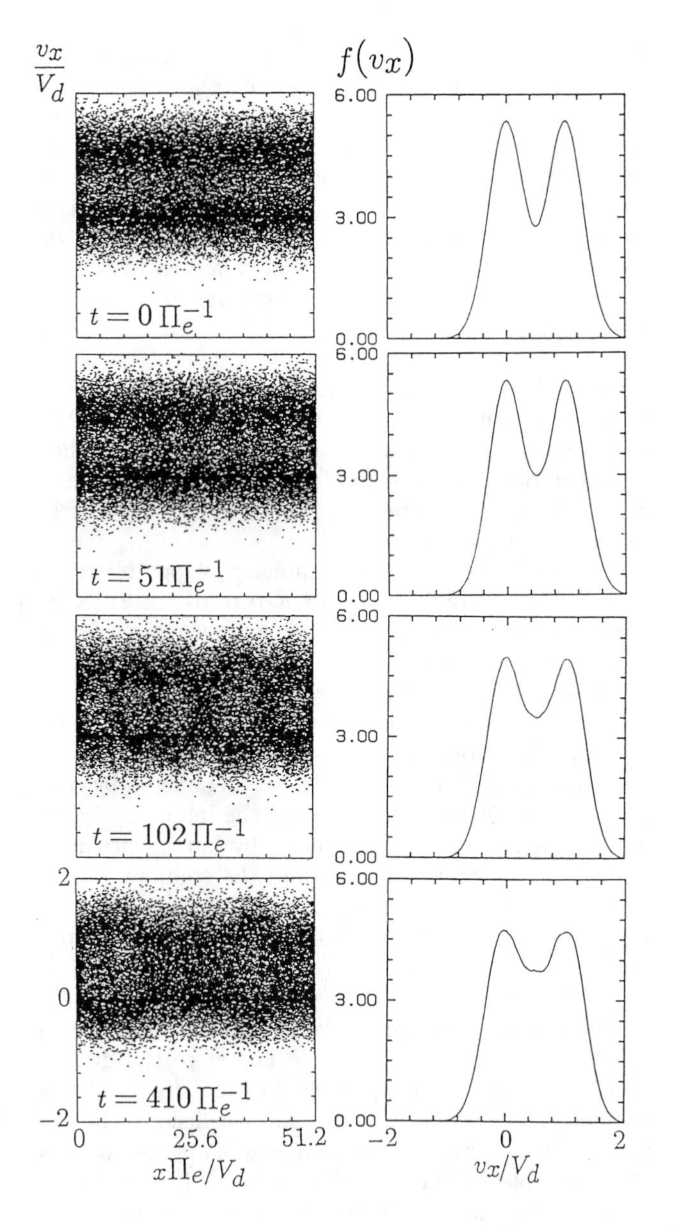

Figure 68. Same as Figure 67 but for the warm bistream instability [*Omura et al.*, 1996].

8. SUMMARY AND DISCUSSION

In this article, we have given an overview of plasma wave research by GEOTAIL for a period from September, 1992 to December, 1996 for about 4 years. Though the materials and subjects dealt in this paper have been piecemeal published by the members of the GEOTAIL PWI team, it may be timely and useful to give their overview in a unified manner. Therefore, we decided to include most of the subjects relating to plasma wave data even though they have already been published in previous publications. Not to mention we attempted to provide an overview of the plasma wave characteristics existing in geospace.

Chapter 2 was prepared to give a background of linear dispersion characteristics of plasma waves in a homogeneous magnetized plasma, though not always useful in inhomogeneous plasmas, nor for waves involving nonlinear dynamics. However, readers who are nonspecialist in plasma wave research may find it helpful in the discussion of various normal modes in a unified manner.

Chapter 3 was dedicated to description of electromagnetic(EM) waves found in geospace. They are very rich in their frequency spectra and could carry information from distant plasmas, sometimes providing a tool for remote sensing. The EM waves dealt there are; a special whistler with a large dispersion which has never been

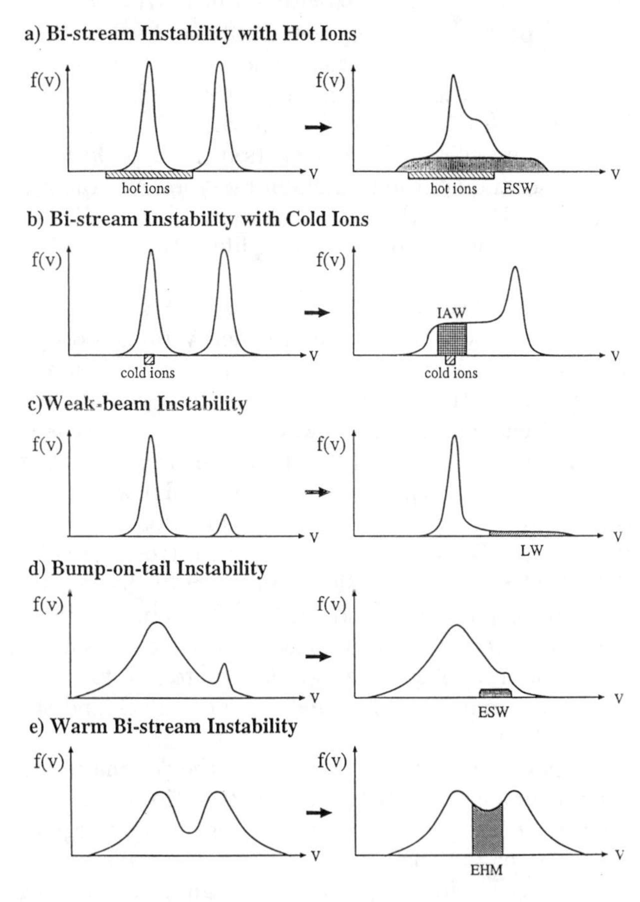

Figure 69. Schematic illustration of the electron velocity distribution functions (left) at $t = 0$ as the initial condition and (right) at $t = 409.6$ as the final equilibrium state [*Omura et al.*, 1996].

observed in the Earth's magnetosphere before; in situ observation of phase and intensity observation of chorus; lion roars in the magnetosheath; magnetic noise bursts(MNB) in the magnetosheath; monochromatic whistler wave packet in the solar wind; trapped and escaping continuum radiation(CR) including enhanced CR showing time-varying discrete spectrum thus providing information on substorm-related electron injection into the plasmasphere; auroral kilometric radiation(AKR) and auroral myriametric radiation(AMR); type III solar burst; and $2f_p$ EM radiation in the upstream region. Most of these waves are well-known waves in plasmas, but show very interesting features in geospace. GEOTAIL PWI added further detailed information of these waves to what has been known by the previous spacecraft research, especially of phase (or waveform) information for waves below 4kHz. Among them, the interesting and yet puzzling features are the short life and monochromaticity of the MNB and of the whistler packet in the magnetosheath and solar wind. They will provide an open problem for future study. Unfortunately the maximum frequency which the waveform capture (WFC) receiver onboard GEOTAIL can cover is only 4kHz, which prevents us from performing the phase study of the continuum radiation and the $2f_p$ emissions. Finally, the statistical study of the AKR is a result of full utilization of the long life of the GEOTAIL spacecraft.

Chapter 4 dealt with the electrostatic (ES) waves observed by GEOTAIL. As the frequency range observable by the PWI receivers is much higher than the ion characteristic frequencies on most of the orbits, the waves discussed are electron waves for most of the cases. The ES waves discussed in Chapter 4 are; Langmuir waves; electron cyclotron harmonics (ECH) waves; ion acoustic (or ion acoustic-like) waves; broadband electrostatic noise (BEN); and narrowband electrostatic noise (NEN). Concerning the BEN, a new fact of the solitary and bipolar pulse wave form of the BEN's in the PSBL was discovered. It is also pointed out that the similar solitary and bipolar pulse wave forms are also found in the ES waves in the solar wind, the bow shock and the magnetosheath.

Chapter 5 was prepared to regard the plasma waves as a tool or a measure for region identification of geospace. First wave features were described for different regions in geospace; the solar wind and the upstream region; bow shock; dayside magnetopause; magnetosheath; magnetopause boundary; tail lobe; plasma sheet and plasma sheet boundary layer(PSBL). Then the knowledge of the characteristic features inherent to these individual regions in geospace was used to study the geospace global structure. Such attempt was carried out for the first time by GEOTAIL database and turns out effective to provide a statistical view of plasma structure in geospace. It is well recognized that the geospace is dynamically changing occasionally deviating from the time-averaged stationary model. The statistical mapping of the electron density and the MNB intensity clearly showed that the distant tail is not to be described by the static and stationary model of the tail but vividly in motion bringing the magnetosheath into the central portion around the Sun-Earth line in the deep tail.

Chapter 6 described wave phenomena associated with transient geophysical or solar phenomena, such as CME, CIR and plasmoid passage. Though the data on these transient occasions are limited in number, they have provided a new and interesting problem which researchers can tackle in the future.

Chapter 7 describes the summary of our efforts to explain the newly found ESW nature of the BEN via particle model computer experiments. A series of computer experiments has confirmed the interpretation of the observed bipolar pulse waveforms of BEN in terms of particle trapping and subsequent formation of the solitary BGK potentials. It has been stressed that the combination of theory and observation assisted by computer experiments (to supplement the inability of dealing the highly nonlinear problems by human brains) is a powerful method to understand the complex plasma processes taking place in geospace. The discovery of the ESW nature of the BEN by GEOTAIL and its interpretation by the nonlinear theory/simulation are one such successful case.

The geospace plasmas beyond the top of the ionosphere are almost collisionless. Therefore the dissipation in various physical processes cannot be attributed to inter-particle collisions. Instead of collisions, the particle scattering and diffusion by waves are supposed to play the role of effective dissipation. This has been recognized for long time but the observational proof of wave contribution to large-scale geophysical phenomena such as collisionless shocks and magnetic reconnection has not been found explicitly. Not to mention, plasma waves are found in the vicinity of the bow shock as shown in previous chapters, and waves such as ESW and Langmuir waves are found in the PSBL which are thought to be connected to the X-point of the reconnection [*Scarf et al.*, 1984]. However, little has been investigated on the action of plasma waves onto the macroscopic instabilities or interactions between the different plasmas (such as between the solar wind and the magnetospheric plasma). This is partly due to the lack of the detailed information of plasma waves in a short time scale to disclose the wave phase or waveform for better understanding of the nonlinear nature of these waves. The feedback or role of plasma waves to macroscopic instabilities through resistivity or diffusion will need substantial intensity of plasma waves. Such large amplitude waves are not free from the nonlinear process. The GEOTAIL PWI made a tiny but quantum step to the quantitative understanding of such nonlinear processes, especially by the WFC data. However, it only opened the door to the steep road to the final destination. It is still very difficult to understand the above-mentioned problems because at the present moment, the time resolution of particle measurement on its velocity distribution function is too coarse to be compared with the fast phenomena such as those disclosed by WFC data. However, we are encouraged to study such problem by looking at the rich (but very fast) variation of plasma waves in those regions as bow shocks and its vicinity. By performing a proper and sound computer experiment, we could step forward with those valuable data and, in return, can predict the expected particle distribution. Even with the lack of the fast resolution of the velocity distribution function $f(v)$, we could compare the averaged velocity distribution function from the simulation based on the wave observation with the observed slow-time scale $f(v)$. Thus much work should be, and could be done with the existing GEOTAIL data. Finally we hope this article can contribute to the space community and provide a guide to the plasma wave paradise in geospace.

Acknowledgments. The authors would like to express their sincere thanks to A. Nishida for his invitation to this volume, and for his insight of suggesting the participation of the plasma wave team on the GEOTAIL spacecraft. Without his leadership of the spacecraft, the current successful scientific results would not have been obtained. We also would like to thank other GEOTAIL Team Members for their collaboration and kind data provision to the plasma wave team. The contribution by the members of the Plasma Wave Team cannot be acknowledged fully. We thank them for their unselfish contribution, from the preparation phase of our instrument to their scientific contribution through collaborations and the resultant publications and thesis works. Special thanks in this regard are due to R. R. Anderson, K. Hashimoto, M. Tsutsui, I. Kimura, H. Usui, S. Yagitani, Y-L. Zhang, T. Murata and Y. Kasaba. Help and contributions from the graduate course students of Kyoto University and Kanazawa University were also substantial in their handling the enormous raw data set and in the development of the data analysis software. Finally we would like to express our thanks to our students for their help of producing figures in this paper, especially to S. Kudo, T. Miyake, D. Morikawa, N. Miki, and other students at Kyoto University.

REFERENCES

Acuña, M. H., K. W. Ogilvie, D. N. Baker, S. A. Curtis, D. H. Fairfield and W. H. Mish, The global geospace science program and its investigations, *Space Science Reviews*, *71*, 5, 1995. Akhiezer, A. I., Plasma electrodynamics, Pergamon Press Ltd., New York, 1975.

Akimoto, K. and N. Omidi, The generation of broadband electrostatic noise by an ion beam in the magnetotail, *Geophys. Res. Lett.*, *13*, 97, 1986.

Akimoto, K., Y. Omura and H. Matsumoto, Rapid generation of Langmuir wave packets during electron beam-plasma instabilities, *Phys. of Plasmas*, *3*, 2559, 1996.

Anderson, K. A., Measurements of the bow shock particles far upstream from the Earth, *J. Geophys. Res.*, *86,* 4445, 1981.

Ashour-Abdalla, M., C. F. Kennel, and W. Liversey, A parametric study of electron multiharmonic instabilities in the magnetosphere, *J. Geophys. Res.*, *84*, 6540, 1979.

Ashour-Abdalla, M. and H. Okuda, Theory and simulations of broadband electrostatic noise in the geomagnetic tail, *J. Geophys. Res.*, *91*, 6833, 1986.

Ashour-Abdalla, M., R. L. Richard, and F. V. Coroniti, A simulation study of electron hole distributions, *Geophys. Res. Lett.*, *16*, 1137, 1989.

Belmont, G., D. Fontaine, and P. Canu, Are equatorial electron cyclotron waves responsible for diffuse auroral precipitation ?, *J. Geophys. Res.*, *88*, 9163, 1983.

Bernstein, I. B., Waves in a Plasma in a Magnetic Field, *Phys. Rev.*, *109*, 10, 1958.

Borda de Agua., L., Y. Omura, H. Matsumoto, and A. L. Brinca, Competing processes of plasma wave instabilities driven by an anisotropic electron beam: linear results and 2-D particle simulations. *J. Geophys. Res.*, *101*, 15475, 1996.

Brown. L. W, The galactic radio spectrum between 130kHz and 2600kHz, *Astrophys. J.*, *180*, 359, 1973.

Budden, K. G., The propagation of radio waves: the theory of radio waves of low power in the ionosphere and magnetosphere 1. Radio wave propagation, Cambridge University Press, 1985.

Burinskaya, T. M. and C. -V. Meister, Contribution of the ion-beam acoustic instability to the generation of broadband electrostatic noise in the earth's magnetotail, *Planet. Space Sci.*, *37*, 145, 1989.

Burinskaya, T. M. and C. -V. Meister, Contribution of the ion-beam acoustic instability to the generation of broadband electrostatic noise -Three dimensional quasilinear approach-, *Planet. Space Sci.*, *38*, 695, 1990.

Calvert, W., The auroral plasma cavity, *Geophys. Res. Lett.*, *8*, 919, 1981.

Coroniti, F. V., and M. Ashour-Abdalla, Electron velocity space hole modes and narrowband electrostatic noise in the distant tail, *Geophys. Res. Lett.*, *16*, 747, 1989.

Coroniti, F. V., E. W. Greenstadt, B. T. Tsurutani, E. J. Smith, R. D. Zwickl, and J. T. Gosling, Plasma waves in the distant geomagnetic tail: ISEE 3, *J. Geophys. Res.*, *95*, 20,977, 1990.

Dunckel, N., and R. A. Helliwell, Whistler-mode emissions on the OGO 1 satellite, *J. Geophys. Res.*, *74*, 6371, 1969.

Data catalogue 1995, Division of Integrated studies for substorms, STE laboratory, Japan.

Dusenbery, P. B. and L. R. Lyons, The generation of electrostatic noise in the plasma sheet boundary layer, *J. Geophys. Res.*, *90*, 10935, 1985.

Dusenbery, P. B., Generation of broadband noise in the magnetotail by the beam acoustic instability, *J. Geophys. Res.*, *91*, 12005, 1986.

Eckersley, T. L., Note on musical atmospheric disturbances, *Phil. Mag.*, *49*, 1250, 1925.

Ellis, G. R. A., On the propagation of whistling atmospherics, *J. A. T. P.*, *8*, 338, 1956.

Filbert, P. C. and P. J. Kellogg, Observations of low frequency radio emissions in the Earth's magnetosphere, *J. Geophys. Res.*, *94*, 8867, 1989.

Fredricks, R. W., and F. L. Scarf, Recent studies of magnetospheric electric field emissions above the electron gyrofrequency, *J. Geophys. Res.*, *78*, 310, 1973.

Fried, B. D., and S. D. Conte, *The Plasma Dispersion Function*, Academic, San Diego, Calif., 1961.

Gallagher, D. L. and D. A. Gurnett, Auroral kilometric radiation: Time-averaged source location, *J. Geophys. Res.*, *84*, 6501, 1979.

Ginzberg, V. L. and V. V. Zheleznyakov, On the possible mechanism of sporadic radio emission (radiation in an isotropic plasma), *Soviet Astron. A. J.*, *2*, 653, 1958.

Gosling, J. T., Coronal mass ejections and magnetic flux ropes in interplanetary space, in *Physics of Magnetic Flux Ropes, Geophys. Monogr. 58*, edited by C. T. Russell, E. R. Priest, and L. C. Lee, AGU, Washington D.C., 343, 1990.

Gosling, J. T., et al., Mass ejections from the Sun, A view from Skylab, *J. Geophys. Res.*, *79*, 4581, 1974.

Gosling, J., T., A. J. Hundhausen, and S. J. Bame, Solar wind stream evolution at large heliospheric distances: Experimental demonstration and the test of a model, *J. Geophys. Res.*, 81, 2111, 1976.

Gough, M. P., Nonthermal continuum emissions associated with electron injections : Remote plasmapause sounding, *Plant. Space Sci.*, *30*, 657, 1982.

Grabbe, C. L., Auroral kilometric radiation: A theoretical review, *Reviews of Geophysics and space physics*, *19*, 627, 1981.

Grabbe, C. L., Numerical study of the spectrum of broadband electrostatic noise in the magnetotail, *J. Geophys. Res.*, *92*, 1185, 1987.

Grabbe, C. L., Wave propagation effects of broadband electrostatic noise in the magnetotail, *J. Geophys. Res.*, *94*, 17299, 1989.

Green, J. L., D. A. Gurnett and S. D. Shawhan, The angular distribution of auroral kilometric radiation, *J. Geophys. Res.*, *82*, 1825, 1977.

Greenstadt, E. W., G. K. Crawford, R. J. Strangeway, S. L. Moses, and F. V. Coroniti, Spatial distribution of electron plasma oscillations in the Earth's foreshock at ISEE 3, *J. Geophys. Res.*, *100*, 19933, 1995.

Gurnett, D. A., and R. R. Shaw, Electromagnetic radiation trapped in the magnetosphere above the plasma frequency, *J. Geophys. Res.*, *78*, 8136, 1973.

Gurnett, D. A., The Earth as a radio source: Terrestrial kilometric radiation, *J. Geophys. Res., 79*, 4227, 1974.

Gurnett, D. A., The earth as a radio source: The nonthermal continuum, *J. Geophys. Res.*, *80*, 2751, 1975.

Gurnett, D. A., L. A. Frank, and R. P. Lepping, Plasma waves in the distant magnetotail, *J. Geophys. Res.*, *81*, 6059, 1976.

Gurnett, D. A., and R. R. Anderson, Plasma wave electric fields in the solar wind: initial results from Helios 1, *J. Geophys. Res.*, *82*, 632, 1977.

Gurnett, D. A., R. R. Anderson, F. L. Scarf and W. S. Kurth, The heliocentric radial variation of plasma oscillations associated with type III radio bursts, *J. Geophys. Res.*. *83*, 4147, 1978.

Gurnett, D. A., W. S. Kurth, and F. L. Scarf, The structure of the Jovian magnetotail from plasma wave observations, *Geophys. Res. Lett.*, *7*, 53, 1980.

Gurnett, D. A., Plasma waves and instabilities, in *Collisionless shocks, Geophysical Monograph 35,* ed. by B. T. Tsurutani and R. G. Stone, 207, 1985.

Gurnett, D. A., J. E. Maggs, D. L. Gallagher, W. S. Kurth, and F. L. Scarf, Parametric interaction and spatial collapse of beam-driven Langmuir waves in the solar wind, *J. Geophys. Res.*, *86*, 8833, 1981.

Gurnett, D. A., G. B. Hospodarsky, W. S. Kurth, D. J. Williams, and S. J. Bolton, Fine structure of Langmuir waves produced by a solar electron event, *J. Geophys. Res.*, *98*, 5631, 1993.

Hammond, C. M., G. K. Crawford, J. T. Gosling, H. Kojima, J. L. Phillips, H. Matsumoto, A. Balogh, L. A. Frank, S. Kokubun and T. Yamamoto, Latitudinal structure of a coronal mass ejection iferred from Ulysses and Geotail observations, *Geophys. Res. Lett.*, *22*, 1169, 1995.

Hashimoto, K., H. Matsumoto, H. Kojima, T. Murata, I. Nagano, T. Okada, K. Tsuruda and T. Iyemori, Auroral myriametric radiation observed by GEOTAIL, *Geophys. Res. Lett.*, *21*, 2927, 1994.

Helliwell, R. A., J. H. Crary, J. H. Pope, and R. L. Smith, The "nose" whistler - a new high latitude phenomenon, *J. Geophys. Res.*, *61*, 139, 1956.

Helliwell, R. A., Whistlers and related ionospheric phenomena, *Stanford University Press*, 1965.

Hilgers, A., H. De, Feraudy, and D. Le, Queau, Measurement of the direction of the auroral kilometric radiation electric field inside the sources with the viking satellite, *J. Geophys, Res.*, *97*, 8381, 1992.

Hoang, S., J. Fainberg, J. -L. Steinberg, R. G. Stone, and R. H. Zwickl, The $2f_p$ circumterrestrial radio radiation as seen from ISEE3, *J. Geophys. Res.*, *86*, 4531, 1981.

Horne, R. B., Path-integrated growth of electrostatic waves: The generation of terrestrial myriametric radiation, *J. Geophys. Res.*, *94*, 8895, 1989.

Horne, R. B., Narrow-band structure and amplitude of terrestrial myriametric radiation, *J. Geophys. Res.*, *95*, 3925, 1990.

Holzer, Robert, E., Malcolm, G. Mcleod, and Edward, J. Smith, Preliminary results from the Ogo 1 search coil magnetometer: boundary positions and magnetic noise spectra, *J. Geophys. Res.*, *71*, 1481, 1966.

Howe, H. C., and J. H. Binsack, Explorer 33 and 35 plasma observations of magnetosheath flow, *J. Geophys. Res.*, *77*, 3334, 1972.

Huba, J. D., N. T. Gladd, and K. Papadopoulos, Lower-hybrid-drift wave turbulence in the distant magnetotail, *J. Geophys, Res.*, *83*, 5217, 1978.

Hubbard, R. F. and T. J. Birmingham, Electrostatic emissions between electron gyroharmonics in the outer magnetosphere, *J. Geophys. Res.*, *83*, 4837, 1978.

Isted, G. A., and G. Millington, The "dawn chorus" in radio observation, *Nature*, *180*, 716, 1957.

Jones, D., Source of terrestrial nonthermal radiation, *Nature*, *260*, 686, 1976.

Jones, D., Latitudinal beaming of planetary radio emissions, *Nature*, *288*, 225, 1980.

Jones, D., Beaming of terrestrial myriametric radiation, *Adv. Space. Res.*, *1*, 373, 1981.

Jones, D., First remote sensing of the plasmapause by terrestrial myriametric radiation, *Nature*, *294*, 728, 1981.

Jones, D., Terrestrial myriametric radiation from the earth's plasmapause, *Planet. Space. Sci.*, *30*, 399, 1982.

Kaiser, M. L. and J. K. Alexander, Relationship between auroral substorms and the occurrence of terrestrial kilometric radiation, *Geophys. Res. Lett.*, *82*, 5283, 1977.

Kaiser, M. L., M. D. Desch, W. M. Farrell, J.-L. Steinberg, and M. J. Reiner, LF band terrestrial radio bursts observed by WIND/WAVES, *Geophys. Res. Lett.*, *23*, 1287, 1996.

Kasaba, Y., Study of radio waves in geospace via spacecraft observations and numerical simulations, Ph. D. thesis (supervised by H. Matsumoto), Kyoto University, 1997a.

Kasaba, Y., H. Matsumoto, and R. R. Anderson, GEOTAIL observation of $2f_p$ emission around the terrestrial foreshock region, *Adv. Space Res.*, *20*, 699, 1997b.

Kasaba, Y., H. Matsumoto, K. Hashimoto, R. R. Anderson, J.-L. Bougeret, M. L. Kaiser, X. Y. Wu, and I. Nagano, Remote sensing of the plasmapause during substorm: GEOTAIL observation of the continuum enhancement, submitted to *J. Geophys. Res.*, 1997c.

Kawano, H., A. Nishida, M. Fujimoto, T.Mukai, S. Kokubun, T. Yamamoto, T. Terasawa, M. Hirahara, Y. Saito, S. Machida, K. Yumoto, H. Matsumoto and T. Murata, A quasi-stagnant plasmoid observed with Geotail on October 15, 1993, *J. Geomag. Geoelectr.*, *48*, 525, 1996.

Kellogg Paul J., S. J. Monson, K. Goetz, R. L. Howard, J.-L. Bougeret, and M. L. Kaiser, Early wind observations of bow shock and foreshock waves, *Geophys. Res. Lett.*, *23*, 1243, 1996.

Kennel, C. F., F. V. Scarf, R. W. Fredricks, J. H. Mcgehee, and F. V. Coroniti, VLF electric field observations in the magnetosphere, *J. Geophys. Res.*, *75*, 6136, 1970.

Kennel, C. F., and M. Ashour-Abdalla, Electrostatic waves and the strong diffusion of magnetospheric electrons, in *Magnetospheric Plasma Physics*, edited by A. Nishida, pp.245, Center for Academic Publications, Tokyo, Japan, 1982.

Kojima, H., H. Matsumoto, T. Miyatake, I. Nagano, A. Fujita, L. A. Frank, T. Mukai, W. R. Paterson, Y. Saito, S. Machida and R. R. Anderson, Radiation between electrostatic solitary waves and hot plasma flow in the plasma sheet boundary layer: GEOTAIL observation, *Geophys. Res. Lett.*, *21*, 2919, 1994.

Kojima, H., H. Matsumoto, S. Chikuba, S. Horiyama, M. Ashour-Abdalla and R. R. Anderson, GEOTAIL waveform observations of broadband/narrowband electrostatic noise in the distant tail, *J. Geophys. Res.*, *102*, 14439, 1997a.

Kojima, H., H. Furuya, H. Usui, and H. Matsumoto, Modulated electron plasma waves observed in the tail lobe: GEOTAIL waveform observations, *Geophys. Res. Lett.*, *24*, 3049, 1997.

Koons, H. C., and J. F. Fennell, Fine structure in electrostatic emission bands between electron gyrofrequency harmonics, *J. Geophys. Res.*, *89*, 3015, 1984.

Kundu, M. R., Solar Radio Astronomy, Interscience Publishers, New York, 1965.

Kurth, W. S., D. A. Gurnett, and R. R. Anderson, Escaping nonthermal continuum radiation, *J. Geophys. Res.*, *86*, 5519, 1981.

Kurth, W. S., Continuum radiation in planetary magnetospheres, Proceedings of the 3rd International Workshop on Radio Emissions from Planetary Magnetospheres in Graz, Austria, September 2-4, 1991.

Lacombe, C., C. C. Harvey, S. Hoang, A. Mangeney, J. -L. Steinberg, and D. Burgess, ISEE observations of emission at twice the solar wind plasma frequency, *Ann. Geophys.*, *6*, 113, 1988.

Lyons, L. R., Electron diffusion driven by magnetospheric electrostati waves, *J. Geophys. Res.*, *79*, 575, 1974.

Machida, S., T. Mukai, Y. Saito, T. Obara, T. Yamamoto, A. Nishida, M. Hirahara, T. Terasawa, and S. Kokubun, GEOTAIL low energy particle and magnetic field observations of a plasmoid at X_{GSM}=$-142R_E$, *Geophys. Res. Lett.*, *21*, 2995, 1994.

Matsumoto, H., Theoretical studies on whistler mode wave-particle interactions in the magnetospheric plasma, Ph. D. Thesis, Kyoto University, 1972.

Matsumoto, H., K. Hashimoto, and I. Kimura, Two types of phase bunching in the whistler mode wave-particle interaction, *J. Geomag. Geoelectr.*, *26*, 55, 1974

Matsumoto, H., I. Nagano, R. R. Anderson, H. Kojima,, K. Hashimoto, M. Tsutsui, T. Okada, I. Kimura, Y. Omura and M. Okada, Plasma wave observations with GEOTAIL spacecraft, *J. Geomag. Geoeletr.*, *46*, 59, 1994a.

Matsumoto, H., H. Kojima, T. Miyake, Y. Omura, M. Okada, I. Nagano and M. Tsutsui, Electrostatic Solitary Waves (ESW) in the magnetotail: BEN wave forms observed by GEOTAIL, *Geophys. Res. Lett.*, *21*, 2915, 1994b.

Matsumoto, H. and H. Usui, Intense bursts of electron cyclotron harmonic waves near the dayside magnetopause observed by Geotail, *Geophys. Res. Lett.*, *24*, 49, 1997.

Matsumoto, H., H. Kojima, Y. Kasaba, T. Miyake, R. R. Anderson and T. Mukai, Plasma waves in the upstream and bow shock regions observed by GEOTAIL, *Advanced Space Research*, *20*, 683, 1997.

Melrose, D. B., A theory for the nonthermal radio continua in the terrestrial and Jovian magnetospheres, *J. Geophys. Res.*, *86*, 30, 1981.

Mukai, T., M. Hirahara, S. Machida, Y. Saito, T. Terasawa, and A. Nishida, GEOTAIL observation of cold ion streams in the medium distance magnetotail lobe in the course of a substorm, *Geophys. Res. Lett.*, *21*, 1023, 1994.

Murata, T., Study of Magnetotail dynamics via Computer Experiments and Spacecraft Observations, Ph. D. Thesis (supervised by H. Matsumoto), Kyoto University, 1995.

Murata, T., H. Matsumoto, H. Kojima, A. Fujita, T. Nagai, T. Yamamoto and R. R. Anderson, Estimation of tail reconnection lines by AKR onsets and plasmoid entries observed with GEOTAIL spacecraft, *Geophys. Res. Lett.*, *22*, 1849, 1995.

Murata, T., H. Matsumoto, H. Kojima and T. Iyemori, Correlations of AKR index with Kp and Dst indices, *Proceedings of the NIPR Symposium on Upper Atomosphere Physics*, *10*, 64, 1997.

Nagano, I., S. Yagitani, H. Kojima, Y. Kakehi, T. Shiozaki, H. Matsumoto, K. Hashimoto, T.Okada, S. Kokubun and

T.Yamamoto, Wave form analysis of the continuum radiation observed by GEOTAIL, *Geophys. Res. Lett.*, *21*, 2911, 1994.

Nagano, I., S. Yagitani, H. Kojima and H. Matsumoto, Analysis of wave normal and Poynting vectors of the chorus emissions observed by GEOTAIL, *J. Geomag. Geoelectr.*, *48*, 299, 1996.

Nagano, I., X.-Y. Wu, S. Yagitani, K. Miyamura, and H. Matsumoto, Unusual whistler-like ELF wave near the magnetopause: GEOTAIL observation and ray-tracing modeling, *J. Geophys. Res.*, *in press*, 1998.

Nunn, D., A Theory of VLF Emissions, *Planet. Space Sci.*, *19*, 1141 , 1971

Nunn, D., A self-consistent theory of triggeren VLF emissions , *Planet. Space Sci.*, *22*, 349, 1974.

Nunn, D., Y. Omura, H. Matsumoto, I. Nagano and S. Yagitani, The numerical simulation of VLF chorus and discrete emissions observed on the Geotail satellite using a Vlasov code, *J. Geophys. Res.*, *102*, 27083, 1997.

Omura, Y., and H. Matsumoto, Computer simulations of basic processes of coherent whistler wave-particle interaction in the magnetosphere, *J. Geophys. Res.*, *87*, 4435, 1982.

Omura, Y., H. Kojima and H. Matsumoto, Computer simulation of electrostatic solitary waves: A nonlinear model of broadband electrostatic noise, *Geophys. Res. Lett.*, *21*, 2923, 1994.

Omura, Y., H. Matsumoto, T. Miyake and H. Kojima, Electron beam instabilities as generation mechanism of electrostatic solitary waves in the magnetotail, *J.Geophys. Res.*, *101*, 2685, 1996.

Onsager, T. G., M. F. Thomsen, R. C. Elphic, J. T. Gosling, R. R. Anderson, and G. Kettmann, Electron generation of electrostatic waves in the plasma sheet boundary layer, *J. Geophys. Res.*, *15*, 15509, 1993.

Oya, H., Conversion of electrostatic plasma waves into electromagnetic waves: numerical calculation of the dispersion relation for all wavelengths, *Radio Sci.*, *6*, 1131, 1971.

Papadopuolos, K., M. L. Goldstein and R. A. Smith, Stabilization of electron streams in type III solar radio bursts, *Astrophys. J.*, *190*, 175, 1974.

Pope, J. H., Diurnal variation in the occurrence of "dawn chorus", *Nature*, *180*, 433, 1957.

Pope, J. H., A high-latitude investigation of the natural very-low-frequency electromagnetic radiation known as chorus, *J. Geophys. Res.*, *68*, 83, 1963.

Reiner, M. J., M. L. Kaiser, J. Fainberg, M. D. Desch and R. G. Stone, 2fp radio emission from the vicinity of the Earth's foreshock: WIND observations, *Geophys. Res. Lett.*, *23*, 1247, 1996.

Reiner, M. J., Y. Kasaba, M. L. Kaiser, H. Matsumoto, I. Nagano and J.-L. Bougeret, Terrestrial 2fp radio source location determined from WIND/GEOTAIL triangulation, *Geophys. Res. Lett.*, *24*, 919, 1997.

Rodriguez, P., Ion waves associated with solar and beam-plasma interaction, *J. Geophys. Res.*, *86*, 1279, 1981.

Roeder, J. L., and H. C. Koons, A survey of electron cyclotron waves in the magnetosphere and the diffuse aurora electron precipitation, *J. Geophys. Res.*, *94*, 2529, 1989.

Russell, Christopher, T., Noise in the geomagnetic tail, *Planet. Space Sci.*, *20*, 1541, 1972.

Russell, C. T., and R. L. McPherron, The magnetotail and substorms, *Space Sci. Rev.*, *15*, 205, 1973.

Scarf, F. L., R. W. Fredricks, L. A. Frank, C. T. Russell, P. J. Coleman, Jr., and M. Neugebauer, Direct correlation of large amplitude waves with suprathermal protons in the upstream solar wind, *J. Geophys. Res.*, *75*, 7316, 1970.

Scarf, F. L., F. V. Coroniti, C. F. Kennel, E. J. Smith, J. A. Slavin, B. T. Tsurutani, S. J. Bame, and W. C. Feldman, Plasma wave spectra near slow mode shocks in the distant tail, *Geophys. Res. Lett.*, *11*, 1050, 1984.

Smith, R. L., and D. L. Carpenter, Extension of nose whistler analysis, *J. Geophys. Res.*, *66*, 2582, 1961.

Smith, R. L., and J. J. Angerami, Magnetospheric properties deduced from OGO 1 observations of ducted and nonducted whistlers, *J. Geophys. Res.*, *73*, 1, 1968.

Smith, E. J., R. E. Holzer, and C.T. Russell, Magnetic emissions in the magnetosheath at frequencies near 100Hz, *J. Geophys, Res.*, *74*, 3027, 1969.

Smith, E. J., A. M. A. Frandsen, and R. E. Holzer, Lion roars in the magnetosheath (abstract), *Eos Trans*, AGU, *52*, 903, 1971.

Smith, E.J., and B.T. Tsurutani, Magnetosheath lion roars, *J. Geophys. Res.*, *81*, 2261, 1976.

Stasiewicz, K., B. Holback, V. Krasnoselskikh, M. Boehm, R. Bostrom, and P. M. Kintner, Parametric instabilities of Langmuir waves observed by Freja, *J. Geophys. Res.*, *101*, 21515, 1996.

Stix, T. H., The Theory of Plasma Waves, McGraw-Hill, New York, 1962.

Stix, T. H., Waves in plasmas, American Institute of Physics, 1992.

Storey, L. R. O., An investigation of whistling atmospheric, *Phil. Tran. Roy. Soc.*, (London), A, *246*, 113, 1953.

Sturrock, P. A., Special characteristics of type II solar radio burst, *Nature*, *192*, 58, 1961.

Summers, D., and R. M. Thorne, A New Tool for Analyzing Microinstabillities in Space Plasmas Modeled by a Generalized Lorentzian *J. Geophys. Res.*, *97*, 16827, 1992.

Summers, D., R. M. Thone, and H. Matsumoto, Evaluation of the modified plasma dispersion function for half-integral indecies, *Phys Plasmas*, *3*, 2496, 1996.

Takahashi, K., S. Kokubun, H. Matsuoka, S. Shiokawa and K. Yumoto, M. Nakamura, H. Kawano, T. Yamamoto, A. Matsuoka, K. Tsuruda, H. Hayakawa, H.Kojima, H. Matsumoto, GEOTAIL observation of magnetosonic Pc 3 waves in the dayside magnetosphere, *Geophys. Res. Lett.*, *21*, 2899, 1994.

Tsurutani, B. T., and G. J. Smith, Two types of magetospheric ELF chorus and their substorm dependences, *J. Geophys. Res.*, *82*, 5112, 1977.

Tsurutani B.T., E.J. Smith,R.R. Anderson, K.W. Oglive, J.D. Scudder, D.N. Baker, and S.J. Bame, Lion roars and nonoscillatory drift mirror waves in the magnetosheath, *J. Geophys. Res.*, *87*, 6060, 1982.

Tsutsui, M., R. J. Strangeway, B. T. Tsurutani, H. Matsumoto, J. L. Phillips, and M. Ashour-Abdalla, Wave mode identification of electrostatic noise observed with ISEE 3 in the deep tail boundary layer, *J. Geophys. Res.*, *96*, 1991.

Tsutsui, M., H. Kojima, I. Nagano, H. Sato, T. Okada, H. Matsumoto, T. Mukai and M. Kawaguchi, Magnetic radiations form harness wires of spacecraft, *IEICE Trans. Commun.*, *E75-B*, 174, 1992.

Tsutsui, M., I. Nagano, H. Kojima, K. Hashimoto, H. Matsumoto, S. Yagitani, and T. Okada, Measurements and analysis of antenna impedance aboard the Geotail spacecraft, *Radio Science*, 32, 1101, 1997.

Usui, H., J. Koizumi, and H. Matsumoto, Statistical study on electron cyclotron harmonic waves observed in the dayside magnetosphere, *Adv. Space Res.*, *20*, 857, 1997.

Vasyliunas, V. H., A survey of low energy electrons in the evening sectors of the magnetosphere with OGO-1 and OGO-3, *J. Geophys. Res.*, *73*, 2839, 1968.

Watts, J. M., Audio frequency electromagnetic hiss recorded at Boulder in 1956, *Geophys. Purae Appl.*, *37*, 169, 1957.

Watts, J. M., An observation of audio-frequency electromagnetic noise during a period of solar disturbance, *J. Geophys. Res.*, *62*, 199, 1957.

Wu, C. S. and L. C. Lee, A theory of the terrestrial kilometric radiation, *Astrophys. J.*, *230*, 621, 1979.

Yagitani, S., I. Nagano, H. Matsumoto, Y. Omura, W.R. Paterson, L. R. Frank and R. R. Anderson, Generation and propagation of chorus emissions observed by GEOTAIL in the dayside magnetosphere, *Proc. of ISAP '96*, Japan, 717, 1996.

Yagitani, S., I. Nagano, H. Takano, H. Matsumoto, Y. Omura, W. R. Paterson, L. A. Frank, and R. R. Anderson, Generation and propagation of chorus emissions in the magnetosphere, *Proceedings of the Fifth International School/symposium for Space Simulations (ISSS-5)*, 68, 1997.

Yamamoto T., K. Shiokawa, and S. Kokubun, Magnetic field structures of the magnetotail as observed by Geotail, *Geophys. Res. Lett.*, *21*, 2875, 1994.

Zhang, Y., H. Matsumoto and H. Kojima, Lion Roars in the Magnetosheath: the GEOTAIL Observations, *accepted for the publication in J. Geophys. Res.*, 1998.

Zhang, Y. L., H. Matsumoto and Y. Omura, Linear and nonlinear interactions of an electron beam with oblique whistler and electrostatic waves in Magnetosphere, *J. Geophys. Res.*, *98*, 21353 , 1993.

H. Kojima, H. Matsumoto, and Y. Omura Radio Atmospheric Science Center, Kyoto University, Uji, Kyoto 611-0011, Japan. (e-mail: matsumot@kurasc.kyoto-u.ac.jp)

I. Nagano, Department of Electrical and Computer Engineering, Kanazawa University, Kodatsuno, Kanazawa 920-0942, Japan.

Multiscale Magnetic Structure of the Distant Tail: Self-Consistent Fractal Approach

Lev M. Zelenyi and Alexander V. Milovanov

Space Research Institute, Moscow, Russia

Gaetano Zimbardo

Dipartimento di Fisica, Università della Calabria, Italy

In this paper, we attempt to describe some general characteristics and the principal structural properties of the magnetic field and plasma turbulence in the distant Earth's magnetotail. We are considering the development of the turbulent magnetic field structures in the tail as an inherent feature of the basic physical processes governing the self-organization of the electric currents and magnetic fields in the distant current sheet. In the framework of the present study, we propose a nontraditional analysis of the structural properties of the turbulence in the tail, based on the general concepts of the fractal geometry. The accent on the fractal geometry made in our study is due to the following important assumption, namely that this turbulence is not of the "smoothened", diffusive type, but has the considerable fine-scale fractal structure in a wide range of spatial scales self-consistently determined by the main dynamical processes in the distant current sheet. We argued that these spatial scales are comparable to the effective Larmor radii of ions having the characteristic energies of $\sim$ 1 keV and migrating through the highly structured turbulent magnetic fields self-consistently generated by the associated cross-tail currents. The key point of our approach is to consider the topology of the (ion) currents across the tail as a highly branched, very complicated percolating network, which maintains, self-consistently, the magnetic field fluctuations in the current sheet. It is shown that the implied percolating structure of the ion currents determines the main characteristics of the Fourier power spectra of the magnetic turbulence in the sheet, which, in turn, affects the plasma particle dynamics itself. We assume that the behavior of the Fourier power spectra in the frequency domain is controlled by the spatial fractal distribution of the turbulent magnetic structures in the current sheet that move away from the Earth with the decelerated solar

New Perspectives on the Earth's Magnetotail
Geophysical Monograph 105

wind velocity. Making use of the implied fractal geometry of the turbulent structures, we could calculate the slopes of the power spectra of the turbulence in the relevant frequency ranges, and compare the theoretical results of our fractal model with the available satellite magnetic turbulence data. In particular, we show that the basic conclusions of our theoretical treatment appear to be in good agreement with the recently reported observational findings from the Geotail survey of plasmas and magnetic fields in the distant Earth's magnetotail.

1. INTRODUCTION

The first direct spacecraft exploration of the distant areas of the Earth's magnetotail was performed in 1966 - 1968 with the Pioneer 7 and 8 spacecrafts at the distances of $\sim 5{\cdot}10^2 - 10^3$ Earth's radii. Much of the attention in the Pioneer plasma experiments had been concentrated on the magnetic field measurements, yielding important information on the magnetotail structural properties [see, e.g., *Ness et al.*, 1967; *Walker et al.*, 1975; *Villante*, 1976; and the references therein]. Already the very existence of a well-defined neutral sheet still preserving most of its near-Earth characteristics at such large distances from the Earth was fairly regarded as a remarkable observational finding. The magnetotail structures customarily detected closer to the Earth (i.e., two lobes separated by a plasma sheet and embedded neutral sheet) were also identifiable in the distant tail data from ISEE 3 spacecraft [see, e.g., *Tsurutani et al.*, 1984]. In the meanwhile, from a series of the magnetic field and plasma measurements it was also recognized that the magnetic structures in the distant tail are much more subject to variability, both spatial and temporal, than those in the near-Earth tail.

Substantial advances in the understanding of the physical properties of the distant tail were achieved recently with the Geotail spacecraft. The Geotail mission not only confirmed the permanent existence of the magnetic configurations in the distant tail with the topological features typical for the near-Earth tail, but, moreover, provided the scientific community with a rich variety of data showing the specific dynamical characteristics of plasmas and magnetic fields in the distant neutral sheet [see, e.g., *Nishida et al.*, 1994; *Yamamoto et al.*, 1994]. These specific characteristics were interpreted as the permanently existing magnetic turbulence, being an *inherent* property of the distant magnetotail.

One of the specific features of the magnetic turbulence in the distant tail as indicated by the Geotail measurements is the *power-law* behavior of the Fourier power density spectra of the magnetic field fluctuations, with the power exponents dependent on the particular frequency range that is analyzed [*Hoshino et al.*, 1994]. In fact, it was found that the power spectrum of the B_z component of the field (below we are using the standard GSM coordinate system) could be well fitted by the power-law function $P(f) \propto f^{-\alpha_1}$ with the slope α_1 ranging between ≈ 0.49 and ≈ 1.48 until the frequency f is less than the characteristic turnover frequency $f_* \approx 0.04$ Hz. As soon as f exceeds the value f_*, the best fit for the spectrum becomes more steep and has the power-law form $P(f) \propto f^{-\alpha_2}$ with the slope α_2 which is unambiguously larger than α_1 and lies between ≈ 1.78 and ≈ 2.43. (These results correspond to the Geotail measurements around one of its most distant apogees, i.e., approximately two hundred Earth's radii.) Thus, the Fourier power density spectrum for the B_z component of the field appears to be a combination of two power-law functions, with a "kink" around the characteristic frequency f_*. As mentioned by *Hoshino et al.* [1994], the possible origin for this kink spectrum may be the turbulent magnetic reconnection where the characteristic scale corresponding to the turnover frequency is given as the reciprocal of the most unstable tearing wavenumber in the current sheet.

The permanent existence of the magnetic field turbulence in the distant tail indicates that the dynamical processes therein may be governed by other physical mechanisms than that in the near-Earth tail. Indeed, the standard approach to understanding of the dynamics of plasmas and magnetic fields in the near-Earth tail implies a sufficient level of smoothness of the magnetic configurations in the plasma sheet, associated with relatively strong magnetic field in the lobes of the tail and in the vicinity of the current sheet. Such an approach is based on the customary treatment of the magnetotail topology as that of a (smoothly stretched) dipole, leading to importance of the quasi-adiabatic behavior of plasma particles [see, e.g., *Ashour-Abdalla et al.*, 1993]. It is, however, clear that this quasi-adiabatic approximation cannot be valid in the relatively distant areas of the tail, where the *local* perturbations in the magnetic field and cross-tail current density might play a decisive role due to the very elongated configuration of the tail at sufficiently large distances away from the Earth (see a schematic illustration on Fig. 1). One could, therefore, expect that these perturbations would strongly influence the magnetic field topology within the plasma sheet, leading to the possible appearance of *fine-scale* magnetic structures responsible for the basic properties of the turbulence in the distant magnetotail.

Thus, we are prepared to conclude that the magnetic and plasma turbulence in the *distant* magnetotail may be *qualitatively* different from the more convential quasilinear turbulence which deals with more smoothened, diffusive magnetic configurations. Rather, the turbulence in the distant tail would more likely involve highly structured, considerably "ragged" magnetic structures self-consistently maintained by the perturbations in the current density which, in turn, are associated with the *strong* nonlinearity of the system. A treatment of such a strongly nonlinear turbulence characterized by self-consistently developed fine-scale magnetic structures would then require exploitation of the mathematical tools that are customarily absent in more convential theoretical studies. In our previous paper [see *Milovanov et al.*, 1996], a self-consistent theoretical treatment of strongly nonlinear, highly structured magnetic turbulence in the distant tail was given in the framework of a basically geometric approach where the dynamical properties of the turbulence were described by using some nonconvential geometric quantities. These quantities were introduced in such a way that the considerable nonlinearity of the turbulence was implicitly present in its *geometry* which was then approximated by a specific construction with the topology of a *fractal* object. In this paper, we proceed along this direction by advocating the *fractal* approach to analyzing magnetic field and plasma turbulence in the distant Earth's magnetotail, with an accent on the intrinsic structural properties of the turbulence self-consistently related to the basic "physical" parameters.

This fractal approach is considered in the next three sections; the first two are devoted to a general discussion of the basic assumptions of this study and concern the estimates of the main physical parameters by order of magnitude, whereas the fourth section deals with a more advanced analysis of turbulent magnetic configurations in the distant magnetotail and actually involves the substantial use of the methods of the fractal geometry. We show that this geometric, fractal approach makes it possible, for instance, to deduce theoretically the slopes of the power spectra for the B_z component of the field in a wide range of frequencies and compare the results obtained with the available Geotail data. Such a comparison is given in the fifth section of this paper.

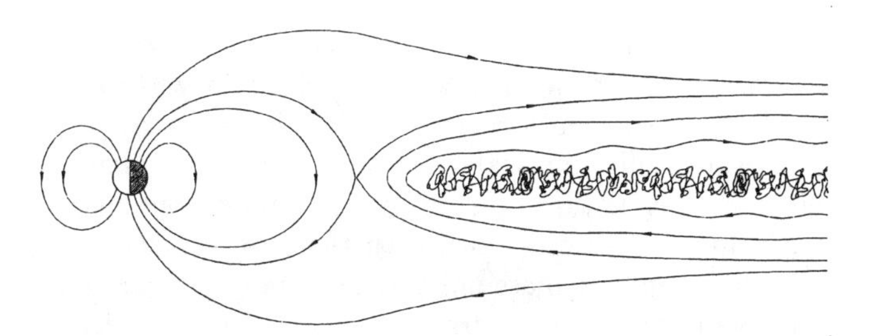

Figure 1. A schematic illustration for the permanently existing magnetic field turbulence in the distant Earth's magnetotail

2. PHYSICAL BACKGROUND AND BASIC EQUATIONS

We begin by remarking that our analysis applies to the stage when the principal geometric structure of the tail magnetic field has already been developed self-consistently, enabling one to bypass the problem of the *primary* formation of the tail. In other words, we imply that all principal transitional processes associated with the formation of the tail itself are already accomplished, and, therefore, the observed turbulence properties are just the *final* result of the self-organization of plasmas and magnetic fields. Consequently, we are considering the existence of the magnetic turbulence in the tail as its fundamental inherent property which controls the statistics of the principal magnetic field and plasma data detected from the satellite experiments.

Having bypassed the problem of the primary formation of the Earth's magnetotail, we still need to specify what would be the characteristic inherent properties of this "already developed" magnetic turbulence in the tail in the self-consistent regime. To meet this necessity, we make the following important assumption. Namely, we assume that the principal property of the self-consistently developed magnetic turbulence in the distant tail is the existence of a range of spatial scales where this turbulence is not the one of a "smeared", diffusive type, but is organized in "clumps" separated by "voids" of various spatial scales. In other words, we imply the existence of the considerable fine-scale structure of the magnetic turbulence in the tail at the range of spatial scales where self-organization of electric currents and magnetic fields necessarily plays a role in formation of the most important characteristics of the turbulence. Let these spatial scales range between some minimal value, a, and some maximum value, ξ. (Numerical estimates for a and ξ for the typical magnetotail plasma parameters will be obtained below. Note that the introduction of the scales a and ξ may have sense only if $\xi \gg a$; this inequality will be verified in what follows. We also mention that the implied existence of the considerable fine-scale structure of the turbulence serves as a signature of its essential strong nonlinearity, in qualitative difference with the more convential, quasilinear approach assuming the spatially "smeared", diffusive nature of the turbulent patterns.) It is clear that the minimal spatial scale, a, could be interpreted as the "smoothness" length of the turbulence, yielding the upper bound for the spatial scales where the turbulence in the tail can be still considered as "smeared" over space. In the meanwhile, the maximum spatial scale, ξ, must be associated with the global topology of the tail which, in its turn, might be self-consistently related to the tearing mode scales.

In fact, the self-consistently developed fine-scale magnetic structures in the plasma sheet could be strongly modulated by the excited tearing modes, having the energy reservoir in the large-scale field reversed geometry of the magnetotail [*Galeev and Zelenyi*, 1975]. (Such a modulation might be treated as an "external" influence on the system coming from these large scales.) Hence, one concludes that the spatial scale ξ may have the same order of magnitude as the (minimum) unstable tearing wavelength in the sheet which we shall denote by λ_*. The quantity λ_* is related to the characteristic current sheet thickness, L, through the widely known estimate $\lambda_* \sim 2\pi L$. (This relation immediately follows from the inequality $kL \leq 1$ which defines the unstable tearing wavenumbers, $k = 2\pi/\lambda$, for a current sheet of thickness L.) Thus, we get, by order of magnitude, $\xi \sim \lambda_* \sim 2\pi L$. (The possible interpretation of the quantity ξ in terms of a correlation length will be given in section 4.) Assuming $2L$ in the distant magnetotail be of the order of $8 \cdot 10^8$ cm (which could be reasonable for a rough estimate) we then find $\xi \sim 2.5 \cdot 10^9$ cm.

This estimate shows that at spatial scales *exceeding* $\xi \sim 2.5 \cdot 10^9$ cm, the topology of the magnetic field in the vicinity of the current sheet would be that of a standard "chain" of magnetic "islands" as discussed in detail by, e.g., *Galeev and Zelenyi* [1977]; whereas at the scales shorter than ξ but longer than the smothness length of the field, $a \ll \xi$, one could expect an evidence for highly structured magnetic configurations with "peaks" and "valleys" in the amplitude of the field, having a typical shape of a mountaneous landscape [see, e.g., *Mandelbrot*, 1983; *Feder*, 1988]. (For the sake of simplicity, we imply the traditional, two-dimensional tearing instability of the current sheet, related to the standard, one-dimensional "chain" structure [see, e.g., *Coppi et al.*, 1966].)

The introduction of the two characteristic spatial scales a and $\xi \gg a$ as implied by our theoretical model, immediately leads to a conclusion that the basic physical mechanisms governing the formation of the magnetic field structures in the distant plasma sheet, are *scale-length dependent.* Indeed, at the spatial scales exceeding $\sim \xi$, the magnetic topology of the plasma sheet is associated with the modulation of the magnetic field configurations by the excited tearing modes, resulting in the appearance of the characteristic "chain" structures. In the intermediate range of scales, i.e., in the range between a and ξ, the basic role in the development of the magnetic structure of the sheet is played by the *self-organization* of currents and magnetic fields in relation with the strong nonlinearity of the magnetic field turbulence and the highly structured, considerably "ragged" shape of the turbulent patterns. After all, at the scales shorter than the smoothness length, a, for this self-consistently developed magnetic field turbulence, the magnetic configurations could be considered as more regular, without having such an internal substructure that might influence significantly the cross-tail current density.

Our further interest is concentrated on a more detailed analysis of the *fine-scale* magnetic structures in the distant current sheet which likely could develop at the spatial scales between a and ξ. Moreover, we intend to restrict our consideration to a treatment of the topology of the $\mathbf{B}_z$ component of the magnetic field in the current sheet as the simplest approach to the problem.

Let us first represent the total current density $\mathbf{j}$ across the sheet as the sum $\mathbf{j} =< \mathbf{j}_y > + \delta\mathbf{j}$. Here, $< \mathbf{j}_y >$ is the average current density in y-direction which produces the basic (i.e., lobe) component of the magnetic field of the tail, $\mathbf{B}_x$; meanwhile, $\delta\mathbf{j}$ is a perturbation responsible for the fluctuating component of the field, $\mathbf{B}_z$. (In order to stress the highly fluctuating nature of the $\mathbf{B}_z$ component, below we denote it as $\delta\mathbf{B}_z$. Note, also, that $\delta\mathbf{j}$ vanishes when averaged over a spatial scale of the order of ξ or more, as it immediately follows from the condition div $\mathbf{j} = 0$.) Thus, we may write

$$\nabla \times \mathbf{B}_x = \frac{4\pi}{c} < \mathbf{j}_y >, \tag{1}$$

$$\nabla \times \delta\mathbf{B}_z = \frac{4\pi}{c} \delta\mathbf{j}. \tag{2}$$

Strictly speaking, the validity of Eqs. (1)–(2) is limited by the appearance of magnetic fluctuations in x- and y-directions, i.e., $\delta\mathbf{B}_x$ and $\delta\mathbf{B}_y$, which could be also induced by the perturbations in current density, $\delta\mathbf{j}$. To simplify the consideration below, we, however, are neglecting these fluctuations in what follows. Thus, hereafter we restict ourselves to a *quasi-two-dimensional* approximation to the magnetic topologies in the current sheet, without due regard for the possible *three-dimensional* effects. Such an approximation would evidently imply that the magnetic fluctuations $\delta\mathbf{B}_z$ would be some functions of the variables x and y, but not of the variable z. Moreover, we assume below the condition $\delta B_z \ll B_x$ to be verified in section 3.

Equation (2) simply says that the currents $\delta\mathbf{j}$, when flowing across the plasma sheet, "avoid" the obstacles of the magnetic field $\delta\mathbf{B}_z$ like streams of a mountaneous river "percolate" through the rocks. This could be easily seen from the explicit form of the vector Eq. (2) written as

$$\delta j_x = \frac{c}{4\pi}\hat{\mathcal{D}}_y\ \delta\mathbf{B}_z \cdot \mathbf{e}_z\,, \tag{3}$$

$$\delta j_y = -\frac{c}{4\pi}\hat{\mathcal{D}}_x\ \delta\mathbf{B}_z \cdot \mathbf{e}_z\,, \tag{4}$$

where $\hat{\mathcal{D}}_x$ and $\hat{\mathcal{D}}_y$ denote differentiations over x and y, respectively, and $\mathbf{e}_z$ is the unit vector in z-direction. In fact, Eqs. (3)–(4) are the standard Hamiltonian equations, $\delta\mathbf{B}_z \cdot \mathbf{e}_z$ being the Hamiltonian function. (Of course, the Hamiltonian form of Eqs. (3)–(4) is the direct consequence of the assumed quasi-two-dimensional geometry of the distant tail.)

The "percolation" of the electric currents through the highly structured, "clumpy" magnetic field $\delta\mathbf{B}_z$ could be viewed geometrically as the formation of a tree-like network schematically illustrated on Fig. 2. The main trunk of this tree is due to the average electric current, $<\mathbf{j}_y>$, and is elongated in y-direction, whereas the numerous branches of this tree originate from the fluctuating component of the current, $\delta\mathbf{j}$, and lie, under the limitations of our quasi-two-dimensional treatment, in the (xy)-plane. The typical size of this tree (i.e., its characteristic scale in the (xy)-plane) is of the order of the unstable tearing mode scale, ξ. In accordance with Eq. (2), the electric currents $\delta\mathbf{j}$ self-consistently generate the fluctuating magnetic field $\delta\mathbf{B}_z$ which could be schematically represented by the "leafs" of this tree, with typical sizes ranging between the smoothness length of the field, a, and the maximum spatial scale comparable with the size of the tree itself, ξ (see Fig. 2). Note that the smallest "leaf" of the tree having, evidently, the typical size $\sim a$, corresponds to an "elementary" magnetic obstacle for the electric currents, $\delta\mathbf{j}$, in the sheet. Under the quasi-two-dimensional approximation, these "elementary" obstacles could be considered as the "elementary" magnetic flux "tubes" carrying the flux $\sim \delta B_z a^2$ in z-direction. (The term "elementary" indicates that the

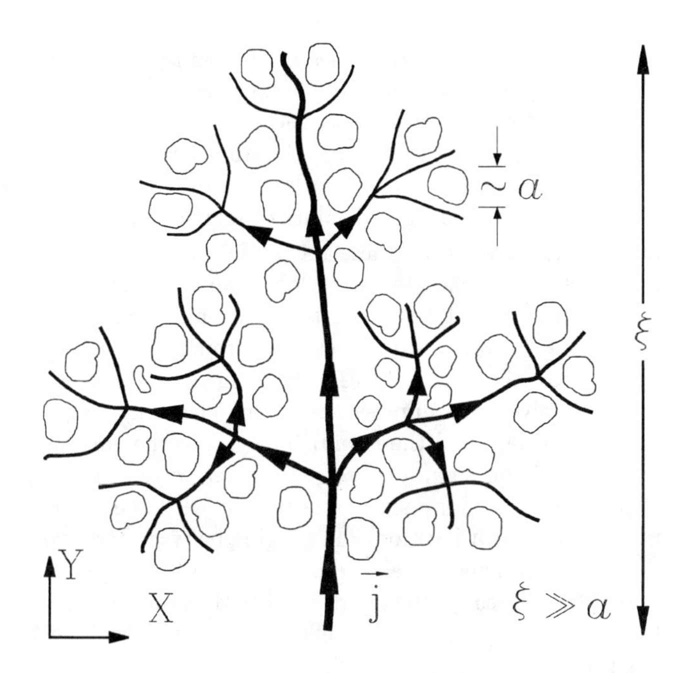

Figure 2. The "sakura model": Percolating structure of the cross-tail electric currents (illustrated as a highly branched "sakura tree"), and the fractal distribution of the self-consistently generated magnetic field fluctuations (shown as the "leafs" of this "tree").

magnetic field of these tubes, $\delta\mathbf{B}_z$, smoothly varies inside the tubes, i.e., within the typical scale of the order of $\sim a$, without having there such an internal substructure that might influence the current flows $\delta\mathbf{j}$ beyond $\sim a$.)

The schematic illustration of the electric currents on Fig. 2 resembles a blossoming sakura tree. (This might suggest to name the model we are considering below, for instance, "sakura model".) Let us also remark that the condition $\operatorname{div}\mathbf{j} = 0$ must be applied. This condition shows that the (xy)-plane is actually covered by many "sakura trees" like the one illustrated on Fig. 2, whose branches are interlaced to form a "sakura jungle" at spatial scales exceeding $\sim \xi$.

3. NUMERICAL ESTIMATES

Before going on with the geometrically accurate treatment of the magnetic topologies in the distant current sheet, we first estimate the basic physical parameters of our model by order of magnitude. We begin with a deeper insight into the basic role played by the smoothness length, a, in plasma dynamics in the sheet. Let r_L be the (effective) plasma particle Larmor radius in the magnetic field $\delta\mathbf{B}_z$, yielding the reciprocal curvature of the particle trajectories in the (xy)-plane. It is clear that plasma particles with Larmor radii $r_L \leq a$ would be strongly affected by the magnetic field $\delta\mathbf{B}_z$, and their contribution into the electric current $\mathbf{j}$ across the sheet is negligible at scale lengths exceeding $\sim a$. (With use of the numerical values of a given below, one may easily verify that these are mainly electrons and cold ions having energies much less than the characteristic energy $W \sim 1$ keV.) On the other hand, hot ions with $r_L \geq \xi$ may easily penetrate through the highly structured magnetic field $\delta\mathbf{B}_z$ existing in the current sheet; however, the number density of such particles is relatively small, and their contribution into the current $\mathbf{j}$ also won't be significant. One, therefore, concludes that the current $\mathbf{j}$ is mostly due to ions with the intermediate values of Larmor radii ranging between the two characteristic length scales a and ξ, i.e., $a \leq r_L \leq \xi$.

It is theoretically important to note that plasma particles with $a \leq r_L \leq \xi$ are effectively "scattered" by the magnetic field $\delta\mathbf{B}_z$ which (as we were speculating above) is organized in an ensemble of "elementary" flux tubes with the typical radii $\sim a$. (In view of Fig. 2, such an ensemble could be interpreted as a set of "leafs" of the tree-like network on the (xy)-plane.) This scattering appears in gradual distortions of the particle trajectories in the (xy)-plane (by the characteristic distortion angle $\psi \sim a/r_L$) when the particles are influenced by the magnetic field of the tubes. During time intervals including many consequent interactions with the "elementary" magnetic tubes, the resulting motion of the plasma particles would be a *diffusion* (i.e., the *random walk*) across the *highly structured* magnetic field of the current sheet, $\delta\mathbf{B}_z$. (We imply that such a diffusion actually takes place not far from the neutral plane, enabling us to neglect the influence of the basic component of the magnetic field, $\mathbf{B}_x$, on the averaged, over z-direction, particle motion.) Now our goal will be to propose a more advanced approach to studying currents and magnetic fields in the distant current sheet considering them in relation with the plasma particle random motion across the self-consistently developed magnetic configurations.

We note, first, that the characteristic distortion angle $\psi \sim a/r_L$ should be of the order of the fluctuations in the current density $\mathbf{j}$, i.e., $\delta j / < j_y > \sim \psi \sim a/r_L$ where δj and $< j_y >$ denote the characteristic values of the vector-functions $\delta\mathbf{j}$ and $< \mathbf{j}_y >$, respectively. (This is due to the essentially fluctuating nature of the magnetic field $\delta\mathbf{B}_z$ at scale lengths exceeding $\sim a$.) From Eqs. (1)–(2) we find that $< j_y > \sim (c/4\pi)(B_x/L)$ and $\delta j \sim (c/4\pi)(\delta B_z/a)$. Taking ratio of these two relations and making use of the explicit expression for the (effective) plasma particle Larmor radius, $r_L \sim vmc/e\delta B_z$, where $v \sim \sqrt{W/m}$ is the characteristic ion velocity and m is the proton mass, we immediately obtain the (self-consistent) "elementary" scale length a yielding the typical size of the smooth, "elementary" magnetic structures in $\delta\mathbf{B}_z$ like magnetic flux tubes:

$$a \sim \sqrt{\frac{vmcL}{eB_x}}. \tag{5}$$

The basic (i.e., lobe) component of the magnetic field, B_x, which appears in Eq. (5), can be estimated from the pressure balance in the magnetotail; assuming $\delta B_z \ll B_x$ one gets $B_x \sim \sqrt{8\pi n W}$ where n is the average plasma number density in the current sheet. The characteristic values of the parameters n and W could be now taken from the survey of plasmas and magnetic fields in the magnetotail with the Geotail mission as reported, e.g., by *Paterson and Frank* [1994]: $n \sim 1$ cm^{-3} and $W \sim 1$ keV. Thus, one finds, numerically, $v \sim 3 \cdot 10^7$ cm·s^{-1}, and $B_x \sim 2 \cdot 10^{-4}$ G. Eq. (5) then yields $a \sim 8 \cdot 10^7$ cm. We finally come to the important result that the ratio of the two scale lengths, $\xi \sim 2.5 \cdot 10^9$ cm and $a \sim 8 \cdot 10^7$ cm, indeed appears to be large, i.e.,

$$\xi/a \sim 30 \gg 1, \tag{6}$$

which substantiates the basic conjecture of our model concerning the possible existence of fine-scale magnetic structures in the distant magnetotail at scale lengths between a and ξ.

To estimate, self-consistently, the level of magnetic perturbations $\delta B_z/B_x$ which are responsible for the development of the fine-scale structure of the tail, we need to introduce the effective plasma conductivity, Σ, associated with the random walk of ions with Larmor radii $a \leq r_L \leq \xi$ on the fine-scale magnetic turbulence. (As mentioned above, the ions having Larmor radii in the range between a and ξ, bring the dominant contribution in the electric current $\mathbf{j}$ across the distant magnetotail.) By order of magnitude, we have $\Sigma \sim ne^2\tau/m$ [*Lifshitz and Pitaevskij*, 1979] where the collisional time τ could be defined as the characteristic diffusion time, i.e., $\tau \sim a^2/\mathcal{D}$, the quantity $\mathcal{D}$ being the plasma particle diffusion coefficient in the (xy)-plane. (Without pretending to be precise, here we assume that the characteristic scale of "voids" between the tubes is of the same order of magnitude as the tube's radii, $\sim a$.) We defer an accurate derivation of the diffusion coefficient $\mathcal{D}$ to the next section and now make use of the rough estimate $\mathcal{D} \sim r_L v/2$ which will be discussed below. Thus, the conductivity Σ becomes $\Sigma \sim 2ne^2a^2/mvr_L$. The average current density $< j_y >$ could be expressed in terms of Σ as $< j_y > \sim \Sigma E_y$ where E_y denotes the characteristic electric field across the magnetotail. For a rough estimate of E_y, we assume a reasonable value of ~ 0.1 mV/m, just yielding the order of magnitude of the electric fields in the distant current sheet. On the other hand, the current density $< j_y >$ may be defined from Eq. (1) to give $< j_y > \sim (c/4\pi)(B_x/L) \sim 1.2 \cdot 10^{-3}$ $\mathrm{g^{1/2}cm^{-1/2}s^{-2}}$. This yields the characteristic ion Larmor radius $r_L \sim 2ne^2a^2E_y/mv < j_y > \sim 1.5 \cdot 10^8$ cm. Hence, $r_L/a \sim 2$, in good agreement with the condition $a \leq r_L \leq \xi$. Finally, the magnetic field δB_z appears to be $\delta B_z \sim vmc/er_L \sim 2 \cdot 10^{-5}$ G. The (relative) level of the magnetic perturbations in the distant magnetotail is, therefore, $\delta B_z/B_x \sim 0.1 \ll 1$, which proves the validity of the quasi-two-dimensional approximation implied by our theoretical model (see Eqs. (1)–(2)). Similar estimate for the perturbation level $\delta B_z/B_x$ that might be reasonable for a quasi-two-dimensional treatment of the magnetic field turbulence in the distant magnetotail could be also deduced numerically from the results of *Veltri et al.* [1998] who studied an effect of the magnetic field fluctuations δB_z on the plasma particle dynamics as a function of $\delta B_z/B_x$.

We also note that the above consideration neglects the possible temporal variations of the magnetic structures in the tail due to the solar wind irregularities. This can be the case when the characteristic time for the substantial solar wind variability, τ_{SW} (which, typically, has the order of magnitude of one or few hours [for a discussion, see, e.g., *Burlaga et al.*, 1989; and the references therein]), exceeds the particle diffusion time at all length scales between a and ξ. Thus, we require $\tau_{SW} \gtrsim \xi^2/\mathcal{D} \sim (\xi/a)^2\tau$, where $\tau \sim a^2/\mathcal{D}$. Taking into account that the ratio $\xi/a \sim 30$ (see Eq. (6)), and the time $\tau \sim a^2/\mathcal{D} \sim 2a^2/r_Lv \sim 2-3$ s, we finally get the condition $\tau_{SW} \gtrsim 30-45$ min to be easily satisfied.

4. SCALING ANALYSIS, FRACTAL DIMENSION, AND INDEX OF CONNECTIVITY

As was already mentioned above, we are considering the highly structured, "clumpy" nature of the magnetic field and plasma turbulence in the distant Earth's magnetotail as its fundamental inherent property which controls the particular statistics of the observed physical parameters. To take the next step, we must find a way to describe on the *quantitative* level this "clumpy" turbulence, in order to develop a framework to studying the turbulence characteristics in an efficient form. Following our previous publication [*Milovanov et al.*, 1996], we proceed here with an approach based on general methods of the *fractal geometry* as the key method of our treatment. Such an approach would involve the *geometric* description of the magnetotail "clumpy" turbulence with use of certain "nonconvential" quantities that define the topology of the turbulent magnetic structures, on one hand, and appear in the self-consistent "physical" equations, on the other.

To introduce these quantities, we need to *assume* that the magnetic field fluctuations, $\delta\mathbf{B}_z$, which are responsible for the "clumpy" character of the turbulence in the range of spatial scales between a and ξ, are not only organized in an ensemble of "elementary" flux tubes with an arbitrary spatial distribution consistent with the highly branched, tree-like topology of the electric currents in the current sheet (see Fig. (2)), but, moreover, that this ensemble reveals the property of the *statistical self-similarity* in the above range of spatial scales. Statistical self-similarity means that each part of the ensemble with a characteristic size between a and ξ could be considered, in the statistical sense, as a reduced scale image of the whole, i.e., the spatial distribution of the "elementary" flux tubes inside a small-scale group of tubes is statistically identical to that of the small-scale groups within a larger group, and so on until one reaches the largest scale length possible, ξ. The assumed statistical self-similarity in the distribution of the magnetic field $\delta\mathbf{B}_z$ appears to be a very general statistical property of the fluctuating magnetic fields in the near-Earth environment [see, e.g., *Burlaga and Klein*, 1986; *Milovanov and Zelenyi*, 1994ab, 1995; *Ohtani et al.*, 1995], and may serve as a manifestation of the *scaling invariance* of the turbulence in the range of scales between a and ξ. Some theoretical substantiation for the possible existence of the statistical self-similarity in the geometry of the field $\delta\mathbf{B}_z$ is proposed below (see subsection 4**D**) where the self-consistent topology of the electric currents in the sheet is approximated by the percolating network at the threshold of percolation (see, also, the schematic illustration on Fig. 2).

Having assumed the property of the statistical self-similarity of the magnetotail turbulence for the scales ranging between a and ξ, we are ready to specify what would be that "nonconvential" geometric quantities which form the basis of application of the fractal geometry in our study. We begin with an introduction of the more widely known quantity which is commonly referred to as the Hausdorff fractal dimension, D [*Mandelbrot*, 1983].

A. Introduction of the Fractal Dimension

Consider the spatial distribution of the "elementary" magnetic flux tubes (each carrying the flux $\sim \delta B_z a^2$) across the current sheet as a statistically self-similar (in the range of scales between a and ξ) ensemble which generates the "clumpy" fine-scale structure of the magnetic turbulence in the sheet in the above range of scales, and which we name below a *fractal* object, where the term "fractal" directly points out the statistical self-similarity of this distribution. Then the Hausdorff fractal dimension, D, is defined as the measure of scaling behavior of the magnetic flux tubes number density in the current sheet. More precisely, let $\rho(\chi)$ denotes the number density of these tubes at a given spatial scale $a \leq \chi \leq \xi$. (We define the function $\rho(\chi)$ as the total number of the tubes contained within a given circle of radius χ, devided by the area of this circle.) For fractal objects, the function $\rho(\chi)$ is well known to have the power-law form $\rho(\chi) \propto \chi^{D-E}$, where E is the topological (integer) dimension of the embedding Euclidean space [*Mandelbrot*, 1983]. (For a fractal distribution of the flux tubes across the neutral *plane*, the topological dimension, E, is, obviously, equal to two, i.e., $E = 2$.) From the introduced scaling behavior of the function $\rho(\chi)$ it is clear that the total number of the tubes contained inside a circle of radius χ is of the order of $\sim \rho(\chi) \cdot \chi^2 \sim \chi^D$ where $\sim \chi^2$ is the total area of the circle. For instance, one may conclude that the total number of "leafs" on a tree shown on Fig. 2, is of the order of $\sim \xi^D$ where D is the associated fractal dimension for the ensemble of the "leafs". Note that, by definition, the Hausdorff fractal dimension, D, cannot exceed the topological dimension, E; in particular, for a (connected) fractal object on a plane ($E = 2$), the quantity D varies between unity and two, i.e., $1 \leq D \leq 2$.

We also draw attention to that fact that an application of fractal geometry to the description of the magnetic structures in the magnetotail would imply that the power exponent, $D - E$, which appears in the scaling behavior of the magnetic flux tubes number density, $\rho(\chi)$, depends on the particular spatial scale χ that is analyzed. To illustrate this, we note that at length scales shorter than a, the magnetic field $\delta \mathbf{B}_z$ is assumed to vary *smoothly* inside a two-dimensional plane area, hence, for the spatial scales $\chi \leq a$, the fractal dimension $D = E = 2$, and, consequently, $\rho(\chi)$ is independent of χ as one could expect for the structureless objects like the "elementary" flux tubes. In the meanwhile, at length scales longer than ξ, the topology of the magnetic field $\delta \mathbf{B}_z$ is that of a one-dimensional chain of "magnetic islands" whose origin is due to the modulation of the fine-scale magnetic fields in the tail by the excited tearing modes; hence, at the scales $\chi \geq \xi$, the fractal dimension D again coincides with the Euclidean dimensionality E which, however, is now equal to unity: $D = E = 1$. As is already clear, in the intermediate range of scales, $a \leq \chi \leq \xi$, the quantity D should describe the statistically self-similar spatial distribution of the "elementary" magnetic flux tubes, leading to the occurrence of the "clumpy" fine-scale structure in $\delta \mathbf{B}_z$. In this case, D may be expected to be a *fraction* $1 < D < 2$.

To find this *fractional* value of the dimension D, we need to perform the scaling analysis of the basic Eq. (2) governing the dynamics of the perturbed currents and magnetic fields in the distant magnetotail. Such an analysis, in turn, requires the derivation of the self-consistent diffusion coefficient, $\mathcal{D}$, for the random walk of plasma particles across the fine-scale magnetic structures in the current sheet. The calculations given below follow a more detailed paper by *Milovanov and Zelenyi* [1995] on plasma diffusion via stochastic magnetic configurations.

A naive approach to the calculation of the diffusion coefficient $\mathcal{D}$ might be based on the widely known Einstein relation between the mean-square displacement of a test particle, $< \delta\chi^2 >$, versus given diffusion time, Δt, i.e., $< \delta\chi^2 >= 2\mathcal{D}\Delta t$. Such an approach, however, has that fatal drawback that it holds only for diffusion in Euclidean spaces; this "Euclidean" approach doesn't take into account the following important property, namely that the diffusion rate of particles via the statistically self-similar, "clumpy" magnetic structures (which we are considering as fractals) would be necessarily dependent on the "microscopic" temporal scale which determines the basic time step of the plasma particle random motion [see, e.g., *Gefen et al.*, 1983]. This actually leads to a necessity to generalize the above Einstein formula, and to introduce a more advanced relation which could be applied to a correct description of transport processes on fractals. Such a relation between the mean-square displacement, $< \delta\chi^2 >$, of a test particle (which experiences random walk on a fractal object), and its diffusion time, Δt, was proposed by *Mandelbrot* [1983] to be $< \delta\chi^2 >= 2\mathcal{D}\theta(\Delta t/\theta)^{2H}$, where the parameter H is arbitrary real number between 0 and 1, i.e., $0 < H < 1$. The case $H = 1/2$ recovers the above Einstein relation $< \delta\chi^2 >= 2\mathcal{D}\Delta t$ for the standard diffusion in Euclidean space, when the diffusion rate is independent of the "microscopic" temporal scale θ. On the contrary, the cases when $H \neq 1/2$ describe the generalized, non-Euclidean diffusion when the "microscopic" scale θ is explicitly present in the description of the particle random walk.

We also note that diffusion processes on fractals can be *persistent* or *antipersistent*, depending on the particular values of the parameter H. The persistent diffusion process is characterized by the values of H ranging between 1/2 and 1, and implies positive correlations between past and future displacements of the test particle. On the contrary, the antipersistent diffusion process is described by the parameter H varying from 0 to 1/2, and leads to negative correlations between past and future particle displacements. In the standard (Euclidean) case of $H = 1/2$, the correlations between past and future displacements are absent on all temporal scales, this being the fundamental property of the Euclidean diffusion. The issue of the *persistence* of the generalized diffusion on fractal objects is discussed in detail by *Feder* [1988].

B. Introduction of the Index of Connectivity

It might be shown that the property for a diffusion process on a fractal to be persistent or antipersistent is related to the topological characteristics of the fractal substrate. To illustrate this, we introduce another, alternative, quantity, σ, instead of the above parameter H, through the simple expression $2 + \sigma = 1/H$. One then immediately concludes that the persistent diffusion on a fractal ($1/2 < H \leq 1$) would formally correspond to the values of σ ranging between -1 and 0, i.e., $-1 \leq \sigma < 0$, whereas the antipersistent diffusion ($0 \leq H < 1/2$) would be described by the parameter σ varying from 0 to $+\infty$, i.e., $0 < \sigma \leq \infty$. The case of the standard Euclidean diffusion (for which the Einstein relation $< \delta\chi^2 >= 2\mathcal{D}\Delta t$ holds) is recovered by the zero value of the quantity σ, i.e., $\sigma = 0$ for $H = 1/2$. Substituting $2+\sigma = 1/H$ into the expression $< \delta\chi^2 >= 2\mathcal{D}\theta(\Delta t/\theta)^{2H}$, one obtains the alternative relation

$$< \delta\chi^2 >= 2\mathcal{D}\theta(\Delta t/\theta)^{2/(2+\sigma)}, \qquad (7)$$

between the mean-square displacement, $< \delta\chi^2 >$, of the test particle on a fractal, and the particle diffusion time, Δt [*Milovanov and Zelenyi*, 1995].

The quantity σ is the topological parameter [*O'Shaughnessy and Procaccia*, 1985] which must be given in addition to the Hausdorff fractal dimension D when describing the geometry of a fractal. Contrary to the parameter D which defines the scaling behavior of the number density of the "elementary" constituents (e.g., "elementary" magnetic flux tubes) of the fractal object, the quantity σ defines how these "elementary" constituents are "glued" together to form the entire fractal structure. (For instance, the quantity σ may take different numerical values even for the fractals having the same Hausdorff dimension D.)

The particular way of "glueing" of the "elementary" constituents into the global geometric configuration might be quantified by the topological concept customarily referred to as "connectivity" [see, e.g., *Nash and Sen*, 1983]. Thus, the quantity σ could be regarded as the measure of connectivity of a fractal, and is called below the "index of connectivity". A detailed consideration of the issue of connectivity for fractal objects is given in the review of *Nakayama et al.* [1994] where the index σ is introduced by considering the structural characteristics of fractal sets. A treatment of the connectivity properties of fractals from the viewpoint of the homotopoic topology is given in the recent papers of *Milovanov* [1997] and *Milovanov and Zimbardo* [1997].

From the definition of the index σ it is clear that this quantity might describe the characteristic "shape" of pathways that connect two given points of a fractal object. This would also mean that the index of connectivity might determine the topology of the possible trajectories for test particles to "walk" from one given point of the fractal object to another, avoiding some inaccessible domains.

In fact, in Euclidean geometry ($\sigma = 0$), the diffusing particles may walk throughout the entire Euclidean space, so that any point of the space could be reached by any particle after sufficiently long diffusion time. In this case, the inaccessible domains are absent, and the diffusion process is described by the standard Einstein relation $< \delta\chi^2 >= 2\mathcal{D}\Delta t$.

The situation is, however, different for the diffusion on fractal objects. Since fractals don't fill the Euclidean space, there always exist the domains of the space that are beyond the fractal substrate, and, therefore, are inaccessible for the particles diffusing on the fractal. Hence, the diffusion rate of these particles would be controled by the particular spatial spreading of the inaccessible domains whose role would appear in the *nonzero* values of the index of connectivity σ.

In the case of *positive* σ, i.e., $\sigma > 0$, the spatial spreading of the inaccessible domains tends to prevent the diffusing particles from a fast penetration deep inside the fractal object, so that the particles would more often return to a vicinity of the starting point than in the case of the standard, Euclidean diffusion (see Fig. 3a). Hence, the diffusion on a fractal having $\sigma > 0$ would be slower (for given diffusion constant, $\mathcal{D}$) than the Euclidean one (see, also, Eq. (7)). This formally recovers the antipersistent diffusion process with the parameter $0 \leq H < 1/2$.

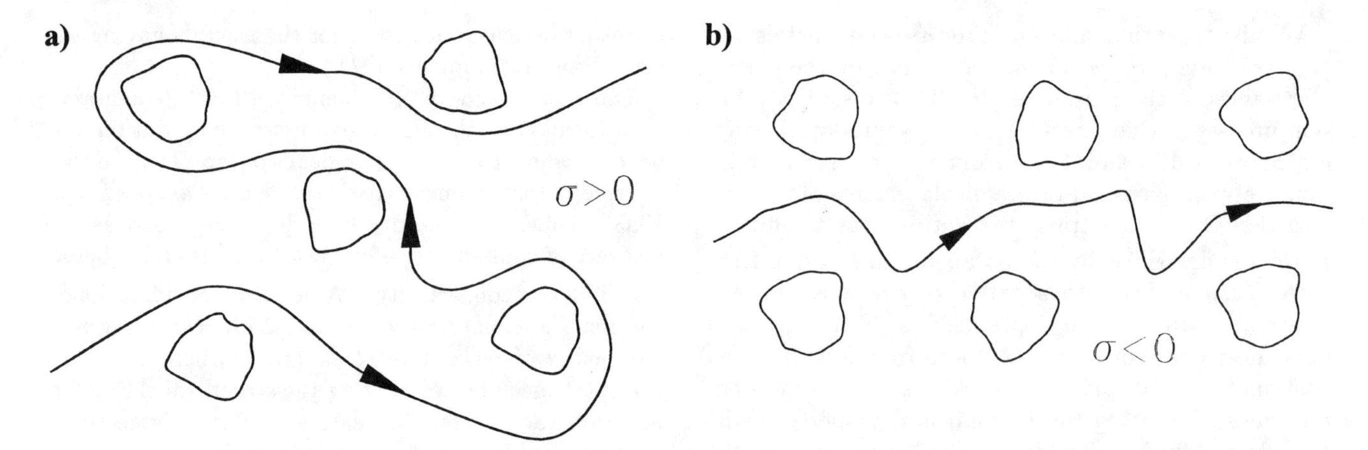

Figure 3. An illustration for the concept of connectivity of fractal objects. Fig. 3a schematically illustrates diffusion on a fractal having $\sigma > 0$ (the antipersistent diffusion process); Fig. 3b shows the opposite case of $\sigma < 0$ (the persistent diffusion process). The domains inaccessible for the diffusing particles, are assumed inside the irregular contours.

On the contrary, the case of *negative* σ, i.e., $-1 \leq \sigma < 0$, assumes such a spreading of the inaccessible domains that stimulates penetration of the diffusing particles into the bulk of the fractal. For instance, the inaccessible domains might provide the particles with "free corridors" where the probability to return towards the starting point is relatively small (see Fig. 3b). It is clear that the diffusion on fractals having $-1 \leq \sigma < 0$ is faster than the Euclidean diffusion (see Eq. (7)), resulting in the persistent diffusion process with $1/2 < H \leq 1$.

A descriptive example of the persistent diffusion might be the limiting case of $\sigma = -1$ which recovers the ballistic regime of the particle propagation. This regime has been analyzed numerically by *Zimbardo and Veltri* [1995] and *Zimbardo et al.* [1995], who recognized the presence of the "free corridors" and ballistic "jumps" (the so-called "Levy flights") in the test particle diffusion through disordered magnetic fields.

Thus, we are coming to a conclusion that the topological characteristics of fractals (quantified by the index of connectivity σ) play the fundamental role in understanding of the transport phenomena on fractal objects. We actually claim that the particular type of the diffusion process on a fractal (persistent or antipersistent) is defined by the intrinsic topological properties of the fractal sets; these properties could be associated with the spatial spreading of the inaccessible domains of the fractal, and determine the sign of the index of connectivity σ (negative for the persistent diffusion, and positive for the antipersistent one). The importance of the index σ for transport phenomena on fractals is discussed in more detail by, e.g., *Gefen et al.* [1983]; *O'Shaughnessy and Procaccia* [1985]; *Havlin and Ben-Avraham* [1987]; *Giona and Roman* [1992]; *Nakayama et al.* [1994]; *Milovanov and Zelenyi* [1995].

C. Scaling Analysis of Basic Equations

We now proceed with the calculation of the diffusion coefficient, $\mathcal{D}$, for plasma particles in the distant current sheet with use of the generalized expression (7). Since the particle diffusion in the sheet is due to the consequent distortions of their trajectories by the magnetic field of the "elementary" flux tubes, the time step θ in Eq. (7) could be estimated as $\theta \sim a/v$ where v is the plasma particle characteristic velocity. (This, of course, implies that the particle Larmor radius r_L is of the order of few a or more, which agrees well with the estimate $r_L/a \sim 2$ obtained above.) Then, it is easy to verify that the mean-square displacement of the particles, $< \delta\chi^2 >$, becomes of the order of their Larmor radius squared, r_L^2, during the time interval $\delta t/\theta \sim \psi^{-1}$ where $\psi \sim a/r_L$ is the characteristic distortion angle of the particle trajectories. Thus, from Eq. (7) one immediately gets

$$\mathcal{D} \sim \frac{1}{2} r_L^2 \frac{v}{a} \left(\frac{a}{r_L}\right)^{2/(2+\sigma)}, \tag{8}$$

yielding the diffusion coefficient for plasma particles with Larmor radii large compared with the "elementary" scale length a. In the standard case of $\sigma = 0$, Eq. (8) is reduced to $\mathcal{D} \sim r_L v/2$ which is a straightforward estimate for the diffusion coefficient expressing the dominant role of the particle (quasi-)Larmor motion. A more general case of $\sigma \neq 0$ might be associated with a diffusion process in presence of the domains "inaccessible" for the plasma particles; in the framework of our approach, such domains might be interpreted as the regions with relatively strong magnetic field fluctuations. (The "inaccessible" domains influence the particle diffusion rate, resulting in the possibilities of the persistent or antipersistent diffusion [see Figs. 3a–3b].)

Since the particle Larmor radius r_L behaves with the magnetic field δB_z as $r_L \propto \delta B_z^{-1}$, from Eq. (8) it follows that the diffusion coefficient $\mathcal{D}$ scales with δB_z as $\mathcal{D} \propto \delta B_z^{-2(1+\sigma)/(2+\sigma)}$. Hence, the current density δj which appears on the right hand side of Eq. (2), scales as $\delta j \sim \psi < j_y > \propto r_L^{-1}\mathcal{D}^{-1} \propto \delta B_z \cdot \delta B_z^{2(1+\sigma)/(2+\sigma)}$ where we took into account that $< j_y > \propto \mathcal{D}^{-1}$.

Let us now turn to the left hand side of Eq. (2). First of all, we draw attention to the important point that in fractal geometries with the nonconvential connectivity $\sigma \neq 0$, the order of differentiation in the operators $\hat{\mathcal{D}}_x$ and $\hat{\mathcal{D}}_y$ (see the explicit, scalar form (3)–(4) of the vector Eq. (2)) becomes *fractional* and is actually equal to $2/(2+\sigma)$ rather than unity [*Zelenyi and Milovanov*, 1996].

The necessity to introduce the *fractional* order of differentiation into the explicit representation of the vector operator $\nabla\times$ through the scalar operators $\hat{\mathcal{D}}_{x,y}$ could be illustrated by the following simple argument. In fact, as soon as the current density, $\delta\mathbf{j}$, is obtained from the magnetic field, $\delta\mathbf{B}_z$, via the *operation of differentiation*, $\nabla\times$ (see Eq. (2)), one must take care of the sufficient level of *smoothness* of the differentiated function, $\delta\mathbf{B}_z$. On the other hand, we have assumed that the field $\delta\mathbf{B}_z$ has the considerable fine-scale structure in a wide range of spatial scales, making it possible to apply the *fractal* geometry approximation to quantify the spatial distribution of $\delta\mathbf{B}_z$. In the framework of such an approximation, the field $\delta\mathbf{B}_z$ becomes a *highly singular* function of spatial variables, which might seem to be in contradiction with the required *smoothness* of the function $\delta\mathbf{B}_z$. Indeed, a formal action of a first-order differential operator on such a field (see Eq. (2)) would yield an essentially diverging function which cannot serve as a correct representation for the current density, $\delta\mathbf{j}$. This contradiction, however, is due to the fact that having approximated the differentiated function, $\delta\mathbf{B}_z$, by a *fractal* distribution, one also needs to apply the same approximation to the differential operators themselves, in order to keep them *consistent* with this distribution. In the exact mathematical theory of fractals, it is proven [see, e.g., *Le Mehaute*, 1991] that the required consistency of the differential operators with the assumed fractal distribution appears in the relevant modification of the *order of differentiation*, which, in our particular case, is given by the value $2/(2+\sigma)$ instead of unity. It can be shown that the action of the differential operator of this modified, fractional order on the field $\delta\mathbf{B}_z$ (which is approximated by the *fractal* spatial distribution discussed above) would yield a nondiverging function, $\delta\mathbf{j}$, with the physically appropriate, *finite* values.

In this paper, we needn't, however, deal with the issue of the fractional differentiation in more detail. For our purposes, it would be sufficient to make use of the only related fact that the above extension of the operators $\hat{\mathcal{D}}_{x,y}$ to fractal geometry provides them with an unusual *scaling behavior*, $\hat{\mathcal{D}}_{x,y} \propto \chi^{-\sigma/(2+\sigma)}$ [*Zelenyi and Milovanov*, 1996], which is the direct consequence of the introduction of the fractional order of differentiation, $2/(2+\sigma)$, into these operators. (Note that in the convential case $\sigma = 0$, the operators $\hat{\mathcal{D}}_{x,y}$ are independent of the length scale χ, as one could expect for the *smooth* distributions (i.e., that without the fractal topology). Note, also, that the "unusual" scaling behavior ($\propto \chi^{-\sigma/(2+\sigma)}$) of the operators $\hat{\mathcal{D}}_{x,y}$ as well as the fractional order of the differentiation over x and y could be regarded as the explicit manifestation of the unusual *topological features* of fractal sets related to the existence of the topological entropy of fractals and nonlocality of the response function in presence of fractal topologies. For more details, see, e.g., *Le Mehaute*, 1991; *Giona and Roman*, 1992; *Milovanov and Zelenyi*, 1995; *Zelenyi and Milovanov*, 1997.)

Thus, we have that the left hand side of Eq. (2) behaves as $\delta B_z \cdot \chi^{-\sigma/(2+\sigma)}$, where the range of scales $a \leq \chi \leq \xi$ is implicitly assumed. Along with the scaling law for the current density, $\delta j \propto \delta B_z \cdot \delta B_z^{2(1+\sigma)/(2+\sigma)}$, this yields the amplitude of the magnetic fluctuations δB_z as a function of the length scale χ, i.e., $\delta B_z \propto \chi^{-\sigma/2(1+\sigma)}$. Hence, the energy density of the magnetic field $\delta\mathbf{B}_z$ scales as $\delta B_z^2 \propto \chi^{-\sigma/(1+\sigma)}$. In the self-consistent regime, this result, however, must coincide with the scaling behavior of the magnetic flux tubes number density, $\rho(\chi) \propto \chi^{D-2}$, as soon as we imply the statistical self-similarity (quantified in terms of the fractal dimension $1 < D < E = 2$) for the distribution of the magnetic flux tubes in the (xy)-plane. This leads to the relation between the fractal dimension D and the exponent σ, i.e.,

$$D = 2 - \frac{\sigma}{1+\sigma} \tag{9}$$

which shows that the self-consistent fractal topology of currents and magnetic fields cannot be arbitrary but imposes a restriction on the possible values of the parameters D and σ.

From Eq. (9) it follows that as σ varies between zero and infinity, the fractal dimension D ranges between its two limiting integer values, i.e., 1 and 2. The convential case of the two-dimensional Euclidean geometry ($D = E = 2$) may be easily obtained from Eq. (9) by setting $\sigma = 0$. In this limit, the electric currents $\delta\mathbf{j}$ responsible for the occurrence of the magnetic perturbations $\delta\mathbf{B}_z$, are organized in such a network that everywhere densely "passes" through each point of a two-dimensional Euclidean plane which, in the meanwhile, is homogeneously (in the statistical sense) covered by the magnetic field fluctuations, $\delta\mathbf{B}_z$. (One could say that the magnetic field fluctuations are "space-filling" in this case.) An increase in the exponent σ implies an appearance of the condensations and rarefactions in the originally statistically homogeneous distribution of the magnetic field fluctuations. It is clear that the electric currents would basically tend to "avoid" the domains where the level of magnetic fluctuations is higher than average, and, therefore, would preferably concentrate along the "corridors" where this level is lower than average. In the self-consistent regime, this would result in the statistically self-similar spatial distribution of the magnetic flux across the (xy)-plane, which we are considering as a fractal pattern having the Hausdorff fractal dimension, $1 < D < 2$, and the nontrivial index of connectivity, $\sigma \neq 0$. (Of course, this statistically self-similar, fractal distribution cannot be already regarded as "space-filling".)

D. Stability Condition

To calculate the numerical values of the fractal dimension, D, and the index of connectivity, σ, in the distant current sheet, we must supplement Eq. (9) with another relation between D and σ, to obtain a set of two independent equations for these two quantities. We choose this other relation from the physically reasonable requirement that the electric currents in the sheet are organized in such a network (having, by the above assumptions, the topology of a fractal "tree" as shown on Fig. 2) that could actually maintain the structure of the current sheet on a *global* scale (i.e., the one comparable with $\xi \sim 2\pi L$ or more). In other words, one must analyze *stability* of the fractal network on Fig. 2, to verify whether this network is really *developed* enough to allow the electric currents effectively flow across the sheet.

To illustrate the stability properties of fractal current networks like that on Fig. 2, let us imagine for a while that this network experiences a structural perturbation which results, for instance, in a more dense covering of the current sheet by the magnetic field fluctuations, $\delta\mathbf{B}_z$, compared with some stable self-consistent covering we are now interested in. Because the "density" of this covering could be characterized by the "elementary" magnetic flux tubes number density, ρ (see subsection 4**A**), this increase in the density of the covering would imply an increase in the Hausdorff fractal dimension, D. But with this increased fractal dimension, the magnetic field fluctuations, $\delta\mathbf{B}_z$, would become less "transparent" for the electric currents which tend to "avoid" the condensations of the magnetic field in the current sheet. (As was mentioned in subsection 4**C**, such condensations of the magnetic field could be treated as the domains "inaccessible" for the plasma particles whose migration across the current sheet we are considering as a diffusion on a fractal object.) The decrease in the "transparency" of the fluctuations would, in turn, result in the relevant decrease in the amplitude of the current fluctuations $\delta\mathbf{j}$ and, moreover, in a change in the topology of the current network itself. The latter would mean, in particular, that the current network becomes less branched than the unperturbed one. "Less branched" means that the modified network is more "coarse-grained" (less "refined"), having, therefore, a *larger* value of the index of connectivity σ. Hence, an original increase in the fractal dimension D causes a consequent increase in the exponent σ. In the self-consistent regime, however, such a re-configuration of the current network immediately influences the magnetic field fluctuations, $\delta\mathbf{B}_z$, as it directly follows from the basic Eq. (2). It is evident that this self-consistent influence acts towards *decreasing* the level of the magnetic field fluctuations, which, for fractal networks, also implies a decrease in the number density of the "elementary" magnetic flux tubes. As is already clear, this would lead to the reconstruction of the unperturbed distribution of the magnetic field and current fluctuations in the sheet towards the *lower* values of the Hausdorff fractal dimension D and the index of connectivity σ.

Thus, we tend to conclude that there exists some stable self-consistent fractal network which must be associated with the particular topology of the magnetic field and current density in the tail. Such a network has the characteristic scale of the order of $\xi \sim 2\pi L \gg a$ (see Eq. (6)) and, therefore, determines the topology of the magnetic field and current density fluctuations on scale lengths far exceeding the "elementary" scale length, a.

The existence of such stable large-scale fractal networks has been clearly established in the framework of the *percolation theory* [see, e.g., *Stauffer*, 1985; *Feder*,

1988]. In fact, it was recognized that the large-scale fractal networks are characteristic of the so-called *threshold* of percolation when the number of the conducting links that form the network first reaches some *critical* value, and the phase transition *insulator* $\rightarrow$ *conductor* occurs in the system.

A remarkable feature of the networks at the threshold of percolation is their substantially *fractal* geometry at the scale lengths ranging between the "elementary" distance, a, and the percolation correlation length, ξ_c [see, e.g., *Isichenko*, 1992; *Nakayama et al.*, 1994]. The scale length ξ_c yields the characteristic size of the fractal network near the percolation threshold; at the point of the percolation transition, this scale length is known to diverge, i.e., $\xi_c/a \rightarrow \infty$ [*Stauffer*, 1985]. This divergence of the percolation correlation length ξ_c is responsible for the anomalous scaling behavior of the physical quantities near the point of the percolation transition, and has a variety of important applications [see, e.g., *Gefen et al.*, 1983].

On the other hand, we have already introduced above the macroscopic scale length, ξ, which was determined as the characteristic size of the electric current network in the magnetotail current sheet. This scale length was associated with the global topology of the tail, leading to the inequality $\xi/a \gg 1$ (see estimate (6)). This suggests an interpretation of the scale length ξ as the *percolation correlation length*, ξ_c, i.e., $\xi \sim \xi_c$, where the formal divergence of the quantity ξ_c is replaced by the approximate estimate $\xi/a \sim 30 \gg 1$, in accordance with Eq. (6).

The use of the estimate $\xi/a \sim 30 \gg 1$ instead of the more formal condition $\xi_c \rightarrow \infty$ would then manifest the restrictions on the spatial scales where the concepts of the percolation theory might be applied. In fact, the percolation correlation length ξ_c defines the characteristic spatial scales where the fractal geometry of percolation crosses over to the Euclidean geometry [*Stauffer*, 1985]. Near the threshold of percolation, the divergence of the correlation length ξ_c could be limited only by some external mechanisms that might influence the structural properties of the system. When applying the percolation theory to the description of the electric currents in the magnetotail current sheet, one must take into account the "external" modulation of the self-consistent fractal geometry of the percolating current networks, that comes from the tearing mode scales, $\sim 2\pi L$ or more. Consequently, at the scales exceeding $\sim 2\pi L$, one might expect an evidence for the tearing instability structures (e.g., magnetic "islands") [*Galeev and Zelenyi*, 1975] whose geometry is clearly Euclidean. This naturally yields the above upper limit of $\sim 2\pi L \sim 30a$ (see Eq. (6)) for the percolation correlation length $\sim \xi_c$.

Application of the percolation theory might easily help substantiate the assumption made in the beginning of section 4. Indeed, it was assumed in section 4 that the electric currents in the magnetotail current sheet have the property of the statistical self-similarity associated with the fractal geometry of the system. This assumption then led us to the additional consideration of the stability properties of the current networks with respect to the structural perturbations. From the viewpoint of the percolation theory, however, the fractal geometry of the networks as well as their topological stability properties are the inherent features of the percolation transition [see, e.g., *Stauffer*, 1985; *Feder*, 1988; *Isichenko*, 1992; *Nakayama et al.*, 1994]. Thus, when considering the topological properties of the self-consistently developed magnetic field and plasma turbulence in the distant magnetotail, we might assume from the very beginning that the electric currents in the magnetotail current sheet are organized in the percolating network which is at the *threshold* of percolation; such an assumption would then immediately guarantee both the applicability of the fractal geometry approximation, and the global structural stability of the system.

Consequently, we come to the following conclusion: The requirement of the global stability of the fractal current network in the distant magnetotail current sheet could be formally reduced to the general topological condition that this network is at the *threshold* of percolation. In the percolation theory, such a topological condition has been first proposed in the conjectural form by *Alexander and Orbach* [1982], and is now widely known as the Alexander – Orbach (AO) conjecture.

In fact, in order to describe the dynamical properties of the percolating fractal networks, *Alexander and Orbach* [1982] introduced the specific combination, $\tilde{d} \equiv 2D/(2+\sigma)$, of the Hausdorff fractal dimension, D, and the index of connectivity, σ, which is commonly referred to as the *spectral fractal dimension*. Then, from a wealth of numerical estimates they noted that the spectral fractal dimension $\tilde{d}$ for the percolating networks at the threshold of percolation was remarkably close to the mean-field value, 4/3, for all embedding Euclidean dimensions E greater than one, even though the parameters D and σ were by no means constant as functions of E. This numerical evidence led them to speculate that the spectral dimension $\tilde{d}$ might be exactly 4/3 for the percolating networks at the threshold for all $E \geq 2$, i.e.,

$$\tilde{d} \equiv \frac{2D}{2+\sigma} = \frac{4}{3}, \quad E \geq 2. \qquad (10)$$

This assertion has come to be known as the AO conjecture. Eq. (10) shows, for instance, that an increase of the Hausdorff fractal dimension D of the percolating network at the threshold of percolation must be followed by the corresponding increase of the index of connectivity σ, for this network to remain at the threshold.

For the relatively high embedding dimensions $E \geq 6$, the general proof of the AO conjecture (10) was found within the mean-field theory [see, e.g., *Havlin and Ben-Avraham*, 1987; *Nakayama et al.*, 1994]. In the lower dimensions $2 \leq E \leq 5$, the mean-field theory couldn't be directly applied, and the validity of the AO conjecture for these E was remaining an open question up to recently. (The case of these low dimensions includes, obviously, the percolation problem on a plane ($E = 2$) which is in the center of attention of our study.) In the attempt to prove the AO conjecture (10) for $2 \leq E \leq 5$, *Milovanov* [1997] proposed special analytical approach based on the general methods of the differential topology. This approach led him to show that the AO conjecture (10) might be improved for $2 \leq E \leq 5$; using the topological arguments, *Milovanov* [1997] proposed the analytical result $\tilde{d} \approx 1.327$ for all $2 \leq E \leq 5$, which is only slightly smaller than the original AO estimate 4/3. This result is based on the most general topological features of fractal geometry of percolation at the threshold.

For the sake of simplicity, we, however, are neglecting below the slight difference between the improved value of $\tilde{d} \approx 1.327$, and the original mean-field estimate, $\tilde{d} = 4/3$. Thus, we are approximately considering Eq. (10) as the stability condition for the magnetic field and current density fractal patterns in the current sheet ($E = 2$), although a more rigorous analytical treatment of the percolation problem on a plane might actually require the use of the improved estimate of $\tilde{d}$.

The stability condition (10) must be given in addition to Eq. (9), to determine the self-consistent values of the fractal dimension D and the index of connectivity σ. Combining Eqs. (9) and (10), one immediately obtains the values of the Hausdorff fractal dimension D and the index of connectivity σ,

$$D \approx 5/3 < 2, \quad \sigma \approx 1/2 > 0. \qquad (11)$$

These findings are the required explicit numerical values of the "nonconvential" geometric parameters that we apply to the self-consistent fractal description of the magnetic field structures in the distant Earth's magnetotail in the range of spatial scales $a \leq \chi \leq \xi$, $\xi \gg a$.

We note that the magnetic field turbulence in the current sheet indeed appears to be not "space-filling" as it immediately follows from the inequality $D < E = 2$. Moreover, we draw attention to the fact that the self-consistent value of the index of connectivity σ is *positive* as one could already expect from the explicit analytical form of Eq. (9). This indicates that the migration of the plasma particles across the current sheet is the antipersistent diffusion process (see Fig. 3a), having, formally, a lower rate than the standard diffusion process in an Euclidean space ($\sigma = 0$).

5. POWER-LAW SPECTRA OF MAGNETIC FIELD FLUCTUATIONS

We now turn to a consideration of the possible behavior of the power spectra of the magnetic field fluctuations in the framework of the proposed fractal approach. Summarizing the results obtained, we first mention that the magnetic field $\delta\mathbf{B}_z$ in the distant magnetotail has different spatial structure depending on the spatial scale χ. Indeed, as $\chi \leq a$, then $\delta\mathbf{B}_z$ smoothly varies within a plane area having the trivial fractal dimension $D = 2$. As $\chi \geq \xi$, then $\delta\mathbf{B}_z$ could be treated as a one-dimensional ($D = 1$) chain of condensations and rarefactions in the z-component of the magnetic field in the current sheet, elongated in x-direction. In the intermediate range of scales, $a \leq \chi \leq \xi$, the magnetic field $\delta\mathbf{B}_z$ is organized in tubes with the substantially fractal spatial distribution, the fractal dimension being $D \approx 5/3$.

Knowing the fractal dimension D, one could immediately calculate the exponent of the Fourier power spectrum α of the magnetic field fluctuations. In fact, we note that the magnetic field structures in the distant tail could be actually considered as moving away from the Earth with the decelerated solar wind velocity, u. The characteristic value of u has the order of magnitude of $\sim$ 200 km$\cdot$s^{-1} as was recently reported by *Paterson and Frank* [1994]. Without regard for the possible dynamical evolution of the magnetic field fluctuations, we now assume that the fine-scale turbulent magnetic structures in the tail are "frozen" ("unchanged") for an observer in the magnetospheric frame of reference, who then maps the characteristic *spatial* scales of the turbulence, χ, into the corresponding *frequency* scales, f, with use of the Doppler relation $f = u/\chi$. (In particular, this immediately yields the two characteristic kink frequencies in the frequency domain, i.e., $f_* = u/\xi$ and

$f_{**} = u/a$, associated with the corresponding changes in the *geometry* of the turbulent field. Using the previously obtained estimates for the characteristic spatial scales ξ and a, one finds, numerically, $f_* \approx 0.01$ Hz and $f_{**} \approx 0.25$ Hz.)

The behavior of the Fourier power density spectra, $P(f)$, in the frequency domain has been widely investigated in relation with the fractal structure of the turbulence [see, e.g., *Berry*, 1979; *Burlaga and Klein*, 1986]. In particular, it has been recognized that the fractal geometry of the turbulence naturally results in the *power-law* behavior of the Fourier spectra, i.e., $P(f) \propto f^{-\alpha}$, with the exponent α dependent on the given fractal characteristics. The most important for our study case when the *temporal* variability of the magnetic field is generated by the fractal *spatial* distribution of the magnetic structures within an outward (plasma) flow having some constant velocity u, was recently analyzed by *Milovanov and Zelenyi* [1994ab]; having introduced the Hausdorff fractal dimension D to characterize the *fractal* spatial distribution of the magnetic structures within the flow, they found a simple expression for the Fourier power exponent α in the frequency domain, i.e.,

$$\alpha = 2D - 1. \tag{12}$$

From Eq. (12) it is evident that the lower the Hausdorff fractal dimension D is, the smaller the exponent of the Fourier power spectrum α would be. (As is already clear, the smaller values of the parameter D are related to the more pronounced scale-length dependence of the density of the given fractal distribution, $\rho(\chi) \propto \chi^{D-E}$.)

Eq. (12) immediately shows that there could exist the following three regimes in the power-law behavior of the Fourier spectrum of the magnetic field fluctuations, associated with the changing fractal spatial structure of the turbulent magnetic field. Indeed, i) the range of relatively large spatial scales $\chi \geq \xi$, where the Hausdorff fractal dimension D is equal to one, is mapped to the lower frequency range $f \leq f_* = u/\xi$, yielding, in this range, the power exponent $\alpha = 1$; ii) the range of the intermediate spatial scales, $a \leq \chi \leq \xi$, where $D \approx 5/3$, is mapped into the intermediate frequency range $f_* \leq f \leq f_{**} = u/a$, the power exponent being $\alpha \approx 7/3$; and iii) the range of relatively small scales, $\chi \leq a$, having $D = 2$, is mapped into the higher frequency range, $f \geq f_{**}$, with the power exponent $\alpha = 3$. Making use of the above numerical estimates of the kink frequencies f_* and f_{**}, one also finds the deduced three regimes in the Fourier power spectrum in an explicit numerical form, i) $\alpha = 1$ for $f \leq 0.01$ Hz, ii) $\alpha \approx 7/3$ for 0.01 Hz $\leq f \leq 0.25$ Hz, and iii) $\alpha = 3$ for $f \geq 0.25$ Hz.

The existence of the first kink, $f_* \approx 0.01$ Hz, separating the two regimes i) and ii), as well as the corresponding values of the power exponents $\alpha_1 = 1$ and $\alpha_2 \approx 7/3$ are in good agreement with the results of *Hoshino et al.* [1994] from the direct Geotail survey of magnetic fields in the distant Earth's magnetotail (see section 1). On the contrary, the possible existence of the second kink, $f_{**} \approx 0.25$ Hz, along with the evidence for the relatively steep component in the power spectrum in the range of frequencies above f_{**} (with the characteristic slope $\alpha = 3$) could be considered as the theoretical prediction of our model. Note that the turnover frequency $f_{**} \approx 0.25$ Hz lies beyond the Nyquist frequency of 0.17 Hz in the Geotail magnetic field data analyzed by *Hoshino et al.* [1994]. Observational substantiation of this prediction might be addressed to further studies of the potentially rich turbulent phenomena in the distant magnetotail.

It is relevant to remark that the existence of similar kinks (for the *lower* frequency ranges) in the power spectra of the magnetospheric turbulence is unambiguously seen in *various* satellite experiments. To be more specific, here we mention the recently reported results by *Ohtani et al.* [1995], who investigated the turbulence data obtained in the near-Earth stretched magnetotail prior to the substorm with the Active Magnetospheric Particle Tracer Explorers (AMPTE). An evidence for a more flat power-law spectrum in the lower frequency range has been recognized, with the characteristic spectral slopes and the turnover frequency f_* close to that reported by *Hoshino et al.* [1994]. Note, however, that the AMPTE experiments were performed at the considerably smaller geocentric distances (around ~ 8.8 Earth's radii) than the Geotail magnetic field experiments discussed by *Hoshino et al.* [1994]. Hence, the observed similarity in the power spectra of the turbulence from Geotail and AMPTE might be considered as a hint that the physical conditions in the considerably *stretched* magnetotail prior to substorms even much *closer* to the Earth could be analogous to that in the *distant* tail, where (according to our theoretical model) the self-organization of currents and magnetic fields plays the dominant role in formation of the principal characteristics of the turbulence. Consequently, the driving mechanisms proposed in our study for the turbulence processes in the *distant* magnetotail might appear to be much more *general*, governing the turbulence phenomena also in the substorm regions characterized by the very stretched topology of the tail. A dynamical scenario for the current sheet disruption during the substorms has been discussed in detail by *Lui et al.* [1988].

By using magnetometer and plasma instrument data of AMPTE obtained during three months of plasma sheet passes, *Bauer et al.* [1995] reported a kink in the power spectra of the magnetic field fluctuations from a computation of the average spectra of the magnetic field vectors in the relevant frequency ranges. The results of *Bauer et al.* [1995] indicate that whereas for frequencies between 0.03 and 2 Hz the power-law behavior with the slopes from ≈ 2 to ≈ 2.5 holds, in the lower frequency range the spectrum is actually more flat, with the slopes clearly less than ≈ 1.5 for frequencies below $\sim 10^{-3}$ Hz. An evidence for the power-law turbulence spectra with the characteristic slopes from 2 to 2.5 in the higher frequency range was also provided by *Russell* [1972] from the OGO 5 plasma sheet data. Similar power-law behavior was recognized by *Borovsky et al.* [1997] from the ISEE 2 Fast Plasma Experiment; having performed the relevant Fourier analysis of the magnetic turbulence data over 1.5 decades of frequency, *Borovsky et al.* [1997] found the mean value of the characteristic spectral slopes to be ≈ 2.2. In addition, the existence of power laws in the magnetic turbulence data has been recently supported by the results of the Interball - 1 mission [*A. Petruckovich*, private communication, 1997]. The obtained spectra of the magnetic field fluctuations as were seen by Interball - 1 have the characteristic shape presented on Fig. 4, with the typical slopes close to the value ≈ 2.35 for frequencies higher than, approximately, $f_* \sim 10^{-2}$ Hz, and with the relatively flat constituent of the spectrum in the frequency range below f_*.

An intriguing point is that the slopes for the magnetic field and velocity fluctuations as indicated by the recent satellite experiments, are not necessarily equal to each other, contrary to what one might expect assuming the standard MHD approach. For instance, the results of *Borovsky et al.* [1997] show the mean slope ≈ 2.2 for the magnetic field data (see the consideration above), whereas the corresponding slope for the velocity data appears to be ≈ 1.5. On the other hand, the theoretical fractal model proposed in our study (note that this model is substantially a non-MHD one) naturally accounts for such a difference. Indeed, it could be shown that the velocity fluctuations might be associated with the Hausdorff fractal dimension $3 - D$ where D is the fractal dimension describing the geometry of the magnetic field patterns in the self-consistent regime. (The fractal dimension $3 - D$ assumes the quasi-two-dimensional geometry of the turbulence and could be defined at the spatial scales exceeding the characteristic ion Larmor radius, r_L. Some topological meaning of the dimension $3 - D$ is discussed in the paper of *Milovanov and Zimbardo* [1997].) Since the fractal dimension D was found to be $\approx 5/3$ (see Eq. (11)), the fractal dimension $3 - D$ becomes, approximately, 4/3. Then from Eq. (12) one immediately obtains the characteristic slope for the velocity fluctuations, $2(3-D)-1 \approx 5/3$, in good agreement with the results of *Borovsky et al.* [1997].

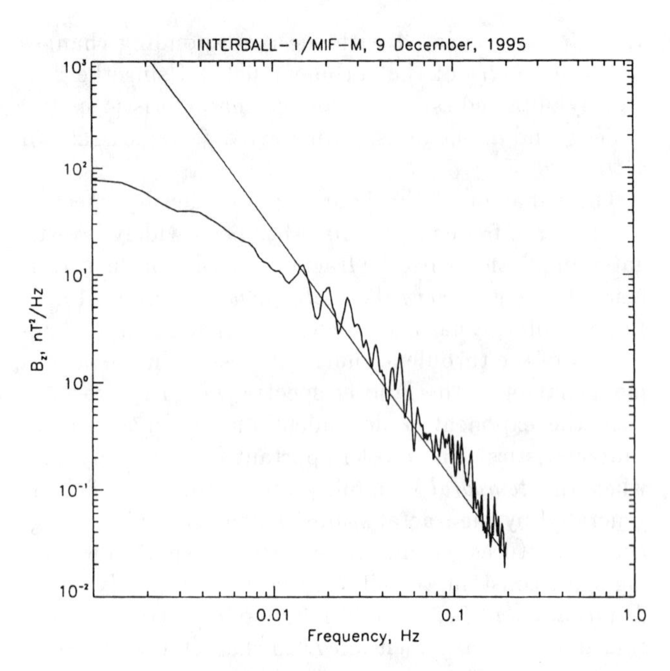

Figure 4. The "kink" power-law spectrum of the magnetic field turbulence from Interball - 1 mission.

In conclusion, we notice that the power-law form of the Fourier spectra could be in many cases associated with the *fractal* nature of the turbulence [see, e.g., *Berry*, 1979; *Burlaga and Klein*, 1986; *Milovanov and Zelenyi*, 1994ab; *Ohtani et al.*, 1995]. Hence, the observed power-law functions might also suggest that the *fractal* geometry of the magnetic field structures (which has been assumed in our study as the characteristic property of the magnetic turbulence in the *distant* magnetotail), could be actually an *inherent* feature of the (strongly nonlinear) turbulent phenomena in the tail associated with the magnetic field and plasma coupling in a wide range of spatial scales.

6. SUMMARY AND CONCLUSIONS

In the framework of the present study, we have proposed a nontraditional, geometric approach to studying magnetic field and plasma turbulence in the distant Earth's magnetotail. This approach is based on an analysis of the fine-scale magnetotail structures that control the basic turbulence properties, with use of such nonconventional quantities as the Hausdorff fractal dimension, D, and the index of connectivity, σ. We have shown that these quantities not only describe the geometry of the turbulent magnetic configurations in the distant tail (we assume this geometry to be *fractal*), but, moreover, appear in the self-consistent "physical" equations governing the self-organization of plasmas and magnetic fields in the distant current sheet. This indeed have made it possible to investigate the main turbulent phenomena in the tail (which, as we argued above, are associated with the strong nonlinearity of the system) by using an efficient geometric description, with the strong "physical" nonlinearity implicitly present in the fractal geometry of the turbulence.

An important point of our study is the assumed existence of a range of scales, $a \leq \chi \leq \xi$ ($a \ll \xi$), where the self-organization of plasmas and magnetic fields plays the decisive role in the formation of the principal characteristics of the turbulence. Such scales χ are comparable to the effective Larmor radius, r_L, of ions having the characteristic energy of $\sim$ 1 keV and migrating through the (highly structured) magnetic field $\delta \mathbf{B}_z$ in the current sheet. We argued that these ions could bring the dominant contribution into the cross-tail current density, and, therefore, are responsible for the self-consistent development of the magnetic configuration of the distant tail. (This self-consistent magnetic configuration includes both the lobe component of the magnetotail, $\mathbf{B}_x$, associated with the *average* ion migration across the tail, and the higly structured, fluctuating component, $\delta \mathbf{B}_z$, related to the fine-scale *perturbations* in the cross-tail current density.)

Having considered the basic fractal properties of the magnetic field and current density structures in the distant current sheet (see section 4), we could deduce self-consistently the main characteristics of the magnetic turbulence spectra as a direct consequence of the assumed fractal geometry of these structures. In particular, we found that the Fourier power spectrum of the z-component of the magnetic field in the current sheet has a *two-kink behavior*, with the kinks around the two characteristic turnover frequencies $f_* \approx 0.01$ Hz and $f_{**} \approx 0.25$ Hz. We argued that the existence of these two kinks could be a result of self-organization of currents and magnetic fields in the distant tail related to the development of fractal topologies in the range of spatial scales between $a \sim 8 \cdot 10^2$ km and $\xi \sim 2.5 \cdot 10^4$ km. Our theoretical conclusions, based on the concepts of the fractal geometry, appear to be in good agreement with the direct Geotail measurements of magnetic fields in the distant magnetotail below the characteristic Nyquist frequency of the data. (We found that this frequency is smaller than the second kink frequency, f_{**}. In the frequency range exceeding f_{**}, our model predicts the possible existence of a more steep constituent of the power spectrum, with a larger slope than that in the intermediate frequency range, $f_* \leq f \leq f_{**}$.) We also note that the possible fractal (i.e., statistically self-similar) nature of the magnetic field fluctuations in the frequency range $f_* \leq f \leq f_{**}$ as it is implied by the present theoretical study, might be an attractive point of the direct examination. Indeed, an analysis on the self-similarity of the magnetic turbulence data have been already accomplished by *Burlaga and Klein* [1986] for the *interplanetary magnetic field turbulence*, and by *Ohtani et al.* [1995] for the magnetic field fluctuations associated with *tail current disruption* during substorms. The results obtained in these studies have revealed a number of interesting properties of the corresponding turbulence data, and, possibly, a comprehensive analysis on the statistical self-similarity for the relevant Geotail magnetic turbulence data might be also advocated.

In our theoretical model, most of the attention has been concentrated on the variability of the z-component of the magnetic field in the tail, without due regard for the fluctuations in x- and y-directions. Of course, the inclusion of the possible three-dimensional effects might provide an explorer with a more detailed understanding of the magnetotail turbulence and with a deeper insight into the magnetic field topologies in the vicinity of the current sheet. In the meanwhile, already a quasi-two-dimensional treatment of the geometry of the turbulence proposed in our study have made it possible to account for the principal turbulence characteristics and for its basic statistical properties at various spatial scales. This might be the starting point for a more exact, three-dimensional consideration of the magnetotail structures, which would be an attractive point of future activity.

Acknowledgments. We gratefully acknowledge very useful discussions with Prof. P. Veltri and Dr. A. Taktakishvili. We would also like to thank Dr. A. Petruckovich for providing the power spectrum of the magnetic field fluctuations from the Interball mission. This work was partly supported by the Russian Foundation of Fundamental Research (projects No 95-02-03998, 96-05-64534, 97-02-16489), by the INTAS grant No 93-2492-ext within the research program of the International Center of Fundamental Physics in Moscow, and by the INTAS grant No 96-2346. In Italy, the work was sponsored by the Italian MURST and CNR, by the Agenzia Spaziale Italiana (ASI), by the Human Capital and Mobility Programme of the European Union, and by the Programma Operativo Plurifondo (POP) of the Regione Calabria.

REFERENCES

Alexander, S., and R.L. Orbach, *J. Phys. (Paris) Lett., 43*, L625, 1982; Orbach, R.L., Fracton dynamics, *Physica D, 38*, 266, 1989.

Ashour-Abdalla, M., L.M. Zelenyi, V. Peroomian, and R.L. Richard, On the structure of the magnetotail current sheet, *Geophys. Res. Lett., 20*, 2019, 1993.

Berry, M.V., Diffractals, *J. Phys. A Math. Gen., 12*, 781, 1979.

Bauer, T.M., W. Baumjohann, R.A. Treumann, N. Sckopke, and H. Luhr, Low-frequency waves in the near-Earth plasma sheet, *J. Geophys. Res., 100*, 9605, 1995.

Borovsky, J.E., R.C. Elphic, H.O. Funsten, and M.F. Thomsen, The Earth's plasma sheet as a laboratory for flow turbulence in high$-\beta$ MHD, *J. Plasma Phys., Dave Montgomery Special Issue, 57*, 1, 1997.

Burlaga, L.F., and L.W. Klein, Fractal structure of the interplanetary magnetic field, *J. Geophys. Res., 91*, 347, 1986.

Burlaga, L.F., W.H. Mish, and D.A. Roberts, Large-scale fluctuations in the solar wind at 1 AU: 1978-1982, *J. Geophys. Res., 94*, 177, 1989.

Coppi, B., G. Laval, and R. Pellat, Dynamics of geomagnetic tail, *Phys. Rev. Lett., 16*, 1207, 1966.

Feder, J., *Fractals*, Plenum Press, New York, 1988.

Galeev, A.A., and L.M. Zelenyi, Nonlinear theory of instability of a diffusive neutral sheet, *Sov. J. JETP, 42*, 450, 1975.

Galeev, A.A., and L.M. Zelenyi, SLAB model of reconnection in collisionless plasma, *Sov. J. Pisma v JETP, 25*, 407, 1977.

Gefen, Y., A. Aharony, and S. Alexander, Anomalous diffusion on percolating clusters, *Phys. Rev. Lett., 50*, 77, 1983.

Giona, M., and H.E. Roman, Fractional diffusion equation for transport phenomena in random media, *Physica A, 185*, 87, 1992.

Havlin, S., and D. Ben-Avraham, Diffusion in disordered media, *Adv. Phys., 36*, 695, 1987.

Hoshino, M., A. Nishida, T. Yamamoto, and S. Kokubun, Turbulent magnetic field in the distant magnetotail: Bottom-up process of plasmoid formation? *Geophys. Res. Lett., 21*, 2935, 1994.

Isichenko, M.B., Percolation, statistical topography, and transport in random media, *Rev. Mod. Phys., 64*, 961, 1992.

Lifshitz, E.M., and L.P. Pitaevskij, *Physical Kinetics*, Nauka, Moscow, 1979. (in Russian).

Le Mehaute, A., *Fractal Geometries: Theory and Applications*, CRC Press, Boca Raton, 1991.

Lui, A.T.Y., R.E. Lopez, S.M. Krimigis, R.W. McEntire, L.J. Zanetti, and T.A. Potemra, A case study of magnetotail current sheet disruption and diversion, *Geophys. Res. Lett., 15*, 721, 1988.

Mandelbrot, B.B., *The Fractal Geometry of Nature*, Freeman, New York, 1983.

Milovanov, A.V., Topological proof for the Alexander-Orbach conjecture, *Phys. Rev. E, 56*, 2437, 1997.

Milovanov, A.V., and L.M. Zelenyi, Fractal clusters in the solar wind, *Adv. Space Res., 14*, (7)123, 1994a.

Milovanov, A.V., and L.M. Zelenyi, Development of fractal structure in the solar wind and distribution of magnetic field in the photosphere, in *Solar System Plasmas in Space and Time. Geophysical Monograph, 84*, edited by J.L. Burch, and J.H. Waite, Jr., pp. 43-52, American Geophysical Union, Washington, DC, 1994b.

Milovanov, A.V., and L.M. Zelenyi, Percolation of a plasma across stochastic magnetic configurations: FLR effects, in *Physics of the Magnetopause. Geophysical Monograph, 90*, edited by P. Song, B.U.O. Sonnerup, and M.F. Thomsen, pp. 357-362, American Geophysical Union, Washington, DC, 1995.

Milovanov, A.V., and L.M. Zelenyi, Fracton excitations in the solar wind and the power-law spectra of the IMF fluctuations, in *Physics of Space Plasma Series, 14*, edited by T. Chang, and J.R. Jasperse, pp. 373-386, Massachusetts Institute of Technology, Cambridge, MA, 1995.

Milovanov, A.V., and G. Zimbardo, Percolation in random scalar fields: Topological aspects and numerical modeling, *Phys. Rev. E*, 1997 (submitted).

Milovanov, A.V., L.M. Zelenyi, and G. Zimbardo, Fractal structures and power law spectra in the distant Earth's magnetotail, *J. Geophys. Res., 101*, 19,903, 1996.

Nakayama, T., K. Yakubo, and R. Orbach, Dynamical properties of fractal networks: scaling, numerical simulations, and physical realizations, *Rev. Mod. Phys., 66*, 381, 1994.

Nash, Ch., and S. Sen, *Topology and Geometry for Physicists*, Academic Press, London, 1983.

Ness, N.F., C.S. Scearce, and S.C. Cantarano, Probable observations of the geomagnetic tail at 10^3 Earth radii by Pioneer 7, *J. Geophys. Res., 72*, 3769, 1967.

Nishida, A., T. Yamamoto, K. Tsuruda, H. Hayakawa, A. Matsuoka, S. Kokubun, M. Nakamura, and K. Maezawa, Structure of the neutral sheet in the distant tail ($x = -210R_e$) in geomagnetically quiet times, *Geophys. Res. Lett., 21*, 2951, 1994.

Ohtani, S., T. Higuchi, A. T. Y. Lui, and K. Takahashi, Magnetic fluctuations associated with tail current disruption: Fractal analysis, *J. Geophys. Res., 100*, 19,135, 1995.

O'Shaughnessy, B., and I. Procaccia, Analytical solutions for diffusion on fractal objects, *Phys. Rev. Lett., 54*, 455, 1985.

Paterson, W.R., and L.A. Frank, Survey of plasma parameters in Earth's distant magnetotail with the GEOTAIL spacecraft, *Geophys. Res. Lett., 21*, 2971, 1994.

Russell, C.T., Noise in the geomagnetic tail, *Planet. Space Sci., 20*, 1541, 1972.

Stauffer, D., *Introduction to Percolation Theory*, Taylor and Francis, London, 1985.

Tsurutani, B.T., J.A. Slavin, E.J. Smith, R. Okida, Magnetic structure of the distant geotail from -60 to -220 R_E: ISEE-3, *Geophys. Res. Lett., 11*, 1, 1984.

Veltri, P., G. Zimbardo, A.L. Taktakishvili, and L.M. Zelenyi, Effect of magnetic turbulence on the ion dynamics in the distant magnetotail, *J. Geophys. Res., 103*, 1998 (in press).

Villante, U., Neutral sheet observations at 1000 R_E, *J. Geophys. Res., 81*, 212, 1976.

Walker, R.C., U. Villante, and A.J. Lazarus, Pioneer 7 observations of plasma flow and field reversal regions in the distant geomagnetic tail, *J. Geophys. Res., 80*, 1238, 1975.

Yamamoto, T., A. Matsuoka, K. Tsuruda, H. Hayakawa, A. Nishida, M. Nakamura, and S. Kokubun, Dense plasmas in the distant magnetotail as observed by GEOTAIL, *Geophys. Res. Lett., 21*, 2879, 1994.

Zelenyi, L.M., and A.V. Milovanov, Large-scale magnetic configurations on the fractal geometry: The force-free approximation, *Russ. J. Astron. Zh., 73*, 805, 1996.

Zelenyi, L.M., and A.V. Milovanov, Dynamical model of the IMF fluctuations: Fracton excitations and power-law spectra, *Russ. J. Geomagnetism i Aeronomia, 37*, 1, 1997.

Zimbardo, G., and P. Veltri, Field line transport in stochastic magnetic fields: Percolation, Levy flights, and non-Gaussian dynamics, *Phys. Rev. E, 51*, 1412, 1995.

Zimbardo, G., P. Veltri, G. Basile, and S. Principato, Anomalous diffusion and Levy random walk of magnetic field lines in three-dimensional turbulence, *Phys. Plasmas, 2*, 2653, 1995.

L. M. Zelenyi and A. V. Milovanov, Department of Space Plasma Physics, Space Research Institute, Profsoyuznaya str. 84/32, 117810 Moscow, Russia. e.mail: lzelenyi@iki.rssi.ru; amilovan@mx.iki.rssi.ru

G. Zimbardo, Dipartimento di Fisica, Università della Calabria, 87030 Arcavacata di Rende, Italy. e.mail: zimbardo@fis.unical.it